COMBUSTION PHYSICS

In the past several decades, combustion has evolved from a scientific discipline that was largely empirical to one that is quantitative and predictive. These advances are characterized by the canonical formulation of the theoretical foundation; the strong interplay between theory, experiment, and computation; and the unified description of the roles of fluid mechanics and chemical kinetics. This graduate-level text incorporates these advances in a comprehensive treatment of the fundamental principles of combustion physics. The presentation emphasizes analytical proficiency and physical insight, with the former achieved through complete, though abbreviated, derivations at different levels of rigor, and the latter through physical interpretations of analytical solutions, experimental observations, and computational simulations. Exercises are designed to strengthen the student's mastery of the theory. Implications of the fundamental knowledge on practical phenomena are discussed whenever appropriate. These distinguishing features provide a solid foundation for an academic program in combustion science and engineering.

Chung K. Law is the Robert H. Goddard Professor of Mechanical and Aerospace Engineering at Princeton University. He obtained his doctorate in engineering physics from the University of California at San Diego in 1973. His research interests are in combustion, propulsion, heat and mass transfer, and issues on energy and the environment. For his research accomplishments, he received the Curtis W. McGraw Research Award of the American Society for Engineering Education (ASEE) in 1984 for outstanding early achievement in research, a silver medal and the Alfred C. Egerton Gold Medal of the Combustion Institute in 1990 and 2006, respectively, the Propellants and Combustion Award, the Energy Systems Award, and the Pendray Aerospace Literature Award of the American Institute of Aeronautics and Astronautics (AIAA) in 1994, 1999, and 2004, respectively, the Heat Transfer Memorial Award, in science, of the American Society of Mechanical Engineers (ASME) in 1997, and several awards for best conference papers. He is an original member of the Highly Cited Researchers database of the Institute for Scientific Information (ISI).

Professor Law is a former president of the Combustion Institute, a Fellow of the AIAA, the ASME, and the American Physical Society (APS), and a member of the U.S. National Academy of Engineering.

COMBUSTION PHYSICS

CHUNG K. LAW

Princeton University

CAMBRIDGE UNIVERSITY PRESS
Cambridge, New York, Melbourne, Madrid, Cape Town,
Singapore, São Paulo, Delhi, Tokyo, Mexico City

Cambridge University Press
32 Avenue of the Americas, New York, NY 10013-2473, USA

www.cambridge.org
Information on this title: www.cambridge.org/9780521154215

First published 2006
Reprinted 2008
First paperback edition 2010
Reprinted 2011 (twice)

A catalog record for this publication is available from the British Library.

Library of Congress Cataloging in Publication Data

Law, Chung K., 1947–
Combustion physics / Chung K. Law.
p. cm.
Includes bibliographical references and index.
ISBN-13: 978-0-521-87052-8 (hardback)
ISBN-10: 0-521-87052-6 (hardback)
1. Combustion, Theory of. 2. Physics. I. Title.
QD516.L27 2006
541′.361 – dc22 2006011565

ISBN 978-0-521-87052-8 Hardback
ISBN 978-0-521-15421-5 Paperback

To my wife Helen
and to our children
Jonathan, Jennifer, and Jeffrey

Contents

Preface

Since the mid-1970s there has been truly significant advancement in combustion science, spurred by the dual societal concerns for energy sufficiency and environmental quality, and enabled by the rapid increase in the sophistication of mathematical analysis, computational simulation, and experimental techniques. Consequently, we have witnessed the evolvement of combustion from a scientific discipline that was largely empirical to one that is quantitative and predictive, leading to its useful applications in combustion-related engineering devices and practices.

This text reflects my desire to incorporate these advances in my lectures on combustion. As a result, its preparation has been guided by the three distinguishing themes characterizing recent developments in combustion research, namely the canonical formulation of the theoretical foundation; the strong interplay between experiment, theory, and computation; and the description of combustion phenomena from the unified viewpoint of fluid mechanics and chemical kinetics.

The text also emphasizes analytical proficiency by presenting complete, albeit abbreviated, derivations that can be followed by the student with a modest effort. Alternate solutions are sometimes presented to demonstrate that a phenomenon can often be analyzed using different approaches and at different levels of rigor. I hope that through this gentle guidance the student can acquire the needed confidence to tackle more difficult problems on his or her own.

This text grew out of the lecture material prepared for a one-year graduate course that I have given at several academic institutions. No prerequisite in mathematics, fluid mechanics, and chemistry is expected apart from the usual undergraduate education in the physical sciences or mechanical, aerospace, or chemical engineering. The text consists of three parts: Chapters 1 through 4 cover the basic components required to describe chemically reacting flows, namely thermodynamics, chemical kinetics, and transport phenomena; Chapters 5 through 10 cover descriptions of the basic combustion phenomena—those of governing equations, nonpremixed and premixed flames, the limit phenomena of ignition, extinction, and flame stabilization, and the aerodynamics of flames; Chapters 11 through 14 cover combustion in the four major classes of flows, namely turbulent, boundary-layer, two-phase, and supersonic flows. Since the amount of material treated in this text is substantial, the instructor may be more selective in the choice of topics. For example, discussion on reaction mechanisms, especially most of Chapter 3, can be omitted if chemistry is not emphasized in the course. Similarly, much of the materials that require extensive mathematical derivations, especially those of Chapter 9, can be omitted if strong mathematical experience is not intended. Furthermore, a one-semester course can be structured by abstracting materials from individual chapters, leaving the rest of the text for the enrichment of individual students.

While a serious attempt was made to make the text comprehensive in its coverage, it is nevertheless inevitable that some important topics were either excluded or inadequately presented. Feedback from readers on possible improvements in future editions will be very much appreciated. Similarly, because of the extensive literature in existence, it is also unavoidable that important references were inadvertently left out. Forbearance of the authors of these articles is requested.

In the preparation of this text I have been ably assisted by many of my present and former graduate students and research associates. In particular, I acknowledge with appreciation the following who have contributed substantially in this effort: John K. Becktold, Beei-Huan Chao, Peck Cho, Suk-Ho Chung, Fokion N. Egolfopoulos, Hong G. Im, Tianfeng Lu, Atsushi Makino, Matei I. Radulescu, Chih-Jen Sung, Hai Wang, Heyang Wang, Shou-Yin Yang, and Delin Zhu. The manuscript was read in part or in whole by Professor Craig T. Bowman of Stanford University, Professor Sau-Hai Lam of Princeton University, and Professor Forman A. Williams of the University of California at San Diego. Their comments have been substantial and most useful, and I thank them sincerely for their collegiality and generosity.

It was by chance that I became a student of Professor Forman A. Williams in the spring of 1970. His influence on my intellectual and professional development has been profound. I am immensely thankful for his mentorship.

I reserve my most heartfelt appreciation for my wife, Helen Kwan-mei, for having transcribed the first drafts of this text, for constantly encouraging me to bring it to fruition, and for her patience and love over the years.

Chung K. Law
Princeton, New Jersey
January 2006

Introduction

This book is about combustion science and technology and, as such, covers not only the basic laws and phenomena related to the physics and chemistry of combustion, but also the implications of the fundamental understanding gained therein to the principles behind the practical combustion phenomena affecting our daily lives. It presents the diverse knowledge required of combustion scientists and engineers, the challenges they face, and the satisfaction they derive in providing the proper linkage between the fundamental and the practical.

In Section 0.1 we identify the major areas of practical combustion phenomena, illustrated by some specific problems of interest. In Section 0.2 we discuss the scientific disciplines comprising the study of combustion, and in Section 0.3 we present the classifications of fundamental combustion phenomena. An overview of the text is given in Section 0.4.

0.1. MAJOR AREAS OF COMBUSTION APPLICATION

It is fair to say that the ability to use fire is an important factor in ushering the dawn of civilization. Today our dependence on the service of fire is almost total, from heating and lighting our homes to powering the various modes of transportation vehicles. Useful as it is, fire can also be menacing and sometimes deadly. Wildland and urban fires cause tremendous loss of property and lives every year; the noxious pollutants from automotive and industrial power plants poison the very environment in which we live; and the use of chemical weapons continues to be an agent of destruction with ever greater efficiency. Combustion is certainly one branch of science that affects almost every aspect of human activities.

Practical combustion problems can be roughly divided into the following five major categories, in each of which we cite some examples of current interest.

Energy and Combustion Devices: Despite the large variety of alternate energy sources available, such as nuclear, solar, wind, hydroelectric, geothermal, and OTEC (ocean thermal energy conversion), chemical energy derived from burning fossil fuels supplies a disproportionately large fraction of the total world energy needs—around 85 percent at present. This trend will continue in the foreseeable future because of its convenience, high-energy density, and the economics.

Combustion energy is mainly used to generate heat and power. Examples of this application are domestic heating, firing of industrial furnaces, and the operation of automotive engines and gas turbines. Hence the design and operation of heat and power devices and engines is closely related to the issue of efficient energy utilization. Because of the importance of transportation vehicles as a major consumer of petroleum fuels and contributor of air pollution, there has been extensive development since the early 1970s for more efficient and cleaner burning internal combustion engines for automobiles. For example, the diesel engine offers substantial advantage over the more widely used gasoline engines, for several reasons. First, even though its combustion cycle efficiency is less than that of the gasoline engine for the same compression ratio, it is more efficient overall because it operates at higher compression ratios. Furthermore, unlike the gasoline engine, which requires highly refined fuels with narrow specifications, the diesel engine is very fuel tolerant. Thus diesel fuel requires less refining than gasoline and, consequently, results in a net saving in processing energy at the refinery stage. This property of fuel tolerance also implies that the diesel engine is a good candidate for the use of unconventional or low-grade fuels. The diesel engine, however, does have the potential disadvantages of being relatively noisier and a heavy emitter of soot and oxides of nitrogen (NO_x); both problems have their origin in its operational principle and therefore require fundamental combustion research. It is nevertheless gratifying to note that much progress has been made recently in alleviating these problems.

An important concept in engine development is that of stratified charge combustion. The basic idea is that the combustion of lean mixtures has the potential of simultaneously increasing the combustion efficiency and reducing the formation of most pollutants. Lean mixtures, however, are hard to ignite. Therefore, the concept of stratified charge combustion is to stratify an overall fuel lean mixture from relatively rich to ultra lean. Since the relatively rich portion can be ignited easier, the hot combustion products so generated can in turn ignite the ultra lean portion of the charge. Thus by combining the merits of high-pressure combustion, direct fuel injection for uniform cylinder-to-cylinder charge distribution and controlled fuel vaporization, spark ignition for controlled ignition event, and stratified charge combustion, there has been considerable development on high-compression-ratio, direct-injection, spark-assisted, stratified charge engines.

In contrast to stratified charge engines, there is also considerable interest in the development of HCCI (homogeneous charge compression ignition) engines. Conceptually, by having reaction taking place homogeneously within the entire engine cylinder, instead of being confined to localized, high-temperature regions constituting the flames, the formation of soot and NO_x can be substantially reduced.

Furthermore, higher compression ratios and hence higher efficiency can be attained with compression ignition.

The fact that improvements in the engine performance can be pursued through the opposite concepts of stratified and homogeneous charges not only demonstrates the complexity of the combustion phenomena underlying such technological processes, but it also highlights the richness of the possible avenues that can be explored for optimization.

Fuels: Combustion needs fuel. Furthermore, the satisfactory operation of different heat and power engines usually depends critically on the compatibility of the fuel used. Examples are the unsuitability of diesel fuel for use in gasoline engines because it is relatively less volatile, and the narrow compositional specifications of gases which can be used in domestic gas stoves in order to maintain flame stabilization by avoiding blowoff and flashback.

The importance of fuel in combustion has been receiving increased interest because of the concern over the shortage and reliability of petroleum supply. Thus "energy crisis" is simply a "fuel crises." Since the world's petroleum supply is projected to be severely depleted within this century, the long term solution for the next few centuries in terms of fossil fuels appears to largely depend on the burning of coal, either through direct utilization or as coal-derived fuels. Two approaches for direct coal utilization are being actively pursued. The first is fluidized-bed combustion, in which air is introduced through the bottom of a bed of coal particles at a sufficiently fast rate such that the particles are levitated, that is, fluidized. This approach has the advantages that the coal particles are in direct contact with the oxidizing air such that their burning rates are maximized, that neutralization of oxides of sulfur (SO_x) can be facilitated by mixing limestone with the coal particles, and that the production of NO_x can be minimized by controlling the fluidization rate. The second approach for direct coal utilization is the burning of coal–water slurries. Here, finely crushed coal particles of sizes ranging between 40–70 μm are mixed in water and sprayed directly into the combustion chamber of industrial furnaces. The advantages are that the physical processes of coal crushing and mixing are less energy expensive than the chemical process of coal liquefaction, and that the slurries can be transported through pipelines and subsequently directly burned in conventional oil-fired combustors. This requires minimum hardware modification, and thereby capital outlay and combustor downtime. Slurries up to 70 percent coal content have been successfully burned.

Oil can also be derived from coal. These coal-derived oils have higher boiling points, wider boiling point ranges, and higher contents of aromatics and nitrogen-containing compounds. Consequently, they tend to produce more soot and NO_x. Various alternate and hybrid fuels have also been formulated. Prominent among these are methanol, ethanol, and mixtures of ethanol with oil. Methanol can be derived from natural gas and coal, while both methanol and ethanol can be produced from biomass. Alcohols have smaller heats of combustion because of the extra oxygen atom in the molecule. However, they have higher knock ratings in gasoline engines and produce less NO_x and soot. Blends of ethanol and gasoline, and methanol and gasoline, have been successfully marketed.

Coal, of course, can also be gasified in the presence of air, with or without steam, to produce a combustible gaseous fuel that consists of hydrogen and carbon monoxide. Coal gasification becomes progressively more attractive as a source of clean fuel with the dwindling supply of natural gas.

Pollution and Health: The major pollutants from combustion are soot, SO_x, NO_x, unburned hydrocarbons (UHC), and carbon monoxide (CO). As just mentioned, soot is expected to be a serious problem with the burning of coal-derived fuels and the large-scale deployment of high-compression engines such as the diesel. Soot not only is unsightly but can also be carcinogenic due to the condensation and thereby presence of carcinogenic liquid combustion products on the particle surface.

The main source of SO_x is from burning coal. When combined with water in the atmosphere, the emitted SO_x forms sulfuric acid and precipitates as acid rain, with devastating effects on aquatic life and soil erosion.

NO_x can be formed from either the N_2 in the atmosphere or the nitrogen atoms in the fuel molecules, with the former produced under high-temperature, intense combustion situations because of the need to dissociate the nominally inert N_2 in the air. Fuel-bound NO_x is less temperature sensitive and could be a major contributor of NO_x emission from burning coal or coal-derived oils. When it reacts with UHC and ozone in the presence of sunlight, NO_x forms smog that is detrimental to the respiratory system.

A problem of potential concern is indoor pollution. With houses being better insulated to conserve energy, the trace pollutants (CO, NO_x, UHC), from such domestic heating devices as the gas stove, furnace, and kerosene heater, may exist at sufficiently high levels as to be injurious to health.

There is also interest in applying combustion technology in the management of municipal, munition, and chemical hazardous wastes through incineration. The problems with burning these wastes are the uncertainty of the toxicity of the combustion intermediates and products and the fact that some of the chemicals are halogenated compounds, which can be resistant to efficient burning because of the scavenging of the crucial hydrogen atom by the halogen radicals in the oxidation process.

A serious, and potentially catastrophic, environmental problem is global warming caused by the increased amount of anthropogenic CO_2 in the atmosphere. Since CO_2 is a by-product of hydrocarbon combustion, suggestions have been made to use hydrogen as the primary fuel source. In the event that hydrogen is derived through the conversion of hydrocarbons, CO_2 is still produced during conversion and needs to be sequestered properly in order to prevent its release into the atmosphere.

A discussion on the adverse effects of combustion on health would not be complete without mentioning the well-established cancer-causing consequence of cigarette smoking, which is simply the slow combustion of tobacco leaves. The knowledge of combustion science has not been sufficiently brought to bear on this problem of immense importance.

Safety: This topic can be divided into three categories, namely fires, explosions, and materials. Fires, both structural and wildland, are costly in terms of human suffering as well as financial loss. Problems of interest include improving fire detection technology and understanding the dynamics of fire propagation in confined spaces such as buildings and aircraft cabins.

Explosions are of concern to safety in mine galleries and grain elevators, as a consequence of LNG (liquefied natural gas) spills or rupturing of pressurized hydrogen storage tanks in urban areas, and in nuclear reactor accidents. In the last example, hydrogen gas is generated and could accumulate in sufficient quantity to cause an explosion. This would in turn rupture the reactor containment structure, causing the release of radioactive gases into the environment.

Since the inhalation of smoke and the toxic products of combustion is a cause of fatality in fires, the choice of materials for structure and decoration is also an important consideration in the overall strategy for fire control.

A strategy toward the prevention of fires and explosions in aircraft and combat vehicles, such as tanks, is the development of fire-safe fuels which, while burning well within the engine, will not catch fire upon spillage. For example, diesel oil emulsified with a small amount of water has been found to be fire resistant.

Defense and Space: The various defense establishments are interested in the formulation of high-energy munitions and propellants; the suppression of combustion instability within jet engines, rockets and guns; signature and detection vulnerability from the exhausts of jet engines and rockets; and measures at preventing explosion of fuel tanks when being penetrated by projectiles. The development of chemical lasers as an intense power source and of hypersonic aircraft up to Mach 25 are also of interest to the national defense.

Since combustion experiments conducted on earth are frequently complicated by the presence of buoyant flows, there has been much interest to conduct these experiments in the weightless environment of a space shuttle or station. The intrusion of buoyancy is particularly problematic when the burning is slow as in the propagation of a flame in a weak mixture, or for long-duration phenomena such as smoldering. The presence of buoyancy can also distort the flame configuration from an otherwise symmetrical one, and hence significantly complicates data reduction as well as theoretical analysis or computational simulation of the phenomenon of interest.

Fire safety is of paramount interest in space exploration. For example, while earthbound smoke detectors of incipient fires are placed at the ceiling of a room in order to capture the buoyancy-driven, upwardly rising smoke, they are clearly inoperative in the weightless space environment. Furthermore, flammability standards established on earth may not have much meaning for the fire safety evaluation of a spacecraft.

Recognizing that the environment within a space craft is artificial anyway, there has been the suggestion of creating an almost fire-proof living environment so that fire hazard ceases to be a concern. This concept is based on the recognition that whereas ignition and combustion intensity depend on the fractional amount of oxygen in the oxidizing gas, human comfort depends only on the absolute amount of oxygen. Furthermore, it is also empirically known that the combustibility of most organic materials decreases drastically with decreasing oxygen concentration. They become hardly flammable when the oxygen concentration is reduced to less than, say, 15 mole percent. Thus if we can reduce the cabin oxygen concentration to half of its value in air, but increase the cabin pressure to two atmospheres, then a comfortable, but fire-proof environment can be created.

0.2. SCIENTIFIC DISCIPLINES COMPRISING COMBUSTION

Combustion is the study of chemically reacting flows with rapid, highly exothermic reactions. It is interdisciplinary in nature, comprising thermodynamics, chemical kinetics, fluid mechanics, and transport phenomena, each of which has the following roles.

Thermodynamics: In combustion processes reactants are converted to products, releasing heat for utilization. The science of thermodynamics allows us to do the bookkeeping on how much chemical energy is converted to thermal energy in such a process, and to determine the thermal and compositional properties of the products when equilibrium is reached. The laws of equilibrium thermodynamics are firmly established, although the thermodynamic properties of many of the reacting species, including large and complex fuels and their reaction intermediates, are still not well determined.

Chemical Kinetics: While thermodynamics links the initial state to the final, equilibrium, state of a reactive mixture, it does not tell us through which path, and for how long, such a transformation takes place. For example, if a particular reaction requires more than an hour to proceed to near completion, we obviously need not take it into consideration when analyzing the cycle performance of an automotive engine. In fact, conclusions based on such equilibrium calculations could be quite erroneous. An example is the calculation of NO_x emission from engines—calculated amounts of NO_x emission based on finite reaction rates far exceed those determined by assuming thermodynamic equilibrium at the exhaust temperature. Since all combustion processes have some finite, characteristic times defining the relevant phenomena, chemical kinetics is needed to prescribe the paths and rates through which reactions take place during such times.

Chemical kinetics is a complex subject, especially for combustion systems in which a myriad of chemical species exist, each of which has the potential of interacting with the rest. As an illustration of its complexity, it may be noted that, at present, only the oxidation mechanisms of such simple fuels as hydrogen and methane can be considered to be reasonably well understood to allow for the prediction of such global combustion characteristic as the burning rate of a laminar flame, provided the pressure is not too high.

Fluid Mechanics: Since the combustion systems we are interested in frequently involve chemical reactions occurring in a flowing medium, knowledge of fluid mechanics is an essential prerequisite for a successful understanding of many combustion phenomena. Here combustion distinguishes itself from being merely a branch of chemistry in that it is chemistry and more. By the same token, a fluid dynamicist cannot satisfactorily describe a combusting flow field without paying adequate attention to the effects of chemical reactions. As an example, the highly localized and exothermic nature of chemical reactions causes significant temperature, and therefore density, variations in a flow, implying that the frequently invoked assumption of constant density in fluid mechanics can be a rather poor one in combustion studies.

Both chemical kinetics and fluid mechanics are major scientific disciplines by themselves. When coupled through combustion, the complexity and richness of the resulting phenomena take on a new dimension, as we shall demonstrate throughout this text.

Transport Phenomena: As just mentioned, in a combustion flow field the chemical reactions frequently occur in highly localized regions of reaction fronts, which are characterized by high temperatures, high product concentrations, and low reactant concentrations. On the other hand, in regions away from these reaction fronts, the temperatures and product concentrations are low while the reactant concentrations are high. The existence of these temperature and concentration gradients will cause the transfer of energy and mass from regions of high values to regions of low values through the molecular process of diffusion. For heat transfer, radiation can also be important. Mechanistically, the existence of diffusive transport is crucial in the sustenance of many types of flames in that it is only through these processes that fresh reactants can be continuously supplied to the flame, while the heat generated there is also being continuously conducted away to heat up and thereby cause ignition of the fresh mixture. Since a reaction can proceed only when its participating species can be brought to the neighborhood of a physical location and remain there for a period of time sufficiently long for the reaction to consummate, strong coupling between transport and chemical kinetics in determining the local reaction rate is to be expected.

0.3. CLASSIFICATIONS OF FUNDAMENTAL COMBUSTION PHENOMENA

In this section, we introduce the various classifications of fundamental combustion phenomena and the terminology usually associated with them.

Premixed versus Nonpremixed Combustion: This is probably the most important classification of combustion phenomena. At the global level, a combustion system frequently consists of two reactants: a fuel and an oxidizer. These two reactants must be brought together and mixed at the molecular level before reaction can take place, as just mentioned. Therefore, the mechanisms of mixing are essential elements in influencing combustion. The requirement of mixedness also implies that at least one of the reactants should be in either the gaseous or the liquid phase so that its molecules can "spread around" those of the other reactant.

Because of the importance of molecular mixedness, combustion systems behave quite differently depending on whether the reactants are initially mixed or not. In a premixed system, the reactants are already well mixed before reaction is initiated. However, in a nonpremixed system the reactants are initially separated and are brought together, through the molecular process of diffusion and the bulk convective motion, to a common region where mixing and subsequently reaction take place. Nonpremixed combustion is also known as "diffusion combustion" because diffusive transport is essential in effecting mixing of the reactants at the molecular level. It is, however, important to recognize that by calling a nonpremixed system diffusional does not imply a premixed system is nondiffusional. The word "diffusional" only indicates the need to bring the reactants together via this transport mechanism. In a premixed system, diffusion is still needed to transport the premixture to—and the thermal energy and the combustion products away from—the reaction region where the reactants are consumed and the thermal energy and combustion products generated.

A Bunsen flame, shown in Figure 0.3.1, provides an illustration of both types of flames. Here as the fuel gas issues from the fuel orifice, air is entrained through the adjustable air intake port and is then mixed with the fuel gas as they travel along the burner tube. The subsequent reaction between the fuel and oxygen in this mixture forms a premixed flame. Assuming that the air flow rate can be manipulated, then the resulting flame can be either fuel rich or fuel lean depending on whether the oxygen or fuel can be completely consumed. If this mixture is fuel lean, then the excess oxygen will remain unreacted after passing through the flame and will be "exhausted" to the environment. However, if it is fuel rich, then after passing through the premixed flame the excess fuel, or rather the fuel-related intermediate species, can further react with oxygen in the ambient air. Since oxygen and the fuel species are initially separated, they need to be transported to a common region where mixing and reactions occur. This results in a nonpremixed flame, at which the outwardly directed fuel species react almost completely with the inwardly directed oxygen. The entire flame ensemble therefore consists of a premixed flame and a nonpremixed flame. Finally, in the event when the air intake port is completely closed, then the burner mixture does not contain any oxygen and, as such, only the nonpremixed flame exists.

It is obvious that one would not find many examples of premixtures in nature because they would have already reacted even if they are only slightly reactive. On

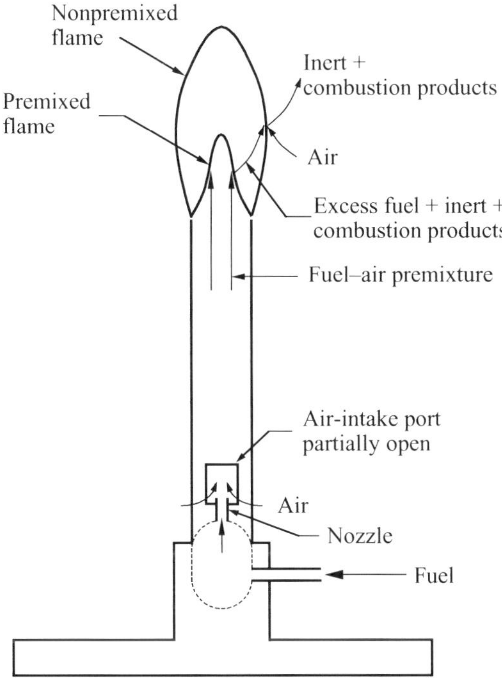

Figure 0.3.1. Schematic of the Bunsen flame.

the other hand, nonpremixed systems abound. Indeed with oxygen in the air as the oxidizer, then all materials that can burn in air are fuels. Examples are fossil deposits such as petroleum and coal, cellulosic materials such as paper and cloth, and metallic substances such as aluminum and magnesium.

Laminar versus Turbulent Combustion: A flame is also characterized by the nature of the flow, whether it is laminar or turbulent. In a laminar flow distinct streamlines exist for the bulk, convective motion, whereas in a turbulent flow such streamlines do not exist such that at any point in space the flow quantities randomly fluctuate in time. The existence of turbulence generally facilitates the coarse mixing process, and therefore has a particularly strong influence on nonpremixed systems in which reactant mixing is essential. The final mixing before reaction can take place, however, must still occur through the molecular diffusion process whether the flow is laminar or turbulent.

Subsonic versus Supersonic Combustion: A second way to characterize combustion according to the nature of the flow is the velocity of the flow, whether it is subsonic or supersonic. In a subsonic flow, the molecular collision processes of diffusion are predominant while reactions also have more time to complete. These are the flames we encounter most frequently in our daily lives, such as the candle flame and the pilot flame. In supersonic combustion the high flow velocity usually renders convective transport to dominate diffusive transport. Reactions also have less time to proceed.

Wave motions involving shocks and rarefactions are likely to be present. Supersonic combustion is usually associated with explosions and supersonic flights.

Homogeneous versus Heterogeneous Combustion: This is among the most confusing terminology in combustion literature. Traditionally, a combustion phenomenon is called homogeneous if both reactants initially exist in the same fluid phase, either gas or liquid. An example is the Bunsen flame just discussed. If the two reactants initially exist in different phases, whether gas–liquid, liquid–solid, or solid–gas, then the combustion is heterogeneous. An example is a coal particle burning in air.

On the other hand, chemists define a heterogeneous reaction as one in which the reactants actually exist in different phases at the location where reaction takes place. Therefore, in the case of coal burning in air, the reaction is heterogeneous when the solid carbon in the coal reacts with the oxygen from air at the particle surface. However, the reaction is homogeneous when there is substantial devolatalization such that the outgassing fuel vapor reacts with oxygen in the gas phase. According to the conventional combustion definition, both modes of burning are called heterogeneous combustion.

Homogeneous-versus-heterogeneous combustion is sometimes also used to designate the uniformity of the mixture. Thus a process is called homogeneous when there is no temperature or concentration gradient in the mixture. An example is the (homogeneous) explosion of a uniform mixture of fuel and air, as in the HCCI engine mentioned earlier. However, if combustion occurs in a gaseous mixture containing fuel vapor pockets produced through, say, vaporization of fuel droplets, then the process is sometimes called heterogeneous.

In order to avoid confusion, we shall be as specific as possible in describing different phenomena. For example, the reaction between vaporized fuel from a coal particle and air will be called "gas-phase reaction," while the reaction between oxygen and solid carbon at the particle surface will be called "surface reaction." This circumvents the uncertainty in designating the former as either a heterogeneous or homogeneous reaction.

0.4. ORGANIZATION OF THE TEXT

The present text aims to give a fairly comprehensive treatment of fundamental combustion phenomena. The next four chapters provide the physical and chemical fundamentals needed to describe combustion processes. Specifically, Chapter 1 discusses equilibrium thermodynamics, which relates the initial and final states of a chemically reacting, multicomponent thermodynamic system, culminating in the calculation of the adiabatic flame temperature. Chapter 2 introduces the general concepts of chemical kinetics and Chapter 3 studies the reaction mechanisms of some practical fuels. Together, these two chapters provide a fairly comprehensive introductory coverage of the chemical aspects of combustion, leading to an appreciation of the complexity of the reaction mechanisms governing hydrocarbon oxidation. In Chapter 4, we study the nonequilibrium diffusive transport of heat, mass, and momentum when nonuniformities in temperature, concentration, and velocity respectively exist in the flow.

In Chapter 5, the general conservation equations for chemically reacting flows are presented, and a simplified, though still fairly general equation system relevant for subsonic combustion is defined. Various formulations needed for subsequent studies are developed, and some important general properties of flames are derived.

In Chapter 6, we start our study of flame phenomena by analyzing the properties of nonpremixed flames in the limit of infinitely fast chemical reaction rates. Some of the formulations introduced in Chapter 5 are now applied to several flame configurations to demonstrate their utility and the canonical nature of the flame responses.

In Chapters 7 and 8, we study the structure of premixed flames for which finite rate kinetics is essential. Chapter 7 first identifies the existence of subsonic deflagration waves and supersonic detonation waves by examining the possible initial and final states of a premixed combustion wave. The basic structure of the one-dimensional deflagration wave, which is commonly called the laminar premixed flame, is then analyzed. The mathematical technique of activation energy asymptotic analysis is introduced via the solution of this important model flame, leading to the derivation of the laminar burning velocity, which is perhaps one of the most important parameters in flame theory. Experimental and computational results on the laminar burning velocity, as well as the flame structure, are then presented and discussed, leading to an appreciation of the unified interpretation of the physical and chemical aspects of the flame structure.

In Chapter 8, the critical phenomena of ignition, extinction, flammability, and flame stabilization are discussed, with emphasis on the mechanisms and criteria governing their occurrence. The influence of residence time, heat loss, and chemical reaction mechanisms are examined.

Having been exposed to the analysis of the reaction zone structure in premixed flames, in Chapter 9 we return to nonpremixed flames to study the asymptotic structure of its reaction zone. A general classification of the nonpremixed flame is presented, and the various flame regimes analyzed, resulting in explicit criteria governing the ignition and extinction of these flames. In particular, we shall show that a special class of the nonpremixed flame turns out to be the premixed flame, and as such provide a unifying formulation and interpretation of premixed and nonpremixed flames.

In Chapter 10, we study the effects of aerodynamics on flames due to flow nonuniformity, flame curvature, and flame–flow unsteadiness, with emphasis on premixed flames because of the richness of their responses. The unequal nature of the diffusivities of heat and species is shown to cause local deficits or surpluses in the total enthalpy and, consequently, modifications of the local flame temperature and propagation velocity. Intrinsic flamefront instabilities as a result of these aerodynamic and diffusional effects are then discussed.

In Chapters 11–14, we study combustion phenomena in the four major classes of fluid flows, namely turbulent flows, boundary-layer flows, two-phase flows, and supersonic flows. For turbulent combustion we shall discuss the general concepts and modeling approaches of turbulent flows and flames, and how concepts of laminar flames can be usefully applied to the understanding of the structure of turbulent flames. For boundary-layer combustion, the concept of flow similitude will be discussed first, followed by analyses of various boundary- and mixing-layer flows that are of particular interest to combustion. The chapter on two-phase combustion will cover various aspects of droplet and particle burning, to be followed by studies on spray combustion. The discussion on supersonic combustion will be conducted in two parts, namely on flows with weak disturbances including sound waves, and on the dynamics and structure of detonation waves.

0.5. LITERATURE SOURCES

The following is a list of books and journals for reference and further study.

Introductory Texts

Bradley, J. 1979. *Flames and Combustion Phenomena*. Chapman and Hall.

Chigier, N. 1981. *Energy, Combustion, and Environment*. McGraw-Hill.

Kanury, A. M. 1975. *Introduction to Combustion Phenomena*. Gordon and Breach.

Strahle, W. C. 1993. *Introduction to Combustion*. Gordon and Breach.

Turns, S. R. 2000. *An Introduction to Combustion: Concepts and Applications*. McGraw-Hill.

Intermediate Texts

Borman, G. L. & Ragland, K. W. 1998. *Combustion Engineering*. McGraw-Hill.

Glassman, I. & Yetter, R. A. 2008. *Combustion*, 4th ed. Academic.

Griffith, J. F. & Barnard, J. A. 1995. *Flame and Combustion*, 3rd ed. CRC Press.

Kee, R. J., Coltrin, M. E. & Glarborg, P. 2003. *Chemically Reacting Flows: Theory and Practice*. Wiley-Interscience.

Kuo, K. K. 2005. *Principles of Combustion*, 2nd ed. John Wiley.

Lewis, B. & von Elbe, G. 1987. *Combustion, Flames, and Explosions of Gases*. Academic.

Liñán, A. & William, F. A. 1993. *Fundamental Aspects of Combustion*. Oxford.

Penner, S. S. 1957. *Chemistry Problems in Jet Propulsion*. Pergamon.

Poinsot, T. & Veynante, D. 2005. *Theoretical and Numerical Combustion*, 2nd ed. Edwards.

Puri, I. K. 1993. *Environmental Implications of Combustion Processes*. CRC Press.

Rosner, D. E. 1986. *Transport Processes in Chemically Reacting Flow Systems*. Butterworth.

Strehlow, R. A. 1984. *Fundamentals of Combustion*. McGraw-Hill.

Toong, T. Y. 1983. *Combustion Dynamics: The Dynamics of Chemically Reacting Fluids*. McGraw-Hill.

Warnatz, J., Maas, U. & Dibble, R. W. 2006. *Combustion: Physical and Chemical Fundamentals, Modeling and Simulation, Experiments, Pollutant Formation*. Springer.

Advanced Texts

Buckmaster, J. D. & Ludford, G. S. S. 1982. *Theory of Laminar Flames*. Cambridge.

Peters, N. 2000. *Turbulent Combustion*. Cambridge.

Williams, F. A. 1985. *Combustion Theory*, 2nd ed. Benjamin-Cummings.
Zel'dovich, Ya. B., Barenblatt, G. I., Librovich, V. B. & Makhviladze, G. M. 1985. *The Mathematical Theory of Combustion and Explosions*. Plenum.

Specialized Texts and Monographs

Fenimore, C. P. 1964. *Chemistry in Premixed Flames*. Pergamon.
Fickett, W. & Davis, W. C. 1979. *Detonations*. University of California.
Fristrom, R. M. & Westenberg, A. A. 1965. *Flame Structure*. McGraw-Hill.
Gaydon, A. G. & Wolfhard, H. G. 1970. *Flames*. Chapman and Hall.
Kohse-Höinghaus, K. & Jeffries, J. B. 2002. *Applied Combustion Diagnostics*. Taylor and Francis.
Lawton, J. & Weinberg F. J. 1969. *Electrical Aspects of Combustion*. Clarendon.
Oran, E. S. & Boris, J. P. 2001. *Numerical Simulation of Reactive Flow*. Cambridge.
Ross, H. 2001. *Microgravity Combustion*. Academic.
Sirignano, W. A. 1999. *Fluid Dynamics and Transport of Droplets and Sprays*. Cambridge.

Journals

Acta Astronautica
AIAA Journal
ASME Transaction: Journal of Heat Transfer
ASME Transaction: Journal of Engineering for Power
Combustion and Flame
Combustion, Explosion, and Shock Waves
Combustion Science and Technology
Combustion Theory and Modeling
International Journal of Chemical Kinetics
International Journal of Heat and Mass Transfer
Journal of Chemical Physics
Journal of Fluid Mechanics
Journal of Physical Chemistry
Journal of Propulsion and Power
Physics of Fluids
Progress in Energy and Combustion Science
Proceedings of the Combustion Institute

1 Chemical Thermodynamics

Chemical thermodynamics is concerned with the description of the equilibrium states of reacting multicomponent systems. Compared to single-component systems in which only thermal equilibrium is required, we are now also interested in chemical equilibrium among all of the components. Since practical combustors are designed to ensure that fuel and air have sufficient residence time to mix, react, and attain thermodynamic equilibrium, global performance parameters such as the heat and power output can frequently be estimated by assuming thermodynamic equilibrium of the combustion products. Thus, the scientific elements of a large part of combustion engineering are covered by the subject of this chapter.

In Section 1.1, we introduce the concept of stoichiometry, which sensitively controls the temperature of a combustion process. In Section 1.2, the criterion for chemical equilibrium is derived and the methodology for calculating the equilibrium composition of a mixture, for given pressure and temperature, is discussed. We then apply this calculation procedure to hydrocarbon–air mixtures as an example in Section 1.3. In Section 1.4, energy conservation is considered, which enables the simultaneous determination of the final composition and temperature of a reactive mixture after equilibrium is established. This final temperature, called the adiabatic flame temperature, T_{ad}, is perhaps the most important parameter of a reactive mixture, indicating not only its potential to deliver heat and power, but also the rates of progress of the various chemical reactions constituting the entire combustion process.

Further exposition of this subject can be found in, for example, Guggenheim (1957), Glasstone (1958), Williams (1985), and Reid, Prausnitz, and Sherwood (1987).

1.1. PRACTICAL REACTANTS AND STOICHIOMETRY

1.1.1. Practical Reactants

For most of the practical combustion devices generating heat and power, the oxidizer is simply the oxygen in air. This somewhat obvious fact underlies the attractiveness of these devices in that the oxidizer not only is "free," it also does not need to be carried along in a transportation vehicle or stored in a power-generation plant.

For practical calculations air can be considered to consist of 21 percent oxygen and 79 percent nitrogen in molar concentrations, implying that for every mole of oxygen there are 3.76 moles of nitrogen. Therefore, we can write

$$\text{Air} = 0.21\text{O}_2 + 0.79\text{N}_2 \quad \text{or} \quad 4.76\text{Air} = \text{O}_2 + 3.76\text{N}_2.$$

Since most of air is nitrogen, which is basically inert as far as the bulk chemical heat release is concerned, the combustion temperature and, hence, intensity are reduced because of the expenditure of thermal energy used to heat it up during the course of burning. Therefore, for applications requiring intense burning, either oxygen-enriched air or even pure oxygen is used.

Fuels can be classified according to their physical states under normal conditions. Representative components of gaseous fuels are hydrogen (H_2), carbon monoxide (CO), and the light hydrocarbons (HC). Liquid fuels are usually the heavier hydrocarbons and alcohols, while solid fuels include carbon, coal, wood, metals, etc. In Chapter 3 the chemical properties of some of these fuels will be discussed.

1.1.2. Stoichiometry

The combustion intensity between a fuel and an oxidizer depends on their relative concentrations. When their concentration ratio is chemically correct in that all the reactants can be totally consumed in the reaction, then the combustion intensity is close to the highest and we call this mode of burning stoichiometric combustion. An example is methane reacting with oxygen, producing only water and carbon dioxide,

$$\text{CH}_4 + 2\text{O}_2 \rightarrow 2\text{H}_2\text{O} + \text{CO}_2.$$

The above reaction is said to be stoichiometrically balanced.

To measure the relative concentrations of fuel and oxidizer in a mixture, we define a fuel–oxidizer ratio, F/O, as the ratio of the mass of fuel to the mass of oxidizer in the mixture. Similarly, a fuel–air ratio, F/A, can also be defined. To indicate the deviation of a mixture's concentration from stoichiometry, an equivalence ratio ϕ is defined as

$$\phi = \frac{(F/O)}{(F/O)_{\text{st}}}, \tag{1.1.1}$$

where the subscript "st" designates the stoichiometric state. Thus $\phi < 1, = 1$, and > 1 respectively correspond to fuel-lean (or simply lean), stoichiometric, and fuel-rich combustion. Note that the oxidizer-to-fuel equivalence ratio is sometimes used, which is simply the reciprocal of the present ϕ.

The definition of ϕ is asymmetrical relative to fuel-lean ($0 < \phi < 1$) and fuel-rich ($1 < \phi < \infty$) situations. Thus, when graphically expressing certain combustion properties as functions of ϕ, the slopes of these curves would be more gentle on the rich side. Caution should therefore be exercised in not interpreting these "slow decay" behaviors as being physically or chemically meaningful because they could simply be consequences of the definition of ϕ. In view of this concern, we introduce a normalized, symmetrical definition,

$$\Phi = \frac{\phi}{1+\phi}, \tag{1.1.2}$$

such that $0 < \Phi < 0.5$, $\Phi = 0.5$, and $0.5 < \Phi < 1$ respectively designate fuel-lean, stoichiometric, and fuel-rich mixtures. In the rest of the text, we shall nevertheless still use ϕ in most of the discussions because it is the representation one finds in the literature.

1.2. CHEMICAL EQUILIBRIUM

1.2.1. First and Second Laws

The first law of thermodynamics states that for a closed system, which is one with a fixed mass, the heat δQ added to the system in an infinitesimal process is used to increase its internal energy by dE and to perform a certain amount of work, δW. Thus

$$\delta Q = dE + \delta W. \tag{1.2.1}$$

Note that E is a property of the system, and hence, dE an exact differential of the process, whereas δQ and δW are path-dependent quantities. Since we are developing the thermodynamics of equilibrium chemical systems, we need consider only the pdV work done by volume change, where p is the pressure, and V is the total volume. Thus Eq. (1.2.1) can be written as

$$\delta Q = dE + pdV. \tag{1.2.2}$$

The second law of thermodynamics states that there exists a quantity S, called the entropy, which has the property that for an infinitesimal process in a closed system,

$$TdS \geq \delta Q, \tag{1.2.3}$$

where T is the temperature. For all natural processes the inequality holds. Equality holds only if the process is reversible.

Relations (1.2.2) and (1.2.3) then imply that

$$dE \leq TdS - pdV. \tag{1.2.4}$$

1.2.2. Thermodynamic Functions

Based on the functional form of (1.2.4), we can define a thermodynamic function E as follows. Since we have assumed that there is only one mode of reversible work, pdV, the state of a single-component thermodynamic system in equilibrium can be completely characterized by two independent variables, say S and V as indicated in (1.2.4). For a multicomponent system, the composition also needs to be specified, say by the number of moles of the ith species, N_i. The same species in different phases is treated as different thermodynamic species by considering phase transition as a chemical reaction. Therefore, we can write, in general, that

$$E = E(S, V, N_i),$$

which can be differentiated to yield

$$dE = \left(\frac{\partial E}{\partial S}\right)_{V,N_i} dS + \left(\frac{\partial E}{\partial V}\right)_{S,N_i} dV + \sum_{i=1}^{N} \left(\frac{\partial E}{\partial N_i}\right)_{S,V,N_{j(j\neq i)}} dN_i. \quad (1.2.5)$$

By comparing Eq. (1.2.5) with (1.2.4), it is clear that the temperature T and the pressure p can be defined as

$$T = \left(\frac{\partial E}{\partial S}\right)_{V,N_i}, \qquad p = -\left(\frac{\partial E}{\partial V}\right)_{S,N_i}.$$

If we further define a chemical potential $\bar{\mu}_i$ as

$$\bar{\mu}_i = \left(\frac{\partial E}{\partial N_i}\right)_{S,V,N_{j(j\neq i)}},$$

then Eq. (1.2.5) can be written as

$$dE = TdS - pdV + \sum_{i=1}^{N} \bar{\mu}_i dN_i, \quad (1.2.6)$$

where the overbar indicates a partial molar quantity. The corresponding symbol without the overbar indicates the same quantity on the basis of per unit mass.

Using Eq. (1.2.6), analogous forms of energy can be obtained for the enthalpy $H = E + pV$, Helmholtz function $A = E - TS$, and Gibbs function $G = H - TS$ as

$$dH = TdS + Vdp + \sum_{i=1}^{N} \bar{\mu}_i dN_i, \quad (1.2.7)$$

$$dA = -SdT - pdV + \sum_{i=1}^{N} \bar{\mu}_i dN_i, \quad (1.2.8)$$

$$dG = -SdT + Vdp + \sum_{i=1}^{N} \bar{\mu}_i dN_i, \quad (1.2.9)$$

with

$$\bar{\mu}_i = \left(\frac{\partial H}{\partial N_i}\right)_{S,p,N_{j(j\neq i)}} = \left(\frac{\partial A}{\partial N_i}\right)_{T,V,N_{j(j\neq i)}} = \left(\frac{\partial G}{\partial N_i}\right)_{T,p,N_{j(j\neq i)}}. \quad (1.2.10)$$

Since the partial molar value of a molar-based extensive property Ψ is defined as $(\partial\Psi/\partial N_i)_{T,p,N_{j(j\neq i)}}$, we readily see that $\bar{\mu}_i$ is simply the partial molar Gibbs function $\bar{g}_i$.

1.2.3. Criterion for Chemical Equilibrium

Comparing (1.2.4) and Eq. (1.2.6), we have

$$\sum_{i=1}^{N} \bar{\mu}_i dN_i \leq 0 \tag{1.2.11}$$

for a closed system. In (1.2.11) the inequality holds for a natural process while the equality applies when the system is in equilibrium. Thus, the equilibrium criterion for a multicomponent system is

$$\sum_{i=1}^{N} \bar{\mu}_i dN_i = 0. \tag{1.2.12}$$

The above criterion is general, and can be similarly derived by considering H, A, or G in the manner above. It is, therefore, not restricted to a constant temperature or a constant pressure process.

Equation (1.2.12) can be made more specific. Consider a chemical reaction given by

$$\sum_{i=1}^{N} \nu_i' \mathrm{M}_i \rightleftharpoons \sum_{i=1}^{N} \nu_i'' \mathrm{M}_i, \tag{1.2.13}$$

where M_i is the chemical symbol for the ith species and ν_i the corresponding molar concentration coefficient. Then, from element conservation, we have

$$\frac{dN_i}{\nu_i'' - \nu_i'} = \frac{dN_j}{\nu_j'' - \nu_j'} = d\lambda, \tag{1.2.14}$$

or

$$dN_i = (\nu_i'' - \nu_i')d\lambda, \tag{1.2.15}$$

where λ is a parameter indicating the progress in reaction.

Substituting Eq. (1.2.15) into Eq. (1.2.12), we have

$$\left[\sum_{i=1}^{N} \bar{\mu}_i (\nu_i'' - \nu_i')\right] d\lambda = 0. \tag{1.2.16}$$

Since $d\lambda$ is arbitrary, the criterion for chemical equilibrium is given by

$$\sum_{i=1}^{N} \bar{\mu}_i (\nu_i'' - \nu_i') = 0, \tag{1.2.17}$$

which shows that at equilibrium the sum of the chemical potentials of all the reactants is equal to that of the products.

1.2.4. Phase Equilibrium

The equilibrium criterion Eq. (1.2.17) derived above can be readily used to describe phase equilibrium. Consider for example a simple transition from phase ℓ to phase g without molecular decomposition, $\mathrm{M}_\ell \to \mathrm{M}_g$. Here we have $\nu_\ell' = 1$, $\nu_g' = 0$, $\nu_\ell'' = 0$, and $\nu_g'' = 1$, where the subscripts ℓ and g can designate any two phases. Equation (1.2.17) then shows that, at phase equilibrium, we have

$$\bar{\mu}_\ell(T, p) = \bar{\mu}_g(T, p), \tag{1.2.18}$$

which implies that

$$d\bar{\mu}_\ell = d\bar{\mu}_g. \tag{1.2.19}$$

Equation (1.2.18) relates the system temperature and pressure at equilibrium.

This phase equilibrium relation can be developed further. Integrating Eq. (1.2.9) through a process in which the size of the system increases while the intensive properties such as T and p remain unchanged, and with $G = 0$ for $N_i = 0$, we obtain

$$G = \sum_{i=1}^{N} \bar{\mu}_i N_i. \tag{1.2.20}$$

Substituting Eq. (1.2.20) back into Eq. (1.2.9) yields the Gibbs–Duhem equation

$$\sum_{i=1}^{N} N_i d\bar{\mu}_i = -SdT + Vdp, \tag{1.2.21}$$

which is a general result. Expressing Eq. (1.2.21) for a single species, and applying it to the two phases, ℓ and g, with the same T and p, we have

$$N_\ell d\bar{\mu}_\ell = -S_\ell dT + V_\ell dp, \qquad N_g d\bar{\mu}_g = -S_g dT + V_g dp. \tag{1.2.22}$$

For phase equilibrium, $d\bar{\mu}_\ell = d\bar{\mu}_g$, and the relations in (1.2.22) yield

$$\frac{dp}{dT} = \frac{\bar{s}_g - \bar{s}_\ell}{\bar{v}_g - \bar{v}_\ell}, \tag{1.2.23}$$

where $\bar{s} = S/N$ and $\bar{v} = V/N$. Equation (1.2.23) is the general Clapeyron relation for phase equilibrium. Since $\bar{g} = \bar{h} - T\bar{s}$, the numerator in Eq. (1.2.23) can be written as

$$\bar{s}_g - \bar{s}_\ell = \frac{\bar{h}_g - \bar{h}_\ell}{T} - \frac{\bar{g}_g - \bar{g}_\ell}{T}. \tag{1.2.24}$$

But $\bar{g}_g = \bar{g}_\ell$ and $(\bar{h}_g - \bar{h}_\ell)$ is simply the heat of transition per mole, $\bar{q}$. Thus, Eq. (1.2.23) becomes

$$\frac{dp}{dT} = \frac{\bar{q}/T}{\bar{v}_g - \bar{v}_\ell}. \tag{1.2.25}$$

If we now specialize ℓ to be a condensed phase and g a gas phase, then $\bar{v}_g \gg \bar{v}_\ell$ because the volume change during gasification is very large, typically by a factor of 10^3 at atmospheric pressure. By further assuming the ideal gas behavior such that $pV_g = N_g R^o T$, where $R^o = 1.987$ cal/mole-K is the universal gas constant, we have

$$\frac{dp}{dT} = \frac{p\bar{q}_v}{R^o T^2}. \tag{1.2.26}$$

Table 1.1. Heats of vaporization and normal boiling temperatures

Formula	Name	$\bar{q}_v^*$ (kcal/mole)	T_b (K)
CCl_4	Carbon Tetrachloride	7.13	349.9
$CHCl_3$	Trichloromethane	6.99	334.3
CH_2Cl_2	Dichloromethane	6.71	313.0
C_4H_{10}	*n*-Butane	5.35	272.7
C_5H_{12}	*n*-Pentane	6.16	309.2
C_6H_{14}	*n*-Hexane	6.90	341.9
C_7H_{16}	*n*-Heptane	7.59	371.6
C_8H_{18}	*n*-Octane	8.22	398.8
C_8H_{18}	Isooctane	7.36	372.4
C_9H_{20}	*n*-Nonane	8.82	424.0
$C_{10}H_{22}$	*n*-Decane	9.26	447.3
$C_{12}H_{26}$	*n*-Dodecane	10.42	489.5
$C_{16}H_{34}$	*n*-Hexadecane	12.24	560.0
$C_{20}H_{42}$	*n*-Eicosane	13.78	617.0
C_6H_6	Benzene	7.34	353.2
C_7H_8	Toluene	7.99	383.8
C_8H_{10}	*o*-Xylene	8.66	417.6
CH_4O	Methanol	8.42	337.7
C_2H_6O	Ethanol	9.22	351.4
C_3H_8O	1-Propanol	9.90	370.3
C_3H_8O	Isopropyl Alcohol	9.52	355.4
$C_4H_{10}O$	1-Butanol	10.35	390.9
C_3H_6O	Acetone	6.95	329.2
H_2O	Water	9.72	373.2

* Measured at T_b.

Source: Lide, D. R. 1990–1991. *CRC Handbook of Chemistry and Physics*, 71th ed., CRC Press, Boca Raton.

Reid, R. C., Prausnitz, J. M. & Poling, B. E. 1987. *The Properties of Gases and Liquids*, 4th ed., McGraw-Hill, New York.

Equation (1.2.26) can be integrated to yield a relation for the vapor pressure p of the substance at temperature T,

$$p = p_{\text{ref}} \exp\left(-\int_T^{T_{\text{ref}}} \frac{\bar{q}_v}{R^o T^2} dT\right), \tag{1.2.27}$$

where p_{ref} is the vapor pressure at a reference temperature T_{ref}, and we have written $\bar{q}_v = \bar{q}$ for subsequent notational purpose, with the subscript v designating vaporization. Equation (1.2.27) is the Clausius–Clapeyron relation for vaporization. By assuming $\bar{q}_v$ = constant, the integral in Eq. (1.2.27) can be readily evaluated, yielding

$$p = p_{\text{ref}} \exp\left[-\frac{\bar{q}_v}{R^o}\left(\frac{1}{T} - \frac{1}{T_{\text{ref}}}\right)\right]. \tag{1.2.28}$$

Frequently the reference state is taken to be that of boiling in the standard state such that T_{ref} is the (normal) boiling point T_b evaluated at $p_{\text{ref}} = 1$ atm. Table 1.1 lists the normal boiling point T_b and the latent heat of vaporization $\bar{q}_v$ at T_b for a number of liquids.

1.2.5. Equilibrium Constants

The criterion for chemical equilibrium, Eq. (1.2.17), can be further developed as follows. For a mixture of ideal gases, the chemical potential of its ith component is given by

$$\bar{\mu}_i(T, p_i) = \bar{\mu}_i^o(T) + R^o T \ln(p_i/p^o), \tag{1.2.29}$$

where $\bar{\mu}_i^o(T) = \bar{g}_i^o(T)$ is the molar standard free energy of i determined at T and a reference pressure p^o. Substituting $\bar{\mu}_i$ into the equilibrium criterion (1.2.17), setting the reference pressure p^o at 1 atmosphere, and expressing p_i in units of atm, Eq. (1.2.17) can be rearranged as

$$\prod_{i=1}^{N} p_i^{(\nu_i''-\nu_i')} = K_p(T), \tag{1.2.30}$$

where

$$K_p(T) = \exp\left\{-\left[\sum_{i=1}^{N}(\nu_i''-\nu_i')\mu_i^o(T)\right]\Big/(R^o T)\right\} \tag{1.2.31}$$

is called the equilibrium constant for partial pressure for the reaction (1.2.13). Equation (1.2.30) relates the mixture concentration, which is proportional to p_i, with the temperature T, at equilibrium. Since $K_p(T)$ is only a function of temperature, it can be tabulated for calculations for a given reaction.

Equation (1.2.31) provides an explicit expression for the evaluation of $K_p(T)$ for a given reaction by simply looking up values of $\mu_i^o(T)$ in tables such as the JANAF (Joint Army–Navy–Air Force) Tables (Stull & Prophet 1971). However, a more fundamental procedure through which $K_p(T)$ can be computed, also by using the JANAF Tables, is to employ the concept of the equilibrium constant for formation. That is, for each of the species M_i in the general reaction scheme (1.2.13), we can write a reaction for the formation of one mole of M_i from its elements $M_{i,j}^o$ in their standard states (e.g., gas, liquid, solid, crystalline); the standard state of an element is the form that is stable at room temperature and atmospheric pressure, such as Ar(g), O_2(g), and graphite C(gr), with $\bar{\mu}_{i,j}^o \equiv 0$. The formation reaction can thus be expressed as

$$\sum_{j=1}^{L} \nu_{i,j}' M_{i,j}^o \rightleftharpoons M_i, \tag{1.2.32}$$

where L is the number of elements. We can then define an equilibrium constant for (1.2.32) as

$$K_{p,i}^o(T) = \exp[-\bar{\mu}_i^o(T)/(R^o T)], \tag{1.2.33}$$

which depends only on the properties of species i and the elements constituting it. Thus $K_p(T)$ for the entire reaction (1.2.13) can be expressed as

$$K_p(T) = \exp\left\{-\left[\sum_{i=1}^{N}(\nu_i'' - \nu_i')\bar{\mu}_i^o(T)\right] \Big/ (R^o T)\right\}$$
$$= \prod_{i=1}^{N}\{\exp[-\bar{\mu}_i^o(T)/(R^o T)]\}^{(\nu_i''-\nu_i')} = \prod_{i=1}^{N}\left[K_{p,i}^o(T)\right]^{(\nu_i''-\nu_i')} \quad (1.2.34)$$

By taking log on both sides of Eq. (1.2.34), we have

$$\log\left[K_p(T)\right] = \sum_{i=1}^{N}(\nu_i'' - \nu_i')\log\left[K_{p,i}^o(T)\right]. \quad (1.2.35)$$

Values of $\log[K_{p,i}^o(T)]$ are listed in the JANAF Tables and are reproduced for some selected species in Table 1.2. Since the JANAF Tables are periodically updated, the most recent edition should be used for quantitative accuracy.

As examples, let us calculate the K_p for the reaction, $CO_2 + H_2 \rightleftharpoons CO + H_2O$, at 1,000 K. From Eq. (1.2.35) and Table 1.2 we have

$$\log(K_p) = -\log\left(K_{p,CO_2}^o\right) - \log\left(K_{p,H_2}^o\right) + \log\left(K_{p,CO}^o\right) + \log\left(K_{p,H_2O}^o\right),$$
$$= -20.680 - 0 + 10.459 + 10.062 = -0.159,$$

which yields $K_p = 0.693$. A similar calculation for the reaction, $CO_2 + C \rightleftharpoons 2CO$, yields $K_p = 1.730$ at $T = 1{,}000$ K.

We now discuss some miscellaneous aspects regarding the equilibrium constant. First, sometimes it is more convenient to work with the concentration c_i (moles per unit volume) instead of the partial pressure p_i. Since $p_i = c_i R^o T$, an equilibrium constant for concentration can be defined as

$$K_c(T) = \prod_{i=1}^{N} c_i^{(\nu_i''-\nu_i')} = K_p(T)/(R^o T)^{\sum_{i=1}^{N}(\nu_i''-\nu_i')}. \quad (1.2.36)$$

The presence of inerts in the system does not affect Eqs. (1.2.31) and (1.2.36). However, it needs to be accounted for when p_i is related to either the molar fraction X_i or the mass fraction Y_i, that is,

$$X_i = \frac{p_i}{p_{\text{total}}} = \frac{p_i}{\sum p_j + p_{\text{inert}}}, \qquad Y_i = \frac{X_i W_i}{\sum X_j W_j + X_{\text{inert}} W_{\text{inert}}},$$

where W_i is the molecular weight of i, and the summation is performed over all chemically active species.

1.2.6. Equilibrium Constants in the Presence of Condensed Phases

The equilibrium constants just derived assume all species are gaseous obeying the ideal gas law. However, it is possible that some of the reacting species may exist both in the gas phase as well as the condensed phase. An example is the formation of carbon particles during the combustion of hydrocarbon fuels. Since the condensed-phase species do not exert any partial pressure, the expression for the equilibrium constant

Table 1.2. Equilibrium constants for formation, $\mathrm{Log}_{10}[K^o_{p,i}(T)]$

T(K)	O	H	OH	H_2O	N	NO	C(g)	CO	CO_2	CH_4
0	$-\infty$	$-\infty$	$-\infty$	∞	$-\infty$	$-\infty$	$-\infty$	∞	∞	∞
100	−126.730	−110.973	−19.438	123.600	−243.615	−46.453	−365.693	62.809	205.645	33.656
200	−61.988	−54.327	−9.350	60.792	−120.422	−22.929	−179.157	33.566	102.922	15.198
298	−40.602	−35.616	−6.005	40.048	−79.812	−15.171	−117.605	24.029	69.095	8.902
300	−40.332	−35.380	−5.963	39.786	−79.301	−15.073	−116.830	23.910	68.670	8.822
400	−29.472	−25.878	−4.265	29.240	−58.713	−11.142	−85.612	19.109	51.540	5.500
500	−22.939	−20.160	−3.246	22.886	−46.344	−8.783	−66.856	16.235	41.260	3.429
600	−18.573	−16.338	−2.568	18.633	−38.087	−7.210	−54.342	14.318	34.405	2.001
700	−15.448	−13.600	−2.085	15.583	−32.182	−6.086	−45.397	12.946	29.506	0.951
800	−13.101	−11.541	−1.724	13.289	−27.749	−5.243	−38.687	11.914	25.830	0.146
900	−11.272	−9.935	−1.444	11.498	−24.297	−4.587	−33.467	11.108	22.970	−0.493
1000	−9.806	−8.647	−1.222	10.062	−21.532	−4.062	−29.291	10.459	20.680	−1.011
1100	−8.606	−7.590	−1.041	8.883	−19.269	−3.633	−25.875	9.926	18.806	−1.440
1200	−7.604	−6.707	−0.890	7.899	−17.380	−3.275	−23.029	9.479	17.243	−1.801
1300	−6.755	−5.959	−0.764	7.064	−15.781	−2.972	−20.621	9.099	15.920	−2.107
1400	−6.027	−5.315	−0.656	6.347	−14.410	−2.712	−18.558	8.771	14.785	−2.372
1500	−5.395	−4.757	−0.563	5.725	−13.220	−2.487	−16.770	8.485	13.801	−2.602
1600	−4.841	−4.267	−0.482	5.180	−12.178	−2.290	−15.207	8.234	12.940	−2.803
1700	−4.353	−3.834	−0.410	4.699	−11.258	−2.116	−13.829	8.011	12.180	−2.981
1800	−3.918	−3.448	−0.347	4.279	−10.440	−1.962	−12.604	7.811	11.504	−3.139
1900	−3.528	−3.103	−0.291	3.886	−9.708	−1.823	−11.508	7.631	10.898	−3.281
2000	−3.177	−2.791	−0.240	3.540	−9.048	−1.699	−10.523	7.469	10.353	−3.408
2100	−2.860	−2.509	−0.195	3.227	−8.451	−1.586	−9.632	7.321	9.860	−3.523
2200	−2.571	−2.252	−0.153	2.942	−7.908	−1.484	−8.823	7.185	9.411	−3.627
2300	−2.307	−2.016	−0.116	2.682	−7.412	−1.391	−8.084	7.061	9.001	−3.722
2400	−2.065	−1.801	−0.082	2.443	−6.957	−1.305	−7.407	6.946	8.625	−3.809
2500	−1.842	−1.602	−0.050	2.224	−6.538	−1.227	−6.785	6.840	8.280	−3.889
2600	−1.636	−1.418	−0.021	2.021	−6.151	−1.154	−6.211	6.741	7.960	−3.962
2700	−1.445	−1.248	0.005	1.833	−5.793	−1.087	−5.680	6.649	7.664	−4.030
2800	−1.268	−1.089	0.030	1.658	−5.460	−1.025	−5.188	6.563	7.388	−4.093
2900	−1.103	−0.942	0.053	1.495	−5.149	−0.967	−4.729	6.483	7.132	−4.152
3000	−0.949	−0.804	0.074	1.343	−4.860	−0.913	−4.302	6.407	6.892	−4.206
3100	−0.805	−0.675	0.094	1.201	−4.589	−0.863	−3.902	6.336	6.668	−4.257
3200	−0.669	−0.554	0.112	1.067	−4.334	−0.815	−3.527	6.269	6.458	−4.304
3300	−0.542	−0.440	0.129	0.942	−4.095	−0.771	−3.176	6.206	6.260	−4.349
3400	−0.422	−0.332	0.145	0.824	−3.870	−0.729	−2.845	6.145	6.074	−4.391
3500	−0.310	−0.231	0.160	0.712	−3.658	−0.690	−2.534	6.088	5.898	−4.430
3600	−0.203	−0.135	0.174	0.607	−3.457	−0.653	−2.240	6.034	5.732	−4.467
3700	−0.102	−0.045	0.188	0.507	−3.268	−0.618	−1.962	5.982	5.574	−4.503
3800	−0.006	0.041	0.200	0.413	−3.088	−0.585	−1.699	5.933	5.425	−4.536
3900	0.084	0.123	0.212	0.323	−2.917	−0.554	−1.449	5.886	5.283	−4.568
4000	0.170	0.200	0.223	0.238	−2.754	−0.524	−1.213	5.841	5.149	−4.598
4100	0.252	0.274	0.234	0.157	−2.600	−0.496	−0.988	5.798	5.020	−4.626
4200	0.331	0.344	0.244	0.079	−2.452	−0.470	−0.774	5.756	4.898	−4.653
4300	0.405	0.411	0.253	0.005	−2.312	−0.444	−0.570	5.717	4.781	−4.679
4400	0.476	0.475	0.262	0.065	−2.178	−0.420	−0.375	5.679	4.670	−4.704
4500	0.544	0.536	0.270	0.133	−2.049	−0.397	−0.189	5.642	4.563	−4.727
4600	0.609	0.595	0.278	0.197	−1.926	−0.375	−0.012	5.607	4.460	−4.750
4700	0.672	0.651	0.286	0.259	−1.808	−0.354	0.158	5.573	4.362	4.772
4800	0.731	0.705	0.293	0.319	−1.696	−0.333	0.321	5.540	4.268	−4.793
4900	0.789	0.756	0.300	0.376	−1.587	−0.314	0.477	5.508	4.178	−4.813
5000	0.844	0.806	0.307	0.430	−1.483	−0.296	0.626	5.477	4.091	−4.832

Sources: JANAF Tables. *Journal of Physical and Chemical Reference Data*, v.3, no.2; v.4, no.1; v.7, no.3; v.11, no.3.

has to be modified. The difference in formulation can be demonstrated simply by the following hypothetical example involving the oxidation of solid carbon,

$$\nu'_{C(s)}C(s) + \nu'_{O_2}O_2 \rightleftharpoons \nu''_{C(g)}C(g) + \nu''_{CO_2}CO_2. \tag{1.2.37}$$

Applying the general criterion for equilibrium, Eq. (1.2.17), we have

$$\nu''_{C(g)}\bar{\mu}_{C(g)} + \nu''_{CO_2}\bar{\mu}_{CO_2} = \nu'_{C(s)}\bar{\mu}_{C(s)} + \nu'_{O_2}\bar{\mu}_{O_2}. \tag{1.2.38}$$

Substituting Eq. (1.2.29) into Eq. (1.2.38) for the gaseous species, we have

$$\begin{aligned} &\ln\left\{[p_{C(g)}]^{\nu''_{C(g)}}[p_{CO_2}]^{\nu''_{CO_2}}/[p_{O_2}]^{\nu'_{O_2}}\right\} \\ &= -\left[\nu''_{C(g)}\bar{\mu}^o_{C(g)} + \nu''_{CO_2}\bar{\mu}^o_{CO_2} - \nu'_{C(s)}\bar{\mu}_{C(s)} - \nu'_{O_2}\bar{\mu}^o_{O_2}\right]\Big/ R^oT, \end{aligned} \tag{1.2.39}$$

where $\bar{\mu}_{C(s)}$ is the chemical potential of solid carbon at the prevailing pressure and temperature of the reaction. Since the chemical potential of a condensed species is quite insensitive to pressure variations, it is customary to simply replace $\bar{\mu}_{C(s)}$ by $\bar{\mu}^o_{C(s)}$. Equation (1.2.39) then becomes

$$\left\{[p_{C(g)}]^{\nu''_{C(g)}}[p_{CO_2}]^{\nu''_{CO_2}}/[p_{O_2}]^{\nu'_{O_2}}\right\} = K'_p(T), \tag{1.2.40}$$

where

$$K'_p(T) = \exp\left\{-\left[\nu''_{C(g)}\bar{\mu}^o_{C(g)} + \nu''_{CO_2}\bar{\mu}^o_{CO_2} - \nu'_{C(s)}\bar{\mu}^o_{C(s)} - \nu'_{O_2}\bar{\mu}^o_{O_2}\right]\Big/ R^oT\right\}. \tag{1.2.41}$$

Thus the final equilibrium expression is similar to Eq. (1.2.30) derived for the totally gaseous system, except now there is no partial pressure for the condensed phase in Eq. (1.2.40).

1.2.7. Multiple Reactions

We now return to our study of chemical equilibrium in general. Our discussion so far has been restricted to a single reaction given by (1.2.13). However, in practically all chemically reacting systems there invariably exists a large number of chemical species and therefore reactions. The generalization of (1.2.13) to a multiple reaction scheme consisting of K reactions is

$$\sum_{i=1}^{N} \nu'_{i,k}M_i \rightleftharpoons \sum_{i}^{N} \nu''_{i,k}M_i, \quad k = 1, 2, \ldots, K. \tag{1.2.42}$$

Each of the reactions in (1.2.42) is described by its own equilibrium relation

$$\prod_{i=1}^{N} p_{i,k}^{(\nu''_{i,k}-\nu'_{i,k})} = K_{p,k}(T), \quad k = 1, 2, \ldots, K. \tag{1.2.43}$$

Thus Eq. (1.2.43) provides K relations.

1.2.8. Element Conservation

In addition to the statement of chemical equilibrium, given by Eq. (1.2.43), element conservation requires that atoms are neither created nor destroyed in chemical reactions. Thus if $\eta_{i,j}$ is the number of atoms of kind j in a molecule of species i, N_i the number of moles of i per unit volume, and $N_{j,0}$ the total number of moles of atom j per unit volume, then we have

$$\sum_{i=1}^{N} \eta_{i,j} N_i = N_{j,0}, \quad j = 1, 2, \ldots, L. \tag{1.2.44}$$

1.2.9. Restricted Equilibrium

For a system with K reactions, N species, and L elements, it is clear that with the $(K + L)$ equations provided by Eqs. (1.2.43) and (1.2.44), concentrations of the N species can be uniquely determined if

$$N = K + L. \tag{1.2.45}$$

If $N > (K + L)$, there are insufficient reactions available and the system is underdetermined. Frequently, however, there are more reactions possible such that $N < (K + L)$. In this case it is necessary to construct out of the original K equations an equivalent set of $K' < K$ equations such that $N = (K' + L)$.

As an example, consider the following hypothetical reaction scheme between oxygen and hydrogen:

$$\begin{aligned} H_2 + O_2 &\rightleftharpoons 2OH \\ 2H_2 + O_2 &\rightleftharpoons 2H_2O \\ H_2 &\rightleftharpoons 2H \\ O_2 &\rightleftharpoons 2O \\ H + OH &\rightleftharpoons H_2 + O. \end{aligned}$$

Here we have $L = 2$ (for O and H), $N = 6$ (for H_2, O_2, H_2O, H, O, and OH), and $K = 5$. Thus $N < (K + L)$. To reduce the number of reactions, we first write down the coefficient matrix of the reactions as

$$\begin{bmatrix} O & H & O_2 & H_2 & OH & H_2O \\ 0 & 0 & -1 & -1 & +2 & 0 \\ 0 & 0 & -1 & -2 & 0 & +2 \\ 0 & +2 & 0 & -1 & 0 & 0 \\ +2 & 0 & -1 & 0 & 0 & 0 \\ +1 & -1 & 0 & +1 & -1 & 0 \end{bmatrix}.$$

The above matrix is less than full rank. It can however be easily reduced to one of rank 4 given by

$$\begin{bmatrix} O & H & O_2 & H_2 & OH & H_2O \\ +2 & 0 & -1 & 0 & 0 & 0 \\ 0 & +2 & 0 & -1 & 0 & 0 \\ +1 & +1 & 0 & 0 & -1 & 0 \\ +1 & +2 & 0 & 0 & 0 & -1 \end{bmatrix}.$$

Thus, an equivalent reaction scheme, with $K' = 4$, is

$$\begin{aligned} O_2 &\rightleftharpoons 2O \\ H_2 &\rightleftharpoons 2H \\ OH &\rightleftharpoons O + H \\ H_2O &\rightleftharpoons O + 2H, \end{aligned} \tag{1.2.46}$$

while element conservation gives

$$\begin{aligned} N_{O,0} &= N_O + N_{OH} + 2N_{O_2} + N_{H_2O} \\ N_{H,0} &= N_H + N_{OH} + 2N_{H_2} + 2N_{H_2O}, \end{aligned} \tag{1.2.47}$$

where $N_{O,0}$ and $N_{H,0}$ are respectively the numbers of moles of oxygen and hydrogen in the mixture. We now have $N = (K' + L)$.

1.3. EQUILIBRIUM COMPOSITION CALCULATIONS

The derivation of the previous sections shows that by using Eq. (1.2.43) for chemical equilibrium and Eq. (1.2.44) for element conservation, we have N equations to determine the N unknown concentrations in a mixture of given temperature, pressure, and concentrations of the individual elements. To be more specific, we shall show in the following how the equilibrium composition of a hydrocarbon–air mixture can be calculated.

1.3.1. Equilibrium Composition of Hydrocarbon–Air Mixtures

To perform an equilibrium composition calculation, we first need to specify the species to be considered. While it is obvious that the more species we include, the more detailed is our knowledge of the composition, the penalty is the excessive extent of calculation. Thus, the calculation should include no more species than necessary. Generally, for energy release calculations it is not necessary to include the minor species whose concentrations are less than, say, 0.1 percent. On the other hand, if we are interested in some specific trace pollutants or radicals, then these minor species also need to be included in the calculation.

As an illustration, let us consider the oxidation of hydrocarbons in air (Penner 1958), which is probably the most important combustion system from the practical point of view. Since the mixture consists of the elements C, H, O, and N, it is often called a CHON system. As a specific example, consider the oxidation of propane in air with an equivalence ratio ϕ, yielding a product composition given by the global reaction

$$\begin{aligned} \phi C_3H_8 + 5(O_2 + 3.76N_2) \rightleftharpoons\ & N_{CO_2}CO_2 + N_{CO}CO + N_{C(g)}C(g) + N_{C(gr)}C(gr) \\ & + N_{H_2O}H_2O + N_{H_2}H_2 + N_H H + N_{OH}OH + N_{O_2}O_2 \\ & + N_O O + N_{NO}NO + N_{N_2}N_2 + N_N N. \end{aligned} \tag{1.3.1}$$

In writing (1.3.1) we have included solid carbon as a combustion product, designated by the specific state, graphite (gr). The possibility that the equilibrium system actually contains solid carbon generally can be assessed beforehand, as will be discussed later. On the other hand, we have not included propane in the product composition because it is mostly reacted and its concentration is extremely low.

Conservation equations for the four elements C, H, O, N are

$$\begin{aligned}
N_{C,0} &= 3\phi = N_{CO_2} + N_{CO} + N_{C(g)} + N_{C(gr)} \\
N_{H,0} &= 8\phi = 2N_{H_2} + 2N_{H_2O} + N_{OH} + N_H \\
N_{O,0} &= 10 = 2N_{O_2} + N_{H_2O} + 2N_{CO_2} + N_{CO} + N_{OH} + N_O + N_{NO} \\
N_{N,0} &= 37.6 = 2N_{N_2} + N_N + N_{NO}.
\end{aligned} \tag{1.3.2}$$

Since there are thirteen unknown species concentrations ($N = 13$) and four elements ($L = 4$), we need nine equilibrium constant relations ($K = 9$). These can be given by the set of nine linearly independent reactions:

$$\begin{aligned}
C(gr) + O_2 &\overset{1}{\rightleftharpoons} CO_2, & p_{CO_2} &= K'_{p,1} p_{O_2} \\
C(gr) + \tfrac{1}{2}O_2 &\overset{2}{\rightleftharpoons} CO, & p_{CO} &= K'_{p,2}\sqrt{p_{O_2}} \\
C(gr) &\overset{3}{\rightleftharpoons} C(g), & p_{C(g)} &= K'_{p,3} \\
H_2 + \tfrac{1}{2}O_2 &\overset{4}{\rightleftharpoons} H_2O, & p_{H_2O} &= K_{p,4} p_{H_2}\sqrt{p_{O_2}} \\
\tfrac{1}{2}H_2 &\overset{5}{\rightleftharpoons} H, & p_H &= K_{p,5}\sqrt{p_{H_2}} \\
\tfrac{1}{2}H_2 + \tfrac{1}{2}O_2 &\overset{6}{\rightleftharpoons} OH, & p_{OH} &= K_{p,6}\sqrt{p_{H_2} p_{O_2}} \\
\tfrac{1}{2}O_2 &\overset{7}{\rightleftharpoons} O, & p_O &= K_{p,7}\sqrt{p_{O_2}} \\
\tfrac{1}{2}N_2 + \tfrac{1}{2}O_2 &\overset{8}{\rightleftharpoons} NO, & p_{NO} &= K_{p,8}\sqrt{p_{N_2} p_{O_2}} \\
\tfrac{1}{2}N_2 &\overset{9}{\rightleftharpoons} N, & p_N &= K_{p,9}\sqrt{p_{N_2}}.
\end{aligned} \tag{1.3.3}$$

To relate the p_is of (1.3.3) with the N_is of (1.3.2), we use the ideal gas relation

$$N_i = N_t(p_i/p_t), \tag{1.3.4}$$

where p_t is the given system pressure, and $p_t = \sum p_i$ and $N_t = \sum N_i$ are to be summed over only the gaseous species. Thus substituting the relations in (1.3.3) into the expressions in (1.3.2) via Eq. (1.3.4), we have four relations to solve for the four unknowns p_{H_2}, p_{O_2}, p_{N_2}, and $N_{C(gr)}$. The solution procedure is to first guess an N_t based on, for example, some stoichiometric relation. Then p_{H_2}, p_{O_2}, p_{N_2}, and $N_{C(gr)}$ can be calculated from (1.3.2) and subsequently the remaining p_is from (1.3.3). Knowing all the p_is, a $p'_t = \sum p_i$ can be calculated. If p'_t is not equal to the given system pressure p_t, then the initial guess N_t is not correct and a new guess can be tried.

After the solution is converged, one final checking is needed. That is, in our calculation we have assumed that solid carbon is present. Thus if the final solution yields a negative value of $N_{C(gr)}$, then it is clear that solid carbon does not exist. In this case we have to repeat the formulation assuming the presence of only gaseous carbon.

An alternate approach is to first assume that solid carbon does not exist. This slightly simplifies the calculation because one less species is involved. If the calculation shows that the partial pressure of the gaseous carbon exceeds the equilibrium vapor pressure of carbon, which is given by the $C(gr) \rightleftharpoons C(g)$ reaction of (1.3.3), then condensation and solid carbon must exist. The calculation should then be carried out by allowing for its presence.

To minimize the uncertainty in guessing, it is judicious to first estimate whether solid carbon is likely to be formed. The most important parameter that controls carbon formation is the equivalence ratio ϕ because it represents the amount of oxygen available to convert carbon to either carbon monoxide or carbon dioxide. Obviously for $\phi < 1$ there exists enough oxygen for complete conversion, implying that carbon, or soot emission, is usually observed only for rich mixtures. To be more precise with the threshold ϕ_C for carbon formation, one can follow an approximate, simple, order of oxidation rule. That is, as oxygen is added to a hydrocarbon system, the oxidation of carbon and hydrogen takes place in a particular sequence in that oxygen is first utilized to convert carbon to CO. Only after all carbon is converted to CO does oxidation of hydrogen to form H_2O occur, which is then followed by the oxidation of CO to CO_2.

Based on this rule, one expects that carbon is formed when the mixture is sufficiently fuel rich such that complete conversion of carbon to CO is not possible. For example, for acetylene (C_2H_2) oxidation, the stoichiometric and threshold reactions are respectively

$$C_2H_2 + 2.5(O_2 + 3.76N_2) \rightarrow 2CO_2 + H_2O + 9.4N_2 \tag{1.3.5}$$

$$C_2H_2 + (O_2 + 3.76N_2) \rightarrow 2CO + H_2 + 3.76N_2. \tag{1.3.6}$$

Thus the threshold ϕ for carbon formation is 2.5. Since this estimation is strictly based on stoichiometry, without regard for the fuel structure and detailed reaction mechanisms, ϕ_C is the maximum ϕ beyond which solid carbon is expected to form in a uniform mixture.

1.3.2. The Major–Minor Species Model

If high accuracy of the product composition is not an issue, then an approximate determination can be accomplished by using a simple yet quite accurate and physically illuminating method based on the concept of major and minor species. Specifically, the species of a hydrocarbon–air mixture in equilibrium can be considered as either a major species or a minor species in terms of its concentration, and therefore the product composition can be calculated in two steps, as illustrated here.

Major Species: In step one we calculate the major species of the mixture, which include CO_2, H_2O, H_2, O_2, and CO. N_2 is of course also a major species, although it does not participate in the reactions to any significant extent. To be more specific, we shall again use propane as the fuel for illustration. We also need to separately discuss fuel-lean and fuel-rich mixtures.

In fuel-lean mixtures we assume the complete conversion of C to CO_2 and H to H_2O, with the excess oxygen remaining as O_2. The chemical reaction is represented by

$$\phi < 1: \quad \phi C_3H_8 + 5(O_2 + 3.76N_2) \rightarrow 3\phi CO_2 + 4\phi H_2O + 5(1-\phi)O_2 + 18.8N_2. \tag{1.3.7}$$

Since the composition of the major species is completely specified through stoichiometry, it does not depend on the temperature and pressure of the mixture.

In fuel-rich mixtures, oxygen is completely reacted but CO and H_2 are now present as the major product species. Thus we can write, in general,

$$\phi > 1: \quad \phi C_3H_8 + 5(O_2 + 3.76N_2) \rightarrow aCO_2 + bCO + cH_2O + dH_2 + 18.8N_2, \tag{1.3.8}$$

where a, b, c, and d are constrained by the element conservation relations,

$$\text{C:} \quad 3\phi = a + b; \qquad \text{H:} \quad 8\phi = 2c + 2d; \qquad \text{O:} \quad 10 = 2a + b + c.$$

Solving the above, we obtain $a = -7\phi + 10 + d$, $b = 10\phi - 10 - d$, and $c = 4\phi - d$. The remaining relation required to uniquely determine d, and thereby a, b, and c, is obtained by assuming chemical equilibrium for the water-gas shift reaction:

$$CO_2 + H_2 \rightleftharpoons CO + H_2O, \qquad K_p(T) = \frac{p_{CO}\,p_{H_2O}}{p_{CO_2}\,p_{H_2}} = \frac{bc}{ad}. \tag{1.3.9}$$

Since K_p increases with increasing temperature, more CO and H_2O are produced at higher temperatures.

With the chemical conversion equations respectively represented by Eqs. (1.3.7) and (1.3.8) for fuel-lean and fuel-rich mixtures, the equilibrium concentrations can be usually calculated accurately as long as ϕ is not too close to unity and temperature is not too high. When $\phi \approx 1$ or the temperature is high, say greater than 2,000 K, dissociation of H_2O and CO_2 is not negligible. This dissociation can be readily taken into account by considering the chemical equilibrium of the following two reactions:

$$CO_2 \rightleftharpoons CO + \frac{1}{2}O_2, \quad K_p = \frac{p_{CO}\sqrt{p_{O_2}}}{p_{CO_2}} \tag{1.3.10}$$

$$H_2O \rightleftharpoons H_2 + \frac{1}{2}O_2, \quad K_p = \frac{p_{H_2}\sqrt{p_{O_2}}}{p_{H_2O}}. \tag{1.3.11}$$

The general equation of chemical conversion should then also include O_2 as a product, given by

$$\phi C_3H_8 + 5(O_2 + 3.76N_2) \rightarrow aCO_2 + bCO + cH_2O + dH_2 + eO_2 + 18.8N_2, \tag{1.3.12}$$

for both fuel-lean and fuel-rich mixtures. Applying the element conservation relations, we have

$$\text{C:} \quad 3\phi = a + b; \qquad \text{H:} \quad 8\phi = 2c + 2d; \qquad \text{O:} \quad 10 = 2a + b + c + 2e.$$

By considering two additional relations given by chemical equilibrium, that is, any two of the three equilibrium constants of Eqs. (1.3.9), (1.3.10), and (1.3.11), we have five equations to solve for the five unknowns (a, b, c, d, and e). The fuel-lean and fuel-rich calculations are of course limiting cases of this complete treatment that includes all the major species.

Minor Species: Basically whatever species that is not included as a major species can be considered as a minor species. Notable examples are OH, O, H, NO, N, etc. For lean mixtures CO is a minor species while for rich mixtures O_2 is a minor species.

To determine the concentrations of the minor species, we assume that the concentrations of the major species remain unaffected by the production of the minor species, and therefore remain at their respective levels determined in the first step. For example, to determine the concentration of NO in the lean combustion of (1.3.7), we write

$$\frac{1}{2}\mathrm{N_2} + \frac{1}{2}\mathrm{O_2} \rightleftharpoons \mathrm{NO}, \qquad K_p(T) = \frac{p_{\mathrm{NO}}}{\sqrt{p_{\mathrm{N_2}}\, p_{\mathrm{O_2}}}}. \tag{1.3.13}$$

Using the stoichiometric coefficients of (1.3.7) and the relation $p_i = X_i p$, we have

$$p_{\mathrm{O_2}} = \frac{5(1-\phi)}{23.8+2\phi}\, p_t, \qquad p_{\mathrm{N_2}} = \frac{18.8}{23.8+2\phi}\, p_t.$$

Thus by specifying T and ϕ, p_{NO} can be determined easily.

For rich mixtures, we do not have O_2 as a major species in the product. But we can write an alternate relation consisting of a nitrogen-containing species and an oxygen containing species, such as

$$\mathrm{H_2O} + \frac{1}{2}\mathrm{N_2} \rightleftharpoons \mathrm{H_2} + \mathrm{NO}, \qquad K_p(T) = \frac{p_{\mathrm{H_2}}\, p_{\mathrm{NO}}}{p_{\mathrm{H_2O}}\sqrt{p_{\mathrm{N_2}}}}. \tag{1.3.14}$$

Since $p_{\mathrm{H_2}}$, $p_{\mathrm{H_2O}}$, and $p_{\mathrm{N_2}}$ have already been determined for the given ϕ and T, p_{NO} can again be calculated.

1.3.3. Computer Solutions

It must be clear by now that although the basic concept of chemical equilibrium is straightforward, the actual calculation can be quite tedious and therefore can best be conducted computationally. If the reaction scheme is sufficiently simple, the iteration algorithm can be either individually written or called from standard subroutine libraries. For more complex mixtures and reaction schemes, computer codes have been developed and are readily available, for example those by Gordon and McBride (1971) and Reynolds (1986). For computational solutions it is more expedient to directly use the equilibrium criterion, Eq. (1.2.12), by minimizing the Gibbs function, $G = \Sigma_i \bar{g}_i N_i$, for the N_is, instead of using the equilibrium constants, K_ps. Since it is so convenient to obtain computer solutions nowadays, hand calculation in the manner described above is rarely performed. It is nevertheless important to understand the underlying principle that leads to the problem formulation and subsequently its solution.

1.4. ENERGY CONSERVATION

During reactions, exchanges in chemical and thermal energy take place. Typically, we are given a cold combustible mixture consisting of reactants and inerts. During the subsequent reaction sequence with net exothermicity, chemical energy is released as the reactant molecules are transformed into the product molecules. This chemical heat release is used to heat the product mixture to the final, adiabatic flame temperature. Since the total energy of the system is conserved, the difference between the initial and final states is simply a rearrangement of the different amounts of thermal and chemical energies in each state. We now study the "accounting" procedure governing such a rearrangement.

1.4.1. Heats of Formation, Reaction, and Combustion

We first need to establish a reference to account for the amount of chemical heat release, or absorption, as different species are transformed. Thus for each species a heat of formation at constant volume ($\bar{e}_i^o$) and a heat of formation at constant pressure ($\bar{h}_i^o$) can be defined as the amount of heat needed to form one mole of the substance from its elements in their standard states, with the reaction taking place in a closed system, either at constant volume or constant pressure, and with the initial and final temperatures, T^o, being the same. The chemical reaction representing such a process is the forward reaction of (1.2.32),

$$\sum_{j=1}^{L} \nu'_{i,j} \mathrm{M}^o_{i,j} \rightarrow \mathrm{M}_i. \tag{1.4.1}$$

By definition, the heat of formation is zero for elements in their standard states.

For a given substance, $\bar{h}_i^o$ is positive if heat is absorbed by the system, and negative if heat is released. As examples, we have

$$\mathrm{C}(gr) + \frac{1}{2}\mathrm{O}_2 \rightarrow \mathrm{CO}, \quad \bar{h}_i^o(T^o = 298.15\ \mathrm{K}) = -26.42\ \mathrm{kcal/mole\ of\ CO}\ ;$$

$$\frac{1}{2}\mathrm{H}_2 + \frac{1}{2}\mathrm{I}_2 \rightarrow \mathrm{HI}, \quad \bar{h}_i^o(T^o = 298.15\ \mathrm{K}) = 6.30\ \mathrm{kcal/mole\ of\ HI}.$$

The negative and positive signs indicate that the formation reactions of CO and HI are exothermic and endothermic respectively. We also note that the reaction $CO + \frac{1}{2}O_2 \rightarrow CO_2$ evolves -67.63 kcal/mole of CO_2. This is not the heat of formation of CO_2 because CO is not an element.

The relation between $\bar{e}_i^o(T^o)$ and $\bar{h}_i^o(T^o)$ is the following. Since $E = H - pV = H - NR^oT$, for reaction (1.4.1) we have

$$\bar{e}_i^o = \bar{h}_i^o - R^oT^o\left(1 - \sum_j \nu'_{i,j}\right), \tag{1.4.2}$$

where the summation is performed over all gaseous elements. For example, for the formation of $CO_2(g)$ from $C(gr) + O_2(g) \rightarrow CO_2(g)$, $(1 - \sum_j \nu'_{i,j})$ is zero while for the $CO(g)$ formation it is $\frac{1}{2}$ according to $C(gr) + \frac{1}{2}O_2(g) \rightarrow CO(g)$. For substances

Table 1.3. Enthalpy of formation, $\bar{h}_i^o(T)$, at 1 atm and 298.15 K

Substance	Formula (state)	$\bar{h}_i^o(T)$ kcal/mol	Substance	Formula (state)	$\bar{h}_i^o(T)$ kcal/mol
Aluminum Oxide	$Al_2O_3(s)$	−400.5	Hydrogen Cyanide	$HCN(g)$	32.3
Diborane	$B_2H_6(g)$	8.5	Formaldehyde	$CH_2O(g)$	−26.0
Boron Oxide	$B_2O_3(s)$	−304.4	Formic Acid	$CH_2O_2(l)$	−101.6
Bromine Atom	$Br(g)$	26.7	Nitromethane	$CH_3NO_2(g)$	−17.9
Bromine	$Br_2(g)$	7.4	Methylnitrate	$CH_3NO_3(g)$	−29.8
Hydrogen Bromide	$HBr(g)$	−8.7	Methane	$CH_4(g)$	−17.8
Calcium Carbonate	$CaCO_3$	−288.5	Methanol	$CH_4O(l)$	−57.1
Calcium Oxide	$CaO(s)$	−151.7	Carbon Monoxide	$CO(g)$	−26.4
Chlorine Atom	$Cl(g)$	29.0	Carbon Dioxide	$CO_2(g)$	−94.0
Hydrogen Chloride	$HCl(g)$	−22.1	Acetylene	$C_2H_2(g)$	54.5
Fluorine Atom	$F(g)$	19.0	Ethylene	$C_2H_4(g)$	12.5
Hydrogen Fluoride	$HF(g)$	−65.3	Acetaldehyde	$C_2H_4O(g)$	−39.7
Iron Oxide	Fe_3O_4	−267.3	Ethylene Oxide	$C_2H_4O(g)$	−12.6
Hydrogen Atom	$H(g)$	52.1	Acetic Acid	$C_2H_4O_2(l)$	−115.8
Iodine Atom	$I(g)$	25.5	Ethane	$C_2H_6(g)$	−20.2
Hydrogen Iodide	$HI(g)$	6.3	Ethanol	$C_2H_6O(l)$	−66.4
Iodine	$I_2(g)$	14.9	Dimethyl Ether	$C_2H_6O(g)$	−44.0
Magnesium Oxide	$MgO(s)$	−143.8	Cyanogen	$C_2N_2(g)$	73.3
Nitrogen Atom	$N(g)$	113.0	Allene	$C_3H_4(g)$	45.5
Ammonia	$NH_3(g)$	−11.0	Propyne	$C_3H_4(g)$	44.2
Nitric Oxide	$NO(g)$	21.6	Propene	$C_3H_6(g)$	4.8
Nitrogen Dioxide	$NO_2(g)$	7.9	Cyclopropane	$C_3H_6(g)$	12.7
Hydroazine	$N_2H_4(g)$	22.8	Acetone	$C_3H_6O(g)$	−51.9
Nitrous Oxide	$N_2O(g)$	19.6	Propylene Oxide	$C_3H_6O(g)$	−22.6
Oxygen Atom	$O(g)$	59.6	Propane	$C_3H_8(g)$	−25.0
Hydroxyl	$OH(g)$	9.3	1,2-Butadiene	$C_4H_6(g)$	38.8
Water	$H_2O(g)$	−57.8	1,3-Butadiene	$C_4H_6(g)$	26.3
Hydrogen Peroxide	$H_2O_2(g)$	−32.6	*n*-Butane	$C_4H_{10}(g)$	−30.0
Ozone	$O_3(g)$	34.1	iso-Butane	$C_4H_{10}(g)$	−32.1
Disilane	$Si_2H_6(g)$	19.2	Diethyl Ether	$C_4H_{10}O(g)$	−60.3
Silane	$SiH_4(g)$	8.2	*n*-Pentane	$C_5H_{12}(g)$	−35.1
Silicon Dioxide	$SiO_2(s)$	−217.7	iso-Pentane	$C_5H_{12}(g)$	−36.7
Sulfur Dioxide	$SO_2(g)$	−70.9	Benzene	$C_6H_6(g)$	19.8
Sulfur Trioxide	$SO_3(g)$	−94.6	Cyclohexane	$C_6H_{12}(g)$	−29.5
Titanium Oxide	$TiO_2(s)$	−225.6	*n*-Hexane	$C_6H_{14}(g)$	−39.9
Graphite	$C(s)$	0.0	Toluene	$C_7H_8(g)$	12.0
Carbon	$C(g)$	171.3	*n*-Heptane	$C_7H_{16}(g)$	−44.9
Carbon Tetrachloride	$CCl_4(g)$	−22.9	*o*-Xylene	$C_8H_{10}(g)$	4.6
Trichloromethane	$CHCl_3(g)$	−24.8	*n*-Octane	$C_8H_{18}(g)$	−49.9
Dichloromethane	$CH_2Cl_2(g)$	−22.9	iso-Octane	$C_8H_{18}(g)$	−53.5
Chloromethane	$CH_3Cl(g)$	−19.6	*n*-Hexadecane	$C_{16}H_{34}(g)$	−89.6

Sources: Lide, D. R. 1990–1991. *CRC Handbook of Chemistry and Physics*, 71st ed., CRC Press, Boca Raton. (Note that $\Delta_f H_i$ is the symbol used for the enthalpy of formation from this source).

of interest to combustion, their heats of formation are usually so large that the work term $R^oT^o(1 - \sum_j \nu'_{i,j})$ is almost negligible. Thus we can assume that $\bar{e}_i^o \simeq \bar{h}_i^o$. Furthermore, if T^o is taken to be 0 K, then $\bar{e}_i^o \equiv \bar{h}_i^o$. Values of $\bar{h}_i^o(T^o)$ for some selected species are listed in Table 1.3.

Table 1.4. Heats of combustion at 25°C and constant pressure

Formula (gas)	Name	$\bar{q}_p$ (kcal/mol)	Lower heating value $[H_2O(g)]$ (cal/g)
Hydrogen	H_2	57.80	28672.3
Methane	CH_4	191.85	11958.7
Methanol	CH_4O	152.55	4760.9
Carbon Monoxide	CO	67.65	2415.2
Acetylene	C_2H_2	2300.40	88348.2
Ethylene	C_2H_4	316.20	11271.2
Ethane	C_2H_6	341.30	11350.3
Allene	C_3H_4	443.25	11063.3
Propyne	C_3H_4	441.95	11030.9
Cyclopropane	C_3H_6	468.25	11127.4
Propane	C_3H_8	488.35	11074.6
1,3-Butadiene	C_4H_6	575.90	10646.7
n-Butane	C_4H_{10}	635.20	10928.5
n-Pentane	C_5H_{12}	781.95	10837.8
Benzene	C_6H_6	757.50	9697.4
Cyclohexane	C_6H_{12}	881.60	10475.1
n-Hexane	C_6H_{14}	929.00	10780.1
Toluene	C_7H_8	901.55	9784.5
n-Heptane	C_7H_{16}	1075.85	10736.6
o-Xylene	C_8H_{10}	1046.00	9852.4
n-Octane	C_8H_{18}	1222.70	10703.8
iso-Octane	C_8H_{18}	1219.10	10672.2
n-Hexadecane	$C_{16}H_{34}$	2397.80	10588.8

Sources: Computed from the enthalpies of formation of Table 1.3.

Knowing $\bar{h}_i^o(T^o)$, we can now define a heat of reaction (at constant pressure) for the general reaction scheme (1.2.13) as

$$\bar{q}_p(T^o) = \sum_{i=1}^{N}(\nu_i'' - \nu_i')\bar{h}_i^o(T^o). \tag{1.4.3}$$

Thus if the heat of formation of the products is less than that of the reactants, then $\bar{q}_p(T^o) < 0$ and we say the reaction is exothermic. Similarly, the reaction is endothermic if $\bar{q}_p(T^o) > 0$.

A special class of the heat of reaction is the heat of combustion (at constant pressure), defined as the amount of heat release when 1 mole of a fuel in its standard state is completely reacted with oxygen to form H_2O, CO_2, and N_2; the need to specify N_2 is due to the possible presence of fuel-bound nitrogen, especially for some explosives (e.g., trinitrotoluene, commonly called TNT). It is also necessary to distinguish whether the product water exists in the gas state, $H_2O(g)$, or the liquid state, $H_2O(\ell)$. Since the former case has less heat release, heats of combustion with water present in gaseous and liquid states are also respectively called lower and higher heating values of the fuel. The difference between them is 10.52 kcal/mole for every mole of H_2O produced. Table 1.4 shows the heats of combustion of various fuels. Note that since reactions of interest to combustion are usually exothermic based

Table 1.5. Mean bond energies (kcal/mole bond)

Bond	Energy	Bond	Energy
Br−Br	46.3	F−F	38.0
C−C	85.5	F−H	136.3
C=C	145.0	H−Br	87.6
C≡C	194.3	H−Cl	103.1
C−Cl	95.0	H−H	104.2
C−H	98.1	H−I	71.3
C−O	86.0	H−O	102.4
C=O (carbon monoxide type)	257.3	I−I	36.1
C=O (carbon dioxide type)	192.0	N−N	225.9
C=O (ketone type)	179.0	O−O	35.0
C=O (aldehyde type)	176.0	O=O	119.0
Cl−Cl	58.1		

Sources: Lide, D. R. 1990–1991. *CRC Handbook of Chemistry and Physics*, 71st ed., CRC Press, Boca Raton.
Souders, Matt & Eshbach, Ovid W. 1975. *Handbook of Engineering Fundamentals*, 3rd ed., Wiley, New York.

on the consumption of fuel, in subsequent references to their heats of combustion we shall simply quote their magnitude, with the understanding that they are negative according to the definition.

For constant volume processes, we simply replace $\bar{h}$ by $\bar{e}$ in this discussion. However, since $\bar{e}_i^o \simeq \bar{h}_i^o$, in most of the subsequent derivations we shall just use the symbol $\bar{q}_c$ to designate the heat of combustion.

1.4.2. Estimation of Heat of Reaction from Bond Energies

The determination of the heat of reaction depends on the availability of the heats of formation. In the event that they are not available, rough estimates of their values can be obtained from the bond energies between the atomic constituents of the reactant and product molecules (McMillen & Golden 1982). The methodology is based on the concept that the energy needed to break a particular type of bond between two atoms is approximately the same regardless of the molecule in which the bond is present. Thus the difference between the sums of the bond energies of the reactants and the products can be approximated as the heat of reaction. Table 1.5 lists the bond energies of a variety of atomic pairs.

As an example, consider the hydrogenation reaction of ethylene to form ethane, $C_2H_4 + H_2 \rightarrow C_2H_6$, or

```
H       H                       H   H
 \     /                        |   |
  C===C    +   H—H   ——>    H—C———C—H
 /     \                        |   |
H       H                       H   H
```

In this reaction one C=C bond and one H—H bond are broken, while one C—C bond and two C—H bonds are created. Thus the net change in the bond energies given by Table 1.5 is (1)(85.5) + (2)(98.1) − (1)(145) − (1)(104.2) = 32.5 kcal/mole. The positive value indicates that the products are more tightly bound and hence less reactive. Thus this reaction is exothermic, with $\bar{q}_p \simeq 32.5$ kcal/mole. Using the values of heats of formation given in Table 1.3, we have $\bar{q}_p = (12.5) - (-20.2) =$ 32.7 kcal/mole. The comparison is close in this case.

1.4.3. Determination of Heat of Reaction from $K_p(T)$

The heat of reaction at temperature T can be determined from $K_p(T)$. Using $\bar{\mu}_i^o(T) = \bar{g}_i^o(T) = \bar{h}_i^o(T) - T\bar{s}_i^o(T)$ in Eq. (1.2.31), we have

$$R^o T \ln K_p(T) = -\sum_{i=1}^{N} (\nu_i'' - \nu_i')\left[\bar{h}_i^o(T) - T\bar{s}_i^o(T)\right]. \tag{1.4.4}$$

Differentiating Eq. (1.4.4) with respect to T, and noting that $d\bar{h}_i^o = Td\bar{s}_i^o + \bar{v}dp = Td\bar{s}_i^o$ for a constant pressure process such that $d\bar{h}_i^o/dT = Td\bar{s}_i^o/dT$, we obtain

$$R^o T \frac{d \ln K_p(T)}{dT} + R^o \ln K_p = \sum_{i=1}^{N} (\nu_i'' - \nu_i')\bar{s}_i^o(T). \tag{1.4.5}$$

Substituting Eq. (1.4.4) into Eq. (1.4.5) yields

$$R^o \frac{d \ln K_p(T)}{d(1/T)} = -\sum_{i=1}^{N} (\nu_i'' - \nu_i')\bar{h}_i^o(T) = -\bar{q}_p(T), \tag{1.4.6}$$

which is the van't Hoff equation. It shows that in a plot of ln K_p versus $(1/T)$, the slope (multiplied by R^o) yields the heat of reaction at T. Furthermore, if the range of temperature change is small, the sensible enthalpy change is also small as compared to the heat of formation. Then the heat of reaction is approximately constant and such a plot is linear within this temperature range.

1.4.4. Sensible Energies and Heat Capacities

Having defined $\bar{h}_i^o(T^o)$ to account for changes in the chemical energy, we now discuss the dependence of thermal energy on temperature. From statistical mechanics, it is known that the internal, or sensible, energy of a pure substance is given by the sum of the energies associated with different modes of excitation of the molecules. That is, by referencing the internal energy level to 0 K, we can write

$$\bar{e}_i^s(T; 0\,K) = \bar{e}_{i,\text{trans}}(T) + \bar{e}_{i,\text{rot}}(T) + \bar{e}_{i,\text{vib}}(T) + \bar{e}_{i,\text{elec}}(T), \tag{1.4.7}$$

where the four terms on the RHS respectively represent the energies associated with translation, rotation, vibration, and electronic excitation. By assigning the same temperature to all modes of excitation in Eq. (1.4.7), it is assumed that equilibrium exists among these modes. Furthermore, translational and rotational equilibria are usually attained readily. Excitation of the vibrational modes depends more strongly on temperature, increasing with increasing temperature. Its attainment of equilibrium usually requires a large number of collisions. For most combustion phenomena, $\bar{e}_{i,\mathrm{elec}}(T)$ is negligible, with the possible exception of strong detonation waves.

The variation of $\bar{e}_i^s$ with temperature is measured by the heat capacity at constant volume, defined as

$$\bar{c}_{v,i}(T) = \left(\frac{\partial \bar{e}_i^s}{\partial T}\right)_v, \tag{1.4.8}$$

which gives the amount of heat needed to raise the temperature of one mole of the substance by 1 K. With the definition of $\bar{c}_{v,i}$, we can now express $\bar{e}_i^s$, with a general reference temperature T^o, as

$$\bar{e}_i^s(T; T^o) = \bar{e}_i^s(T; 0\,\mathrm{K}) - \bar{e}_i^s(T^o; 0\,\mathrm{K}) \tag{1.4.9}$$

$$= \int_{T^o}^{T} \bar{c}_{v,i}\, dT. \tag{1.4.10}$$

Equation (1.4.10) shows the important relation that for a given amount of energy change $\bar{e}_i^s$, the increase in temperature T varies somewhat inversely with the heat capacity $\bar{c}_{v,i}$. That is, the larger the $\bar{c}_{v,i}$, the smaller the $(T - T^o)$, and vice versa. Physically, a larger $\bar{c}_{v,i}$ implies the activation of more modes of excitation of the molecules. Thus the given amount of energy has to be distributed over more excitation modes, leading to less energy for each mode and thereby a smaller temperature increase. Furthermore, $\bar{c}_{v,i}$ assumes higher values for larger molecules and at higher temperatures. For example, while a monatomic gas only has the three translational degrees of freedom ($\bar{c}_{v,i} = \frac{3}{2}R^o$), a diatomic gas has two additional degrees of freedom involving rotation ($\bar{c}_{v,i} = \frac{5}{2}R^o$). At higher temperatures the vibrational degree of freedom is also activated to give a still larger $\bar{c}_{v,i}$. For larger molecules more vibrational degrees of freedom are possible, again leading to a larger $\bar{c}_{v,i}$.

For constant pressure processes, we have the corresponding relations,

$$\bar{h}_i^s(T; 0\,\mathrm{K}) = \bar{e}_i^s(T; 0\,\mathrm{K}) + R^o T \tag{1.4.11}$$

$$\bar{h}_i^s(T; T^o) = \bar{h}_i^s(T; 0\,\mathrm{K}) - \bar{h}_i^s(T^o; 0\,\mathrm{K}) \tag{1.4.12}$$

$$\bar{c}_{p,i}(T) = \left(\frac{\partial \bar{h}_i^s}{\partial T}\right)_p \tag{1.4.13}$$

$$\bar{h}_i^s(T; T^o) = \int_{T^o}^{T} \bar{c}_{p,i}(T)\, dT. \tag{1.4.14}$$

Using Eq. (1.4.11) in Eq. (1.4.13), we obtain

$$\bar{c}_{p,i}(T) = \bar{c}_{v,i}(T) + R^o. \tag{1.4.15}$$

Note that Eqs. (1.4.11) and (1.4.15) apply only to ideal gases.

In the theoretical developments to be presented in the rest of the text, we shall repeatedly encounter the combined term $\bar{c}_p T$. It is clear that any uncertainty or error associated with estimating $\bar{c}_p$ will lead to a corresponding uncertainty or error in T.

Since chemical reaction rates, and consequently the rates of combustion processes, are frequently very temperature sensitive, a quantitatively reliable description of many combustion phenomena requires a correspondingly accurate knowledge of $\bar{c}_p$.

Tables 1.6 and 1.7 provide $\bar{h}_i^s(T; T^o)$ and $\bar{c}_{p,i}(T)$ for selected species of importance to combustion.

1.4.5. Energy Conservation in Adiabatic Chemical Systems

Having defined $\bar{h}_i^o(T^o)$ and $\bar{h}_i^s(T; T^o)$, we can now state that, for a constant pressure process in a closed system, the total enthalpy content per mole of species i at temperature T, $\bar{h}_i(T; T^o)$, is the sum of its heat of formation at T^o, $\bar{h}_i^o(T^o)$, and the sensible heat at T relative to T^o, $\bar{h}_i^s(T; T^o)$, that is,

$$\bar{h}_i(T; T^o) = \bar{h}_i^o(T^o) + \bar{h}_i^s(T; T^o). \tag{1.4.16}$$

Therefore energy conservation for a gas mixture before and after a chemical reaction, respectively designated by subscripts 1 and 2, is

$$\sum_{i=1}^{N} N_{i,1}\bar{h}_i(T_1; T^o) = \sum_{i=1}^{N} N_{i,2}\bar{h}_i(T_2; T^o). \tag{1.4.17}$$

Substituting Eq. (1.4.16) into Eq. (1.4.17), we have

$$\sum_{i=1}^{N} N_{i,1}\bar{h}_i^o(T^o) - \sum_{i=1}^{N} N_{i,2}\bar{h}_i^o(T^o) = \sum_{i=1}^{N} N_{i,2}\bar{h}_i^s(T_2; T^o) - \sum_{i=1}^{N} N_{i,1}\bar{h}_i^s(T_1; T^o). \tag{1.4.18}$$

The LHS side of Eq. (1.4.18) is the chemical heat release at the standard state, and the RHS represents the difference between states 1 and 2 in the total sensible heat relative to T^o. Thus, for an initial composition $N_{i,1}$ and temperature T_1, the unknowns in Eq. (1.4.18) are the final composition $N_{i,2}$ and temperature T_2. T_2 is called the adiabatic flame temperature, designated by T_{ad}, to be discussed next.

1.4.6. Adiabatic Flame Temperature and Equilibrium Composition

If a given uniform mixture with an initial temperature and composition is made to approach chemical equilibrium through an adiabatic, isobaric process at pressure p, then the final temperature is the adiabatic flame temperature, T_{ad}. This quantity is of importance in the study of combustion because it not only indicates the exothermicity and the maximum attainable temperature of this mixture when equilibrium is attained, it also directly affects the reactivity of the various chemical processes including those involving pollutant formation.

Figure 1.4.1 shows the enthalpy–temperature diagram illustrating the principle of the adiabatic flame temperature calculation. We shall conduct the discussion on the basis of per unit mass as mass is conserved in a closed system. It is seen that, instead of proceeding directly from state 1 to state 2 on constant h, which is the total enthalpy of the mixture, the process can be interpreted alternately as having the reactant temperature first changed from T_1 to T^o. At T^o, chemical reaction takes place and releases the chemical heat. This amount of heat is then used to heat the

Table 1.6. Sensible enthalpy, $\bar{h}_i^s(T; T^o = 298.15 \text{ K})$, kcal/mole

T(K)	O_2	O	H_2	H	OH	H_2O	N_2	N	NO	C(gr)	C(g)	CO	CO_2	CH_4
0	−2.075	−1.607	−2.024	−1.481	−2.192	−2.367	−2.072	−1.481	−2.197	−0.251	−1.562	−2.072	−2.238	−2.396
100	−1.381	−1.080	−1.307	−0.984	−1.467	−1.581	−1.379	−0.984	−1.451	−0.237	−0.992	−1.379	−1.543	−1.601
200	−0.685	−0.523	−0.663	−0.488	−0.711	−0.784	−0.683	−0.488	−0.705	−0.159	−0.489	−0.683	−0.816	−0.805
298	0.000	0.000	0.000	0.000	0.000	0.000	0.000	0.000	0.000	0.000	0.000	0.000	0.000	0.000
300	0.013	0.010	0.013	0.009	0.013	0.015	0.013	0.009	0.013	0.004	0.009	0.013	0.016	0.016
400	0.723	0.528	0.707	0.506	0.725	0.825	0.710	0.506	0.727	0.248	0.507	0.711	0.958	0.923
500	1.454	1.038	1.406	1.003	1.432	1.654	1.413	1.003	1.448	0.565	1.004	1.417	1.987	1.960
600	2.209	1.544	2.106	1.500	2.137	2.509	2.126	1.500	2.186	0.942	1.501	2.437	3.087	3.138
700	2.987	2.048	2.808	1.996	2.845	3.390	2.853	1.996	2.942	1.366	1.999	2.873	4.245	4.454
800	3.785	2.550	3.514	2.493	3.556	4.300	3.596	2.493	3.716	1.825	2.495	3.627	5.453	5.897
900	4.599	3.051	4.225	2.990	4.275	5.240	4.355	2.990	4.507	2.312	2.992	4.397	6.702	7.458
1000	5.426	3.552	4.943	3.487	5.004	6.209	5.130	3.487	5.313	2.819	3.489	5.183	7.984	9.125
1100	6.265	4.051	5.669	3.984	5.742	7.210	5.918	3.984	6.131	3.343	3.986	5.983	9.296	10.887
1200	7.113	4.550	6.405	4.480	6.491	8.240	6.718	4.480	6.960	3.881	4.483	6.794	10.632	12.732
1300	7.969	5.049	7.151	4.977	7.252	9.298	7.529	4.977	7.798	4.431	4.980	7.616	11.988	14.652
1400	8.833	5.548	7.907	5.474	8.023	10.384	8.350	5.474	8.644	4.990	5.477	8.446	13.362	16.637
1500	9.703	6.046	8.674	5.971	8.805	11.495	9.179	5.971	9.496	5.558	5.975	9.285	14.750	18.679
1600	10.580	6.544	9.451	6.468	9.596	12.630	10.015	6.468	10.354	6.132	6.472	10.130	16.157	20.772
1700	11.462	7.042	10.238	6.964	10.397	13.787	10.858	6.964	11.217	6.714	6.970	10.980	17.565	22.910
1800	12.350	7.540	11.035	7.461	11.207	14.964	11.706	7.461	12.084	7.301	7.469	11.836	18.987	25.086
1900	13.244	8.038	11.841	7.958	12.024	16.160	12.559	7.958	12.955	7.893	7.968	12.697	20.418	27.298
2000	14.143	8.536	12.656	8.455	12.849	17.373	13.417	8.455	13.829	8.491	8.469	13.561	21.857	29.540
2100	15.048	9.033	13.479	8.952	13.682	18.602	14.279	8.952	14.706	9.093	8.970	14.430	23.303	31.809
2200	15.958	9.531	14.311	9.448	14.520	19.846	15.144	9.449	15.587	9.699	9.472	15.301	24.755	34.103
2300	16.874	10.029	15.150	9.945	15.364	21.103	16.012	9.946	16.469	10.309	9.976	16.175	26.212	36.418
2400	17.795	10.527	15.996	10.442	16.214	22.372	16.883	10.443	17.354	10.924	10.482	17.052	27.674	38.753
2500	18.721	11.025	16.849	10.939	17.069	23.653	17.757	10.941	18.241	11.541	10.988	17.931	29.141	41.106
2600	19.652	11.524	17.709	11.436	17.929	24.945	18.634	11.439	19.129	12.163	11.497	18.813	30.613	43.474
2700	20.588	12.023	18.575	11.932	18.794	26.246	19.512	11.937	20.020	12.788	12.007	19.696	32.088	45.857
2800	21.529	12.522	19.448	12.429	19.662	27.556	20.393	12.436	20.911	13.416	12.519	20.582	33.567	48.253
2900	22.475	13.021	20.326	12.926	20.535	28.875	21.275	12.936	21.805	14.047	13.033	21.469	35.049	50.660
3000	23.426	13.522	21.209	13.423	21.411	30.201	22.159	13.436	22.700	14.682	13.549	22.357	36.535	53.079
3100	24.381	14.022	22.098	13.920	22.291	31.535	23.045	13.938	23.596	15.319	14.067	23.248	38.024	55.507
3200	25.340	14.524	22.993	14.416	23.174	32.876	23.933	14.441	24.493	15.960	14.586	24.139	39.515	57.944
3300	26.303	15.026	23.892	14.913	24.060	34.223	24.821	14.945	25.392	16.603	15.108	25.032	41.010	60.389
3400	27.271	15.528	24.797	15.410	24.949	35.577	25.711	15.451	26.291	17.250	15.631	25.927	42.507	62.842
3500	28.242	16.032	25.706	15.907	25.841	36.936	26.603	15.958	27.192	17.899	16.156	26.822	44.006	65.302
3600	29.217	16.537	26.620	16.404	26.736	38.300	27.496	16.468	28.094	18.551	16.683	27.719	45.508	67.768
3700	30.196	17.042	27.539	16.900	27.633	39.669	28.389	16.980	28.997	19.206	17.212	28.617	47.012	70.241
3800	31.178	17.549	28.463	17.397	28.533	41.043	29.284	17.494	29.900	19.863	17.743	29.516	48.518	72.719
3900	32.163	18.056	29.391	17.894	29.435	42.422	30.180	18.011	30.805	20.524	18.275	30.416	50.027	75.202
4000	33.151	18.565	30.324	18.391	30.339	43.805	31.077	18.531	31.710	21.187	18.809	31.316	51.538	77.690
4100	34.143	19.075	31.261	18.888	31.246	45.192	31.975	19.053	32.616	21.853	19.344	32.218	53.051	80.182
4200	35.137	19.585	32.202	19.384	32.154	46.583	32.874	19.579	33.523	22.521	19.881	33.121	54.566	82.678
4300	36.135	20.097	33.148	19.881	33.065	47.977	33.774	20.109	34.431	23.192	20.419	34.025	56.082	85.179
4400	37.135	20.610	34.098	20.378	33.978	49.375	34.674	20.642	35.340	23.866	20.958	34.930	57.601	87.683
4500	38.138	21.125	35.053	20.875	34.893	50.777	35.576	21.179	36.249	24.542	21.499	35.835	59.122	90.190
4600	39.145	21.640	36.011	21.372	35.809	52.181	36.478	21.720	37.159	25.221	22.041	36.741	60.644	92.701
4700	40.154	22.157	36.974	21.868	36.728	53.589	37.382	22.265	38.070	25.903	22.584	37.649	62.169	95.214
4800	41.166	22.675	37.940	22.365	37.648	55.000	38.286	22.815	38.982	26.587	23.128	38.557	63.695	97.730
4900	42.181	23.194	38.910	22.862	38.571	56.413	39.191	23.369	39.894	27.274	23.674	39.465	65.223	100.249
5000	43.200	23.715	39.884	23.359	39.495	57.829	40.096	23.927	40.807	27.963	24.220	40.375	66.753	102.771

Sources: JANAF Tables. *Journal of Physical and Chemical Reference Data*, v.3, no.2; v.4, no.1; v.7, no.3; v.11, no.3.

Table 1.7. Heat capacities, $\bar{c}_{p,i}(T)$, cal/mol-K

T(K)	O_2	O	H_2	H	OH	H_2O	N_2	N	NO	C(gr)	C(g)	CO	CO_2	CH_4
0	0.000	0.000	0.000	0.000	0.000	0.000	0.000	0.000	0.000	0.000	0.000	0.000	0.000	0.000
100	6.956	5.665	6.729	4.968	7.798	7.961	6.956	4.968	7.721	0.400	5.084	6.956	6.981	7.949
200	6.961	5.433	6.560	4.968	7.356	7.969	6.957	4.968	7.271	1.196	4.996	6.957	7.734	8.001
298	7.021	5.237	6.892	4.968	7.167	8.025	6.961	4.968	7.133	2.036	4.981	6.965	8.874	8.518
300	7.023	5.234	6.895	4.968	7.165	8.027	6.961	4.968	7.132	2.051	4.980	6.965	8.896	8.535
400	7.196	5.134	6.974	4.968	7.087	8.186	6.991	4.968	7.157	2.824	4.975	7.013	9.877	9.680
500	7.431	5.081	6.993	4.968	7.056	8.415	7.070	4.968	7.287	3.495	4.972	7.121	10.666	11.076
600	7.670	5.049	7.009	4.968	7.057	8.676	7.196	4.968	7.466	4.026	4.971	7.276	11.310	12.483
700	7.883	5.029	7.036	4.968	7.090	8.954	7.350	4.968	7.655	4.430	4.970	7.450	11.846	13.813
800	8.062	5.015	7.080	4.968	7.150	9.246	7.513	4.968	7.832	4.739	4.970	7.624	12.293	15.041
900	8.211	5.006	7.142	4.968	7.233	9.547	7.670	4.968	7.988	4.977	4.969	7.786	12.667	16.157
1000	8.334	4.999	7.219	4.968	7.332	9.851	7.815	4.968	8.123	5.165	4.969	7.931	12.980	17.160
1100	8.437	4.994	7.309	4.968	7.439	10.152	7.945	4.968	8.238	5.316	4.969	8.057	13.243	18.052
1200	8.525	4.990	7.407	4.968	7.549	10.444	8.060	4.968	8.336	5.441	4.970	8.168	13.466	18.842
1300	8.601	4.987	7.510	4.968	7.659	10.723	8.161	4.968	8.419	5.546	4.970	8.263	13.656	19.538
1400	8.670	4.984	7.615	4.968	7.766	10.987	8.250	4.968	8.491	5.635	4.972	8.346	13.815	20.150
1500	8.734	4.982	7.719	4.968	7.867	11.233	8.328	4.968	8.552	5.713	4.975	8.417	13.953	20.688
1600	8.795	4.981	7.821	4.968	7.963	11.462	8.396	4.968	8.605	5.782	4.978	8.480	14.074	21.161
1700	8.853	4.979	7.920	4.968	8.053	11.674	8.456	4.968	8.651	5.843	4.983	8.535	14.177	21.579
1800	8.909	4.978	8.016	4.968	8.137	11.869	8.508	4.968	8.692	5.899	4.990	8.583	14.269	21.947
1900	8.965	4.978	8.106	4.968	8.214	12.048	8.555	4.968	8.727	5.950	4.998	8.626	14.352	22.273
2000	9.020	4.978	8.193	4.968	8.286	12.214	8.597	4.969	8.759	5.997	5.008	8.664	14.424	22.562
2100	9.075	4.978	8.275	4.968	8.353	12.366	8.634	4.970	8.788	6.042	5.019	8.698	14.489	22.820
2200	9.129	4.978	8.354	4.968	8.415	12.505	8.668	4.971	8.813	6.083	5.031	8.728	14.547	23.050
2300	9.182	4.980	8.428	4.968	8.473	12.634	8.699	4.972	8.837	6.123	5.045	8.756	14.600	23.256
2400	9.235	4.981	8.499	4.968	8.526	12.753	8.726	4.975	8.858	6.160	5.061	8.781	14.648	23.441
2500	9.287	4.983	8.566	4.968	8.576	12.863	8.751	4.978	8.877	6.196	5.077	8.804	14.692	23.608
2600	9.337	4.986	8.631	4.968	8.622	12.965	8.775	4.982	8.895	6.231	5.094	8.825	14.734	23.758
2700	9.387	4.990	8.692	4.968	8.665	13.059	8.796	4.987	8.912	6.265	5.112	8.844	14.771	23.894
2800	9.435	4.994	8.752	4.968	8.706	13.146	8.815	4.993	8.927	6.297	5.130	8.863	14.807	24.018
2900	9.482	4.999	8.809	4.968	8.744	13.228	8.833	5.001	8.941	6.329	5.149	8.879	14.841	24.131
3000	9.528	5.004	8.864	4.968	8.780	13.304	8.850	5.010	8.955	6.360	5.168	8.895	14.873	24.233
3100	9.572	5.010	8.917	4.968	8.814	13.374	8.866	5.021	8.968	6.391	5.187	8.910	14.902	24.327
3200	9.614	5.017	8.969	4.968	8.846	13.441	8.881	5.034	8.980	6.420	5.206	8.924	14.930	24.413
3300	9.655	5.024	9.020	4.968	8.877	13.503	8.895	5.049	8.991	6.450	5.224	8.937	14.956	24.493
3400	9.694	5.032	9.069	4.968	8.906	13.562	8.908	5.066	9.002	6.478	5.243	8.949	14.982	24.565
3500	9.731	5.041	9.118	4.968	8.933	13.617	8.920	5.085	9.012	6.507	5.261	8.961	15.006	24.633
3600	9.768	5.050	9.165	4.968	8.959	13.669	8.932	5.106	9.022	6.535	5.279	8.973	15.030	24.695
3700	9.802	5.060	9.212	4.968	8.985	13.718	8.944	5.130	9.032	6.563	5.296	8.984	15.053	24.752
3800	9.836	5.070	9.258	4.968	9.009	13.764	8.954	5.155	9.041	6.590	5.313	9.994	15.075	24.806
3900	9.868	5.080	9.304	4.968	9.032	13.808	8.965	5.183	9.050	6.617	5.329	9.004	15.097	24.855
4000	9.900	5.091	9.349	4.968	9.055	13.850	8.975	5.212	8.058	6.644	5.345	9.014	15.119	24.901
4100	9.930	5.102	9.393	4.968	9.076	13.890	8.984	5.244	9.066	6.671	5.360	9.024	15.139	24.944
4200	9.960	5.114	9.437	4.968	9.098	13.927	8.993	5.278	9.074	6.698	5.374	9.033	15.159	24.984
4300	9.990	5.126	9.480	4.968	9.118	13.963	9.002	5.313	9.082	6.724	5.388	9.042	15.179	25.022
4400	10.019	5.137	9.523	4.968	9.138	13.997	9.011	5.351	9.090	6.751	5.401	9.051	15.197	25.057
4500	10.048	5.149	9.564	4.968	9.157	14.030	9.020	5.390	9.097	6.777	5.414	9.059	15.216	25.090
4600	10.077	5.162	9.605	4.968	9.176	14.061	9.028	5.431	9.105	6.803	5.426	9.068	15.234	25.121
4700	10.107	2.174	9.645	4.968	9.195	14.091	9.036	5.473	9.112	6.828	5.437	9.076	15.254	25.150
4800	10.137	5.186	9.684	4.968	9.213	14.120	9.045	5.516	9.119	6.854	5.448	9.084	15.272	25.177
4900	10.168	5.198	9.722	4.968	9.232	14.148	9.053	5.561	9.125	6.880	5.458	9.092	15.290	25.203
5000	10.200	5.210	9.758	4.968	9.249	14.174	9.061	5.607	9.132	6.905	5.468	9.100	15.306	25.227

Sources: JANAF Tables. *Journal of Physical and Chemical Reference Data*, v.3, no.2; v.4, no.1; v.7, no.3; v.11, no.3.

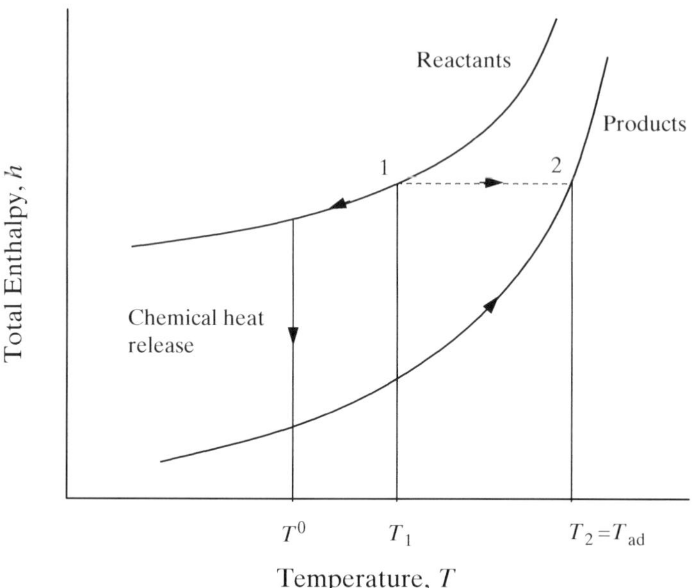

Figure 1.4.1. The principle of energy conservation in the definition of adiabatic flame temperature and composition.

products to the final temperature $T_2 = T_{ad}$. To calculate the adiabatic flame temperature and the associated mixture composition, we simply add the energy conservation relation, Eq. (1.4.18), to the system of equations for chemical equilibrium calculations, as described in Sections 1.2 and 1.3.

The calculation can also be simplified for lean hydrocarbon–air mixtures, with the assumption of the major–minor species, in that the concentrations of the major species are independent of temperature and pressure, and therefore can be determined first. Knowing these concentrations, T_{ad} can be iteratively determined from Eq. (1.4.18) and Table 1.6. For rich mixtures, T_{ad} and the stoichiometric coefficient d need to be simultaneously determined from Eqs. (1.3.9) and (1.4.18). Once T_{ad} and the concentrations of the major species are known, determination of the concentrations of the minor species is straightforward. With the convenience of modern computational capability, such a simplification, with the collateral inaccuracy, is however neither necessary nor worthwhile.

We shall now discuss the dependence of T_{ad} on the fuel concentration, fuel type, and pressure of the fuel–air mixture, obtained from computer solutions. Figures 1.4.2, 1.4.3, and 1.4.4 respectively show T_{ad} as functions of the fuel mole fraction, equivalence ratio ϕ, and normalized equivalence ratio Φ, for various fuel–air mixtures at 1 atm and $T_1 = T^o = 298.15$ K. The vertical bar on each curve in Figure 1.4.2 indicates the stoichiometric fuel concentration. Figure 1.4.3 shows that the most important factor influencing T_{ad} is ϕ, with T_{ad} peaking around $\phi = 1$ and decreasing steadily as the mixture becomes either leaner or richer. This is reasonable because of the need to heat up the excess reactants for off-stoichiometric conditions. It is also to be noted that for rich mixtures the excess "fuel" is actually CO and H_2 instead of the original

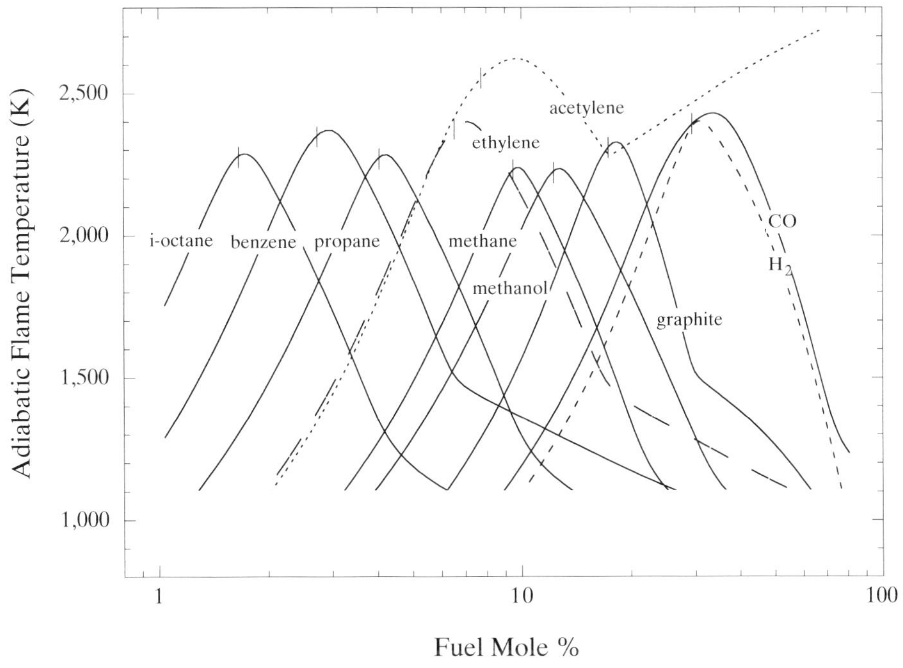

Figure 1.4.2. Adiabatic flame temperature, T_{ad}, as a function of the mole percent of fuel for several fuel–air mixtures at STP. The vertical bars indicate stoichiometric concentrations.

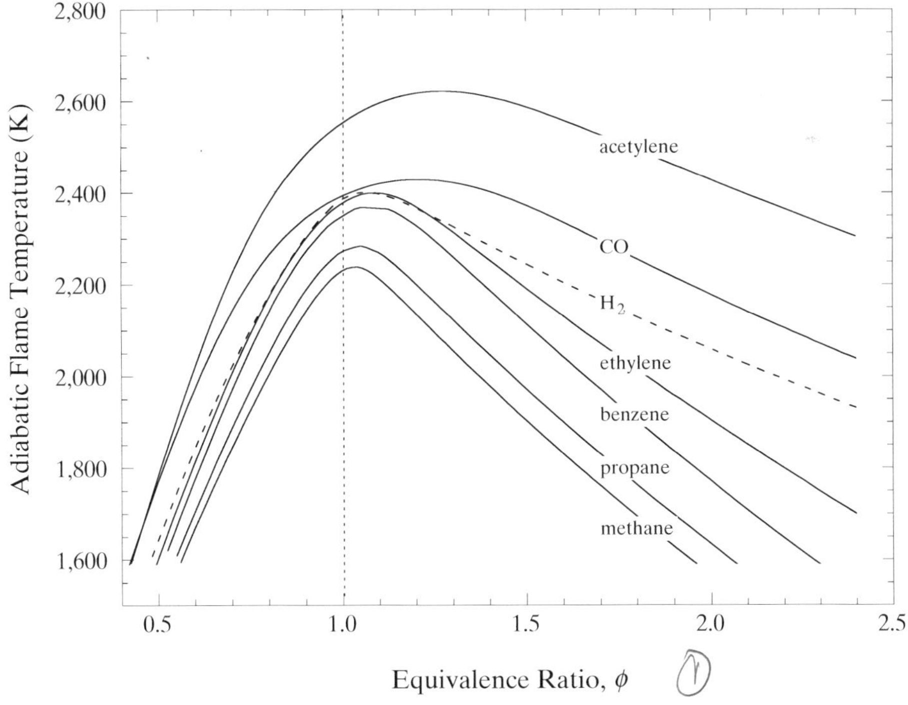

Figure 1.4.3. Adiabatic flame temperature, T_{ad}, as a function of fuel equivalence ratio, ϕ, for several fuel–air mixtures at STP.

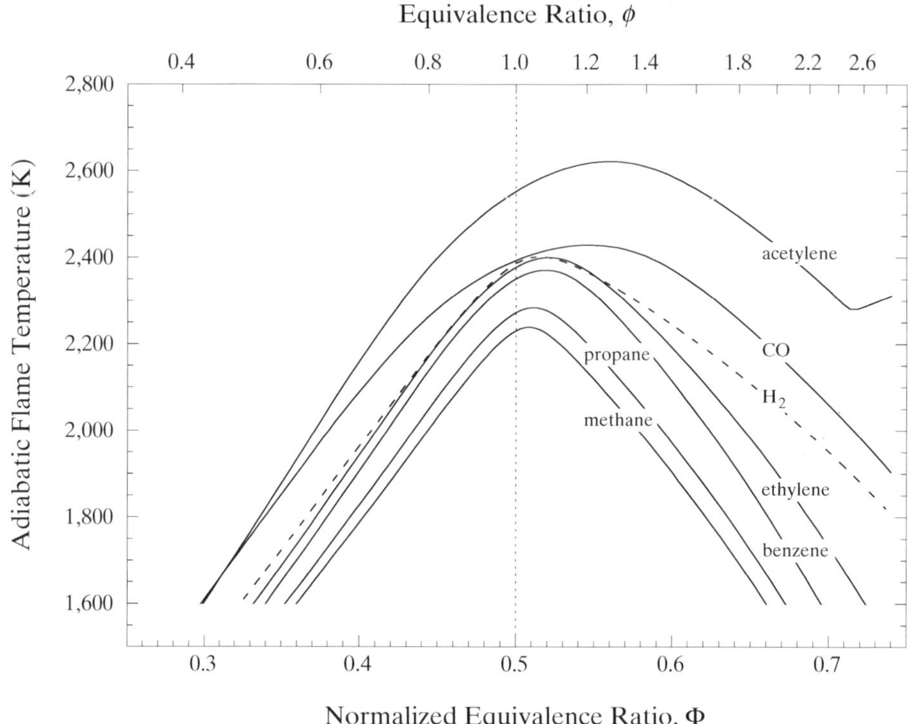

Figure 1.4.4. Adiabatic flame temperature, T_{ad}, as a function of normalized fuel equivalence ratio, Φ, for several fuel–air mixtures at STP.

fuel species. Since the total number of moles of CO and H_2 is larger than that of the original fuel molecule, even it is a larger molecule and, hence, has a larger $\bar{c}_p$, the specific heat of the product based on it being CO and H_2 can be substantially larger. This then leads to a lower T_{ad}.

The parameter that has the next strongest influence on T_{ad} is the C/H ratio of the fuel molecule, with T_{ad} increasing with increasing C/H ratio. There are three factors that need to be considered to explain such a behavior. First, species with large C/H ratios often contain more double and triple C—C bonds, which can hold more potential energy than the single bond. For example, we have shown that the enthalpy of the reaction $C_2H_6 \rightarrow C_2H_4 + H_2$ is endothermic by 32.7 kcal/mol. Since the enthalpy of formation of H_2 is zero, one can then conclude that ethylene stores more energy than ethane and thereby contributes to the higher T_{ad} for ethylene.

Second, we consider the enthalpies of combustion for ethane and ethylene,

$$C_2H_6 + 3.5O_2 \rightarrow 2CO_2 + 3H_2O, \tag{1.4.19}$$

$$C_2H_4 + 3O_2 \rightarrow 2CO_2 + 2H_2O, \tag{1.4.20}$$

which are respectively −341.2 kcal/mole and −316.1 kcal/mole. Thus the oxidation of one mole of ethane is more exothermic than one mole of ethylene, which is reasonable because ethane has one more mole of H_2, and through it offsets the higher bond

energy of ethylene. The third factor, which is contributory, is that the oxidation of one mole of ethylene requires a smaller amount of oxygen (3 versus 3.5 moles), and hence there is less nitrogen in the combustion product that needs to be heated up. Thus the dependence of T_{ad} on the C/H ratio is the net effect of the bond energy, the number of H available per C atom to form H_2O, and the collateral need to heat up the N_2 for the amount of O_2 reacted.

The above dependence is the basis for the development of high-energy-density fuels as explosives or for tactical propulsion. These fuel molecules are usually highly strained, which imply large bond energies and C/H ratios. Nitrogen is also frequently present as an elemental constituent in explosives because of the associated large bond energy. Furthermore, since fuel and oxidizer are present together in these explosives as well as the monopropellants for rocket propulsion, the resulting T_{ad} can be quite high because there is no atmospheric nitrogen that needs to be heated up.

Figures 1.4.2 and 1.4.3 also show that T_{ad} behaves asymmetrically with respect to ϕ in three major aspects, namely the slope is gentler on the rich side of the maximum than that on the lean side, that there are sharp "bends" in the curves for some fuels at very rich compositions, and that it peaks slightly on the rich side of stoichiometry. These behaviors are explained in the following.

First, comparing Figures 1.4.3 and 1.4.4 and, as anticipated earlier, we see that a plot of T_{ad} versus the normalized equivalence ratio Φ yields a profile that is considerably more symmetrical about the maximum T_{ad} concentration as compared to those based on the conventionally defined equivalence ratio, ϕ. Consequently, the gentler slope for T_{ad} on the rich side in a plot involving ϕ, as compared to the steeper slope on the lean side, is largely a consequence of the asymmetrical definition of ϕ, and as such has no bearing on the nature of lean-versus-rich chemical equilibria. Thus there is merit in examining lean-versus-rich exothermicities of a thermodynamic system based on Φ.

We next consider the results (Figure 1.4.2) that, for very fuel-rich mixtures of fuels with large C/H ratios (e.g., benzene, ethylene, graphite), the T_{ad} curves exhibit sharp bends and long tails. The same tail would appear in Figures 1.4.3 and 1.4.4 if the abscissas were extended to larger values. The appearance of this tail is due to the fact that the available oxygen is not stoichiometrically sufficient to convert all carbon to CO, hence leading to the formation of graphite. This can be seen in Figure 1.4.5 where the variations of T_{ad} and the major product (CO, CO_2, graphite) concentrations are plotted as functions of the fuel content for ethylene–air mixtures.

We further note that, unlike many other fuels, acetylene is unique in that its T_{ad} at the tail part of rich stoichiometry increases with increasing fuel concentration. This is because acetylene is an exceptionally energetic fuel, with its heat of formation considerably larger than most other fuels on a per carbon atom basis. It can therefore be converted to solid carbon and H_2 spontaneously through a highly exothermic reaction,

$$C_2H_2 \rightarrow 2C(gr) + H_2, \quad \bar{q}_p = -54.2 \text{ kcal/mol}. \tag{1.4.21}$$

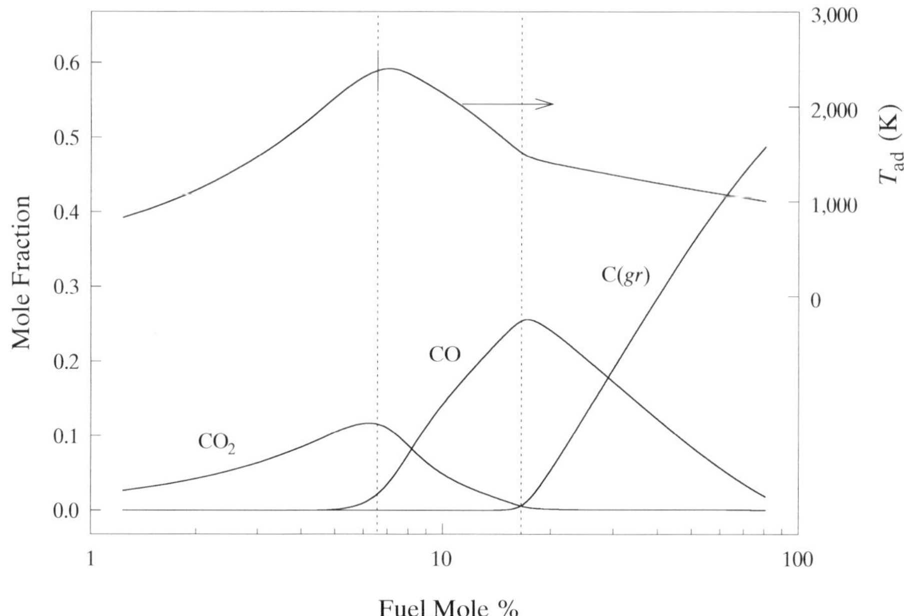

Figure 1.4.5. Equilibrium combustion products CO_2, CO and graphite, C(*gr*), and adiabatic flame temperature, T_{ad}, of ethylene–air mixtures as a function of the mole percent of fuel.

The increase in T_{ad} at high acetylene concentrations is therefore a consequence of this exothermic pyrolysis process. This explains why acetylene can explode by itself, without the presence of oxygen.

We now discuss the shifting of the maximum T_{ad} to the rich side of stoichiometry (Law, Makino & Lu 2006). To identify the cause of this shifting, in Figures 1.4.6–1.4.8 we show for methane–air mixtures the T_{ad}, the heat release per unit mass of the mixture q_p, the specific heat of the burned mixture $c_{p,2}$, and the major and minor species concentrations. Normalized equivalence ratio Φ is used so as to eliminate the asymmetrical effect due to the definition of ϕ.

We first consider the situation of frozen equilibrium when dissociation of the products CO_2 and H_2O is suppressed as defined by Eqs. (1.3.7)–(1.3.9) for the determination of the major species concentrations. The result that both T_{ad} and q_p peak at $\Phi = 0.5$, as shown in Figure 1.4.6, is caused by the need to heat up the excess reactants for off-stoichiometric concentrations, as just discussed. Since the shifting does not exist for frozen equilibrium, it is therefore reasonable to expect that product dissociation is the cause of the shifting. This possibility is further supported by noting that the heat release, which depends on the extent of dissociation, is also rich shifted.

Concerning the role of the product specific heat, $c_{p,2}$, Figure 1.4.6 shows that it monotonically increases from lean to rich mixtures for both frozen and dissociation equilibria. Thus product dissociation does not qualitatively affect the lean-to-rich increasing trend of $c_{p,2}$. This increasing trend is caused by the increased number of moles of product species formed per unit mass of the mixture reacted, especially those of H_2 and CO, as the mixture becomes progressively richer.

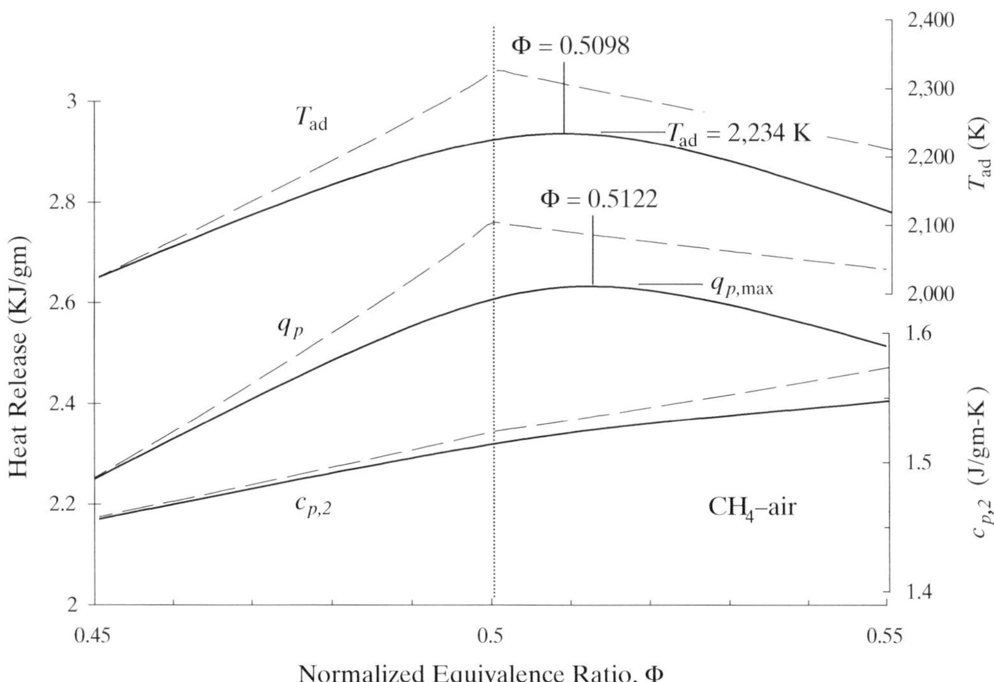

Figure 1.4.6. Adiabatic flame temperature, T_{ad}, heat of combustion, q_p, and product specific heat, $c_{p,2}$, as functions of Φ for methane–air mixtures at *STP* (– – Frozen equilibrium; — Dissociation equilibrium).

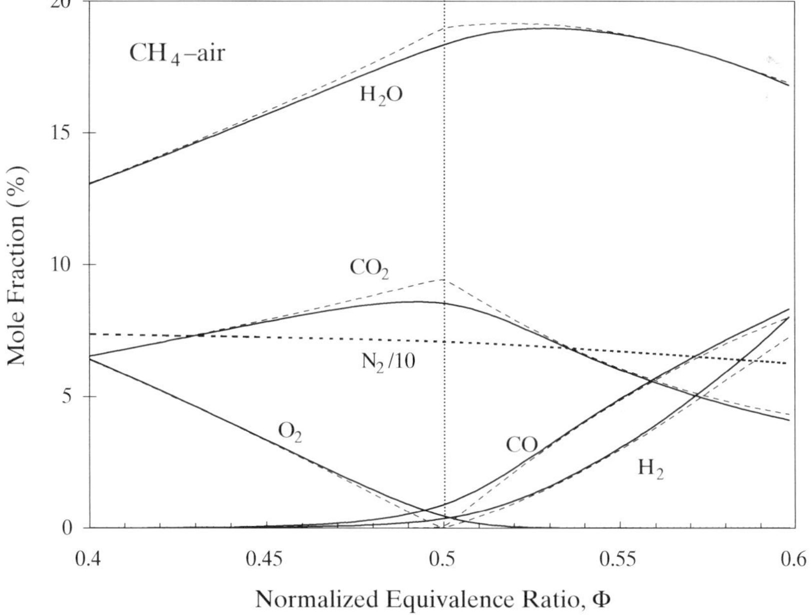

Figure 1.4.7. Concentrations of major product species as functions of Φ for methane–air mixtures at STP (– – Frozen equilibrium; — Dissociation equilibrium).

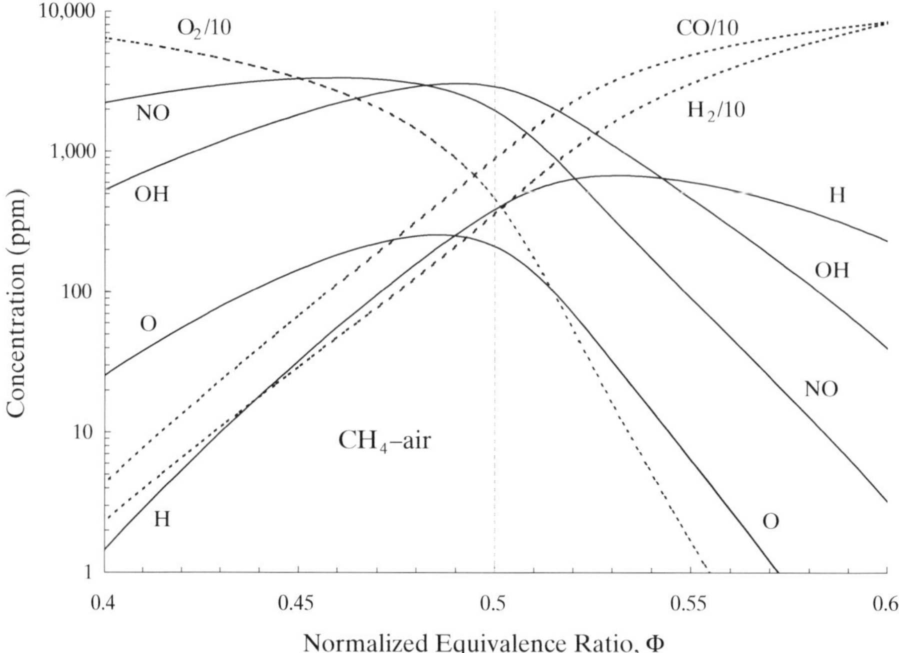

Figure 1.4.8. Concentrations of minor product species as functions of Φ for methane–air mixtures at STP.

The fact that $c_{p,2}$ increases from lean to rich then implies that it tends to cause the peak in T_{ad} to shift to the lean side. Indeed, we see that while q_p peaks at $\Phi = 0.5122$, T_{ad} is lean-shifted to $\Phi = 0.5098$ by $c_{p,2}$. This point can be quantified further by noting that since $T_{\text{ad}} \sim q_p/c_{p,2}$, we have $\delta T_{\text{ad}}/T_{\text{ad}} \approx \delta q_p/q_p - \delta c_{p,2}/c_{p,2}$. Thus in Figure 1.4.9 we have plotted the individual difference terms in this relation for small departures from $\Phi = 0.5$. The result shows that, for such small departures, the influences are linearized and the difference, $\delta q_p/q_p - \delta c_{p,2}/c_{p,2}$, is indeed almost equal to $\delta T_{\text{ad}}/T_{\text{ad}}$. We therefore conclude that the rich shifting of T_{ad} is caused by the corresponding rich shifting of the maximum heat release instead of the influence of $c_{p,2}$, which actually causes a lean shifting.

The final question is why the shifting is to the rich instead of the lean side. This can be demonstrated by considering only CO oxidation for simplicity, with an ϵ amount of dissociation,

$$\phi \text{CO} + \frac{1}{2}(\text{O}_2 + 3.76\text{N}_2) \rightarrow a(1-\epsilon)\text{CO}_2 + (b + a\epsilon)\text{CO} + \left(c + \frac{a}{2}\epsilon\right)\text{O}_2 + 1.88\text{N}_2, \tag{1.4.22}$$

where $a = \phi$, $b = 0$, and $c = (1-\phi)/2$ for lean mixtures and $a = 1$, $b = \phi - 1$, and $c = 0$ for rich mixtures. Using the definition of the equilibrium constant K_p, we have

$$K_p = \frac{p_{\text{co}}\sqrt{p_{\text{O}_2}}}{p_{\text{co}_2}} = \frac{(b + a\epsilon)}{a(1-\epsilon)}\sqrt{\frac{(c + \frac{a}{2}\epsilon)p_t}{a + b + c + \frac{a}{2}\epsilon + 1.88}}, \tag{1.4.23}$$

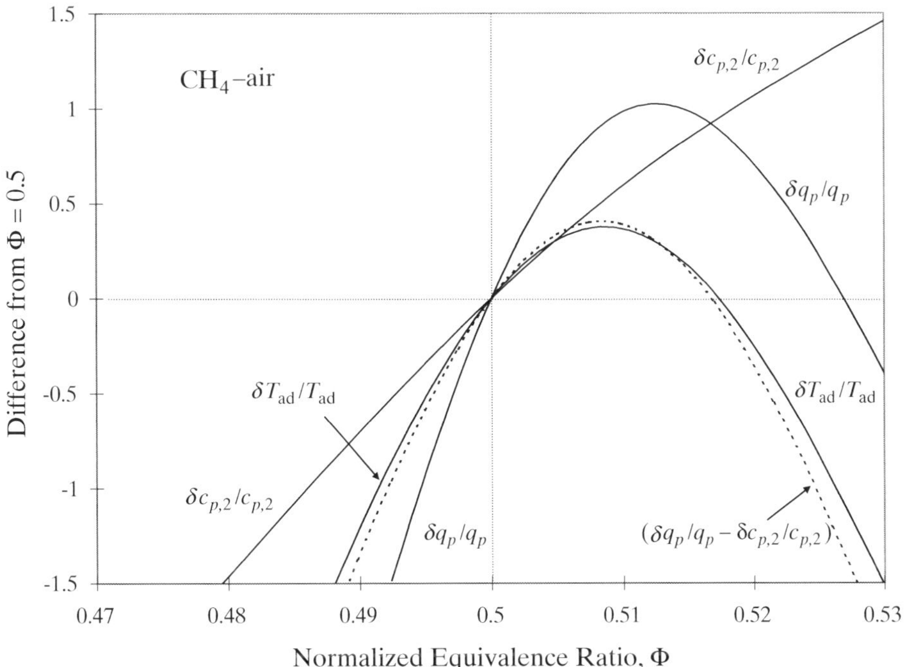

Figure 1.4.9. Difference terms for q_p, $c_{p,2}$, T_{ad}, and $(q_p - c_{p,2})$, referenced to the stoichiometric state as function of Φ for methane–air mixtures at STP.

which, for $\epsilon \ll 1$, yields $K_p \sim \epsilon\sqrt{p_t}$ and $K_p \sim \sqrt{\epsilon p_t}$ for lean and rich mixtures respectively. Consequently, for the same T_{ad} and hence $K_p(T_{ad})$, we have $\epsilon_{lean} = \sqrt{\epsilon_{rich}}$. Since $\epsilon < 1$, we have proven that $\epsilon_{lean} > \epsilon_{rich}$. The reason that there is a greater amount of dissociation on the lean side is because the dissociation of one mole of CO_2 results in one mole of CO but only half mole of O_2. Thus the lean mixtures can accommodate more O_2 from dissociation than the rich mixtures can accommodate CO.

It can also be readily demonstrated that the same conclusion holds for H_2 oxidation and hence H_2O dissociation into H_2 and $\frac{1}{2}O_2$. Combining the results of CO and H_2 oxidation, it is then obvious that the observed rich shifting for hydrocarbons is due to the dissociation of the products CO_2 and H_2O.

Since product dissociation is the cause for the shifting, and since the extent of dissociation is reduced with increasing pressure, Figure 1.4.10 shows that the shiftings in both q_p and T_{ad} are diminished as the mixture pressure increases from 1 atm to 100 atm. Indeed, we retrieve the frozen equilibrium limit as $p \rightarrow \infty$.

The reduced amount of dissociation with increasing pressure also implies that T_{ad} is higher at higher pressures, with the effect most prominent around stoichiometry where dissociation is most severe. Figure 1.4.11 shows this increasing trend of T_{ad} on pressure for the stoichiometric fuel–air mixtures of Figure 1.4.3. The pressure effect is quite modest, although it is not negligible in view of the temperature sensitivity of chemical reactions.

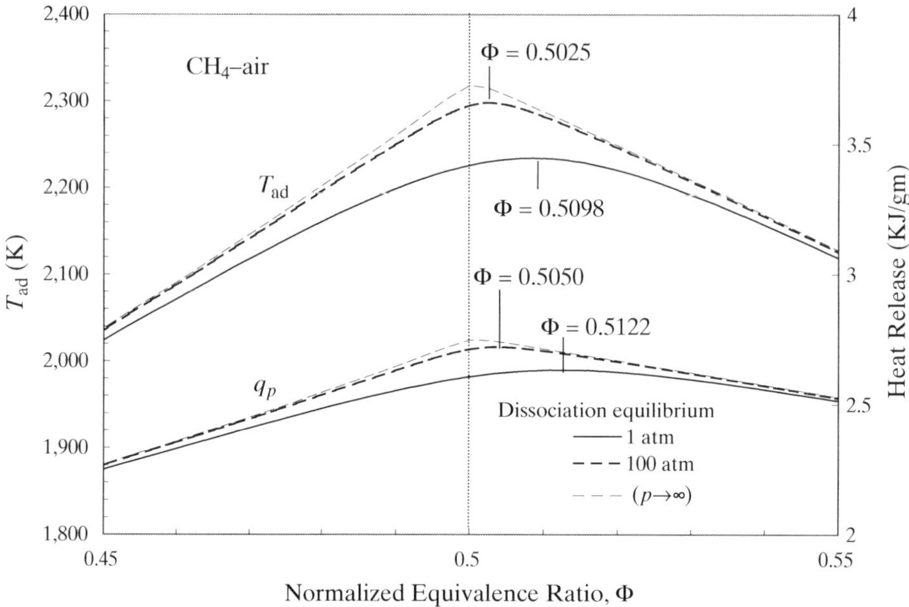

Figure 1.4.10. T_{ad} and q_p as functions of Φ for methane–air mixtures at 298.15 K and 1 and 100 atm.

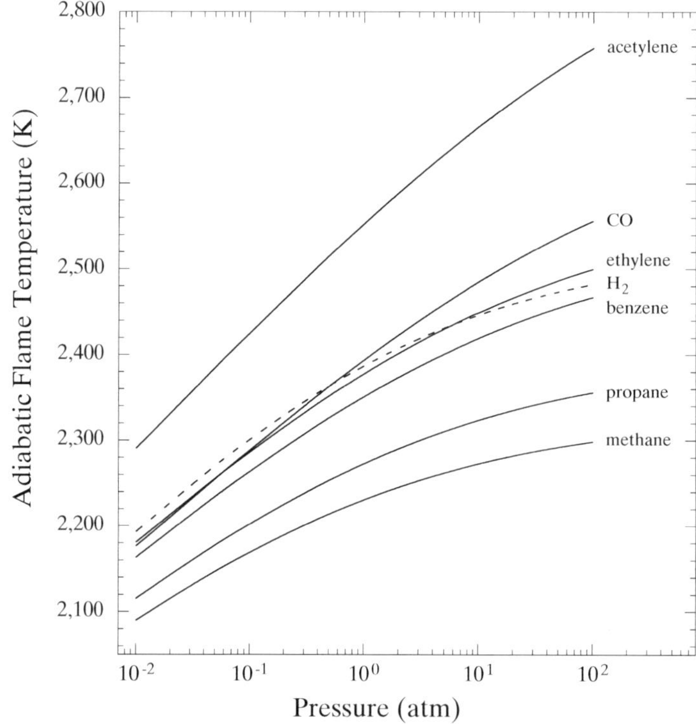

Figure 1.4.11. T_{ad} as function of pressure for various stoichiometric fuel–air mixtures at 298.15 K.

Referring back to Figures 1.4.7 and 1.4.8 for the dependence of the product composition on ϕ, we see that while CO and H_2, and O_2, are the major species for rich and lean mixtures respectively, they become the minor species for lean and rich mixtures. This substantiates our earlier discussion in Section 1.3.2 on the major–minor species model. It is also of interest to note that while the formation of NO is highly temperature sensitive (see Chapter 3), and that T_{ad} peaks on the rich side, the NO concentration actually peaks on the lean side of stoichiometric. The reason is that its formation requires oxygen, whose concentration increases rapidly as the mixture becomes leaner. The net response thus reflects the combined needs for high temperature and high oxygen concentration.

PROBLEMS

1. Show that one does not need to distinguish whether the fuel–oxidizer equivalence ratio is on molar or mass bases. Furthermore, the fuel–air equivalence ratio and fuel–oxygen equivalence ratio are the same.

2. An alcohol with the chemical formula C_mH_nOH reacts with air at an equivalence ratio ϕ. Assuming the reactions of C and H with O_2 produce only the major product species, write down the chemical reactions for $\phi = 1$, < 1, and > 1.

3. (a) For a general hydrocarbon fuel C_mH_n, derive the threshold equivalence ratio ϕ_C for carbon formation. Plot ϕ_C as a function of the carbon-to-hydrogen molar ratio.

   (b) Similarly, derive ϕ_C for a general alcohol C_mH_nOH.

   (c) Plot ϕ_C as a function of the carbon number m, up to $m = 20$, for alkanes and alkane-based alcohols.

   (d) Make appropriate observations regarding the tendency to soot in the above results.

4. The chemical reaction

$$H_2 + \frac{1}{2}O_2 \rightleftharpoons H_2O$$

   can be alternately expressed as

$$2H_2 + O_2 \rightleftharpoons 2H_2O.$$

   What is the relation between their equilibrium constants K_ps and K_cs?

5. The equilibrium constant for the formation of C_2H_5OH, CO_2, and H_2O are respectively K_1, K_2, and K_3. What are the equilibrium constants, K_p (if K_is are K_ps) and K_c (if K_is are K_cs), for the oxidative reaction of C_2H_5OH assuming only major product species?

6. Using the major–minor species model, calculate the concentrations of CO_2, H_2O, O_2, CO, and H_2 for methane–air mixtures at 1,000 K and 1 atm pressure, with $\phi = 0.6, 0.8, 1.0, 1.2$, and 1.5. Tabulate and plot your results.

7. For the following fuels burning in air, write down the stoichiometric reactions, the respective fuel–air ratios (F/A), and the heats of combustion per mole ($\bar{q}_c$) and per gm (q_c). Assume the product water is in the gaseous state. Fuels: CO, H_2, CH_4, C_7H_{16}, $C_{16}H_{34}$, C_6H_6, CH_3OH, $C_7H_{15}OH$.

What can you say about the q_cs for (a) CH_4, C_7H_{16}, $C_{16}H_{34}$, and (b) C_7H_{16}, CH_3OH, $C_7H_{15}OH$?

8. Compare the heats of combustion of n-octane, aluminum, and boron on molar, mass, and volume bases; their respective densities are 0.703 g/cm^3, 2.70 g/cm^3, and 2.34 g/cm^3. The complete oxidation product of Al and B are respectively Al_2O_3 and B_2O_3. These values indicate the potential of boron as a high-energy-density fuel, provided complete combustion can be achieved.

9. By knowing the heats of formation $\bar{h}_i^o(T^o)$ for the components of a general reaction, what is the heat of reaction $\bar{q}_p(T^1)$ evaluated at a temperature $T^1 \neq T^o$? Calculate $\bar{q}_p$ for the complete oxidation of CO at 1,000 K. Compare the $\bar{q}_p$ (1,000 K) here with the $\bar{q}_c$ (298.15 K) in Problem (7).

10. A stoichiometric amount of CO and O_2 initially at 298.15 K react adiabatically and isobarically at pressure p.

(a) If the extent of dissociation of CO_2 is ϵ, derive an expression for K_p in terms of ϵ and p.

(b) For $p = 6$ atm, calculate ϵ and the adiabatic flame temperature T_{ad}.

(c) Repeat (b), but use air instead of O_2 as the oxidizer. What can you say about the differences in the answers here as compared with those in (b)?

11. Using the major–minor species model, calculate the adiabatic flame temperatures of methane–air mixtures at 1 atm pressure, with initial temperature of 298.15 K and $\phi = 0.6$ and 0.8.

2 Chemical Kinetics

In Chapter 1, we discussed the equilibrium states of a thermochemical system. We have shown that by knowing the initial state of a reactive mixture, we can determine its final state after chemical and thermal equilibria have been established. However, the equilibrium calculations are not able to answer such relevant questions as: how does the mixture get from the initial state to the final state, and how long does it take to do so? Obviously if a particular reaction proceeds exceedingly slowly compared to other physical or chemical processes of interest, it is likely that this reaction could be either irrelevant to the system's behavior or not of controlling importance in being the rate-limiting step in the system's evolution.

In this chapter, we first present the phenomenological law describing the general dependence of reaction rates on reactant concentrations and temperature. We then discuss multistep reactions and some approximation techniques used to simplify the representation of these reactions.

In Section 2.2, we present the specific functional dependence of the reaction rate on temperature—the Arrhenius law. We then derive and discuss three theories of reaction rates. The collision theory is based on the kinetic theory of gases, counting the frequency of molecular collisions that are energetic enough to cause the colliding molecules to react. The transition state theory examines the activated state of molecules and derives the reaction rate by considering the characteristic times and energies associated with their transition from activated to reacted states. Section 2.3 presents the unimolecular reaction theory, which examines the dynamics of collision energization of an isolated molecule and its subsequent de-energization and reaction.

In Section 2.4, we discuss chain reaction mechanisms and how these mechanisms can affect the reaction rate in qualitatively different manners. This is followed by Section 2.5 on experimental and computational techniques used to study reaction rates and mechanisms.

Detailed exposition on the basics of chemical kinetics can be found in Benson (1960), Laidler (1965), Williams (1985), Pilling and Seakins (1995), and Gardiner (1999). The oxidation mechanisms of specific fuel systems involving hydrogen, carbon monoxide, and various hydrocarbons, and the formation and destruction of pollutants will be studied in Chapter 3.

2.1. PHENOMENOLOGICAL LAW OF REACTION RATES

2.1.1. The Law of Mass Action

For a single, forward chemical reaction represented by

$$\sum_{i=1}^{N} \nu_i' \mathrm{M}_i \overset{k_f}{\rightarrow} \sum_{i=1}^{N} \nu_i'' \mathrm{M}_i, \tag{2.1.1}$$

the rate of change in the molar concentration c_i (moles per unit volume) of species i,

$$\hat{\omega}_i = \frac{dc_i}{dt}, \tag{2.1.2}$$

is uniquely related to $\hat{\omega}_j$ of species j by

$$\frac{\hat{\omega}_i}{\nu_i'' - \nu_i'} = \frac{\hat{\omega}_j}{\nu_j'' - \nu_j'} = \omega, \tag{2.1.3}$$

where the subscript f in k_f designates the (forward) direction of the reaction. Since ω (moles per unit volume per second) is species independent, it can be defined as the reaction rate of (2.1.1). Then the phenomenological law of mass action states that ω is proportional to the product of the concentrations of the reactants, or

$$\omega = k_f(T) \prod_{i=1}^{N} c_i^{\nu_i'}, \tag{2.1.4}$$

where the proportionality factor $k_f(T)$, to be specified later, is called the specific reaction rate constant and is primarily a function of temperature.

Equation (2.1.4) is based on the microscopic viewpoint that the reaction rate is proportional to the collision frequency, which in turn is proportional to the product of the species concentrations. Thus implicit in Eq. (2.1.4) is the requirement that reaction (2.1.1) is an elementary one in which the reaction actually takes place at the molecular level.

As an example of an elementary reaction, we have $\mathrm{H} + \mathrm{HO_2} \rightarrow \mathrm{OH} + \mathrm{OH}$, which could be an important step in the oxidation of hydrogen and hydrocarbons, as will be discussed in Chapter 3. For this reaction we can write

$$\omega = -\frac{d[\mathrm{H}]}{dt} = -\frac{d[\mathrm{HO_2}]}{dt} = \frac{1}{2}\frac{d[\mathrm{OH}]}{dt},$$

while, from the law of mass action, the reaction rate is given by

$$\omega = k_f[\mathrm{H}][\mathrm{HO_2}].$$

2.1.2. Reversible Reactions

Associated with every forward reaction (2.1.1) is the corresponding backward reaction

$$\sum_{i=1}^{N} \nu_i'' \mathrm{M}_i \overset{k_b}{\rightarrow} \sum_{i=1}^{N} \nu_i' \mathrm{M}_i . \tag{2.1.5}$$

Therefore the net reaction rate in the presence of both forward and backward reactions is $\hat{\omega}_i = \hat{\omega}_{i,f} + \hat{\omega}_{i,b} = (\nu_i'' - \nu_i')(\omega_f - \omega_b) = (\nu_i'' - \nu_i')\omega$, with ω given by

$$\omega = k_f \prod_{i=1}^{N} c_i^{\nu_i'} - k_b \prod_{i=1}^{N} c_i^{\nu_i''} . \tag{2.1.6}$$

Only one of k_f and k_b needs to be determined. This is because at equilibrium, $\omega \equiv 0$, the rate of forward reaction is balanced by that of backward reaction, and Eq. (2.1.6) yields

$$\frac{k_f}{k_b} = \prod_{i=1}^{N} c_i^{(\nu_i'' - \nu_i')} . \tag{2.1.7}$$

By definition Eq. (2.1.7) is just the equilibrium constant K_c as given by Eq. (1.2.36),

$$\frac{k_f}{k_b} = K_c . \tag{2.1.8}$$

Substituting Eq. (2.1.8) into Eq. (2.1.6), we have

$$\omega = k_f \left(\prod_{i=1}^{N} c_i^{\nu_i'} - K_c^{-1} \prod_{i=1}^{N} c_i^{\nu_i''} \right) . \tag{2.1.9}$$

Since K_c can be usually determined to a much greater accuracy than k_b, Eq. (2.1.9) is preferred over Eq. (2.1.6).

In many instances the backward reaction occurs at a much slower rate than the forward reaction, that is, $\omega_f \gg \omega_b$. For example, the backward reaction of $H + O_2 \rightarrow OH + O$ involves the collision between two radicals. Since concentrations of radicals are much lower than those of the starting reactants, the frequency of collision for the backward reaction is much lower than that of the forward reaction, which involves only one radical species. There are also situations in which the products formed are continuously eliminated from the reaction region such that $c_{\text{products}} \simeq 0$. Furthermore, the progress of the backward reaction can be substantially slowed if it is endothermic or has a high energy barrier, as will be shown in section 2.2.1. One can then approximate Eq. (2.1.6) by

$$\omega \simeq k_f \prod_{i}^{N} c_i^{\nu_i'} . \tag{2.1.10}$$

Such a reaction is called an irreversible reaction. The approximation that $\omega_f \gg \omega_b$ obviously breaks down as the reaction approaches equilibrium, at which the forward and backward reaction rates are equal. However, at this stage, an accurate description of these reactions is probably not important anyway.

2.1.3. Multistep Reactions

There are seldom reactions in which the original reactants interact with each other in a single step, at the molecular level, and produce the final products as shown in (2.1.1). For example, the representation of stoichiometric methane oxidation by

$$CH_4 + 2O_2 \rightarrow CO_2 + 2H_2O \tag{2.1.11}$$

is a gross simplification. There are actually a large number of intermediate elementary steps with intermediate species involved before the final formation of products. The products also consist of many more species than just CO_2 and H_2O, as we have seen in the last chapter. Therefore if there are K such intermediate reactions involved,

$$\sum_{i=1}^{N} \nu'_{i,k}\mathrm{M}_i \underset{k_{k,b}}{\overset{k_{k,f}}{\rightleftharpoons}} \sum_{i=1}^{N} \nu''_{i,k}\mathrm{M}_i, \quad k = 1, 2, \ldots, K, \tag{2.1.12}$$

then the generalized law of mass action is

$$\omega_k = k_{k,f}\prod_{i=1}^{N} c_i^{\nu'_{i,k}} - k_{k,b}\prod_{i=1}^{N} c_i^{\nu''_{i,k}}, \quad k = 1, 2, \ldots, K, \tag{2.1.13}$$

such that

$$\hat{\omega}_i = \sum_{k=1}^{K}(\nu''_{i,k} - \nu'_{i,k})\omega_k. \tag{2.1.14}$$

It may be noted that in the above the subscript for $\hat{\omega}$ is for species i, whereas that for ω is for reaction k.

Obviously if the K elementary steps of (2.1.12) are identified and their associated reaction rate constants are known, the rates of production and destruction of any species can be precisely determined. However, this is frequently not the case due to the difficulty to identify the elementary steps as well as their reaction rate constants. Furthermore, even if these elementary steps are known precisely, the task of solving a combustion flow field by including all reactions is an extremely taxing one, even with the most powerful computers. Therefore, various approximation techniques have been introduced to facilitate the solution procedure. In the following we discuss the concepts involved with several of these approximations.

2.1.4. Steady-State Approximation

During the course of a complex chemical reaction scheme leading to the conversion of the reactants to the products, reaction intermediates are produced. Some of these intermediates are chain carriers, which play a crucial role in the propagation of the overall scheme because they provide linkage between the individual reactions. The individual reactions in which these intermediates participate frequently proceed at rapid rates, although the concentrations of these intermediates and thereby their net rates of change are quite low. In other words, the consumption and regeneration of the intermediates occur at rapid, but approximately equal rates such that their concentrations can be considered to remain constant. Thus if i is such an intermediate, and if we express its reaction rate, Eq. (2.1.14), by

$$\hat{\omega}_i = \frac{dc_i}{dt} = \hat{\omega}_i^+ - \hat{\omega}_i^-, \tag{2.1.15}$$

where $\hat{\omega}_i^+$ and $\hat{\omega}_i^-$ respectively represent the rates of all the generation and consumption reactions, then the steady-state approximation assumes

$$\left|\frac{dc_i}{dt}\right| \ll \left(\hat{\omega}_i^+, \hat{\omega}_i^-\right), \tag{2.1.16}$$

such that

$$\hat{\omega}_i^+ = \hat{\omega}_i^-. \tag{2.1.17}$$

This approximation therefore reduces the solution of a differential equation, Eq. (2.1.15), to that of an algebraic one, Eq. (2.1.17), for the concentration c_i, which can then be used in the overall reaction scheme. The algebraic expression obtained from Eq. (2.1.17), however, could be implicit in c_i. Consequently, the determination of c_i might involve algebraic iterations, which could still be computationally taxing, especially when convergence is not readily achievable.

It is important to note that the steady-state approximation does not imply $dc_i/dt = 0$. Rather it simply requires that the magnitudes of $\hat{w}_i^+$ and $\hat{w}_i^-$ are both large compared with that of dc_i/dt. Furthermore, the identification of the intermediates to which the steady-state approximation is applicable is not always obvious. The approximation may also just hold locally, either spatially or during a transient evolution. Thus it may be necessary to check the validity of the approximation by evaluating dc_i/dt by using results from the approximate solution, and assess the adequacy of (2.1.16).

2.1.5. Partial Equilibrium Approximation

Partial equilibrium approximation assumes that the forward and backward rates of a reaction k are much larger than the net reaction ω_k such that we can set $\omega_k \approx 0$ in Eq. (2.1.13), yielding

$$k_{k,f} \prod_{i=1}^{N} c_i^{\nu'_{i,k}} \simeq k_{k,b} \prod_{i=1}^{N} c_i^{\nu''_{i,k}}. \tag{2.1.18}$$

It is important to note that ω_k is small only when it is compared to those of the forward and backward reactions in Eq. (2.1.13). It is, however, not necessarily small when compared to $\hat{\omega}_i$ in Eq. (2.1.14). Therefore while Eq. (2.1.18) is obtained by setting $\omega_k = 0$, it does not imply that ω_k is necessarily small in Eq. (2.1.14), and we can directly apply $\omega_k = 0$.

A systematic procedure to apply the partial equilibrium approximation is the following. If this approximation is applied to, say, reaction $k = 1$, then we identify a species $i = 1$ and solve for c_1 in terms of the other concentrations c_i, $i \neq 1$, from $\omega_1 = 0$ and hence Eq. (2.1.18). However, ω_1 does not need to vanish when compared to $\hat{\omega}_i$. Thus we express $\hat{\omega}_i$ in Eq. (2.1.14) as

$$\hat{\omega}_i = \frac{dc_i}{dt} = (\nu''_{i,1} - \nu'_{i,1})\omega_1 + \sum_{k=2}^{K} \hat{\omega}_{i,k}, \quad i = 1, 2, \ldots, N, \tag{2.1.19}$$

which becomes, for $i = 1$,

$$\hat{\omega}_1 = \frac{dc_1}{dt} = (\nu''_{1,1} - \nu'_{1,1})\omega_1 + \sum_{k=2}^{K} \hat{\omega}_{1,k}. \tag{2.1.20}$$

Eliminating ω_1 between Eqs. (2.1.19) and (2.1.20) results in

$$\frac{dc_i}{dt} = \sum_{k=2}^{K} \hat{\omega}_{i,k} + \left(\frac{\nu''_{i,1} - \nu'_{i,1}}{\nu''_{1,1} - \nu'_{1,1}}\right)\left[\frac{dc_1}{dt} - \sum_{k=2}^{K} \hat{\omega}_{1,k}\right], \quad i = 2, \ldots, N. \tag{2.1.21}$$

Since c_1 is known as a function of c_i, $i \neq 1$, we have therefore reduced the number of equations that need to be solved from N to $(N - 1)$.

2.1.6. Approximations by Global and Semiglobal Reactions

The steady-state and partial-equilibrium approximations are systematic reductions of a complete kinetic scheme to a simpler one involving a smaller number of reactions and unsteady species. The final results, of course, still depend on the kinetic constants of the original elementary steps. The limiting case of such a reduction would be a lumped, one-step overall reaction, in the manner of, say, Eq. (2.1.11), involving the starting reactants. Such a representation is called a global reaction. It is obvious that, instead of the global reaction, more chemical information can be retained by reducing the complete mechanism to several steps, involving the starting reactants and a few important species participating in the overall reaction scheme. Such a reduced mechanism is said to involve semiglobal reactions.

Although the above reduction procedure is systematic, it can be algebraically tedious and unwieldy, especially because the algebraic relations for c_i are frequently nonlinear and cannot be explicitly expressed, as just mentioned. Furthermore, the range of applicability of the reduced mechanism also becomes progressively restricted as more species are assumed to be in steady state and more reactions are assumed to be in partial equilibrium. Thus an alternate approach is simply to represent the detailed scheme by a few postulated reaction steps involving some important species, with the kinetic constants empirically determined. The simplest representation is the one-step overall reaction involving the starting reactants,

$$\text{Fuel} + \text{Oxidizer} \xrightarrow{k} \text{Products}, \tag{2.1.22}$$

with the reaction rate given by

$$\omega = k \prod_{i=1}^{N} c_i^{n_i}, \tag{2.1.23}$$

where n_i, the exponent to c_i, is called the reaction order with respect to the ith species. It is to be distinguished from the stoichiometric coefficient ν'_i used in, say, Eq. (2.1.4). We shall further address their differences in the next section.

It is obvious that not all multistep reactions can be approximated by a one-step overall reaction. Frequently it is found that the bulk behavior of a complex reaction scheme can only be satisfactorily described by reactions involving two, three, or even more steps, constituting an empirical semiglobal reaction mechanism. The range of applicability of such empirical global and semiglobal reaction mechanisms depends on the particular combustion phenomenon, as well as the range of the parameters characterizing the phenomenon, through which the empirical constants are determined.

2.1.7. Reaction Order and Molecularity

The molecularity of a reaction is the number of atoms or molecules that interact with each other at the molecular level, leading to the completion of the reaction. Thus for the elementary reaction (2.1.1), ν_i' is the molecularity of the reaction with respect to species i, and $\sum_{i=1}^{N} \nu_i'$ is the overall molecularity of the reaction.

Since the overall molecularity represents the number of molecules that collide within their molecular range of interaction, for a very brief period of time when such an interaction is effective, and because of the dilute nature of gases, the higher the molecularity the less frequent is the event. Indeed, most elementary reactions have overall molecularity of two, representing two-body collisions. However, reactions with overall molecularity of three are important ones for the recombination of two colliding molecules, for example, $H + O_2 + M \rightarrow HO_2 + M$, because from the consideration of simultaneous momentum and energy conservation, a third body, M, is needed to carry away the excess energy. For large, complex molecules the excess energy can be absorbed by the various vibrational modes of the combined molecules and a third body may not be needed. Active radicals are frequently rendered inactive through such recombination reactions. Reactions with overall molecularity of one are also possible. For example, a molecule can become excited when irradiated by a light source of a particular frequency and subsequently dissociate into two molecules. A chemically active molecule can also become deactivated upon collision with a solid surface.

The order of a reaction, n_i, indicates the influence of the concentration of the ith reactant on the reaction rate. As such, the experimentally determined stoichiometric coefficients based on concentration variations are reaction orders instead of reaction molecularities. When the reaction under study is indeed elementary, then the reaction orders determined are the respective molecularities of the reaction. However, when the species whose concentrations are measured are the starting and final species of a global or semiglobal reaction, then the reaction orders represent the net effects of the molecularities of the individual elementary reactions. In such cases the reaction orders do not need to be integers. Furthermore, since reactions with molecularities of 1 and 2 can either promote or retard the progress of a global reaction, while reactions with molecularities of 3 are usually recombinatory in nature and hence retarding, the overall reaction order of a global reaction should be no larger than 2. Indeed, it is not unreasonable to anticipate that the overall reaction order can actually be negative for situations in which three-body termination reactions dominate. We shall encounter instances of negative n_i in our later studies.

A higher order reaction may exhibit a lower order behavior when some of the reactants are present in large quantities such that their concentrations change very little during the reaction. A prominent example is the unimolecular reaction to be studied in Section 2.3, for which the reaction order is 1 at high pressures but becomes 2 at low pressures. The measured reaction order is sometimes called the pseudo-molecularity of the reaction. This potential loss of sensitivity at large concentrations also forms the basis of the isolation method in determining the reaction orders of individual components of a complex reaction scheme. That is, by keeping all but one of the reactants in high concentrations, the reaction order of the lean component can be approximately identified as the apparent overall reaction order, which can be easily measured. This concept is also useful in modeling premixed combustion phenomena because the concentration of the stoichiometrically abundant reactant can be assumed to be constant during the course of reaction.

Since reaction order describes both elementary and global reactions, it will be used from now on in specifying all reactions.

2.2. THEORIES OF REACTION RATES: BASIC CONCEPTS

2.2.1. The Arrhenius Law

The specific reaction rate constant $k(T)$ gives the functional dependence of the reaction rate on temperature. For an elementary reaction the Arrhenius law states that

$$\frac{d\ln k(T)}{dT} = \frac{E_a}{R^o T^2}, \tag{2.2.1}$$

where E_a is called the activation energy of the reaction, having the unit of cal/mole or joule/mole. If E_a is a constant with respect to temperature, integrating Eq. (2.2.1) yields

$$k(T) = Ae^{-E_a/R^o T}, \tag{2.2.2}$$

where A is called the frequency factor or the preexponential factor. We note that since R^o is a constant, it is convenient to define a new quantity,

$$T_a = \frac{E_a}{R^o},$$

and call it the activation temperature of the reaction. Furthermore, following tradition, we shall also use the uppercase letter to designate molar quantities.

For constant values of A and E_a, a plot of $\ln k(T)$ versus $1/T$ exhibits a linear relationship, with A and E_a respectively determined from the intercept and slope of such a plot. However, for many elementary reactions A is found to be temperature dependent over an extended temperature range. A modified Arrhenius equation can

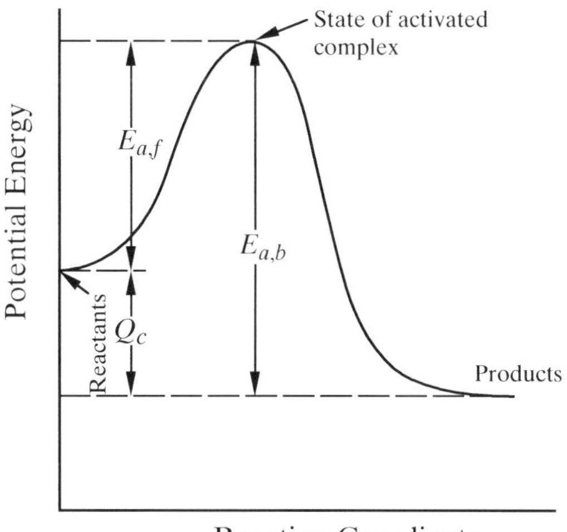

Figure 2.2.1. Potential energy diagram showing the concept of reaction activation.

thus be introduced by expressing $A = A(T) = BT^{\alpha}$ such that

$$k(T) = BT^{\alpha}e^{-E_a/R^oT}, \tag{2.2.3}$$

where B is a constant, and α is the temperature exponent.

The parameters in Eq. (2.2.3) are usually determined by a numerical fitting of the experimental data. When reporting such kinetic rate expressions, it is essential to specify the temperature range over which the fitting is performed. Likewise, when using such kinetic expressions, caution should be exercised when extending beyond their respective ranges of fitting.

For a one-step global reaction or multistep semiglobal reaction scheme, the description of the kinetic rate expression may be more complex than that for elementary reactions. Frequently, a numerical fitting of the experimental data according to Eq. (2.2.2) or Eq. (2.2.3), with $\alpha \neq 0$, constitutes an attempt to force a detailed reaction mechanism to conform to a substantially simplified scheme. Since each elementary reaction has its own activation energy that dictates the temperature range over which the reaction is most effective, the existence of a finite, discernible curvature may imply the need for a more detailed description of the kinetic mechanism. Therefore if it is necessary to use such global or semiglobal schemes, then the curve may need to be approximated by a series of segments according to Eq. (2.2.2) or Eq. (2.2.3), each of which has its own kinetic constants of frequency factor, temperature exponent, and activation energy, as well as the temperature range over which it is valid.

2.2.2. The Activation Energy

The sensitivity of a chemical reaction to temperature variation depends critically on the activation energy, which represents the minimum energy the colliding molecules must possess for the reaction to be possible. Figure 2.2.1 shows the situation for a forward exothermic reaction, in which the reactants must pass through a highly energized, activated complex stage before they are converted to products. It is indicated that the energy difference between the activated complex and the reactants is equal to $E_{a,f}$. The factor $\exp(-E_{a,f}/R^oT)$ then physically represents the fraction of the collisions having an energy higher than $E_{a,f}$. For a given T, the Arrhenius factor decreases rapidly with increasing $E_{a,f}$. The larger the $E_{a,f}$, the more sensitive is the reaction to temperature variations.

Figure 2.2.1 also shows that

$$E_{a,f} + Q_c = E_{a,b}, \tag{2.2.4}$$

where Q_c is the heat release (per mole) in the reaction, and $E_{a,b}$ is the activation energy for the backward reaction. It is clear that the backward reaction for an exothermic process is a slower process because of the larger activation energy $E_{a,b}$. An alternate interpretation is that since the backward reaction is endothermic, $E_{a,b}$ has to be at least as large as the heat absorbed in the reaction. Therefore it is slow except at high temperatures.

The largeness of the activation energy is measured by the Arrhenius number, defined as

$$Ar = \frac{E_a}{R^o T_{\max}} = \frac{T_a}{T_{\max}}, \tag{2.2.5}$$

where $T_{\max}$ is a reference, maximum temperature in the flow field, for example, the flame temperature. Thus Ar is a measure of the ratio of the activation energy to the maximum thermal energy in the flow. To obtain an estimate of the magnitude of Ar, if we let E_a to assume values around 20 to 60 kcal/mole for noncatalyzed, overall reactions of interest to combustion, and $T_{\max}$ to be between 500 K and 3,000 K, then Ar is between 4 and 60. Thus Ar can be considered to be a large number in studies of many combustion phenomena.

Reaction characteristics can be profoundly affected by the sensitive nature of the Arrhenius factor for large activation energy reactions. Figure 2.2.2 shows a normalized plot of

$$\frac{\exp(-T_a/T)}{\exp(-T_a/T_{\max})} = \exp\left[Ar\left(1 - \frac{T_{\max}}{T}\right)\right]$$

versus $T/T_{\max}$. It is seen that the maximum reaction rate is attained at the maximum temperature. Furthermore, for small Ar the reaction rate increases progressively from $T = 0$ to $T_{\max}$. However, for large values of Ar, the reaction rate is suppressed until T approaches $T_{\max}$, or $T/T_{\max} \to 1$, at which it increases rapidly. This is a crucial feature of reactions of interest to combustion.

To further demonstrate the importance of the concept of large activation energy reactions, consider the first-order reaction of a homogeneous mixture whose reaction rate is

$$\frac{dc}{dt} = -Bce^{-E_a/R^oT}. \tag{2.2.6}$$

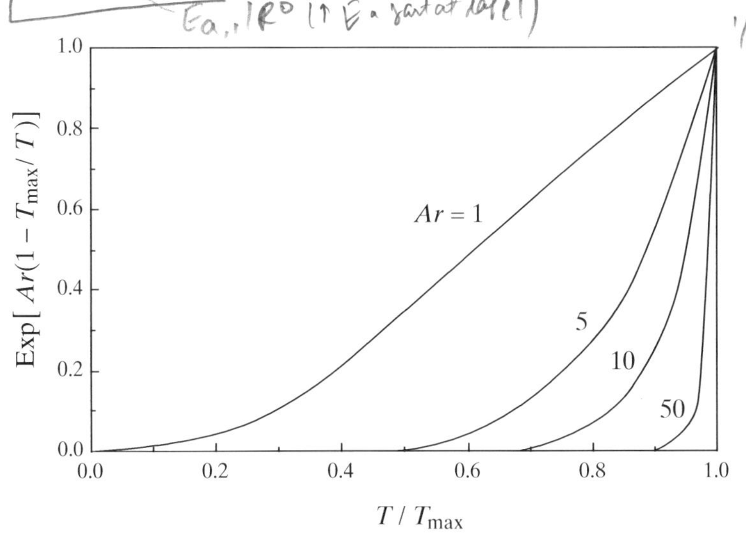

Figure 2.2.2. Normalized Arrhenius factor versus normalized temperature, demonstrating the sensitivity of the Arrhenius factor with the Arrhenius number Ar.

Figure 2.2.3 shows that $(-dc/dt)$ is a highly peaked function of time: the initial rapid rise is due to the sensitive variation of the Arrhenius factor in response to the increase in temperature, while the subsequent decline is due to depletion of the reactant $(c \to 0)$ as the reaction is completed. This implies that a reaction with large activation energy is confined to either a very small time interval (e.g., a rapid thermal explosion) or a very small spatial volume (e.g., a thin flame). This concept is useful in understanding and analyzing combustion phenomena and flame structures. Equation (2.2.6) will be studied in more detail in connection with the thermal explosion problem in Section 8.1.2.

It is also important to recognize that there are many reactions whose activation energies are zero, and hence are not temperature sensitive. In fact some of them exhibit a weak negative temperature dependence. Prominent among them are the

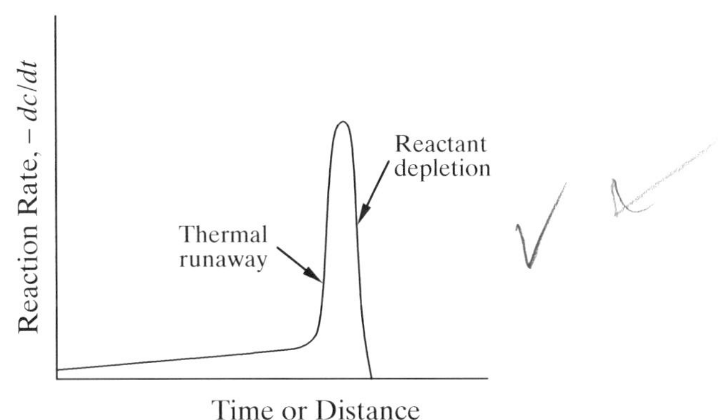

Figure 2.2.3. Typical reaction rate profile for large activation energy reactions.

three-body recombination reactions, mentioned earlier, in which the requirement for the consummation of the reaction is the presence of the third body to carry away the excess energy instead of the availability of the activation energy to initiate the reaction. Frequently these reactions are highly exothermic, and hence are the sources of thermal energy in the flow.

2.2.3. Collision Theory of Reaction Rates

The collision theory of reaction rates equates the reaction rate with the rate of molecular collision having a collision energy exceeding the activation energy. Thus if we assume that the collision energy necessary to effect molecular change is derived from the relative translational energy between the colliding molecules, that the gas is sufficiently dilute such that only two-body collisions are of importance, and that the equilibrium Maxwell velocity distribution function can still be used for the highly transient process of chemical reaction, then the reaction rate can be determined by summing over all possible collisions satisfying the requirement of minimum collision energy.

A detailed derivation of this theory allowing for three-dimensional collision dynamics, and the subsequent comparison with experimental results (Fowler & Guggenheim 1939), show that the effective collision energy is the component of the relative translational energy along the line of centers instead of the total relative translational energy. Physically this component of the relative motion represents a head-on collision situation, while the other two components normal to the line of centers only affect the dynamics of the center of mass, and therefore are not effective in chemical transformation.

In view of the above considerations, we shall present a simplified derivation involving only one-dimensional, head-on collisions. Thus if we assume that within unit volume there are n_i molecules of species i and n_j molecules of species j, and that these molecules are rigid, nonattracting spheres of masses m_i and m_j, and diameters σ_i and σ_j, respectively, then within unit time a collision between a particle of i and a particle of j would lie within a cylinder of volume $\pi\sigma_{i,j}^2\overline{V}_{i,j}$ (Figure 2.2.4), where $\sigma_{i,j} = (\sigma_i + \sigma_j)/2$ is the maximum separation distance between the centers of the molecules for collision to be possible, and $\overline{V}_{i,j}$ is their relative velocity. For a gas at temperature T, the molecules have a spectrum of velocities given by the Maxwell velocity distribution function, which has an average velocity

$$\overline{V}_{i,j} = \left(\frac{8k^oT}{\pi m_{i,j}}\right)^{1/2}, \tag{2.2.7}$$

where k^o is the Boltzmann constant and $m_{i,j} = m_i m_j/(m_i + m_j)$ the reduced mass.

Summing over n_i and n_j, the total number of collisions per unit volume per unit time between all the molecules of i and j is

$$Z_{i,j} = \pi\sigma_{i,j}^2 n_i n_j \left(\frac{8k^oT}{\pi m_{i,j}}\right)^{1/2}. \tag{2.2.8}$$

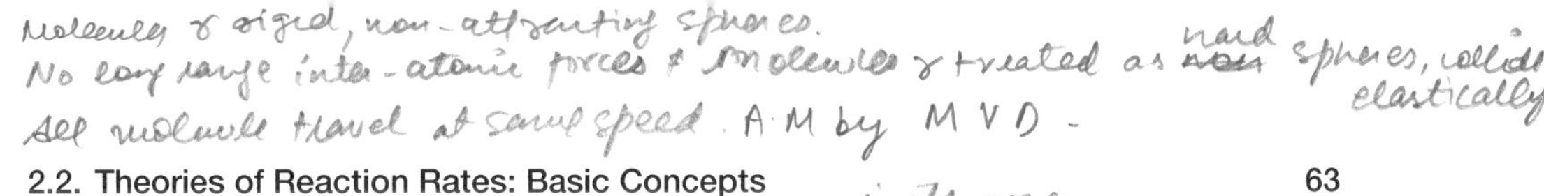

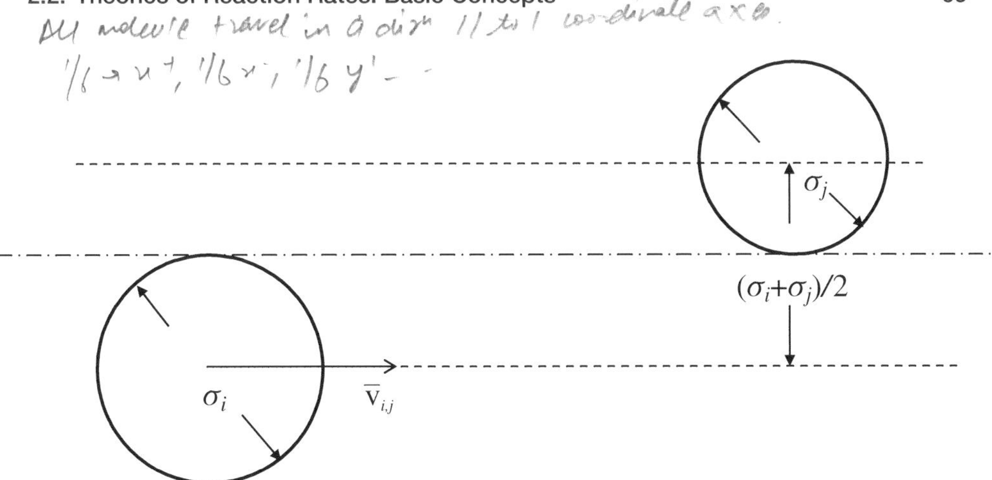

Figure 2.2.4. Schematic showing the collision volume swept by molecules of diameters σ_i and σ_j with a relative velocity $\overline{V}_{i,j}$.

Only a fraction of $Z_{i,j}$ involves collisions that are sufficiently energetic to effect a reaction. The Boltzmann distribution has the property that the number of molecules n^* that possess an amount of energy in excess of E^* is

$$\frac{n^*}{n} = e^{-E^*/R^oT}. \tag{2.2.9}$$

Thus if we define

$$Z^*_{i,j} = \pi\sigma^2_{i,j} n^*_i n^*_j \left(\frac{8k^oT}{\pi m_{i,j}}\right)^{1/2} \tag{2.2.10}$$

as the collision frequency involving molecules of i and j having energies in excess of E^*_i and E^*_j respectively, and set $(E^*_i + E^*_j)$ to E_a, which is the minimum energy to effect reaction, we have

$$Z^*_{i,j} = Z_{i,j}e^{-E_a/R^oT}. \tag{2.2.11}$$

Identifying

$$Z^*_{i,j} = -\frac{dn_i}{dt} = -\frac{dn_j}{dt}, \tag{2.2.12}$$

and noting that $c_i = n_i/A^o$, where A^o is Avogadro's number, 6.022×10^{23}, the reaction rate $\hat{\omega}_i = \omega$ of Eq. (2.1.2) is simply

$$\begin{aligned}\omega &= -\frac{dc_i}{dt} = -\frac{dc_j}{dt} \\ &= A^o\sigma^2_{i,j}\left(\frac{8\pi k^oT}{m_{i,j}}\right)^{1/2} c_ic_je^{-E_a/R^oT}.\end{aligned} \tag{2.2.13}$$

Comparing Eq. (2.2.13) with the reaction rate expressions given by the law of mass action, Eq. (2.1.4), and the Arrhenius law, Eq. (2.2.2), we have

$$\omega = A(T)c_ic_je^{-E_a/R^oT}, \tag{2.2.14}$$

with the frequency factor $A(T)$ identified as

$$A(T) = A^o \sigma_{i,j}^2 \left(\frac{8\pi k^o T}{m_{i,j}} \right)^{1/2}, \tag{2.2.15}$$

which is a function of T. We have therefore successfully identified results from the collision theory with those of the phenomenological law of mass action and the Arrhenius law, with a temperature-dependent frequency factor characterized by $\alpha = 1/2$.

As an example, consider the reaction involving hydrogen iodide, $2HI \rightarrow H_2 + I_2$, at $T = 600$ K (Fowler & Guggenheim 1939). Using $\sigma_{HI} = 3.5 \times 10^{-8}$ cm, $m_{HI} \simeq (127\ \text{g/mole})/A^o$, $k^o = 1.38 \times 10^{-16}$ erg/K, and $A^o = 6.02 \times 10^{23}$/mole, Eq. (2.2.15) gives $A = 5.1 \times 10^{13}$ cm^3/mole-sec. The experimental value is 5.0×10^{13} cm^3/mole-sec. While this close agreement is somewhat fortuitous considering the various assumptions made in the formulation and the difficulty in obtaining kinetic data of high accuracy, it does illustrate the reasonable predictive capability of this simple theory.

There are, however, experimentally determined frequency factors that are considerably smaller than the values given by Eq. (2.2.15), say, by a factor of 10^4 to even 10^8. Such slow reaction rates cannot be explained by the simple theory just presented. Mechanistically, the collision and subsequent consummation of reaction between two molecules involve more than just the transfer of translational energies. Since polyatomic molecules also possess rotational and vibrational motion, the state of the molecules is highly nonequilibrium during collision, with active energy transfer among these various modes of excitation. Furthermore, since the structure of polyatomic molecules is not spherically symmetric, the collision efficiency should also depend on the relative geometrical orientation of the molecules upon collision. For example, the efficiency in causing dissociation of a diatomic molecule by a colliding particle obviously depends on whether the collision occurs along or perpendicular to the line of centers of the atoms. These nonequilibrium and configurational influences have been accounted for by introducing a steric factor ψ such that the frequency factor is expressed as

$$A \rightarrow Z\psi. \tag{2.2.16}$$

The steric factor cannot be determined from the simple collision theory.

We shall next present the transition state theory, which has greater capability to describe the reaction rate, including the effect represented by the steric factor.

2.2.4. Transition State Theory of Reaction Rates

The transition state theory of reaction rates examines the state of the activated complex and its influence on the reaction rate. First it is recognized that upon collision between molecules of the reactants with a collision energy in excess of E_a, a highly energized, unstable molecule called the activated complex is formed. This activated complex possesses a number of vibrational modes, the bonding of one of which is particularly weak. Thus during one outward vibration the complex breaks up to form two product molecules. The reaction scheme is given by

$$\sum_{i=1}^{N} \nu_i' \mathrm{R}_i \underset{k_{1,b}}{\overset{k_{1,f}}{\rightleftharpoons}} \mathrm{R}^{\ddagger} \tag{2.2.17}$$

$$\mathrm{R}^{\ddagger} \overset{k_2}{\rightarrow} \sum_{i=1}^{N} \nu_i'' \mathrm{P}_i, \tag{2.2.18}$$

which respectively represent the forward and backward reactions between the reactants and the complex, and the decomposition of the complex to the product species. The backward reaction for (2.2.18) is assumed to be slow and therefore does not need to be included.

If we follow the rate of change of a typical reactant species R_i, then the reaction rates for R_i and $\mathrm{R}^{\ddagger}$ are respectively given by

$$\frac{dc_{\mathrm{R}_i}}{dt} = -\nu_i' k_{1,f} \prod_{j=1}^{N} c_j^{\nu_j'} + \nu_i' k_{1,b} c_{\mathrm{R}^{\ddagger}} \tag{2.2.19}$$

$$\frac{dc_{\mathrm{R}^{\ddagger}}}{dt} = k_{1,f} \prod_{j=1}^{N} c_j^{\nu_j'} - k_{1,b} c_{\mathrm{R}^{\ddagger}} - k_2 c_{\mathrm{R}^{\ddagger}}. \tag{2.2.20}$$

The theory makes two assumptions, namely partial equilibrium exists between the reactants and the activated complex, and the complex is a steady-state species. Thus invoking the partial equilibrium assumption for reaction (2.2.17) by setting $dc_{\mathrm{R}_i}/dt = 0$ in Eq. (2.2.19), we obtain

$$c_{\mathrm{R}^{\ddagger}} = K_c^{\ddagger} \prod_{j=1}^{N} c_j^{\nu_j'}, \tag{2.2.21}$$

where $K_c^{\ddagger}$ is the equilibrium constant of (2.2.17). To apply the steady-state assumption for $\mathrm{R}^{\ddagger}$, we first add Eqs. (2.2.19) and (2.2.20) to yield

$$\frac{dc_{\mathrm{R}_i}}{dt} + \nu_i' \frac{dc_{\mathrm{R}^{\ddagger}}}{dt} = -\nu_i' k_2 c_{\mathrm{R}^{\ddagger}}. \tag{2.2.22}$$

We then assume that the rate of change of the complex is much smaller than that of the reactant, $dc_{\mathrm{R}^{\ddagger}}/dt \ll dc_{\mathrm{R}_i}/dt$. Consequently Eq. (2.2.22) becomes

$$\frac{dc_{\mathrm{R}_i}}{dt} \approx -\nu_i' k_2 c_{\mathrm{R}^{\ddagger}} = -\nu_i' k_2 K_c^{\ddagger} \prod_{j=1}^{N} c_j^{\nu_j'}, \tag{2.2.23}$$

which gives the net rate of reaction for c_{R_i}. It is important to note that, in accordance with the earlier discussion on the partial equilibrium assumption, while we have set $dc_{\mathrm{R}_i}/dt = 0$ to obtain Eq. (2.2.21) on the basis of partial equilibrium because it is much smaller than either the forward or backward reaction rates in Eq. (2.2.19), it is however still much larger than $dc_{\mathrm{R}^{\ddagger}}/dt$, which is therefore in turn set to zero in Eq. (2.2.22).

Thus if we represent the reaction rate of c_{R_i} by an overall reaction based on the reactants R_i,

$$\frac{dc_{R_i}}{dt} = -\nu_i' k \prod_{j=1}^{N} c_j^{\nu_j'}, \tag{2.2.24}$$

then a comparison between Eqs. (2.2.23) and (2.2.24) yields an expression for the overall reaction rate constant k as

$$k = k_2 K_c^{\ddagger}. \tag{2.2.25}$$

A simplified expression for k_2 can be obtained by noting that for the decay process we expect the vibrational energy of the particular bond that ruptures is provided by the translational energy of the two colliding molecules. From quantum theory the energy associated with a vibrational mode of frequency ν is $h^o\nu$, where h^o is Planck's constant. Furthermore, because each molecule has $\frac{1}{2}k^oT$ amount of translational energy, the two colliding molecules would have a total energy of k^oT. Thus equating the vibrational and translational energies we have

$$\nu = \frac{k^oT}{h^o}, \tag{2.2.26}$$

which is called the universal frequency factor. If we further assume that rupturing of the bond occurs rapidly, within a few vibrations, then the vibrational frequency can be approximately identified as the characteristic decay rate for the activated complex,

$$k_2 \approx \nu. \tag{2.2.27}$$

Thus Eq. (2.2.25) yields the specific rate constant of the transition state theory as

$$k = \left(\frac{k^oT}{h^o}\right) K_c^{\ddagger}. \tag{2.2.28}$$

Equation (2.2.28) can be further developed by relating $K_c^{\ddagger}$ to $K_p^{\ddagger}$ as

$$\begin{aligned} K_c^{\ddagger} &= K_p^{\ddagger}(R^oT)^{-(1-\sum_{i=1}^{N}\nu_i')} \\ &= \exp\left[-\left(G^{o\ddagger} - \sum_{i=1}^{N}\nu_i' G_i^o\right) \Big/ R^oT\right](R^oT)^{-(1-n)}, \end{aligned} \tag{2.2.29}$$

where $G^{o\ddagger}$ and G_i^o are the Gibbs' free energies of the activated complex and reactant i, respectively, and $n = \sum_{i=1}^{N}\nu_i'$ is the overall reaction order. Writing $\Delta G^{o\ddagger} = G^{o\ddagger} - \sum_{i=1}^{N}\nu_i' G_i^o$ and recognizing that

$$\Delta G^{o\ddagger} = \Delta H^{o\ddagger} - T\Delta S^{o\ddagger}, \tag{2.2.30}$$

where $\Delta H^{o\ddagger}$ and $\Delta S^{o\ddagger}$ are the enthalpy and entropy of activation, respectively, Eq. (2.2.28) can be written as

$$k = \left(\frac{k^oT}{h^o}\right)(R^oT)^{n-1}\exp\left(\Delta S^{o\ddagger}/R^o\right)\exp\left(-\Delta H^{o\ddagger}/R^oT\right). \tag{2.2.31}$$

Comparing Eq. (2.2.31) with Eq. (2.2.2), which can be expressed as

$$k = Z\psi e^{-E_a/R^oT} \tag{2.2.32}$$

through the use of (2.2.16), the activation energy is identified as

$$E_a = \Delta H^{o^{\ddagger}}, \tag{2.2.33}$$

which is the change in the heat content between the reactants and the activated complex. Furthermore, Z and ψ are respectively given by

$$Z = \left(\frac{k^oT}{h^o}\right)(R^oT)^{n-1} \tag{2.2.34}$$

$$\psi = \exp\left(\Delta S^{o^{\ddagger}}/R^o\right). \tag{2.2.35}$$

The steric factor ψ thus depends on the entropy change in the formation of the activated complex. Since the change in entropy from translational to vibrational motion is negative, $\psi \ll 1$ and the preexponential factor is reduced.

An evaluation of ψ requires a knowledge of the entropy of the activated complex, which is generally not available. However, if we assume that the entropy of the activated complex is close to that of the products, then the change in entropy may be identified with that between the reactants and the products. Predictions based on this assumption agree well with the experimental data of those reactions that fail to conform with the simple collision theory.

2.3. THEORIES OF REACTION RATES: UNIMOLECULAR REACTIONS

In this section, we discuss theories specially developed for unimolecular reaction rates. A unimolecular reaction describes a chemical process in which a reactant undergoes an isomerization or decomposition process. The reaction,

$$\mathrm{R} \xrightarrow{k} \mathrm{P}, \tag{2.3.1}$$

indicates such a transformation in terms of R and P. Since the reactant of a unimolecular reaction must acquire sufficient energy before the reaction can take place, it requires intermolecular energy transfer through collision with another molecule. Consequently, a unimolecular reaction is really second order in nature, exhibiting a pseudo first-order behavior only under certain conditions. To demonstrate this behavior, an arbitrary species M is added to the reaction equation, as in

$$\mathrm{R} + \mathrm{M} \rightleftharpoons \mathrm{P} + \mathrm{M}. \tag{2.3.2}$$

Unimolecular reactions are of critical importance to combustion. For example, during methane ignition the reaction $CH_4 + M \rightleftharpoons CH_3 + H + M$ is often the initiation reaction of radical chain processes. Furthermore, the reverse of a unimolecular decomposition reaction is precisely a radical-radical or radical-molecule recombination reaction as mentioned previously, with M being the third body. These recombination

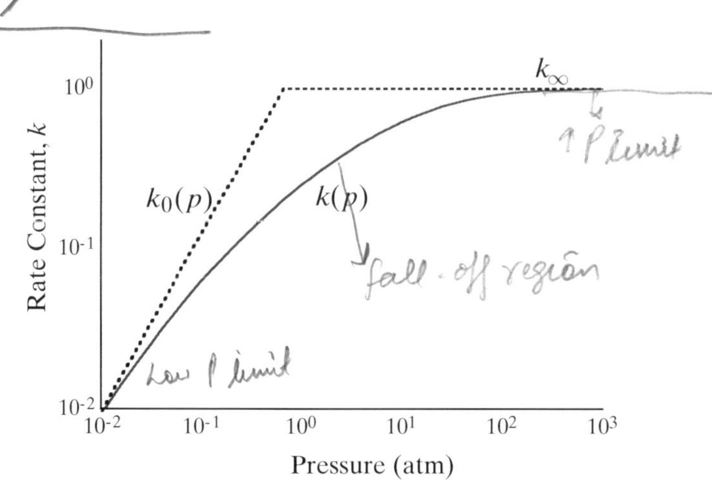

Figure 2.3.1. Characteristic variation of a unimolacular reaction rate constant with pressure.

reactions tend to slow down the overall reaction process by removing the radicals, and thus play important roles in combustion.

The specific rate constant of a unimolecular reaction is both temperature and pressure dependent. Figure 2.3.1 illustrates the variation of k as a function of pressure at a fixed temperature. It is seen that while k is a constant at high pressures, it falls off at lower pressures where it becomes proportional to pressure and hence the total gas molar density, $c = p/R^oT$. That is, with decreasing pressure, reaction (2.3.2) undergoes transition from first order toward second order. The rate constants at the two limiting pressures are called the high- and low-pressure limit rate constants, k_∞ and k_o, respectively. In the following we shall present two theories that describe such a transition.

2.3.1. Lindemann Theory

The characteristic fall-off behavior of a unimolecular rate constant can be explained phenomenologically by the Lindemann theory. It states that since stable molecules cannot spontaneously break up into products in the manner of (2.3.1), the reactant R must attain sufficient energy to undergo reaction. The process of forming an energized molecule, R*, involves collisions between R and another molecule M. The energized molecule R* may undergo a unimolecular reaction to form products, or it may undergo de-energization by collision with other molecules. These three distinct processes can be depicted as

$$\mathrm{R} + \mathrm{M} \underset{k_{1,b}}{\overset{k_{1,f}}{\rightleftharpoons}} \mathrm{R}^* + \mathrm{M} \tag{2.3.3}$$

$$\mathrm{R}^* \overset{k_2}{\rightarrow} \mathrm{P}. \tag{2.3.4}$$

The reaction rates of R and R* are respectively given by

$$\frac{dc_\mathrm{R}}{dt} = -k_{1,f}c_\mathrm{R}c_\mathrm{M} + k_{1,b}c_{\mathrm{R}^*}c_\mathrm{M} \tag{2.3.5}$$

$$\frac{dc_{\mathrm{R}^*}}{dt} = k_{1,f}c_\mathrm{R}c_\mathrm{M} - k_{1,b}c_{\mathrm{R}^*}c_\mathrm{M} - k_2c_{\mathrm{R}^*}. \tag{2.3.6}$$

Note that the reaction system (2.3.3) and (2.3.4) is analogous to that of (2.2.17) and (2.2.18) for the transition-state theory, except the presence of M is distinguished.

Invoking the steady-state assumption for c_{R^*} by neglecting dc_{R^*}/dt in Eq. (2.3.6), that is, by setting $dc_{R^*}/dt = 0$, we have

$$c_{R^*} = \frac{k_{1,f} c_R c_M}{k_{1,b} c_M + k_2}. \tag{2.3.7}$$

Putting Eq. (2.3.7) into Eq. (2.3.5) and rearranging, we obtain

$$\frac{dc_R}{dt} = -\frac{k_2(k_{1,f}/k_{1,b})}{1 + k_2/(k_{1,b} c_M)} c_R, \tag{2.3.8}$$

such that the rate constant of reaction (2.3.1) can be identified as

$$k = \frac{k_2(k_{1,f}/k_{1,b})}{1 + k_2/(k_{1,b} c_M)}. \tag{2.3.9}$$

Since any molecule can participate in the de-energization process, and if it is assumed that only a single collision is needed to de-energize R*, c_M in Eq. (2.3.9) is equal to the total molar gas density and hence is proportional to the total pressure. If the collision efficiencies of all species are the same, then c_M can be replaced by p through $p = c_M R^o T$. We further note that contrary to the formulation of the transition state theory, partial equilibrium is not assumed for reaction (2.3.3). In this way the second-order nature of the reaction can be captured.

Equation (2.3.9) shows that at the high-pressure limit, $c_M \to \infty$, and we have

$$k_\infty = k_2(k_{1,f}/k_{1,b}), \tag{2.3.10}$$

which demonstrates that k_∞ is independent of pressure. Under this condition, a unimolecular reaction, (2.3.1), has a reaction order of one. At the low-pressure limit, $c_M \to 0$, and we have

$$k_0 = k_{1,f} c_M, \tag{2.3.11}$$

which shows that k_0 is proportional to c_M and thus pressure at a given temperature. Reaction (2.3.1) therefore becomes second order.

Having identified k_∞ and k_0, the rate constant k can be expressed as

$$\frac{1}{k} = \frac{1}{k_\infty} + \frac{1}{k_0}. \tag{2.3.12}$$

A useful parameter indicating the pressure range over which the transition between the low- and high-pressure limits is most prominent is the transition pressure, $p_{1/2}$, at which $k/k_\infty = 1/2$, or equivalently $k_0(p) = k_\infty$. From Eqs. (2.3.11) and (2.3.12) we have

$$p_{1/2} \sim c_{M,1/2} = \frac{k_\infty}{k_{1,f}}. \tag{2.3.13}$$

The Lindemann theory qualitatively explains the variation of unimolecular reaction rate constant as a function of pressure. Specifically, Eq. (2.3.12) shows that $1/k$ varies linearly with $1/k_0$ such that a plot of $1/k$ versus $1/c_M$ should yield a straight line in the low-pressure limit. However, experiments showed that such a plot yields a nonlinear relationship. Furthermore, the experimental $p_{1/2}$ can be several orders smaller than the calculated value obtained by using the experimental k_∞ and the collision theory to estimate the $k_{1,f}$ (Robinson & Holbrook 1972), indicating that $k_{1,f}$ is actually faster. This would lead to a larger value of k_0 at a lower c_M according to Eq. (2.3.11), which in turn facilitates $k \to k_\infty$ as shown in Eq. (2.3.12). These weaknesses are addressed by higher order theories to be discussed next.

2.3.2. Rice–Ramsperger–Kassel (RRK) Theory

The Lindemann mechanism has been further developed by Hinshelwood, Slater, Rice, Ramsperger, and Kassel, attaining a physically satisfactory formulation through the RRK theory (Robinson & Holbrook 1972; Pilling & Seakins 1995). This theory incorporates two major improvements. First, the Lindemann theory considers only the relative kinetic energy along the line of centers in effecting reaction upon collision, and hence does not account for the additional internal energy, E, the colliding molecules possess through the vibrational modes of excitation. Inclusion of this additional source of energy, which also implies that all the rate constants should be functions of E, is expected to significantly increase the rate of activation upon collision. Hinshelwood has thus extended the analysis to a hypothetical molecule with s equivalent harmonic oscillators with the same frequency ν, and found that the total rate of activation is modified from the result of hard-sphere collision by the factor $(E_0/k^oT)^{s-1}/(s-1)!$, where E_0 is the internal energy of the molecule and can be approximately considered to be the energy needed for activation (see Figure 2.2.1). Since $E_0 \gg k^oT$, this factor is much larger than unity and as such leads to an increased theoretical values for $k_{1,f}$. This then largely resolves the major weakness of the Lindemann theory. Further developments along this direction through the RRK theory have included the internal energy in the decomposition rate of $R^* \to P$.

The second improvement recognizes that since the energized molecules have a distribution of internal energies, their contribution should be appropriately weighted and summed. In particular, recognizing that $k_{1,f}/k_{1,b}$ is just the equilibrium constant for reaction (2.3.3), and since $k_{1,b}$ is basically not energy dependent, we can define a distribution function $f(E)$ such that $f(E)dE$ is the number of molecules of R having an energy between E and $E + dE$, and is given by

$$\frac{dk_{1,f}}{k_{1,b}} = f(E)dE. \tag{2.3.14}$$

Thus we can express Eq. (2.3.9) in differential form as

$$dk(E) = \frac{k_2(E)\, f(E)}{1 + k_2(E)/[k_{1,b}(E)c_M]}\, dE, \tag{2.3.15}$$

which upon integration over the energy range from E_0, the minimum energy that can lead to reaction, to infinity,

$$k = \int_{E_0}^{\infty} \frac{k_2(E)\ f(E)}{1 + k_2(E)/[k_{1,b}(E)c_M]}\, dE, \tag{2.3.16}$$

yields the specific rate constant, k. Its high- and low-pressure limits are

$$k_\infty = \int_{E_0}^{\infty} k_2(E) f(E) dE, \qquad k_0 = c_M \int_{E_0}^{\infty} k_{1,b}(E) f(E) dE, \tag{2.3.17}$$

which are the generalized expressions for Eqs. (2.3.10) and (2.3.11) respectively.

The RRK theory has been further developed to the Quantum–Rice–Ramsperger–Kassel (QRRK) theory, and the Rice–Ramsperger–Kassel–Marcus (RRKM) theory, in which the energy levels are quantized. With subsequent developments (Troe 1979; Troe 1983; Gilbert et al. 1983; Dean 1985; Gilbert & Smith 1990), the fall-off behavior for some simple unimolecular reactions is now generally predictive. However, an accurate evaluation of $k_2(E)$, and thus the unimolecular reaction rate constant, requires accurate knowledge of the microscopic properties of the reactant and the energized molecule. These properties include the chemical structure, the potential energy surface, the vibrational frequencies, and moments of inertia. It is still a challenging goal to predict these properties accurately. In practice, these theories are used mainly for extrapolation of available experimental data to the entire temperature and pressure domain.

2.3.3. Representation of Unimolecular Reaction Rate Constants

In the previous discussion, we have shown why a unimolecular reaction rate constant is dependent on pressure. We shall now introduce a simple formula to represent the pressure dependence of the rate constant. Since the rate constants of many unimolecular reactions critical to combustion are in the pressure fall-off region, an effective representation of such rate constants is an important element of combustion kinetics.

The most widely used representation of unimolecular reaction rate constant is Troe's fall-off formula (Troe 1983), which is an extension of Lindemann's treatment. Troe's treatment corrects the Lindemann fall-off by a pressure-dependent factor called the broadening factor F. Specifically, Eq. (2.3.12) is modified to the form

$$k(T, p) = \left[\frac{1}{k_0(T, p)} + \frac{1}{k_\infty(T)}\right]^{-1} F(T, p). \tag{2.3.18}$$

It is usually sufficiently accurate to express $k_\infty(T)$ and k_0/c_M by the Arrhenius expression (2.2.2) or the modified Arrhenius expression (2.2.3). However, there is little theoretical development for the broadening factor $F(T, p)$. In practice, F is usually expressed empirically as

$$\log_{10} F(T, p) = \frac{\log_{10} F_c(T)}{1 + \left\{\frac{\log_{10}[k_0(T,p)/k_\infty(T)]+c}{N-d\{\log_{10}[k_0(T,p)/k_\infty(T)]+c\}}\right\}^2}, \tag{2.3.19}$$

where c, N, and d are empirical expressions or constants

$$c = -0.4 - 0.67\log_{10} F_c(T)$$
$$N = 0.75 - 1.27\log_{10} F_c(T)$$
$$d = 0.14,$$

and F_c is called the central broadening factor, defined as $F_c = 2k_{k_0/k_\infty=1}/k_\infty$ and, in practice, fitted empirically as a function of temperature

$$F_c(T) = (1-a)e^{-T/T^{***}} + ae^{-T/T^*} + e^{-T^{**}/T}.$$

In the above equation, a, T^{***}, T^*, and T^{**} are fitting parameters.

2.3.4. Chemically Activated Reactions

We have shown that an isolated molecule cannot undergo chemical transformation on its own. It must attain sufficient energy by collision with other molecules, and only then can the energized molecule undergo reaction, via its activated state, leading to products. This collisional activation process is called thermal energization because the reactant molecule is not chemically changed upon collision, which only imparts a certain amount of energy to it. There is another type of activation process, called chemical activation, through which a chemical transformation occurs when an energized molecule is produced. Consider the following reaction scheme:

$$\begin{array}{c} \mathrm{A + B} \underset{k_{1,b}}{\overset{k_{1,f}}{\rightleftharpoons}} \mathrm{(AB)^*} \overset{k_2}{\rightarrow} \mathrm{C + D} \\ + \\ \mathrm{M} \\ k_3 \downarrow \\ \mathrm{(AB)} \end{array} \tag{2.3.20}$$

In the above scheme, the "hot" $(AB)^*$ molecule is produced through the chemical combination of A and B, and possesses their combined translational, rotational, and vibrational energies. Additionally, if $A + B \rightarrow (AB)^*$ is exothermic, the reaction heat is also transformed into the vibrational energy in $(AB)^*$. This "hot" molecule may dissociate into $C + D$, or it may decompose back to $A + B$, or it may be stabilized to (AB) by collision with a third molecule, M. The net bimolecular process,

$$\mathrm{A + B} \overset{k}{\rightarrow} \mathrm{C + D}, \tag{2.3.21}$$

is called a chemically activated reaction.

The dynamics of a chemically activated reaction can be described satisfactorily by the transition state theory if the lifetime of $(AB)^*$ is short. That is, the decomposition of $(AB)^*$ into $C + D$ is much faster than the characteristic time scale of collision stabilization, $(AB)^* + M \rightarrow (AB) + M$. However, this is usually not the case when medium or large molecules are involved. An alternate treatment, based on unimolecular reaction theories, is devised to describe the rate of this type of reactions.

It is first noted that even though a chemically energized molecule $(AB)^*$ acquires energy differently from a thermally energized molecule, their physical natures are the same. Thus the dynamics of a chemically activated reaction can be treated by unimolecular reaction rate theories. In doing so, we assume that the concentrations of C and D are sufficiently low such that the production of $(AB)^*$ from C + D is negligible. Following the Lindemann analysis, we write the reaction rates for the production of $(AB)^*$, C, and D as

$$\frac{dc_{(AB)^*}}{dt} = k_{1,f}c_A c_B - (k_{1,b} + k_2 + k_3 c_M)c_{(AB)^*} \tag{2.3.22}$$

$$\frac{dc_C}{dt} = \frac{dc_D}{dt} = k_2 c_{(AB)^*}. \tag{2.3.23}$$

Invoking the steady-state assumption by setting $dc_{(AB)^*}/dt = 0$, we have

$$c_{(AB)^*} = \frac{k_{1,f}}{k_{1,b} + k_2 + k_3 c_M} c_A c_B. \tag{2.3.24}$$

Putting Eq. (2.3.24) into Eq. (2.3.23), we obtain

$$\frac{dc_C}{dt} = \frac{dc_D}{dt} = \frac{k_2 k_{1,f}}{k_{1,b} + k_2 + k_3 c_M} c_A c_B, \tag{2.3.25}$$

which yields the rate constant for reaction (2.3.21) as

$$k = \frac{k_2 k_{1,f}}{k_{1,b} + k_2 + k_3 c_M}. \tag{2.3.26}$$

If we assume that $(AB)^*$ can be stabilized by a single collision with M, k_3 is then the collision frequency factor between $(AB)^*$ and M,

$$k_3 = \pi \sigma^2_{(AB)^*,M} \left(\frac{8k^o T}{\pi m_{(AB)^*,M}} \right)^{1/2}. \tag{2.3.27}$$

Furthermore, if the breakup of $(AB)^*$ to C and D occurs within one molecular vibration, then k_2 is given by

$$k_2 = \frac{k^o T}{h^o}. \tag{2.3.28}$$

An important consequence of the above derivation is that Eq. (2.3.26) contains the molar density term, c_M, which is directly proportional to the system pressure. The rate constant k of an apparent second-order bimolecular reaction is now pressure dependent. Furthermore, as $c_M \to \infty$, Eq. (2.3.26) becomes

$$k = \frac{k_2 k_{1,f}}{k_3 c_M}. \tag{2.3.29}$$

Consequently the rate constant is inversely proportional to pressure and the net reaction is first order. However, at the low pressure limit where $c_M \to 0$, the $k_3 c_M$ term in Eq. (2.3.26) becomes insignificant. Then the rate constant is independent of pressure and the net reaction is second order. Extension of Eq. (2.3.26) to include the dependence on internal energy in the manner of the RRK theory can be readily conducted.

Chemically activated reactions are usually encountered for medium- to large-size reactants. For example, the important methyl self-reaction

$$
\begin{aligned}
CH_3 + CH_3 &\rightarrow C_2H_6 \\
&\rightarrow C_2H_5 + H,
\end{aligned}
$$

and the CO oxidation reaction

$$
\begin{aligned}
CO + OH &\rightarrow HOCO \\
&\rightarrow CO_2 + H
\end{aligned}
$$

belong to this class of reactions. The existence of these reactions adds yet another dimension of difficulty in the identification of elementary reaction steps and the determination of their rate constants.

2.4. CHAIN REACTION MECHANISMS

In most reaction mechanisms there are present small concentrations of highly reactive intermediate species that participate and are regenerated in a sequence of reactions. It is through the participation of these species, called the chain carriers, that the reactants are converted to the products. Chain reactions can be further classified into straight-chain and branched-chain reactions, which are now separately discussed.

2.4.1. Straight-Chain Reactions: The Hydrogen–Halogen System

We consider straight-chain reactions via the example of the hydrogen–halogen system, which has a direct reaction path of

$$H_2 + X_2 \stackrel{k_0}{\rightarrow} 2HX, \tag{X0}$$

with a production rate of HX given by

$$\frac{d[HX]}{dt} = 2k_0[H_2][X_2], \tag{2.4.1}$$

where X_2 is the halogen molecule F_2, Cl_2, Br_2, or I_2. It has been found, however, that the reaction between H_2 and X_2 to produce HX actually follows a more complex scheme consisting of the following five major steps:

$X_2 + M \stackrel{k_{1,f}}{\rightarrow} X + X + M$	Chain initiation	(X1f)
$X + H_2 \stackrel{k_{2,f}}{\rightarrow} HX + H$	Chain carrying	(X2f)
$H + X_2 \stackrel{k_{3,f}}{\rightarrow} HX + X$	Chain carrying	(X3f)
$X + X + M \stackrel{k_{1,b}}{\rightarrow} X_2 + M$	Chain termination	(X1b)
$H + HX \stackrel{k_{2,b}}{\rightarrow} X + H_2$	Chain carrying.	(X2b)

In designating the specific elementary reactions of a halogen, X_2, we have adopted the practice of prefixing the reaction number of a reaction mechanism by a symbol for the distinguishing reactant. For example, we shall use H for reactions involved in hydrogen oxidation, M for methane oxidation, C_2 for the C_2 hydrocarbon species, and so on. This provides a measure of consistency in subsequent discussions.

In the above halogen–hydrogen mechanism, reaction (X1*f*) is the chain initiation step through which a halogen molecule is dissociated into two halogen atoms. Once some X is formed through (X1*f*), the production of HX is effected by reactions (X2*f*) and (X3*f*) through the reaction intermediates, also called chain carriers, H and X. These reactions are called chain-carrying steps because in each of them the destruction of one chain carrier leads to the creation of a new chain carrier such that the number of chain carriers in each step remains unchanged. These two reactions form a closed sequence in that at the end of the reaction sequence, (X3*f*), a new X is produced that can start the process all over again from (X2*f*). Indeed, the net result of (X2*f*) and (X3*f*) is the same as that of the direct path (X0).

Reactions (X1*b*) and (X2*b*) are simply the backward reactions of (X1*f*) and (X2*f*) respectively. Reaction (X1*b*) is a chain-termination step in which two X atoms recombine through collision with the third body M. Reaction (X2*b*) is a chain-carrying step, although it is inhibitive to the net production rate of HX. The backward reaction of (X3*f*) is highly endothermic and therefore not important.

The rates of concentration variation of the five components are given by

$$\frac{d[\mathrm{H_2}]}{dt} = -k_{2,f}[\mathrm{X}][\mathrm{H_2}] + k_{2,b}[\mathrm{H}][\mathrm{HX}] \tag{2.4.2}$$

$$\frac{d[\mathrm{X_2}]}{dt} = -k_{1,f}[\mathrm{X_2}][\mathrm{M}] - k_{3,f}[\mathrm{H}][\mathrm{X_2}] + k_{1,b}[\mathrm{X}]^2[\mathrm{M}] \tag{2.4.3}$$

$$\frac{d[\mathrm{H}]}{dt} = k_{2,f}[\mathrm{X}][\mathrm{H_2}] - k_{3,f}[\mathrm{H}][\mathrm{X_2}] - k_{2,b}[\mathrm{H}][\mathrm{HX}] \tag{2.4.4}$$

$$\begin{aligned}\frac{d[\mathrm{X}]}{dt} &= 2k_{1,f}[\mathrm{X_2}][\mathrm{M}] - k_{2,f}[\mathrm{X}][\mathrm{H_2}] + k_{3,f}[\mathrm{H}][\mathrm{X_2}] \\ &\quad + k_{2,b}[\mathrm{H}][\mathrm{HX}] - 2k_{1,b}[\mathrm{X}]^2[\mathrm{M}]\end{aligned} \tag{2.4.5}$$

$$\frac{d[\mathrm{HX}]}{dt} = k_{2,f}[\mathrm{X}][\mathrm{H_2}] + k_{3,f}[\mathrm{H}][\mathrm{X_2}] - k_{2,b}[\mathrm{H}][\mathrm{HX}]. \tag{2.4.6}$$

Invoking the steady-state assumption by setting

$$\frac{d[\mathrm{H}]}{dt} = 0 \quad \text{and} \quad \frac{d[\mathrm{X}]}{dt} = 0 \tag{2.4.7}$$

in Eqs. (2.4.4) and (2.4.5), solving for [H] and [X], and substituting them into Eq. (2.4.6) yields

$$\frac{d[\mathrm{HX}]}{dt} = \frac{2k_{2,f}(k_{1,f}/k_{1,b})^{1/2}[\mathrm{H_2}][\mathrm{X_2}]^{1/2}}{1 + (k_{2,b}/k_{3,f})[\mathrm{HX}]/[\mathrm{X_2}]}. \tag{2.4.8}$$

Table 2.1. Heats of reaction of the various reaction steps for the halogen–hydrogen systems at 300 K in kcal/mole

Reaction		F	Cl	Br	I
$H_2 + X_2 \rightarrow 2HX$	(X0)	−129.7	−44.0	−17.4	−2.8
$X_2 + M \rightarrow X + X + M$	(X1*f*)	37.8	57.8	53.6	36.4
$X + H_2 \rightarrow HX + H$	(X2*f*)	−31.6	1.2	16.7	32.5
$H + X_2 \rightarrow HX + X$	(X3*f*)	−98.1	−45.2	−34.1	−35.3

Comparing Eq. (2.4.8) with Eq. (2.4.1) for the direct path reaction (X0), it is seen that these two rate expressions are quite different. For example, while these rates still depend on $[H_2]$ to the first power, the dependence on $[X_2]$ is more involved for the chain mechanism. The chain mechanism also correctly predicts the inhibitive effect of increasing product concentration on its own production rate, in that $d[HX]/dt$ varies inversely with [HX].

We next discuss the influence of the heats of reaction of the individual steps, shown in Table 2.1, on their relative efficiencies in the chain process. This also explains the differences in the observed behavior of the various halogens.

First we note that for all the halogens the net heat change, represented by the first equation in Table 2.1, is exothermic. Thus halogen–hydrogen reactions are basically self-sustaining. The intensity of the reactions, however, varies significantly from the highly exothermic, violent HF system to the weakly exothermic, mild HI system.

The initiation reactions for all halogens, (X1*f*), are endothermic. Therefore the halogen–hydrogen reactions are not spontaneous and an ignition stimulus is needed. We also note that the heats of initiation reactions are much smaller than the value of 104.2 kcal required to dissociate H_2 to two H through $H_2 + M \rightarrow 2H + M$. Therefore initiation of the chain cycle is favored by first dissociating X_2 and going through the chain-carrying steps (X2*f*) and (X3*f*), instead of first producing H from H_2 and then going through the chain-carrying steps (X3*f*) and (X2*f*).

Specializing to the individual halogens, we see that both of the basic chain-carrying F–H_2 and H–F_2 steps of the F_2–H_2 system are highly exothermic, implying that the overall reaction should proceed rapidly even though its initiation reaction is endothermic by 37.8 kcal/mole. For the chlorine case the Cl–H_2 reaction is slightly endothermic. Thus while Cl_2–H_2 mixtures are stable in the dark, they become reactive in the presence of light. In the case of bromine, the overall reaction intensity is further weakened because of the moderate endothermicity of the Br–H_2 reaction. The H–Br_2 recombination reaction is however sufficiently exothermic to make it an important step in the overall chain mechanism. Finally, for the case of iodine, the endothermicity of the I–H_2 step not only is very high but is also about the same value as the exothermicity of the H–I_2 step. Thus the chain mechanism is not favored and the direct, bimolecular H_2–I_2 reaction is the observed one.

2.4.2. Branched-Chain Reactions

In straight-chain reactions there is no net production of chain carries for each cycle of reactions. There are, however, reactions in which there is a net generation of chain carriers. This leads to an extremely rapid rate of the overall reaction, which may eventually culminate into an explosion. A well-known example is the chain cycle in the hydrogen–oxygen reaction scheme,

$$\mathrm{H + O_2 \rightarrow OH + O} \qquad \text{Chain branching} \tag{H1}$$

$$\mathrm{O + H_2 \rightarrow OH + H} \qquad \text{Chain branching} \tag{H2}$$

$$\mathrm{OH + H_2 \rightarrow H_2O + H} \qquad \text{Chain carrying.} \tag{H3}$$

In this reaction sequence the O and H atoms and the OH radical are the chain carriers. It is seen that in each of (H1) and (H2) the presence of one chain carrier as a reactant results in two chain carriers. Therefore they are called chain-branching steps. The net reaction of the chain cycle (H1) to (H3) is

$$\mathrm{3H_2 + O_2 \rightarrow 2H_2O + 2H},$$

which shows that two H atoms are generated per cycle. It should also be noted that since the chemical reactivities of different chain carriers are necessarily different, the net reactivity of, say, a chain-carrying reaction can be either weakened or strengthened through the exchange of chain carries of different reactivities. A well-known example is the reaction

$$\mathrm{H + O_2 + M \rightarrow HO_2 + M} \tag{H9}$$

considered previously in which a very active carrier H is consumed, producing a very inactive carrier HO_2. Thus under certain situations (H9) is considered to be terminating instead of carrying. Another example, with a milder weakening of reactivity, is the reaction $CH_4 + H \rightarrow CH_3 + H_2$, in which the reactive H atom is exchanged for the less reactive radical CH_3.

The general behavior of an explosive mixture in response to changes in pressure and temperature can be illustrated by using the following representative scheme (Strehlow 1984):

$$n\mathrm{R} \overset{k_1}{\rightarrow} \mathrm{C} \qquad \text{Initiation} \tag{2.4.9}$$

$$\mathrm{R + C} \overset{k_2}{\rightarrow} a\mathrm{C + P} \qquad \text{Chain-branching cycle} \tag{2.4.10}$$

$$\mathrm{C + R + R} \overset{k_g}{\rightarrow} \mathrm{P} \qquad \text{Gas termination} \tag{2.4.11}$$

$$\mathrm{C} \overset{k_w}{\rightarrow} \mathrm{P} \qquad \text{Surface termination.} \tag{2.4.12}$$

In the above C is the chain carrier, $a > 1$ is the multiplication factor in the chain-branching cycle, and P represents the stable products formed through the respective reactions. The gas termination reaction is a three-body process while that of surface termination is a one-body process. The rate of production of C is then given by

$$\frac{d[\mathrm{C}]}{dt} = k_1[\mathrm{R}]^n + (a-1)k_2[\mathrm{R}][\mathrm{C}] - k_g[\mathrm{R}]^2[\mathrm{C}] - k_w[\mathrm{C}], \tag{2.4.13}$$

which can be rearranged to show

$$\frac{d[\mathrm{C}]}{dt} = k_1[\mathrm{R}]^n + k_2[\mathrm{R}](a - a_c)[\mathrm{C}], \tag{2.4.14}$$

where

$$a_c = 1 + \frac{k_g[\mathrm{R}]^2 + k_w}{k_2[\mathrm{R}]}. \tag{2.4.15}$$

Equation (2.4.14) shows that [C] varies exponentially with time, growing for $(a - a_c) > 0$ and decaying otherwise. Consequently the condition for the occurrence of branched-chain explosion corresponds to the situation of

$$a \geq a_c, \tag{2.4.16}$$

where a_c is the critical multiplication factor at which the mixture becomes explosive.

Equations (2.4.15) and (2.4.16) show that explosion is favored for small a_c, which corresponds to situations of fast chain-branching reactions (large k_2) and/or slow chain-termination reactions (small k_g and k_w), as is physically reasonable. Furthermore, since [R] is proportional to the system pressure p, we have

$$\begin{aligned} a_c &\to 1 + \frac{k_w}{k_2[\mathrm{R}]} \to \infty, \quad \text{as } p \to 0, \\ a_c &\to 1 + \frac{k_g[\mathrm{R}]}{k_2} \to \infty, \quad \text{as } p \to \infty, \end{aligned} \tag{2.4.17}$$

which shows that explosion is not possible at either very low or very high pressures. The reason is that as the gas density decreases with $p \to 0$, the chain cycle becomes less efficient because it requires the collision between two molecules. The wall termination reaction, however, depends only on the concentration of the chain carrier and therefore becomes more efficient. The net effect is the tendency to inhibit explosion. Similarly, as $p \to \infty$, the increase in density favors the three-body gas termination reaction as compared to the two-body chain-branching reaction, hence again inhibiting explosion.

Since the creation of chain carriers requires energy, the chain-branching reaction is endothermic and is more sensitive to temperature variation in that it is characterized by a large activation energy. On the other hand, the gas and wall termination reactions are not temperature sensitive, implying that k_g and k_w are nearly constants and therefore the associated activation energies can be taken to be identically zero. Thus with increasing temperature, a_c decreases and the gas becomes more explosive. We again note that the termination reactions are highly exothermic because of the release of energy from the activated radicals.

The above behavior is represented by a C-shaped explosion limit curve shown in Figure 2.4.1. It demonstrates that, at a given temperature, continuously increasing the pressure of an initially nonexplosive gas will cause it to become first explosive and then nonexplosive again. Increasing temperature widens the range of explosivity. In Chapter 3 we shall show that, for the hydrogen–oxygen system, with further increase in pressure the mixture will become explosive again due to the reactivation of HO_2.

The chain ignition mechanism discussed here is somewhat different from the thermal ignition process to be studied in a later topic. Presently the requirement for ignition is the presence of some chain carriers that multiply through the branched-chain steps, leading to an explosive acceleration of the overall reaction rate. For

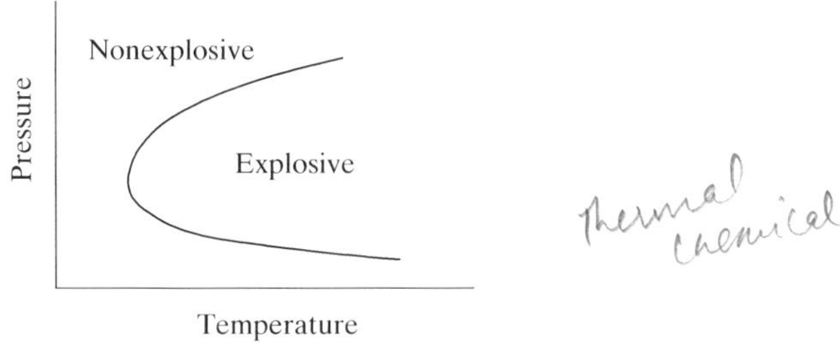

Figure 2.4.1. *C*-shaped pressure–temperature explosion limits due to chain mechanisms and wall termination.

the thermal ignition mechanism a critical mass of the reactive mixture has to be heated to a sufficiently high temperature such that the rate of chemical heat generation exceeds that of heat loss through various transport mechanisms. The net heat accumulated further enhances the reaction rate and eventually leads to a thermal runaway situation. We do note, however, that the chain and thermal mechanisms are usually related in that the creation of chain carriers through the endothermic chain-branching reactions requires energy and therefore the presence of heating. Furthermore, the rapid increase in the radical concentration will lead to the initiation of some highly exothermic, chain-terminating reactions and hence the eventual thermal runaway.

2.4.3. Flame Inhibitors

Flame inhibitors are most commonly used as extinguishing agents. Of these, the most effective are the halogenated species that inhibit the highly explosive, branched-chain H_2–O_2 reactions (H1)–(H3) by catalyzing the recombination of hydrogen atoms into relatively nonreactive radicals and molecules. This reduces the available radical pool and lowers the overall rate of chain branching. The inhibition reactions are primarily (X3*f*) and (X2*b*), through which the H atoms are deactivated by forming the H_2 and HX molecules and the less active X radicals.

The halogen radical initially may be present in the form of a halogen acid (e.g., HI, HBr, HCl) or a halogenated hydrocarbon (e.g., CH_3Br). Several halogen atoms can also be associated with one carbon atom forming, say, CF_3Br. Dissociation of the halogens from the carrier atoms (carbon or hydrogen) is much easier and therefore has a much greater rate than hydrogen atom abstraction from the carbon atoms. This can be demonstrated by comparing the bond energies of carbon–halogen to that of carbon–hydrogen:

Species	Bond energy (kcal/mole)
H_3C–H	104
H_3C–Cl	84
H_3C–Br	70
H_3C–I	56

The relative ease of iodine abstraction makes it the most effective inhibitor listed.

When the inhibitor is a halogenated hydrocarbon, the fuel content of the inhibitor molecule can also affect its efficiency. For example, addition of CH_3X to lean flames will provide more fuel in the form of CH_3, which tends to increase the flame temperature and hence the reaction rate, while at the same time the halogen radical tends to reduce the reaction rate. In rich mixtures both effects tend to reduce the reaction rate.

In the case of CF_3Br, both F and Br can serve as the inhibiting radical and therefore the molecule is an effective inhibitor. However, while inhibition through Br basically follows reactions (X3*f*) and (X2*b*), the hydroflouric acid (HF) produced through the reaction between F and H is virtually chemically inert due to its very high bond energy (140 kcal/mole). Thus HF does not participate in the recombination reaction (X2*b*). Nevertheless, since the formation of HF still removes a hydrogen atom from the radical pool, and since CF_3Br has a large heat capacity, which tends to lower the mixture's flame temperature, it is still an effective inhibitor.

2.5. EXPERIMENTAL AND COMPUTATIONAL TECHNIQUES

In order to characterize the gross properties and detailed reaction steps of a given fuel–air mixture, various experimental techniques have been developed and are briefly discussed in the following.

Rapid Compression Machines: Here the fuel–air mixture is initially contained in a cylindrical or square chamber with a piston forming one of the end walls. At time = 0 the piston is impulsively driven into the chamber and is locked at a preset location. Thus the mixture is instantaneously heated and compressed to high temperature and pressure. By using this technique one can determine whether a mixture at a given temperature and pressure is ignitable, and the associated ignition delay if ignition does occur. The temporal variation of the supposedly uniform mixture, temperature, and composition can also be studied using various probing and sampling techniques.

Turbulent Flow Reactors: In a flow reactor small amounts of fuel and oxidizer are rapidly mixed with and heated by a hot, inert flow as the highly diluted reacting mixture flows steadily along a tube. The flow in the tube is turbulent such that the flow velocity is relatively uniform over the tube's cross-section. The tube is heated to maintain isothermicity. The gas is probed and sampled along the flow to determine the stable species during the course of reaction.

This technique is useful for the study of pyrolytic and oxidative reactions in the 1,000–1,500 K range. The reaction zone can be spread out to about one meter in length using relatively low subsonic flows such that reactions with characteristic times as short as 3 msec can be studied. The main drawback of the plug flow reactor technique is the time needed to mix the reactants with the hot inert carrier gas, which renders uncertain the initial time for the reaction to commence.

Shock Tubes: In a shock tube a diaphragm initially separates a high and a low pressure region. By instantaneously bursting the diaphragm a shock wave is generated which propagates into the low pressure region. Thus if the low pressure region consists of the reactants, then upon passage of the shock wave they are instantly heated and pressurized to high values. Further heating and pressurization can be achieved by working with the stagnant region behind the reflected shock. The advantages of this technique are the instantaneous nature of heating and the achievable high temperature, up to 5,000 K. The available time for reaction is limited by the arrival of the rarefaction waves behind the contact surface.

Well-Stirred Reactors (WSR): In a well-stirred, or perfectly stirred, reactor such as the Longwell jet-stirred reactor (Longwell & Weiss 1955), reactants are centrally injected into a spherical chamber with a high injection velocity through a perforated tube. These reactants then burn intensely and the products exit through ports at the outer wall of the chamber. Thus by assuming that mixing is instant such that conditions are uniform within the chamber, and by controlling the flow rate of the mixture and thereby the residence time of the reaction, the reaction intensity of such a uniform mixture can be controlled. An overall reaction rate can be determined by measuring the temperature within the chamber and the concentrations of the products. The states of ignition and extinction, when the reacting mixture is subjected to different flow rates, can also be determined.

Laminar Flames: The technique basically involves establishing a stationary premixed or nonpremixed flame and spatially probing the temperature and reactant concentration profiles. The thickness of the flame can be stretched by conducting the experiment under low pressures so that finer resolution can be achieved. Unlike the previous technique in which the reactions take place in homogeneous media, in a flame environment steep temperature and concentration gradients exist. The significant temperature change across a flame implies a corresponding shift in the dominant temperature-sensitive reactions. Furthermore, the progress of reactions is intimately coupled to species diffusion, which can bring radicals produced in, say, the high temperature region of the flame to a low temperature region, thereby facilitating the reactions there. Obviously these effects must be accounted for and "subtracted out" before the chemical information can be identified.

Strictly speaking, compared to the homogeneous systems, laminar flames are not considered to be chemically clean and well-defined for kinetics studies. They are, however, useful as validation checks in that, assuming the transport aspects of the flame are accurately described, any chemical reaction mechanism which cannot reproduce the structure and global response of a flame under diverse conditions obviously is deficient in some aspects.

PROBLEMS

1. For the direct hydrogen–iodine reaction

$$H_2 + I_2 \underset{k_b}{\overset{k_f}{\rightleftharpoons}} 2HI$$

show that

$$[H_2] = [I_2] = c_0 - \tfrac{1}{2}[HI]$$

for a mixture whose concentrations at time $t = 0$ are $[H_2] = [I_2] = c_0$ and $[HI] = 0$. Then show that the instantaneous concentration of HI is given by

$$[HI] = \frac{2c_0\{\exp(4c_0\sqrt{k_f k_b}t) - 1\}}{(2\sqrt{k_b/k_f} - 1) + (2\sqrt{k_b/k_f} + 1)\exp(4c_0\sqrt{k_f k_b}t)}.$$

2. Consider the following chain reaction process in the production of NO and NO_2:

$$NO_2 \underset{h^o v}{\overset{k_1}{\rightarrow}} NO + O$$

$$O + O_2 + M \overset{k_2}{\rightarrow} O_3 + M$$

$$NO + O_3 \overset{k_3}{\rightarrow} NO_2 + O_2.$$

By using the steady-state assumption, show that the concentration of ozone, a major component of the photochemical smog, is determined by the ratio of $[NO_2]$ to [NO].

3. Consider the reaction mechanism for hydrogen and chlorine:

$$Cl_2 + M \rightarrow Cl + Cl + M, \quad (R1)$$

$$Cl + H_2 \rightarrow HCl + H, \quad (R2)$$

$$H + Cl_2 \rightarrow HCl + Cl, \quad (R3)$$

$$Cl + Cl + M \rightarrow Cl_2 + M. \quad (R4)$$

(a) With appropriate steady-state assumptions, show that

$$\frac{d[HCl]}{dt} = 2k_2\sqrt{\frac{k_1}{k_4}}\,[Cl_2]^{0.5}\,[H_2].$$

(b) If NCl_3 is added to the mixture as an inhibitor, the following reaction needs to be considered:

$$NCl_3 + Cl \rightarrow NCl_2 + Cl_2. \quad (R5)$$

What are the rate expressions in the limits of fast and slow rates of this reaction?

4. Consider the Lindemann mechanism for the following association reactions:

$$A + B \underset{k_d}{\overset{k_a}{\rightleftharpoons}} C^*$$

$$C^* + M \overset{k_s}{\rightarrow} C + M,$$

where

$$\frac{dC}{dt} = k_a[A][B].$$

Derive the association rate constant, k_a, in the low and high pressure limits using the steady-state assumption for C^*.

5. NH_3 is used as an agent for NO reduction in the temperature range of about 1,000 K. The mechanism for NO reduction by NH_3 contains the following reaction steps:

$$NH_3 + OH \rightarrow NH_2 + H_2O \tag{R1}$$

$$NH_2 + NO \rightarrow N_2 + H + OH \tag{R2}$$

$$NH_2 + NO \rightarrow N_2 + H_2O \tag{R3}$$

$$H + O_2 \rightarrow O + OH \tag{R4}$$

$$O + H_2O \rightarrow OH + OH. \tag{R5}$$

Show that the above process runs away when the (branching) ratio of the rate of the branching reaction (R2) over that of the termination reaction (R3) exceeds the critical value of $\frac{1}{3}$. Assume that NH_2, H, and O are in steady state.

3 Oxidation Mechanisms of Fuels

In Chapter 2 we introduced the principles of the chemical kinetics of gas-phase reactions. In particular, we discussed the fundamental dependence of reaction rates on temperature, pressure, and reactant concentrations. We studied the concept of multistep reactions, and then categorized the reaction mechanisms and showed how they can affect the reaction rate in qualitatively different manners.

In this chapter, we discuss the oxidation mechanisms of specific fuel systems involving hydrogen, carbon monoxide, and various hydrocarbons. The formation of pollutants will also be covered. In particular, the reaction pathways leading to fuel consumption, the formation and destruction of intermediate species, and the final product formation are discussed in a qualitative manner. We shall see that numerous elementary chemical reactions are involved in the conversion of reactants to products, and that even for a given fuel these reactions often play different roles in different combustion environments. The intricate paths followed by hydrocarbon oxidation illustrate the complexity of chemical kinetics in combustion. Nevertheless, in spite of such apparent complexity, we shall also show that there appears to be only a finite number of reactions that exert significant influence in a combustion process, providing the possibility that the reaction mechanisms of fuel oxidation can be largely understood.

While most of our discussion will be conducted in a general manner, frequently it is necessary to discuss systems with specific fuels and combustion intermediate species. Therefore, in Section 3.1 we introduce the nomenclature and properties of common fuels and some of the important species that are involved in their oxidation. In Section 3.2 we study the oxidation mechanisms of hydrogen and carbon monoxide. This is followed by discussion of the high-temperature oxidation mechanisms of methane in Section 3.3, C_2 hydrocarbons in Section 3.4, alcohols in Section 3.5, higher aliphatic hydrocarbons in Section 3.6, and aromatics in Section 3.7. Hydrocarbon oxidation at low to intermediate temperatures is discussed in Section 3.8. In Section 3.9, we study the chemistry of pollutant formation. In Sections 3.10–3.12, we discuss the approaches toward reducing a large reaction mechanism to smaller ones suitable for further analytical and computational studies.

Further coverage on fuels chemistry can be found in Lewis and von Elbe (1987), Glassman (1996), Miller (1996), Lindstedt (1998), Gardiner (1999), Warnatz, Maas, and Dibble (2001), Simmie (2003), and Miller, Pilling, and Troe (2005).

3.1. PRACTICAL FUELS

Fuels can be classified according to their physical states under normal conditions. Representative components of gaseous fuels are hydrogen (H_2), carbon monoxide (CO), and the light hydrocarbons (HC). Liquid fuels are usually the heavier hydrocarbons and alcohols, and solid fuels include carbon, coal, wood, metals, solid propellants, and so on. Among all materials that can be used as fuel, hydrocarbons make up the bulk of the fuel supply simply because the complete oxidation of a hydrocarbon is usually accompanied by an appreciable amount of heat release. For example, the stoichiometric oxidation of one $-CH_2-$ group in a typical aliphatic fuel molecule,

$$-CH_2- + 1.5O_2 \rightarrow H_2O + CO_2,$$

releases about 156 kcal amount of heat per mole of CH_2 consumed. The extent of exothermicity represents one of the highest energy densities in the discipline of chemistry.

As we discussed in Chapter 2, the conversion of a hydrocarbon fuel to products during combustion is rarely achieved in one step. Instead, it takes place through many stages of reaction, during which a variety of intermediates are produced. Some of them are radicals (e.g., H, O, and OH), which are extremely reactive due to the presence of unpaired electrons, and are short-lived during combustion. To understand the mechanism of hydrocarbon oxidation, it is necessary to first introduce the nomenclature and molecular structures of some of the important hydrocarbon fuels and the intermediates during their combustion.

Alkanes (Paraffins): The molecules have open-chain, single-bond, saturated structures with the general chemical formula C_mH_{2m+2}. The smallest alkane compound is methane (CH_4), which is the major component of natural gas.

Alkanes can be further classified as normal alkanes with a straight-chain structure (e.g., *n*-butane) and iso-alkanes with one or more branched chains (e.g., *i*-butane, and 2,2,4-trimethylpentane, also known as iso-octane, which is a desirable gasoline fuel for its anti-knock property).

```
    H   H   H   H
    |   |   |   |
H — C — C — C — C — H
    |   |   |   |
    H   H   H   H
```

n-butane

```
    H   H   H
    |   |   |
H — C — C — C — H
    |   |   |
    H   M   H
```

i-butane

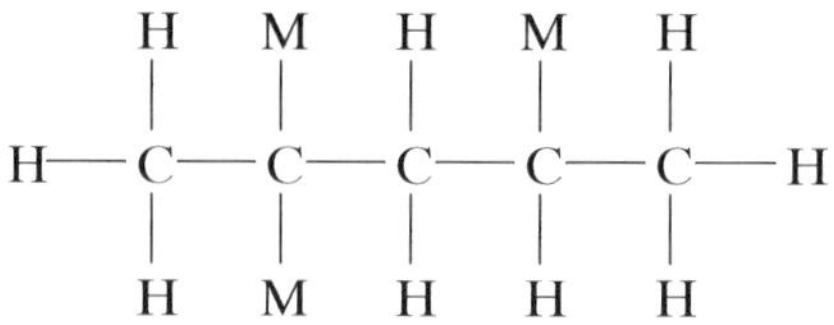

```
      H   M   H   M   H
      |   |   |   |   |
  H — C — C — C — C — C — H           M:  H — C — H
      |   |   |   |   |                       |
      H   M   H   H   H                       H
```

2,2,4-trimethylpentane

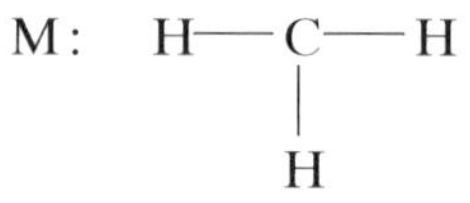

methyl group

The molecular structures shown above indicate only the bonding structures of the molecules. They do not, however, reflect their true, three-dimensional geometric structure. In fact, the carbon atoms in an aliphatic compound are not connected in a linear fashion, but are zigzagged, with a C—C—C angle of about 109 degrees. The C—C bonds in an aliphatic compound can rotate nearly freely, allowing for a cyclic structure to exist.

An alkyl radical is produced by breaking a C—H bond in an alkane molecule. For example, the methyl radical shown above is derived from breaking one C—H bond in methane. Because the reactivity of a radical species depends on where the unpaired electron is in a molecule, it is necessary to distinguish the radical isomers. For example, two isomers can be derived from *n*-butane, that is, 1-butyl and 2-butyl,

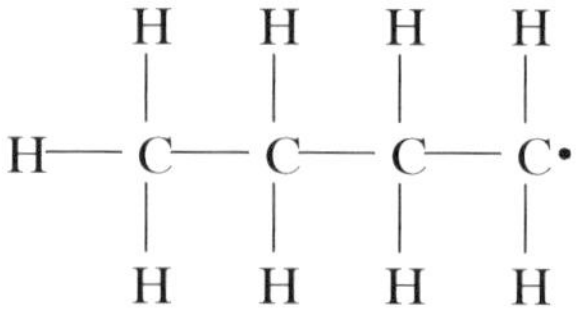

```
      H   H   H   H                  H   H       H
      |   |   |   |                  |   |   •   |
  H — C — C — C — C•             H — C — C — C — C — H
      |   |   |   |                  |   |   |   |
      H   H   H   H                  H   H   H   H
```

1-butyl

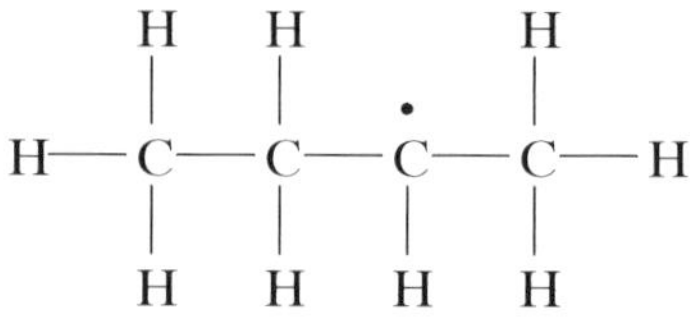

2-butyl

where the dot specifies the position of the unpaired electron, with the numerical number before the compound name indicating this position closest to the end of the chain. With the increase in the molecular size of the parent alkane, and depending on the number of branched chains, the number of possible radical isomers can increase rapidly. The alkyl radicals are of great importance in the oxidation process of alkanes.

Cycloalkanes (Cyclanes): These are single-bond, saturated, ring structured compounds with the general formula C_mH_{2m} (e.g., cyclopropane and cyclohexane). The cyclanes are almost as reactive as the linear alkanes. Large compounds, say those larger than C_{10}, are unstable and do not exist naturally.

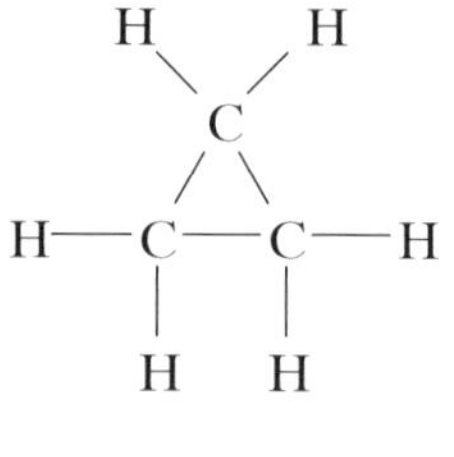

```
                                           H2
      H     H                              C
        \  /                            /     \
         C                         H2C         CH2
        / \                          |           |
  H — C — C — H                    H2C         CH2
      |   |                             \     /
      H   H                                C
                                           H2
```

cyclopropane

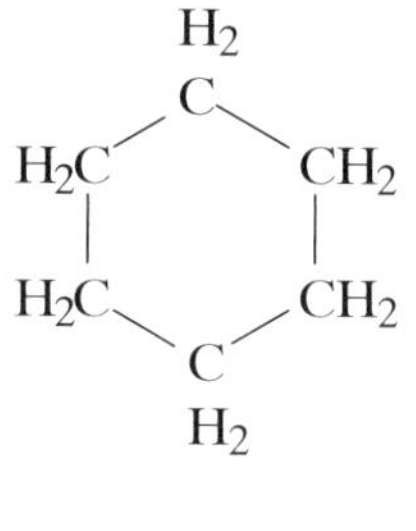

cyclohexane

Alkenes (Olefins): The molecules have open-chain structures with one double, C=C, bond, and with the general chemical formula C_mH_{2m}. The simplest olefin compounds include, for example, ethene, which is commonly called ethylene,

```
H       H
 \     /
  C==C
 /     \
H       H
```

ethene

and the butene isomers,

```
    H   H   H    H              H                   H
    |   |   |   /               |                   |
H---C---C---C==C            H---C---C===C---C---H
    |   |       \               |   |   |   |
    H   H        H              H   H   H   H
```

1-butene 2-butene

Unlike a single C–C bond, a double C=C bond is rigid and does not allow rotation about the bond. Consequently, the double-bonded C atoms form a rigid planar structure along with its immediate neighboring atoms.

Similar to the alkanes, an alkene compound also has its corresponding radicals. The simplest alkene-derived radical is vinyl,

```
H       H
 \     /
  C==C•
 /
H
```

vinyl

Polyolefins: These are open-chain, highly unsaturated compounds with more than one double bond, and with the general formula $C_mH_{2m-2(n-1)}$, where n is the number of double bonds. The simplest polyolefin is 1,3-butadiene,

```
H       H
 \     /
  C==C        H
 /     \     /
H       C==C
       /     \
      H       H
```

1,3-butadiene

Alkynes (Acetylenes): The molecules have open-chain structures with one triple, C≡C, bond, and with the general chemical formula C_mH_{2m-2}. The simplest acetylenic compounds are ethyne, which is commonly called acetylene, and propyne,

H—C≡C—H

acetylene

M—C≡C—H

propyne

Aromatics: These belong to the ring-structured, planar or nearly planar hydrocarbon family, with the ring containing mostly six carbon atoms. The simplest aromatic compounds are benzene (C_6H_6)

benzene

and toluene, the latter obtained by substituting an H by a methyl group in the benzene structure. The above ring structure can be written in either of the following simplified forms:

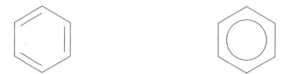

Although unsaturated, the double bonds in benzene are not distributed within the ring in an alternate fashion, but are rather distributed evenly over the entire ring structure. Such a behavior is called resonant stabilization.

Larger aromatics can be classified into two categories, those with condensed and uncondensed ring structures. Condensed ring structures have one or more C—C bonds shared by aromatic rings, and the uncondensed compounds have two or more individual aromatic rings loosely connected by C—C bonds. For example, the simplest condensed and uncondensed polycyclic aromatic compounds are naphthalene ($C_{10}H_8$) and biphenyl ($C_{12}H_{10}$), respectively:

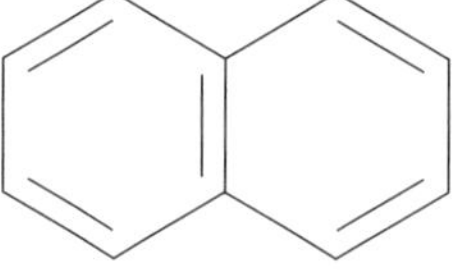

naphthalene

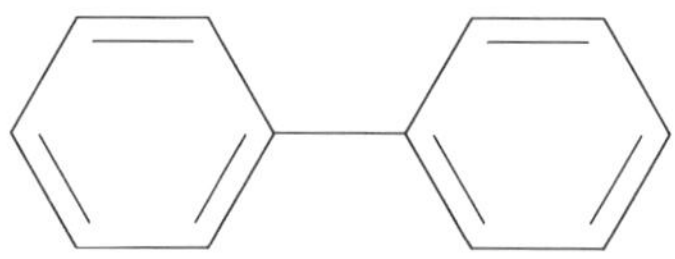

biphenyl

Alcohols: The general formula is ROH, where R is a general functional group attached to the hydroxyl group OH. Examples are methyl alcohol (CH_3OH) and ethyl alcohol (C_2H_5OH), also known as methanol and ethanol, respectively. Compared to hydrocarbons, RH, alcohols have lower heats of combustion per unit mass because a fuel containing oxygen in its molecular structure is carrying unnecessary mass for oxidation in air.

Other Fuels: There are, of course, other important fuels for specialized application. Examples are ammonium perchlorate (NH_4ClO_4) for solid propellant rockets, hydrazine (N_2H_4) for liquid propellant rockets, hydrogen and fluorine for chemical lasers, and all substances that can burn in air as hazardous materials causing fires and explosions.

3.2. OXIDATION OF HYDROGEN AND CARBON MONOXIDE

A basic understanding of the oxidation mechanisms of hydrogen and carbon monoxide is important for two reasons. First, hydrogen and carbon monoxide are major sources of fuel themselves. Second, their oxidation mechanisms are subsets of those of hydrocarbons, which are mainly made up of hydrogen and carbon. We shall therefore separately discuss their oxidation mechanisms. The complete reaction mechanism for hydrogen and CO oxidation is shown in Table 3.1, which also lists the reaction numbers as well as the rate constants defined in Eq. (2.2.3). A reverse reaction will be indicated by a negative sign in the reaction number.

3.2.1. Explosion Limits of Hydrogen–Oxygen Mixtures

In Chapter 2, we studied the role of the chain mechanism in the chemical runaway of a homogeneous mixture when it is also subjected to wall deactivation. We shall now consider the hydrogen–oxygen system, which exhibits a *Z*-shaped, pressure–temperature explosion-limit boundary, shown in Figure 3.2.1 for the response of such a mixture in a heated and pressurized chamber. It is seen that whereas explosion is always possible at high temperatures, the response is highly nonmonotonic at moderate temperatures in that, by steadily increasing the pressure from a low value, a nonexplosive mixture would first become explosive, then nonexplosive, and finally explosive again. The phenomenon of interest here, involving the *Z*-shaped response, is of a general nature in that similar responses have also been observed, both experimentally and computationally, for hydrogen–oxygen mixtures in other systems, for example the ignition temperature of a uniform flow for a given pressure and residence time.

Since initiation reactions involve only the reactant species, there are three possible initiation reactions for the present H_2–O_2 system, namely the dissociation of H_2, the dissociation of O_2, and the reaction between H_2 and O_2, as in

$$H_2 + M \rightarrow H + H + M \qquad \text{(H5)}$$

$$O_2 + M \rightarrow O + O + M \qquad \text{(−H6)}$$

$$H_2 + O_2 \rightarrow HO_2 + H. \qquad \text{(−H10)}$$

Table 3.1. Oxidation of H_2–CO mixtures

No.	Reaction	B[cm, mol, s]	α	E_a(kcal/mol)
	H_2–O_2 Chain Reactions			
(1)	$H + O_2 \rightleftharpoons O + OH$	1.9×10^{14}	0	16.44
(2)	$O + H_2 \rightleftharpoons H + OH$	5.1×10^{04}	2.67	6.29
(3)	$OH + H_2 \rightleftharpoons H + H_2O$	2.1×10^{08}	1.51	3.43
(4)	$O + H_2O \rightleftharpoons OH + OH$	3.0×10^{06}	2.02	13.40
	H_2–O_2 Dissociation/Recombination			
(5)	$H_2 + M \rightleftharpoons H + H + M$	4.6×10^{19}	−1.40	104.38
(6)	$O + O + M \rightleftharpoons O_2 + M$	6.2×10^{15}	−0.50	0
(7)	$O + H + M \rightleftharpoons OH + M$	4.7×10^{18}	−1.0	0
(8)	$H + OH + M \rightleftharpoons H_2O + M$	2.2×10^{22}	−2.0	0
	Formation and Consumption of HO_2			
(9)	$H + O_2 + M \rightleftharpoons HO_2 + M$	6.2×10^{19}	−1.42	0
(10)	$HO_2 + H \rightleftharpoons H_2 + O_2$	6.6×10^{13}	0	2.13
(11)	$HO_2 + H \rightleftharpoons OH + OH$	1.7×10^{14}	0	0.87
(12)	$HO_2 + O \rightleftharpoons OH + O_2$	1.7×10^{13}	0	−0.40
(13)	$HO_2 + OH \rightleftharpoons H_2O + O_2$	1.9×10^{16}	−1.00	0
	Formation and Consumption of H_2O_2			
(14)	$HO_2 + HO_2 \rightleftharpoons H_2O_2 + O_2$	4.2×10^{14}	0	11.98
		1.3×10^{11}	0	−1.629
(15)	$H_2O_2 + M \rightleftharpoons OH + OH + M$	1.2×10^{17}	0	45.50
(16)	$H_2O_2 + H \rightleftharpoons H_2O + OH$	1.0×10^{13}	0	3.59
(17)	$H_2O_2 + H \rightleftharpoons H_2 + HO_2$	4.8×10^{13}	0	7.95
(18)	$H_2O_2 + O \rightleftharpoons OH + HO_2$	9.5×10^{06}	2.0	3.97
(19)	$H_2O_2 + OH \rightleftharpoons H_2O + HO_2$	1.0×10^{12}	0	0
		5.8×10^{14}	0	9.56
	Oxidation of CO			
(1)	$CO + O + M \rightleftharpoons CO_2 + M$	2.5×10^{13}	0	−4.54
(2)	$CO + O_2 \rightleftharpoons CO_2 + O$	2.5×10^{12}	0	47.69
(3)	$CO + OH \rightleftharpoons CO_2 + H$	1.5×10^{07}	1.3	−0.765
(4)	$CO + HO_2 \rightleftharpoons CO_2 + OH$	6.0×10^{13}	0	22.95

Source: Kim, T. J., Yetter, R. A. & Dryer, F. 1994. New results on moist CO oxidation: high-pressure, high-temperature experiments, and comprehensive kinetic modeling. *Proc. Combust. Inst.* **25**, 759–766.

The endothermicities of these three reactions are respectively 104, 118, and 55 kcal/mole. Since the activation energy of a dissociation reaction is roughly equal to its endothermicity, reaction (−H10) is the most important initiation reaction under almost all conditions. Reaction (H5) may also contribute to initiation, but only at high temperatures. Reaction (−H6) is usually not preferred because oxygen has a larger dissociation energy than hydrogen.

With the production of H from either (H5) or (−H10), the following chain reaction is initiated:

$$H + O_2 \rightarrow O + OH \quad \text{(H1)}$$

$$O + H_2 \rightarrow H + OH \quad \text{(H2)}$$

$$OH + H_2 \rightarrow H + H_2O. \quad \text{(H3)}$$

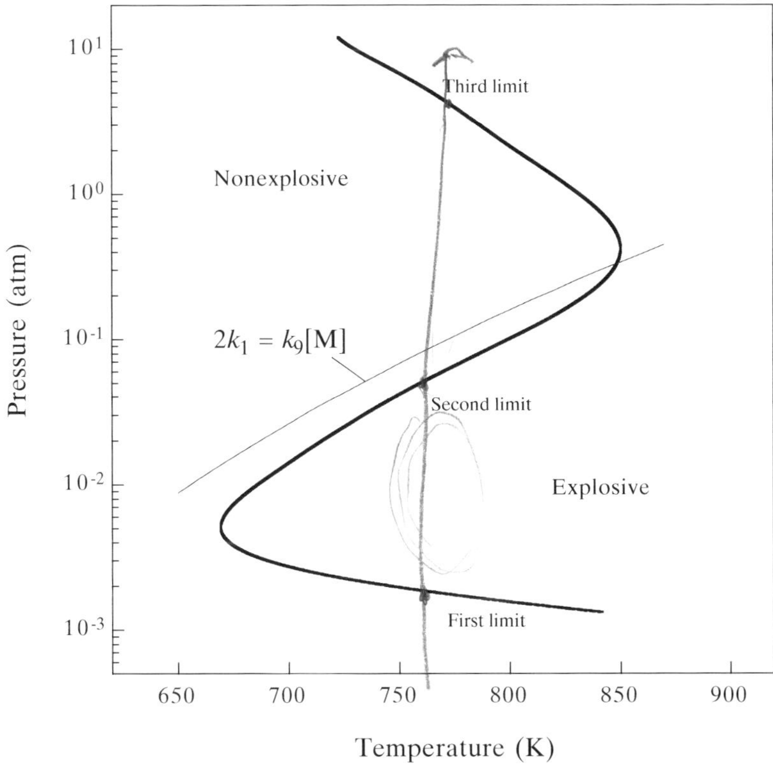

Figure 3.2.1. Explosion limits of a stoichiometric H_2-O_2 mixture.

The reverse reactions for (H1) and (H2) are not important at this stage because they involve the collisions between two radical species whose concentrations are very low. The reverse reaction of (H3) is also not important because of the low concentration of the product species, H_2O, during the initiation stage of explosion.

For sufficiently low temperatures and pressures, explosion is not possible even with the addition of some H or OH. This is because the key chain-branching step (H1) is endothermic by 17 kcal/mole and is therefore not favored at low temperatures. Furthermore, at low pressures these active species either rapidly diffuse to the chamber wall where they are destroyed,

$$\text{H, O, OH} \rightarrow \text{wall destruction},$$

or react too slowly relative to an imposed finite residence time.

As pressure increases, collision becomes more frequent and reactions are facilitated. As the first explosion limit is crossed, the rate of branching becomes overwhelmingly fast relative to either the rate of removal at the wall or the finite residence time, and explosion occurs.

By further increasing pressure, the three-body reaction,

$$H + O_2 + M \rightarrow HO_2 + M, \tag{H9}$$

becomes more frequent and will eventually replace (H1) as the dominant reaction between H and O_2. Being relatively inactive, the HO_2 radicals can survive many collisions and eventually will either diffuse to the wall where they are destroyed, or exhausted in a flow of finite residence time. Reaction (H9) is therefore an effective termination step of the radical chain process. As a result, the chain sequence (H1)–(H3) is broken.

The second explosion limit is therefore determined by the competition between the growth, (H1)–(H3), and destruction, (H9), of the H atom. To determine the p-T dependence of this limit, we write the rate equations for the H, O, and OH radicals as

$$\frac{d[\mathrm{H}]}{dt} = -k_1[\mathrm{H}][\mathrm{O}_2] + k_2[\mathrm{O}][\mathrm{H}_2] + k_3[\mathrm{OH}][\mathrm{H}_2] - k_9[\mathrm{H}][\mathrm{O}_2][\mathrm{M}] \quad (3.2.1)$$

$$\frac{d[\mathrm{O}]}{dt} = k_1[\mathrm{H}][\mathrm{O}_2] - k_2[\mathrm{O}][\mathrm{H}_2] \quad (3.2.2)$$

$$\frac{d[\mathrm{OH}]}{dt} = k_1[\mathrm{H}][\mathrm{O}_2] + k_2[\mathrm{O}][\mathrm{H}_2] - k_3[\mathrm{OH}][\mathrm{H}_2]. \quad (3.2.3)$$

Assuming steady state for the O and OH radicals, that is, $d[\mathrm{O}]/dt = 0$, and $d[\mathrm{OH}]/dt = 0$, we obtain

$$k_2[\mathrm{O}][\mathrm{H}_2] = k_1[\mathrm{H}][\mathrm{O}_2]$$

$$k_3[\mathrm{OH}][\mathrm{H}_2] = k_1[\mathrm{H}][\mathrm{O}_2] + k_2[\mathrm{O}][\mathrm{H}_2] = 2k_1[\mathrm{H}][\mathrm{O}_2].$$

Putting the above expressions into Eq. (3.2.1), we have

$$\frac{d[\mathrm{H}]}{dt} = 2k_1[\mathrm{H}][\mathrm{O}_2] - k_9[\mathrm{H}][\mathrm{O}_2][\mathrm{M}] \quad (3.2.4a)$$

$$= (2k_1 - k_9[\mathrm{M}])[\mathrm{H}][\mathrm{O}_2], \quad (3.2.4b)$$

which is simply

$$\frac{d[\mathrm{H}]}{dt} = 2\omega_1 - \omega_9. \quad (3.2.5)$$

Following the same discussion as that for branched-chain explosions in Section 2.4.2, Eq. (3.2.4b) shows that [H] varies exponentially with time, growing for $(2k_1 - k_9[\mathrm{M}]) > 0$ and decaying otherwise. This then implies that the second limit is given by the neutral condition,

$$2k_1 = k_9[\mathrm{M}]. \quad (3.2.6)$$

Since $p = [\mathrm{M}]R^\circ T$, and if the fall-off parameters of all third body species in (H9) are the same, then the pressure and temperature relationship of the second limit is explicitly given by

$$p = \frac{2k_1}{k_9} R^\circ T. \quad (3.2.7)$$

Based on the rate constant expressions listed in Table 3.1, we obtain the p–T relationship shown in Figure 3.2.1. It is seen that the second explosion limit is well described by this relation, which is called the crossover temperature.

It is to be noted that although Eq. (3.2.6) corresponds to the condition $d[\mathrm{H}]/dt \equiv 0$, it is not a consequence of the steady-state assumption for [H]. Indeed, while the steady-state assumption for [O] and [OH] holds over extensive regimes in p and T, the transition boundary $d[\mathrm{H}]/dt \equiv 0$ holds only along the crossover temperature line.

As the third explosion limit is crossed, with further increase in pressure, the concentration of HO_2 becomes higher. The reactions

$$HO_2 + H_2 \rightarrow H_2O_2 + H \tag{$-$H17}$$

$$H_2O_2 + M \rightarrow OH + OH + M \tag{H15}$$

are more frequent and overtake the stability of HO_2. Therefore, instead of having HO_2 lost either to the wall or through the flow, it reacts with H_2 and generates the active species, H, O, and OH, to again induce explosion.

At higher temperatures, say about 900 K, more radicals are produced and reactions between them become important. The HO_2 radical can either react with itself,

$$HO_2 + HO_2 \rightarrow H_2O_2 + O_2, \tag{H14}$$

followed by (H15), or with the H and O radicals,

$$HO_2 + H \rightarrow OH + OH \tag{H11}$$

$$HO_2 + O \rightarrow OH + O_2. \tag{H12}$$

In this situation, reaction (H9) is a part of the chain propagation process, and thus explosion will always occur.

In Table 3.1 the H_2—O_2 chain reaction sequence is given in Nos. 1–4, the H_2—O_2 recombination and dissociation reactions in Nos. 5–8, the formation and consumption of the hydroperoxyl radical, HO_2, in Nos. 9–13, and the chemistry of hydrogen peroxide, H_2O_2, in Nos. 14–19.

We also note from Table 3.1 that some reactions have negative activation energies. This often occurs for reactions without intrinsic energy barriers such as those involving two radicals. The negative activation energy then implies that it is easier for a cold radical to combine with another cold radical than is for two hot radicals to combine. In the latter case, the combined kinetic energy of the two hot radicals are so large that the colliding radicals may just fly by without forming a bond.

Table 3.1 further shows that reaction (H14) has two sets of reaction constants. This reflects the fact that depending on the initial point of contact, the reaction may proceed by two different routes. One of the routes dominates at low temperatures while the other dominates at high temperatures. For this reason, the overall rate constant is the sum of these two elementary reactions.

3.2.2. Carbon Monoxide Oxidation

It is difficult to ignite and sustain a dry $CO{-}O_2$ flame because the direct reaction between CO and O_2,

$$CO + O_2 \rightarrow CO_2 + O, \tag{CO2}$$

has a high activation energy (48 kcal/mol) and therefore is a very slow process even at high temperatures. Furthermore, the O atom produced does not lead to any rapid chain-branching reactions (Lewis & von Elbe 1987). However, in the presence of even a small quantity of hydrogen, say 20 ppm, OH radicals are formed through reactions (H1–H3, H5, and –H10). Then the formation of CO_2 via the reaction

$$CO + OH \rightarrow CO_2 + H \tag{CO3}$$

becomes the dominant path for CO oxidation. The H atoms produced in (CO3) feed the chain-branching reactions (H1)–(H3), and thereby accelerate the CO oxidation rate.

In moist air, water can also catalyze CO oxidation through

$$O_2 + M \rightarrow O + O + M, \tag{−H6}$$

followed by

$$O + H_2O \rightarrow OH + OH, \tag{H4}$$

which provides the OH radicals needed by (CO3).

At high pressures, the reaction

$$CO + HO_2 \rightarrow CO_2 + OH \tag{CO4}$$

provides another route of CO to CO_2 conversion. Thus CO oxidation can be modeled by adding mainly (CO3) and (CO4) to the hydrogen–oxygen mechanism, as shown in Table 3.1.

Since trace amounts (ppm) of moisture are invariably present in practical systems and laboratory experiments, the relevant oxidation mechanism to use in understanding the response of these systems is that of wet instead of dry CO. Indeed, sometimes it maybe useful in experiments to deliberately add a small amount of hydrogen or water in the mixture, exceeding the background moisture contamination, so as to accurately define the composition of the mixture.

3.2.3. Initiation Reactions in Flames

The mechanisms of fuel oxidation in flames, including those of hydrogen and carbon monoxide, are quite different from those of ignition of homogeneous reactive mixtures discussed above because the radical pool concentration in a flame is always much larger than that generated during the induction period of homogeneous ignition. In addition, the initiation chemistry responsible for homogeneous ignition is often less important for flames because there are always abundant radicals that can back diffuse from the high-temperature flame region to the colder, unburned regions of fuel and oxidizer, resulting in a different initiation chemistry. For example, for hydrogen flames, the dominant initiation reaction is the $H + O_2$ branching reaction (H1), because of the abundance and mobility of the highly reactive H atom, instead of reactions (H5) and (−H10) for homogeneous mixtures. The global activation energy is, of course, also affected.

3.3. OXIDATION OF METHANE

3.3.1. General Considerations of Hydrocarbon Oxidation

Recently, considerable progress has been made in our understanding of the detailed reaction mechanisms of hydrocarbon oxidation. The general consensus is that the two most important reactions in a combustion process are

$$H + O_2 \rightarrow O + OH \tag{H1}$$

$$CO + OH \rightarrow CO_2 + H, \tag{CO3}$$

which do not involve the specific hydrocarbon fuel at all. Reaction (CO3) is responsible almost exclusively for converting CO to CO_2. It also generates some of the H atoms needed by (H1).

In addition to the similarity of the most important reactions for hydrocarbon fuels, it is recognized that although the initial fuel breakdown is fuel specific, its rate is in general too fast to limit the overall rate of combustion. Furthermore, the initial fuel breakdown always leads to C_1, C_2, and C_3 fragments. A comprehensive understanding of hydrocarbon oxidation must then be strongly hierarchical in that mechanisms for the oxidation of complex fuels contain within them the submechanisms of simpler molecules. The key to the understanding of the oxidation mechanism of hydrocarbon fuels, therefore, must start from an understanding of the simpler hydrocarbon oxidation mechanisms. Methane, being the lightest hydrocarbon fuel, is naturally the starting point.

Before proceeding to the discussion of methane oxidation, it is important to recognize that there exist diverse ranges of conditions in which a combustion process can take place. The dominant reaction mechanisms may thus vary substantially, depending on the local thermodynamic states of the combustion process. In particular, as we have just shown in the previous section, (H1) has a sufficiently large activation energy whereas (H9) has none. Therefore (H1) is the dominant chain-branching step at high temperatures. However, in the intermediate- to low-temperature regimes, the reaction chemistry is often dominated by that of the HO_2 radicals produced through (H9) as well as the initiation reactions such as (−H10) for hydrogen oxidation and (M2) for methane oxidation, which will be discussed in the following section. It is therefore necessary to distinguish the reaction mechanisms in the high-temperature regime (usually above 1,100 K at atmospheric pressure) from those in the low- to intermediate-temperature regimes.

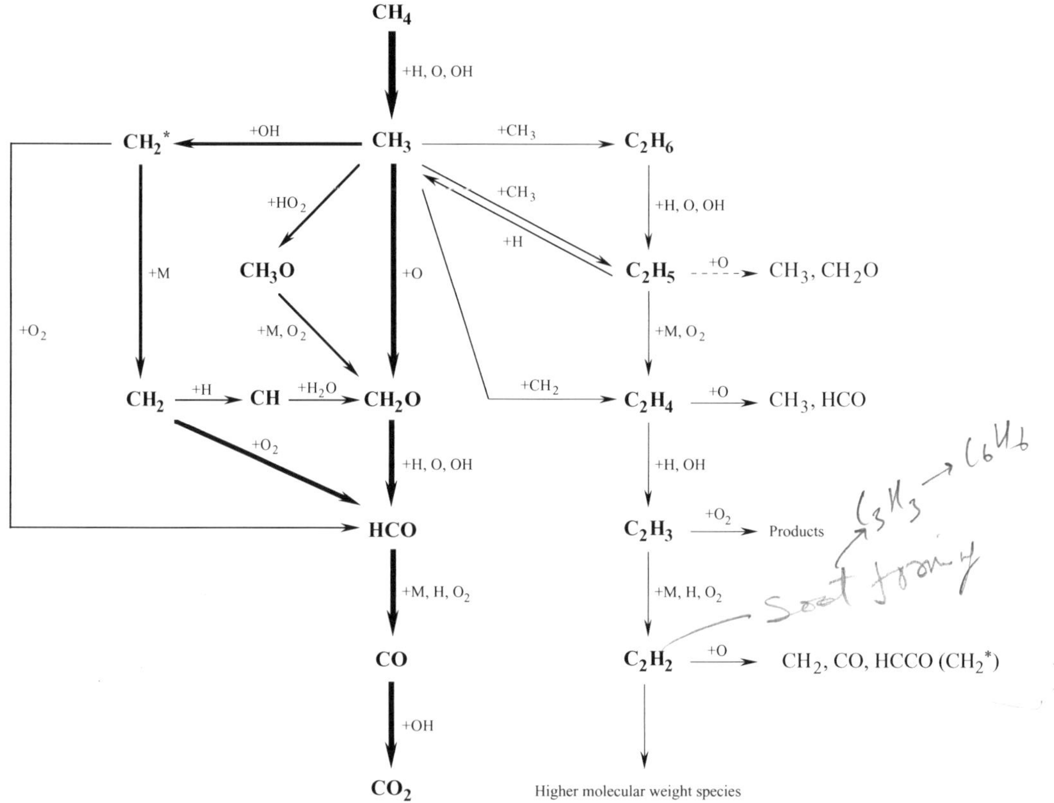

Figure 3.3.1. Reaction pathways in methane flames (adapted from Warnatz 1981).

In this section we study the ignition of methane and the methane flame chemistry, including a detailed discussion of the formaldehyde (CH_2O) oxidation mechanism, which is an integral part of the methane oxidation process.

The reaction pathways for the oxidation involving methane flames are depicted in Figure 3.3.1, with the thickness of the arrows indicating the relative importance of individual pathways. Much of the following discussion on methane autoignition is also included in this diagram.

3.3.2. Methane Autoignition

For methane, the ignition delay time τ, also called the induction time, has been empirically correlated by

$$\tau = 2.5 \times 10^{-15} \exp(26700/T)[CH_4]^{0.32}[O_2]^{-1.02},$$

where τ is in seconds (s) and $[CH_4]$ and $[O_2]$ are respectively the molar concentrations (mole/cm^3) of methane and oxygen in the reactant mixture. This correlation represents shock tube measurements (Tsuboi & Wagner 1975) in argon over the temperature range of 1,200 to 2,100 K, and shows that τ decreases with increasing temperature and pressure. The inhibitive effect of methane in the global consideration, as indicated by the negative reaction order for methane, is particularly worth noting.

The preignition chemistry of methane is dictated by the rate of radical accumulation. The chain reaction is initiated primarily by the following two reactions:

$$CH_4 + M \rightleftharpoons CH_3 + H + M \qquad \text{(M1)}$$

$$CH_4 + O_2 \rightleftharpoons CH_3 + HO_2. \qquad \text{(M2)}$$

Since (M1) is a large activation-energy, unimolecular reaction, it may be favored over (M2) only at high temperatures. Thus if (M1) is the dominant initiation step, the reactions that follow are

$$H + O_2 \rightleftharpoons OH + O \qquad \text{(H1)}$$

$$CH_4 + (H, O, OH) \rightleftharpoons CH_3 + (H_2, OH, H_2O). \qquad \text{(M3, M4, M5)}$$

Reaction (M3) plays an inhibiting role in the ignition process because it competes with the chain-branching step (H1) for H and converts the active H atoms to the less active CH_3 radicals. This is the primary reason for the positive dependence of the induction time τ on the methane concentration. The initiation reaction (M1) can also play the role of inhibiting ignition in that when the radical pool begins to build up, its backward reaction becomes increasingly fast, leading to increased radical chain termination.

When (M2) is the dominant initiation step, the following reactions may participate in further radical growth:

$$CH_4 + HO_2 \rightleftharpoons CH_3 + H_2O_2 \qquad \text{(M6)}$$

$$H_2O_2 + M \rightleftharpoons OH + OH + M. \qquad \text{(H15)}$$

The similarity between (M6) and (−H17) is to be noted.

The methyl radicals produced in the above reactions subsequently react with molecular oxygen through two reaction channels,

$$CH_3 + O_2 \rightleftharpoons CH_3O + O \qquad \text{(M7a)}$$

$$\rightleftharpoons CH_2O + OH. \qquad \text{(M7b)}$$

The branching ratio, that is, the relative contributions of the two channels to the overall rate constant, appears to be crucial to methane ignition because channel *b* is a chain-carrying step, producing formaldehyde, CH_2O, whereas channel *a* leads to chain branching.

Close to ignition, the reaction

$$CH_3 + HO_2 \rightleftharpoons CH_3O + OH \qquad \text{(M8)}$$

becomes the dominant step for methyl oxidation. The CH_3O radical, produced from (M7a) and (M8), is highly active. It is converted to formaldehyde rapidly via

$$CH_3O + M \rightleftharpoons CH_2O + H + M \qquad \text{(M9)}$$

$$CH_3O + O_2 \rightleftharpoons CH_2O + HO_2. \qquad \text{(M10)}$$

The formaldehyde produced in (M9) and (M10) subsequently reacts with OH and O_2,

$$CH_2O + OH \rightleftharpoons HCO + H_2O \tag{M11}$$

$$CH_2O + O_2 \rightleftharpoons HCO + HO_2, \tag{M12}$$

producing the highly active formyl radical, HCO, which is consumed by reactions similar to the destruction of the CH_3O radical,

$$HCO + M \rightleftharpoons H + CO + M \tag{M13}$$

$$HCO + O_2 \rightleftharpoons CO + HO_2. \tag{M14}$$

The further conversion of CO to CO_2 via reaction (CO3) occurs after ignition. Although the final temperature of the burned mixture depends on the conversion of CO to CO_2, the induction time is usually insensitive to (CO3).

It is therefore clear that through the above sequence of reactions we have progressively reduced the hydrogen content of the original fuel molecule, CH_4, to CH_3 and CH_3O, then to CH_2O, HCO, and finally to CO. The hydrogen atoms abstracted at different stages are eventually oxidized to form H_2O, while the CO that is formed through (M14) at the end of the sequence is oxidized to form CO_2 through the CO oxidation mechanism described earlier. The hierarchical nature of the oxidation process in terms of the progressive reduction of the parent hydrocarbon fuel species to the constituents of the H_2 and CO oxidation systems, through which the final products H_2O and CO_2 are formed, is particularly noteworthy.

The above mechanism describes the route through which the size of the original fuel molecule is reduced. There is, however, another route through which the size of the fuel molecule is increased. Specifically, upon the formation of the methyl radical CH_3, it can self react through

$$CH_3 + CH_3 + M \rightleftharpoons C_2H_6 + M \tag{M15}$$

$$CH_3 + CH_3 \rightleftharpoons C_2H_5 + H. \tag{M16}$$

Reaction (M15) is a radical termination step, which inhibits ignition, while reaction (M16) is chain carrying, leading to the formation of the ethyl radical C_2H_5 and the highly active H atoms. Since (M15) tends to be more important at higher pressures because it is a third-order radical–radical recombination reaction, it has also shown strong sensitivity for ignition near atmospheric pressure.

It is important to note that through reactions such as (M15) and (M16), species containing two carbon atoms are generated even though the original fuel molecule, methane, consists of only one carbon atom. These larger molecules will subsequently undergo oxidation themselves. Thus the oxidation mechanism of methane must include not only those of the smaller species such as CH_2O and CO, but also those of the larger, C_2 species. The influence of C_2 chemistry is particularly important for rich mixtures because of the abundance of CH_3 radicals.

By extending this concept further, we can anticipate that, in principle, the C_2H_5 radicals produced through, say, (M16) can self react to generate the even larger molecule C_4H_{10}, while the recombination between the CH_3 and C_2H_5 radicals produces C_3H_8. Consequently the oxidation mechanism of CH_4, and indeed that of any hydrocarbon, will necessarily involve the oxidation of all hydrocarbons of larger molecules. In reality, the concentrations of these larger molecules, except those produced through the first generation of recombination, are so small that their effects on the overall reaction response are negligible.

3.3.3. Methane Flames

Similar to the discussion on the H_2 and CO flames, the main difference between the autoignition and flame chemistry is the abundance of the radicals H, O, and OH that are produced at the flame and back diffuse upstream. Consequently, the destruction of methane is achieved mainly through H abstraction by H, O, and OH, that is, reactions (M3)–(M5), producing the methyl radical. The methyl radical is then consumed mainly by the O atom through

$$CH_3 + O \rightleftharpoons CH_2O + H, \tag{M17}$$

and by HO_2 through (M8). Upon further hydrogen abstraction of CH_2O by H,

$$CH_2O + H \rightarrow HCO + H_2, \tag{M18}$$

and by OH through (M11), the formyl radical is produced. It then decomposes rapidly to yield CO and H, or undergoes the H-abstraction by H and O_2.

Some CH_3 radicals will also react with the OH radical to yield the highly active, singlet methylene radical (CH_2^*),

$$CH_3 + OH \rightleftharpoons CH_2^* + H_2O. \tag{M19}$$

The CH_2^* radical has no unpaired electrons, but it has an empty orbital which makes it a highly energetic and active species. Most of the CH_2^* are de-energized by collision with other molecules to the more stable triplet CH_2 radical, with two unpaired electrons,

$$CH_2^* + M \rightarrow CH_2 + M. \tag{M20}$$

The CH_2^* radical can also react with O_2 to provide secondary chain branching, producing the highly active H and OH radicals,

$$CH_2^* + O_2 \rightarrow CO + H + OH \tag{M21}$$

to speed up the overall oxidation.

Even if reaction (M20) is the dominant CH_2^* removal channel, the resulting triplet CH_2 radicals are still highly active. They react readily with O_2 to provide secondary chain branching,

$$CH_2 + O_2 \rightarrow HCO + OH. \tag{M22}$$

Some of the CH_2 will form the CH radical through the H-abstraction reaction,

$$CH_2 + H \rightleftharpoons CH + H_2. \quad \text{(M23)}$$

The CH radicals are quickly consumed by H_2O or O_2,

$$CH + H_2O \rightarrow CH_2O + H, \quad \text{(M24)}$$

$$CH + O_2 \rightarrow HCO + O. \quad \text{(M25)}$$

In rich mixtures, the relative abundance of CH_3 enhances their recombination to yield C_2 species via reactions (M15) and (M16) and via the fast reaction of CH_3 and CH_2,

$$CH_3 + CH_2 \rightleftharpoons C_2H_4 + H. \quad \text{(M26)}$$

These C_2 species either undergo further oxidation or may survive the attack by oxygenated species, eventually leading to acetylene according to the reaction steps shown in Figure 3.3.1. The production of acetylene is crucial to the formation of soot, as will be discussed later.

3.4. OXIDATION OF C_2 HYDROCARBONS

We shall now discuss the oxidation mechanisms of the C_2 hydrocarbons, namely ethane (C_2H_6), ethylene (C_2H_4), and acetylene (C_2H_2). We first note that not only is ethane a major intermediate during the rich combustion of methane, as shown above, it is also the second most important constituent of natural gas. Similar to ethane, ethylene and acetylene are fuels by themselves. Furthermore, they are the major intermediates of ethane and higher hydrocarbon oxidation. We shall discuss the current understanding of the oxidation mechanisms of the three C_2 hydrocarbons in high-temperature flames.

The oxidation of ethane in flames starts from the H-abstraction of C_2H_6 by H, O, and OH, producing the ethyl radical,

$$C_2H_6 + (H, O, OH) \rightleftharpoons C_2H_5 + (H_2, OH, H_2O). \quad \text{(C}_2\text{1)}$$

The ethyl radical is not very stable. It reacts rapidly with H and O_2, or decomposes to ethylene and an H atom,

$$C_2H_5 + (H, O_2) \rightleftharpoons C_2H_4 + (H_2, HO_2) \quad \text{(C}_2\text{2)}$$

$$C_2H_5 + M \rightleftharpoons C_2H_4 + H + M, \quad \text{(C}_2\text{3)}$$

or it reacts with O_2 to produce acetaldehyde (CH_3CHO),

$$C_2H_5 + O_2 \rightarrow CH_3CHO + OH. \quad \text{(C}_2\text{4)}$$

Acetaldehyde subsequently reacts with H, O, and OH, producing the CH_3CO radical, followed by its unimolecular decomposition, leading to CH_3 and CO,

$$CH_3CHO + (H, O, OH) \rightleftharpoons CH_3CO + (H_2, OH, H_2O) \quad \text{(C}_2\text{5)}$$

$$CH_3CO + M \rightleftharpoons CH_3 + CO + M. \quad \text{(C}_2\text{6)}$$

Under fuel lean conditions, the ethyl radical may also react with the O atom to produce CH_3 and CH_2O,

$$C_2H_5 + O \rightarrow CH_3 + CH_2O. \tag{C$_2$7}$$

The oxidation mechanism of ethylene is different from that of the alkanes in that it does not necessarily require H-abstraction reactions before it can be oxidized. In fact, the double bond in ethylene is susceptible to direct O and OH attack. Specifically, when ethylene reacts with the O atom, its double bond breaks readily, yielding CH_3 and HCO,

$$C_2H_4 + O \rightarrow CH_3 + HCO. \tag{C$_2$8}$$

Because the products of the above reaction are radicals, reaction ($C_2$8) provides a secondary chain-branching step which may significantly speed up the oxidation process.

For fuel-rich mixtures, ethylene may survive the attack by the O atom. The H-abstraction by H and OH becomes an important source of ethylene consumption, producing the vinyl (C_2H_3) radical,

$$C_2H_4 + (H, OH) \rightleftharpoons C_2H_3 + (H_2, H_2O). \tag{C$_2$9}$$

Similar to the ethyl radical, vinyl is also quite reactive. It reacts primarily with O_2, producing CH_2O and HCO, and CH_2CHO and O, via

$$C_2H_3 + O_2 \rightarrow CH_2O + HCO \tag{C$_2$10a}$$

$$\rightarrow CH_2CHO + O. \tag{C$_2$10b}$$

Subsequent reactions of CH_2CHO lead to the formation of C_1 radical species. The vinyl radicals may also undergo the following three reactions, all of which lead to the formation of acetylene,

$$C_2H_3 + (H, O_2) \rightarrow C_2H_2 + (H_2, HO_2) \tag{C$_2$11}$$

$$C_2H_3 + M \rightleftharpoons C_2H_2 + H + M. \tag{C$_2$12}$$

Figure 3.3.1 shows that acetylene is a major intermediate during the fuel-rich combustion of methane. It is also a major product of incomplete, fuel-rich combustion, which can be understood by examining the thermodynamics of the conversion of a fuel molecule to acetylene. Consider the conversion of a typical $-CH_2-$ group in an aliphatic hydrocarbon to acetylene via

$$-CH_2- \rightleftharpoons \frac{1}{2}C_2H_2 + \frac{1}{2}H_2.$$

The process is typically endothermic by, say, 32 kcal/mol at 1,600 K. Then why is acetylene favored? The answer lies in the entropy change of the process. The release of H_2 generates a significant amount of entropy, equal to 30.8 cal/mol-K at the same temperature. The resulting Gibbs' function value is then quite negative, $\Delta G = 32 - 30.8 \times 10^{-3} \times 1{,}600 = -17$ kcal/mol, favoring equilibrium toward the product side. The corresponding equilibrium constant, $K_p \approx 200$, is substantially greater than unity, thus explaining the tendency of acetylene production. Of course, the accumulation of acetylene as an incomplete combustion product requires that oxygen is so deficient that the residual hydrocarbons remain unoxidized.

Acetylene, having an enthalpy of formation of 54 kcal/mol at the standard state, is a highly energetic and dangerous fuel. Even in the absence of oxygen, it may undergo spontaneous polymerization. When used as a fuel, acetylene represents a compound with the highest energy density among the common fuels, with an adiabatic flame temperature considerably higher than those of other fuels at the same stoichiometry, as shown in Chapter 1.

Similar to ethylene, acetylene does not necessarily require H-abstraction to initiate the oxidation process. This, in part, is due to the high C—H bond energy in acetylene, equal to 131 kcal/mol, which is substantially larger than the C—H bond energy in alkanes (e.g., 101 kcal/mol in ethane). In addition, the reaction between acetylene and O is rapid,

$$C_2H_2 + O \rightarrow CH_2 + CO \quad (C_213a)$$

$$\rightarrow HCCO + H. \quad (C_213b)$$

The ketenyl radical, HCCO, is extremely active, reacting readily with the H atom to produce the singlet methylene radical,

$$HCCO + H \rightarrow CH_2^* + CO. \quad (C_214)$$

In fuel-rich mixtures, acetylene may also combine with the H atom to produce the vinyl radical. If there is oxygen available, the vinyl radical can be oxidized very quickly via reactions (C_210a) and (C_210b).

The reaction for ethane, ethylene and acetylene combustion has now proceeded to the stage that all the species (e.g., CH_3, CH_2s, CH_2O, HCO, and CO) are those that participate in the oxidation mechanism of methane, and hence can be subsequently described by the CH_4—O_2 chemistry.

3.5. OXIDATION OF ALCOHOLS

The oxidation of alcohols in flames starts with H-abstraction by H, O, and OH. For example, in methanol flames the initial attack of CH_3OH occurs via the following reactions:

$$CH_3OH + (H, O, OH) \rightarrow CH_2OH + (H_2, OH, H_2O). \quad (A1)$$

These reactions yield the hydroxymethyl radical, which readily dissociates or reacts with molecular oxygen to form formaldehyde,

$$CH_2OH + M \rightarrow CH_2O + H + M \quad (A2)$$

$$CH_2OH + O_2 \rightarrow CH_2O + HO_2. \quad (A3)$$

The abstraction of an H atom from CH_3OH may also lead to the formation of the methoxyl radical, CH_3O,

$$CH_3OH + (H, O, OH) \rightarrow CH_3O + (H_2, OH, H_2O). \quad (A4)$$

When active radical species are not available for the attack of methanol, the initiation of the oxidation process is achieved through

$$CH_3OH + M \rightarrow CH_3 + OH + M. \tag{A5}$$

A less important channel may also contribute to the initiation, producing the singlet methylene radical, CH_2^*,

$$CH_3OH + M \rightarrow CH_2^* + H_2O + M. \tag{A6}$$

A chain reaction then follows via (M21), the reactions (A1)–(A5), (M9), (M10), and most importantly, (H1).

For ethanol oxidation in flames, it first undergoes the H-abstraction reactions according to

$$C_2H_5OH + (H, O, OH) \rightarrow CH_3CHOH + (H_2, OH, H_2O). \tag{A7}$$

Similar to the hydroxymethyl radical, the CH_3CHOH radical undergoes either dissociation or reaction with molecular oxygen to produce acetaldehyde,

$$CH_3CHOH + M \rightarrow CH_3CHO + H + M \tag{A8}$$

$$CH_3CHOH + O_2 \rightarrow CH_3CHO + HO_2. \tag{A9}$$

The H-abstraction from CH_3CHO via (C_25) yields CH_3CO, which readily dissociates to CH_3 and CO through (C_26).

In the absence of active radical species, ethanol may dissociate when exposed to high temperatures through breaking of the C—C bond, which is the weakest among all bonds,

$$C_2H_5OH + M \rightarrow CH_3 + CH_2OH + M. \tag{A10}$$

It may also react with O_2 through

$$C_2H_5OH + O_2 \rightarrow CH_3CHOH + HO_2. \tag{A11}$$

Depending on the ethanol-to-oxygen ratio, different amounts of ethylene are produced among the intermediate species. The ratio of ethylene to acetaldehyde increases with increasing equivalence ratio. The methyl radical, which is produced from CH_3CO dissociation, recombines to form ethane, which is subsequently converted to ethylene via reactions (C_21), (C_22), and (C_23).

3.6. HIGH-TEMPERATURE OXIDATION OF HIGHER ALIPHATIC FUELS

We shall limit our discussion here to the oxidation mechanism of higher aliphatic compounds under high-temperature combustion conditions. It will be shown that the mechanistic features of the oxidation process are similar among higher aliphatic hydrocarbons, and that the submechanisms of H_2—CO—O_2, CH_4—O_2 and C_2H_x—O_2 discussed in previous sections are the building blocks for the higher aliphatic hydrocarbon combustion. A general discussion of aliphatic fuel oxidation at low- to intermediate-temperature regime, specifically the cool flame phenomenon and the temperature dependence of the oxidation rates, will be given in Section 3.8.

Before we proceed to elaborate on the oxidation mechanism of the higher aliphatic fuels, it is necessary to devote some discussion here on the chemical bond energy and the β-scission rule, which is one of the foundations upon which the oxidation mechanisms of aliphatics are understood.

3.6.1. The β-Scission Rule

In Chapter 1, we have defined the bond energy, or more precisely, the bond dissociation energy, BDE, as the enthalpy associated with the breaking of a chemical bond (McMillen & Golden 1982). Since bond breaking is usually endothermic, the activation energy of such a reaction is then equal to the BDE plus the energy barrier of the reverse bond association reaction. Hence, a low activation energy, and thus, a large rate constant requires necessarily, although not sufficiently, a small BDE. In other words, the chemical bonds in a molecule that have smaller BDEs are more susceptible to dissociation than those with larger BDEs.

The β-scission rule states that, for a radical species, the bonds that will break are those one removed from the radical site because they have the lowest bond energies amongst all the bonds. Physically, the unpaired electron at the radical site strengthens the adjacent bonds, rendering the bonds next to them most susceptible to break. Taking the primary propyl radical (n-C_3H_7) as an example, the bond dissociation energies, BDE (kcal/mol), for all the bonds are shown in the following diagram:

$$\begin{array}{ccccccc} & & H & & H & & H \\ & & | & & | & & | \\ H & \overset{\sim 100}{—} & C & \overset{24}{—} & C & \overset{99}{—} & C\bullet \\ & & | & & |\,34 & & |\,>100 \\ & & H & & H & & H \end{array}$$

Clearly, the β C—C bond has the lowest BDE, 24 kcal/mol, and is the most preferred bond to break. The resulting products are ethylene (C_2H_4) and methyl (CH_3). The second least stable bond is the β C—H bond, which has a BDE of 34 kcal/mol. The breaking of the β C-H leads to propene (C_3H_6) and an H atom.

Another example is the n-hexyl radical,

$$\begin{array}{ccccccccccccc} & & H & & H & & H & & H & & H & & H \\ & & | & & | & & | & & | & & | & & | \\ H & — & C & \overset{\sim 90}{—} & C & \overset{\sim 90}{—} & C & \overset{\sim 90}{—} & C & \overset{22}{—} & C & \overset{99}{—} & C\bullet \\ & & |\,\sim 100 & & |\,\sim 100 & & |\,\sim 100 & & |\,\sim 100 & & |\,36 & & |\,>100 \\ & & H & & H & & H & & H & & H & & H \end{array}$$

Again, the weakest bond, with a BDE of 22 kcal/mol, is the β C—C bond. The most probable products of decomposition are then ethylene (C_2H_4) and n-butyl radical

(n-C_4H_9). The second weakest bond, with a BDE of 36 kcal/mol, is the β C—H bond. In this case, the decomposition yields 1-haxene (C_6H_{12}) and an H atom.

It may also be noted that, while the C—C bond breaking appears to be more probable based on bond energy considerations, C—H bond dissociation reactions usually have a larger A factor than C—C bond dissociation reactions. Considering the influences of both activation energy and A factor on the rate constant, we can generally assume that whereas the C—C bond dissociation is preferred at low temperatures, the C—H bond breaking prevails at high temperatures.

For radicals without a β C—C bond, the β C—H bonds are always the most susceptible to dissociation, for example,

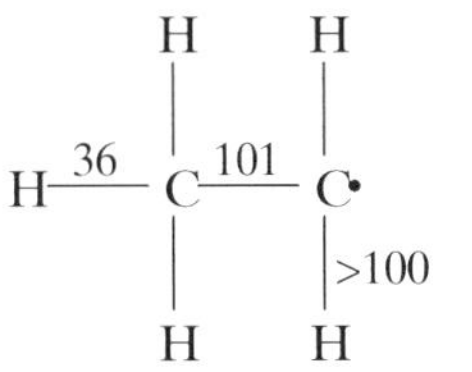

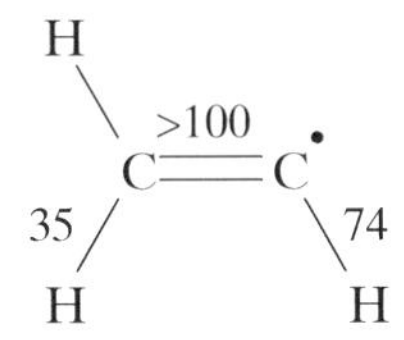

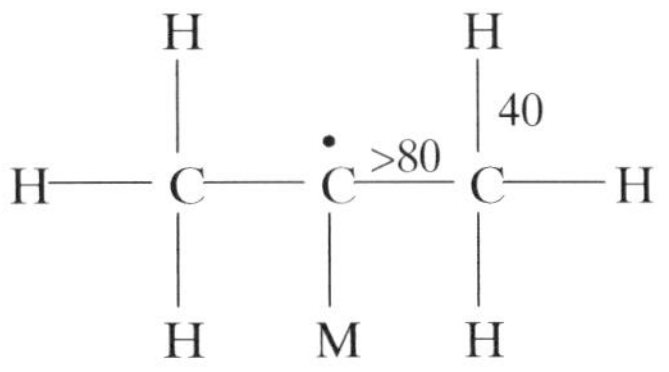

```
     H    H           H                                H          H
     |    |            \   >100    •                   |          | 40
H-36-C-101-C•           C=====C                   H----C----C->80-C----H
     |    |>100     35 /       \ 74                    |    |     |
     H    H           H         H                      H    M     H
```

This explains why the dissociation reactions of ethyl (C_2H_5) and vinyl (C_2H_3), producing C_2H_4 + H through ($C_2$3) and C_2H_2 + H through ($C_2$11), respectively, are the dominant decomposition channels of ethyl and vinyl, as discussed in Section 3.4.

For many radical species derived from unsaturated hydrocarbons, the β C—H bonds may actually be weaker than the β C—C bonds. Consider the 1,3-butadien-1-yl (n-C_4H_5) radical,

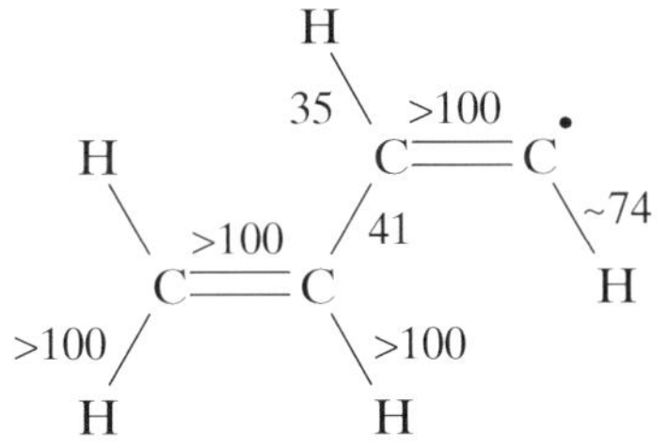

```
                     H
                  35  \  >100    •
          H            C=====C
           \  >100    / 41    \ ~74
            C=====C            H
      >100 /       \ >100
          H         H
```

The most probable products of dissociation are vinylacetylene (C_4H_4) and the H atom. In general, the β C—H bond is the weakest if the γ bond, that is, two removed from the radical site, is a double or triple bond.

The above consideration, however, does not apply to cyclic species. An example is the phenyl (C_6H_5) radical as shown here:

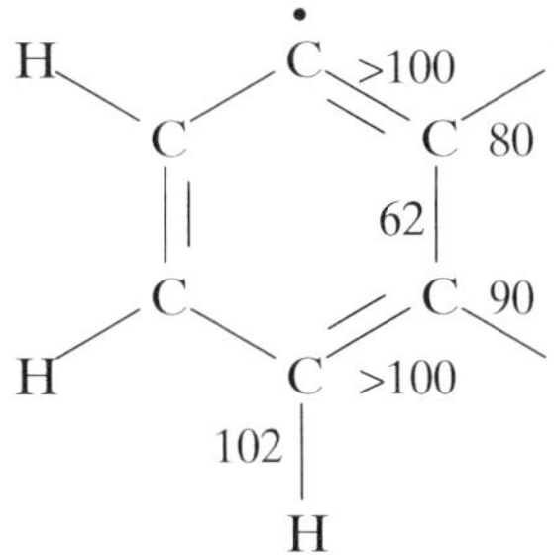

```
                 •
     H          C  >100     H
       \     /     \\     /
         C            C  80
         ||        62 |
         C            C  90
       /    \     //     \
     H          C  >100    H
           102  |
                H
```

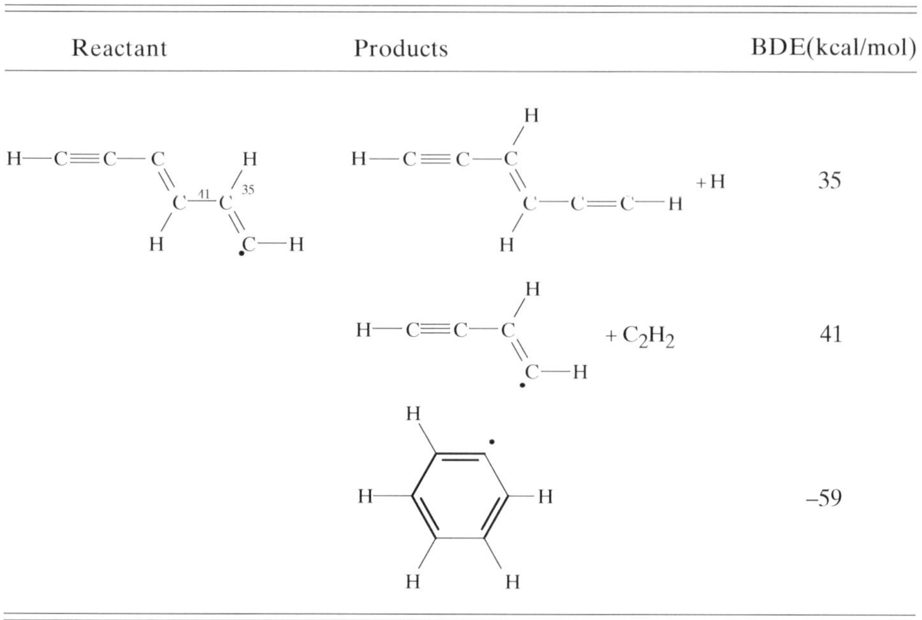

Reactant	Products	BDE(kcal/mol)
	$+H$	35
	$+C_2H_2$	41
		–59

Figure 3.6.1. Example illustrating an exception to the β-scission rule.

The β C—C bond, with a BDE value of 62 kcal/mol, is apparently the most susceptible to decomposition.

The examples given above illustrate the simplicity and reliability of the β-scission rule. One should, however, exercise caution when dealing with the decomposition mechanism of a large radical species. Figure 3.6.1 illustrates that for certain radical species the β-scission process itself is less probable than the isomerization and the formation of a new bond. As the diagram shows, the β-scission rule states that the reaction should proceed with $C_6H_4 + H$ and $C_4H_3 + C_2H_2$ as the most probable products. However, since the isomerization of C_6H_5, leading to the cyclic phenyl radical, is actually exothermic by 59 kcal/mol, it is more favorable than the β-scission reactions.

3.6.2. Oxidation Mechanisms

Similar to methane and ethane oxidation, the oxidation of higher aliphatic compounds in a homogeneous mixture of fuel and oxidant is initiated by a bond-breaking process when such a mixture is exposed to a high temperature environment. For a straight-chain aliphatic fuel, the initial bond breaking occurs most likely between one of the C—C bonds.

In flames, the initial attack is through the H-abstraction reactions,

$$RH + (H, O, OH) \rightarrow R + (H_2, OH, H_2O), \tag{3.6.1}$$

Table 3.2. Comparison of bond energies among primary, secondary, and tertiary C—H bonds in aliphatic compounds

Type of C—H Bond	Compound	BDE (kcal/mol)
primary	C_2H_6	101.1 ± 0.4
secondary	C_3H_8	98.6 ± 0.4
	n-C_4H_8	98.3 ± 0.5
tertiary	i-C_4H_{10}	96.4 ± 0.4

where R is a function group (e.g., C_3H_7). The rate of the abstraction reaction depends somewhat on the type of C—H bonds in the aliphatic compound. Consider the C—H bonds in 2-methylbutane,

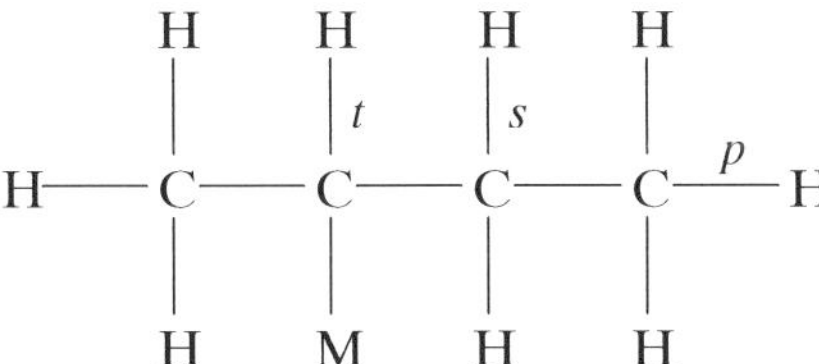

Depending on the neighboring environments of the C atom, the C—H bonds can be classified into three types, primary, secondary, and tertiary, which are denoted by "*p*," "*s*," and "*t*," respectively. While a primary C—H bond is one whose C atom is bonded with one C atom and three H atoms, a secondary C—H bond is one whose C atom has two bonded C atoms and two H atoms. A tertiary C—H bond involves a C atom that is bonded to three other C atoms. Thus, while ethane has only primary C—H bonds, propane has six primary and two secondary C—H bonds.

Compared to the primary C—H bond, the secondary and tertiary C—H bonds are usually weaker (Table 3.2). The activation energies of H abstraction from the primary C—H bonds tend to be larger than those from the secondary and tertiary C—H bonds. On the other hand, H-abstraction from the primary positions usually has a larger A factor than the secondary and tertiary H-abstractions. As a result, the production of primary, secondary, and tertiary radicals by H-abstraction at high temperatures tends to proceed at about the same rate, within a few factors.

Obviously, the attack on the p, s, and t C—H bonds by H-abstraction yields different reaction products. For example, while the H-abstraction of propane at the primary C—H bonds yields the n-propyl radical,

$$C_3H_8 + (H, O, OH) \rightarrow n\text{-}C_3H_7 + (H_2, OH, H_2O), \quad (C_31)$$

the attack of the secondary C—H bond produces the i-propyl radical,

$$C_3H_8 + (H, O, OH) \rightarrow i\text{-}C_3H_7 + (H_2, OH, H_2O). \quad (C_32)$$

Then the further oxidation of the propyl radical bifurcates, with the first through the β scission of the n-propyl radical to form ethylene and a methyl radical,

$$n\text{-}C_3H_7 \rightarrow C_2H_4 + CH_3, \tag{C$_3$3}$$

and the second via the dissociation of the i-propyl radical, yielding propene (C_3H_6) and the H atom,

$$i\text{-}C_3H_7 \rightarrow C_3H_6 + H. \tag{C$_3$4}$$

The oxidation mechanism of propene is in principle similar to that of ethylene, in that propene is oxidized by the O atom to break the C=C bond,

$$C_3H_6 + O \rightarrow C_2H_5 + HCO \tag{C$_3$5a}$$

$$\rightarrow CH_2CO + CH_3 + H, \tag{C$_3$5b}$$

or it may undergo the H-abstraction reactions, producing the allyl radical (C_3H_5). The allyl radicals are further oxidized by O, OH, O_2, and HO_2, producing oxygenated or nonoxygenated C_1 and C_2 species. Now the reaction has proceeded to the stage where all the species are those participating in the reactions of the $H_2{-}CO{-}O_2$, $CH_4{-}O_2$, and $C_2H_x{-}O_2$ systems, it can be subsequently described by the mechanisms already discussed.

For higher, straight-chain alkanes, upon H abstraction, the primary alkyl radicals will decompose sequentially to several ethylene molecules and the ethyl or methyl radicals, depending on whether the fuel molecule contains an even or odd number of C atoms. For example, while the decomposition of the n-hexyl radical (n-C_6H_{13}) follows

$$n\text{-}C_6H_{13} \rightarrow C_2H_4 + n\text{-}C_4H_9 \tag{C$_6$1}$$

$$n\text{-}C_4H_9 \rightarrow C_2H_4 + C_2H_5, \tag{C$_4$1}$$

the n-pentyl radical (n-C_5H_{11}) decomposes according to

$$n\text{-}C_5H_{11} \rightarrow C_2H_4 + n\text{-}C_3H_7 \tag{C$_5$1}$$

$$n\text{-}C_3H_7 \rightarrow C_2H_4 + CH_3. \tag{C$_3$3}$$

Again, subsequent reactions can now be described by the systems involving methane, ethane, ethylene, CO, and hydrogen.

The β-scission reactions of secondary alkyl radicals are more complex than those of the primary alkyl radicals. The products depend on the radical position and size of the fuel molecule. Nonetheless, they must include an alkene (propene, butene, etc.) and methyl or ethyl as the β-scission products. For example, while 2-pentyl (2-C_5H_{11}) decomposes to propene and ethyl,

$$CH_3\text{-}CH_2\text{-}CH_2\text{-}CH\text{-}CH_3 \rightarrow C_2H_5 + C_3H_6, \tag{C$_5$2}$$

the 3-pentyl (3-C_5H_{11}) radical decomposes to 1-butene and methyl,

$$CH_3\text{-}CH_2\text{-}CH\text{-}CH_2\text{-}CH_3 \rightarrow CH_3 + 1\text{-}C_4H_8. \tag{C$_5$3}$$

The resulting propene reacts readily with the O atom according to reaction steps ($C_3$5a) and ($C_3$5b). For 1-butene, further H-abstraction reactions, followed by the β-scission process, may lead to a variety of products, including 1,3-butadiene (1,3-C_4H_6), 1,2-butadiene (1,2-C_4H_6), ethylene, and the vinyl radical. The butadiene species may either react with the O atom and the OH radical, or undergo reaction with the H atoms, producing C_1, C_2, and C_3 fragments.

3.7. OXIDATION OF AROMATICS

The importance of gaining a basic understanding of the oxidation mechanisms of aromatics is appreciated by recognizing that modern gasoline fuels contain a substantial amount of aromatics, and that aromatics are generally thought as the precursor to soot. In this section, we shall focus on the oxidation mechanisms of benzene and alkylbenzene (Brezinsky 1986; Dagaut, Pengloan & Ristori 2002), recognizing that the oxidation mechanisms of polycyclic aromatic hydrocarbons are not at all well understood at present.

A significant difference in the oxidation mechanisms of aliphatics and aromatics is the rate of the disintegration of the initial fuel molecules. Specifically, the initial decomposition of the alkyl radical is relatively rapid at high temperatures, such that the rate-limiting steps during the combustion of aliphatic fuels are mostly the reactions of the C_1 to C_3 species. This is not the case with aromatics. The disintegration of the aromatic ring is often slow, and sometimes constitutes as a bottleneck. The reason can be easily understood by comparing the β C—C bond energies in phenyl (62 kcal/mol) with an alkyl radical like n-hexyl (22 kcal/mol). As a result, the phenyl radical does not decompose as readily as the alkyl radical. Its persistence in large concentrations in benzene combustion allows direct oxidation of the ring structure, as depicted in Figure 3.7.1 and discussed in the following.

Like the alkanes, the oxidation of benzene in flames can start with the H-abstraction reactions,

$$C_6H_6 + (H, OH) \rightarrow C_6H_5 + (H_2, H_2O). \quad \text{(B1)}$$

The phenyl radical tends to combine with the H atom, reproducing benzene,

$$C_6H_5 + H + M \rightarrow C_6H_6 + M. \quad \text{(B2)}$$

The remaining phenyl radicals react with molecular oxygen to form mainly the phenoxyl (C_6H_5O) radicals, and to a lesser extent, benzoquinone ($C_6H_4O_2$),

$$C_6H_5 + O_2 \rightarrow C_6H_5O + O \quad \text{(B3a)}$$

$$\rightarrow C_6H_4O_2 + OH. \quad \text{(B3b)}$$

In addition to the above reactions, benzene can react directly with the O atom, leading to the formation of a CO bond,

$$C_6H_6 + O \rightarrow C_6H_5O + H. \quad \text{(B4)}$$

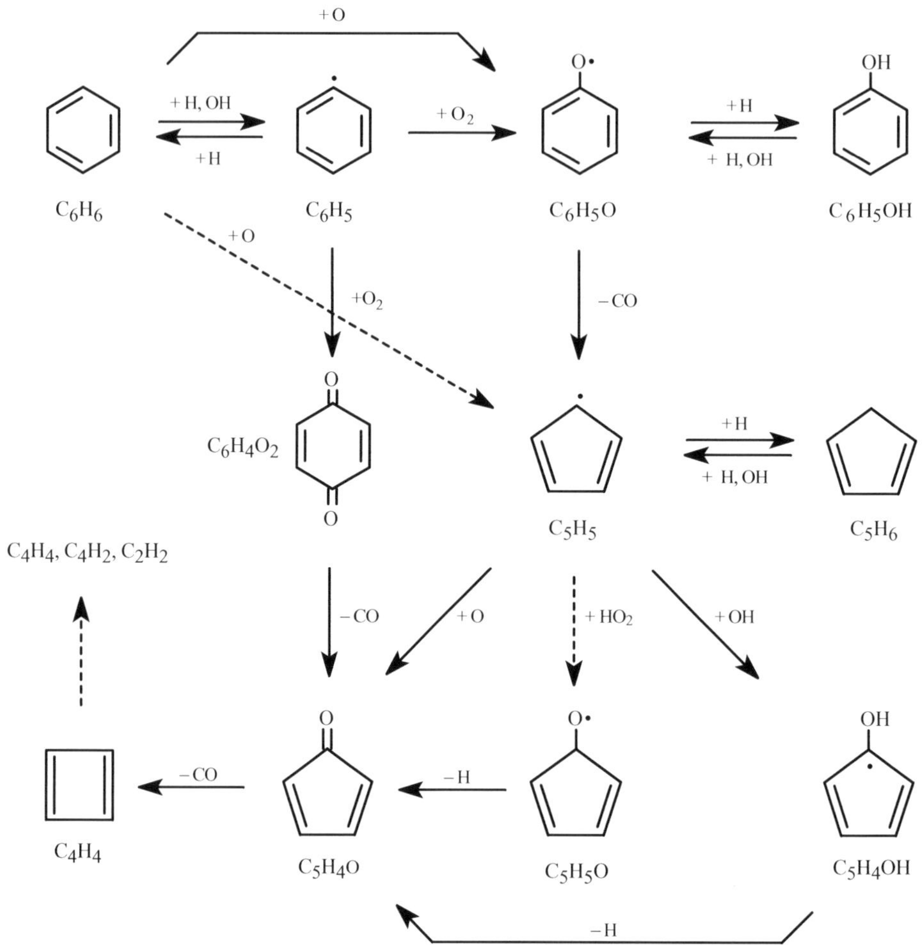

Figure 3.7.1. Reaction pathways of high-temperature benzene oxidation.

Like phenyl, the phenoxyl radical tends to recombine with the H atom to form phenol, which has been observed to exist in large quantity in both flow-reactor experiments and in laminar premixed flames,

$$C_6H_5O + H + M \rightarrow C_6H_5OH + M. \qquad \text{(B5)}$$

Competing with the above radical–radical combination channel is the thermal decomposition of the phenoxyl radical, which expels CO and leads to the cyclopentadienyl radical (C_5H_5),

$$C_6H_5O + M \rightarrow C_5H_5 + CO + M. \qquad \text{(B6)}$$

The cyclopentadienyl radical can then combine with an H atom to form cyclopentadiene (C_5H_6),

$$C_5H_5 + H + M \rightarrow C_5H_6 + M. \qquad \text{(B7)}$$

The further oxidation pathways of C_5H_5 and C_5H_6 remain unclear as at what stage the transition from the cyclic to the noncyclic structure takes place. The reactions of C_5H_5 and O, OH, and HO_2 via the following reaction sequence

$$C_5H_5 + O \rightarrow C_5H_4O + H \tag{B8}$$

$$C_5H_5 + OH \rightarrow C_5H_4OH + H \tag{B9}$$

$$C_5H_4OH + M \rightarrow C_5H_4O + H + M \tag{B10}$$

$$C_5H_5 + HO_2 \rightarrow C_5H_5O + OH \tag{B11}$$

$$C_5H_4O + M \rightarrow 2C_2H_2 + CO + M \tag{B12}$$

$$C_5H_5O + M \rightarrow C_4H_5 + CO + M \tag{B13}$$

$$C_4H_5 + M \rightarrow C_2H_2 + C_2H_3 + M, \tag{B14}$$

appear to explain the species profile measured in flow reactor studies of benzene oxidation. Recent studies suggest that the disintegration of the ring structure occurs with cyclopentadienone (C_5H_4O) thermal decomposition, leading to cyclobutadiene (c-C_4H_4), which subsequently dissociates to vinylacetylene (C_4H_4), diacetylene (C_4H_2), and acetylene,

$$C_5H_4O + M \rightarrow c\text{-}C_4H_4 + CO + M \tag{B15}$$

$$c\text{-}C_4H_4 + M \rightarrow (C_4H_4, C_4H_2) + (H_2, 2C_2H_2) + M. \tag{B16}$$

The C_5H_5O radical appears to be highly unstable. It tends to expel an H atom to form C_5H_4O,

$$C_5H_5O + M \rightarrow C_5H_4O + H + M. \tag{B17}$$

The benzoquinone produced from (B3a) presumably decomposes by expelling a CO,

$$C_6H_4O_2 + M \rightarrow C_5H_4O + CO + M, \tag{B18}$$

although it can also react with the H, O, and OH radical species.

Recent studies also suggest that other reaction channels exist, even during the initial attack of the benzene molecule. For example, it is shown that the reaction between benzene and the O atom can lead directly to the cyclopentadienyl and formyl radicals,

$$C_6H_6 + O \rightarrow C_5H_5 + HCO. \tag{B19}$$

Likewise, similar reactions are possible with phenol and cyclopentadiene.

It is generally accepted that the oxidation of the alkyl benzene compounds starts with the disintegration of the alkyl functional group. For example, the initial oxidation of toluene ($C_6H_5CH_3$) in flames appears to proceed through the following steps, which eventually lead to the formation of the intermediate phenyl radical:

$$C_6H_5CH_3 + (H, OH) \rightarrow C_6H_5CH_2 + (H_2, H_2O) \quad \text{(B20)}$$

$$C_6H_5CH_2 + H + M \rightarrow C_6H_5CH_3 + M \quad \text{(B21)}$$

$$C_6H_5CH_2 + O \rightarrow C_6H_5CHO + H \quad \text{(B22)}$$

$$C_6H_5CH_2 + HO_2 \rightarrow C_6H_5CHO + H + OH \quad \text{(B23)}$$

$$C_6H_5CHO + (H, OH, O_2) \rightarrow C_6H_5CO + (H_2, H_2O, HO_2) \quad \text{(B24)}$$

$$C_6H_5CO + M \rightarrow C_6H_5 + CO + M \quad \text{(B25)}$$

$$C_6H_5CH_3 + H \rightarrow C_6H_5 + CH_4. \quad \text{(B26)}$$

The destruction of the phenyl radicals follows that of the benzene oxidation mechanism.

3.8. HYDROCARBON OXIDATION AT LOW TO INTERMEDIATE TEMPERATURES

So far we have discussed the oxidation mechanisms of fuels during high-temperature combustion. In this section, we shall shift our focus to hydrocarbon oxidation in the low- to intermediate-temperature regime (<1,000 K). The chemistry in this regime is relevant to ignition, including such practical combustion phenomenon as engine knock.

Similar to the H_2—O_2 system, homogeneous mixtures of aliphatic hydrocarbon and oxygen also exhibit explosion limits. The pressure–temperature limits at explosion are, however, quite different from that of hydrogen. Figure 3.8.1 shows a typical p–T diagram, which characterizes the ignition boundaries of methane, ethane, and propane. The explosion behavior of higher aliphatic hydrocarbons resembles that of propane.

In general, the homogeneous ignition temperature decreases with increasing aliphatic hydrocarbon size, indicating that the tendency of explosion is higher for higher hydrocarbons. This behavior can be qualitatively understood by examining the C—H bond dissociation energy of methane, ethane, and the higher aliphatic hydrocarbons. That is, the C—H bonds in methane are 4 kcal/mol stronger than those of ethane, whereas the C—H bonds in ethane are 2–3 kcal/mol stronger (see Table 3.2) than the secondary C—H bonds, which exist only in propane and higher aliphatic hydrocarbons. Since the explosion process depends on the initial attack of the hydrocarbon molecules by O_2 and consequently by active radicals, the difference in bond energy is translated into the difference in the activation energy of these reactions. As a result, it is more difficult to abstract an H atom from methane than from ethane, and from ethane than from the higher hydrocarbons. A higher temperature is thus needed to make a methane–oxygen mixture explosive.

As demonstrated in Figure 3.8.1, the explosion boundary varies from the simple p–T relationship in methane oxidation, to more complex behaviors in higher hydrocarbon oxidation. For methane, the explosion temperature varies smoothly

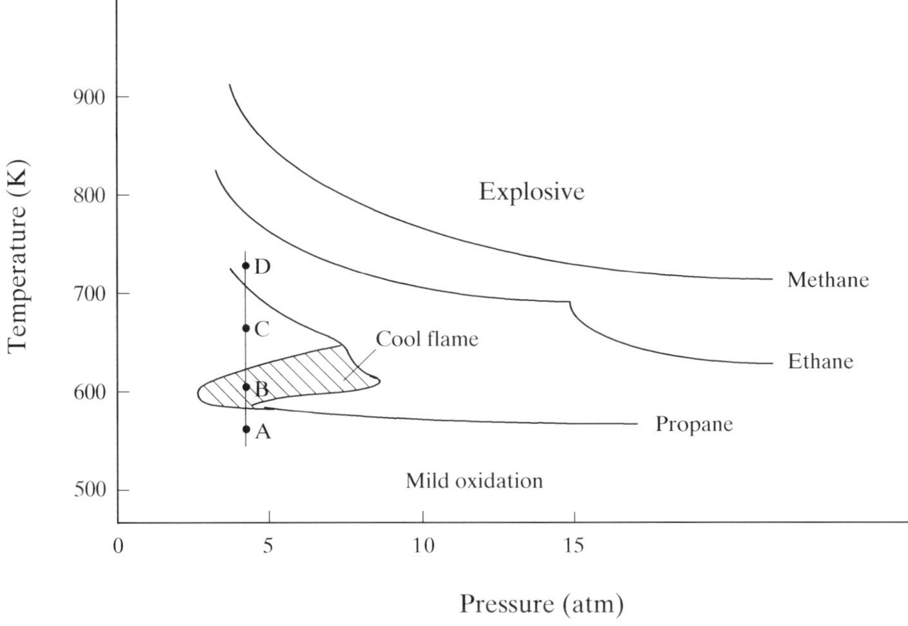

Figure 3.8.1. Explosion limits of methane, ethane, and propane.

and decreases monotonically with increasing pressure. In the case of ethane, the explosion temperature first shows a slow decrease as pressure increases, then a sharp decline with further increase in pressure.

For higher aliphatic hydrocarbons, the explosion p–T diagram is more complex. For a fixed pressure of a few atmospheres, say along ABCD, the following is observed with increasing temperature. First, there is no observable change in the mixture at A, where the temperature is below $T \approx 530$ K. This low-temperature regime is one of chemical synthesis that produces oxygen-containing organic molecules. Oxidation in this regime can only occur through the agency of initiators or catalysts; without them the homogeneous oxidation rates are negligibly slow. From the combustion point of view, this low-temperature regime is not important.

With increasing temperature, we arrive at point B enclosed by a peninsula, which lies between 570 and 670 K. In this peninsular regime, one observes pale blue emissions characteristic of peroxides and formaldehyde, as compared to the green C_2 and blue-violet CH emissions characteristic of the high-temperature regime. The emission of such chemiluminescence occurs without having violent temperature and/or pressure rises seen in an explosion process. This phenomenon is commonly called the "cool flame" (Benson 1981; Lewis & von Elbe 1987). Cool flames can appear either in the upstream region of a premixed flame, or as precursors to the onset of explosion in a homogeneous mixture. Their reaction rates are generally much lower than those of high-temperature oxidation, and the reactions consume only 5–10 percent of the hydrocarbons. Cool flames may also occur in a periodic manner. That is, during the passage of a cool flame, the temperature can be raised by 100–200 K. The increase in temperature, however, rapidly slows down the reaction. With simultaneous

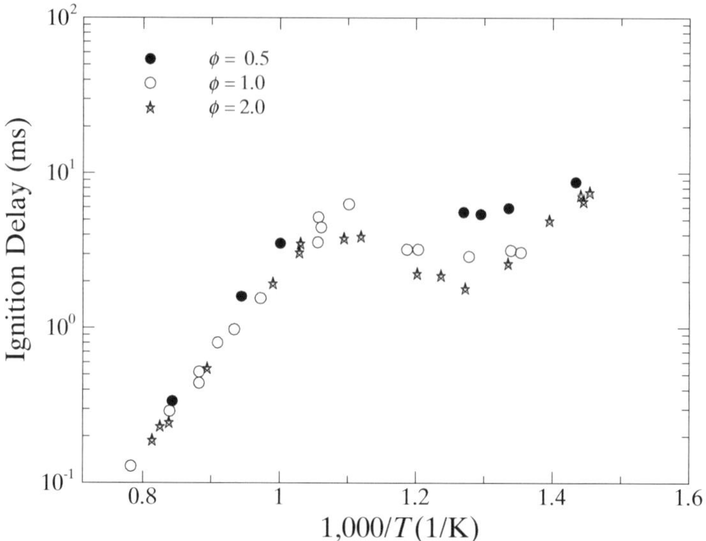

Figure 3.8.2. 1.79 percent *n*-heptane oxidation behind reflected shock waves in 80 percent N_2, pressure: 13.5 bar (adapted from Ciezki & Adomeit 1993).

heat loss, say from the wall of the vessel, the mixture is then cooled. This continues until a sufficiently low temperature is reached such that the reaction is again facilitated. The cycle thus repeats itself.

Consistent with the periodic behavior of the cool flame is that when the system temperature is increased from B to C, the reactions slow down until a complete stop. Such a temperature dependence is often called the "negative" temperature coefficient (NTC) for the reaction rate. With a further increase in temperature from C to D at around 700 K, the explosion is preceded by a narrow regime with a bright blue flame.

Figure 3.8.2 shows experimental data (Ciezki & Adomeit 1993) on the ignition delay of heptane–oxygen–nitrogen mixtures behind reflected shock waves, demonstrating the existence and influence of the NTC regime.

The cool flame behavior, and the associated negative temperature coefficient for the reaction rate, are consequences of the following mechanism. Initiation comes from the reaction between fuel and molecular oxygen,

$$RH + O_2 \rightarrow R + HO_2, \tag{R1}$$

producing R, which further reacts with oxygen according to

$$R + O_2 + M \rightleftharpoons RO_2 + M \tag{R2}$$

$$R + O_2 \rightarrow \text{olefin} + HO_2 \tag{R3}$$

$$RO_2 + RH \rightarrow RO_2H + R \tag{R4}$$

$$RO_2H + M \rightarrow RO + OH + M \tag{R5}$$

$$RH + (OH, HO_2, RO) \rightarrow R + (H_2O, H_2O_2, ROH). \tag{R6}$$

Thus it is seen that the reaction between R and O_2 can proceed in two paths. The first is a radical propagation path, involving an exothermic step (R2) with 39 kcal/mol energy release, followed by (R4), which generates the radicals R to feed (R2). The RO_2H decomposes readily to two radical species, RO and OH, and therefore produces a degenerate chain branching. This path is responsible for the oxidation process in a cool flame. In the second path involving (R3), an olefin forms with 9 kcal/mol heat release. The HO_2 reacts with RH to form H_2O_2, which is a metastable species below 750 K, and hence does not contribute to chain branching.

When the temperature is increased to above 600 K, reaction (R3) becomes increasingly fast, but (R2) slows down because k_{R2} tends to decrease with increasing temperature, and its reverse reaction also becomes progressively more important. As a result, the amount of RO_2 that can be fed to (R4) decreases rapidly. The production rate of RO_2H decreases to the extent that the subsequent chain-branching step (R5) stops, and the overall reaction will eventually shut itself off. This explains the negative temperature coefficient for the reaction rate.

If the mixture can be heated to about 700 K, then the branching reaction

$$H_2O_2 + M \rightarrow OH + OH + M \qquad \text{(H15)}$$

becomes important, generating large quantities of OH radicals to initiate reactions in the intermediate-temperature regime.

3.9. CHEMISTRY OF POLLUTANT FORMATION

It is estimated that over 90 percent of the air pollutants are generated from burning fossil fuels. A basic understanding of the mechanism of pollutant formation is then a first step toward developing rational strategies for pollutant reduction.

There are three principal pollutants that are produced from fossil fuel combustion, namely oxides of nitrogen, oxides of sulfur, and soot. Oxides of nitrogen consist of nitric oxide (NO) and nitrogen dioxide (NO_2), which are collectively referred to as NO_x, and nitrous oxide (N_2O). The nitrogen atom needed for their formation comes from either the molecular nitrogen in the oxidizing gas, particularly air, or the fuel. Since most processed gas and liquid petroleum fuels have practically no nitrogen content, their formation from the burning of these fuels then primarily depends on the ease with which N can be abstracted from the tightly bound N_2. On the other hand, fuel-bound nitrogen could be present in abundance in unprocessed fossil fuels, such as coal and even some natural gas, and could constitute a significant source for their formation.

As mentioned in the Introduction, nitrogen oxides are directly responsible for the photochemical smog; sulfur oxides are related to acid rain; and soot is one of the major sources of small atmospheric aerosols in urban areas. Unburnt intermediate species during combustion, including carbon monoxide and formaldehyde, and unburnt hydrocarbons are also pollutants; the mechanisms of their formation and destruction have already been discussed in this chapter. Furthermore, since the source of sulfur in SO_x is mostly fuel-bound, the minimization of SO_x emission can be basically achieved at the stage of fuel processing and by flue gas scrubbing. This section therefore focuses on understanding the reaction paths involved in the formation and control of nitrogen oxides and soot. Reviews on the formation and control of nitrogen oxides can be found in Miller and Bowman (1989) and Bowman (1992, 2000), and those on soot properties are given in Haynes and Wagner (1981), Kennedy (1997), and Richter and Howard (2000). In particular, the following discussion on the formation and removal of oxides of nitrogen follow closely that of Bowman (2000).

3.9.1. Oxides of Nitrogen

3.9.1.1. Mechanisms of NO Formation: Nitric oxide can be formed from atmospheric N_2 through three mechanisms, namely the thermal NO, the prompt NO, and the N_2O mechanisms. The thermal NO mechanism consists of three reactions:

$$N_2 + O \rightarrow NO + N \tag{N1}$$

$$O_2 + N \rightarrow NO + O \tag{N2}$$

$$N + OH \rightarrow NO + H. \tag{N3}$$

This sequence is referred to as the Zel'dovich mechanism. The rate-controlling reaction for this mechanism is (N1), which requires the breaking of the tight N_2 bond and as such is favored in high-temperature gases. The formation is also weakly dependent on the availability of O_2. Consequently NO emission from engines typically peaks slightly on the lean side of stoichiometry, as shown in Figure 1.4.8 for the equilibrium product composition of methane–air mixtures. As a rough guideline, thermal NO formation is usually considered to be unimportant at temperatures below 1,800 K.

The prompt NO mechanism was proposed when it was observed that there is substantial NO formation in the upstream, colder part of premixed hydrocarbon flames, where the O atom concentration is relatively low, and the Zel'dovich mechanism cannot fully explain the NO production. It was also observed that the flamefront NO formation tends to increase as the unburned mixture becomes fuel rich. These evidences led Fenimore (1971) to propose that the reactions between N_2 and the hydrocarbon radical species are responsible for the formation of NO in the colder part of the flame. This prompt NO mechanism has now been extended to include a large set of reactions. The main sequence of reactions involve the initial reaction of N_2 with CH and CH_2, producing NCN, HCN (hydrogen cyanide) and the H and NH radicals,

$$N_2 + CH \rightarrow NCN + H \tag{N4}$$

$$N_2 + CH_2 \rightarrow HCN + NH. \tag{N5}$$

The HCN and NH formed can undergo further reactions (Bowman 1973) which ultimately lead to the formation of N and consequently NO via (N2) and (N3),

$$\mathrm{HCN + O \rightarrow NCO + H} \tag{N6}$$

$$\mathrm{HCN + (H, OH) \rightarrow CN + (H_2, H_2O)} \tag{N7}$$

$$\mathrm{NCO + H \rightarrow NH + CO} \tag{N8}$$

$$\mathrm{NH + (H, OH) \rightarrow N + (H_2, H_2O)} \tag{N9}$$

$$\mathrm{CN + O \rightarrow N + CO,} \tag{N10}$$

while the oxidation of NCN by O, OH, and O_2 also leads to the formation of NO (Moskaleva & Lin 2000). Since the concentrations of O, CH, and CH_2 tend to increase with increasing temperature, the formation of prompt NO still increases with increasing flame temperature.

Formation of NO through the N_2O route follows the sequence

$$\mathrm{N_2 + O + M \rightarrow N_2O + M} \tag{N11}$$

$$\mathrm{N_2O + O \rightarrow NO + NO.} \tag{N12}$$

Since this sequence requires the O atom and the three-body recombination reaction (N11), the formation of NO is favored with increasing air concentration and pressure.

For NO formation from fuel-bound nitrogen, the nitrogen-containing fuel species are usually aromatic and polyaromatic compounds with one or more nitrogen atoms. The nitrogen-containing species that evolve from fuel pyrolysis are mostly HCN and to a lesser extent ammonia (NH_3). Consequently the NO formation mechanism follows those involving HCN and NH_3. This is discussed in detail in Miller and Bowman (1989).

3.9.1.2. NO Control: Since the formation of NO frequently requires high temperatures, an obvious strategy towards reducing NO formation is to reduce the combustion temperature. This can be achieved by increasing the inert content of the combustion mixture, for example through exhaust gas recirculation (EGR), or reducing the extent of nonpremixed, stoichiometric, burning (see Chapter 6) through faster vaporization of the fuel droplets or better mixing of the fuel and oxidizer streams. Water injection has also been found effective because its presence not only lowers the flame temperature, but it also removes the O atom via

$$\mathrm{O + H_2O \rightarrow OH + OH,} \tag{H4}$$

thereby reducing the extent of the rate-limiting reaction (N1) in the Zel'dovich mechanism.

The formation of NO can also be reduced through staged combustion. In this process burning first occurs in a fuel-rich stage, typically with $\phi \approx 1.4$, such that NO formation is reduced due to the reduced temperature and oxygen concentration.

In the second stage, air is added to consume the fuel. Although the addition of oxygen and the associated increase in the combustion temperature tend to favor NO formation, the residence time is now limited and as such the overall NO formation is reduced. Furthermore, excess air can be introduced during the second stage to limit the temperature.

Additional techniques have been developed to remove the NO formed during active combustion. The first involves adding sufficient fuel to the combustion products to render it fuel rich, which promotes the "reburning" of NO through the following sequence:

$$CH + NO \rightarrow HCN + O \tag{N13}$$

$$CH + NO \rightarrow HCO + N \tag{N14}$$

$$CH_3 + NO \rightarrow HCN + H_2O \tag{N15}$$

$$CH_3 + NO \rightarrow H_2CN + OH \tag{N16}$$

$$HCCO + NO \rightarrow HCN + CO_2 \tag{N17}$$

$$HCCO + NO \rightarrow HCNO + CO \tag{N18}$$

$$N + NO \rightarrow N_2 + O. \tag{−N1}$$

The HCN formed proceeds to react with the O atom through (N6), followed by (N7)-(N10). The N atoms produced from (N9) and (N10) feed reaction (−N1) to complete the reburn sequence. Additional air is then introduced to complete the consumption of the fuel.

The second technique, referred to as Selective Non-Catalytic Reduction (SNCR), involves injecting a nitrogen-containing additive into the combustion products, and thereby initiating a sequence of reactions that chemically reduces the NO to other species. For example, through the Thermal DeNO$_x$ process, NO is reacted away by injecting ammonia within a temperature window of 1,100–1,400 K, leading to the reactions

$$NH_2 + NO \rightarrow NNH + OH \tag{N19}$$

$$NH_2 + NO \rightarrow N_2 + H_2O, \tag{N20}$$

with the NNH produced either decompose through

$$NNH \rightarrow N_2 + H, \tag{N21}$$

or react with O_2 through

$$NNH + O_2 \rightarrow N_2 + HO_2. \tag{N22}$$

A second SNCR process is that of RAPRENOx, in which cyanuric acid is injected into the combustion products. The resulting reaction sequence involves first sublimation of the cyanuric acid

$$(HOCN)_3 \rightarrow 3HNCO \tag{N23}$$

$$HNCO + OH \rightarrow NCO + H_2O \tag{N24}$$

$$HNCO + OH \rightarrow NH_2 + CO_2 \tag{N25}$$

$$NCO + NO \rightarrow N_2O + CO. \tag{N26}$$

Thus NO is converted to N_2O, which is not regulated but is nevertheless a greenhouse gas. Its reaction mechanism is discussed next.

3.9.1.3. Formation and Removal of NO_2 and N_2O: The principal formation step of NO_2 is

$$NO + RO_2 \rightarrow NO_2 + RO, \tag{N27}$$

where RO_2 is a peroxy radical. The NO_2 formed, however, can be rapidly converted back to NO through

$$NO_2 + H \rightarrow NO + OH \tag{N28}$$

$$NO_2 + O \rightarrow NO + O_2, \tag{N29}$$

especially in the high-temperature flame environment containing an abundance of H and O radicals.

Nitrous oxide is formed through (N11), as mentioned earlier, and is removed through (N12) and

$$N_2O + H \rightarrow N_2 + OH \tag{N30}$$

$$N_2O + O \rightarrow N_2 + O_2. \tag{N31}$$

3.9.2. Soot Formation

Formation of particulate carbon, or soot, is a phenomenon often observed in the combustion of hydrocarbons. The yellowish luminescence observed in a wood or coal fire is caused by the thermal radiation of soot in flames. The smoke emitted from the exhaust of an automobile or from a household chimney is also a result of dispersion of soot particles in the burned gas. Soot formed in a boiler flame provides the necessary radiation source for heat transfer, while airborne carbon particulates from combustion are known to be carcinogenic.

Soot is not a uniquely defined chemical substance in that it contains mostly carbon, with up to 10 percent (mol) of hydrogen. The atomic C/H ratio is about 8 to 1. The mass density of soot is about 2 g/cm^3. When soot is extracted with organic solvents, highly condensed polycyclic aromatic hydrocarbons (PAHs) are found in the solvents.

Electron microscopy studies show that a soot "particle" often consists of chain-like aggregates of nearly spherical particles. These spherical particles, called the primary particles, may contain between 10^5 to 10^6 carbon atoms, and usually have diameters between 20 to 50 nm. X-ray diffraction shows that the primary soot particles are

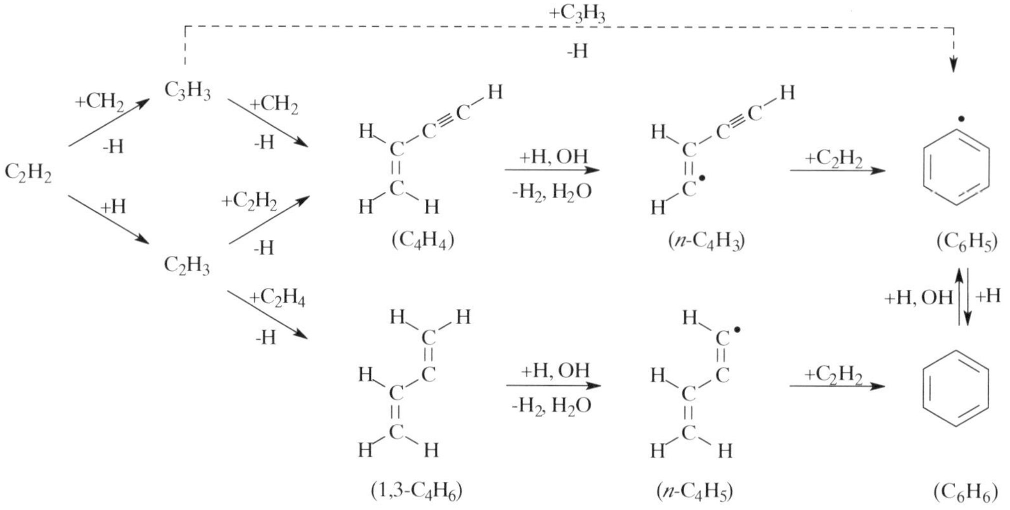

Figure 3.9.1. Selected pathways of benzene formation in hydrocarbon combustion.

made up of a large number of randomly arranged grains. Each grain consists of 5 to 10 nearly parallel planes arranged in a turbostractic fashion. Each layer is between 1 to 2 nm in dimension and contains on the order of 50 carbon atoms. The inner layer spacing is about 0.35 nm, which is of the same order as that of graphite.

It is well accepted that the physical and chemical coalescence of PAHs is responsible for the inception of soot. (Frenklach et al. 1984; Frenklach & Warnatz 1987). The growth of soot particles to the size observed in combustion exhaust is caused by the coagulation of smaller incipient soot particles, by PAH surface condensation, and by surface reactions between soot and gaseous species like acetylene. Soot produced from flames may be oxidized by OH, O, and O_2 before the combustion gas reaches the exhaust. Addition of steam and carbon dioxide to the combusting gas can also enhance soot oxidation, possibly achieved by the increase in OH concentration as a result of increased concentrations of H_2O and through the direct reaction between C and CO_2 to produce CO.

Because PAHs are the precursors to soot, a basic understanding of the mechanism of soot formation must start with that of PAH formation. It is known that acetylene forms in large concentrations in fuel-rich combustion, and its polymerization is thought to be responsible for the formation of PAHs. In particular, the first aromatic ring may be produced from nonaromatic species in the reaction sequence depicted in Figure 3.9.1. It is seen that in addition to the importance of acetylene during the formation of the first aromatic ring, that is, benzene and phenyl, the H atom also plays a critical role in that it activates/deactivates the radical species from which the first aromatic ring forms. This is also the case during the growth of the aromatic ring to PAHs, as will be seen later.

When burning long-chain aliphatic fuels, methylene radicals are produced from the reaction between C_2H_2 and the O atom, that is, reactions ($C_2$13a), ($C_2$13b), and ($C_2$14), thus their presence is directly related to acetylene. Only in fuel-rich

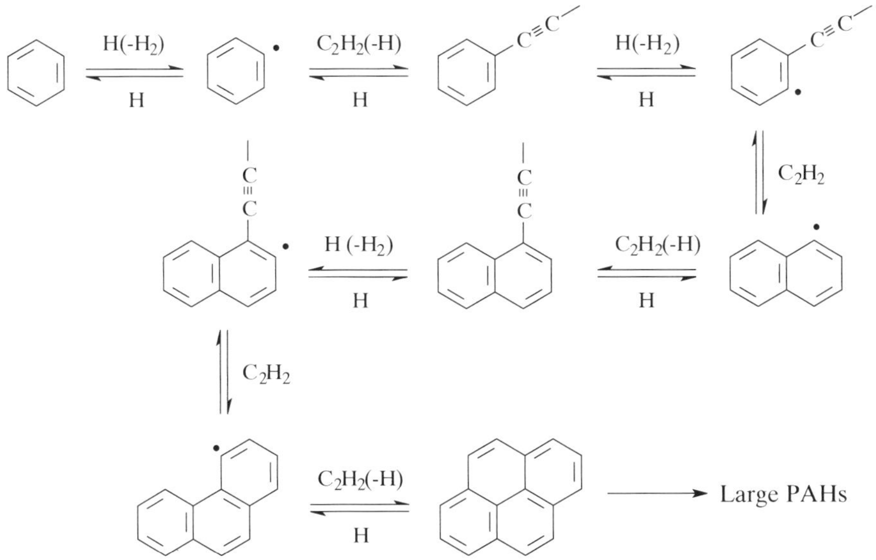

Figure 3.9.2. The H-abstraction–C_2H_2-addition (HACA) mechanism of polycyclic aromatic hydrocarbon formation.

methane or natural gas flames are methylene radicals produced from the direct reaction between CH_3 and OH through (M19), which enhances the propargyl (C_3H_3) recombination path to benzene.

The further growth of the aromatics is thought to proceed through the H-abstraction—C_2H_2-addition (HACA) mechanisms, as shown in Figure 3.9.2. In this mechanism, the addition of acetylene to an aromatic radical, like phenyl, leads to either the bonding of an ethynyl ($-C_2H$) group with the aromatic ring, or the formation of an additional condensed aromatic ring. Depending on the neighboring ring structure, the newly formed ring can either be a radical, which can grow readily with acetylene, or it may be a molecular species. The latter will have to be "activated" through the H-abstraction reaction to produce a PAH radical species, before it can undergo the further growth reaction with acetylene.

If the concentrations of aromatic species are sufficiently large, PAH growth through the direct ring-ring condensation is also possible. For example, benzene and phenyl can react to form biphenyl. Through the H-abstraction reaction, a biphenyl radical forms and can react with acetylene to form the three-ring phenanthrene, or it can react with benzene to form a four ring aromatic species. Such a reaction sequence is shown in Figure 3.9.3.

The competition between the HACA mechanism and the aromatic condensation mechanism is largely determined by the ratio of acetylene to benzene. If the concentration of acetylene is substantially larger than that of benzene, the HACA mechanism dominates. However, if the acetylene concentration is about equal to that of benzene, as in the very early stage of a premixed benzene flame, then the aromatic–aromatic condensation mechanism may prevail.

The reaction pathways leading to PAH formation and growth as depicted in the three diagrams just discussed is highly reversible. When the temperature exceeds

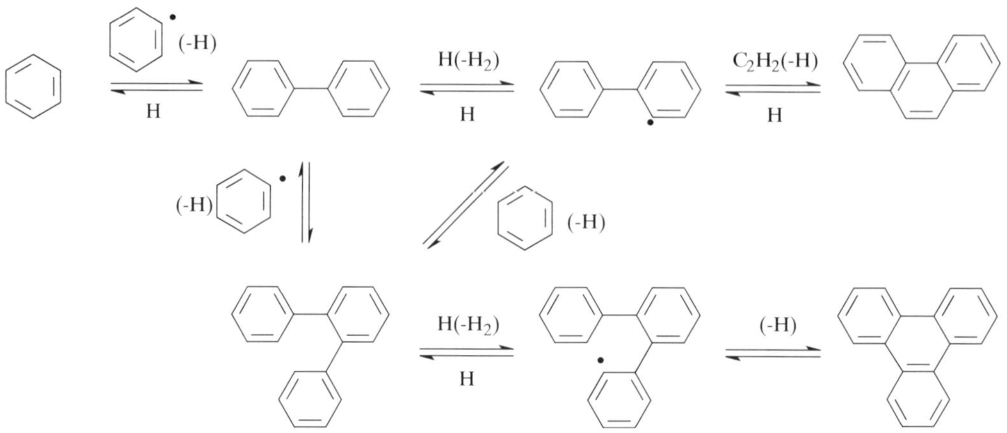

Figure 3.9.3. An alternate polycyclic aromatic hydrocarbon growth mechanism.

around 1,800 K, some of these reactions may proceed in the reverse direction in favor of the reactants. Hence, the same reactions that are responsible for PAH formation and growth also cause PAHs to thermally decompose at high temperatures. In fact, the reduction of PAH concentrations in the post-burning region of a premixed flame is caused by the thermal decomposition of the PAHs following the reverse of the reaction pathways shown in these diagrams, and to a lesser extent, due to oxidation.

When PAHs grow to the size of pyrene (a four-ringed PAH) or larger, they may be able to condense onto each other upon collision and form small clusters. These clusters can continue to react with acetylene following the same mechanism of PAH growth, or they may coagulate to form larger clusters. These chemical and physical processes eventually lead to the formation of soot particles. When the PAH concentrations are sufficiently high, surface condensation may become a major source of soot mass growth. Detailed kinetic models formulated on the basis of these physical and chemical processes can predict reasonably well soot production in laminar premixed and nonpremixed flames of simple hydrocarbons such as acetylene and ethylene.

3.10. MECHANISM DEVELOPMENT AND REDUCTION

3.10.1. Postulated Semiglobal Mechanisms

Recognizing the complexity of a detailed reaction mechanism and the intricacy of fuel oxidation kinetics, rational modeling and simulation of combustion phenomena are invariably faced with the need for simpler but chemically realistic mechanisms. An early attempt toward achieving this goal is to extend the concept of the one-step overall reaction between reactants and products, with constant kinetic parameters, by postulating some global and semiglobal reactions characterized by additional major intermediates and empirically determined kinetic parameters. Since the approach is basically empirical, it does not require knowledge of the detailed mechanism.

The simplest semiglobal mechanism is, of course, the one-step overall reaction, such as that for methane oxidation,

$$CH_4 + 2O_2 \rightarrow CO_2 + 2H_2O.$$

Chemical information can be introduced into its rate expression by using empirically extracted kinetic parameters of frequency factor, activation energy, and reaction order based on results from some experimental phenomena, such as ignition delay times and propagation speeds of laminar flames. It was however found (Westbrook & Dryer 1981, 1984) that such a scheme cannot satisfactorily describe the flame propagation data from fuel-lean to fuel-rich conditions. A major weakness of this global reaction step is the neglect of CO, because in typical hydrocarbon flames large amounts of CO and H_2 may exist in equilibrium with CO_2 and H_2O, while CO oxidation is also a rather slow process. Thus a two-step mechanism can be postulated to account for the influence of CO oxidation, as

$$CH_4 + \frac{3}{2}O_2 \rightarrow CO + 2H_2O$$

$$CO + \frac{1}{2}O_2 \rightarrow CO_2.$$

Note that even though at the elementary level CO is oxidized by reacting with OH in the manner of (CO3), at the global level its oxidation is represented by the second step of this two-step mechanism because the net result of the chain cycle is a reduction in the concentrations of CO and O_2.

A logical extension of the two-step mechanism is the three-step mechanism, allowing for the presence and thereby separate oxidation of H_2,

$$CH_4 + \frac{1}{2}O_2 \rightarrow CO + 2H_2$$

$$CO + \frac{1}{2}O_2 \rightarrow CO_2$$

$$H_2 + \frac{1}{2}O_2 \rightarrow H_2O.$$

For higher hydrocarbons, for example those of n-alkanes, a four-step mechanism has been proposed, involving the initial decomposition of C_mH_{2m+2} to C_2H_4 and H_2, the subsequent oxidation of C_2H_4 to CO and H_2, and the eventual oxidation of CO and H_2. The specific reaction steps are

$$C_mH_{2m+2} \rightarrow \frac{m}{2}C_2H_4 + H_2$$

$$C_2H_4 + O_2 \rightarrow 2CO + 2H_2$$

$$CO + \frac{1}{2}O_2 \rightarrow CO_2$$

$$H_2 + \frac{1}{2}O_2 \rightarrow H_2O.$$

Postulated mechanisms determined in the above manner have enjoyed considerable usage because of their simplicity. However, since the amount of empirical kinetic information used in their determination is quite limited, it is reasonable to expect that their range of applicability is correspondingly limited. Thus there is the need to develop simplified mechanisms from the detailed mechanisms.

3.10.2. Need for Comprehensiveness and Reduction

Before discussing mechanism reduction, it is important to first recognize that the overriding requirement for accurate and reliable description of combustion phenomena is that the detailed mechanism developed must be accurate and comprehensive itself. Specifically, a comprehensively developed detailed mechanism is expected to be able to describe all kinds of combustion phenomena over all possible ranges of the thermodynamic parameters of the system such as the temperature, pressure, and reactant composition. It should also be hierarchical in terms of the fuel structure. For example, since H_2 is an intermediate in the oxidation of a hydrocarbon, the H_2 oxidation mechanism must constitute a submechanism of a hydrocarbon mechanism. Consequently a comprehensive mechanism for the hydrocarbon must degenerate to that for H_2 when all elementary reactions not related to H_2 oxidation are stripped away.

Since the size of a mechanism is proportional to its comprehensiveness, it is sometimes desirable to aim for local comprehensiveness, described by substantially smaller mechanisms that are applicable only to a particular fuel and a restricted range of combustion parameters and phenomena. For example, mechanisms can be developed for methane oxidation at atmospheric pressure for application in furnaces and boilers utilizing natural gas, or for blends of heptane and iso-octane at high-pressures for application in internal combustion engines burning gasoline.

The need to reduce detailed mechanisms arises from three considerations. First, calculations using these mechanisms are usually computationally demanding because of the large number of species and reactions involved, as mentioned earlier. Second, different reactions have vastly different time scales because of the Arrhenius kinetics. This renders the system of conservation equations computationally stiff. Third, since variations of these species and reactions are highly coupled, it is difficult to identify the dominant reaction species and pathways through straightforward inspection.

In the next two sections we shall first discuss the procedure of mechanism reduction, using the hydrogen–oxygen system as an example. We shall then present several theories on mechanism reduction. The reader may wish to revisit this topic after having studied the text materials up to Chapter 8, as several phenomena mentioned in the following are covered in the intervening chapters.

3.11. SYSTEMATIC REDUCTION: THE HYDROGEN–OXYGEN SYSTEM

Systematic mechanism reduction is frequently conducted at two levels of approximation. We start with a detailed mechanism which may consist of tens to hundreds of species and hundreds to thousands of elementary reactions. The size of this detailed mechanism is first reduced by eliminating species and reactions that have negligible influence on the phenomena of interest, resulting in a so-called skeletal mechanism. Elimination of these components can be performed through either experience, which could be rather taxing for a large mechanism, or by a systematic assessment of the effects of varying individual species and reactions on the response of certain combustion phenomena.

Based on the skeletal mechanism, we then apply quasi-steady-state (QSS) approximations to certain species and partial equilibrium assumptions to certain reactions, resulting in the final reduced mechanism with a minimal number of rate equations to solve. Computationally, it is usually more desirable to eliminate species than reactions because this directly reduces the number of species conservation equations that needs to be solved. The individual reactions in the reduced mechanism are now lumped ones, with their respective reaction rates dependent on those of the elementary ones constituting the skeletal mechanism. Clearly, the reduced mechanism can be rendered arbitrarily small, down to the one-step overall reaction, when enough species are put into QSS and enough reactions assumed to be in partial equilibrium. This, however, is achieved at the expense of the fidelity of the mechanism.

The accuracy of the skeletal mechanism needs to be assessed by comparing its calculated responses with those of the detailed mechanism, while the accuracy of the reduced mechanism is assessed by comparing its calculated results with those of the skeletal mechanism. It is important to preserve and validate comprehensiveness at each level of the reduction.

We now illustrate the procedure of systematic reduction by using the simple hydrogen–oxygen system.

3.11.1. Reduction to Skeletal Mechanisms

Methods with varying degrees of rigor have been developed to identify the unimportant species and reactions that can be eliminated from a detailed mechanism. The most commonly adopted one is sensitivity analysis, which can be used to identify the unimportant reactions.

As will be more formally stated in the next section, the sensitivity of a system response parameter y with respect to the perturbation of the reaction rate constant k of a reaction is defined as $\partial y/\partial k$. Thus its lognormal form, $\partial \ln y/\partial \ln k$, measures the relative error induced by the removal of this reaction. Reactions with sensitivity smaller than certain specified values can be considered to be unimportant and hence neglected.

A sensitivity analysis of the individual reactions of the hydrogen–oxygen mechanism of Table 3.1, using the ignition delay time as the response parameter y and for the range of thermodynamic states mapped, as indicated in Figure 3.11.1, yields the

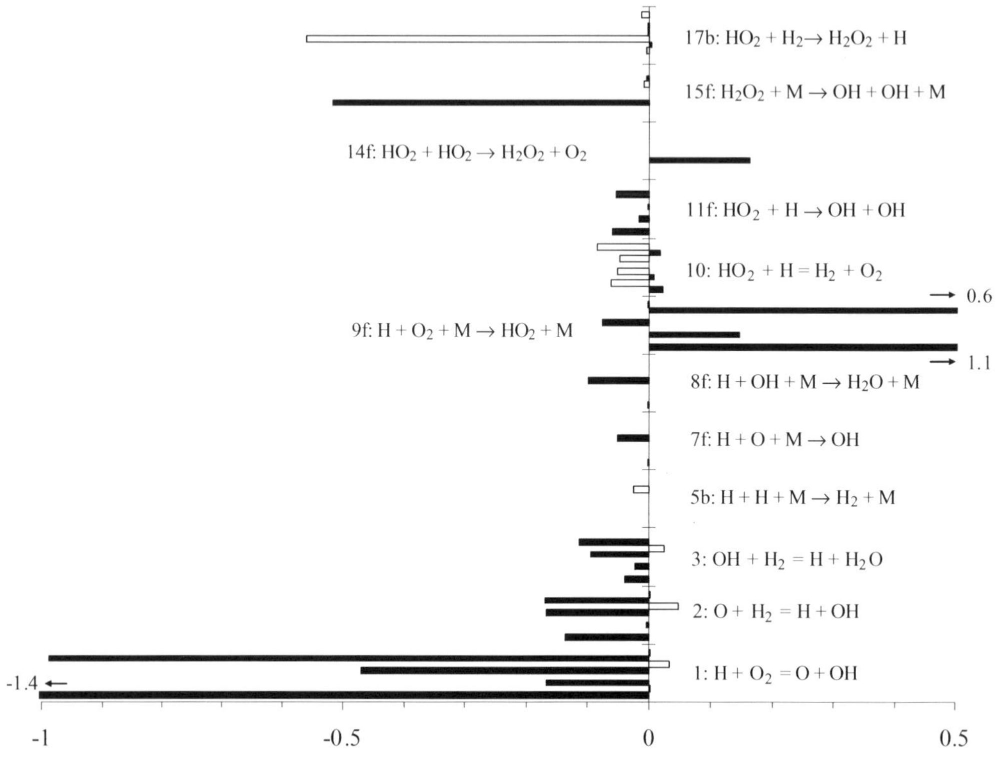

Figure 3.11.1. Lognormal sensitivities of ignition time with respect to the rates of each elementary reaction for stoichiometric H_2–air mixture under different pressure and initial temperatures. The solid bars indicate forward reactions and the open ones the reverse. The sequence of the data series pairs from bottom is: (1,000 K, 1 atm), (1,000 K, 30 atm), (1,500 K, 1 atm), and (1,500 K, 30 atm). Reactions with negligible sensitivities are not shown.

following sixteen-step skeletal mechanism:

$$H + O_2 \rightleftharpoons O + OH \qquad (H1, -H1)$$

$$O + H_2 \rightleftharpoons H + OH \qquad (H2, -H2)$$

$$OH + H_2 \rightleftharpoons H + H_2O \qquad (H3, -H3)$$

$$H + H + M \rightarrow H_2 + M \qquad (-H5)$$

$$H + O + M \rightarrow OH + M \qquad (H7)$$

$$H + OH + M \rightarrow H_2O + M \qquad (H8)$$

$$H + O_2 + M \rightarrow HO_2 + M \qquad (H9)$$

$$HO_2 + H \rightleftharpoons H_2 + O_2 \qquad (H10, -H10)$$

$$HO_2 + H \rightarrow OH + OH \qquad (H11)$$

$$HO_2 + HO_2 \rightarrow H_2O_2 + O_2 \qquad (H14)$$

$$H_2O_2 + M \rightarrow OH + OH + M \qquad (H15)$$

$$HO_2 + H_2 \rightarrow H_2O_2 + H. \qquad (-H17)$$

It is seen that all eight species (H_2, O_2, H_2O, H, O, OH, HO_2, H_2O_2) of the detailed mechanism appear in the skeletal mechanism, indicating that none of them is uniformly negligible in all possible ignition conditions. These eight species are described by eight kinetic rate equations,

$$L(\mathrm{H_2}) = -\omega_2 - \omega_3 + \omega_{5b} + \omega_{10} - \omega_{17b} \tag{3.11.1}$$

$$L(\mathrm{O_2}) = -\omega_1 - \omega_{9f} + \omega_{10} + \omega_{14f} \tag{3.11.2}$$

$$L(\mathrm{H_2O}) = \omega_3 + \omega_{8f} \tag{3.11.3}$$

$$L(\mathrm{H}) = -\omega_1 + \omega_2 + \omega_3 - 2\omega_{5b} - \omega_{7f} - \omega_{8f} - \omega_{9f} - \omega_{10} - \omega_{11f} + \omega_{17b} \tag{3.11.4}$$

$$L(\mathrm{O}) = \omega_1 - \omega_2 - \omega_{7f} \tag{3.11.5}$$

$$L(\mathrm{OH}) = \omega_1 + \omega_2 - \omega_3 + \omega_{7f} - \omega_{8f} + 2\omega_{11f} + 2\omega_{15f} \tag{3.11.6}$$

$$L(\mathrm{HO_2}) = \omega_{9f} - \omega_{10} - \omega_{11f} - 2\omega_{14f} - \omega_{17b} \tag{3.11.7}$$

$$L(\mathrm{H_2O_2}) = \omega_{14f} - \omega_{15f} + \omega_{17b}, \tag{3.11.8}$$

which can be expressed in matrix form as

$$\mathbf{L}_{8\times 1} = \mathbf{s}_{8\times 12} \cdot \boldsymbol{\omega}_{12\times 1}, \tag{3.11.9}$$

where

$$\mathbf{L}_{8\times 1} = \left(L(\mathrm{H_2}) \quad L(\mathrm{O_2}) \quad L(\mathrm{H_2O}) \quad L(\mathrm{H}) \quad L(\mathrm{O}) \quad L(\mathrm{OH}) \quad L(\mathrm{HO_2}) \quad L(\mathrm{H_2O_2})\right)^T \tag{3.11.10}$$

$$\mathbf{s}_{8\times 12} = \left|\begin{array}{cccccccccccc} 0 & -1 & -1 & 1 & 0 & 0 & 0 & 1 & 0 & 0 & 0 & -1 \\ -1 & 0 & 0 & 0 & 0 & 0 & -1 & 1 & 0 & 1 & 0 & 0 \\ 0 & 0 & 1 & 0 & 0 & 1 & 0 & 0 & 0 & 0 & 0 & 0 \\ -1 & 1 & 1 & -2 & -1 & -1 & -1 & -1 & -1 & 0 & 0 & 1 \\ 1 & -1 & 0 & 0 & -1 & 0 & 0 & 0 & 0 & 0 & 0 & 0 \\ 1 & 1 & -1 & 0 & 1 & -1 & 0 & 0 & 2 & 0 & 2 & 0 \\ 0 & 0 & 0 & 0 & 0 & 0 & 1 & -1 & -1 & -2 & 0 & -1 \\ 0 & 0 & 0 & 0 & 0 & 0 & 0 & 0 & 0 & 1 & -1 & 1 \end{array}\right| \tag{3.11.11}$$

$$\boldsymbol{\omega}_{12\times 1} = \left(\omega_1 \quad \omega_2 \quad \omega_3 \quad \omega_{5b} \quad \omega_{7f} \quad \omega_{8f} \quad \omega_{9f} \quad \omega_{10} \quad \omega_{11f} \quad \omega_{14f} \quad \omega_{15f} \quad \omega_{17b}\right)^T \tag{3.11.12}$$

and $\mathbf{L}(\cdot) = d[\cdot]/dt$. Specifically, $\mathbf{L}_{8\times 1}$ is the production rate vector of the species, $\mathbf{s}_{8\times 12}$ the stoichiometric coefficient matrix, and $\boldsymbol{\omega}_{12\times 1}$ the reaction rate vector of the elementary reactions.

At this stage partial equilibrium assumptions can be applied to some of the reactions in the skeletal mechanism for which both the forward and backward reactions are important, while QSS assumptions can be applied to some of the species by setting the corresponding rate equations, (11.1)–(11.8), to zero. These approximations will reduce the number of rate equations that needs to be solved. However, since the size of the skeletal mechanism is still quite large for convenient algebraic manipulation, we shall perform a systematic lumping to derive a mechanism that is chemically equivalent but is more compact and informative, thereby facilitating further analysis.

3.11.2. Linearly Independent Representation

Since the rank of the stoichiometric coefficient matrix of the skeletal mechanism can be at most equal to the number of species, K, the number of linearly independent expressions for the reaction rates can be no more than K. Furthermore, for each of the M elements, we also have an independent relation for element conservation. Consequently, there are only $(K - M)$ linearly independent expressions to describe the reactions. In other words, a mechanism with a multitude of elementary reactions can be reduced to a set of $(K - M)$, semiglobal reactions, each of which is a linear combination of the elementary reactions with a reaction rate that is also a linear combination of the reaction rates of the elementary reactions.

Since $L = 8$ and $M = 2$ for the hydrogen–oxygen system, there are six linearly independent semiglobal reactions. While the choice of this reaction set is not unique, it is physically illuminating to select these reactions such that they mimic the functions of some key groups of the elementary reactions. Thus, for example, we choose the following six semiglobal reactions that bear resemblance to some key hydrogen–oxygen elementary reactions:

Chain initiation:	$H_2 + O_2 \rightleftharpoons H + HO_2$	(I)
Main branching:	$H + O_2 \rightleftharpoons O + OH$	(II)
Main heat release:	$H_2 + OH \rightleftharpoons H_2O + H$	(III)
Chain termination:	$H + O_2 \rightleftharpoons HO_2$	(IV)
Secondary chain termination:	$HO_2 + HO_2 \rightleftharpoons H_2O_2 + O_2$	(V)
Secondary chain branching:	$H_2O_2 \rightleftharpoons OH + OH.$	(VI)

The production rates of the species then become

$$L(H_2) = -\Omega_I - \Omega_{III} \tag{3.11.13}$$

$$L(O_2) = -\Omega_I - \Omega_{II} - \Omega_{IV} + \Omega_V \tag{3.11.14}$$

$$L(H_2O) = \Omega_{III} \tag{3.11.15}$$

$$L(H) = \Omega_I - \Omega_{II} + \Omega_{III} - \Omega_{IV} \tag{3.11.16}$$

$$L(O) = \Omega_{II} \tag{3.11.17}$$

$$L(OH) = \Omega_{II} - \Omega_{III} + 2\Omega_{VI} \tag{3.11.18}$$

$$L(HO_2) = \Omega_I + \Omega_{IV} - 2\Omega_V \tag{3.11.19}$$

$$L(H_2O_2) = \Omega_V - \Omega_{VI}, \tag{3.11.20}$$

where Ω_{I} to Ω_{VI} are the semiglobal reaction rates. Since the production rate of this semiglobal mechanism should be equal to that of the skeletal mechanism, we have

$$\mathbf{S}_{8\times 6}\boldsymbol{\Omega}_{6\times 1} = \mathbf{s}_{8\times 12}\boldsymbol{\omega}_{12\times 1}, \tag{3.11.21}$$

where $\mathbf{S}_{8\times 6}$ is the stoichiometric coefficient matrix for the semiglobal reactions and $\boldsymbol{\Omega}_{6\times 1}$ is the corresponding reaction rate vector. The problem can thus be solved by multiplying the transpose of $\mathbf{S}_{8\times 6}$ on both sides of (3.11.21),

$$\mathbf{S}^T_{6\times 8}\mathbf{S}_{8\times 6}\boldsymbol{\Omega}_{6\times 1} = \mathbf{S}^T_{6\times 8}\mathbf{s}_{8\times 12}\boldsymbol{\omega}_{12\times 1}, \tag{3.11.22}$$

such that

$$\boldsymbol{\Omega}_{6\times 1} = (\mathbf{S}^T_{6\times 8}\mathbf{S}_{8\times 6})^{-1}\mathbf{S}^T_{6\times 8}\mathbf{s}_{8\times 12}\boldsymbol{\omega}_{12\times 1}. \tag{3.11.23}$$

The solution of (3.11.23) yields the reaction rates of I through VI, as

$$\Omega_{\text{I}} = \omega_2 - \omega_{5b} - \omega_{8f} - \omega_{10} + \omega_{17b} \tag{3.11.24}$$

$$\Omega_{\text{II}} = \omega_1 - \omega_2 - \omega_{7f} \tag{3.11.25}$$

$$\Omega_{\text{III}} = \omega_3 + \omega_{8f} \tag{3.11.26}$$

$$\Omega_{\text{IV}} = \omega_2 + \omega_{5b} + 2\omega_{7f} + \omega_{8f} + \omega_{9f} + \omega_{11f} \tag{3.11.27}$$

$$\Omega_{\text{V}} = \omega_2 + \omega_{7f} + \omega_{11f} + \omega_{14f} + \omega_{17b} \tag{3.11.28}$$

$$\Omega_{\text{VI}} = \omega_2 + \omega_{7f} + \omega_{11f} + \omega_{15f}. \tag{3.11.29}$$

It is clear that the rate expressions based on the lumped reactions, Eqs. (3.11.13)–(3.11.20), are simpler than those based on the elementary reactions, Eqs. (3.11.1)–(3.11.8). Although not necessary, this procedure facilitates their manipulation for further reduction, discussed next.

3.11.3. Reduction through QSS Assumption

We now proceed to reduce the size of the lumped skeletal mechanism derived above by applying the QSS assumption to appropriate species. All the assumptions made, except that for H, have been found (Lu, Law & Ju 2003) to be acceptable for the conditions that led to the development of the skeletal mechanism.

3.11.3.1. Four-Step Reduced Mechanism: For the sake of demonstration let us first assume that HO_2 and H_2O_2 are QSS species. Such an assumption could lead to inaccuracies in the HO_2 and H_2O_2 chemistry, especially behavior related to the third limit.

Thus from Eqs. (3.11.19) and (3.11.20), we have

$$L(HO_2) = \Omega_{\text{I}} + \Omega_{\text{IV}} - 2\Omega_{\text{V}} = 0 \tag{3.11.30}$$

$$L(H_2O_2) = \Omega_{\text{V}} - \Omega_{\text{VI}} = 0. \tag{3.11.31}$$

These relations can then be used to reduce the number of linearly independent rate equations (3.11.24)–(3.11.29) from six to four. The reaction rates for the individual species, Eqs. (3.11.13)–(3.11.18), can be manipulated to assume the forms

$$\begin{aligned} L(\mathrm{H_2}) &= -\Omega_{\mathrm{I}} - \Omega_{\mathrm{III}} = -(\Omega_{\mathrm{I}} - \Omega_{\mathrm{V}}) - (\Omega_{\mathrm{III}} + \Omega_{\mathrm{V}}) \\ &= -\Omega_{\mathrm{I'}} - \Omega_{\mathrm{III'}} + \Omega_{\mathrm{IV'}} \end{aligned} \tag{3.11.32}$$

$$\begin{aligned} L(\mathrm{O_2}) &= -\Omega_{\mathrm{I}} - \Omega_{\mathrm{II}} - \Omega_{\mathrm{IV}} + \Omega_{\mathrm{V}} \\ &= -\Omega_{\mathrm{II}} - \Omega_{\mathrm{V}} + (-\Omega_{\mathrm{I}} - \Omega_{\mathrm{IV}} + 2\Omega_{\mathrm{V}}) \\ &= -\Omega_{\mathrm{II'}} + \Omega_{\mathrm{IV'}} \end{aligned} \tag{3.11.33}$$

$$L(\mathrm{H_2O}) = \Omega_{\mathrm{III}} = \Omega_{\mathrm{III'}} \tag{3.11.34}$$

$$\begin{aligned} L(\mathrm{H}) &= \Omega_{\mathrm{I}} - \Omega_{\mathrm{II}} + \Omega_{\mathrm{III}} - \Omega_{\mathrm{IV}} + (\Omega_{\mathrm{I}} + \Omega_{\mathrm{IV}} - 2\Omega_{\mathrm{V}}) \\ &= 2(\Omega_{\mathrm{I}} - \Omega_{\mathrm{V}}) - \Omega_{\mathrm{II}} + \Omega_{\mathrm{III}} = 2\Omega_{\mathrm{I'}} - \Omega_{\mathrm{II'}} + \Omega_{\mathrm{III'}} \end{aligned} \tag{3.11.35}$$

$$L(\mathrm{O}) = \Omega_{\mathrm{II}} = \Omega_{\mathrm{II'}} \tag{3.11.36}$$

$$L(\mathrm{OH}) = \Omega_{\mathrm{II}} - \Omega_{\mathrm{III}} + 2\Omega_{\mathrm{VI}} = \Omega_{\mathrm{II'}} - \Omega_{\mathrm{III'}} - 2\Omega_{\mathrm{IV'}} \tag{3.11.37}$$

with the four semiglobal reactions given by

$$\mathrm{H_2} \rightleftharpoons \mathrm{H} + \mathrm{H} \tag{I'}$$

$$\mathrm{H} + \mathrm{O_2} \rightleftharpoons \mathrm{O} + \mathrm{OH} \tag{II'}$$

$$\mathrm{H_2} + \mathrm{OH} \rightleftharpoons \mathrm{H_2O} + \mathrm{H} \tag{III'}$$

$$\mathrm{OH} + \mathrm{OH} \rightleftharpoons \mathrm{H_2} + \mathrm{O_2} \tag{IV'}$$

and the semiglobal reaction rates being $\Omega_{\mathrm{I'}} = \Omega_{\mathrm{I}} - \Omega_{\mathrm{V}}$, $\Omega_{\mathrm{II'}} = \Omega_{\mathrm{II}}$, $\Omega_{\mathrm{III'}} = \Omega_{\mathrm{III}}$, and $\Omega_{\mathrm{IV'}} = -\Omega_{\mathrm{IV}}$ respectively. The steady-state concentrations of $\mathrm{HO_2}$ and $\mathrm{H_2O_2}$ are obtained by solving the algebraic equations (3.11.30) and (3.11.31).

3.11.3.2. Two-Step Reduced Mechanism: We now assume that all the other radicals except H are in steady state. The accuracy of the mechanism, for the ignition phenomena of interest here, should still be quite satisfactory even at this level of approximation. Thus, we set

$$L(\mathrm{O}) = \Omega_{\mathrm{II''}} = 0 \tag{3.11.38}$$

$$L(\mathrm{OH}) = \Omega_{\mathrm{II''}} - \Omega_{\mathrm{III''}} - 2\Omega_{\mathrm{IV''}} = 0. \tag{3.11.39}$$

Substitution of (3.11.38) and (3.11.39) into (3.11.32)–(3.11.37) yields

$$L(\mathrm{H_2}) = -\Omega_{\mathrm{I'}} - \Omega_{\mathrm{III'}} + \Omega_{\mathrm{IV'}} = -\Omega_{\mathrm{I'}} - 1.5\Omega_{\mathrm{III'}} = -\Omega_{\mathrm{I'}} - 3\Omega_{\mathrm{II''}} \tag{3.11.40}$$

$$L(\mathrm{O_2}) = -\Omega_{\mathrm{II'}} + \Omega_{\mathrm{IV'}} = -0.5\Omega_{\mathrm{III'}} = -\Omega_{\mathrm{II''}} \tag{3.11.41}$$

$$L(\mathrm{H_2O}) = \Omega_{\mathrm{III'}} = 2\Omega_{\mathrm{II''}} \tag{3.11.42}$$

$$L(\mathrm{H}) = 2\Omega_{\mathrm{I'}} - \Omega_{\mathrm{II'}} + \Omega_{\mathrm{III'}} = 2\Omega_{\mathrm{I'}} + \Omega_{\mathrm{III'}} = 2\Omega_{\mathrm{I''}} + 2\Omega_{\mathrm{II''}} \tag{3.11.43}$$

with the semiglobal reactions being

$$H_2 \rightleftharpoons H + H \tag{I''}$$

$$3H + O_2 \rightleftharpoons 2H_2O + 2H \tag{II''}$$

and the corresponding reaction rates given by: $\Omega_{I''} = \Omega_{I'}$ and $\Omega_{II''} = 0.5\Omega_{III'}$. The concentrations of the QSS species, O, OH, HO_2, and H_2O_2 are solved from the algebraic equations (3.11.30), (3.11.31), (3.11.38), and (3.11.39).

3.11.3.3. One–Step Overall Reaction: If we finally assume that H is in steady state,

$$L(H) = 2\Omega_{I''} + 2\Omega_{II''} = 0, \tag{3.11.44}$$

Eqs. (3.11.40)–(3.11.42) then become

$$L(H_2) = -\Omega_{I''} - 3\Omega_{II''} = 2\Omega_{I''} = -2\Omega_{1\text{-step}} \tag{3.11.45}$$

$$L(O_2) = -\Omega_{II''} = \Omega_{I''} = -\Omega_{1\text{-step}} \tag{3.11.46}$$

$$L(H_2O) = 2\Omega_{II''} = -2\Omega_{I''} = 2\Omega_{1\text{-step}} \tag{3.11.47}$$

and the mechanism is reduced to one step,

One-step overall reaction: $2H_2 + O_2 \rightleftharpoons 2H_2O$

with the global reaction rate given by $\Omega_{1\text{-step}} = -\Omega_{I''}$.

It is quite apparent that although reduced mechanisms can be derived manually step by step using the above procedure, the algebra will quickly become too unwieldy for the more complex hydrocarbon oxidation mechanisms. Systematic procedures by matrix manipulation (Chen 1988; Lu 2004) have been developed for the reduction of detailed mechanisms with any number of QSS species. Consequently a skeletal mechanism can be reduced to any degree, with specified species. Indeed, the procedure can be applied to hydrocarbon oxidation with postulated semiglobal mechanisms discussed in the previous section, extending their comprehensiveness but at the expense of the need for the internal iteration to solve for the QSS species. That is, in addition to the inaccuracies imparted to the reduced mechanisms with the QSS and partial equilibrium assumptions, another collateral cost is the need to solve for the concentrations of the species that participated in these assumptions from a set of algebraic equations. Frequently these relations are not algebraically explicit such that their solution requires iteration, which may also experience difficulty in convergence and hence can significantly complicate and prolong the solution process. Alternate approaches, such as truncation (Peters 1991) through which certain terms in an algebraically nonlinear expression are either dropped or approximated so as to make the solution tractable, have been applied, but at the expense of incurring unquantifiable inaccuracies to the solution. Iteration can also be avoided through tabulation (Pope 1997), which has been found to be quite useful for mechanisms that are not too large.

3.12. THEORIES OF MECHANISM REDUCTION

Various theories and methods have been developed for mechanism reduction (Lu & Law 2009), each with its merits and restrictions. These methods are mentioned here, to be followed by a more detailed discussion of three of them.

Concerning reduction of detailed mechanism to skeletal mechanisms, useful methods for the elimination of reactions include sensitivity and Jacobian analysis (Turanyi 1990; Tomlin et al. 1992; Tomlin, Turanyi & Pilling 1997), detailed reduction (Wang & Frenklach 1991), and computational singular perturbation (CSP) (Massias et al. 1999a, 1999b). A method involving the elimination of species was developed based on the theory of directed relation graph (DRG) (Lu & Law 2005), and has been demonstrated to be particularly suitable for the reduction of very large mechanisms.

Subsequent reduction employing partial equilibrium and QSS assumptions have employed methods such as reaction rate analysis (Peters & Kee 1987; Chen 1988; Smooke 1991; Ju & Niioka 1994; Sung et al. 1998) and lifetime analysis (Lovas, Nilsson & Mauss 2000). More rigorous methods include those of intrinsic low-dimensional manifolds (ILDM) (Maas & Pope 1992) and CSP (Lam 1993, 1994; Lam & Goussis 1988; Goussis & Lam 1992). It is noted that ILDM and CSP are essentially the same, except CSP undergoes an additional step of refinement.

We shall now discuss, in more detail, the methods of sensitivity analysis, directed relation graph, and computational singular perturbation.

3.12.1. Sensitivity Analysis

As demonstrated in Section 3.11.1, sensitivity analysis can be used to eliminate unimportant reactions from detailed mechanisms. At the simplest level of analysis, a global sensitivity can be computed by evaluating the change Δy in a global combustion parameter y, such as the laminar flame speed or the ignition delay to be discussed in later chapters, due to a small perturbation Δk in the reaction rate constant k, or more specifically the preexponential factor A. The method is rather straightforward but can be cumbersome and quite time consuming. This was the method used in obtaining the sensitivity shown in Figure 3.11.1.

A more efficient analysis can be conducted by first expressing the system of rate equations as

$$\frac{d\mathbf{y}}{dt} = \mathbf{g}(\mathbf{y}; \mathbf{a}), \tag{3.12.1}$$

where $\mathbf{y}$ is the dependent variable vector that in our case consists of the reaction scalars such as the temperature and the concentrations of species, $\mathbf{g}(\mathbf{y}; \mathbf{a})$ is the production rate term, and $\mathbf{a}$ is the vector for the preexponential factor A. The sensitivity matrix $\mathbf{E}$ is then defined as

$$\mathbf{E} = \frac{\partial \mathbf{y}}{\partial \mathbf{a}}. \tag{3.12.2}$$

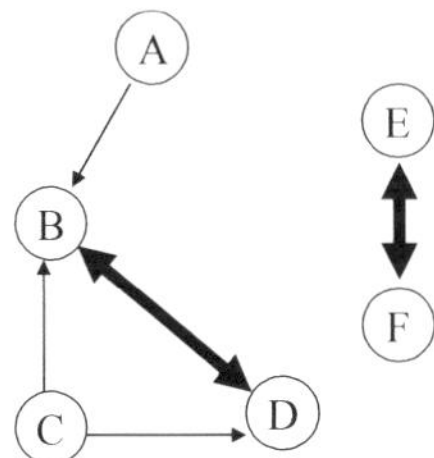

Figure 3.12.1. A directed relation graph showing typical relations of the species.

Taking the time derivative of (3.12.2), we get

$$\frac{\partial \mathbf{E}}{\partial t} = \frac{\partial}{\partial t}\left(\frac{\partial \mathbf{y}}{\partial \mathbf{a}}\right) = \frac{\partial}{\partial \mathbf{a}}\left(\frac{\partial \mathbf{y}}{\partial t}\right) = \frac{\partial \mathbf{g}(\mathbf{y}, \mathbf{a})}{\partial \mathbf{a}} = \frac{\partial \mathbf{g}}{\partial \mathbf{a}} + \frac{\partial \mathbf{g}}{\partial \mathbf{y}} \cdot \frac{\partial \mathbf{y}}{\partial \mathbf{a}} = \frac{\partial \mathbf{g}}{\partial \mathbf{a}} + \mathbf{J} \cdot \frac{\partial \mathbf{y}}{\partial \mathbf{a}}, \quad (3.12.3)$$

where $\mathbf{J} = \partial \mathbf{g}/\partial \mathbf{y}$ is the Jacobian matrix of (3.12.1). Equation (3.12.3) is to be solved together with Eq. (3.12.1) for $\mathbf{y}$ and $\mathbf{E}$, from which the lognormal sensitivity $\partial \ln \mathbf{y}/\partial \ln \mathbf{a}$ can be readily evaluated. It is noted that although sensitivity analysis is simple to apply, the computation is typically time consuming due to the large number of equations to be solved.

3.12.2. Theory of Directed Relation Graph

While it is relatively straightforward to identify the unimportant elementary reactions through, say, sensitivity analysis, it is much more involved to identify and eliminate species of negligible importance. This difficulty is mostly due to the indirect as well as direct couplings of the species. In the presence of such couplings, removal of a species from a detailed mechanism could result in significant error in predicting the concentration of another species, through their participation with a group of third party species, even though they never appear together in any of the elementary reactions. It is therefore necessary to identify and keep the group of species that are either directly or indirectly strongly coupled with a species.

The method of directed relation graph has been exploited to abstract the couplings between the species (Lu & Law 2005). A DRG is a graph in which each vertex designates a species and each directed edge represents the requirement relation of one species to another. Figure 3.12.1 shows the typical relations between the species in a DRG. The arrow shows the direction of dependence of one species on another, with its width indicating the strength of the dependence. Thus, for a species A that has to be kept, it is seen that A depends on B, but B does not depend on A. Furthermore, since species B and D are strongly coupled, they form the dependent set {B, D} of A, and have to be kept in order to correctly predict A. However, species C, E, and F can be eliminated because they are not required by either A or any species in the dependent set of A. Furthermore, species within the strongly coupled groups, {B, D} and {E, F}, should be either kept or eliminated together.

The requirement relation of one species to another can be quantified as

$$r_{AB} \equiv \frac{\sum_{k=1,K} |\nu_{A,i}\omega_k \delta(B)_k|}{\sum_{k=1,K} |\nu_{A,k}\omega_k|} \tag{3.12.4}$$

$$\delta(B)_k = \begin{cases} 1, & \text{if the } k\text{th elementary reaction involves species B,} \\ 0, & \text{otherwise} \end{cases} \tag{3.12.5}$$

where $\nu_{A,k}$ is the stoichiometric coefficient of species A in the kth elementary reaction, ω_k its production rate, and K the number of reactions. Thus r_{AB} measures the relative error in the production rate of species A due to the removal of species B. If r_{AB} is large, the removal of species B from the skeletal mechanism could induce significant error in the evaluation of the production rate of species A, that is, species A has strong dependence on species B. By comparing r_{AB} with a small threshold value, the weak dependence between species can be truncated and a DRG can be formed. It is clear that the size of the skeletal mechanism varies inversely with the threshold value, which measures the accuracy and comprehensiveness of the mechanism. Thus the specification of the threshold value is a compromise between the size and accuracy of the resulting skeletal mechanism.

Once a DRG is constructed for a single reaction state and for a particular application such as ignition delay and well-stirred reactor (WSR), some important species A, such as the fuel, can be selected as the starting species for a graph search, which subsequently identifies all the species required by A either directly or indirectly. These species then constitute the species set of this local, subskeletal mechanism, valid for the particular reaction state, application, and accuracy threshold. All the other species, as well as all the elementary reactions that do not involve any of the participating species, can thus be eliminated from the mechanism.

To obtain a skeletal mechanism that is valid over a wide range of parameters such as pressure, temperature, equivalence ratio, and residence time, and for different applications, subskeletal mechanisms are developed by sampling many reaction states within the parametric ranges of these applications. The final skeletal mechanism is then the union of all these subskeletal mechanisms, for a given accuracy threshold, ϵ.

It is also reasonable to expect that strongly coupled groups frequently exist in large mechanisms, and as such intragroup couplings are strong while intergroup couplings are relatively weak (Figure 3.12.1). Consequently the number of species in the skeletal mechanism would jump abruptly around certain values of ϵ across which a group of strongly coupled species is eliminated. This property facilitates selection of the desirable skeletal mechanism, which should be both sufficiently small and comprehensive. Thus reduction of the mechanism size is most efficient for values of ϵ slightly larger than those where the jumps occur.

Figure 3.12.2 demonstrates the occurrence of such jumps for the reduction of a detailed ethylene mechanism consisting of 70 species and 463 elementary reactions (Qin et al. 2000; Lu & Law 2005). Mechanisms of this size are considered to be moderately large. It is seen that the number of species in the skeletal mechanism determined

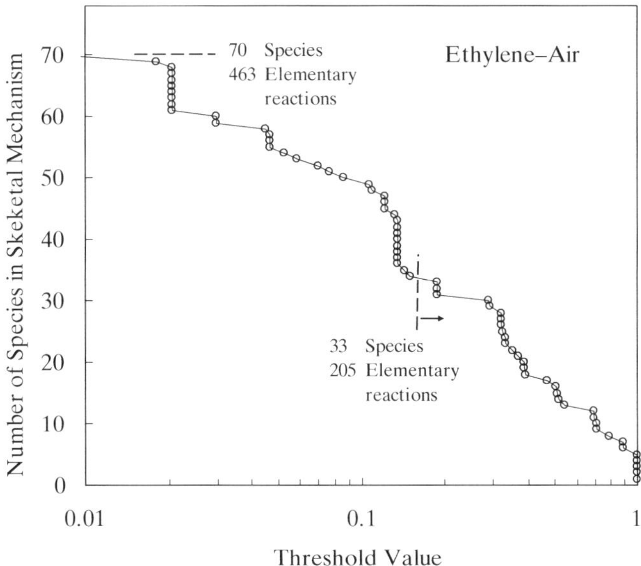

Figure 3.12.2. Dependence of the number of species of the skeletal mechanism on the specified threshold value used to truncate the weak relations of the species. The jumps in the number of species show the existence of strongly coupled groups.

from DGR decreases abruptly around the threshold values of $\epsilon = 0.02$ and $\epsilon = 0.12$. For the two jumps observed, it is reasonable to select the second jump because the mechanism for the first jump is still too large even though it has a higher accuracy. This results in a skeletal mechanism consisting of 205 elementary reactions and 33 species, which are: H_2, H, O, O_2, OH, H_2O, HO_2, H_2O_2, C, CH, CH_2, CH_2^*, CH_3, CH_4, CO, CO_2, HCO, CH_2O, CH_2OH, CH_3O, CH_3OH, C_2H_2, C_2H_3, C_2H_4, C_2H_5, HCCO, CH_2CO, CH_2CHO, n-C_3H_7, C_3H_6, α-C_3H_5, Ar, and N_2. The skeletal mechanism can then be readily constructed by identifying the 205 elementary reactions of the detailed mechanism that consist of one or more of the above 33 species. Thus a skeletal mechanism is defined by simply listing the participating species, as is done here.

3.12.3. Theory of Computational Singular Perturbation

Computational singular perturbation is a systematic method to remove the short time scales in the general reacting system of Eq. (3.12.1),

$$\frac{d\mathbf{y}}{dt} = \mathbf{g}(\mathbf{y}). \tag{3.12.6}$$

By taking the time derivative of (3.12.6), we obtain

$$\frac{d\mathbf{g}}{dt} = \mathbf{J} \cdot \mathbf{g}, \tag{3.12.7}$$

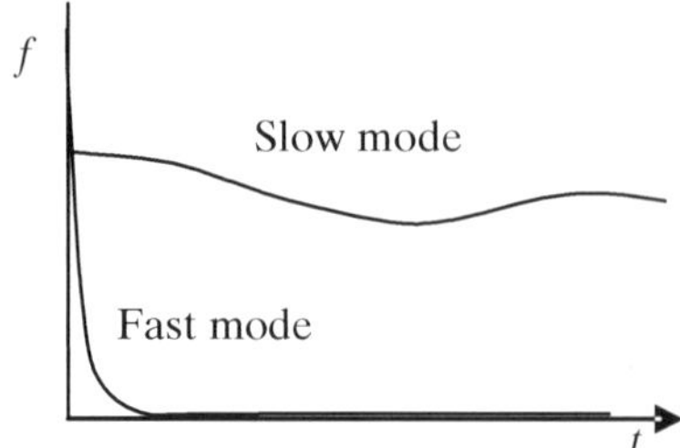

Figure 3.12.3. Schematic showing typical evolution of fast and slow modes obtained from CSP analysis.

where $\mathbf{J} = d\mathbf{g}/d\mathbf{y}$ is the time-dependent Jacobian matrix. By using the decomposition $\mathbf{J} = \mathbf{A}\mathbf{\Lambda}\mathbf{B}$, where $\mathbf{B} = \mathbf{A}^{-1}$, Eq. (3.12.7) can be written as

$$\frac{d\mathbf{f}}{dt} = \mathbf{\Lambda} \cdot \mathbf{f} \tag{3.12.8a}$$

$$\mathbf{f} \equiv \mathbf{B} \cdot \mathbf{g} \tag{3.12.8b}$$

$$\mathbf{\Lambda} = \left(\frac{d\mathbf{B}}{dt} + \mathbf{B} \cdot \mathbf{J}\right) \cdot \mathbf{A}, \tag{3.12.8c}$$

where the components of $\mathbf{f}$ are called the modes of the system, which are linear combinations of the components of $\mathbf{g}$. If the Jacobian matrix is time independent, $\mathbf{\Lambda}$ can be either a diagonal matrix or in Jordan form obtained through eigenvalue decomposition. These modes are then decoupled, resulting in rate equations for the individual modes as

$$\frac{df_i}{dt} = \lambda_i f_i \tag{3.12.9}$$

such that $f_i \sim \exp(\lambda_i t)$, where the $\lambda_i s$ are the elements of $\mathbf{\Lambda}$, and $1/|\ \lambda_i\ |$ yields the time scales of the modes. Figure 3.12.3 schematically shows the rapid decay of f_i for the fast modes relative to the slow modes.

In general, however, $\mathbf{J}$ is time dependent. An iterative procedure in the CSP theory can be applied to separate the fast and slow spaces, and the modes in the refined fast space vanish after a short transient period, that is, $\mathbf{f}_{\text{fast}} = 0$. This set of algebraic equations can therefore be employed to reduce the stiffness and number of differential equations.

The CSP theory can also be applied to identify the QSS species for the QSS-based reduction procedure discussed earlier. Specifically, from Eq. (3.12.8b) the production rate of the ith species can be expressed as

$$\frac{dy_i}{dt} = g_i = \mathbf{a}_i \cdot \mathbf{f} = \mathbf{a}_{i,\text{fast}} \cdot \mathbf{f}_{\text{fast}} + \mathbf{a}_{i,\text{slow}} \cdot \mathbf{f}_{\text{slow}}, \tag{3.12.10}$$

where $\mathbf{a}_i = (\mathbf{a}_{i,\text{fast}}, \mathbf{a}_{i,\text{slow}})$ is the ith row of the matrix $\mathbf{A}$, and $\mathbf{f} = (\mathbf{f}_{\text{fast}}^T, \mathbf{f}_{\text{slow}}^T)^T$. It is seen that if $\mathbf{a}_{i,\text{slow}} \approx 0$ such that $g_i \approx \mathbf{a}_{i,\text{fast}} \cdot \mathbf{f}_{\text{fast}}$, then it can be assumed that $dy_i/dt = g_i = 0$ because $\mathbf{f}_{\text{fast}} = 0$, and the species is in steady state.

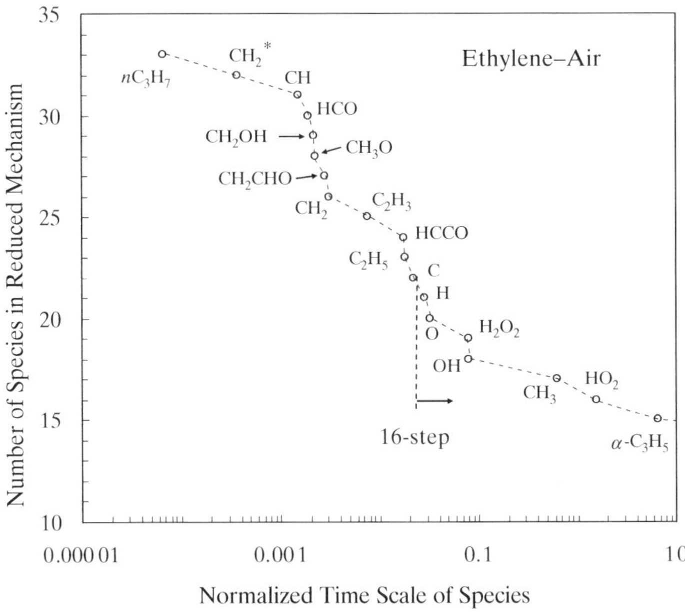

Figure 3.12.4. Number of species in the reduced mechanism as a function of the normalized cutoff time scale of the species.

Using the CSP theory and an appropriate cutoff time scale for the fast species, twelve QSS species, given by C, CH, CH_2, CH_2^*, HCO, CH_2OH, CH_3O, C_2H_3, C_2H_5, HCCO, CH_2CHO, and n-C_3H_7, have been identified (Figure 3.12.4) from the skeletal mechanism developed in the previous section. Since there are five elements (C, H, O, N, Ar), the reduced mechanism should consist of sixteen semiglobal steps, which can be constructed by using the systematic matrix operations (Chen 1988; Lu 2004).

3.12.4. Mechanism Validation

Comprehensive validation is a crucial step towards developing skeletal and reduced mechanisms. Thus to demonstrate the accuracy of the mechanisms derived above, we first compare the results of WSR and autoignition with those of the detailed mechanism, noting that WSR operations encounter extensive temperature ranges covering both ignition and extinction, while autoignition is relevant to the low- to intermediate-temperature ranges. Figures 3.12.5a and 3.12.5b then respectively show the calculated temperature profiles and the ignition times with the detailed, skeletal, and reduced mechanisms for extensive ranges of the system parameters for the ethylene–air system considered in Figures 3.12.2 and 3.12.4. It is seen that the skeletal and reduced mechanisms simulate well the response of the detailed mechanism, exhibiting minimal deviations. Furthermore, although these comparisons are only for the stoichiometric mixture, it has been found that the same degree of agreement also exists for very lean and rich mixtures.

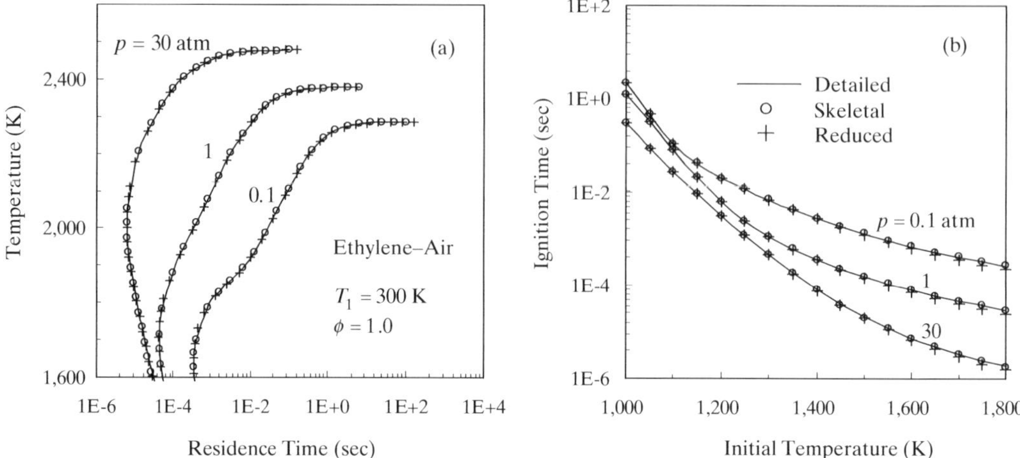

Figure 3.12.5. Comparison of (a) temperature on the residence time of WSR, and (b) autoignition time on the initial temperature under constant pressure and enthalpy, between the detailed, skeletal, and reduced mechanisms for stoichiometric ethylene–air mixtures under various pressures.

Since WSR and autoignition are homogeneous systems and are used in mechanism reduction, it is necessary to extend the validation to systems that involve diffusive transport and that are also not part of the reduction process. Figures 3.12.6 and 3.12.7 respectively compare, over extensive ranges of pressure and equivalence ratio, the calculated laminar burning velocities of premixtures and the ignition temperatures of nonpremixed counterflow obtained by using the detailed, skeletal, and reduced mechanisms; the related phenomena will be studied in Chapters 7 and 8 respectively. Very close agreement is again observed.

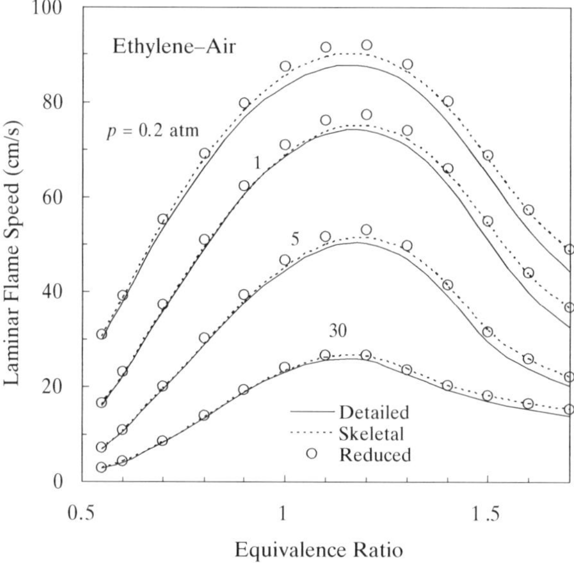

Figure 3.12.6. Comparison of the laminar flame speed on the equivalence ratio between detailed, skeletal, and reduced mechanisms for ethylene–air flames at various pressures.

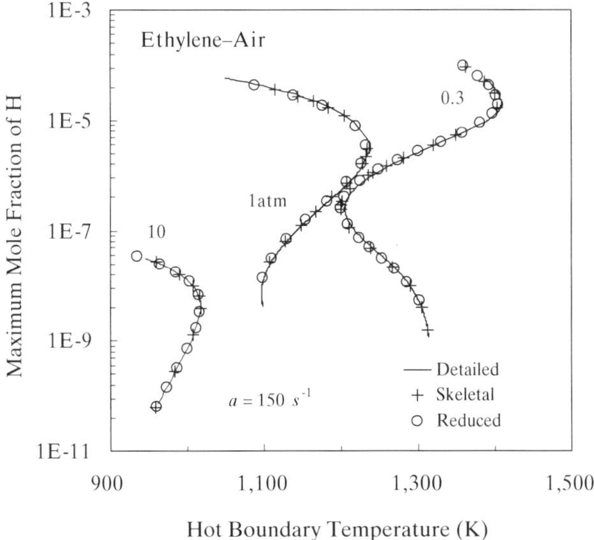

Figure 3.12.7. Comparison of maximum mole fraction of the H radical on the hot boundary temperature of counterflow ignition between detailed, skeletal, and reduced mechanisms for ethylene–air mixtures at various pressures.

PROBLEMS

1. For a hydrogen–oxygen mixture undergoing homogeneous explosion in the second-limit regime, show that the maximum [H] production rate in terms of pressure variations is given by the relation

$$4k_1 = 3k_9[\mathrm{M}].$$

Sketch/plot this relation as well as the second limit relation in the same graph. Discuss the results.

2. For a hydrogen–oxygen mixture that is either highly enriched with hydrogen or at higher temperatures, the generation of the hydrogen atom is increased such that the reactions

$$\mathrm{HO_2 + H \rightarrow H_2 + O_2} \tag{H10}$$

$$\mathrm{HO_2 + H \rightarrow OH + OH} \tag{H11}$$

are facilitated. The second limit is consequently modified to (Zheng & Law 2004):

$$2k_1 = \frac{2k_{10}}{k_{10} + k_{11}} k_9[\mathrm{M}],$$

which has been referred to as the extended second limit. Derive this expression. Experimentally (Mueller et al. 1999) it has been found that, for a given temperature, this extended second limit is situated at a slightly higher pressure than that of the second limit. Discuss implications of the above results.

3. Based on the Zel'dovich mechanism, show that the rate of NO formation can be expressed as

$$\frac{d[\mathrm{NO}]}{dt} = 2k_{\mathrm{N}1} K_c [\mathrm{O}_2]_{\mathrm{eq}}^{1/2} [\mathrm{N}_2],$$

where K_c is the equilibrium constant for concentration for the reaction $\frac{1}{2}\mathrm{O}_2 = \mathrm{O}$ and $[\mathrm{O}_2]_{\mathrm{eq}}$ is the associated equilibrium O_2 concentration. State all assumptions made in deriving the above expression.

4. Construct the reaction pathway diagram for methanol oxidation in the manner of Figure 3.3.1 for methane, using the reaction pathway information provided in the text. Terminate your branches wherever they can be connected to the methane diagram.

5. A semiglobal four-step mechanism has been constructed for methane oxidation:

$$\begin{aligned} &\text{I.} \quad \mathrm{CH_4 + 2H + H_2O \rightleftharpoons CO + 4H_2},\\ &\text{II.} \quad \mathrm{CO + H_2O \rightleftharpoons CO_2 + H_2},\\ &\text{III.} \quad \mathrm{H + H + M \rightleftharpoons H_2 + M},\\ &\text{IV.} \quad \mathrm{O_2 + 3H_2 \rightleftharpoons 2H + 2H_2O}, \end{aligned}$$

with the global reaction rates $\omega_{\mathrm{I}}, \omega_{\mathrm{II}}, \omega_{\mathrm{III}}, \omega_{\mathrm{IV}}$, respectively.

(a) Derive a reduced mechanism with three global steps and the corresponding reaction rates, by assuming that H is in steady state.

(b) Can you derive a two-step reduced mechanism by further assuming that reaction II is in partial equilibrium?

6. Explain why the multistep reaction rate for CO,

$$\frac{d[\mathrm{CO}]}{dt} \sim [\mathrm{CO}][\mathrm{H_2O}]^{0.5}[\mathrm{O_2}]^{0.25},$$

involves the concentration of water vapor to the 0.5 power, even though $\mathrm{H_2O}$ does not participate in the overall reaction.

7. Use CSP to analyze the following reaction mechanism with three species, F, R, and P, and two reversible reactions:

$$\mathrm{F} \rightleftharpoons \mathrm{R}, \quad \omega_1 = 1 \times (c_{\mathrm{F}} - c_{\mathrm{R}}) \qquad \text{(R1)}$$

$$\mathrm{R} \rightleftharpoons \mathrm{P}, \quad \omega_2 = 10^3 \times (c_{\mathrm{R}} - c_{\mathrm{P}}). \qquad \text{(R2)}$$

Compare the information provided by the CSP solution with that from the classical partial equilibrium assumption. Explain why the leading order solution from the partial equilibrium assumption should not be directly substituted back to the ODEs to compute the species production rates.

4 Transport Phenomena

When the molecules in any region of a fluid medium possess an excess of energy, concentration, or momentum, such that gradients of these properties exist in the neighborhood of this region, the system will attempt to restore spatial uniformity by transporting the relevant property in the direction of the deficient region. The transport occurs even in the absence of any bulk motion in this direction. As an illustration, consider a body of stagnant gas situated between two parallel plates as shown in Figure 4.1.1a. If plate A is suddenly raised to a temperature $T_A > T_B$, then the region around plate B also will be heated soon. Therefore there can be a net transfer of heat from A to B despite the lack of any bulk fluid movement. Similarly if there is initially a higher concentration of a species at A relative to its concentration at B, then its molecules will slowly migrate from A to B, as shown in Figure 4.1.1b. Finally, if plate A is impulsively started to move in a direction parallel to itself (Figure 4.1.1c), then plane B will soon feel the motion and, if unrestrained, will tend to be dragged along. Note that in the last case even though there is a bulk flow in the direction parallel to the plates, the transport occurs in the direction normal to the plates in which there is no motion.

The phenomena we have described are known respectively as heat conduction, mass diffusion, and viscous motion. The fundamental physical mechanism responsible for the transport is the incessant collision between the gas molecules, which are in continuous random motion. Through these collisions macroscopic nonuniformities in the fluid medium are evened out. We shall call this mode of transport as diffusional transport whether it is for mass, momentum, or energy.

Diffusional transport is to be distinguished from another important transport mechanism, namely convection, in which the relevant property is "carried along" by the bulk movement within the fluid medium. The existence and intensity of convective transport are described by the fluid dynamics of the system. Depending on the direction of the motion, convection can either facilitate or retard diffusional transport, which is always present whenever nonuniformities exist in the flow field. Furthermore, since the rate of convective transport depends only on the fluid motion, it is the same for the transport of mass, momentum, and energy. However, the diffusional

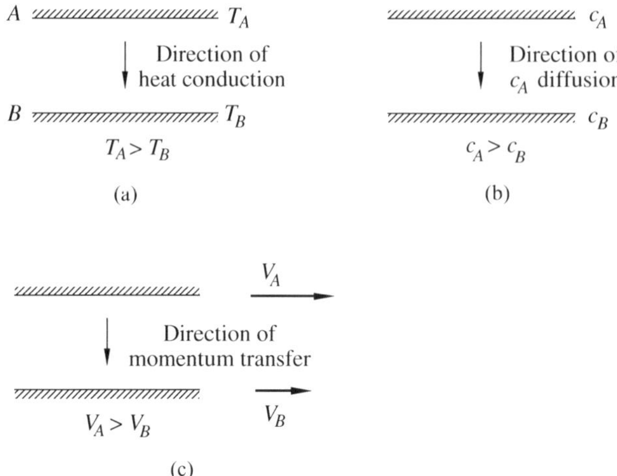

Figure 4.1.1. Schematic showing the diffusion of (a) heat, (b) mass, and (c) momentum.

transport of mass, momentum, and energy occurs at different rates, depending on the respective property gradients and diffusion coefficients, as will be defined later.

Frequently mass diffusion can also generate convection. This can be illustrated by covering the bottom plate in Figure 4.1.1b with water and by making the upper surface porous and completely dry. Water will continuously diffuse from the bottom surface to the upper surface until it is completely dried. Thus there is a net transfer of mass in the upward direction due to diffusion. This convective motion is called Stefan flow.

Diffusion is essential to many combustion processes, especially for subsonic flows. Consider an active reaction region, which has low reactant concentrations and high temperature and product concentrations. Away from this reaction region the reactant concentrations are high whereas the temperature and product concentrations are low. Figure 5.5.1 shows the situations for premixed and nonpremixed flames that will be studied later. Therefore, in general, through the concentration gradients the reactants are continuously supplied to the reaction region to sustain reaction; through the temperature gradients the reaction heat generated is used to preheat the reactants to the reaction temperature as they are transported to the reaction region; and through the product concentration gradients the products can be continuously "drained" away. It is obvious that if products were allowed to accumulate in the reaction region, reactants would be displaced and reactions would terminate rapidly.

The kinetic theory of dilute gases (Hirschfelder, Curtiss & Bird 1954; Vincenti & Kruger 1965; Williams 1985) provides a statistical description of a gas which is not in equilibrium in the spatial distribution of either its concentration, momentum, or temperature. It relates the collisional dynamics at the molecular level to the macroscopic phenomena of diffusion, yielding explicit description of the diffusional transport rates. In the following we shall first present an elementary derivation of

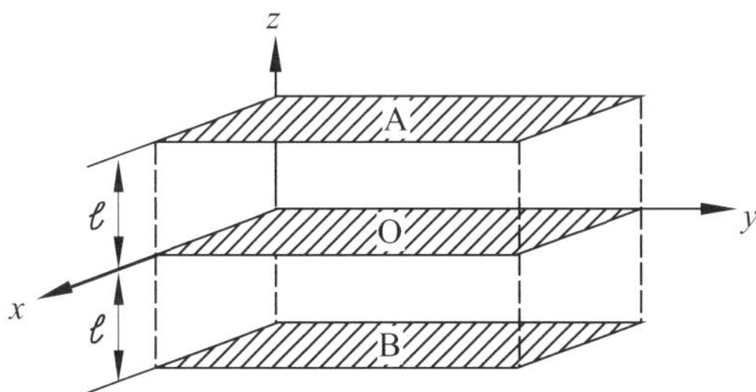

Figure 4.1.2. Schematic showing the concept of diffusive transport in the presence of spatial gradient.

these transport rates by adopting the simplest possible description of the molecular structure, collisional dynamics, and statistics. The derivation not only identifies the dominant physical mechanisms and the important parameters, but it also yields results that are surprisingly close to those of the formal analysis. Following this we shall present results from the kinetic theory of gases, which allow for estimations of the transport coefficients of gases with different molecular structures and collisional dynamics. The focus of this chapter is on the transport coefficients; conservation equations governing the transport of various properties are discussed in the next chapter.

4.1. PHENOMENOLOGICAL DERIVATION OF DIFFUSION COEFFICIENTS

4.1.1. Derivation

In this derivation, we aim to determine the rate of transport of a property Ψ of a gas through molecular collisions in the same manner as the derivation for the collision theory of reaction rates in Chapter 2. The gas is sufficiently dilute that only binary collisions are of importance. The two colliding molecules have masses m_i, m_j and diameters σ_i, σ_j respectively. If we further assume that transport occurs only in, say,  the z-direction as shown in Figure 4.1.2, then the property values at planes (ℓ) and $(-\ell)$ can be related to the value Ψ_0 at plane 0 through

$$\Psi(+\ell) = \Psi_0 + \ell(d\Psi/dz)_0, \qquad \Psi(-\ell) = \Psi_0 - \ell(d\Psi/dz)_0, \tag{4.1.1}$$

where ℓ is a small distance along z. Using Eq. (4.1.1), the net property flux across plane 0 is

$$F_\Psi = -\overline{V}_{i,j}[\Psi(+\ell) - \Psi(-\ell)] \approx -2\overline{V}_{i,j}\ell(d\Psi/dz)_0, \tag{4.1.2}$$

where $\overline{V}_{i,j}$ is the characteristic collision velocity in the z–direction between planes 0^+ and 0^- given by Eq. (2.2.7), and the negative sign indicates that the direction of transport is opposite to that of the property gradient.

If we next assume that complete exchange and equalization of properties between molecules occur within one collision, then ℓ can be identified as the molecular mean free path given by

$$\ell = \frac{1}{n\pi\sigma_{i,j}^2}, \tag{4.1.3}$$

where n is the total number density of the molecules.

Substituting Eqs. (4.1.3) and (2.2.7) into Eq. (4.1.2), we have

$$F_\Psi = -2\left(\frac{8k^o T}{\pi m_{i,j}}\right)^{1/2}\left(\frac{1}{n\pi\sigma_{i,j}^2}\right)\left(\frac{d\Psi}{dz}\right), \tag{4.1.4}$$

where we have omitted the subscript 0 in $(d\Psi/dz)$ because it is an arbitrary reference plane anyway.

Specializing now to the transport of species, momentum, and internal energy, we have respectively $\Psi = n_i m_{i,j}$, $nm_{i,j}v_y$, and $nm_{i,j}e(T)$. Substituting these values into Eq. (4.1.4), and recognizing that while n_i varies with z during species transport, n is approximately constant, and that $de/dz = (de/dT)(dT/dz) = c_v(dT/dz)$, we have

$$F_{\text{species } i} = -D_{i,j}\frac{d\rho_i}{dz}, \quad F_{\text{momentum}} = -\mu_{i,j}\frac{dv_y}{dz}, \quad F_{\text{energy}} = -\lambda_{i,j}\frac{dT}{dz}, \tag{4.1.5}$$

where we have defined:

Mass Diffusion Coefficient:

$$D_{i,j} = \frac{2(8m_{i,j}k^o T/\pi)^{1/2}}{\rho(\pi\sigma_{i,j}^2)} \tag{4.1.6}$$

Viscosity Coefficient:

$$\mu_{i,j} = \frac{2(8m_{i,j}k^o T/\pi)^{1/2}}{\pi\sigma_{i,j}^2} \tag{4.1.7}$$

Thermal Conductivity Coefficient:

$$\lambda_{i,j} = \frac{2(8m_{i,j}k^o T/\pi)^{1/2}c_v}{\pi\sigma_{i,j}^2}. \tag{4.1.8}$$

The equations in (4.1.5) can be generalized to their respective vector forms by substituting (d/dz) by the gradient operator ∇. They are respectively known as Fick's law of mass diffusion, Newton's law of viscosity, and Fourier's law of heat conduction. These laws state that the diffusional flux of a transported quantity is proportional to its gradient.

In the following discussion, we shall treat μ and λ as the average properties of an effectively one-component gas, with $m_{i,j} = m/2$. For mass diffusion, the identities of the diffusing species are essential and the subscripts i, j need to be retained.

4.1.2. Discussion on Diffusion Coefficients

First it is significant to note that, for the ideal gases treated herein, μ and λ are independent of the mass density ρ of the gas. This interesting phenomenon is due to the fact that if we, say, double the number of molecules between the plates, then even though there are now twice as many molecules available for momentum transport, the mean free path of each molecule is halved such that the transfer is only half as effective. Therefore, the net rate of transfer is unchanged.

The mass diffusion coefficient $D_{i,j}$ does depend on density through $D_{i,j} \sim \rho^{-1}$. However, the product $\rho D_{i,j}$ is independent of density. Indeed, had we used (ρ_i/ρ) instead of ρ_i in the definition for $F_{\text{species } i}$ in (4.1.5), the appropriate coefficient that emerges would be $\rho D_{i,j}$. It will be further demonstrated in later studies that the relevant parameter controlling the diffusion process in many mass transport problems is $\rho D_{i,j}$ instead of $D_{i,j}$.

It is seen that $\rho D_{i,j}$, μ, and λ are also independent of pressure, which implies that the diffusion rates of mass, momentum, and energy are correspondingly pressure insensitive. These coefficients, however, do depend on temperature through the factor $T^{1/2}$ in the present simplified formulation. Results from more rigorous formulation show a stronger temperature dependence. However, if quantitative accuracy is not essential for a particular situation, then it is convenient in analytical modeling to assume that they are constants.

4.1.3. Characteristic Diffusion Rates and Nondimensional Numbers

Associated with each diffusion process is the characteristic rate of spreading, which has the dimension of length²/sec because we are dealing with the surface of the "sphere of influence." For mass diffusion this is simply $D_{i,j}$. For viscous spreading the relevant coefficient is the kinematic viscosity,

$$\nu = \frac{\mu}{\rho}, \tag{4.1.9}$$

while for thermal conduction we have the thermal diffusivities,

$$\alpha_h = \frac{\lambda}{c_p \rho}, \tag{4.1.10}$$

where c_p is the appropriate specific heat to use. Note that $D_{i,j}$, ν, and α_h all vary with ρ^{-1}, and hence the system pressure.

Taking the ratios of these characteristic spreading rates, we can define a Schmidt number

$$Sc_{i,j} = \frac{\nu}{D_{i,j}} = \frac{\mu}{\rho D_{i,j}}, \tag{4.1.11}$$

which is a measure of the relative influence of viscosity to mass diffusion, a Prandtl number

$$Pr = \frac{\nu}{\alpha_h} = \frac{\mu c_p}{\lambda}, \tag{4.1.12}$$

which is a measure of the relative influence of viscosity to thermal diffusion, and a Lewis number

$$Le_{i,j} = \frac{\alpha_h}{D_{i,j}} = \frac{Sc_{i,j}}{Pr} = \frac{\lambda}{c_p \rho D_{i,j}}, \tag{4.1.13}$$

which is a measure of the relative influence of thermal to mass diffusion. These numbers assume values close to unity, and are almost constants, being quite insensitive to variations in temperature. When these numbers do assume the value of unity, the conservation equations describing the transport of heat, mass, and momentum bear great similarity to each other, leading to the existence of conserved quantities and substantial facilitation of analysis, as will be demonstrated in Chapter 5.

4.1.4. Second-Order Diffusion

The above discussion was concerned with heat diffusion in the presence of a temperature gradient and mass diffusion in the presence of a concentration gradient. However, it has been observed, as well as predicted by the kinetic theory of gases, that heat diffusion can occur in the presence of a concentration gradient while mass diffusion can occur in the presence of a temperature gradient. These second-order processes are respectively known as Dufour and Soret effects.

The possible existence of second-order diffusion can be appreciated by recognizing that the presence of a temperature gradient in a mixture of uniform concentration renders the molecules in the high temperature region to have a larger random velocity than the molecules in the low temperature region. A diffusive flux then results as the mixture attempts to achieve equilibrium by having the light molecules migrating to the hot region so as to increase its random kinetic energy, and the heavy molecules migrating to the cold region to reduce this energy.

The second-order diffusion processes are represented by the thermal diffusion coefficients, $D_{T,i}$, which are generally much smaller than the first order, Fickian, diffusion coefficients. Prominent exceptions that are of particular interest to combustion are the counter-gradient diffusion of the low-molecular-weight hydrogen and gradient diffusion of the "high-molecular-weight" soot particles in the presence of steep temperature gradients around flames. The Dufour effect is usually quite small such that it is negligible.

Species diffusion can also occur in the presence of body forces and pressure gradients. These effects are again usually small unless these forces are extremely large.

4.2. SOME USEFUL RESULTS FROM KINETIC THEORY OF GASES

4.2.1. General Concepts

The above phenomenological derivation adopts a simplified description of the molecular velocity and collision frequency, without considering the energetics of collision and the distribution of the molecular velocities. The statistical aspects of particle

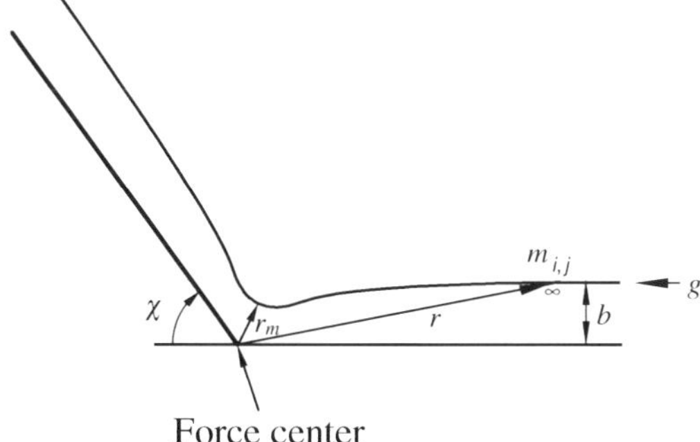

Figure 4.2.1. Binary particle collision in a central force field in the force-center coordinate.

collision are properly accounted for in the kinetic theory of gases. Binary collision is still assumed, implying that the gas is sufficiently dilute such that collisions involving three or more molecules are rare and their effects negligible. Furthermore, the collision is assumed to be elastic and inert, without internal excitation or change of the molecular structure.

In the simplest analysis, the nature of the molecular structure and the collision dynamics are assumed to be described by a central force field represented by a force potential $\phi(r)$, where r is the radial distance away from the molecule. Thus strictly speaking the theory requires that the molecules are point masses in that the collisional distances are much larger than the sizes of the molecules. The spherically symmetric force field also requires that the gas is monatomic such that the molecules actually do not have any structure. Violation of this requirement is especially serious for polar molecules with dipole moments (e.g., water). However, as we shall see, these nonidealities can be approximately accounted for through the use of special forms of force potentials and the associated force constants.

Thus once $\phi(r)$ is given, the deflection angle $\chi(b, g)$ for a given collision (Figure 4.2.1) can be shown from classical mechanics (Goldstein 1980) to be

$$\chi(b, g) = \pi - 2b \int_{r_m}^{\infty} \left(1 - \frac{\phi}{m_{i,j} g^2/2} - \frac{b^2}{r^2}\right)^{-1/2} \frac{dr}{r^2}, \tag{4.2.1}$$

where g is the relative approach velocity, b the impact parameter, which is the distance of closest approach if there were no potential, and r_m the minimum approach distance given by

$$\phi(r_m) = \frac{m_{i,j} g^2}{2} \left(1 - \frac{b^2}{r_m^2}\right).$$

Summing over all possible collision events characterized by g and b, the fluxes of transport can be described by the collision integrals $\Omega_{i,j}^{(k,\ell)}$ defined as

$$\Omega_{i,j}^{(k,\ell)} = \left(\frac{2\pi k^o T}{m_{i,j}}\right)^{1/2} \int_0^{\infty} \int_0^{\infty} e^{-\hat{g}^2} \hat{g}^{(2\ell+3)} \left[1 - (\cos\chi)^k\right] b\, db\, d\hat{g}, \tag{4.2.2}$$

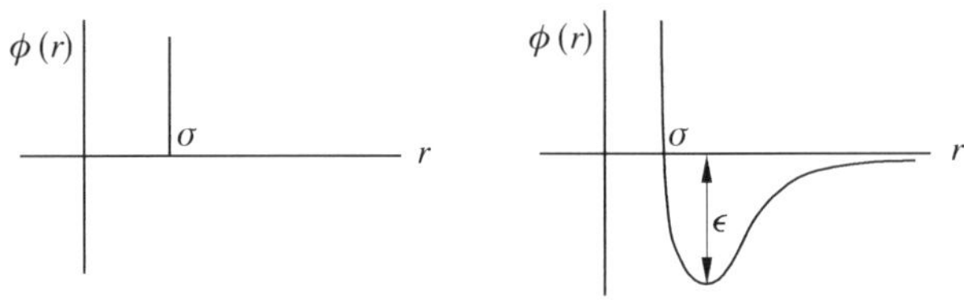

(a) Hard-sphere potential
$\phi(r) = \infty \quad r < \sigma$
$\phi(r) = 0 \quad r > \sigma$

(b) Lennard–Jones 6–12 potential
$\phi(r) = 4\epsilon\,[(\frac{\sigma}{r})^{12} - (\frac{\sigma}{r})^{6}]$

Figure 4.2.2. Force potential for (a) hard-sphere potential, and (b) Lennard–Jones 6–12 potential.

where $\hat{g} = g(m_{i,j}/2k^oT)^{1/2}$. Combinations of different integers k and ℓ indicate different modes of transport.

4.2.2. Collision Potentials and Integrals

An evaluation of the integral in $\Omega_{i,j}^{(k,\ell)}$ requires a knowledge of the force potential $\phi(r)$. The force potential for the hard sphere collision, shown in Figure 4.2.2a, is given by

$$\begin{aligned} \phi(r) &= \infty \qquad r < \sigma \\ &= 0 \qquad r > \sigma, \end{aligned} \tag{4.2.3}$$

where σ is the sphere diameter.

A more realistic, frequently used collision model for nonpolar gases is the Lennard–Jones 6–12 potential, given by

$$\phi(r) = 4\epsilon\left[\left(\frac{\sigma}{r}\right)^{12} - \left(\frac{\sigma}{r}\right)^{6}\right], \tag{4.2.4}$$

and shown in Figure 4.2.2b. The potential is the combination of a sixth-power attraction and twelfth-power repulsion. The sixth-power attractive force represents the induced-dipole interaction between two nonpolar molecules, while the twelfth-power repulsive force arises when molecules are close to each other such that their outer electronic shells interact repulsively. Here ϵ is the characteristic collision energy and σ can be interpreted as the effective molecular size.

Now if the collision integral $\Omega_{i,j}^{(k,\ell)}$ is nondimensionalized by that of the hard sphere, and the temperature by ϵ/k^o, then we can define

$$\begin{aligned} \Omega_{i,j}^{(k,\ell)*}(T^*) &= \Omega_{i,j}^{(k,\ell)} \Big/ \left[\Omega_{i,j}^{(k,\ell)}\right]_{\text{hard sphere}} \\ T^* &= k^oT/\epsilon_{i,j}. \end{aligned} \tag{4.2.5}$$

Thus, for a given potential, the nondimensional $\Omega_{i,j}^{(k,\ell)*}$ is only a function of T^* and can be tabulated for different combinations of (k, ℓ).

The above discussions are for nonpolar gases. For polar gases the Stockmayer potential,

$$\phi(r) = 4\epsilon\left[\left(\frac{\sigma}{r}\right)^{12} - \left(\frac{\sigma}{r}\right)^{6}\right] - \frac{d_i d_j}{r^3}\zeta, \tag{4.2.6}$$

is frequently used, where

$$\zeta = 2\cos\theta_i \cos\theta_j - \sin\theta_i \sin\theta_j \cos\beta,$$

d_i and d_j are the dipole moments of the two interacting molecules, θ_i and θ_j the angles of inclination of the axes of the two dipoles to the line joining the centers of these molecules, and β the azimuthal angle between them. Integrating over θ_i, θ_j, and β, it is found that the nondimensional collision integrals now depend on T^* as well as an additional parameter

$$\delta^* = \frac{1}{2}(d^*)^2 = \frac{1}{2}\frac{d^2}{\epsilon\sigma^3}. \tag{4.2.7}$$

Tables 4.1 and 4.2 respectively list the collision integrals $\Omega^{(1,1)*}(T^*;\delta^*)$ and $\Omega^{(1,2)*}(T^*;\delta^*)$ needed to calculate the coefficients of binary diffusion, viscosity, and thermal conductivity to be specified later; the case of $\delta^* = 0$ corresponds to nonpolar gases. Note that in expressing Ω we have omitted the indices i and j.

The effects of different molecules are accounted for by defining effective force constants

$$\sigma_{i,j} = \frac{1}{2}(\sigma_i + \sigma_j), \qquad \epsilon_{i,j} = \sqrt{\epsilon_i \epsilon_j}, \qquad d_{i,j} = \sqrt{d_i d_j}, \tag{4.2.8}$$

which hold for either polar–polar or nonpolar–nonpolar interactions. The force constants for the collision between a polar gas (p) and a nonpolar gas (n) are given by

$$\sigma_{n,p} = \frac{1}{2}(\sigma_n + \sigma_p)\xi^{-1/6}, \qquad \epsilon = \sqrt{\epsilon_n \epsilon_p}\xi^2, \qquad \delta^* = \frac{1}{2}d_p^{*2}, \tag{4.2.9}$$

where

$$\xi = \left[1 + \frac{1}{4}\alpha_n^* d_p^* \sqrt{\frac{\epsilon_p}{\epsilon_n}}\right], \qquad \alpha_n^* = \frac{\alpha_n}{\sigma_n^3}, \qquad d_p^{*2} = \frac{d_p^2}{\epsilon_p \sigma_p^3}, \tag{4.2.10}$$

and α_n is the polarizability of the nonpolar molecule, (σ_n, ϵ_n) the Lennard–Jones force constants, and (σ_p, ϵ_p) the Stockmayer force constants.

Table 4.3 lists the potential parameters for nonpolar and polar gases often encountered in combustion. These values have been evaluated by fitting experimentally determined transport properties.

Table 4.1. Collision integral $\Omega^{(1,1)*}(T^*;\delta^*)$

T^*	$\delta^*=0$	0.25	0.50	0.75	1.0	1.5	2.0	2.5
0.1	4.0079	4.002	4.665	5.521	6.454	8.214	9.824	11.31
0.2	3.1300	3.164	3.355	3.721	4.198	5.230	6.225	7.160
0.3	2.6494	2.657	2.770	3.002	3.319	4.054	4.785	5.483
0.4	2.3144	2.320	2.402	2.572	2.812	3.386	3.972	4.539
0.5	2.0661	2.073	2.140	2.278	2.472	2.946	3.437	3.918
0.6	1.8767	1.885	1.944	2.060	2.225	2.628	3.054	3.474
0.7	1.7293	1.738	1.791	1.893	2.036	2.388	2.763	3.137
0.8	1.6122	1.622	1.670	1.760	1.886	2.198	2.535	2.872
0.9	1.5175	1.527	1.572	1.653	1.765	2.044	2.349	2.657
1.0	1.4398	1.450	1.490	1.564	1.665	1.917	2.196	2.478
1.2	1.3204	1.330	1.364	1.425	1.509	1.720	1.956	2.199
1.4	1.2336	1.242	1.272	1.324	1.394	1.573	1.777	1.990
1.6	1.1679	1.176	1.202	1.246	1.306	1.461	1.639	1.827
1.8	1.1166	1.124	1.146	1.185	1.237	1.372	1.530	1.698
2.0	1.0753	1.082	1.102	1.135	1.181	1.300	1.441	1.592
2.5	1.0006	1.005	1.020	1.046	1.080	1.170	1.278	1.397
3.0	0.95003	0.9538	0.9656	0.9852	1.012	1.082	1.168	1.265
3.5	0.91311	0.9162	0.9256	0.9413	0.9626	1.019	1.090	1.170
4.0	0.88453	0.8871	0.8948	0.9076	0.9252	0.9721	1.031	1.098
5.0	0.84277	0.8446	0.8501	0.8592	0.8716	0.9053	0.9483	0.9984
6.0	0.81287	0.8142	0.8183	0.8251	0.8344	0.8598	0.8927	0.9316
7.0	0.78976	0.7908	0.7940	0.7993	0.8066	0.8265	0.8526	0.8836
8.0	0.77111	0.7720	0.7745	0.7788	0.7846	0.8007	0.8219	0.8474
9.0	0.75553	0.7562	0.7584	0.7619	0.7667	0.7800	0.7976	0.8189
10.0	0.74220	0.7428	0.7446	0.7475	0.7515	0.7627	0.7776	0.7957
12.0	0.72022	0.7206	0.7220	0.7241	0.7271	0.7354	0.7464	0.7600
14.0	0.70254	0.7029	0.7039	0.7055	0.7078	0.7142	0.7228	0.7334
16.0	0.68776	0.6880	0.6880	0.6901	0.6919	0.6970	0.7040	0.7125
18.0	0.67510	0.6753	0.6760	0.6770	0.6785	0.6827	0.6884	0.6955
20.0	0.66405	0.6642	0.6648	0.6657	0.6669	0.6704	0.6752	0.6811
25.0	0.64136	0.6415	0.6418	0.6425	0.6433	0.6457	0.6490	0.6531
30.0	0.62350	0.6236	0.6239	0.6243	0.6249	0.6267	0.6291	0.6321
35.0	0.60882	0.6089	0.6091	0.6094	0.6099	0.6112	0.6131	0.6154
40.0	0.59640	0.5964	0.5966	0.5969	0.5972	0.5983	0.5998	0.6017
50.0	0.57626	0.5763	0.5764	0.5766	0.5768	0.5775	0.5785	0.5798
75.0	0.54146	0.5415	0.5416	0.5416	0.5418	0.5421	0.5424	0.5429
100.0	0.51803	0.5181	0.5182	0.5184	0.5184	0.5185	0.5186	0.5187

Finally, for nonpolar gases ($\delta^*=0$), $\Omega^{(1,1)*}(T^*)$ and $\Omega^{(2,2)*}(T^*)$ have been fitted (Monchick & Mason 1961) to excellent accuracy by the formulas

$$\Omega^{(1,1)*}(T^*) = 1.069(T^*)^{-0.1580} + 0.3445e^{-0.6537T^*} + 1.556e^{-2.099T^*} + 1.976e^{-6.488T^*}, \tag{4.2.11}$$

$$\Omega^{(2,2)*}(T^*) = 1.155(T^*)^{-0.1462} + 0.3945e^{-0.6672T^*} + 2.05e^{-2.168T^*}, \tag{4.2.12}$$

valid in the range $0.30 \le T^* \le 400$.

Table 4.2. Collision integral $\Omega^{(2,2)*}(T^*; \delta^*)$

T^*	$\delta^* = 0$	0.25	0.50	0.75	1.0	1.5	2.0	2.5
0.1	4.1005	4.266	4.833	5.742	6.729	8.624	10.34	11.89
0.2	3.2626	3.305	3.516	3.914	4.433	5.570	6.637	7.618
0.3	2.8399	2.836	2.936	3.168	3.511	4.329	5.126	5.874
0.4	2.5310	2.522	2.586	2.749	3.004	3.640	4.282	4.895
0.5	2.2837	2.277	2.329	2.460	2.665	3.187	3.727	4.249
0.6	2.0838	2.081	2.130	2.243	2.417	2.862	3.329	3.786
0.7	1.9220	1.924	1.970	2.072	2.225	2.614	3.028	3.435
0.8	1.7902	1.795	1.840	1.934	2.070	2.417	2.788	3.156
0.9	1.6823	1.689	1.733	1.820	1.944	2.258	2.596	2.933
1.0	1.5929	1.601	1.644	1.725	1.838	2.124	2.435	2.746
1.2	1.4551	1.465	1.504	1.574	1.670	1.913	2.181	2.451
1.4	1.3551	1.365	1.400	1.461	1.544	1.754	1.989	2.228
1.6	1.2800	1.289	1.321	1.374	1.447	1.630	1.838	2.053
1.8	1.2219	1.231	1.259	1.306	1.370	1.532	1.718	1.912
2.0	1.1757	1.184	1.209	1.251	1.307	1.451	1.618	1.795
2.5	1.0933	1.100	1.119	1.150	1.193	1.304	1.435	1.578
3.0	1.0388	1.044	1.059	1.083	1.117	1.204	1.310	1.428
3.5	0.99963	1.004	1.016	1.035	1.062	1.133	1.220	1.319
4.0	0.96988	0.9732	0.9830	0.9991	1.021	1.079	1.153	1.236
5.0	0.92676	0.9291	0.9360	0.9473	0.9628	1.005	1.058	1.121
6.0	0.89616	0.8979	0.9030	0.9114	0.9230	0.9545	0.9955	1.044
7.0	0.87272	0.8741	0.8780	0.8845	0.8935	0.9181	0.9505	0.9893
8.0	0.85379	0.8549	0.8580	0.8632	0.8703	0.8901	0.9164	0.9482
9.0	0.83795	0.8388	0.8414	0.8456	0.8515	0.8678	0.8895	0.9160
10.0	0.82435	0.8251	0.8273	0.8308	0.8356	0.8493	0.8676	0.8901
12.0	0.80184	0.8024	0.8039	0.8065	0.8101	0.8201	0.8337	0.8504
14.0	0.78363	0.7840	0.7852	0.7872	0.7899	0.7976	0.8081	0.8212
16.0	0.76834	0.7687	0.7696	0.7712	0.7733	0.7794	0.7878	0.7983
18.0	0.75518	0.7554	0.7562	0.7575	0.7592	0.7642	0.7711	0.7797
20.0	0.74364	0.7438	0.7445	0.7455	0.7470	0.7512	0.7569	0.7642
25.0	0.71982	0.7200	0.7204	0.7211	0.7221	0.7250	0.7289	0.7339
30.0	0.70097	0.7011	0.7014	0.7019	0.7026	0.7047	0.7076	0.7112
35.0	0.68545	0.6855	0.6858	0.6861	0.6867	0.6883	0.6905	0.6932
40.0	0.67232	0.6724	0.6726	0.6728	0.6733	0.6745	0.6762	0.6784
50.0	0.65099	0.6510	0.6512	0.6513	0.6516	0.6524	0.6534	0.6546
75.0	0.61397	0.6141	0.6143	0.6145	0.6147	0.6148	0.6148	0.6147
100.0	0.58870	0.5889	0.5894	0.5900	0.5903	0.5901	0.5895	0.5885

4.2.3. Transport Coefficients

Knowing the collision integrals, the various transport coefficients can be calculated. Thus the viscosity coefficient for a single component gas is given by

$$\mu_i = \frac{5}{16} \frac{\sqrt{\pi m_i k^o T}}{\pi \sigma_i^2 \Omega_{i,i}^{(2,2)*}(T^*; \delta^*)}. \tag{4.2.13}$$

Expressing Eq. (4.2.13) in practical units, we have

$$\mu_i = 2.6693 \times 10^{-6} \frac{(W_i T)^{1/2}}{\sigma_i^2 \Omega_{i,i}^{(2,2)*}(T^*; \delta^*)} \quad \text{(kg/m-s)}, \tag{4.2.14}$$

Table 4.3. Potential parameters for stable and unstable species often encountered in combustion processes

Species	n	ϵ/k^o (K)	σ (Å)	d (debyes)	δ^*	α (Å^3)	$\alpha_n^* \times 10^2$
Ar	0	136.500	3.330	0.000	0.000	0.000	0.000
CH	1	80.000	2.750	0.000	0.000	0.000	0.000
CH_2	1	144.000	3.800	0.000	0.000	0.000	0.000
CH_2CO	2	436.000	3.970	0.000	0.000	0.000	0.000
CH_2O	2	498.000	3.590	0.000	0.000	0.000	0.000
CH_3	1	144.000	3.800	0.000	0.000	0.000	0.000
CH_3CHO	2	436.000	3.970	0.000	0.000	0.000	0.000
CH_3CO	2	436.000	3.970	0.000	0.000	0.000	0.000
CH_3O	2	417.000	3.690	1.700	0.500	0.000	0.000
CH_4	2	141.000	3.746	0.000	0.000	2.600	4.946
CH_4O	2	417.000	3.690	1.700	0.500	0.000	0.000
C_2H	1	209.000	4.100	0.000	0.000	0.000	0.000
C_2H_2	1	209.000	4.100	0.000	0.000	0.000	0.000
C_2H_2OH	2	224.700	4.162	0.000	0.000	0.000	0.000
C_2H_3	1	209.000	4.100	0.000	0.000	0.000	0.000
C_2H_4	2	280.000	3.971	0.000	0.000	0.000	0.000
C_2H_5	2	252.300	4.302	0.000	0.000	0.000	0.000
C_2H_6	2	252.300	4.302	0.000	0.000	0.000	0.000
C_3H_2	2	209.000	4.100	0.000	0.000	0.000	0.000
C_3H_3	1	252.000	4.760	0.000	0.000	0.000	0.000
C_3H_4	1	252.000	4.760	0.000	0.000	0.000	0.000
C_3H_6	2	266.800	4.982	0.000	0.000	0.000	0.000
C_3H_7	2	266.800	4.982	0.000	0.000	0.000	0.000
C_3H_8	2	266.800	4.982	0.000	0.000	0.000	0.000
C_4H	1	357.000	5.180	0.000	0.000	0.000	0.000
C_4H_2	1	357.000	5.180	0.000	0.000	0.000	0.000
C_4H_2OH	2	224.700	4.162	0.000	0.000	0.000	0.000
C_3H_3	1	357.000	5.180	0.000	0.000	0.000	0.000
C_4H_4	1	357.000	5.180	0.000	0.000	0.000	0.000
C_4H_8	2	357.000	5.176	0.000	0.000	0.000	0.000
C_4H_9	2	357.000	5.176	0.000	0.000	0.000	0.000
CO	1	98.100	3.650	0.000	0.000	1.950	4.010
CO_2	1	244.000	3.763	0.000	0.000	2.650	4.973
F	0	80.000	2.750	0.000	0.000	0.000	0.000
F_2	1	125.700	3.301	0.000	0.000	1.600	4.448
H	0	145.000	2.050	0.000	0.000	0.000	0.000
H_2	1	38.000	2.920	0.000	0.000	0.790	3.173
H_2O	2	572.400	2.605	1.844	1.217	0.000	0.000
H_2O_2	2	107.400	3.458	0.000	0.000	0.000	0.000
He	0	10.200	2.576	0.000	0.000	0.000	0.000
HF	1	352.000	2.490	1.730	1.995	0.000	0.000
HCCO	2	150.000	2.500	0.000	0.000	0.000	0.000
HCO	1	498.000	3.590	0.000	0.000	0.000	0.000
HNO	1	116.700	3.492	0.000	0.000	0.000	0.000
HNNO	2	232.400	3.828	0.000	0.000	0.000	0.000
HO_2	1	107.400	3.458	0.000	0.000	0.000	0.000
N	0	71.400	3.298	0.000	0.000	0.000	0.000

Species	n	ϵ/k^o (K)	σ (Å)	d (debyes)	δ^*	α (Å^3)	$\alpha_n^* \times 10^2$
N_2	1	97.530	3.621	0.000	0.000	1.760	3.707
N_2H_2	2	71.400	3.798	0.000	0.000	0.000	0.000
N_2H_3	2	200.000	3.900	0.000	0.000	0.000	0.000
N_2H_4	2	205.000	4.230	0.000	0.000	4.260	5.628
N_2O	1	232.400	3.828	0.000	0.000	0.000	0.000
NH	1	80.000	2.650	0.000	0.000	0.000	0.000
NH_2	2	80.000	2.650	0.000	0.000	2/260	12.144
NH_3	2	481.000	2.920	1.470	0.653	0.000	0.000
NNH	2	71.400	3.798	0.000	0.000	0.000	0.000
NO	1	97.530	3.621	0.000	0.000	1,760	3.707
NO_2	2	200.000	3.500	0.000	0.000	0.000	0.000
O	0	80.000	2.750	0.000	0.000	0.000	0.000
O_2	1	107.400	3.458	0.000	0.000	1.600	3.869
O_3	2	180.000	4.100	0.000	0.000	0.000	0.000
OH	1	80.000	2.750	0.000	0.000	0.000	0.000

Nomenclature of Table:

$n = 0$: Monatomic molecule
$n = 1$: Linear molecule
$n = 2$: Nonlinear molecule
ϵ/k^o : Lennard–Jones potential well depth
σ : Lennard–Jones collison diameter
d : Dipole moment
α : Molecular polarizability

where W_i is the molecular weight and T and σ_i are respectively expressed in units of K and Å.

The thermal conductivity coefficient is related to μ_i through

$$\lambda_{i,\text{mono}} = \frac{15}{4}\frac{k^o}{m_i}\mu_i \tag{4.2.15a}$$

$$= \frac{5}{2}c_{v,i,\text{mono}}\mu_i \tag{4.2.15b}$$

where we have used the subscript "mono" to indicate that the relation is only for a monatomic gas because the theory assumes hard sphere collision.

It is of interest to note that Eqs. (4.1.7) and (4.1.8) yield

$$\lambda_i = c_{v,i}\mu_i, \tag{4.2.16}$$

which is different from Eq. (4.2.15b). The reason for this difference is that in deriving Eqs. (4.1.7) and (4.1.8), and hence Eq. (4.2.16), we have drawn analogy between transfers in momentum and internal energy, without considering the collision dynamics. On the other hand, Eq. (4.2.15b) is derived based on the collision dynamics of two structureless hard spheres, without considering the internal energy. Thus the energy transferred is that associated with translation. Consequently, the specific heat in Eq. (4.2.15b) is that for the translational degrees of freedom, $\frac{3}{2}(k^\circ/m_i)$, while the specific heat in Eq. (4.2.16) is that associated with the internal degrees of freedom, which is $c_v - \frac{3}{2}(k^o/m_i)$. Thus λ_i can be approximated as the sum of Eqs. (4.2.15b) and (4.2.16), with the appropriate specific heats used for the two different modes of energy transfer,

$$\lambda_i = \frac{5}{2}\left(\frac{3}{2}\frac{k^o}{m_i}\right)\mu_i + \left(c_{v,i} - \frac{3}{2}\frac{k^o}{m_i}\right)\mu_i, \tag{4.2.17}$$

which can be written as

$$\lambda_i = \frac{9\gamma - 5}{4} c_{v,i}\mu_i \tag{4.2.18a}$$

$$= \lambda_{i,\text{mono}}\left[\frac{3}{5} + \frac{4}{15(\gamma - 1)}\right] \tag{4.2.18b}$$

by using the relation $c_{p,i} - c_{v,i} = k^o/m_i$. Equation (4.2.18) is the Eucken formula for polyatomic gases. It is fairly accurate for most diatomic gases at moderate densities, but is a poor representation for larger molecules.

An alternative expression for λ_i is the Hirschfelder's formula,

$$\lambda_i = \lambda_{i,\text{mono}}\left(0.115 + 0.354\frac{\gamma}{\gamma - 1}\right), \tag{4.2.19}$$

which has been found to be quite accurate over a large range of T^* for a number of realistic potential functions.

Using the Eucken formula, we also have

$$Pr = \frac{4\gamma}{9\gamma - 5}. \tag{4.2.20}$$

Thus $Pr = \frac{2}{3}$ and 0.74 for monatomic ($\gamma = \frac{5}{3}$) and diatomic ($\gamma = \frac{7}{5}$) gases respectively.

The binary diffusion coefficient between species i and j is given by

$$D_{i,j} = \frac{3}{16}\frac{\sqrt{2\pi(k^oT)^3/m_{i,j}}}{p\pi\sigma_{i,j}^2\Omega_{i,j}^{(1,1)*}(T^*, \delta^*)}, \tag{4.2.21}$$

which becomes, in practical units,

$$D_{i,j} = 1.8583 \times 10^{-7}\frac{\left[T^3(W_i + W_j)/W_iW_j\right]^{1/2}}{p\sigma_{i,j}^2\Omega_{i,j}^{(1,1)*}(T^*; \delta^*)}(\text{m}^2/\text{s}), \tag{4.2.22}$$

where p is the pressure expressed in atmospheres.

For multicomponent mixtures the viscosity and thermal conductivity coefficients are respectively given by the approximate expressions

$$\mu_{\text{mix}} = \sum_{i=1}^{N} \frac{\mu_i}{1 + \frac{1}{X_i}\sum_{j \neq i} X_j \Phi_{i,j}} \tag{4.2.23}$$

$$\lambda_{\text{mix}} = \sum_{i=1}^{N} \frac{\lambda_i}{1 + \frac{1.065}{X_i}\sum_{j \neq i} X_j \Phi_{i,j}}. \tag{4.2.24}$$

where

$$\Phi_{i,j} = \frac{1}{\sqrt{8}}\left(1 + \frac{W_i}{W_j}\right)^{-1/2}\left[1 + \left(\frac{\mu_i}{\mu_j}\right)^{1/2}\left(\frac{W_j}{W_i}\right)^{1/4}\right]^2. \tag{4.2.25}$$

Finally, the diffusion coefficient of a species i with a very dilute concentration in a given mixture can be approximated by

$$D_i \approx \frac{1 - Y_i}{\sum_{j \neq i} \frac{X_j}{D_{i,j}}}. \tag{4.2.26}$$

The complete diffusion equation is discussed in Chapter 5.

PROBLEMS

1. Review the problem of two-body collision in a central force field discussed in text books on classical mechanics (e.g., Goldstein 1980). Then derive Eq. (4.2.1).

2. Show that for a hard sphere potential (Hirschfelder, Curtiss & Bird 1954, pp. 523–546):

$$\begin{aligned} \chi(b, g) &= 2\cos^{-1}(b/\sigma) \qquad & b \leq \sigma \\ &= 0 & b \geq \sigma, \end{aligned}$$

$$Q^{(k)} = \int_0^\infty \left[1 - (\cos\chi)^k\right] b\, db = \left[1 - \frac{1}{2}\frac{1 + (-1)^k}{1 + k}\right]\frac{\sigma^2}{2}$$

$$\Omega_{i,j}^{(k,\ell)} = \left(\frac{2\pi k^o T}{m_{i,j}}\right)^{1/2} \frac{(\ell + 1)!}{2} Q^{(k)}.$$

3. For propane–air mixtures at 1 atm pressure and with an initial temperature 298.15 K, calculate the following for $\phi = 0.55$ and 2.25.

   (a) The adiabatic flame temperatures and the species molar fractions, using the major-minor species model and accounting for only the major species.

   (b) The pure component transport coefficients μ_i, λ_i, the binary diffusion coefficients $D_{i,j}$, the mixture transport coefficients $\mu_{\text{mix}}, \lambda_{\text{mix}}$, and the Prandtl number Pr, for both the initial and final states.

(c) The diffusion coefficient D_i of each species in the mixture, assuming it is a trace component.

(d) The $Sc_{i,j}$ and $Le_{i,j}$ between the deficient reactant (i.e., fuel for $\phi < 1$ and oxygen for $\phi > 1$) and the abundant species of the initial mixture.

In (b), you need to use the heat capacities for propane, which are tabulated as follows (T in K, c_p in cal/mol-K)*:

T	298.15	300	400	500	600	700	800	900
c_p	17.627	17.725	22.669	27.040	30.884	34.247	37.174	39.713
T	1000	1100	1200	1300	1400	1500		
c_p	41.908	43.807	45.455	46.897	48.181	49.352		

* Source: Barin, I. 1989. *Thermochemical Data of Pure Substances*, Part 1, VCH.

5 Conservation Equations

The dynamics and thermodynamics of a chemically reacting flow are governed by the conservation laws of mass, momentum, energy, and the concentration of the individual species. In this chapter, we shall first present a derivation of these conservation equations based on control volume considerations. We shall then derive a simplified form of these equations describing only those effects which are of predominant importance in most of the subsonic combustion phenomena to be studied later. Some useful concepts and analytical techniques for combustion modeling will be discussed and several important nondimensional numbers will be introduced. Whenever possible, we shall adhere to the nomenclature of Williams (1985) for consistency and ease of referencing. A summary of the symbols is given at the end of this chapter.

Further discussion emphasizing on the mathematical aspects of combustion theory can be found in Buckmaster and Ludford (1982) and Williams (1985).

5.1. CONTROL VOLUME DERIVATION

To derive the various conservation equations, we first take a control volume that is at rest with respect to an inertia reference frame. It has a volume V and a control surface **S**, with a unit normal vector **n**, as shown in Figure 5.1.1. Within this control volume a flow element of velocity **v** passes through. This bulk, mass-weighted velocity **v** is the resultant of the individual velocities $\mathbf{v}_i$ of the various species. Thus, by definition,

$$\sum \rho_i \mathbf{v}_i = \rho \mathbf{v}. \tag{5.1.1}$$

The difference between $\mathbf{v}_i$ and **v** is then the molecular diffusion velocity

$$\mathbf{V}_i = \mathbf{v}_i - \mathbf{v}. \tag{5.1.2}$$

Multiplying Eq. (5.1.2) by ρ_i and summing over i, we have

$$\sum \rho_i \mathbf{V}_i = \sum \rho_i \mathbf{v}_i - \rho \mathbf{v} = 0, \tag{5.1.3}$$

as should be the case. Furthermore, since $Y_i = \rho_i/\rho$, the above results can also be expressed as $\mathbf{v} = \sum Y_i \mathbf{v}_i$ and $\sum Y_i \mathbf{V}_i = 0$.

Consider an extensive fluid property Ψ whose magnitude depends on the size of the control volume V, and its corresponding intensive quantity ψ, which is the "density" of Ψ per unit volume of the fluid. The rate of change of Ψ is then given by the sum of the temporal change of Ψ within V and the loss/gain of Ψ through fluxes across the surface of V. That is,

$$\frac{\delta \Psi}{\delta t} = \frac{\partial}{\partial t}\int_V \psi dV + \int_S \psi(\mathbf{v}\cdot\mathbf{n})dS. \tag{5.1.4}$$

Using the divergence theorem, we can write

$$\int_S \psi(\mathbf{v}\cdot\mathbf{n})dS = \int_V (\nabla\cdot\psi\mathbf{v})dV,$$

which, when substituted into Eq. (5.1.4), yields the general equation describing the rate of change of Ψ as

$$\frac{\delta \Psi}{\delta t} = \int_V \left(\frac{\partial \psi}{\partial t} + \nabla\cdot\psi\mathbf{v}\right)dV. \tag{5.1.5}$$

In the following we shall first apply Eq. (5.1.5) to derive the general conservation equations for mass, individual species concentration, momentum, and energy. Following this we shall also derive the conservation conditions across an interface.

5.1.1. Conservation of Total Mass

If the fluid property Ψ is its total mass m, then ψ is the mass density ρ. Thus Eq. (5.1.5) becomes

$$\frac{\delta m}{\delta t} = \int_V \left(\frac{\partial \rho}{\partial t} + \nabla\cdot\rho\mathbf{v}\right)dV. \tag{5.1.6}$$

Since matter is neither created nor destroyed in a chemical system, $\delta m/\delta t = 0$ along a fluid element. Furthermore, since the control volume is arbitrary, Eq. (5.1.6) implies

$$\boxed{\frac{\partial \rho}{\partial t} + \nabla\cdot(\rho\mathbf{v}) = 0.} \tag{5.1.7}$$

Equation (5.1.7) is the conservation equation for total mass, commonly known as the continuity equation.

5.1.2. Conservation of Individual Species

If Ψ is the mass m_i of the ith species, then ψ is its partial density ρ_i. Equation (5.1.5) then becomes

$$\frac{\delta m_i}{\delta t} = \int_V \left(\frac{\partial \rho_i}{\partial t} + \nabla\cdot\rho_i\mathbf{v}\right)dV. \tag{5.1.8}$$

There are two sources which can lead to a change in m_i. The first is volumetric in nature, caused by the presence of chemical reaction as represented by the rate of

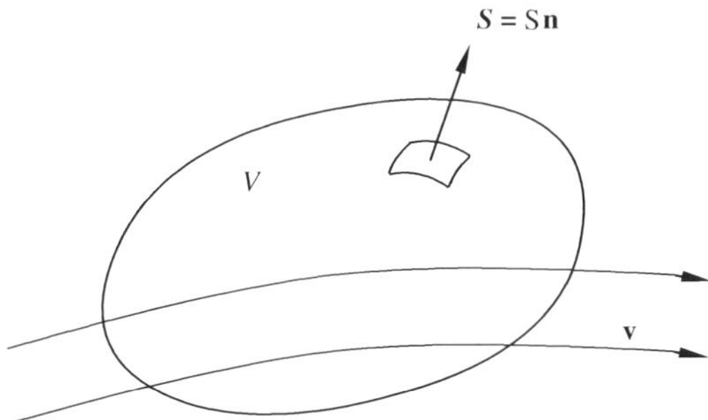

Figure 5.1.1. Control volume for the derivation of the conservation equations.

production of i per unit volume, w_i. The second is a surface process, due to diffusion across the control surface when spatial nonuniformity exists in the concentration of i. This diffusive transport is effected through molecular collision and its magnitude is proportional to the mass flux $\rho_i \mathbf{V}_i$ of the molecular random motion. Thus

$$\frac{\delta m_i}{\delta t} = \int_V w_i dV - \int_S (\rho_i \mathbf{V}_i \cdot \mathbf{n}) dS = \int_V (w_i - \nabla \cdot \rho_i \mathbf{V}_i) dV. \tag{5.1.9}$$

The negative sign for the diffusion term indicates the fact that since $d\mathbf{S}$ is pointed outward (Figure 5.1.1), an outwardly directed $\mathbf{V}_i$ represents a net loss of mass by the control volume. Also note that w_i, which is the mass reaction rate (gm/cm^3-sec), is related to the molar reaction rate $\hat{\omega}_i$ (mole/cm^3-sec) introduced in Chapter 2 through $w_i = W_i \hat{\omega}_i$. Equating (5.1.8) and (5.1.9) yields

$$\frac{\partial \rho_i}{\partial t} + \nabla \cdot [\rho_i (\mathbf{v} + \mathbf{V}_i)] = w_i, \quad i = 1, 2, \ldots, N. \tag{5.1.10}$$

Equation (5.1.10) is the conservation equation for the ith species. Summing Eq. (5.1.10) over the N species, and noting that $\sum_{i=1}^{N} w_i = 0$, we retrieve Eq. (5.1.7). Thus only N of the $(N+1)$ equations given by Eqs. (5.1.7) and (5.1.10) are independent.

Using $Y_i = \rho_i / \rho$, Eq. (5.1.10) can be expressed in two alternate forms as

$$\frac{\partial (\rho Y_i)}{\partial t} + \nabla \cdot [\rho (\mathbf{v} + \mathbf{V}_i) Y_i] = w_i \tag{5.1.11}$$

$$\rho \frac{D Y_i}{Dt} = -\nabla \cdot (\rho \mathbf{V}_i Y_i) + w_i, \tag{5.1.12}$$

where

$$\frac{D}{Dt}(\cdot) = \frac{\partial}{\partial t}(\cdot) + \mathbf{v} \cdot \nabla (\cdot) \tag{5.1.13}$$

is the material derivative, and we have used the continuity relation Eq. (5.1.7). Since density variation is an essential feature in combustion, it is frequently more illuminating to use the density-weighted material derivative, $\rho D(\cdot)/Dt$, as shown in Eq. (5.1.12).

5.1.3. Conservation of Momentum

If Ψ is the momentum $\mathbf{M}$ of the flow, then ψ is the momentum flux $\rho\mathbf{v}$. Thus Eq. (5.1.5) becomes

$$\frac{\delta \mathbf{M}}{\delta t} = \int_V \left(\frac{\partial \rho \mathbf{v}}{\partial t} + \nabla \cdot \rho \mathbf{v}\mathbf{v} \right) dV. \tag{5.1.14}$$

Newton's second law of motion states that the force acting on a system is equal to the rate of change of its momentum. The force can be further divided into a surface force, represented by the stress tensor $\mathbf{P}$, and a volumetric force $\mathbf{f}_i$, frequently called body force, which represents all of the external forces acting on unit mass of the ith species. Thus,

$$\frac{\delta \mathbf{M}}{\delta t} = -\int_S (\mathbf{P} \cdot \mathbf{n})\, dS + \sum_{i=1}^{N} \int_V \rho_i \mathbf{f}_i dV = \int_V \left(-\nabla \cdot \mathbf{P} + \sum_{i=1}^{N} \rho_i \mathbf{f}_i \right) dV. \tag{5.1.15}$$

The negative sign for the stress tensor term indicates the convention that when $\mathbf{P}$ is in the same direction as $d\mathbf{S}$, the system is exerting force on its surrounding. Equating (5.1.14) and (5.1.15), we have

$$\frac{\partial (\rho \mathbf{v})}{\partial t} + \nabla \cdot \rho \mathbf{v}\mathbf{v} = -\nabla \cdot \mathbf{P} + \rho \sum_{i=1}^{N} Y_i \mathbf{f}_i, \tag{5.1.16}$$

which can also be expressed as

$$\rho \frac{D\mathbf{v}}{Dt} = -\nabla \cdot \mathbf{P} + \rho \sum_{i=1}^{N} Y_i \mathbf{f}_i \tag{5.1.17}$$

by using Eq. (5.1.7).

5.1.4. Conservation of Energy

If Ψ is the total internal energy E of the system, which includes the chemical, sensible, and flow kinetic energies, then ψ is $(\rho e + \rho v^2/2)$ because e contains both the sensible and chemical energies. Thus Eq. (5.1.5) becomes

$$\frac{\delta E}{\delta t} = \int_V \left[\frac{\partial \rho (e + v^2/2)}{\partial t} + \nabla \cdot \rho \mathbf{v}(e + v^2/2) \right] dV. \tag{5.1.18}$$

The internal energy of the system can be changed by three sources. The first source, Q, is due to the energy flux $\mathbf{q}$ incident at the boundary of the system,

$$Q = -\int_S (\mathbf{q} \cdot \mathbf{n}) dS = -\int_V \nabla \cdot \mathbf{q} dV. \tag{5.1.19}$$

The second source is the work done on the system by the surface force $\mathbf{F_S}$, or

$$W_S = \int_S \mathbf{v} \cdot d\mathbf{F_S} = -\int_S \mathbf{v} \cdot (\mathbf{P} \cdot \mathbf{n}) dS, \tag{5.1.20}$$

where the negative sign indicates that if $\mathbf{v}$ is pointed in the same outward direction as the surface force $d\mathbf{F}_S$, then work is being done by the system. This term can be further rearranged to show

$$W_S = -\int_S \mathbf{n} \cdot (\mathbf{v} \cdot \mathbf{P}) dS = -\int_V \nabla \cdot (\mathbf{v} \cdot \mathbf{P}) dV. \tag{5.1.21}$$

The third source is the work done by the body forces $\mathbf{F}_{V,i}$ on the various species moving at $\mathbf{v}_i$. Thus

$$\begin{aligned} W_V &= \sum_{i=1}^{N} \int_V \mathbf{v}_i \cdot d\mathbf{F}_{V,i} = \sum_{i=1}^{N} \int \mathbf{v}_i \cdot (\rho_i \mathbf{f}_i) dV \\ &= \sum_{i=1}^{N} \int_V (\mathbf{v} + \mathbf{V}_i) \cdot (\rho_i \mathbf{f}_i) dV. \end{aligned} \tag{5.1.22}$$

Consequently we have

$$\frac{\delta E}{\delta t} = Q + W_S + W_V, \tag{5.1.23}$$

or

$$\begin{aligned} \frac{\partial \rho(e + v^2/2)}{\partial t} + \nabla \cdot \rho \mathbf{v}(e + v^2/2) &= \rho \frac{D(e + v^2/2)}{Dt} \\ &= -\nabla \cdot \mathbf{q} - \nabla \cdot (\mathbf{v} \cdot \mathbf{P}) + \sum_{i=1}^{N} (\mathbf{v} + \mathbf{V}_i) \cdot (\rho_i \mathbf{f}_i). \end{aligned} \tag{5.1.24}$$

Equation (5.1.24) can be cast in a somewhat simpler form. If we take the scalar product of Eq. (5.1.17) with $\mathbf{v}$, we have

$$\mathbf{v} \cdot \rho \frac{D\mathbf{v}}{Dt} = \rho \frac{D}{Dt}\left(\frac{v^2}{2}\right) = -\mathbf{v} \cdot (\nabla \cdot \mathbf{P}) + \rho \mathbf{v} \cdot \sum_{i=1}^{N} Y_i \mathbf{f}_i. \tag{5.1.25}$$

Subtracting Eq. (5.1.25) from Eq. (5.1.24) results

$$\rho \frac{De}{Dt} = \frac{\partial(\rho e)}{\partial t} + \nabla \cdot (\rho \mathbf{v} e) = -\nabla \cdot \mathbf{q} - \mathbf{P} : \nabla \mathbf{v} + \rho \sum_{i=1}^{N} Y_i \mathbf{f}_i \cdot \mathbf{V}_i, \tag{5.1.26}$$

where we have used the relation $\nabla \cdot (\mathbf{v} \cdot \mathbf{P}) = \mathbf{v} \cdot (\nabla \cdot \mathbf{P}) + \mathbf{P} : \nabla \mathbf{v}$. The operation symbol (:) means that the tensor is to be contracted twice.

This completes our derivation of the conservation equations for chemically reacting flows, given by Eqs. (5.1.7), (5.1.12), (5.1.17), and (5.1.26) for the conservation of total mass, individual species concentration, momentum, and energy respectively. These equations, however, are not complete until we have specified the constitutive relations for the pressure tensor $\mathbf{P}$, the diffusion velocity $\mathbf{V}_i$, the heat flux $\mathbf{q}$, and the reaction rate w_i. These will be presented in Section 5.2.2. Before doing so, we shall first apply the above conservation relations to derive the general conservation conditions across an interface. These relations are frequently needed to serve as the

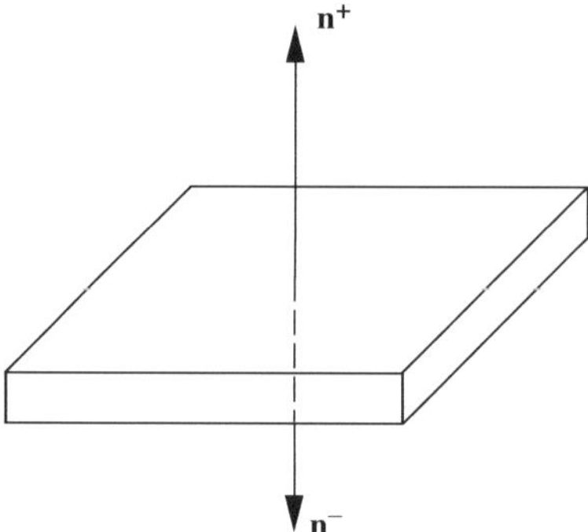

Figure 5.1.2. Control volume for the derivation of conservation relations across an interface.

boundary or matching conditions when analyzing processes occurring in different flow regions or media.

5.1.5. Conservation Relations across an Interface

Consider the control volume of Figure 5.1.1 to be a thin slab sandwiching a control surface $\mathbf{S}_I$, which has a velocity $\mathbf{v}_I$ (Figure 5.1.2). The unit normal vectors of the two surfaces of the slab are $\mathbf{n}^+$ and $\mathbf{n}^-$, with $\mathbf{n}^- \to -\mathbf{n}^+$ as the thickness and thereby the mass of the slab approach zero. The integral conservation equations derived previously of course still hold within this control volume. However, since we are now referencing changes in properties across the interface, it is more convenient to consider these changes in the reference frame at the interface. Thus, the only modification necessary is to replace $\mathbf{v}$ by $(\mathbf{v} - \mathbf{v}_I)$ for the quantities within the surface integrals describing their transport due to the mass flux $\rho\mathbf{v}$.

With the above considerations, the conservation equations derived previously can be evaluated in the limit of vanishing slab thickness such that the integration volume $V \to 0$ while the integration surface $\mathbf{S}$ degenerates to $(\mathbf{S}_I^+ + \mathbf{S}_I^-)$. Thus, using the rate of change equation (5.1.4) and appropriate source terms, the interfacial conservation relations for total mass, individual species concentrations, momentum, and energy are respectively given by

$$\int_{S_I} [\rho^+(\mathbf{v}^+ - \mathbf{v}_I) - \rho^-(\mathbf{v}^- - \mathbf{v}_I)] \cdot \mathbf{n}^+ dS = -\lim_{V\to 0}\left[\frac{\partial}{\partial t}\int_V \rho dV\right], \tag{5.1.27}$$

$$\int_{S_I} [\rho^+ Y_i^+(\mathbf{v}^+ + \mathbf{V}_i^+ - \mathbf{v}_I) - \rho^- Y_i^-(\mathbf{v}^- + \mathbf{V}_i^- - \mathbf{v}_I)] \cdot \mathbf{n}^+ dS$$
$$= \lim_{V\to 0}\left[\int_V w_i dV - \frac{\partial}{\partial t}\int_V \rho Y_i dV\right], \quad i = 1, \ldots, N, \tag{5.1.28}$$

$$\int_{S_I} \{\rho^+\mathbf{v}^+[(\mathbf{v}^+ - \mathbf{v}_I)\cdot\mathbf{n}^+] - \rho^-\mathbf{v}^-[(\mathbf{v}^- - \mathbf{v}_I)\cdot\mathbf{n}^+] + (\mathbf{P}^+ - \mathbf{P}^-)\cdot\mathbf{n}^+\} dS$$
$$= \lim_{V\to 0}\left[\int_V \rho \sum_{i=1}^{N} Y_i \mathbf{f}_i dV - \frac{\partial}{\partial t}\int_V \rho\mathbf{v} dV\right], \tag{5.1.29}$$

and

$$\int_{S_I} \left\{ \rho^+ \left[e^+ + \frac{(v^+)^2}{2} \right] (\mathbf{v}^+ - \mathbf{v}_I) - \rho^- \left[e^- + \frac{(v^-)^2}{2} \right] (\mathbf{v}^- - \mathbf{v}_I) \right.$$
$$\left. + (\mathbf{q}^+ - \mathbf{q}^-) + (\mathbf{v}^+ \cdot \mathbf{P}^+ - \mathbf{v}^- \cdot \mathbf{P}^-) \right\} \cdot \mathbf{n}^+ dS$$
$$= \lim_{V \to 0} \left[\int_V \rho \sum_{i=1}^{N} Y_i \mathbf{f}_i \cdot (\mathbf{v} + \mathbf{V}_i) dV - \frac{\partial}{\partial t} \int_V \rho \left(e + \frac{v^2}{2} \right) dV \right], \quad (5.1.30)$$

where Eq. (5.1.27) has been used in deriving Eq. (5.1.29). The volume integrals in the above relations vanish in the absence of source or sink at the interface, leading to the corresponding vanishing of the integrands in the surface integrals and consequently the differential conservation relations for the fluxes in crossing the interface.

5.2. GOVERNING EQUATIONS

5.2.1. Conservation Equations

For ease of referencing we summarize in the following the conservation equations derived above.

Overall Continuity:

$$\frac{\partial \rho}{\partial t} + \nabla \cdot (\rho \mathbf{v}) = 0 \quad (5.2.1)$$

Continuity of Species:

$$\rho \frac{DY_i}{Dt} = w_i - \nabla \cdot (\rho Y_i \mathbf{V}_i), \quad i = 1, \ldots, N \quad (5.2.2)$$

Momentum:

$$\rho \frac{D\mathbf{v}}{Dt} = -\nabla \cdot \mathbf{P} + \rho \sum_{i=1}^{N} Y_i \mathbf{f}_i \quad (5.2.3)$$

Energy:

$$\rho \frac{De}{Dt} = -\nabla \cdot \mathbf{q} - \mathbf{P} : (\nabla \mathbf{v}) + \rho \sum_{i=1}^{N} Y_i \mathbf{f}_i \cdot \mathbf{V}_i. \quad (5.2.4)$$

5.2.2. Constitutive Relations

Derivation of the constitutive relations specifying the diffusion velocity $\mathbf{V}_i$, the pressure tensor $\mathbf{P}$, and the heat flux vector $\mathbf{q}$ can be found in, for example, Hirschfelder, Curtiss, and Bird (1954), and Williams (1985), while the reaction rate is stated in Chapter 2 through the law of mass action.

Diffusion Velocity $\mathbf{V}_i$:

$$\nabla X_i = \sum_{j=1}^{N}\left(\frac{X_i X_j}{D_{i,j}}\right)(\mathbf{V}_j - \mathbf{V}_i) + (Y_i - X_i)\left(\frac{\nabla p}{p}\right) + \left(\frac{\rho}{p}\right)\sum_{j=1}^{N} Y_i Y_j(\mathbf{f}_i - \mathbf{f}_j)$$
$$+ \sum_{j=1}^{N}\left[\left(\frac{X_i X_j}{\rho D_{i,j}}\right)\left(\frac{D_{T,j}}{Y_j} - \frac{D_{T,i}}{Y_i}\right)\right]\left(\frac{\nabla T}{T}\right), \quad i = 1, \ldots, N. \tag{5.2.5}$$

Pressure Tensor $\mathbf{P}$:

$$\mathbf{P} = \left[p + \left(\frac{2}{3}\mu - \kappa\right)(\nabla \cdot \mathbf{v})\right]\mathbf{U} - \mu[(\nabla \mathbf{v}) + (\nabla \mathbf{v})^T], \tag{5.2.6}$$

where $\mathbf{U}$ is the unit tensor and the superscript T denotes transpose of the tensor.

Heat Flux Vector $\mathbf{q}$:

$$\mathbf{q} = -\lambda \nabla T + \rho \sum_{i=1}^{N} h_i Y_i \mathbf{V}_i + R^o T \sum_{i=1}^{N}\sum_{j=1}^{N}\left(\frac{X_j D_{T,i}}{W_i D_{i,j}}\right)(\mathbf{V}_i - \mathbf{V}_j) + \mathbf{q}_R, \tag{5.2.7}$$

where $\mathbf{q}_R$ is the radiant heat flux vector.

Species Reaction Rate:

$$w_i = W_i \sum_{k=1}^{K}(\nu''_{i,k} - \nu'_{i,k}) B_k T^{\alpha_k} \exp(-E_{a,k}/R^o T) \prod_{j=1}^{N} c_j^{\nu'_{j,k}}, \quad i = 1, \ldots, N, \tag{5.2.8}$$

where $w_i = W_i \hat{\omega}_i$ with $\hat{\omega}_i$ specified in Section 2.1.1, and $c_j = \rho_j / W_j = Y_j \rho / W_j = (X_j p / R^o T)$.

Although the above expressions for $\mathbf{V}_i$, $\mathbf{P}$, and $\mathbf{q}$ were not derived, it is important to recognize the physical meaning of these fluxes. The four terms on the RHS of Eq. (5.2.5) respectively show that mass diffusion can be effected by Fickian diffusion through the concentration gradient, diffusion in the presence of pressure gradient ∇p and of body force $\mathbf{f}_i$, and the second-order, Soret diffusion in the presence of temperature gradient. The efficiencies of the concentration and Soret diffusions depend on their respective diffusion coefficients $D_{i,j}$ and $D_{T,i}$. Among these four processes, concentration diffusion dominates in most situations of physical interest. Pressure diffusion could be important under exceptionally large values of ∇p, or for species i whose molecular weight is substantially different from those of the other species such that Y_i is very different from X_i. Body force diffusion could be relevant when considering electromagnetic forces associated with ions and electrons in gas mixtures. When gravity is the only body force, then $\mathbf{f}_i = \mathbf{f}_j = \mathbf{g}$ and there is no net body force diffusion, where $\mathbf{g}$ is the gravity acceleration vector.

The meaning of the pressure tensor $\mathbf{P}$, especially for its role in momentum conservation Eq. (5.2.3), is explained in elementary fluid mechanics texts and is not repeated here. Suffice to note that the bulk viscosity coefficient κ is frequently neglected but is not necessarily negligible as the density changes are large in combustion processes. Regarding the influence of buoyancy on the fluid motion (Law & Faeth 1994; Ronney 1998; Ross 2001), clearly it is inherently an important process for most of the combustion processes on earth because in the active burning region the temperature is relatively high and density low, whereas in regions away from the flame the temperature is low and density high. Therefore the fluid elements in the burning region tend to rise relative to its surrounding, and produce a net natural convective motion. This can change the burning rate, for example in increasing it by enhancing the rate of oxidizer transfer to the burning region. In fundamental combustion studies buoyancy is sometimes considered a "nuisance" because it could distort the flame shape from symmetry based on which theoretical predictions are frequently made.

It has also been suggested that electromagnetic forces could be significant under certain situations (Lawton & Weinberg 1969). Obviously ions and electrons are produced during reactions such that the flame can be manipulated by externally applied electric fields. Examples are the inducement of ionic winds and of flamefront instability in the form of ridges over the surface of a Bunsen flame. The stabilization characteristics of Bunsen flames, to be studied in Chapter 8, have been found to be affected (Calcote & Pease 1951). However, while the effects of charged particles on the bulk flow have not been adequately studied, they are believed to be small.

The significant change in density within a reacting flow also renders it inadequate to adopt the constant density assumption commonly used in nonreacting flows of low Mach numbers, recognizing nevertheless that the density variation here arises from large temperature nonuniformities instead of the high-speed nature of the flow, which renders the variation of the specific volume with pressure significant.

Concerning the heat flux $\mathbf{q}$ in Eq. (5.2.7), the first term on the RHS represents the well-known conduction heat transfer in the presence of a temperature gradient. The second term is the transfer of heat through mass diffusion due to the different heat contents of the various species. This term vanishes when the specific heats of different species are the same. The third term represents second-order diffusion, the Dufour effect, accounting for the transfer of heat in the presence of mass diffusion, especially due to the concentration gradient. Unlike the second term, which also is caused by mass diffusion, this term exists even if the specific heats of the species are the same.

The radiation heat transfer vector, $\mathbf{q}_R$, can be an extremely complicated integral function accounting for the radiation effect in all directions (Vincenti & Kruger 1965; Tien & Lee 1982; Sarofim 1986; Viskanta & Mengüc 1987). It depends on the gas temperature as well as the molecular structure because the efficiency of molecular radiation absorption and emission is sensitive to the wavelength of the radiation. It is through this term that the vast field of radiation is coupled to combustion. This heat transfer mode is important for flames with heavy soot loading, and because of its long-range effect in large-scale phenomena such as flames in furnaces as well as building and wildland fires. Sometimes the radiative heat loss from a sooty flame can be so substantial that the flame temperature is significantly reduced. Furthermore, since radiative heat loss is inherent to a flame, there exist situations in which it becomes the only loss mechanism experienced by the flame when all external loss mechanisms are eliminated, and hence constitutes the mechanism responsible for the extinction for such near-adiabatic flames.

5.2.3. Auxiliary Relations

If we let Y_i, ρ, T, and $\mathbf{v}$ be the $N+5$ dependent variables in Eqs. (5.2.1) through (5.2.4), then the other variables are related to them through the following auxiliary relations.

Ideal Gas Equation of State:

$$p = \rho R^o T / \sum_{i=1}^{N} X_i W_i = \rho R^o T \sum_{i=1}^{N} \frac{Y_i}{W_i} = \frac{\rho R^o T}{\overline{W}}, \tag{5.2.9}$$

where $\overline{W} = \sum X_i W_i = (\sum Y_i / W_i)^{-1}$ is the average molecular weight of the mixture.

Energy–Enthalpy Relation:

$$h = \sum_{i=1}^{N} Y_i h_i = e + p/\rho. \tag{5.2.10}$$

The Caloric Equation of State:

$$h_i = h_i^o(T^o) + h_i^s(T; T^o), \quad i = 1, \ldots, N, \tag{5.2.11}$$

where

$$h_i^s(T; T^o) = \int_{T^o}^{T} c_{p,i} dT. \tag{5.2.12}$$

Conversion between Molar and Mass Fractions: The relation between the molar fraction X_i and mass fraction Y_i is

$$X_i = \frac{Y_i / W_i}{\sum_{j=1}^{N} Y_j / W_j}, \qquad Y_i = \frac{X_i W_i}{\sum_{j=1}^{N} X_j W_j}. \tag{5.2.13}$$

In combustion problems it is frequently necessary to convert between X_i and Y_i. The reason is that while chemical reactions are mole-based processes in that their rates vary with the molecular concentrations of the mixture, convection is a mass-based process, describing the inertia and motion of the bulk flow according to Newton's law of motion. The diffusion process is a mixed one as can be seen by the presence of both X_i and Y_i in the definition of the diffusion velocity given by Eq. (5.2.5). However, if concentration diffusion is the dominant mode of diffusion such that ∇X_i is balanced by only the first term on the RHS of Eq. (5.2.5), then the diffusion velocity $\mathbf{V}_i$ is again a mole-based quantity.

Because of the complexities involved in converting between X_i and Y_i, a constant average molecular weight $\overline{W}$ is frequently used in combustion analysis, as shown in Eq. (5.2.9). Furthermore, X_i and Y_i are now related through $X_i W_i = Y_i \overline{W}$. This assumption is reasonable for hydrocarbon–air mixtures because the abundance of nitrogen makes $\overline{W}$ to be less sensitive to compositional variations of the mixture as combustion proceeds. It is however inadequate for, say, a nonpremixed system of air and a very light fuel such as hydrogen or a very large fuel such as a large hydrocarbon or a polymer. The dominance of nitrogen also becomes progressively weaker as a premixture becomes richer in its fuel concentration.

Finally, since we will be working with intensive properties from now on, whenever possible we will use the upper and lower case letters to respectively designate thermodynamic quantities on molar and mass bases, as in H_i, $C_{p,i}$, Q_c, and h_i, $c_{p,i}$, q_c. Recall that in Chapter 1 we used the upper case letter (e.g., H) to designate extensive quantities, and the lower case letter with overbar (e.g., $\bar{h}$) to designate partial molar quantities.

5.2.4. Some Useful Approximations

5.2.4.1. Diffusion Velocity, $\mathbf{V}_i$*:* In the absence of pressure gradient, body-force, and second-order diffusion, Eq. (5.2.5) simplifies to the Stefan–Maxwell equation,

$$\nabla \ln X_i = \sum_{j=1}^{N} \left(\frac{X_j}{D_{i,j}} \right) (\mathbf{V}_j - \mathbf{V}_i), \qquad i = 1, 2, \ldots, N, \tag{5.2.14}$$

which shows that $\mathbf{V}_i$ not only is implicitly expressed in terms of the concentration gradient, it is also coupled to all the $\mathbf{V}_j$s. It is therefore desirable to have an explicit expression for $\mathbf{V}_i$ so that it can be readily substituted into the conservation equations. The simplest expression is obtained by assuming that the binary diffusion coefficients of all pairs of species are equal, that is, $D_{i,j} = D$. Then Eq. (5.2.14) becomes

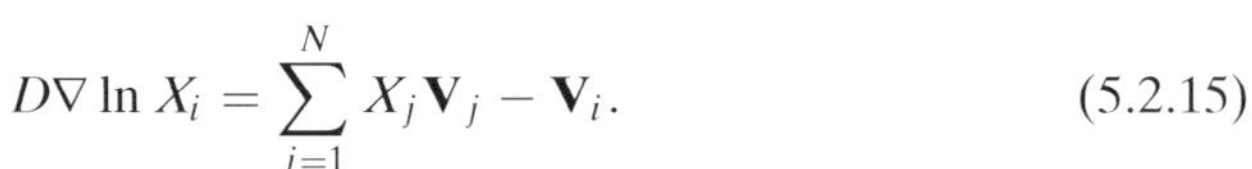

$$D \nabla \ln X_i = \sum_{j=1}^{N} X_j \mathbf{V}_j - \mathbf{V}_i. \tag{5.2.15}$$

Multiplying Eq. (5.2.15) by Y_i, summing over i, and noting that $\sum_{i=1}^{N} Y_i \mathbf{V}_i = 0$, we have

$$\sum_{j=1}^{N} X_j \mathbf{V}_j = D \sum_{j=1}^{N} Y_j \nabla \ln X_j.$$

Substituting this relation into Eq. (5.2.15), and converting X_i to Y_i by using Eq. (5.2.13), it can be shown that

$$\mathbf{V}_i = -D \nabla \ln Y_i, \tag{5.2.16}$$

which is Fick's law of mass diffusion.

The assumption of equal binary diffusion coefficient can be quite a restrictive one because of the significant difference in the molecular weights and structure between the different gaseous components in a combustion environment. Compared to the assumption of an average molecular weight used in, say, the equation of state, this assumption can now qualitatively affect the combustion behavior because it controls the motion and hence concentration of the individual species, and because of the dominant influence of diffusion in flames.

An alternate simplification for the mass diffusion velocity can be obtained by assuming that all but one of the N species in the mixture exist in trace quantities. Thus diffusion of any ith of the $(N-1)$ species is mainly governed by its interaction with the Nth, abundant species. Then the diffusion velocity of the ith species can be approximated by

$$\mathbf{V}_i = -D_{i,N}\nabla \ln Y_i. \tag{5.2.17}$$

Equation (5.2.17) is quite a reasonable assumption for fuels burning in air because of the abundance of nitrogen in the mixture relative to fuel, oxygen, and the combustion intermediates and products, as mentioned earlier.

5.2.4.2. Constant Transport Coefficients: Together with the assumption of equal binary diffusion coefficient, in combustion modeling it is also frequently assumed that in the gas phase the specific heat c_p, the thermal conductivity coefficient λ, and the product ρD (or $\rho^2 D$ in boundary-layer flows discussed in Chapter 12) are constants. Realistically these transport properties are moderately to fairly strong functions of both temperature and species concentrations. For example, the gas temperature can vary from a few hundred degrees at the freestream locations of the fuel and oxidizer to 2,000–3,000 K at the flame. The gas mixture can also consist of a variety of species, such as polar molecules like water, nonpolar molecules like oxygen and carbon dioxide, and large and long molecules like branched and normal paraffins, while their molecular weights can vary from 1 for the hydrogen atom, to 18 for water, to several hundred for the large hydrocarbons. Thus although considerable analytical simplification can be achieved with these constant property assumptions, quantitative accuracy is frequently compromised.

5.2.4.3. Isobaric Assumption: Most combustion phenomena take place in low-speed subsonic flows. For these flows the description of the spatial pressure variation is considerably simplified. This can be demonstrated by considering the inviscid momentum equation in one dimension,

$$\rho u \frac{du}{dx} = -\frac{dp}{dx}. \tag{5.2.18}$$

Nondimensionalizing p by p_o, ρ by ρ_o, and u by u_o, we have

$$\frac{\rho_o u_o^2}{p_o}\left(\hat{\rho}\hat{u}\frac{d\hat{u}}{dx}\right) = -\frac{d\hat{p}}{dx}, \tag{5.2.19}$$

where quantities with "^" are nondimensional. Since $\gamma p_o/\rho_o = a_o^2$, where a_o is a reference speed of sound, and with the Mach number given by $M = u/a$, Eq. (5.2.19) becomes

$$\gamma M_o^2 \left(\hat{\rho}\hat{u}\frac{d\hat{u}}{dx} \right) = -\frac{d\hat{p}}{dx}. \tag{5.2.20}$$

Since the terms in Eq. (5.2.20) must balance each other, and since $M_o^2 \ll 1$ for low-speed subsonic flames, Eq. (5.2.20) implies that a small, $O(M_o^2)$, change in the pressure gradient, $d\hat{p}/dx$, will produce an $O(1)$ change in the inertia force, $\hat{\rho}\hat{u}d\hat{u}/dx$, and vice versa. This concept can be demonstrated more clearly by expressing $\hat{p}$ as the sum of a background pressure $\hat{p}_o$ and a dynamic pressure $\hat{p}_1$,

$$\hat{p}(x,t) = \hat{p}_o(x,t) + \hat{p}_1(x,t), \tag{5.2.21}$$

where $\hat{p}_o = O(1)$ and $\hat{p}_1 = O(M_o^2)$. Substituting Eq. (5.2.21) into Eq. (5.2.20), and equating terms of equal order, we have

$$O(1): \quad \frac{d\hat{p}_o}{dx} = 0 \tag{5.2.22}$$

$$O(M_o^2): \quad \gamma M_o^2 \left(\hat{\rho}\hat{u}\frac{d\hat{u}}{dx} \right) = -\frac{d\hat{p}_1}{dx}. \tag{5.2.23}$$

We therefore readily conclude from Eq. (5.2.22) that

$$\nabla \hat{p}_o = 0, \quad \text{or} \quad p_o = p_o(t). \tag{5.2.24}$$

There are several implications of the above results. First, because $\nabla\hat{p} \approx \nabla\hat{p}_1 = O(M_o^2)$, it can be neglected when compared to the spatial variations of other quantities such as temperature, density, and concentrations. Consequently $\hat{p}(\mathbf{x},t) \approx \hat{p}(t)$. Furthermore, in an open system that does not have any temporal pressure variation forced upon it, then the temporal influence of any pressure wave generated by the combustion process is $\partial p/\partial t \sim O(M_o^2)$ and therefore can again be neglected. However, in a closed system in which either a temporal pressure variation is imposed, as within a piston-driven, reciprocating internal combustion engine, or the combustion-generated pressure waves actively interact with the combustion chamber environment, as with the amplification of pressure waves within the "acoustic cavity" of a rocket motor, then the $\partial p/\partial t = dp/dt$ term needs to be retained.

In the momentum equation, we must retain the spatial pressure variation, with $p(\mathbf{x},t)$, because it is the pressure gradient that drives the flow, as shown in Eq. (5.2.23).

Regardless of the nature of $\nabla\hat{p}$, the fact that $\hat{p}_1 \ll \hat{p}_o$ leads to a significant simplification of the equation of state, as

$$\rho RT = p(x,t) \approx p_o(t), \tag{5.2.25}$$

which implies that ρT varies only with time but not with space, if the gas constant R is approximately constant. Thus $\hat{p}_o$ can be identified as the thermodynamic pressure of the system.

The above result is referred to as the isobaric or low Mach number approximation.

5.3. A SIMPLIFIED DIFFUSION-CONTROLLED SYSTEM

5.3.1. Assumptions

In this section we shall formulate and describe a simplified system that consists of the four essential processes constituting most subsonic combustion phenomena, namely: (a) the unsteady terms describing the time variation of the system, (b) the diffusion terms because of the existence of strong temperature and concentration gradients in flames, (c) the convection terms which describe the fluid mechanical aspects of the system, and (d) the chemical reaction terms because, otherwise, we would not be studying combustion! Consequently we shall neglect body forces, radiation heat transfer, and all modes of diffusion except that due to concentration gradients.

Furthermore, we also assume that in low-speed subsonic flows viscous heating is much weaker than the heat involved with other modes of heat transfer as well as the chemical heat release. Thus $\mathbf{P}:(\nabla\mathbf{v}) = p\mathbf{U}:(\nabla\mathbf{v}) = p\nabla\cdot\mathbf{v}$ in Eq. (5.2.4). This is frequently justifiable as a first approximation for low speed, subsonic flows. In supersonic flows, however, the extent of conversion of the flow kinetic energy to thermal energy can be quite substantial when the flow is slowed down through viscous action within boundary layers.

5.3.2. Derivation

We shall now derive the simplified forms of the energy and species equations. We are not concerned with the momentum equation because momentum is assumed to be conserved during molecular collision leading to chemical reaction. Chemical reaction terms therefore do not explicitly appear in the momentum equation, which is identical to that of chemically nonreacting flows. Indirectly, momentum transport is strongly affected by the presence of reactions in the flow field, through density variation, because of the large changes in temperature and species composition as a consequence of localized heat release and chemical transformation, as mentioned earlier.

With the above discussion, Eq. (5.1.26) for energy conservation is simplified to

$$\frac{\partial}{\partial t}(\rho e) + \nabla\cdot(\rho\mathbf{v}e) = -\nabla\cdot\mathbf{q} - p(\nabla\cdot\mathbf{v}), \tag{5.3.1}$$

where

$$\mathbf{q} = -\lambda\nabla T + \rho\sum_{i=1}^{N} h_i Y_i \mathbf{V}_i, \tag{5.3.2}$$

and $\mathbf{V}_i$ is given implicitly by Eq. (5.2.14). Substituting $e = h - p/\rho$ and $\mathbf{q}$ into Eq. (5.3.1) yields

$$\frac{\partial}{\partial t}\left(\rho\sum_{i=1}^{N} Y_i h_i\right) + \nabla\cdot\left[\rho\sum_{i=1}^{N} Y_i h_i(\mathbf{v}+\mathbf{V}_i) - \lambda\nabla T\right] = \frac{dp}{dt}, \tag{5.3.3}$$

where we have invoked the isobaric assumption. Species conservation is still given by Eq. (5.1.11),

$$\frac{\partial}{\partial t}(\rho Y_i) + \nabla \cdot [\rho Y_i(\mathbf{v} + \mathbf{V}_i)] = w_i, \quad i = 1, \ldots, N. \tag{5.3.4}$$

Using Eq. (5.3.4), energy conservation can be expressed in an alternate form in which the influence of reaction is explicitly displayed. Substituting $h_i = h_i^o + h_i^s$ in Eq. (5.3.3), multiplying Eq. (5.3.4) by h_i^o, summing over i, and subtracting the resulting expression from Eq. (5.3.3), we obtain

$$\frac{\partial}{\partial t}(\rho h^s) + \nabla \cdot \left(\rho \mathbf{v} h^s + \rho \sum_{i=1}^{N} Y_i h_i^s \mathbf{V}_i - \lambda \nabla T\right) = \frac{dp}{dt} - \sum_{i=1}^{N} h_i^o w_i, \tag{5.3.5}$$

where $h^s = \sum_{i=1}^{N} Y_i h_i^s$. The third term on the LHS of Eq. (5.3.5), $\sum Y_i h_i^s \mathbf{V}_i$, shows that the molecular transfer of energy is effected not only through the usual thermal conduction term, $\lambda \nabla T$, but also through the imbalances in the enthalpy fluxes of different species. These imbalances can be the results of either different diffusivities, which affect $\mathbf{V}_i$, or different specific heats, which affect h_i^s. Thus two formulations can be pursued, depending on the assumptions regarding the diffusivities and the specific heats.

5.3.2.1. Distinct Specific Heat Formulation: Here we keep $c_{p,i}$ distinct but let $D_{i,j} \equiv D$. Noting that

$$\nabla h^s = \nabla \sum_{i=1}^{N} Y_i h_i^s = \sum_{i=1}^{N} h_i^s \nabla Y_i + \sum_{i=1}^{N} Y_i \nabla h_i^s, \tag{5.3.6}$$

$$\sum_{i=1}^{N} Y_i \nabla h_i^s = \sum_{i=1}^{N} Y_i \nabla \int^{T} c_{p,i} dT = \sum_{i=1}^{N} Y_i c_{p,i} \nabla T = c_p \nabla T, \tag{5.3.7}$$

where $c_p = \sum_{i=1}^{N} Y_i c_{p,i}$, Eq. (5.3.5) becomes

$$\frac{\partial}{\partial t}(\rho h^s) + \nabla \cdot \left[\rho \mathbf{v} h^s - \rho D \nabla h^s + \lambda \left(\frac{1}{Le} - 1\right) \nabla T\right] = \frac{dp}{dt} - \sum_{i=1}^{N} h_i^o w_i. \tag{5.3.8}$$

The equivalent expression for energy conservation with the total enthalpy h as the dependent variable can be similarly obtained, from Eq. (5.3.3), as

$$\frac{\partial}{\partial t}(\rho h) + \nabla \cdot \left[\rho \mathbf{v} h - \rho D \nabla h + \lambda \left(\frac{1}{Le} - 1\right) \nabla T\right] = \frac{dp}{dt}. \tag{5.3.9}$$

For species conservation, using Eqs. (5.2.16) in Eq. (5.3.4) yields

$$\frac{\partial}{\partial t}(\rho Y_i) + \nabla \cdot (\rho \mathbf{v} Y_i - \rho D \nabla Y_i) = w_i, \quad i = 1, 2, \ldots, N. \tag{5.3.10}$$

Equations (5.3.8) or (5.3.9), and Eq. (5.3.10) are the final simplified conservation relations for energy and species for this formulation. The species equation shows the four dominant terms of interest, namely the unsteady term, the convection term as indicated by the bulk flow velocity $\mathbf{v}$, the diffusive term as indicated by the mass diffusion coefficient D, and the reaction term w_i. It may also be noted that the convection and diffusion terms are respectively described by first and second order differentials, and the negative sign in front of the diffusion term indicates that diffusive transport occurs in the direction of decreasing concentration.

5.3.2.2. Distinct Diffusivity Formulation: Here we keep $D_{i,j} = D_{i,N}$ distinct, but let $c_{p,i} \equiv c_p$. Since now $h_i^s \equiv h^s$, the third term in Eq. (5.3.5) vanishes because $\sum Y_i \mathbf{V}_i \equiv 0$. Equation (5.3.5) then becomes

$$\frac{\partial}{\partial t}(\rho h^s) + \nabla \cdot (\rho \mathbf{v} h^s - \lambda \nabla T) = \frac{dp}{dt} - \sum_{i=1}^{N} h_i^o w_i, \tag{5.3.11}$$

which can be alternately written as

$$\frac{\partial}{\partial t}(\rho h^s) + \nabla \cdot [\rho \mathbf{v} h^s - (\lambda/c_p)\nabla h^s] = \frac{dp}{dt} - \sum_{i=1}^{N} h_i^o w_i \tag{5.3.12}$$

by using $\nabla h^s = c_p \nabla T$ from Eq. (5.3.7). Total energy and individual species conservations are now respectively given by

Sensible energy is not conserved

$$\frac{\partial}{\partial t}(\rho h) + \nabla \cdot \left[\rho \mathbf{v} h - \lambda \nabla T - \sum_{i=1}^{N} h_i^o(\rho D_i)\nabla Y_i\right] = \frac{dp}{dt} \tag{5.3.13}$$

$$\frac{\partial}{\partial t}(\rho Y_i) + \nabla \cdot (\rho \mathbf{v} Y_i - \rho D_i \nabla Y_i) = w_i, \quad i = 1, 2, \ldots, N, \tag{5.3.14}$$

where we have written $D_i = D_{i,N}$ for simplicity. Equations (5.3.12) or (5.3.13), and Eq. (5.3.14) are the final governing equations for this formulation.

It may be noted that in the above relations $\mathbf{v}$ and D (or D_i) do not appear separately but are rather grouped with ρ through $\rho\mathbf{v}$ and ρD. Thus the relevant quantities representing convective intensity and mixture mass diffusivity are the mass flux $\rho\mathbf{v}$ and the density-weighted mass diffusivity ρD. Furthermore, λ/c_p is also the density-weighted thermal diffusivity $\rho(\lambda/c_p\rho)$. As discussed in Chapter 4, these density-weighted diffusivities are insensitive to pressure variations and are only moderately sensitive to temperature variations ($\sim T^{0.7}$).

5.4. CONSERVED SCALAR FORMULATIONS

A major difficulty in the solution of chemically reacting flows is the presence of the reaction term, which not only is nonlinear but also couples the energy and species equations. However, recognizing that the concentrations of the various reactive species Y_i and the system enthalpy are related through stoichiometry, it is reasonable to expect that under suitable situations these quantities can be stoichiometrically combined such that the resulting term is again not affected by chemical reactions in the flow. Such a combined quantity is called a conserved scalar or coupling function.

An obvious example of a conserved scalar is the total energy of a homogeneous mixture, which is the sum of the stoichiometrically scaled sensible energy and chemical energy as represented by the fuel concentration, as discussed in Chapter 1. In the presence of diffusive transport in a nonhomogeneous mixture, different flow scalars will diffuse at different rates and it is not at all clear if such conserved scalars still exist. In this section we will identify several of them and the separate requirements for each of them to exist.

5.4.1. Coupling Function Formulation

This is perhaps the most frequently used conserved scalar formulation, which is sometimes also referred to as the Shvab–Zel'dovich formulation. While the basic concept of the coupling function formulation is applicable to a general reaction scheme, it is most useful for the one-step overall reaction,

$$\sum_{i=1}^{N} \nu_i' \mathrm{M}_i \to \sum_{i=1}^{N} \nu_i'' \mathrm{M}_i. \tag{5.4.1}$$

This reaction has a species-independent reaction rate ω given by Eq. (2.1.4), and is related to w_i of Eq. (5.2.8) through

$$\omega = \frac{w_i}{W_i(\nu_i'' - \nu_i')}, \tag{5.4.2}$$

where

$$\omega = BT^\alpha \exp(-E_a/R^o T) \prod_{j=1}^{N} c_j^{\nu_j'}. \tag{5.4.3}$$

Specializing to the distinct specific heat formulation for illustration, we substitute Eq. (5.4.2) into Eq. (5.3.10) to yield

$$\frac{\partial}{\partial t}(\rho Y_i) + \nabla \cdot [\rho \mathbf{v} Y_i - (\rho D)\nabla Y_i] = W_i(\nu_i'' - \nu_i')\omega. \tag{5.4.4}$$

We now define a stoichiometrically weighted mass fraction as

$$\tilde{Y}_i = \frac{1}{\sigma_{i,n}}\left(\frac{Y_i}{Y_{n,B}}\right), \tag{5.4.5}$$

where

$$\sigma_{i,n} = \frac{W_i(\nu_i'' - \nu_i')}{W_n(\nu_n'' - \nu_n')}$$

is defined in general as the stoichiometric mass ratio of species i to an as yet unspecified reference species n, and the subscript B is a boundary location, say the freestream, at which Y_n is known. Equation (5.4.4) can then be written as

$$L_D(\tilde{Y}_i) = w_n, \quad i = 1, \ldots, N, \tag{5.4.6}$$

where $w_n = [W_n(\nu_n'' - \nu_n')/Y_{n,B}]\omega$, and

$$L_D(\cdot) = \left[\frac{\partial \rho}{\partial t} + \nabla \cdot (\rho \mathbf{v} - \rho D \nabla)\right](\cdot) \tag{5.4.7}$$

designates an operator whose diffusion term is characterized by ρD.

Since Eq. (5.4.6) consists of N equations, we can eliminate the complex and nonlinear reaction term w_n from $(N-1)$ of them by subtracting any jth equation from an ith equation to yield

$$L_D(\tilde{Y}_i - \tilde{Y}_j) = 0. \tag{5.4.8}$$

Thus, if we define a species coupling function

$$\beta_{i,j} = \tilde{Y}_i - \tilde{Y}_j, \tag{5.4.9}$$

then Eq. (5.4.8) is expressed as

$$L_D(\beta_{i,j}) = 0. \tag{5.4.10}$$

The reaction term w_n can also be eliminated from the energy conservation equation by further assuming that the mixture Lewis number, $Le = \lambda / c_p \rho D$, is unity. Then Eq. (5.3.8) is simplified to

$$L(h^s) = \frac{dp}{dt} - \sum_{i=1}^{N} h_i^o w_i, \tag{5.4.11}$$

where we have removed the subscript D from the operator $L_D(\cdot)$ because the diffusion coefficient can now be either ρD or λ / c_p as a consequence of the $Le = 1$ assumption. Substituting Eq. (5.4.2) into Eq. (5.4.11) yields

$$L(h^s) = \frac{dp}{dt} - \left[\sum_{i=1}^{N} h_i^o W_i(\nu_i'' - \nu_i')\right]\omega. \tag{5.4.12}$$

If we now define stoichiometrically weighted nondimensional enthalpy and temperature as

$$\tilde{h}^s = \frac{h^s}{Y_{n,B} q_{c,n}}, \qquad \tilde{T} = \frac{c_p T}{Y_{n,B} q_{c,n}}, \tag{5.4.13}$$

where

$$q_{c,n} = \frac{\sum_{k=1}^{N} h_k^o W_k(\nu_k'' - \nu_k')}{W_n(\nu_n'' - \nu_n')}$$

is the chemical heat release per unit mass of species n reacted, then Eq. (5.4.12) becomes

$$L(\tilde{h}^s) = \left(1 - \frac{1}{\gamma}\right)\frac{d\tilde{p}}{dt} - w_n, \tag{5.4.14}$$

where $\tilde{p} = \rho \tilde{T}$ and we have also used the ideal gas equation of state to relate p, ρ, and T. Thus, a species–enthalpy coupling function can be defined as

$$\beta_i = \tilde{h}^s + \tilde{Y}_i, \tag{5.4.15}$$

which satisfies

$$L(\beta_i) = \left(1 - \frac{1}{\gamma}\right)\frac{d\tilde{p}}{dt}. \tag{5.4.16}$$

Equation (5.4.16) is somewhat more complex than Eq. (5.4.10) because of the inhomogeneous term describing the variation of $\tilde{p}$ with time. However, in a steady-state situation, Eq. (5.4.16) simplifies to the homogeneous equation

$$L_{s.s.}(\beta_i) = [\nabla \cdot (\rho \mathbf{v} - \rho D \nabla)]\,\beta_i = 0. \tag{5.4.17}$$

Another possible simplification of Eq. (5.4.16) can be obtained for flames and combustion processes occurring in an open environment of constant pressure, such that $dp/dt \simeq 0$ and consequently

$$L(\beta_i) = 0. \tag{5.4.18}$$

The coupling function formulation simplifies, but does not eliminate, the chemical aspect of the problem. That is, for a combustion system governed by $(N+1)$ equations representing conservation of energy and the N chemically active species, the reaction term w_n is eliminated from all but one of them. For example, for an open flame, the system is described by Eq. (5.4.18) for the coupling functions β_i together with Eq. (5.4.14), given by

$$L(\tilde{h}^s) = -w_n. \tag{5.4.19}$$

Thus the chemical information is still contained in Eq. (5.4.19), as it should be.

The coupling functions $\beta_{i,j}$ and β_i are called conserved scalars because they are not affected by chemical reactions in the flow field. Physically, the existence of conserved scalars can be explained as follows. First we note that $\beta_{i,j}$ and β_i are conserved quantities in a static, nondiffusive medium because they simply represent the stoichiometric relations between the creation and destruction of the various species and the corresponding change in the sensible enthalpy of the mixture. Next, these relations should still hold in a convective medium because in the frame of reference of the flow the medium is again static. In other words, all scalar quantities are carried along by the flow at the same rate. Finally, consider the effect of diffusion. Since heat and different species have different diffusivities, their respective diffusive fluxes are transported with different efficiencies and consequently $\beta_{i,j}$ and β_i cease to be conserved. However, by making the equal diffusivity and unity Lewis number assumptions, we are requiring that these diffusive fluxes be transported with the same efficiency, hence preserving the conserved nature of $\beta_{i,j}$ and β_i. Thus the assumption of equal diffusivities is essential in the existence of coupling functions.

Since a coupling function, or conserved scalar, is not affected by the chemical reaction in the flow field, the concentration of any inert species is by definition a conserved scalar in chemically reacting flows, satisfying differential operators even more general than $L(\cdot)$. Experimentally, sometimes it has been found useful to determine some of the flow field properties by tracking the response of an inert species such as nitrogen.

It is also important to recognize that coupling functions, being functions of space and time, are in general not necessarily constants of the flow. The terminology "conserved scalar" only implies that the quantity is conserved in a chemical reaction. On the other hand, there are situations in which they are indeed constants in the flow, as for the one-dimensional planar premixed flame to be studied in Chapter 7.

Finally, for later use we shall write down the various parameters for a specific one-step reaction scheme, namely one between a fuel species F and an oxidizer species O, given by

$$\nu'_F F + \nu'_O O \rightarrow \nu''_P P, \tag{5.4.20}$$

with $q_{c,F}$ amount of heat release per unit mass of fuel consumed. If we let the reference species n be the fuel, then by definition

$$\frac{\sum_{k=1}^{N} h_k^o W_k(\nu''_k - \nu'_k)}{W_F(\nu''_F - \nu'_F)} = q_c, \qquad \frac{W_i(\nu''_i - \nu'_i)}{W_F(\nu''_F - \nu'_F)} = \sigma_i, \tag{5.4.21}$$

and $\tilde{Y}_i = Y_i/(\sigma_i Y_{F,B})$, where $q_c \equiv q_{c,F}$, $\sigma_i \equiv \sigma_{i,F}$ is the stoichiometric mass ratio of the ith species to the fuel, and $Y_{F,B}$ the fuel mass fraction in the freestream. Thus the various stoichiometrically scaled parameters are

$$\tilde{h}^s = \frac{h^s}{Y_{F,B}q_c}, \quad \tilde{T} = \frac{c_p T}{Y_{F,B}q_c}, \quad \tilde{Y}_F = \frac{Y_F}{Y_{F,B}}, \quad \tilde{Y}_O = \frac{Y_O}{\sigma_O Y_{F,B}}, \tag{5.4.22}$$

where $Y_{F,B}q_c/c_p$ is the increase in the temperature from the freestream value upon complete consumption of $Y_{F,B}$ amount of fuel in a homogeneous mixture.

5.4.2. Local Coupling Function Formulation

The major drawback of the coupling function formulation is the need to assume equal diffusivities for all species as well as heat, when both $\beta_{i,j}$ and β_i need to be considered. This results in equal Lewis numbers which also have the special value of unity. While our discussion on transport phenomena shows that the unity Lewis number assumption is a reasonable one as far as its magnitude is concerned, we shall demonstrate in due course that there exist some important combustion phenomena, such as flamefront instability, that can only be explained by the fact that Le deviates from unity, that is, diffusion of heat and the various species occur with different diffusivities. Thus, in combustion modeling, caution should be exercised when invoking the unity Lewis number assumption lest some crucial phenomena are unknowingly suppressed.

Fortunately, many combustion problems can be studied, through rational approximation, without assuming $Le = 1$ or using the coupling function formulation. The reason being that the main advantage of the coupling function formulation is the ability to eliminate the reaction terms from the conservation equations which govern the entire flow field. However, as discussed in Chapter 2, active chemical reactions are frequently confined in narrow spatial regions because of the large-activation-energy Arrhenius kinetics involved. Thus much of the flow field is essentially chemically nonreactive and therefore can be described by the governing equations with $w_i = 0$ but $Le_i \neq 1$. Furthermore, as will be shown in later chapters, in the narrow reaction region the diffusion term frequently dominates over the convection and transient terms because of the steep spatial gradients associated with the rapid changes in the flow properties. Thus Eqs. (5.3.12) and (5.3.14) can be approximated by

$$\nabla \cdot [(\lambda/c_p)\nabla h^s] = \sum_{i=1}^{N} h_i^o w_i, \tag{5.4.23}$$

$$\nabla \cdot (\rho D_i \nabla Y_i) = -w_i, \tag{5.4.24}$$

in the reaction region, where we have used the distinct diffusivity formulation to capture the role of preferential diffusion. By further assuming that λ/c_p and ρD_i are constants, Eqs. (5.4.23) and (5.4.24) become

$$\nabla^2 \tilde{h}^s = (c_p/\lambda) w_n, \tag{5.4.25}$$

$$\nabla^2 (\tilde{Y}_i/Le_i) = -(c_p/\lambda) w_n, \tag{5.4.26}$$

where $Le_i = \lambda/(c_p \rho D_i)$. Thus addition of Eqs. (5.4.25) and (5.4.26) yields the Laplace equation

$$\nabla^2 \beta_i = 0, \tag{5.4.27}$$

where

$$\beta_i = \tilde{h}^s + \frac{\tilde{Y}_i}{Le_i} \tag{5.4.28}$$

is the local coupling function which is valid only in the reaction region. Equation (5.4.28) shows that $\tilde{Y}_i$ is now to be scaled by Le_i in the local coupling function. Thus for $Le_i > 1$, that is, mass diffusivity being smaller than thermal diffusivity, β_i suffers a net reduction in the concentration of i. The converse holds when $Le_i < 1$. This result is physically reasonable.

The expression of β_i as given by Eq. (5.4.28) demonstrates the possibility that a thermodynamic quantity, such as the total enthalpy of a mixture as represented by $(\tilde{h}^s + \tilde{Y}_i)$, can be affected by the nonequilibrium, diffusive transport processes when the thermal and species diffusivities differ from each other. Such an influence can therefore modify the local concentrations of a mixture from its freestream values, and through it the local flame temperature from the adiabatic flame temperature, which is defined by the freestream values.

5.4.3. Near-Equidiffusion Formulation

Since the Lewis numbers of many gaseous reactants are indeed close to unity, it is possible to approximately account for Lewis number effects by expanding the governing equations around $Le_i = 1$. To do this, we again assume that λ/c_p and ρD_i are constants. Then Eqs. (5.3.12) and (5.3.14) can be written, for an open flame, as

$$\frac{\partial}{\partial t}(\rho \tilde{h}^s) + \nabla \cdot [\rho \mathbf{v} \tilde{h}^s - (\lambda/c_p)\nabla \tilde{h}^s] = -w_n, \tag{5.4.29}$$

$$\frac{\partial}{\partial t}(\rho \tilde{Y}_i) + \nabla \cdot \left[\rho \mathbf{v} \tilde{Y}_i - \frac{(\lambda/c_p)}{Le_i}\nabla \tilde{Y}_i\right] = w_n. \tag{5.4.30}$$

Expressing Eq. (5.4.30) as

$$L_\lambda(\tilde{Y}_i) = w_n + \left(\frac{1}{Le_i} - 1\right)\nabla \cdot \left[(\lambda/c_p)\nabla \tilde{Y}_i\right], \tag{5.4.31}$$

where

$$L_\lambda(\cdot) = \left\{\frac{\partial}{\partial t}\rho + \nabla \cdot [\rho \mathbf{v} - (\lambda/c_p)\nabla]\right\}(\cdot), \tag{5.4.32}$$

and adding Eqs. (5.4.29) and (5.4.31) yields

$$L_\lambda(\beta_i) = \left(\frac{1}{Le_i} - 1\right)\nabla \cdot \left[(\lambda/c_p)\nabla \tilde{Y}_i\right], \tag{5.4.33}$$

with $\beta_i = \tilde{h}^s + \tilde{Y}_i$. Equation (5.4.33) clearly demonstrates that nonequidiffusion plays the role of a sink/source for the conserved scalar β_i. For $Le_i \approx 1$, solution can be sought in series form as in, say, $\tilde{Y}_i = \tilde{Y}_{i,0} + \delta_i \tilde{Y}_{i,1} + O(\delta_i^2)$, with

$$\left|\frac{1}{Le_i} - 1\right| = \delta_i \ll 1$$

being the small parameter of expansion. Thus expanding Eq. (5.4.33), we have, to the first two orders,

$$L_\lambda(\beta_{i,0}) = 0, \tag{5.4.34}$$

$$L_\lambda(\beta_{i,1}) = \nabla \cdot [(\lambda/c_p)\nabla \tilde{Y}_{i,0}]. \tag{5.4.35}$$

The nonequidiffusion effects are therefore captured at the first-order solution, for $\beta_{i,1}$.

5.4.4. Element Conservation Formulation

All our discussion so far has been based on tracking the fate of the individual species as chemical reactions evolve. The coupling functions so identified are based on chemical stoichiometry. Since elements are conserved in chemical reactions, it is logical to investigate relations governing the conservation of elements in chemically reacting flows.

In a mixture consisting of N species of Y_i made up of L elements, we can define an element mass fraction Z_k of element k as

$$Z_k = \sum_{i=1}^{N} \mu_{i,k} Y_i, \quad k = 1, 2, \ldots, L, \tag{5.4.36}$$

where $\mu_{i,k}$ is the mass fraction of the kth element in the ith species. For example, for the elements carbon and oxygen in carbon dioxide, we have $\mu_{CO_2,C} = 12/44$ and $\mu_{CO_2,O} = 32/44$. If we now multiply Eq. (5.3.10) by $\mu_{i,k}$, sum over all i, and require that elements be conserved in chemical reactions such that

$$\sum_{i=1}^{N} \mu_{i,k} w_i = 0, \quad k = 1, 2, \ldots, L, \tag{5.4.37}$$

then we have

$$L_D(Z_k) = 0, \quad k = 1, 2, \ldots, L. \tag{5.4.38}$$

Thus the element mass fraction Z_k is a conserved scalar, provided a single diffusion coefficient D is used.

The advantage of using Z_k instead of the coupling function $\beta_{i,j}$ is that Eq. (5.4.38) is independent of the specific chemical reactions that take place in the mixture, while the identification of $\beta_{i,j}$ requires the specification of the reaction scheme, which is restricted to a one-step overall reaction in the case we used for demonstration. The restriction of using Z_k is that in a chemically reacting mixture the number of elements is usually much smaller than the number of species ($L < N$), implying that a solution of Z_k is frequently not sufficient to solve for all the species concentrations.

It is, however, still advantageous to use Eq. (5.4.38) to supplement the solution of Eq. (5.3.10). That is, instead of directly solving for all the Y_is from the N species equations of Eq. (5.3.10), the extent of solution is reduced by solving the L element conservation equations and the $(N - L)$ species conservation equations. For a chemically simple mixture consisting of only a few species, the element conservation formulation can be quite useful. An example is the system consisting of only a hydrocarbon fuel, oxygen, and water and carbon dioxide as the products. Here $N = 4$, while $L = 3$, representing hydrogen, carbon and oxygen.

When employing the element conservation formulation, it is more advantageous to use the total energy conservation, Eq. (5.3.9), because it is also independent of the specific reaction scheme adopted.

5.4.5. Mixture Fraction Formulation

Similar to the element conservation formulation, the mixture fraction formulation is another derivative of the coupling function formulation and as such is also restricted by the assumptions of a single diffusion coefficient and unity Lewis number. The mixture fraction variable so defined has been found to be especially suitable for studying problems of nonpremixed turbulent flames (Bilger 1980; Peters 2000).

Let the flow consists of a fuel (F) stream and an oxidizer (O) stream. These streams have uniform properties at their respective upstream boundaries, which will be designated by the subscripts B^- and B^+. We thus set $Y_{F,B^+} = 0$ and $Y_{O,B^-} = 0$. We then define a mixture fraction Z in terms of the fuel–oxidizer coupling function $\beta_{F,O}$ as

$$Z = \frac{\beta_{F,O} - \beta_{F,O,B^-}}{\beta_{F,O,B^+} - \beta_{F,O,B^-}}, \tag{5.4.39}$$

such that $Z = 0$ at the fuel boundary and $Z = 1$ at the oxidizer boundary. Since Z varies linearly with $\beta_{F,O}$, which satisfies $L_D(\beta_{F,O}) = 0$ as given by Eq. (5.4.10), it is clear that Z must also be given by

$$L_D(Z) = 0. \tag{5.4.40}$$

Furthermore, since Z varies between 0 and 1, and carries the physical meaning of the relative amounts of fuel and oxidizer, it is sometimes more illuminating to use Z as the independent variable to indicate the progress in mixing and reaction. For example, consider another coupling function for an ith species and fuel, $\beta_{i,F} = \tilde{Y}_i - \tilde{Y}_F$. Since $\beta_{i,F}$ also satisfies

$$L_D(\beta_{i,F}) = 0, \tag{5.4.41}$$

a possible solution of $\beta_{i,F}$ is the linear relation with Z,

$$\beta_{i,F} = c_{1,i} + c_{2,i} Z, \tag{5.4.42}$$

which satisfies both Eqs. (5.4.40) and (5.4.41). Applying the boundary conditions of $\tilde{Y}_i$ and $\tilde{Y}_F$ at $Z = 0$ and 1, we obtain

$$\tilde{Y}_i - \tilde{Y}_F = (\tilde{Y}_{i,B^-} - \tilde{Y}_{F,B^-}) + (\tilde{Y}_{F,B^-} + \tilde{Y}_{i,B^+} - \tilde{Y}_{i,B^-})Z. \tag{5.4.43}$$

Taking $i = O$, Eq. (5.4.43) becomes

$$\tilde{Y}_O - \tilde{Y}_F = -\tilde{Y}_{F,B^-} + (\tilde{Y}_{F,B^-} + \tilde{Y}_{O,B^+})Z. \tag{5.4.44}$$

Similarly, with the further assumption of unity Lewis number, the fuel–enthalpy and oxidizer–enthalpy coupling functions are expressed as

$$\tilde{h}^s + \tilde{Y}_F = (\tilde{h}^s_{B^-} + \tilde{Y}_{F,B^-}) + (\tilde{h}^s_{B^+} - \tilde{h}^s_{B^-} - \tilde{Y}_{F,B^-})Z, \tag{5.4.45}$$

$$\tilde{h}^s + \tilde{Y}_O = \tilde{h}^s_{B^-} + (\tilde{h}^s_{B^+} - \tilde{h}^s_{B^-} + \tilde{Y}_{O,B^+})Z. \tag{5.4.46}$$

In the absence of chemical reaction, the flow is frozen and we have $L(\tilde{h}^s) \equiv 0$, or simply $L(\tilde{T}) \equiv 0$. Similarly $L(\tilde{Y}_i) \equiv 0$ for all i. Thus $\tilde{T}$ and $\tilde{Y}_i$ are all linearly related to Z, given by

$$\tilde{T} = \tilde{T}_{B^-} + (\tilde{T}_{B^+} - \tilde{T}_{B^-})Z, \tag{5.4.47}$$

$$\tilde{Y}_F = \tilde{Y}_{F,B^-}(1 - Z), \quad \tilde{Y}_O = \tilde{Y}_{O,B^+} Z. \tag{5.4.48}$$

This corresponds to the situation of pure mixing.

The utility of the mixture fraction formulation is that variations of the combustion response are studied in the Z-space instead of the physical space. Since Z varies between (0,1) from the fuel to the oxidizer streams, such a variation provides a more direct indication of the dependence of flame characteristics on the reactant concentrations. For example, we have just shown that a conserved scalar would vary linearly with Z. Furthermore, since Z is configurational independent, results obtained from different flame configurations (e.g., spherical flame versus jet flame) can be systematically and meaningfully compared by using Z as the independent variable.

To study the influence of chemistry, we consider the energy equation $L(\tilde{h}^s) = -w_n$, that is,

$$\rho \frac{\partial \tilde{h}^s}{\partial t} + \rho \mathbf{v} \cdot \nabla \tilde{h}^s - \nabla \cdot (\rho D \nabla \tilde{h}^s) = -w_n. \qquad (5.4.49)$$

To incorporate Z as one of the coordinates, we consider an orthogonal coordinate system in which Z is such a coordinate while the other two coordinates, $X = x$ and $Y = y$, are distances along surfaces of constant Z. Thus we need to perform the coordinate transformation $(t, x, y, z) \rightarrow (\tau, X, Y, Z)$, with $\tau \equiv t$.

We first note that

$$\frac{\partial}{\partial t} = \frac{\partial}{\partial \tau} + \frac{\partial Z}{\partial t}\frac{\partial}{\partial Z}, \qquad \frac{\partial}{\partial x} = \frac{\partial}{\partial X} + \frac{\partial Z}{\partial x}\frac{\partial}{\partial Z}, \qquad \frac{\partial}{\partial z} = \frac{\partial Z}{\partial z}\frac{\partial}{\partial Z},$$

$$\begin{aligned} \frac{\partial}{\partial x}\left(\rho D \frac{\partial}{\partial x}\right)\tilde{h}^s &= \left(\frac{\partial}{\partial X} + \frac{\partial Z}{\partial x}\frac{\partial}{\partial Z}\right)\left[\rho D \left(\frac{\partial}{\partial X} + \frac{\partial Z}{\partial x}\frac{\partial}{\partial Z}\right)\tilde{h}^s\right] \\ &= \frac{\partial}{\partial X}\left(\rho D \frac{\partial}{\partial X}\right)\tilde{h}^s + \left(\frac{\partial Z}{\partial x}\right)\left[\frac{\partial}{\partial X}\left(\rho D \frac{\partial}{\partial Z}\right) + \frac{\partial}{\partial Z}\left(\rho D \frac{\partial}{\partial X}\right)\right]\tilde{h}^s \\ &\quad + \rho D \left(\frac{\partial Z}{\partial x}\right)^2 \left(\frac{\partial^2}{\partial Z^2}\right)\tilde{h}^s + \left(\frac{\partial \tilde{h}^s}{\partial Z}\right)\frac{\partial}{\partial x}\left(\rho D \frac{\partial Z}{\partial x}\right), \end{aligned} \qquad (5.4.50)$$

$$\begin{aligned} \frac{\partial}{\partial z}\left(\rho D \frac{\partial}{\partial z}\right)\tilde{h}^s &= \left(\frac{\partial Z}{\partial z}\frac{\partial}{\partial Z}\right)\left[\rho D \left(\frac{\partial Z}{\partial z}\frac{\partial}{\partial Z}\right)\tilde{h}^s\right] \\ &= \rho D \left(\frac{\partial Z}{\partial z}\right)^2 \left(\frac{\partial^2}{\partial Z^2}\right)\tilde{h}^s + \left(\frac{\partial \tilde{h}^s}{\partial Z}\right)\frac{\partial}{\partial z}\left(\rho D \frac{\partial Z}{\partial z}\right), \end{aligned} \qquad (5.4.51)$$

with the differentials involving y being analogous to those involving x. Combining the second terms on the RHS of Eq. (5.4.50) for the X and Y directions, and if we define

$$\nabla_t = \left(\frac{\partial}{\partial X}, \ \frac{\partial}{\partial Y}, \ 0\right)$$

as the two-dimensional tangential gradient operator in X and Y, then the combined second term can be written as

$$(\nabla Z) \cdot \left[\nabla_t \left(\rho D \frac{\partial}{\partial Z}\right) + \frac{\partial}{\partial Z}(\rho D \nabla_t)\right]\tilde{h}^s \quad .$$

However, since ∇_t is over the surface of constant Z while ∇Z is perpendicular to it, their dot product vanishes identically. Substituting the above results into Eq. (5.4.49), and using the conserved scalar equation (5.4.40),

$$\rho \frac{\partial Z}{\partial t} + \rho \mathbf{v} \cdot \nabla Z - \nabla \cdot (\rho D \nabla Z) = 0, \qquad (5.4.52)$$

it can be readily shown that energy conservation in the new coordinate system is given by

$$\rho \frac{\partial \tilde{h}^s}{\partial t} + \rho \mathbf{v}_t \cdot \nabla_t \tilde{h}^s = -w_n + \rho D |\nabla Z|^2 \frac{\partial^2 \tilde{h}^s}{\partial Z^2} + \nabla_t \cdot (\rho D \nabla_t \tilde{h}^s), \qquad (5.4.53)$$

where $\mathbf{v}_t$ is the two-dimensional velocity vector in the X and Y directions. Equation (5.4.53) is useful for the analysis of flame structures in which the variations of temperature and concentrations occur predominantly in the Z direction.

Finally, we note that a mixture fraction on the basis of the element mass fraction Z_k can also be defined as

$$Z = \frac{Z_k - Z_{k,B^+}}{Z_{k,B^-} - Z_{k,B^+}}. \tag{5.4.54}$$

5.4.6. Progress Variable Formulation

The analogue of the mixture fraction for premixed flames is the progress variable c. In this formulation we consider an unreacted premixed stream, originating from the upstream boundary and consisting of a fuel and an oxidizer with concentrations $Y_{F,u}$ and $Y_{O,u}$, where the subscript u designates the upstream unburned state, and we shall also use the subscript b to designate the downstream burned state. This premixture traverses the reaction zone and continuously converts the fuel and oxidizer species to a product species with concentration Y_P. Upon complete reaction at the downstream boundary of the flame, the product concentration is $Y_{p,b}$. Thus a progress variable c can be defined as

$$Y_p = cY_{P,b}, \tag{5.4.55}$$

such that $c = 0$ and 1 for the completely unreacted and reacted states respectively. The temperature T is then readily related to Y_P and hence c through total energy conservation. Substituting Y_p into Eq. (5.4.6), the progress variable is governed by

$$L_D(c) = \frac{w_n}{Y_{P,b}}. \tag{5.4.56}$$

5.5. REACTION-SHEET FORMULATION

5.5.1. Jump Relations for Coupling Functions

Since chemical reactions characterized by high activation energies are spatially confined to very thin regions, it is sometimes possible to collapse this reaction region into an infinitesimally thin reaction sheet, which then simply serves as a sink for the reactants and a source for chemical heat release and the combustion products. The flow in regions bounded away from both sides of this reaction surface is chemically nonreactive and can be separately solved as such. These solutions are then matched at the reaction sheet, where conservation requirements are imposed relating changes in the values as well as the gradients of temperature and species concentrations in crossing it. These requirements, called jump relations, are derived in the following.

We consider a general wrinkled reaction zone whose thickness is much smaller than its radius of curvature. The surface can thus be treated as being locally planar, with properties varying predominantly in the direction normal to it. Using the distinct diffusivity formulation, from Eqs. (5.3.12) and (5.3.14) we have

$$\frac{\partial}{\partial t}(\rho \tilde{h}^s) + \frac{\partial}{\partial n}\left(\rho u \tilde{h}^s - (\lambda/c_p)\frac{\partial \tilde{h}^s}{\partial n}\right) = \left(1 - \frac{1}{\gamma}\right)\frac{d\tilde{p}}{dt} - w_n, \tag{5.5.1}$$

$$\frac{\partial}{\partial t}(\rho \tilde{Y}_i) + \frac{\partial}{\partial n}\left(\rho u \tilde{Y}_i - \rho D_i \frac{\partial \tilde{Y}_i}{\partial n}\right) = w_n, \tag{5.5.2}$$

$$\frac{\partial}{\partial t}(\rho \tilde{Y}_j) + \frac{\partial}{\partial n}\left(\rho u \tilde{Y}_j - \rho D_j \frac{\partial \tilde{Y}_j}{\partial n}\right) = w_n, \tag{5.5.3}$$

where n is the coordinate normal to the flame surface, and we have applied the stoichiometric scaling for h^s and Y_i. By subtracting Eq. (5.5.3) from Eq. (5.5.2), we obtain

$$\frac{\partial}{\partial t}\left[\rho(\tilde{Y}_i - \tilde{Y}_j)\right] + \frac{\partial}{\partial n}\left[\rho u(\tilde{Y}_i - \tilde{Y}_j) - \left(\rho D_i \frac{\partial \tilde{Y}_i}{\partial n} - \rho D_j \frac{\partial \tilde{Y}_j}{\partial n}\right)\right] = 0. \tag{5.5.4}$$

Integrating Eq. (5.5.4) across the reaction zone, bounded between $n_f^- < n < n_f^+$, where n_f is the location of reaction sheet, we have

$$\int_{n_f^-}^{n_f^+} \frac{\partial}{\partial t}\left[\rho(\tilde{Y}_i - \tilde{Y}_j)\right] dn + \left[\rho u(\tilde{Y}_i - \tilde{Y}_j)\right]_{n_f^-}^{n_f^+} - \left[\rho D_i \frac{\partial \tilde{Y}_i}{\partial n} - \rho D_j \frac{\partial \tilde{Y}_j}{\partial n}\right]_{n_f^-}^{n_f^+} = 0. \tag{5.5.5}$$

In the limit of an infinitesimally thin reaction zone, and if the reaction sheet does not have any temporal discontinuity across it, the first term in Eq. (5.5.5) vanishes because $(n_f^+ - n_f^-) \to 0$. The second term also vanishes if we assume the values of the individual quantities, ρu, $\tilde{Y}_i$, and $\tilde{Y}_j$ are continuous across the reaction sheet such that

$$(\rho u)^+ = (\rho u)^-, \quad Y_i^+ = Y_i^-, \quad T^+ = T^-, \quad \text{etc.} \tag{5.5.6}$$

Consequently Eq. (5.5.5) becomes

$$\left[\rho D_i \frac{\partial \tilde{Y}_i}{\partial n}\right]_{n_f^-}^{n_f^+} = \left[\rho D_j \frac{\partial \tilde{Y}_j}{\partial n}\right]_{n_f^-}^{n_f^+}, \tag{5.5.7}$$

which, when expressed in Y_i, is

$$\left[\rho D_i \frac{\partial Y_i}{\partial n}\right]_{n_f^-}^{n_f^+} = \sigma_{i,j}\left[\rho D_j \frac{\partial Y_j}{\partial n}\right]_{n_f^-}^{n_f^+}. \tag{5.5.8}$$

Equation (5.5.8) clearly shows that, in crossing the reaction sheet, the change in the diffusive flux of the ith species is stoichiometrically proportional to the change in the diffusive flux of the jth species, which is physically reasonable. It relates the extent of the discontinuous changes in the concentration gradients in crossing the reaction sheet. Similar derivation involving Eqs. (5.5.1) and (5.5.2) leads to

$$\left(\frac{q_c}{\sigma_i}\right)\left[\rho D_i \frac{\partial Y_i}{\partial n}\right]_{n_f^-}^{n_f^+} = -\left[\lambda \frac{\partial T}{\partial n}\right]_{n_f^-}^{n_f^+}, \tag{5.5.9}$$

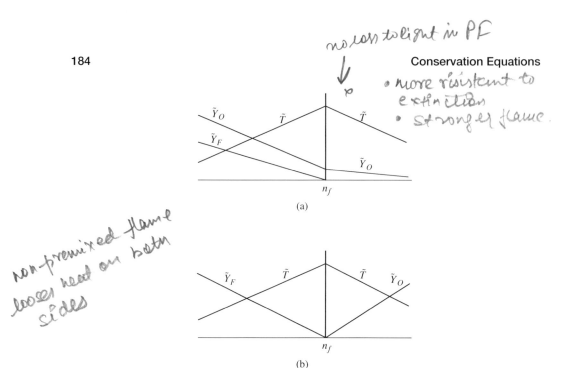

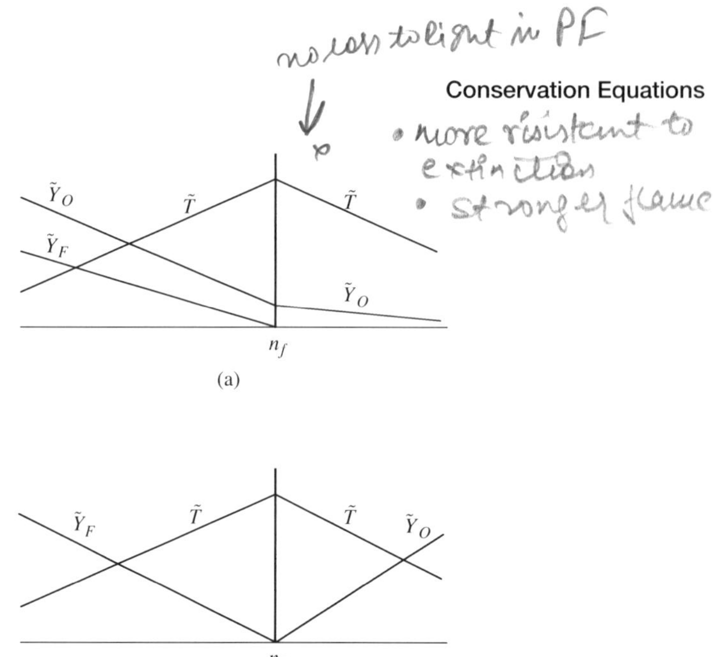

Figure 5.5.1. Flame structure in the vicinity of the reaction sheet for (a) premixed flame, (b) nonpremixed flame.

which shows that the amount of heat generated from the reaction is equal to the heat conducted away to both sides of the reaction zone.

Writing the above results in vector form, we have

$$\left[\rho D_i \mathbf{n} \cdot \nabla \tilde{Y}_i\right]_{n_f^-}^{n_f^+} = \left[\rho D_j \mathbf{n} \cdot \nabla \tilde{Y}_j\right]_{n_f^-}^{n_f^+} \tag{5.5.10}$$

and

$$\left[\rho D_i \mathbf{n} \cdot \nabla \tilde{Y}_i\right]_{n_f^-}^{n_f^+} = -\left[(\lambda/c_p)\mathbf{n} \cdot \nabla \tilde{h}^s\right]_{n_f^-}^{n_f^+}, \tag{5.5.11}$$

where $\mathbf{n}$ is the unit normal vector of the reaction surface.

An additional jump relation can be derived from one of Eqs. (5.5.1) to (5.5.3), accounting for the reaction term as a source/sink of the various reaction quantities. This will be discussed in Chapter 9.

The reaction sheet analyzed is a weak discontinuity in that only the gradients are discontinuous. As a comparison, the detonation wave to be studied later is a strong discontinuity in that the values of the flow variables are discontinuous across the wave surface.

Up to now we have neither specified the nature of the flame nor imposed the states of i and j at the reaction sheet. Thus the above results are applicable to all reacting species and to both premixed and nonpremixed flames. If we now refer i and j to fuel (F) and oxidizer (O) respectively, then the jump relations Eqs. (5.5.10) and (5.5.11) can be explicitly written for premixed and nonpremixed flames, as follows.

5.5.1.1. Premixed Flames: In a premixed flame fuel and oxidizer are supplied from the same side, say the negative side (Figure 5.5.1a). Thus if we assume that the fuel is the deficient species, then its concentration at the reaction sheet is $\tilde{Y}_{F,f} = 0$ and the concentration and energy jump relations are respectively given by

$$-\left[\rho D_F \mathbf{n} \cdot \nabla \tilde{Y}_F\right]_{n_f^-} = \left[\rho D_O \mathbf{n} \cdot \nabla \tilde{Y}_O\right]_{n_f^-}^{n_f^+}, \tag{5.5.12}$$

$$\left[\rho D_F \mathbf{n} \cdot \nabla \tilde{Y}_F\right]_{n_f^-} = \left[(\lambda/c_p)\mathbf{n} \cdot \nabla \tilde{h}^s\right]_{n_f^+} - \left[(\lambda/c_p)\mathbf{n} \cdot \nabla \tilde{h}^s\right]_{n_f^-}. \tag{5.5.13}$$

In the case that the downstream flow is adiabatic, then $\mathbf{n} \cdot \nabla \tilde{h}^s$ identically vanishes at n_f^+.

5.5.1.2. Nonpremixed Flames: In most situations involving nonpremixed flames, we shall simply specify $\tilde{Y}_{F,f} = \tilde{Y}_{O,f} = 0$ and thereby suppress leakage of the fuel and oxidizer across the reaction sheet. Thus if fuel and oxidizer are supplied from the negative and positive sides of the flame respectively (Figure 5.5.1b), then $(\mathbf{n} \cdot \nabla \tilde{Y}_F)$ and $(\mathbf{n} \cdot \nabla \tilde{Y}_O)$ vanish on the positive and negative sides respectively. Equation (5.5.10) becomes

$$\left[\rho D_F \mathbf{n} \cdot \nabla \tilde{Y}_F\right]_{n_f^-} = -\left[\rho D_O \mathbf{n} \cdot \nabla \tilde{Y}_O\right]_{n_f^+}, \tag{5.5.14}$$

while the energy jump relation is the same as Eq. (5.5.13). The negative sign in Eq. (5.5.14) illustrates that fuel and oxidizer are transported toward each other.

5.5.2. Adiabatic Flame Temperature

We can also derive a fairly general expression for the temperature of an unsteady three-dimensional reaction sheet, assuming equal density-weighted diffusivities, unity Lewis number, and the absence of heat loss and temporal pressure variation.

With the above assumptions, Eqs. (5.3.12) and (5.3.14) can be written for each side of the reaction sheet, with $\mathbf{x} < \mathbf{x}_f$ and $\mathbf{x} > \mathbf{x}_f$, as

$$\frac{\partial}{\partial t}(\rho \tilde{h}^s) + \nabla \cdot \left(\rho \mathbf{v} \tilde{h}^s - \rho D \nabla \tilde{h}^s\right) = 0, \tag{5.5.16}$$

$$\frac{\partial}{\partial t}(\rho \tilde{Y}_i) + \nabla \cdot \left(\rho \mathbf{v} \tilde{Y}_i - \rho D \nabla \tilde{Y}_i\right) = 0, \quad i = O, F. \tag{5.5.17}$$

A solution that satisfies Eqs. (5.5.16) and (5.5.17) on each side of the reaction zone is given by

$$\tilde{h}^s = c_{1,i} + c_{2,i} \tilde{Y}_i, \tag{5.5.18}$$

where $c_{1,i}$ and $c_{2,i}$ are constants. We now separately consider premixed and nonpremixed flames.

5.5.2.1. Premixed Flames: Here the temperature distribution on the reactant side of the reaction sheet, for a fuel-deficient mixture, is

$$(\tilde{h}^s)^- = c_{1,F} + c_{2,F} \tilde{Y}_F. \tag{5.5.19}$$

Evaluating Eq. (5.5.19) at the upstream boundary, where $(\tilde{h}^s)^- = \tilde{h}^s_u$ and $\tilde{Y}_F = \tilde{Y}_{F,u}$, and also at the reaction sheet where $(\tilde{h}^s)^- = \tilde{h}^s_f$ and $\tilde{Y}_F = 0$, $c_{1,F}$ and $c_{2,F}$ can be determined, giving

$$(\tilde{h}^s)^- = \tilde{h}^s_f - \frac{\left(\tilde{h}^s_f - \tilde{h}^s_u\right)}{\tilde{Y}_{F,u}} \tilde{Y}_F. \tag{5.5.20}$$

Substituting $(\tilde{h}^s)^-$ into the jump relation (5.5.13), it is seen that, for an adiabatic downstream, the flame temperature is

$$\tilde{h}^s_f = \tilde{h}^s_u + \tilde{Y}_{F,u}, \tag{5.5.21}$$

which, in dimensional form, is

$$h^s_f = h^s_u + Y_{F,u} q_c. \tag{5.5.22}$$

Equation (5.5.22) shows that the sensible enthalpy at the flame is the sum of the sensible and chemical enthalpies of the freestream. Thus T_f given by h^s_f is simply the adiabatic flame temperature, $T_{\rm ad}$, of the mixture. Alternatively, Eq. (5.5.22) can be interpreted as the statement that the heat release $Y_{F,u} q_c$ from burning $Y_{F,u}$ amount of fuel, per unit mass of the unburned mixture, which consists of the reactants and the inert, is used to increase the temperature of the mixture from T_u to T_f.

5.5.2.2. Nonpremixed Flames: Here the temperature distribution on the fuel and oxidizer sides of the reaction sheet can be respectively written as

$$(\tilde{h}^s)^- = c_{1,F} + c_{2,F} \tilde{Y}_F, \qquad (\tilde{h}^s)^+ = c_{1,O} + c_{2,O} \tilde{Y}_O. \tag{5.5.23}$$

Evaluation of $(\tilde{h}^s)^-$ is the same as that for the premixed flame, yielding

$$(\tilde{h}^s)^- = \tilde{h}^s_f - \frac{\left(\tilde{h}^s_f - \tilde{h}^s_{B^-}\right)}{\tilde{Y}_{F,B^-}} \tilde{Y}_F, \tag{5.5.24}$$

while a similar evaluation for $(\tilde{h}^s)^+$, with $(\tilde{h}^s)^+ = \tilde{h}^s_{B^+}$ at $\tilde{Y}_O = \tilde{Y}_{O,B^+}$, gives

$$(\tilde{h}^s)^+ = \tilde{h}^s_f - \frac{\left(\tilde{h}^s_f - \tilde{h}^s_{B^+}\right)}{\tilde{Y}_{O,B^+}} \tilde{Y}_O. \tag{5.5.25}$$

Substituting $(\tilde{h}^s)^\pm$ into the jump relations (5.5.13) and (5.5.14), we obtain

$$\frac{\tilde{h}^s_f - \tilde{h}^s_{B^-}}{\tilde{Y}_{F,B^-}} + \frac{\tilde{h}^s_f - \tilde{h}^s_{B^+}}{\tilde{Y}_{O,B^+}} = 1, \tag{5.5.26}$$

which can be expressed in dimensional form as

$$(Y_{F,B^-}) q_c = \left(h^s_f - h^s_{B^-}\right) + \left(\frac{Y_{F,B^-}}{Y_{O,B^+}} \sigma_O\right) \left(h^s_f - h^s_{B^+}\right). \tag{5.5.27}$$

Thus analogous to the interpretation of Eq. (5.5.22), Eq. (5.5.27) states that the heat release $(Y_{F,B^-}) q_c$ from burning Y_{F,B^-} amount of fuel in unit mass of the fuel mixture is used to heat this fuel mixture from T_{B^-} to T_f, and

$$\phi^* = \frac{Y_{F,B^-}}{Y_{O,B^+}}\sigma_O \tag{5.5.28}$$

amount of the oxidizer mixture from T_{B^+} to T_f. The fact that the burning of unit mass of the fuel mixture requires ϕ^* amount of the oxidizer mixture can be understood by recognizing that the reaction of Y_{F,B^-} amount of fuel stoichiometrically requires $\sigma_O Y_{F,B^-}$ amount of oxidizer, which corresponds to $\sigma_O Y_{F,B^-}/Y_{O,B^+}$ amount of the oxidizer mixture because it also consists of an inert. The T_f so determined is therefore the adiabatic flame temperature for the stoichiometric reaction between the fuel and oxidizer mixtures.

The parameter ϕ^* so identified can be interpreted as

$$\phi^* = \frac{\text{Available fuel/Stoichiometric fuel requirement}}{\text{Available oxidizer/Stoichiometric oxidizer requirement}}. \tag{5.5.29}$$

That is, ϕ^* is the ratio of the available fuel to the available oxidizer, each measured with reference to the stoichiometric requirement. Thus in analogy to premixed systems, we can interpret ϕ^* as the fuel-to-oxidizer equivalence ratio for nonpremixed systems. Furthermore, we can also define a normalized equivalence ratio for nonpremixed burning as

$$\Phi^* = \frac{\phi^*}{1+\phi^*}. \tag{5.5.30}$$

The significance of this parameter will be demonstrated in Chapter 6 on nonpremixed flames.

5.6. FURTHER DEVELOPMENT OF THE SIMPLIFIED DIFFUSION-CONTROLLED SYSTEM

5.6.1. Conservation Equations

In subsequent chapters, we shall frequently use the distinct diffusivity formulation and the one-step overall reaction given by (5.4.1) for illustration. The governing equations (5.3.12) and (5.3.14), when expressed in the stoichiometrically scaled variables, are

$$\frac{\partial}{\partial t}(\rho\tilde{h}^s) + \nabla\cdot\left[\rho\mathbf{v}\tilde{h}^s - (\lambda/c_p)\nabla\tilde{h}^s\right] = -w_F, \tag{5.6.1}$$

$$\frac{\partial}{\partial t}(\rho\tilde{Y}_i) + \nabla\cdot\left[\rho\mathbf{v}\tilde{Y}_i - \frac{1}{Le_i}(\lambda/c_p)\nabla\tilde{Y}_i\right] = w_F, \tag{5.6.2}$$

where we have used the fuel as the reference species, and have set $dp/dt = 0$ for simplicity. For the fuel–oxidizer reaction scheme of (5.4.20), the reaction rate w_F is

$$\begin{aligned} w_F &= -B\left(\frac{\nu_F' W_F}{Y_{F,B}}\right) c_O^{\nu_O'} c_F^{\nu_F'} T^\alpha e^{-E_a/R^oT} \\ &= -B_C \tilde{Y}_O^{\nu_O'} \tilde{Y}_F^{\nu_F'} \tilde{T}^{\alpha-(\nu_F'+\nu_O')} e^{-\tilde{T}_a/\tilde{T}}, \end{aligned} \tag{5.6.3}$$

where

$$B_C = B\left(\frac{\nu'_F W_F}{Y_{F,B}}\right)\left[\frac{\sigma_O(p\bar{W})/(R^o W_O)}{(q_c/c_p)}\right]^{\nu'_O}\left[\frac{p\bar{W}/(R^o W_F)}{(q_c/c_p)}\right]^{\nu'_F}(Y_{F,B}q_c/c_p)^{\alpha}, \quad (5.6.4)$$

and the relation $c_i = \rho Y_i/W_i = (p\bar{W}Y_i)/(R^o T W_i)$ has been used. B_C is thus a density-weighted collision rate of the reaction, having the unit gm/cm^3-sec. It may also be noted that since w_F varies more sensitively with the Arrhenius factor than with $\tilde{T}^{\alpha-(\nu'_F+\nu'_O)}$, and since chemical reaction is frequently confined to the region of the highest temperature characterized by the temperature $T_{\max}$, this term can be approximated by $\tilde{T}_{\max}^{\alpha-(\nu'_F+\nu'_O)}$ and hence absorbed in the definition of B_C in Eq. (5.6.4), as will be done from now on.

5.6.2. Nondimensional Numbers

Using ℓ_o as a reference length scale and ρ_o a reference density, we define

$$\tilde{\mathbf{x}} = \frac{\mathbf{x}}{\ell_o}, \qquad \tilde{t} = \frac{\lambda/c_p\rho_o}{\ell_o^2}t, \qquad \tilde{\mathbf{v}} = \frac{\ell_o}{\lambda/c_p\rho_o}\mathbf{v}. \quad (5.6.5)$$

Equations (5.6.1) and (5.6.2) then become

$$\left[\frac{\partial\tilde{\rho}}{\partial\tilde{t}} + \tilde{\nabla}\cdot(\tilde{\rho}\tilde{\mathbf{v}}) - \tilde{\nabla}^2\right]\tilde{h}^s = -\tilde{w}_F, \quad (5.6.6)$$

$$\left[\frac{\partial\tilde{\rho}}{\partial\tilde{t}} + \tilde{\nabla}\cdot(\tilde{\rho}\tilde{\mathbf{v}}) - \frac{1}{Le_i}\tilde{\nabla}^2\right]\tilde{Y}_i = \tilde{w}_F, \quad (5.6.7)$$

where

$$\tilde{w}_F = -Da_C\tilde{Y}_O^{\nu'_O}\tilde{Y}_F^{\nu'_F}e^{-E_a/R^oT}, \quad (5.6.8)$$

and we have defined a collision Damköhler number

$$Da_C = \frac{\ell_o^2 B_C}{\lambda/c_p}, \quad (5.6.9a)$$

which can also be expressed as

$$Da_C = \frac{\ell_o^2/(\lambda/c_p\rho_o)}{(\rho_o/B_C)}. \quad (5.6.9b)$$

The numerator of Da_C in Eq. (5.6.9b) is a characteristic diffusion time while the denominator is a characteristic collision time. Thus the collision Damköhler number represents the ratio of these two characteristic times,

$$Da_C = \frac{\text{Characteristic diffusion time}}{\text{Characteristic collision time}}. \quad (5.6.10)$$

In certain problems a reference velocity v_o instead of the reference length scale ℓ_o is given. Then in (5.6.5) we have instead $\tilde{\mathbf{v}} = \mathbf{v}/v_o$ such that $\ell_o = (\lambda/c_p\rho_o)/v_o$ and the numerator of Da_C in Eq. (5.6.9b) becomes $(\lambda/c_p\rho_o)/v_o^2$. Since both λ/c_p and B depend on the collision processes between molecules, Da_C/ℓ_o^2 can be further expressed in terms of the fundamental parameters characterizing such collisions and described by the kinetic theory of gases.

In a totally nondiffusive system such as the homogeneous flow mentioned in Chapter 2,

$$Da_C = \frac{\text{Characteristic flow time}}{\text{Characteristic collision time}} \tag{5.6.11}$$

in which the characteristic flow time is simply ℓ_o/v_o.

From our studies on chemical kinetics we know that not every collision results in reaction. Thus a more realistic measure of the time needed for reaction to consummate should include the Arrhenius factor, $\exp(-E_a/R^oT)$, which gives the probability of reaction from collisions. Furthermore, because of the largeness of the activation energy, we know reaction is concentrated in the region of maximum temperature, $T_{\max}$, which is frequently the flame temperature. Thus the Arrhenius factor should be evaluated at $T_{\max}$, giving $\exp(-E_a/R^oT_{\max}) = \exp(-Ar)$, where $Ar = E_a/R^oT_{\max}$ is the Arrhenius number defined in Chapter 2. We can now define a reaction Damköhler number as

$$Da = Da_C \exp(-Ar), \tag{5.6.12}$$

which has the physical significance of

$$Da = \frac{\text{Characteristic flow time or diffusion time}}{\text{Characteristic reaction time}} \tag{5.6.13}$$

Since Da is more relevant than Da_C, the reaction rate given by Eq. (5.6.8) can now be expressed as

$$\tilde{w}_F = -Da\tilde{Y}_O^{\nu'_O}\tilde{Y}_F^{\nu'_F} \exp\left[Ar\left(1 - \frac{\tilde{T}_{\max}}{\tilde{T}}\right)\right]. \tag{5.6.14}$$

The temperature-sensitive nature of the factor $\exp[Ar(1 - \tilde{T}_{\max}/\tilde{T})]$ in Eq. (5.6.14) has already been demonstrated in Figure 2.2.2. Specifically, for large values of Ar, the Arrhenius factor and thereby the reaction rate are exponentially small as long as $\tilde{T}$ is not too close to $\tilde{T}_{\max}$. It assumes an $O(1)$ value only when $\tilde{T}$ is sufficiently close to $\tilde{T}_{\max}$ such that $Ar(1 - \tilde{T}_{\max}/\tilde{T}) = O(1)$, which is equivalent to

$$\tilde{T}_{\max} - \tilde{T} = O(Ze^{-1}), \tag{5.6.15}$$

where we have defined a Zel'dovich number as

$$Ze = \frac{Ar}{\tilde{T}_{\max}} = \frac{\tilde{T}_a}{\tilde{T}^2_{\max}}. \tag{5.6.16}$$

Thus for large Zel'dovich number reactions the reaction zone is concentrated within a narrow region over which the temperature deviates from $\tilde{T}_{\max}$ only by $O(Ze^{-1})$. As $Ze \to \infty$, this region degenerates to a reaction sheet.

For a premixed flame, $T_{\max} \equiv T_b = T_u + Y_{F,u}q_c/c_p$ such that $\tilde{T}_b = 1/(1-\alpha)$ and $\tilde{T}_u = \alpha/(1-\alpha)$, where $\alpha = T_u/T_b \approx \rho_b/\rho_u$ is the upstream-to-downstream temperature ratio and $1/\alpha$ can be interpreted as the thermal expansion ratio across the flame. Then

$$Ze = (1 - \alpha)Ar. \tag{5.6.17}$$

Thus the implication that $Ar \gg 1$ corresponds to $Ze \gg 1$ simultaneously requires $\alpha \ll 1$. Since the upstream must be sufficiently cold to freeze the reaction, $Ze \approx Ar \gg 1$ is usually satisfied. A similar consideration can be extended to nonpremixed flames.

In summary, we have identified three nondimensional numbers that are of particular relevance to combustion phenomena. The first is the Lewis number, *Le*, which measures the relative rates of thermal conduction to mass diffusion. Deviation of *Le* from unity implies a local nonconservation of the total enthalpy, and can lead to flame temperatures which are either smaller or larger than the adiabatic flame temperature. This, in turn, affects the reaction rate in an Arrhenius manner.

The second is the Damköhler number, *Da*, which measures the residence time available for a chemical reaction of certain rate to proceed. Thus $Da \to 0$ for a chemically frozen situation because the reaction time is excessively long relative to the flow time available for the reaction to consummate. At the other extreme, $Da \to \infty$ in an equilibrium flow because reaction is completed instantly with vanishing reaction time. Finite values of *Da* indicate flow situations in which reaction occurs with finite rate. In particular, we must have $Da = O(1)$ in the reaction region of a flame.

The third is the Zel'dovich number, *Ze*, which measures the combined effects of the temperature sensitivity of the reaction, through $Ar = \tilde{T}_a/\tilde{T}_{\max}$, and the extent of heat release relative to the initial content of the mixture's sensible energy, through $(1 - \alpha)$. The Zel'dovich number assumes large values only when both the activation energy and chemical heat release are sufficiently large, which is usually satisfied for reactions of interest to combustion and for typical burning situations. A large *Ze* is responsible for the spatially or temporally localized nature of reaction regions or periods in combustion phenomena.

NOMENCLATURE

a	Speed of sound
Ar	Arrhenius number
B_C	Collision frequency factor
B_k	Constant in the frequency factor for the kth reaction
c	Progress variable
c_i	Molar concentration of species i (moles per unit volume)
$c_{p,i}$	Specific heat at constant pressure for species i
Da	Reaction Damköhler number
Da_C	Collision Damköhler number
$D_{i,j}$	Binary diffusion coefficient for species i and j

$D_{T,i}$ Thermal diffusion coefficient for species i
e Specific internal energy of the gas mixture
$E_{a,k}$ Activation energy for the kth reaction
$\mathbf{f}_i$ External body force per unit mass of species i
$\mathbf{g}$ Gravity acceleration vector
h_i Specific enthalpy of species i
h_i^o Specific standard heat of formation for species i at temperature T^o
h_i^s Specific sensible enthalpy relative to T^o for species i
K Total number of chemical reactions
L Total number of elements
Le Lewis number
M Mach number
$\mathbf{n}$ Unit normal vector
N Total number of chemical species
p Hydrostatic pressure
$\mathbf{P}$ Pressure tensor
$\mathbf{q}$ Heat flux vector
q_c Heat reaction per unit mass of fuel
$\mathbf{q}_R$ Radiant heat flux vector
R^o Universal gas constant
T Temperature
T_a Activation temperature
T^o Standard, reference temperature
$\mathbf{v}$ Mass-averaged velocity of the gas mixture
$\mathbf{V}_i$ Diffusion velocity of species i
w_i Mass production rate of species i (mass per unit volume per unit time)
$\overline{W}$ Average molecular weight
W_i Molecular weight of species i
X_i Mole fraction of species i
Y_i Mass fraction of species i
Z Mixture fraction
Ze Zel'dovich number
Z_k Mass fraction of element k
α Ratio of unburnt-to-burnt temperature
α_k Temperature exponent of the frequency factor for the kth reaction
β_i Coupling function for enthalpy and species i
$\beta_{i,j}$ Coupling function for species i and j
γ Specific heat ratio
κ Bulk viscosity coefficient
λ Thermal conductivity
μ Viscosity coefficient
$\nu'_{i,k}$ Stoichiometric coefficient for species i appearing as a reactant in reaction k
$\nu''_{i,k}$ Stoichiometric coefficient for species i appearing as a product in reaction k

ρ	Density
σ_i	Stoichiometric mass ratio of species i to fuel
ϕ	Fuel-to-oxidizer equivalence ratio
ϕ^*	Equivalence ratio defined for nonpremixed flames
Φ	Normalized equivalence ratio
ω	Species-independent reaction rate
ψ	Ψ per unit volume
Ψ	A general fluid property

Subscripts and Superscripts

F, O	Fuel, oxidizer
u, b	Unburned and burned states
f	Flame; reaction zone
$\sim$	Stoichiometrically weighted quantities
B	Flow boundary
$+, -$	Nonreactive regions separated by a reaction sheet

PROBLEMS

1. Consider a mixture of N species with the composition that, except for the Nth species, all the $(N-1)$ species exist in trace amounts such that not only $X_i \ll X_N$ and $Y_i \ll Y_N$ for $i \neq N$, but sums of the quantities weighted by either X_i or Y_i are also much smaller than the corresponding quantities involving N. Starting from the simplified diffusion relation Eq. (5.2.14),

$$\nabla X_i = \sum_{j=1}^{N} \left(\frac{X_i X_j}{D_{i,j}} \right) (\mathbf{V}_j - \mathbf{V}_i), \tag{5.P.1}$$

show that the diffusion velocity of the ith species can be approximated by

$$\mathbf{V}_i \approx -D_{i,N} \nabla \ln X_i \approx -D_{i,N} \nabla \ln Y_i. \tag{5.P.2}$$

2. Let us reconsider Problem 1, but be more precise with the assumption needed to arrive at the diffusion velocity $\mathbf{V}_i$ given by Eq. (5.P.2). Let's assume for simplicity that the concentration of the trace species i is proportional to that of N such that $X_i \approx b_i X_N$ and $Y_i \approx b_i Y_N$, where $b_i \ll 1$ and is a constant. Show that Eq. (5.P.2) holds if

$$\sum_{j \neq N} b_j \ll 1 \tag{5.P.3}$$

$$\sum_{j \neq N} b_j \left(\frac{D_{i,N} - D_{j,N}}{D_{i,j}} + \frac{D_{j,N}}{D_{i,N}} \right) \ll 1. \tag{5.P.4}$$

In deriving the above you may need to use Eq. (5.P.2). Note that more accurate representation for X_i and Y_i would be $X_i \approx a_i + b_i X_N$ and $Y_i = a_i' + b_i' Y_N$. The final expressions would become quite involved.

3. Starting from Eq. (5.2.14), show that the general solution for $\mathbf{V}$ is

$$\mathbf{V} = \mathbf{A}^{-1}\mathbf{B}, \tag{5.P.5}$$

where $\mathbf{V}$ and $\mathbf{B}$ are the column vectors $\{\mathbf{V}_i\}$ and $\{\nabla \ln X_i\}$ respectively, and $\mathbf{A}$ is the matrix with the diagonal and off-diagonal elements being

$$a_{i,i} = -\sum_{j \neq i} X_j / D_{i,j}, \quad a_{i,j} = X_j / D_{i,j}.$$

4. For a stoichiometric amount of methane–air mixture calculate the mass fractions and stoichiometrically weighted mass fractions (with the fuel being the reference species and $Y_{F,B} = 1$) of O_2 and CH_4 before reaction, and of CO_2 and H_2O after complete reaction without dissociation. What can you say about $\tilde{Y}_i$ and the small value of Y_{CH_4} relative to Y_{O_2} in terms of the suitability of hydrocarbons as transportation fuels?

5. (a) Starting from Eqs. (5.3.8) and (5.3.10), derive the appropriate coupling functions for a two-step reaction scheme

$$\sum_{i=1}^{N} \nu'_{i,1}\mathrm{M}_i \xrightarrow{k_1} \sum_{i=1}^{N} \nu''_{i,1}\mathrm{M}_i$$

$$\sum_{i=1}^{N} \nu'_{i,2}\mathrm{M}_i \xrightarrow{k_2} \sum_{i=1}^{N} \nu''_{i,2}\mathrm{M}_i$$

in an open system. Assume $Le = 1$.

(b) Write down all the possible coupling functions for the following scheme involving chlorine–hydrogen reactions:

$$\mathrm{Cl} + \mathrm{H_2} \rightarrow \mathrm{HCl} + \mathrm{H}$$

$$\mathrm{H} + \mathrm{Cl_2} \rightarrow \mathrm{HCl} + \mathrm{Cl}.$$

(c) Generalize the above results to a system of N species and K reactions. What is the relation between N and K?

6. Show that the flame-sheet temperature for $Le^{\pm} = 1$ but $\lambda^+ \neq \lambda^-$ and $(\rho D)^+ \neq (\rho D)^-$ is still the adiabatic flame temperature.

6 Laminar Nonpremixed Flames

Either by nature or design, in most combustion systems fuel and oxidizer are initially spatially separated. If the subsequent mixing between them is not sufficiently fast before chemical reaction is initiated, then the mixing and reaction will take place only in thin reaction zones that separate them. Examples are a wood panel on fire, an oil spray burning in a furnace, a candle flame, and the sparks (i.e., burning metal particles) generated when a metal surface is abraded.

The structure of a nonpremixed flame therefore consists of three zones, with a reaction zone separating a fuel-rich zone and an oxidizer-rich zone. Figure 6.1.1a shows a typical configuration for the model problem to be studied in the next section. As the fuel and oxidizer are transported toward each other, through diffusion as well as whatever convective motion the system may have, they become heated and eventually meet and mix within the reaction zone. Reaction between them subsequently takes place rapidly. The combustion products together with the heat of combustion are then transported away from the reaction zone in both directions. Since reaction occurs at a finite rate and the reaction zone has a finite thickness, complete reaction cannot be accomplished. Small amounts of fuel and oxidizer invariably leak through the reaction zone, as shown in Figure 6.1.1b.

It is, however, frequently useful to assume that the reaction occurs infinitely fast and thereby is confined to a reaction sheet. Fuel and oxidizer are each confined to their respective regions of supply, and attain vanishing concentrations at the reaction sheet. Consequently no leakage occurs. The reaction sheet then acts as a sink for the reactants and a source of combustion heat and products. The situation is shown in Figure 6.1.1c.

Since we have assumed an infinitely fast reaction rate, the relevant factors controlling the combustion phenomena here are the stoichiometric rates of transport of fuel and oxidizer to the reaction sheet, which also determine the heat release rate. Thus in this limit combustion is controlled by the slower process of diffusion as compared to reaction, and its characteristics can therefore be predicted without knowing the specific reaction mechanism and rate except the stoichiometry. It is for this reason that the flame is conventionally called a diffusion flame. We shall, however, purposely

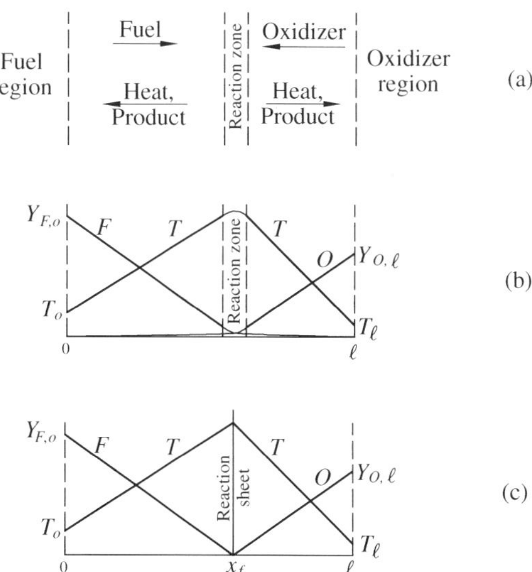

Figure 6.1.1. Structure of the nonpremixed flame: (a) physical configuration of a one-dimensional, purely diffusive system; (b) temperature and concentration profiles with finite flame thickness and reactant leakage; (c) temperature and concentration profiles with reaction-sheet assumption.

avoid using this terminology because calling a nonpremixed flame a diffusion flame could convey the erroneous impression that a premixed flame does not require diffusion.

In the presence of finite-rate kinetics, the reaction sheet is broadened, the reactants leak through it, and the burning rate is reduced. This reduction, however, is small and burning is still diffusion controlled. When leakage and the reduction in flame temperature become relatively severe, extinction occurs. Therefore, as long as the flame is burning steadily, the bulk combustion characteristics can be satisfactorily approximated by the reaction-sheet solution.

In this chapter, we shall therefore confine our study of nonpremixed combustion to the reaction-sheet limit. This will be illustrated by four problems, namely a chambered flame involving one-dimensional, planar, diffusive transport (Section 6.1); the classical Burke–Schumann flame as a simplified representation of jet combustion (Section 6.2); droplet vaporization and combustion (Section 6.4); and the counterflow flame (Section 6.5). Through separate analysis of these configurationally different nonpremixed flames, we shall demonstrate that they share some fundamental features such as the flame temperature and reaction-sheet location. In Section 6.3, the concept of Stefan flow, which arises from the gasification of condensed fuels, is introduced via the examples of a one-dimensional chambered flow and droplet vaporization and condensation.

The critical phenomenon of extinction of nonpremixed flames will be analyzed in Chapter 9. The problem of droplet and particle combustion is of sufficient fundamental and practical interest that it will be further studied via two-phase combustion in Chapter 13.

Before further discussion, we shall clarify the usage of the term "flame." From the fundamental viewpoint, a flame is a region within which the nonequilibrium processes of diffusion and reaction take place, and as such should consist of the fuel and oxidizer diffusion zones as well as the reaction zone shown in Figure 6.1.1. Conventionally, however, the word "flame" has been used to indicate either the reaction zone or more loosely the narrow, high temperature region in the flow. This has the potential of causing confusion in, for example, specifying the thickness of laminar "flamelets" in a turbulent flame brush. Thus in the following discussion, whenever possible, we shall be specific in distinguishing the reaction zone from the entire flame structure. At the same time, however, we shall also bow to tradition and sometimes use the word "flame" to designate a narrow high temperature region, which invariably also includes the reaction zone. Such designations are quite obvious and should not cause confusion.

6.1. THE ONE-DIMENSIONAL CHAMBERED FLAME

In this problem, we have a chamber with a constant cross-sectional area and a constant pressure throughout (Figure 6.1.1a). Two porous walls, situated at $x = 0$ and $x = \ell$, maintain constant concentrations and temperatures according to

$$\begin{aligned} x = 0:&\quad Y_F = Y_{F,o},\ Y_O = 0,\ Y_P = 0,\ T = T_o \\ x = \ell:&\quad Y_F = 0,\ Y_O = Y_{O,\ell},\ Y_P = 0,\ T = T_\ell. \end{aligned} \tag{6.1.1}$$

For simplicity we shall also assume that there is no net flow through the chamber. Fuel and oxidizer diffuse toward each other and react at a reaction-sheet located at x_f. The burning is steady and the combustion products are continuously eliminated through the walls. We aim to determine the combustion characteristics such as the temperature and species profiles, the temperature and location of the reaction sheet, and the consumption rates of the reactants. Note that since the reaction-sheet temperature is the highest in the flow field, it will be simply called the flame temperature.

The problem will be solved using the coupling function formulation, the reaction-sheet formulation, the mixture-fraction formulation, and the element conservation formulation so as to demonstrate the usefulness of these four procedures.

6.1.1. Coupling Function Formulation

Since there is no net mass flow, we have $u = 0$. By further assuming $Le_i = 1$, then Eqs. (5.4.27) and (5.4.25) are reduced to their respective one-dimensional forms,

$$\frac{d^2 \beta_i}{dx^2} = 0 \tag{6.1.2}$$

$$(\lambda/c_p)\frac{d^2 \tilde{T}}{dx^2} = w_F, \tag{6.1.3}$$

where $\beta_i = \tilde{T} + \tilde{Y}_i$ and we have assumed $c_p = \text{constant}$.

The solution for Eq. (6.1.2) is straightforward, given by

$$\beta_i = c_{1,i} + c_{2,i}x. \tag{6.1.4}$$

The constants $c_{1,i}$ and $c_{2,i}$ can be determined by applying the boundary conditions in (6.1.1). Thus for $i = F, O$, we have

$$\beta_F = \tilde{T} + \tilde{Y}_F = (\tilde{Y}_{F,o} + \tilde{T}_o) + (\tilde{T}_\ell - \tilde{T}_o - \tilde{Y}_{F,o})\tilde{x} \tag{6.1.5}$$

$$\beta_O = \tilde{T} + \tilde{Y}_O = \tilde{T}_o + (\tilde{T}_\ell - \tilde{T}_o + \tilde{Y}_{O,\ell})\tilde{x}, \tag{6.1.6}$$

where $\tilde{x} = x/\ell$. The above solutions are general, being valid for any w_F. We now apply the reaction-sheet assumption. Since there is no reactant leakage, we have

$$\tilde{Y}_F = 0, \quad \tilde{x}_f \le \tilde{x} \le 1; \quad \tilde{Y}_O = 0, \quad 0 \le \tilde{x} \le \tilde{x}_f. \tag{6.1.7}$$

Applying Eq. (6.1.7) to the expressions for β_F and β_O, we readily obtain the temperature profile as

$$\tilde{T}^- = \tilde{T}_o + (\tilde{T}_\ell - \tilde{T}_o + \tilde{Y}_{O,\ell})\tilde{x}, \qquad 0 \le \tilde{x} \le \tilde{x}_f \tag{6.1.8}$$

$$\tilde{T}^+ = (\tilde{T}_o + \tilde{Y}_{F,o}) + (\tilde{T}_\ell - \tilde{T}_o - \tilde{Y}_{F,o})\tilde{x}, \quad \tilde{x}_f \le \tilde{x} \le 1, \tag{6.1.9}$$

where $\tilde{T}^-$ and $\tilde{T}^+$ respectively represent the temperature distributions in the fuel and oxidizer sides of the reaction sheet. Knowing $\tilde{T}$, we can now substitute $\tilde{T}^-$ into β_F for $\tilde{Y}_F$ and $\tilde{T}^+$ into β_O for $\tilde{Y}_O$, giving

$$\tilde{Y}_F = \tilde{Y}_{F,o} - (\tilde{Y}_{F,o} + \tilde{Y}_{O,\ell})\tilde{x}, \quad 0 \le \tilde{x} \le \tilde{x}_f \tag{6.1.10}$$

$$\tilde{Y}_O = -\tilde{Y}_{F,o} + (\tilde{Y}_{F,o} + \tilde{Y}_{O,\ell})\tilde{x}, \quad \tilde{x}_f \le \tilde{x} \le 1. \tag{6.1.11}$$

The reaction sheet location $\tilde{x}_f$ is then given by setting, say, $\tilde{Y}_F = 0$ in Eq. (6.1.10), yielding

$$\tilde{x}_f = \frac{\tilde{Y}_{F,o}}{\tilde{Y}_{F,o} + \tilde{Y}_{O,\ell}} = \frac{1}{1 + \tilde{Y}_{O,\ell}}, \tag{6.1.12}$$

in which we have noted that $\tilde{Y}_{F,o} \equiv 1$ because $\tilde{Y}_F$ is defined as $Y_F/Y_{F,o}$. We nevertheless shall leave it in symbol form in most of the following derivations because of its direct physical meaning. Equation (6.1.12) shows that in order to achieve stoichiometry, the reaction sheet tends to be situated closer to the boundary at which the reactant concentration is stoichiometrically lower; that is, $\tilde{x}_f \to 0$ for $\tilde{Y}_{F,o} \ll \tilde{Y}_{O,\ell}$, and $\tilde{x}_f \to 1$ for $\tilde{Y}_{F,o} \gg \tilde{Y}_{O,\ell}$. Through this adjustment a steeper gradient and thereby faster diffusion rate can be achieved to compensate for the lower concentration. By the same reasoning, increasing the concentration at a boundary will cause the reaction sheet to recede from it. For $\tilde{Y}_{F,o} = \tilde{Y}_{O,\ell}$, $\tilde{x}_f = 1/2$; that is, the reaction sheet is situated midway between the boundaries.

Recognizing that $\tilde{Y}_{F,o}/\tilde{Y}_{O,\ell}$ is simply the fuel-to-oxidizer equivalence ratio for the nonpremixed system, as identified in Eq. (5.5.28), thus in terms of ϕ^*,

Eq. (6.1.12) becomes

$$\tilde{x}_f = \frac{\phi^*}{1+\phi^*}, \tag{6.1.13}$$

which shows that the scaled reaction sheet location depends only on the equivalence ratio ϕ^*. Furthermore, in terms of the normalized equivalence ratio $\Phi = \phi/(1+\phi)$ given by Eq. (5.5.30), $\tilde{x}_f$ is simply

$$\tilde{x}_f = \Phi^*, \tag{6.1.14}$$

which shows that the reaction sheet location varies linearly with Φ^*. This again demonstrates that Φ, instead of ϕ, is the fundamental parameter representing the equivalence ratio.

The flame temperature can be found by evaluating either Eq. (6.1.8) or Eq. (6.1.9) at the reaction sheet, and by using Eq. (6.1.14), yielding

$$\tilde{T}_f - \tilde{T}_o = (\tilde{T}_\ell - \tilde{T}_o + \tilde{Y}_{O,\ell})\Phi^*. \tag{6.1.15}$$

The flame temperature can also be expressed as

$$\frac{\tilde{T}_f - \tilde{T}_o}{\tilde{Y}_{F,o}} + \frac{\tilde{T}_f - \tilde{T}_\ell}{\tilde{Y}_{O,\ell}} = 1, \tag{6.1.16}$$

which is simply Eq. (5.5.26), previously derived for the flame temperature of a general nonpremixed reaction sheet for $Le = 1$, as should be the case.

The final quantity of interest is the mass flux of the fuel, which indicates the fuel consumption rate. By definition,

$$f_F = -(\lambda/c_p)\left(\frac{dY_F}{dx}\right)_{x_f}, \tag{6.1.17}$$

and we also have $f_O = \sigma_O f_F$. Substituting Eq. (6.1.10) into Eq. (6.1.17), we get

$$f_F = \frac{\lambda/c_p}{\ell}\frac{Y_{F,o}}{\Phi^*}. \tag{6.1.18}$$

The maximum burning rate occurs when $Y_{F,o} = Y_{O,\ell} = 1$, or $f_{F,\max} = (\lambda/c_p\ell)(1+\sigma_O^{-1})$, as is reasonable to expect.

We have thus determined most of the combustion characteristics of interest. The expediency with which the coupling function formulation can lead to the final solution is evident. It may also be noted that since the above solution is determined based on the conservation of coupling functions, we need not even separately use the energy conservation relation (6.1.3). Once the reaction-sheet assumption is invoked, w_F in Eq. (6.1.3) is suppressed to be zero in regions bounded away from the reaction sheet.

Frequently the problem is considered to be solved at this stage. However, if one is also interested in the product and inert distributions, they can be easily determined. Thus, using Eq. (6.1.2) for the product coupling function β_P, and applying the boundary conditions that $\tilde{Y}_P = Y_P = 0$ at $x = 0$ and ℓ, we find

$$\beta_P = \tilde{T} + \tilde{Y}_P = \tilde{T}_o + (\tilde{T}_\ell - \tilde{T}_o)\tilde{x}. \tag{6.1.19}$$

With $\tilde{T}^{\mp}$ given by Eqs. (6.1.8) and (6.1.9), the product distribution $\tilde{Y}_P^{\mp}$ can be explicitly expressed.

For the concentration of the inert, since it is a conserved scalar, Y_I satisfies

$$\frac{d^2 Y_I}{d\tilde{x}^2} = 0, \quad 0 \le \tilde{x} \le 1, \tag{6.1.20}$$

throughout the flow field. The solution is

$$Y_I = Y_{I,o} + (Y_{I,\ell} - Y_{I,o})\tilde{x}. \tag{6.1.21}$$

Furthermore, since $Y_{I,\ell} = 1 - Y_{O,\ell}$, $Y_{I,o} = 1 - Y_{F,o}$, Eq. (6.1.21) can be written as

$$Y_I = (1 - Y_{F,o}) + (Y_{F,o} - Y_{O,\ell})\tilde{x}. \tag{6.1.22}$$

The problem is solved completely. The species and temperature profiles are shown in Figure 6.1.1c. Note that in the absence of convection, all these functions vary linearly in the purely diffusing regions in planar flows.

6.1.2. Reaction-Sheet Formulation

The main drawback of the coupling function formulation is that it cannot be applied to systems with both convection and nonunity Lewis numbers because, as discussed in Chapter 5, coupling functions cannot be formed. For these systems the reaction-sheet formulation may be adopted.

To demonstrate the methodology, we shall solve the same problem with the understanding that the approach is really meant for the more complex situation of $u \neq 0$ and $Le_i \neq 1$.

The methodology involves first separately solving the concentration and temperature distributions in the two nonreactive regions separated by the reaction sheet. These solutions are then matched at the reaction sheet through the jump relations of (5.5.13) and (5.5.14).

Thus the governing equations become

$$0 \le \tilde{x} < \tilde{x}_f: \quad \frac{d^2 \tilde{Y}_F}{d\tilde{x}^2} = 0, \quad \frac{d^2 \tilde{T}^-}{d\tilde{x}^2} = 0, \tag{6.1.23}$$

$$\tilde{x}_f < \tilde{x} \le 1: \quad \frac{d^2 \tilde{Y}_O}{d\tilde{x}^2} = 0, \quad \frac{d^2 \tilde{T}^+}{d\tilde{x}^2} = 0. \tag{6.1.24}$$

Solving Eqs. (6.1.23) and (6.1.24) subject to the boundary conditions at the reaction sheet ($\tilde{T}^- = \tilde{T}^+ = \tilde{T}_f$, $\tilde{Y}_F = \tilde{Y}_O = 0$ at $\tilde{x} = \tilde{x}_f$) and the respective porous walls, we have

$$0 \le \tilde{x} \le \tilde{x}_f: \quad \tilde{Y}_F = \tilde{Y}_{F,o}\left(1 - \frac{\tilde{x}}{\tilde{x}_f}\right), \tag{6.1.25}$$

$$\tilde{T}^- = \tilde{T}_o + (\tilde{T}_f - \tilde{T}_o)\left(\frac{\tilde{x}}{\tilde{x}_f}\right), \tag{6.1.26}$$

$$\tilde{x}_f \le \tilde{x} \le 1: \qquad \tilde{Y}_O = \tilde{Y}_{O,\ell}\frac{(\tilde{x}-\tilde{x}_f)}{(1-\tilde{x}_f)}, \tag{6.1.27}$$

$$\tilde{T}^+ = \tilde{T}_f - (\tilde{T}_f - \tilde{T}_\ell)\frac{(\tilde{x}-\tilde{x}_f)}{(1-\tilde{x}_f)}. \tag{6.1.28}$$

Equations (6.1.25)–(6.1.28) are expressed in terms of $\tilde{x}_f$ and $\tilde{T}_f$, which are to be determined from the jump relations across the reaction sheet. With $Le_i = 1$, these relations are

$$\left(\frac{d\tilde{Y}_F}{d\tilde{x}}\right)_{\tilde{x}_f^-} = -\left(\frac{d\tilde{Y}_O}{d\tilde{x}}\right)_{\tilde{x}_f^+} \tag{6.1.29}$$

$$\left(\frac{d\tilde{Y}_F}{d\tilde{x}}\right)_{\tilde{x}_f^-} = \left(\frac{d\tilde{T}^+}{d\tilde{x}}\right)_{\tilde{x}_f^+} - \left(\frac{d\tilde{T}^-}{d\tilde{x}}\right)_{\tilde{x}_f^-}. \tag{6.1.30}$$

Thus substituting Eqs. (6.1.25) and (6.1.27) into Eq. (6.1.29), we readily obtain $\tilde{x}_f$ as given by Eq. (6.1.12). Further using Eqs. (6.1.25), (6.1.26), (6.1.28) and $\tilde{x}$ in Eq. (6.1.30), we obtain Eq. (6.1.16) for $\tilde{T}_f$.

To determine the product distribution, we note that Y_P separately satisfies $d^2Y_p^{\mp}/d\tilde{x}^2 = 0$ for the two nonreactive regions. Hence, its solution is

$$Y_P^- = Y_{P,f}\left(\frac{\tilde{x}}{\tilde{x}_f}\right), \qquad 0 \le \tilde{x} \le \tilde{x}_f, \tag{6.1.31}$$

$$Y_P^+ = Y_{P,f}\frac{(1-\tilde{x})}{(1-\tilde{x}_f)}, \qquad \tilde{x}_f \le \tilde{x} \le 1. \tag{6.1.32}$$

Equations (6.1.31) and (6.1.32) depend on the product concentration at the reaction sheet, $Y_{P,f}$, which can be determined by analyzing the inert concentration and overall mass conservation. Thus evaluating Eq. (6.1.22) at the reaction sheet, we have

$$Y_{I,f} = (1 - Y_{F,o}) + (Y_{F,o} - Y_{O,\ell})\tilde{x}_f, \tag{6.1.33}$$

which leads to $Y_{P,f} = 1 - Y_{I,f}$. It may be cautioned that in summing over the mass fractions, Y_i instead of its stoichiometry-weighted value $\tilde{Y}_i$ should be used.

6.1.3. Mixture Fraction Formulation

Using the formulation of Section 5.4.5, and identifying the boundaries B^- and B^+ as those corresponding to F and O, Eq. (5.4.44) readily yields the reaction-sheet location in Z-space, with $\tilde{Y}_{O,f} = \tilde{Y}_{F,f} = 0$, as

$$Z_f = \frac{\tilde{Y}_{F,o}}{\tilde{Y}_{F,o} + \tilde{Y}_{O,\ell}} = \Phi^*, \tag{6.1.34}$$

which is simply $\tilde{x}_f$ given by Eq. (6.1.14). Further evaluating Eq. (5.4.46) at the flame, we have the expression for the flame temperature $\tilde{T}_f$,

$$\tilde{T}_f - \tilde{T}_o = (\tilde{T}_\ell - \tilde{T}_o + \tilde{Y}_{O,\ell})Z_f, \tag{6.1.35}$$

which is just Eq. (6.1.15) by using Eq. (6.1.34) for Z_f.

The fuel and oxidizer concentrations are simply obtained by respectively setting $\tilde{Y}_O$ and $\tilde{Y}_F$ to zero in Eq. (5.4.44),

$$\tilde{Y}_F = \tilde{Y}_{F,o} - (\tilde{Y}_{F,o} + \tilde{Y}_{O,\ell})Z = \tilde{Y}_{F,o}\left(1 - \frac{Z}{Z_f}\right), \quad Z < Z_f \tag{6.1.36}$$

$$\tilde{Y}_O = -\tilde{Y}_{F,o} + (\tilde{Y}_{F,o} + \tilde{Y}_{O,\ell})Z = \tilde{Y}_{O,\ell}\left(\frac{Z - Z_f}{1 - Z_f}\right), \quad Z > Z_f, \tag{6.1.37}$$

while the temperature profile obtained by setting $\tilde{Y}_F$ and $\tilde{Y}_O$ to zero in Eqs. (5.4.45) and (5.4.46) respectively,

$$\tilde{T}^- = \tilde{T}_o + (\tilde{T}_\ell - \tilde{T}_o + \tilde{Y}_{O,\ell})Z = \tilde{T}_f - (\tilde{T}_f - \tilde{T}_o)\left(1 - \frac{Z}{Z_f}\right), \quad Z < Z_f, \tag{6.1.38}$$

$$\tilde{T}^+ = (\tilde{T}_o + \tilde{Y}_{F,o}) + (\tilde{T}_\ell - \tilde{T}_o - \tilde{Y}_{F,o})Z = \tilde{T}_f - (\tilde{T}_f - \tilde{T}_\ell)\left(\frac{Z - Z_f}{1 - Z_f}\right), \quad Z > Z_f. \tag{6.1.39}$$

The above results clearly demonstrate that $\tilde{T}_f$ by itself, and the flame location and all the scalar profiles in Z-space, are configurational independent and are therefore general properties of nonpremixed flames, as discussed in Section 5.4.5.

To relate the above results to the specific flame configuration of interest, we need to solve $L_D(Z) = 0$ of Eq. (5.4.40). For the present case it is simply

$$\frac{d^2 Z}{d\tilde{x}^2} = 0. \tag{6.1.40}$$

For $Z(0) = 0$ and $Z(1) = 1$, the solution of Eq. (6.1.40) is

$$Z = \tilde{x}. \tag{6.1.41}$$

Thus, for this problem, the mixture fraction Z is simply given by the physical coordinate $\tilde{x}$. This is because all conserved scalars vary linearly with $\tilde{x}$ for this simple problem. It also explains why Z_f of Eq. (6.1.34) is identical to $\tilde{x}_f$ given by Eq. (6.1.14).

6.1.4. Element Conservation Formulation

To demonstrate this approach, we first have to specify the fuel–oxidizer system. The simplest system is that of hydrogen reacting with oxygen to produce water,

$$2H_2 + O_2 \rightarrow 2H_2O. \tag{6.1.42}$$

Here we have three species:

$$i = 1, 2, 3 \quad \text{for} \quad H_2, O_2, H_2O,$$

and two elements:

$$k = 1, 2 \quad \text{for} \quad H, O.$$

Thus the element mass fraction Z_k of element k can be expressed in terms of Y_i and the coefficients $\mu_{i,k}$ as

$$Z_1 = \mu_{1,1} Y_1 + \mu_{3,1} Y_3, \tag{6.1.43}$$

$$Z_2 = \mu_{2,2} Y_2 + \mu_{3,2} Y_3. \tag{6.1.44}$$

The Z_ks are governed by Eq. (5.4.38),

$$\frac{d^2 Z_k}{d\tilde{x}^2} = 0, \tag{6.1.45}$$

with the boundary conditions,

$$\tilde{x} = 0: \quad Z_1 = \mu_{1,1} Y_{1,o}, \quad Z_2 = 0, \quad T = T_o \tag{6.1.46}$$

$$\tilde{x} = 1: \quad Z_1 = 0, \quad Z_2 = \mu_{2,2} Y_{2,\ell}, \quad T = T_\ell. \tag{6.1.47}$$

The solution of Eq. (6.1.45) is then

$$Z_1 = \mu_{1,1} Y_{1,o}(1 - \tilde{x}), \tag{6.1.48}$$

$$Z_2 = \mu_{2,2} Y_{2,\ell} \tilde{x}, \tag{6.1.49}$$

which holds throughout the flow field. At the reaction sheet, we have $Y_{1,f} = Y_{2,f} = 0$, which, when substituted into Eqs. (6.1.43), (6.1.44), (6.1.48), and (6.1.49), yields

$$\tilde{x}_f = \frac{(\mu_{1,1}/\mu_{3,1}) Y_{1,o}}{(\mu_{1,1}/\mu_{3,1}) Y_{1,o} + (\mu_{2,2}/\mu_{3,2}) Y_{2,\ell}} \tag{6.1.50}$$

and an expression for $Y_{3,f}$. Recognizing from (6.1.42) that $(\mu_{1,1}/\mu_{3,1})$=$(\nu_3'' W_3/\nu_1' W_1)$, and that $i = 1, 2$ are respectively the fuel (F) and oxidizer (O) of the present system, we retrieve from Eq. (6.1.50) the flame location expression of Eq. (6.1.14).

Finally, energy conservation is given by

$$\frac{d^2 h}{d\tilde{x}^2} = 0. \tag{6.1.51}$$

Using the definition $h = \sum_{i=1}^N Y_i h_i = c_p T + \sum_{i=1}^N Y_i h_i^o$, the boundary conditions $h_o = c_p T_o + Y_{1,o} h_1^o$, $h_\ell = c_p T_\ell + Y_{2,\ell} h_2^o$, and the definition of the heat of combustion,

$$q_c = h_1^o + \left(\frac{\mu_{1,1}}{\mu_{3,1}}\right)\left(\frac{\mu_{3,2}}{\mu_{2,2}}\right) h_2^o - \left(\frac{\mu_{1,1}}{\mu_{3,1}}\right) h_3^o, \tag{6.1.52}$$

the flame temperature expression, Eq. (6.1.16), can be readily derived.

6.2. THE BURKE–SCHUMANN FLAME

In 1928 Burke and Schumann presented the first detailed analysis of nonpremixed flames. The situation studied is the steady-state coaxial flow of a fuel gas issuing into an oxidizing gas, as shown schematically in Figure 6.2.1 and as photographic images in Figure 6.2.2. It is obvious on physical grounds that the resulting flame can be either closed or open at its tip, depending on the ratio of the inner to the outer

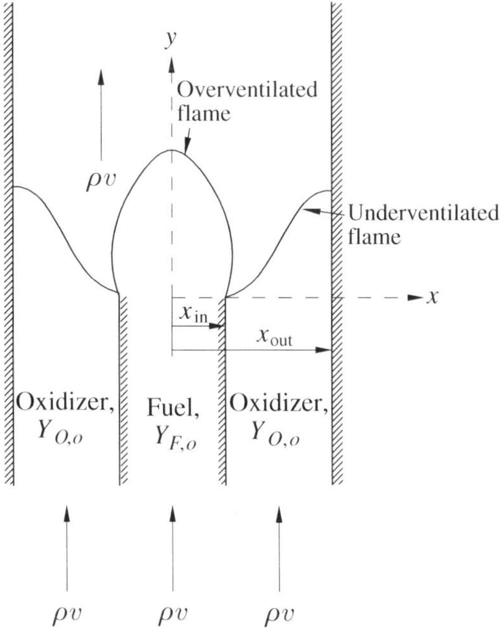

Figure 6.2.1. Schematic of the Burke–Schumann flame, with overventilated and underventilated configurations.

tube diameters, the fuel and oxidizer concentrations in their respective streams, and the fuel–oxidizer stoichiometric ratio. These flames are respectively known as overventilated (Figure 6.2.2a, c, d) and underventilated (Figure 6.2.2b) flames, depending on whether the rate of the oxidizer supply is stoichiometrically more or less than that of the fuel.

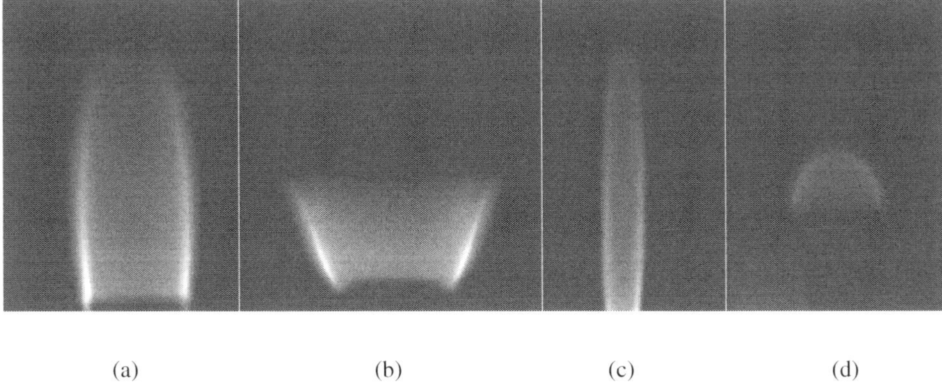

(a) (b) (c) (d)

Figure 6.2.2. Photographic images of the various Burke–Schumann flame configurations, obtained by issuing diluted methane into coflowing air of equal momentum: (a) overventilated flame with moderate $Pe(\sim 50)$, (b) underventilated flame with moderate $Pe(\sim 50)$, (c) overventilated flame with high $Pe(\sim 200)$, and overventilated flame with low $Pe(\sim 1)$. Inner tube diameter is 0.635 cm for (a) to (c), and 0.25 cm for (d). Outer tube diameter is 4 cm (Courtesy: S. W. Yoo).

For simplicity in illustration, we shall study the two-dimensional slot geometry instead of the usual cylindrical geometry. There is no loss in physical insight with the use of the mathematically simpler rectangular configuration.

The problem is governed in general by diffusion and convection in the streamwise (y) and transverse (x) directions (Figure 6.2.1). To simplify the analysis, we first assume that the mass flux is in the streamwise direction everywhere, implying that the x-flux, ρu, is zero. This is a reasonable assumption because, with the flow bounded by the outer walls, movement in the x-direction is restricted, being mainly induced by thermal expansion. Thus for the present steady flow, continuity given by Eq. (5.2.1) becomes

$$\frac{\partial}{\partial y}(\rho v) = 0. \tag{6.2.1}$$

Equation (6.2.1) shows that ρv is only a function of x. For further simplification, such a dependence is also suppressed by assuming that

$$\rho v = \text{constant}. \tag{6.2.2}$$

Equation (6.2.2) is probably the most restrictive assumption in the analysis. It limits the analysis to situations in which the freestream momentum fluxes of the oxidizer and fuel flows are the same.

In the original Burke–Schumann formulation diffusion in the streamwise direction is neglected. The analysis is therefore not applicable to flows with small velocities for which streamwise convection can be of the same order of, if not smaller than, streamwise diffusion. From the analytical viewpoint this restrictive assumption is not necessary and therefore will not be made in our analysis (Chung & Law 1984a). It must, however, also be recognized that since the present analysis neglects body force, it still becomes inaccurate for very small flow rates for which buoyancy effects can dominate the flow. Furthermore, there is also an upper limit for the flow rate at which the flow becomes turbulent. The transition Reynolds number is around 2,300 for tube flows.

Thus following the same approach as that of the previous section, the coupling function $\beta_i = \tilde{T} + \tilde{Y}_i$ for $Le_i = 1$ is governed by

$$\frac{\partial(\rho v \beta_i)}{\partial y} - \frac{\partial}{\partial x}\left[(\lambda/c_p)\frac{\partial \beta_i}{\partial x}\right] - \frac{\partial}{\partial y}\left[(\lambda/c_p)\frac{\partial \beta_i}{\partial y}\right] = 0. \tag{6.2.3}$$

Because ρv = constant and if we assume (λ/c_p) = constant, then Eq. (6.2.3) becomes

$$Pe\frac{\partial \beta_i}{\partial \tilde{y}} - \left(\frac{\partial^2 \beta_i}{\partial \tilde{x}^2} + \frac{\partial^2 \beta_i}{\partial \tilde{y}^2}\right) = 0, \tag{6.2.4}$$

where $\tilde{x} = x/x_{\text{out}}$, $\tilde{y} = y/x_{\text{out}}$, and

$$Pe = \frac{\rho v x_{\text{out}}}{\lambda/c_p}$$

is the Peclet number of the flow, measuring the relative intensities of convective to diffusive transports.

The boundary conditions for Eq. (6.2.4) are

$$0 < \tilde{x} < \tilde{x}_{\text{in}}, \quad \tilde{y} = 0: \quad \tilde{Y}_F = \tilde{Y}_{F,o}, \quad \tilde{T} = \tilde{T}_o \tag{6.2.5}$$

$$\tilde{x}_{\text{in}} < \tilde{x} < 1, \quad \tilde{y} = 0: \quad \tilde{Y}_O = \tilde{Y}_{O,o}, \quad \tilde{T} = \tilde{T}_o \tag{6.2.6}$$

$$\tilde{x} = 0, \quad \tilde{y} > 0: \quad \partial(\cdot)/\partial\tilde{x} = 0 \text{ (symmetry)} \tag{6.2.7}$$

$$\tilde{x} = 1, \quad \tilde{y} > 0: \quad \partial(\cdot)/\partial\tilde{x} = 0 \text{ (adiabatic wall)} \tag{6.2.8}$$

$$\tilde{y} \to \infty: \quad \text{boundedness.} \tag{6.2.9}$$

In writing Eq. (6.2.8) it is assumed that the wall is adiabatic and nonpermeable such that $\partial\tilde{T}/\partial\tilde{x} = 0$ and $\partial\tilde{Y}_i/\partial\tilde{x} = 0$ respectively. Furthermore, by specifying the temperature and reactant concentrations at the slot exit, we are suppressing back diffusion into the slot. This can be achieved experimentally by placing a porous plate at the slot exit.

The solution will be sought using the method of separation of variables. Thus let

$$\beta_i(\tilde{x}, \tilde{y}) = \Theta(\tilde{x})\Phi(\tilde{y}), \tag{6.2.10}$$

which, upon substitution into Eq. (6.2.4), yields

$$\frac{Pe\Phi' - \Phi''}{\Phi} = \frac{\ddot{\Theta}}{\Theta}, \tag{6.2.11}$$

where $\Phi' = d\Phi/d\tilde{y}$, $\dot{\Theta} = d\Theta/d\tilde{x}$, etc. For simplicity of notation we have omitted the subscript i for Θ and Φ and the quantities associated with them.

Since the LHS of Eq. (6.2.11) is only a function of $\tilde{y}$ while the RHS only a function of $\tilde{x}$, they can be equal only if they are both equal to a constant, $-k^2$. Thus $\Theta(\tilde{x})$ and $\Phi(\tilde{y})$ individually satisfy the following equations:

$$\ddot{\Theta} + k^2\Theta = 0 \tag{6.2.12}$$

$$\Phi'' - Pe\Phi' - k^2\Phi = 0, \tag{6.2.13}$$

which have the solutions

$$\begin{aligned} \Theta &= c_1 + c_2\tilde{x}, & k &= 0 \\ &= c_1 \cos k\tilde{x} + c_3 \sin k\tilde{x}, & k &\neq 0, \end{aligned} \tag{6.2.14}$$

$$\Phi = d_1 \exp\left\{\left[Pe + \sqrt{Pe^2 + 4k^2}\right]\tilde{y}/2\right\} + d_2 \exp\left\{\left[Pe - \sqrt{Pe^2 + 4k^2}\right]\tilde{y}/2\right\}. \tag{6.2.15}$$

Since $c_2 = c_3 = 0$ due to symmetry at $\tilde{x} = 0$, and $d_1 = 0$ due to boundedness as $\tilde{y} \to \infty$, β_i assumes the functional form

$$\beta_i = c(\cos k\tilde{x}) \exp\left\{\left[Pe - \sqrt{Pe^2 + 4k^2}\right]\tilde{y}/2\right\}. \tag{6.2.16}$$

Applying Eq. (6.2.8) to Eq. (6.2.16), we have $\sin k = 0$, which implies $k = n\pi$, $n = 0, 1, 2, \ldots$. Therefore β_i becomes

$$\beta_i = c_0 + \sum_{n=1}^{\infty} c_n(\cos n\pi\tilde{x}) \exp\left\{\left[Pe - \sqrt{Pe^2 + 4\pi^2 n^2}\right]\tilde{y}/2\right\}. \tag{6.2.17}$$

To determine the coefficients c_n, we note that at $\tilde{y} = 0$,

$$\beta_i = c_0 + \sum_{n=1}^{\infty} c_n \cos n\pi\tilde{x}, \quad i = O, F, \tag{6.2.18}$$

while the boundary conditions at $\tilde{y} = 0$, Eqs. (6.2.5) and (6.2.6), give

$$\begin{aligned} \beta_F &= \tilde{T}_o + \tilde{Y}_{F,o}, \qquad && 0 < \tilde{x} < \tilde{x}_{\text{in}} \\ &= \tilde{T}_o, && \tilde{x}_{\text{in}} < \tilde{x} < 1, \end{aligned} \tag{6.2.19}$$

$$\begin{aligned} \beta_O &= \tilde{T}_o, \qquad && 0 < \tilde{x} < \tilde{x}_{\text{in}} \\ &= \tilde{T}_o + \tilde{Y}_{O,o}, && \tilde{x}_{\text{in}} < \tilde{x} < 1. \end{aligned} \tag{6.2.20}$$

Hence, first equating and integrating β_F in Eqs. (6.2.18) and (6.2.19) from $\tilde{x} = 0$ to 1, we have

$$\begin{aligned} \int_0^1 \beta_F d\tilde{x} &= \int_0^{\tilde{x}_{\text{in}}} (\tilde{T}_o + \tilde{Y}_{F,o}) d\tilde{x} + \int_{\tilde{x}_{\text{in}}}^1 \tilde{T}_o d\tilde{x} \\ &= \int_0^1 c_0 d\tilde{x} + \sum_{n=1}^{\infty} c_n \int_0^1 \cos n\pi\tilde{x} d\tilde{x}, \end{aligned}$$

which gives

$$c_0 = \tilde{T}_o + \tilde{x}_{\text{in}} \tilde{Y}_{F,o}, \tag{6.2.21}$$

where we have used the result $\int_0^1 \cos n\pi\tilde{x} d\tilde{x} = 0$. Next, we multiply β_F by $\cos m\pi\tilde{x}$ and integrate it from $\tilde{x} = 0$ to 1 to obtain

$$\begin{aligned} \int_0^1 \beta_F \cos m\pi\tilde{x} d\tilde{x} &= (\tilde{T}_o + \tilde{Y}_{F,o}) \int_0^{\tilde{x}_{\text{in}}} \cos m\pi\tilde{x} d\tilde{x} + \tilde{T}_o \int_{\tilde{x}_{\text{in}}}^1 \cos m\pi\tilde{x} d\tilde{x} \\ &= c_0 \int_0^1 \cos m\pi\tilde{x} d\tilde{x} + \sum_{n=1}^{\infty} c_n \int_0^1 \cos m\pi\tilde{x} \cos n\pi\tilde{x} d\tilde{x}, \end{aligned}$$

which gives

$$c_n = 2\tilde{Y}_{F,o} \left(\frac{\sin n\pi\tilde{x}_{\text{in}}}{n\pi} \right), \tag{6.2.22}$$

where we have used

$$\begin{aligned} \int_0^1 \cos m\pi\tilde{x} \cos n\pi\tilde{x} d\tilde{x} &= 0, \qquad && m \neq n \\ &= 1/2, && m = n. \end{aligned}$$

Thus $\beta_F = \tilde{T} + \tilde{Y}_F$ is finally given by

$$\beta_F = \tilde{T}_o + \tilde{x}_{\text{in}} \tilde{Y}_{F,o} + 2\tilde{Y}_{F,o} \sum_{n=1}^{\infty} G_n(\tilde{x}, \tilde{y}; Pe), \tag{6.2.23}$$

where

$$G_n(\tilde{x}, \tilde{y}; Pe) = \left(\frac{\sin n\pi \tilde{x}_{\text{in}}}{n\pi}\right)(\cos n\pi \tilde{x}) \exp\left\{\left[Pe - \sqrt{Pe^2 + 4\pi^2 n^2}\right]\tilde{y}/2\right\}. \quad (6.2.24)$$

Similarly it can be shown that $\beta_O = \tilde{T} + \tilde{Y}_O$ is given by

$$\beta_O = \tilde{T}_o + (1 - \tilde{x}_{\text{in}})\tilde{Y}_{O,o} - 2\tilde{Y}_{O,o} \sum_{n=1}^{\infty} G_n(\tilde{x}, \tilde{y}; Pe). \quad (6.2.25)$$

Evaluating β_F and β_O at the reaction sheet, we obtain

$$\tilde{T}_f = \tilde{T}_o + \tilde{Y}_{F,o}\left[\tilde{x}_{\text{in}} + 2\sum_{n=1}^{\infty} G_n(\tilde{x}_f, \tilde{y}_f; Pe)\right] \quad (6.2.26)$$

$$\tilde{T}_f = \tilde{T}_o + \tilde{Y}_{O,o}\left[(1 - \tilde{x}_{\text{in}}) - 2\sum_{n=1}^{\infty} G_n(\tilde{x}_f, \tilde{y}_f; Pe)\right]. \quad (6.2.27)$$

From Eqs. (6.2.26) and (6.2.27), the reaction sheet location and temperature are given by

$$(1 - \tilde{x}_{\text{in}}) - 2\sum_{n=1}^{\infty} G_n(\tilde{x}_f, \tilde{y}_f; Pe) = \Phi^* \quad (6.2.28)$$

$$(\tilde{T}_f - \tilde{T}_o) = \tilde{Y}_{O,o}\Phi^*, \quad (6.2.29)$$

where $\phi^* = \tilde{Y}_{F,o}/\tilde{Y}_{O,o}$. Equation (6.2.28) shows that the relation defining the reaction sheet location and the physical aspects of the problem, namely $\tilde{x}_{\text{in}}$ and Pe, is again given by the stoichiometry parameter Φ^*, in conformity with Eq. (6.1.14) for the chambered flame. Furthermore the flame temperature expression also is in the same form as Eq. (6.1.15), recognizing that the freestream temperatures for the fuel and oxidizer are the same for the present problem.

Equation (6.2.28) shows that the transition between the underventilated and overventilated flames, when $\tilde{y}_f \to \infty$ and therefore $G_n \to 0$, occurs at

$$1 - \tilde{x}_{\text{in}} = \Phi^*, \quad (6.2.30)$$

which provides a unique relation between the dimension of the burner, the stoichiometry of the fuel–oxidizer mixture, and the freestream concentrations, as anticipated earlier. Specifically, it shows that for reduced (increased) fuel concentration, a larger (smaller) inner slot is needed to achieve tip opening.

The flame temperature can also be expressed as

$$\frac{\tilde{T}_f - \tilde{T}_o}{\tilde{Y}_{F,o}} + \frac{\tilde{T}_f - \tilde{T}_o}{\tilde{Y}_{O,o}} = 1, \quad (6.2.31)$$

yielding the equivalent of Eq. (6.1.16), as it should. Thus the flame temperature is again the adiabatic flame temperature, and is independent of the nature of the flow field characterized by Pe. As mentioned in Chapter 5, this result holds for general, three-dimensional reaction sheets.

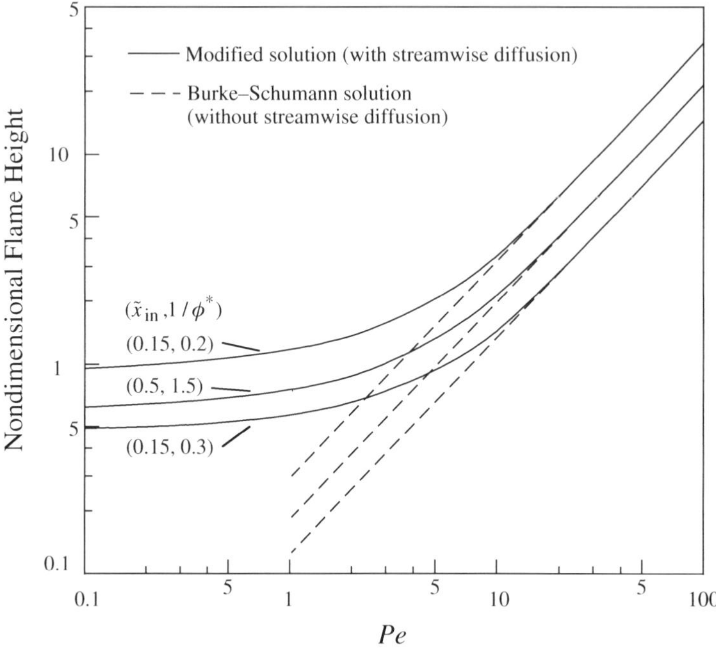

Figure 6.2.3. Variation of the flame height with the Peclet number for the Burke–Schumann flame with and without streamwise diffusion.

Figure 6.2.3 shows the variation of the nondimensional flame height, $\tilde{y}_f(\tilde{x}_f = 0)$, with Pe for an overventilated flame and various values of $\tilde{x}_{in}$ and ϕ^*. It is seen that, for large values of Pe, convection dominates over diffusion in the streamwise direction such that the flame height increases linearly with Pe. Since $Pe \sim \rho v/(\lambda/c_p)$, the flame height increases linearly with the mass flux ρv, but decreases with increasing diffusivity, $\lambda/c_p = \rho D$. Furthermore, since ρD is pressure insensitive, the flame height is also pressure insensitive. This behavior is basically that of the original Burke–Schumann flame, shown as the dashed lines in Figure 6.2.3, as should be the case. However, with decreasing Pe the flame height asymptotes to a constant value, signifying the progressive dominance of streamwise diffusion.

Figures 6.2.2c and 6.2.2d show photographs of overventilated Burke–Schumann flames with high and low Pe flows respectively. Specifically, the somewhat hemispherical flame shape of the low-Pe flame suggests the nearly equal importance of streamwise and transverse diffusion when streamwise convection is very weak, while the highly elongated shape of the high-Pe flame indicates the dominance of convection in the streamwise direction.

6.3. CONDENSED FUEL VAPORIZATION AND THE STEFAN FLOW

For the Burke–Schumann flame the convective motion is specified. Convection is usually induced by externally applied pressure gradients, and is controlled by the fluid mechanical aspects of the problem. However, as mentioned in Chapter 4, there

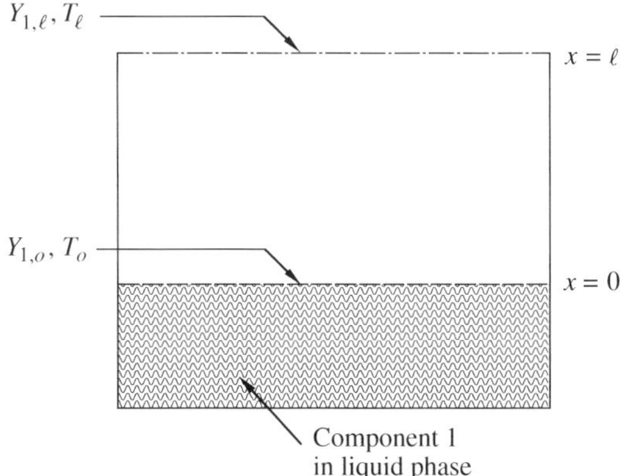

Figure 6.3.1. Schematic of vaporization from a one-dimensional chamber.

is another kind of convection, called the Stefan flow, which is internally generated and can be present even in the absence of any externally imposed flow. Such a flow frequently arises as a consequence of the gasification of a condensed fuel. Here a fuel vapor source is present at the surface of the condensed fuel, and the ambience, or the flame, which is located away from the surface, is typically low in the fuel vapor concentration and thus represents a sink for the fuel vapor. The fuel vapor then continuously diffuses from the surface to either the ambience or the flame. Since the consequence of this diffusion is a net transport of mass, a convective motion, that is, the Stefan flow, is induced. This motion can be quite significant and has to be accounted for, especially for rapid rates of vaporization when the condensed fuel is volatile, or when it is placed in a hot environment or undergoes combustion. This phenomenon is also particularly relevant for nonpremixed combustion because practical fuels are frequently present in the condensed phase, for example fuel droplets and coal particles.

To demonstrate this phenomenon we consider the following chemically nonreactive example. Here a pool of pure liquid, say water, designated by $i = 1$, undergoes vaporization in an open container (Figure 6.3.1), with its height ℓ fixed. Its vapor concentration at the surface is $Y_{1,o}$. There is a constant breeze over the container such that at its edge the gas composition is the same as that of the environment, consisting of $Y_{1,\ell}$ of species 1 and $Y_{2,\ell} = 1 - Y_{1,\ell}$ of a noncondensable species, say air. We aim to determine the mass vaporization flux, $f = (\rho u)_o$.

From continuity, $d(\rho u)/dx = 0$, we have

$$\rho u = f = \text{constant}, \tag{6.3.1}$$

which shows that this internally generated convection is a constant, and as such directly yields the mass vaporization flux.

From species conservation, Eq. (5.3.14), we have

$$f\frac{dY_i}{dx} - \frac{d}{dx}\left(\rho D\frac{dY_i}{dx}\right) = 0. \tag{6.3.2}$$

In writing Eq. (6.3.2) we have used Y_i instead of $\tilde{Y}_i$ because the stoichiometry proportionality constant between them is irrelevant for the present nonreactive case. Furthermore, since we now have a two-component system, the binary diffusion coefficient is uniquely defined, that is, $D_i \equiv D$.

Integrating Eq. (6.3.2) once, we have

$$fY_i - \rho D\frac{dY_i}{dx} = f_i, \tag{6.3.3}$$

where the constant of integration is simply the constant mass flux of species i, f_i, which is the sum of its convective and diffusive fluxes. Since species 2 is assumed to be noncondensable, its net mass flux vanishes, implying

$$f_2 \equiv 0 \tag{6.3.4}$$

such that

$$fY_2 - \rho D\frac{dY_2}{dx} = 0. \tag{6.3.5}$$

Applying the relation $Y_1 + Y_2 = 1$ into Eq. (6.3.5) then readily yields

$$fY_1 - \rho D\frac{dY_1}{dx} = f = f_1, \tag{6.3.6}$$

which shows that the net mass flux of the system is that of species 1.

Integrating Eq. (6.3.6), assuming that $\rho D =$ constant, and applying the boundary condition $Y_1(0) = Y_{1,o}$ and $Y_1(\ell) = Y_{1,\ell}$ yields the vaporization rate,

$$f = \frac{\rho D}{\ell}\ln\left(1 + B_{m,v}\right), \tag{6.3.7}$$

where we have defined a mass transfer number for vaporization,

$$B_{m,v} = \frac{Y_{1,o} - Y_{1,\ell}}{1 - Y_{1,o}}, \tag{6.3.8}$$

and the subscript v designates vaporization. For the $\ln(\cdot)$ term to be defined, we require $B_{m,v} > -1$ or $Y_{1,\ell} < 1$, which is always satisfied. Equation (6.3.7) shows that the vaporization flux f increases with increasing mass diffusivity D and decreasing diffusion distance ℓ, as should be the case. Furthermore, for $Y_{1,o} > Y_{1,\ell}$, we have $B_{m,v} > 0$ and therefore $f > 0$; this is the vaporization case. On the other hand, when $Y_{1,o} < Y_{1,\ell}$, we have $B_{m,v} < 0$ and therefore $f < 0$; this is the condensation case.

It is significant to note the logarithmic dependence of the mass flux for the present Stefan flow problem. For the special case of $Y_{1,o} \ll 1$ and $Y_{1,\ell} \ll 1$, we have $|B_{m,v}| \ll 1$. Thus expanding Eq. (6.3.7) we obtain

$$f \approx \frac{\rho D}{\ell}B_{m,v}, \tag{6.3.9}$$

which shows a linear dependence of f on $B_{m,v}$ for slow rates of vaporization and condensation. This result can be similarly obtained by carrying out the above derivation without the convection term in the governing equation (6.3.2).

It is perhaps somewhat surprising that we seem to be able to determine the liquid vaporization rate without considering energy conservation, even though it is physically obvious that heat transfer must be involved to effect vaporization. The linkage to energy conservation is the vapor concentration at the surface, $Y_{1,o}$, which is not known a priori and has to be determined through energy conservation and certain information regarding the phase change process. Physically, in order to ensure that the vapor concentration at the surface is maintained at $Y_{1,o}$ in the presence of its continuous transport to the ambience, the liquid needs to be continuously gasified. This gasification is effected through heat transfer from the ambience to the surface.

Heat transfer is described by the energy conservation equation (5.3.12), which, for constant c_p, is

$$fc_p\frac{dT}{dx} - \frac{d}{dx}\left(\lambda\frac{dT}{dx}\right) = 0. \tag{6.3.10}$$

Equation (6.3.10) is in the same form as the species conservation relation, Eq. (6.3.2). Integrating it once yields

$$fc_pT - \lambda\frac{dT}{dx} = \text{constant}. \tag{6.3.11}$$

An independent statement regarding the vaporization process at the liquid surface can also be made. If we assume the container is well insulated such that all the heat conducted to the liquid surface is used to vaporize a certain amount of the liquid, and if q_v is the latent heat of vaporization for unit mass of liquid, then

$$\lambda\left(\frac{dT}{dx}\right)_0 = fq_v, \tag{6.3.12}$$

where q_v is the specific latent heat of vaporization. Evaluating Eq. (6.3.11) by using Eq. (6.3.12), we have

$$fc_p(T - T_o) - \lambda\frac{dT}{dx} = -fq_v. \tag{6.3.13}$$

Integrating Eq. (6.3.13), assuming λ = constant, and applying the boundary conditions $T(0) = T_o$ and $T(\ell) = T_\ell$ yields

$$f = \frac{\lambda/c_p}{\ell}\ln(1 + B_{h,v}), \tag{6.3.14}$$

where $B_{h,v}$ is a heat transfer number for vaporization, defined as

$$B_{h,v} = \frac{c_p(T_\ell - T_o)}{q_v}. \tag{6.3.15}$$

Equation (6.3.15) shows that vaporization occurs when $T_\ell > T_o$ such that there is a continuous supply of heat from the environment to the liquid in order to sustain vaporization; on the other hand condensation occurs when $T_\ell < T_o$ such that the condensation heat generated is continuously dissipated to the ambience. The heat transfer number represents the ratio of the "driving potential" for gasification, namely the enthalpy difference between the two boundaries, to the "resistance" to gasification, which is the latent heat of vaporization.

We again note the logarithmic variation of f. Thus, except for very slow rates of vaporization, an increase in the ambient temperature, relative to the surface temperature, does not lead to a corresponding fractional increase in the gasification rate. In fact, the influence becomes progressively weaker at higher temperatures.

We have therefore derived an alternate expression for the mass vaporization flux by considering energy conservation alone. In equilibrium the two expressions for f, Eqs. (6.3.7) and (6.3.14), must agree. Furthermore if for simplicity we assume unity Lewis number, $\lambda/(c_p \rho D) = 1$, then we have $B_{m,v} = B_{h,v}$, or

$$\frac{Y_{1,o} - Y_{1,\ell}}{1 - Y_{1,o}} = \frac{c_p(T_\ell - T_o)}{q_v} = B_{h,v}. \tag{6.3.16}$$

Thus $Y_{1,o}$ can be solved as

$$Y_{1,o}(T_o) = \frac{Y_{1,\ell} + B_{h,v}}{1 + B_{h,v}}, \tag{6.3.17}$$

where $B_{h,v} = B_{h,v}(T_o)$ is given by Eq. (6.3.15).

Since the container is insulated, the liquid temperature T_o is not given but is rather a parameter to be determined. To do so we need the final piece of information to fully describe the system. One may realize that so far we have not said anything about the volatility of the liquid. Obviously rubbing alcohol would vaporize much faster than lubricating oil! Now, if we assume that the phase change reaction

$$\text{M(liquid)} \rightleftharpoons \text{M(gas)} \tag{6.3.18}$$

at the surface occurs so fast, as compared to the gas-phase mass diffusion rate, and that the vapor concentration at the surface is saturated, then equilibrium vaporization occurs such that a definite relation exists between the surface temperature and vapor concentration. The most frequently used relation is the Clausius–Clapeyron relation, Eq. (1.2.28), developed for ideal gases,

$$p_i(T) = p_n \exp\left[\frac{Q_v}{R^o}\left(\frac{1}{T_{b,n}} - \frac{1}{T}\right)\right], \tag{6.3.19}$$

where p_i is the partial pressure of the vaporizing liquid, $T_{b,n}$ the boiling point of the liquid at the reference pressure p_n, $Q_v(\equiv \bar{q}_v)$ the molar latent heat of vaporization, and $T \equiv T_o$ for the present problem. Usually p_n is the normal atmospheric pressure such that $T_{b,n}$ is the normal boiling point. Since $X_i = p_i/p$, where p is the prevailing pressure, Eq. (6.3.19) can also be expressed as

$$X_i(T) = \left(\frac{p_n}{p}\right) \exp\left[\frac{Q_v}{R^o}\left(\frac{1}{T_{b,n}} - \frac{1}{T}\right)\right]. \tag{6.3.20}$$

Equation (6.3.20) provides an additional relation between the surface concentration and the surface temperature, $X_{1,o}(T_o)$. Together with Eq. (6.3.17), T_o and $X_{1,o}$ can be iteratively determined. The value of T_o so determined is called the wet-bulb

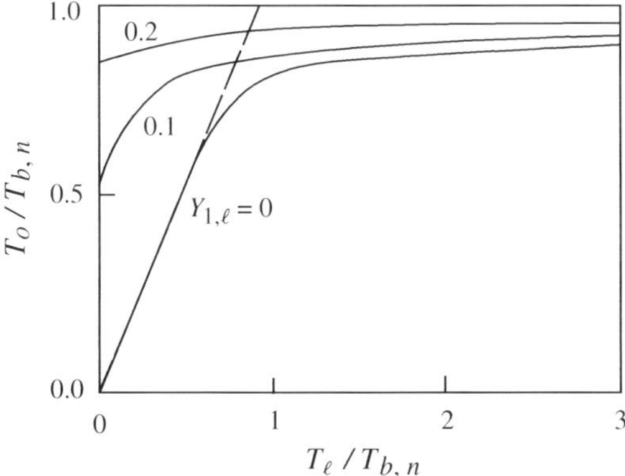

Figure 6.3.2. Normalized wet-bulb temperature of water as a function of the normalized ambient temperature and water vapor concentration.

temperature (Figure 6.3.2). It increases with the ambient temperature, pressure, and vapor concentration $Y_{1,\ell}$, and asymptotically approaches the boiling point of the liquid with increasing ambient temperature.

The problem just studied exemplifies the situation of simultaneous heat and mass transfer, which are coupled through some thermodynamic or chemical processes dependent on both temperature and species concentrations.

6.4. DROPLET VAPORIZATION AND COMBUSTION

6.4.1. Phenomenology

There are many technological processes in which it is desired to gasify and possibly also combust a given mass of liquid in a gaseous medium at either a very fast or a controlled rate. Since the heat and mass exchange rates between the liquid mass and the gaseous medium increase with increasing interfacial area, a standard technique to increase the overall gasification rate is to disperse the liquid mass into an ensemble of liquid fragments such that the total surface area of the fragments is much greater than that of the original liquid mass. The act of dispersal is called spraying or atomization. Since these fragments are typically very small, they contract under surface tension to form droplets that are nearly spherical in shape. Therefore in order to understand the overall spray behavior, it is necessary to study how a single droplet exchanges heat, mass, and momentum with its gaseous medium.

Droplet vaporization and combustion are rich in problems related to fluid dynamics and heat and mass transfer. For example, in realistic situations a nonradial velocity usually exists between the droplet and the ambience. This can be caused either by the inertia of the droplet acquired during spraying, or by its slower response, compared to the gas, to changes in the flow velocity and configuration. Even in a stagnant environment, buoyancy alone can induce a relative velocity. Such a relative velocity, when coupled to the Stefan flow resulting from surface gasification, can lead to interesting considerations regarding the net fuel gasification rate, the drag experienced by the droplet, and the generation of internal circulatory motion within the droplet.

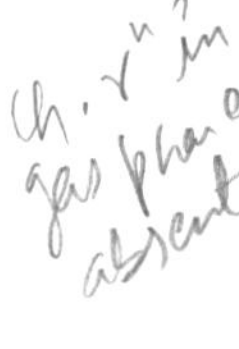

Since droplet gasification is basically a two-phase flow problem, a complete analysis will involve the four interacting processes consisting of liquid-phase transport, gas-phase transport, phase change at liquid–gas interface, and chemical reactions in the gas phase. The last component is absent for pure vaporization. In this section we shall confine ourselves to study only the simplest possible situation of droplet gasification, namely the d^2-law of droplet vaporization and combustion. In this situation the droplet is treated as a constant source of single-component fuel vapor, implying that we do not need to be concerned with the heat and mass transfer processes within the droplet interior. Furthermore, external forced and natural convection is absent such that spherical symmetry prevails. Finally, because of the significant density disparity between liquid and gas, the liquid possesses great inertia such that its properties at the droplet surface, for example, the regression rate as well as the temperature and species concentrations in more complicated situations, change at rates much slower than those of the gas-phase transport processes. Since the ambience is also assumed to be constant, the gas-phase processes can therefore be treated as steady, with the boundary variations occurring at longer time scales. This is called the quasi-steady assumption, which is frequently invoked in problems involving gasification and combustion of condensed-phase materials.

The d^2-law model embodies much of the essential physics governing droplet gasification and yields rough estimates of the gasification rate. Formulation of this model and the experimental verifications were first reported by Spalding (1953), Godsave (1953), Goldsmith and Penner (1954), and Wise, Lorell, and Wood (1955). Further developments on droplet combustion can be found in Faeth (1977), Law (1982), Law and Law (1993), and Sirignano (1999). In Chapter 13, we shall revisit droplet combustion and study such topics as modifications of the d^2-law, multicomponent droplet combustion, droplet dynamics and convection effects, high-pressure combustion, and droplet interaction.

6.4.2. d^2-Law of Droplet Vaporization

The problem to be studied is shown in Figure 6.4.1. Here a droplet of radius r_s vaporizes in an environment of temperature T_∞ and mass fraction $Y_{1,\infty}$ of the vaporizing species 1, which we shall designate as the fuel; the rest is air. The vaporization process is basically the same as that of the previous section. Thus in the case of vaporization the relatively cold droplet receives heat from the hot ambience and gasifies. The gasified fuel is transported to the ambience, which has a lower concentration of the fuel vapor. The transport is through both diffusion and Stefan convection, causing the continuous "shrinking" of the droplet. The reverse holds for condensation. We aim to determine the vaporization or condensation rate.

Analysis of the problem is analogous to the previous one, except now the system is spherically symmetric and the cross-sectional area of the flow field, $4\pi r^2$, continuously increases in the flow direction, where r is the radial coordinate. We have also

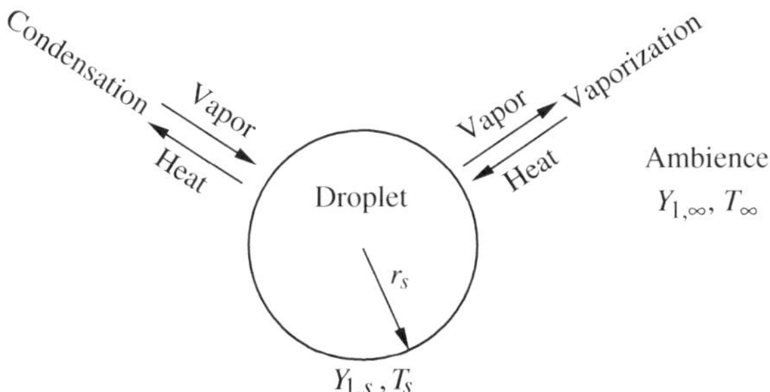

Figure 6.4.1. Schematic showing spherically symmetric droplet vaporization.

changed the subscript for the surface from o to s, and that for the outer boundary from ℓ to ∞. The liquid surface is now situated at a finite location, r_s.

Continuity, $d(r^2 \rho u)/dr = 0$, yields the constant mass flow rate

$$m_v = 4\pi r^2 \rho u, \tag{6.4.1}$$

where u is the radial velocity.

Taking a slightly different approach from that of the last section, integrating twice the fuel conservation equation,

$$\frac{d}{dr}\left(r^2 \rho u Y_1 - \rho D r^2 \frac{dY_1}{dr}\right) = 0, \tag{6.4.2}$$

subject to the boundary conditions $Y_1(r_s) = Y_{1,s}$ and $Y_1(\infty) = Y_{1,\infty}$ yields

$$Y_1 = Y_{1,s} + (Y_{1,\infty} - Y_{1,s})\left[\frac{e^{\tilde{m}_v(1-\tilde{r}^{-1})} - 1}{e^{\tilde{m}_v} - 1}\right], \tag{6.4.3}$$

where we have defined the nondimensional quantities $\tilde{m}_v = m_v/[4\pi(\lambda/c_p)r_s]$ and $\tilde{r} = r/r_s$.

Applying Y_1 given by Eq. (6.4.3) to the condition that the mass flow rate of component 1 constitutes the total mass flow rate,

$$\tilde{m}_v Y_1 - \tilde{r}^2 \frac{dY_1}{d\tilde{r}} = \tilde{m}_v, \tag{6.4.4}$$

the mass vaporization rate is given by

$$\tilde{m}_v = \ln(1 + B_{m,v}), \tag{6.4.5}$$

where the mass transfer number is identical to Eq. (6.3.8),

$$B_{m,v} = \frac{Y_{1,s} - Y_{1,\infty}}{1 - Y_{1,s}}. \tag{6.4.6}$$

This shows that the mass transfer number so identified is a universal parameter independent of the geometry of the system.

Working with energy conservation it can also be shown that

$$\tilde{m}_v = \ln(1 + B_{h,v}), \tag{6.4.7}$$

where the heat transfer number is identical to Eq. (6.3.15),

$$B_{h,v} = \frac{c_p(T_\infty - T_s)}{q_v}. \tag{6.4.8}$$

Since $B_{m,v}$ and $B_{h,v}$ are system-independent, the droplet temperature T_s is identical to that determined in the previous problem and as such is also system independent.

Expressing Eq. (6.4.7) in dimensional form,

$$m_v = 4\pi(\lambda/c_p)r_s \ln(1 + B_{h,v}), \tag{6.4.9}$$

we see that the mass vaporization rate increases linearly with the droplet radius r_s. Furthermore, the mass flux at the surface, given by

$$f_{v,s} = \frac{m_v}{4\pi r_s^2} = \frac{\lambda/c_p}{r_s} \ln(1 + B_{h,v}), \tag{6.4.10}$$

has the same form as Eq. (6.3.14) by identifying r_s and ℓ as the characteristic lengths for the respective problems.

Finally, we need to determine the instantaneous droplet size, r_s. By definition m_v is the rate of change of the droplet mass,

$$m_v = -\frac{d}{dt}\left(\frac{4}{3}\pi r_s^3 \rho_\ell\right), \tag{6.4.11}$$

where ρ_ℓ is the liquid density. Assuming $\rho_\ell =$ constant, Eq. (6.4.11) can be expressed as

$$m_v = -2\pi \rho_\ell r_s \frac{dr_s^2}{dt}. \tag{6.4.12}$$

Equating Eqs. (6.4.9) and (6.4.12), we have

$$\frac{dr_s^2}{dt} = -\frac{2(\lambda/c_p)}{\rho_\ell} \ln(1 + B_{h,v}). \tag{6.4.13}$$

Noting that the quantity on the RHS of Eq. (6.4.13) is independent of the droplet size, a vaporization rate constant can be defined as

$$K_v = \frac{2(\lambda/c_p)}{\rho_\ell} \ln(1 + B_{h,v}), \tag{6.4.14}$$

which is also called the surface regression rate when it is multiplied by 4π. Equation (6.4.13) can be readily integrated, using the initial condition that $r_s = r_{s,o}$ at $t = 0$, yielding the instantaneous droplet size

$$r_s^2 = r_{s,o}^2 - K_v t. \tag{6.4.15}$$

Equation (6.4.15) shows that the square of the droplet radius decreases linearly with time (Figure 6.4.2). This is the d^2-law of droplet vaporization, where d stands for the droplet diameter. This result is physically reasonable because the phenomenon of interest is a spherically symmetric, diffusion-controlled process in which quantities

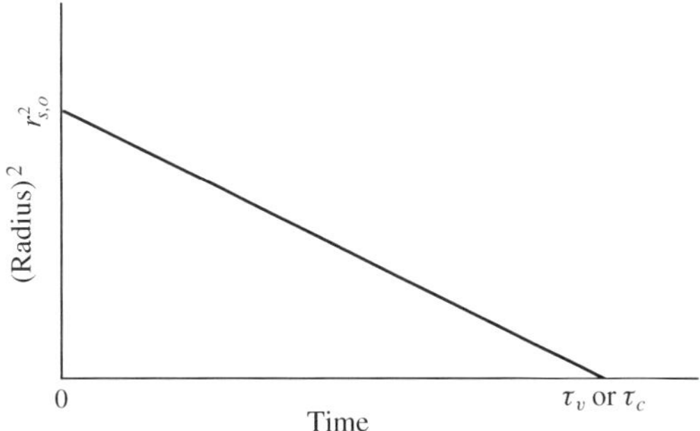

Figure 6.4.2. d^2-Law behavior of droplet vaporization and combustion.

vary with the surface area of the propagating sphere of influence. The d^2-variation has been repeatedly shown through experimentation to be largely correct.

The time τ_v needed to completely vaporize a droplet of initial size $r_{s,o}$ is obtained by setting $r_s = 0$ in Eq. (6.4.15),

$$\tau_v = \frac{r_{s,o}^2}{K_v}, \tag{6.4.16}$$

which illustrates the importance of fine atomization in that the time to achieve complete vaporization decreases quadratically with the initial droplet size.

6.4.3. d^2-Law of Droplet Combustion

Figure 6.4.3 shows the spherically symmetric droplet combustion process. The droplet gasification mechanism is basically the same as that of droplet vaporization, except now the heat source is the flame instead of the ambience. The flame also serves as the sink for the outwardly transported fuel vapor and inwardly transported oxidizer gas. Figure 6.4.4 shows representative temperature and concentration profiles.

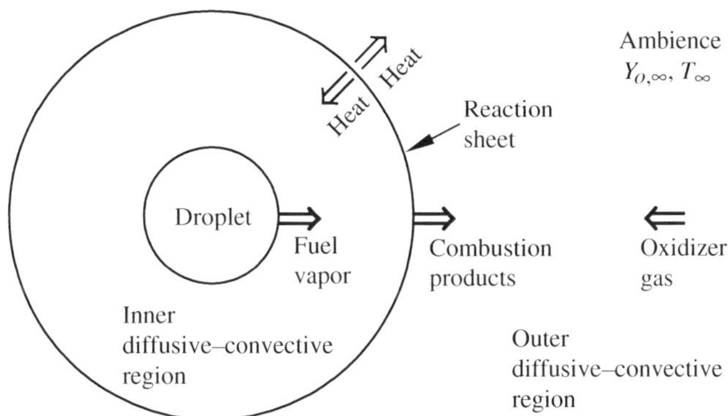

Figure 6.4.3. Schematic showing spherically symmetric droplet combustion.

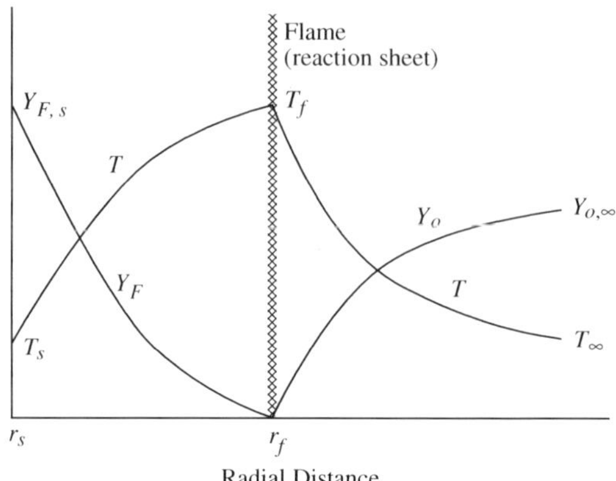

Figure 6.4.4. Concentration and temperature profiles for the reaction-sheet combustion of a droplet.

Continuity yields the constant mass burning rate m_c, given by

$$m_c = 4\pi r^2 \rho u. \tag{6.4.17}$$

Conservation of coupling function $\beta_i = \tilde{T} + \tilde{Y}_i$ is

$$\frac{d}{dr}\left[r^2 \rho u \beta_i - (\lambda/c_p) r^2 \frac{d\beta_i}{dr}\right] = 0, \quad i = O, F. \tag{6.4.18}$$

Assuming $(\lambda/c_p) =$ constant, and defining $\tilde{m}_c = m_c/[4\pi(\lambda/c_p) r_s]$, Eq. (6.4.18) becomes

$$\frac{d}{d\tilde{r}}\left(\tilde{m}_c \beta_i - \tilde{r}^2 \frac{d\beta_i}{d\tilde{r}}\right) = 0, \quad i = O, F, \tag{6.4.19}$$

whose first and second integrals are

$$\tilde{r}^2 \frac{d\beta_i}{d\tilde{r}} = -c_{1,i} + \tilde{m}_c \beta_i \tag{6.4.20}$$

$$\beta_i(\tilde{r}) = \frac{c_{1,i}}{\tilde{m}_c} + c_{2,i} \exp\left(-\frac{\tilde{m}_c}{\tilde{r}}\right), \tag{6.4.21}$$

where $c_{1,i}$ and $c_{2,i}$ are the integration constants to be evaluated by applying the boundary conditions,

$$\tilde{r} \to \infty: \quad \tilde{Y}_O = \tilde{Y}_{O,\infty}, \quad \tilde{Y}_F = 0, \quad \tilde{T} = \tilde{T}_\infty, \tag{6.4.22}$$

$$\tilde{r} = 1: \quad \tilde{m}_c \tilde{Y}_{O,s} - \left(\frac{d\tilde{Y}_O}{d\tilde{r}}\right)_1 = 0, \quad \tilde{m}_c \tilde{Y}_{F,s} - \left(\frac{d\tilde{Y}_F}{d\tilde{r}}\right)_1 = \tilde{m}_c,$$

$$\left(\frac{d\tilde{T}}{d\tilde{r}}\right)_1 = \tilde{m}_c \tilde{q}_v, \quad \tilde{T} = \tilde{T}_s, \tag{6.4.23}$$

where $\tilde{q}_v = q_v/q_c$, $\tilde{Y}_F \equiv Y_F$, and $\tilde{Y}_O = Y_O/\sigma_O$. The first three relations in (6.4.23) respectively state that, at the droplet surface, the net oxidizer convective–diffusive transport vanishes because there is no oxidizer penetration into the liquid, the net fuel vapor transport is the fuel gasification rate, and the heat conduction from the flame is used to effect fuel gasification.

Applying the boundary conditions of (6.4.22) and (6.4.23) to Eqs. (6.4.20) and (6.4.21) respectively, for $i = O, F$, we obtain

$$\beta_O = \tilde{T} + \tilde{Y}_O = (\tilde{T}_s - \tilde{q}_v) + \{[\tilde{T}_\infty - (\tilde{T}_s - \tilde{q}_v)] + \tilde{Y}_{O,\infty}\} e^{-\tilde{m}_c/\tilde{r}}, \quad (6.4.24)$$

$$\beta_F = \tilde{T} + \tilde{Y}_F = [1 + (\tilde{T}_s - \tilde{q}_v)] + \{\tilde{T}_\infty - [1 + (\tilde{T}_s - \tilde{q}_v)]\} e^{-\tilde{m}_c/\tilde{r}}. \quad (6.4.25)$$

Equations (6.4.24) and (6.4.25) show that $(\tilde{T}_s - \tilde{q}_v)$ appears as a group, providing an indication of the energy levels of the problem. In particular, it implies that the need for gasification is equivalent to lowering the droplet enthalpy $\tilde{T}_s$ by the latent heat of gasification $\tilde{q}_v$. This is physically reasonable.

To determine the mass burning rate $\tilde{m}_c$, the reaction sheet standoff ratio $\tilde{r}_f$, and the flame temperature $\tilde{T}_f$, we now apply the reaction sheet requirements,

$$\tilde{Y}_O(\tilde{r}_f) = 0, \qquad \tilde{Y}_F(\tilde{r}_f) = 0. \quad (6.4.26)$$

Furthermore, as a consequence of $\tilde{Y}_O(\tilde{r}_f) = 0$, the relation

$$\tilde{Y}_O(1) = 0 \quad (6.4.27)$$

obviously also holds because if there is no leakage of the oxidizer across the reaction sheet, the oxidizer concentration in the inner region to the reaction sheet, in particular that at the droplet surface, must also vanish. Applying Eqs. (6.4.26) and (6.4.27) to Eqs. (6.4.24) and (6.4.25), we obtain three relations from which $\tilde{m}_c$, $\tilde{r}_f$, and $\tilde{T}_f$ can be solved as

$$\tilde{m}_c = \ln(1 + B_{h,c}), \quad (6.4.28)$$

$$\tilde{r}_f = \frac{\tilde{m}_c}{\ln(1 + \tilde{Y}_{O,\infty})} = 1 + \frac{\ln[1 + (\tilde{T}_f - \tilde{T}_s)/\tilde{q}_v]}{\ln(1 + \tilde{Y}_{O,\infty})}, \quad (6.4.29)$$

and

$$q_c = q_v + c_p(T_f - T_s) + c_p(T_f - T_\infty)\left(\sigma_O + \frac{1 - Y_{O,\infty}}{Y_{O,\infty}}\sigma_O\right). \quad (6.4.30)$$

In the above,

$$B_{h,c} = \frac{(\tilde{T}_\infty - \tilde{T}_s) + \tilde{Y}_{O,\infty}}{\tilde{q}_v} = \frac{c_p(T_\infty - T_s) + (Y_{O,\infty}/\sigma_O)q_c}{q_v} \quad (6.4.31)$$

is the heat transfer number for combustion. Compared to the heat transfer number for pure vaporization, $B_{h,v}$ given by Eq. (6.3.15), we see that the driving potential for gasification now consists of an additional source, $(Y_{O,\infty}/\sigma_O)q_c$, representing chemical heat release. For combustion of practical fuels, the chemical contribution is usually much larger than the thermal contribution, $c_p(T_\infty - T_s)$, especially when the environment is cold such that T_∞ is close to T_s. Numerical values of $B_{h,c}$ typically range between 1 and 10.

The flame temperature is deliberately expressed in the dimensional, implicit form of Eq. (6.4.30) in order to demonstrate the fact that it is again the stoichiometric adiabatic flame temperature of the system. Specifically, Eq. (6.4.30) shows that the heat release q_c by unit mass of fuel is equal to the amount needed to first gasify it and then heat it from the droplet temperature T_s to the flame temperature T_f, plus the amount needed to heat the stoichiometric, σ_O unit of oxidizer and the remaining $[(1 - Y_{O,\infty})/Y_{O,\infty}]\sigma_O$ unit of inert from the ambient temperature T_∞ to the flame temperature T_f. This is equivalent to the interpretation of Eq. (5.5.27), now extended to include condensed phase gasification.

The above expressions are defined to within $\tilde{T}_s$. Following the same procedure as discussed in Section 6.3 for pure vaporization, $\tilde{T}_s$ can be determined by evaluating Eq. (6.4.25) at $\tilde{r} = 1$ to yield an expression for $Y_{F,s} = Y_{F,s}(\tilde{T}_s)$,

$$\tilde{Y}_{F,s} = 1 - \tilde{q}_v + \left[\tilde{T}_\infty - (1 + \tilde{T}_s - \tilde{q}_v)\right] e^{-\tilde{m}_c}. \tag{6.4.32}$$

From Eq. (6.4.32), a mass transfer number $B_{m,c}$ can be defined through $\tilde{m}_c = \ln(1 + B_{m,c})$ such that by equating $B_{h,c}$ and $B_{m,c}$, and by using the Clausius–Clapeyron relation, $\tilde{T}_s$ can be iteratively solved.

An accurate knowledge of $\tilde{T}_s$, however, is frequently not necessary in the evaluation of the bulk combustion parameters $\tilde{m}_c$, $\tilde{r}_f$, and $\tilde{T}_f$ because the enthalpy contribution from T_s is usually much smaller than the chemical source term, as just mentioned. Furthermore, realizing that the droplet is expected to be close to its boiling state under the situation of intense heating during steady burning, it is then frequently adequate to assume that

$$T_s = T_b, \tag{6.4.33}$$

where T_b is the liquid's boiling point under the prevailing pressure. Using Eq. (6.4.33), the bulk combustion parameters can be obtained through straightforward evaluation. It is, however, also important to recognize that the adoption of Eq. (6.4.33) falsifies the phase-change process and thereby necessitates the abandonment of the proper phase change description, for example the Clausius–Clapeyron relation, and all of its physical implications. Indeed, because of the presence of species other than fuel vapor at the droplet surface, for example nitrogen from the air and the combustion products generated at the reaction zone, the state of boiling can never be attained for the droplet, theoretically as well as in realistic situations.

It is also of interest to note that the present results specialize to those of pure vaporization, in an environment free of fuel vapor, by simply setting $Y_{O,\infty} = 0$ in Eqs. (6.4.28) to (6.4.31). This yields the result that the heat source, namely the reaction zone, is now simply the ambience, with $\tilde{r}_f \to \infty$ and $\tilde{T}_f = \tilde{T}_\infty$.

It is appropriate to recognize at this point the similarity between droplet vaporization and droplet combustion. Apart from the gas-phase reactions, the gasification process at the droplet surface is qualitatively the same in both cases. Thus during combustion, the droplet simply perceives the flame as a hot "ambience" located at $\tilde{r}_f$. Consequently, understanding gained from studying droplet combustion frequently can be applied to the modeling of droplet vaporization. Indeed, from an experimental design point of view, droplet vaporization in a high temperature environment can be usefully simulated by studying droplet burning in a cold environment. The flame now conveniently serves as a high-temperature, constant-pressure "chamber" within which vaporization takes place.

It is reasonable to ask whether the flame location and temperature for the present problem still conform to the canonical expressions for the chambered as well as the Burke–Schumann flames, given by Eqs. (6.1.14) and (6.1.15) for the former and Eqs. (6.2.28) and (6.2.29) for the latter. This is indeed the case and can be expediently demonstrated by using the set of alternate boundary conditions at $\tilde{r} = 1$, as $\tilde{Y}_O = \tilde{Y}_{O,s}$, $\tilde{Y}_F = \tilde{Y}_{F,s} = 1$, and $\tilde{T} = \tilde{T}_s$. This yields the alternate expressions for the coupling functions,

$$\tilde{T} + \tilde{Y}_O = (\tilde{T}_\infty + \tilde{Y}_{O,\infty}) - \left[(\tilde{T}_\infty + \tilde{Y}_{O,\infty}) - (\tilde{T}_s + \tilde{Y}_{O,s})\right] \frac{1 - e^{-\tilde{m}_c/\tilde{r}}}{1 - e^{-\tilde{m}_c}}, \tag{6.4.34}$$

$$\tilde{T} + \tilde{Y}_F = \tilde{T}_\infty - [\tilde{T}_\infty - (\tilde{T}_s + \tilde{Y}_{F,s})] \frac{1 - e^{-\tilde{m}_c/\tilde{r}}}{1 - e^{-\tilde{m}_c}}. \tag{6.4.35}$$

Evaluating the above at the reaction sheet, for which $\tilde{Y}_{O,s} = 0$, we obtain

$$\frac{e^{-\tilde{m}_c/\tilde{r}_f} - e^{-\tilde{m}_c}}{1 - e^{-\tilde{m}_c}} = \Phi^*, \tag{6.4.36}$$

$$\tilde{T}_f - \tilde{T}_s = (\tilde{T}_\infty - \tilde{T}_s + \tilde{Y}_{O,\infty})\Phi^*, \tag{6.4.37}$$

where $\phi^* = \tilde{Y}_{F,s}/\tilde{Y}_{O,\infty} = 1/\tilde{Y}_{O,\infty}$. The above results are determined to within $Y_{F,s}$, which is given by Eq. (6.4.32). The burning rate $\tilde{m}_c$ is also readily given by evaluating Eq. (6.4.34) at $\tilde{r} = 1$.

To relate $\tilde{m}_c$ to the rate of decrease of the droplet size, by definition the droplet gasification rate $\tilde{m}_v$ is given by Eq. (6.4.11). If we now assume that the instantaneous rate of fuel gasification at the droplet surface is equal to that of fuel consumption at the reaction sheet, or

$$m_v = m_c, \tag{6.4.38}$$

and since in dimensional form

$$m_c = 4\pi(\lambda/c_p)r_s \ln(1 + B_{h,c}), \tag{6.4.39}$$

equating Eqs. (6.4.11) and (6.4.39) yields

$$\frac{dr_s^2}{dt} = -K_c, \tag{6.4.40}$$

where

$$K_c = \frac{2(\lambda/c_p)}{\rho_\ell} \ln(1 + B_{h,c}) \tag{6.4.41}$$

is the droplet burning rate constant.

Integrating Eq. (6.4.40) with the initial condition $r_s(t=0) = r_{s,o}$ yields

$$r_s^2 = r_{s,o}^2 - K_c t, \tag{6.4.42}$$

which is the analogue of the d^2-law for droplet vaporization given in Eq. (6.4.15) and shown in Figure 6.4.2. Setting $t = \tau_c$ for $r_s = 0$ at complete burnout, we obtain the total time for combustion as

$$\tau_c = \frac{r_{s,o}^2}{K_c}. \tag{6.4.43}$$

It is also of interest to note that since $(\lambda/c_p)_g \sim (\rho D)_g$, and since $B_c = O(1 \sim 10)$ such that $\ln(1 + B_{h,c}) = O(1)$, Eq. (6.4.41) shows that

$$\frac{K_c}{D_g} \sim \frac{\rho_g}{\rho_\ell}, \tag{6.4.44}$$

where the subscript g designates gas-phase property. Under atmospheric pressure $(\rho_g/\rho_\ell) = O(10^{-3} \sim 10^{-2})$, therefore the rate of surface regression is much slower than that of gas-phase diffusion. This is in agreement with the assumption of gas-phase quasi-steadiness, as it should be. Furthermore, if we take $D_g = O(10^0\ \text{cm}^2/\text{sec})$, then $K_c = O(10^{-3} \sim 10^{-2}\ \text{cm}^2/\text{sec})$, which is the typical order of the burning rate constants for fuel droplets determined experimentally. For the pure vaporization case, the above estimates need to be modified by the factor $\ln(1 + B_{h,v}) \approx B_{h,v}$ because it is usually less than unity.

Summarizing the above results, the d^2-law states that, during quasi-steady droplet combustion, the droplet surface regression rate, the reaction sheet standoff ratio (r_f/r_s), and the flame temperature (T_f), remain as constants, and that T_f is also the stoichiometric adiabatic flame temperature of the fuel–oxidizer system.

6.4.4. Experimental Results on Single-Component Droplet Combustion

We shall briefly discuss the experimental observations on the combustion characteristics of single-component droplets, and show that certain aspects can be substantially different from predictions of the d^2-law. Discussion of the experimental methodologies as well as detailed explanations of these differences will be conducted in Chapter 13.

Figures 6.4.5 and 6.4.6 show the experimental data (Law, Chung & Srinivasan 1980) on the temporal variations of the square of the normalized droplet radius, $R_s^2 = (r_s/r_{s,o})^2$, the nondimensional flame radius, $R_f = (r_f/r_{s,o})$, and the flamefront standoff ratio, $\tilde{r}_f = (r_f/r_s)$, for the spark-ignited, nearly spherically symmetric burning of an octane droplet in air and pure oxygen environments, respectively. The experiments were conducted under reduced pressure and thereby reduced buoyancy situations. The flame location was taken to be the midpoint of the luminous zone.

The experimental results show that after ignition a period exists during which the burning rate is very slow as indicated by the almost lack of variation in R_s^2 in Figure 6.4.5. This period is generally quite short, spanning about 5 to 10 percent of

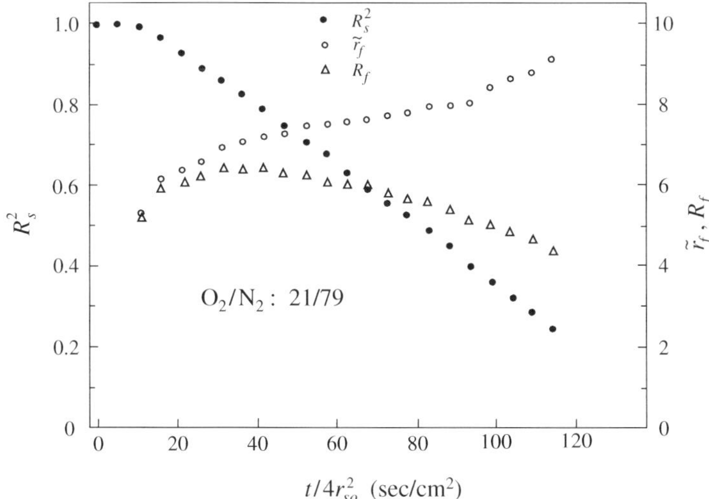

Figure 6.4.5. Experimental measurements of the droplet and flame sizes for an octane droplet burning in the air environment at 0.1–0.15 atm pressure (Law, Chung & Srinivasan 1980).

the droplet lifetime depending on the fuel boiling point and the ignition duration. After this initial period R_s^2 varies almost linearly with time. In the pure oxygen environment it is so short that it falls within the period during which the droplet has not resumed its spherical shape from the disturbance caused by the spark discharge.

The flamefront standoff ratio exhibits two distinct behaviors. At low oxygen concentrations, $\tilde{r}_f$ continuously increases, with the increase being actually faster towards the end of the droplet lifetime. For high oxygen concentrations, the increase

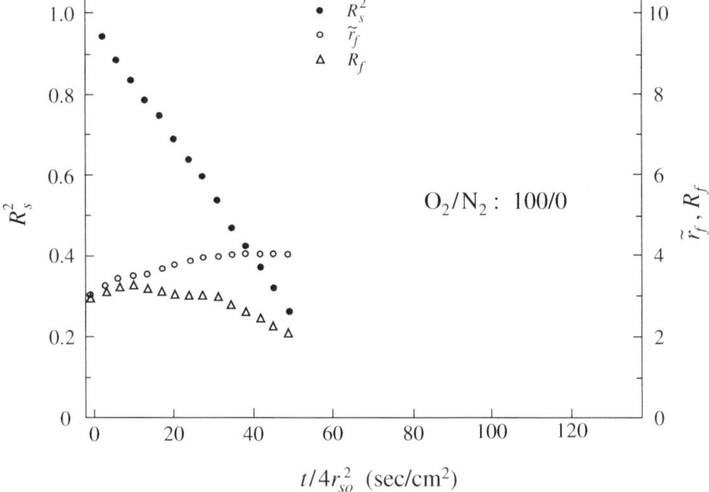

Figure 6.4.6. Experimental measurements of the droplet and flame sizes for an octane droplet burning in the pure oxygen environment at 0.1–0.15 atm pressure (Law, Chung & Srinivasan 1980).

instead levels off. The actual flame size, R_f, first increases and then decreases for both cases.

The experimental results show a number of inadequacies of the d^2-law that, however, have all been satisfactorily explained. The presence of the short, initial period during which the droplet size hardly changes, indicating very slow rate of droplet gasification, signifies the need to heat up the initially cold droplet to close to the liquid boiling point for steady burning. Thus much of the heat arriving the droplet surface is used for droplet heating instead of liquid gasification. This feature is not captured by the d^2-law which does not describe influences due to initial conditions except the initial droplet size.

The movement of the flamefront is also a consequence of suppressing the initial condition. Here the d^2-law prescribes that upon ignition, the flame instantaneously assumes its quasi-steady value of $\tilde{r}_f$. However, in order to support a flame of this size, considerable amount of fuel vapor needs to be present in the inner region to the flame in accordance to the d^2-law fuel vapor concentration profile. This amount of fuel vapor is not present initially and therefore needs to be gradually built up, leading to the corresponding gradual increase in the flame size. In a low-oxygen environment the amount of fuel vapor that needs to be accumulated is large because of the larger flame size; hence the continuous increase of $\tilde{r}_f$ as observed recognizing, of course, that r_s also continuously decreases. By the same reasoning $\tilde{r}_f$ levels off in a high-oxygen environment because of the smaller $\tilde{r}_f$.

A corollary of this fuel vapor accumulation phenomenon is that because part of the fuel gasified is accumulated in the inner region to the flame instead of being instantaneously reacted, m_v is actually not equal to m_c. Thus the heat release rate from the droplet flame cannot be directly related to the droplet size in the manner of Eq. (6.4.42). This can have significant implications in the modeling of spray combustion.

The d^2-law prediction of the flamefront standoff ratio has also been found to be much larger than experimentally observed value. For example, for an alkane droplet burning in air, the observed $\tilde{r}_f$ is less than 10, whereas the predicted value is around 40. Such a large discrepancy has been found to be caused by the unity Lewis number assumption in the d^2-law. A much smaller flame size is predicted by allowing for unequal rates of heat and mass diffusion, especially for the slow diffusion rate of the large fuel molecules in the inner region. Physically, it is reasonable to expect that a slower fuel diffusion rate would lead to the flame located closer to the fuel region.

6.5. THE COUNTERFLOW FLAME

The various flame configurations that we have studied actually cannot be readily established in the laboratory. First, attempt has been made only recently to establish a chambered flame (Lo Jacono et al. 2005). Furthermore, the Burke–Schumann flame and the droplet flame are both susceptible to buoyancy effects that can be minimized only under very restricted conditions. The Burke–Schumann flame is also affected

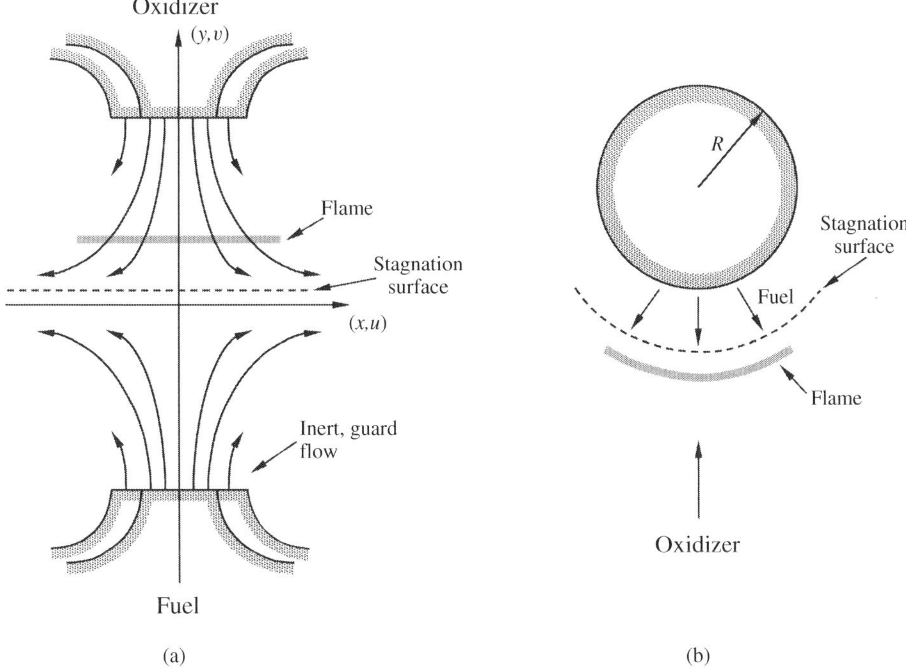

Figure 6.5.1. Schematics showing counterflow flame generated by: (a) opposing nozzles and (b) porous burner in uniform flow.

by the stabilization process at the burner rim. The flame, which has actually been extensively used in the study of nonpremixed as well as premixed flames, is one situated in a counterflow, which can be generated by impinging a uniform oxidizer jet against a uniform fuel jet, as shown in Figure 6.5.1a. An alternate arrangement is to eject the fuel stream from a porous cylinder or sphere, of radius R, which is immersed in a uniform flow of the oxidizer gas (Figure 6.5.1b). Since such a flow is usually dominated by forced convection, buoyancy effect is small and the resulting flame is quite steady. This is especially the case for the porous burner arrangement because of the presence of a rigid boundary in the flow. Stability can be further facilitated by adding a guard flow of inert gas (Figure 6.5.1a) and adjusting its flow velocity. This guard flow also isolates the reactant jets from the environment.

There are three additional desirable characteristics of a counterflow. First, the flow velocity along the centerline near the stagnation region varies linearly with distance. The flow can therefore be characterized by a single parameter, namely its velocity gradient a, which constitutes the local strain rate. Second, due to symmetry the flame is either planar or with a well-defined curvature, as in Figures 6.5.1a and 6.5.1b respectively. Thus detailed experimental and computational resolution of the flame structure can be conducted in the direction normal to it along the centerline. Third, the inverse of the velocity gradient, $1/a$, represents a characteristic flow time, which, when compared to the characteristic reaction time, yields the system Damköhler number.

The counterflow and the associated velocity gradient can be described at various levels of rigor (Kee, Coltrin & Glarborg 2003). The simplest is to assume that the flow is potential, with the oxidizer and fuel streams approaching from $\pm\infty$ and having a single strain rate a throughout the flow. However, in the laboratory the flows are usually generated by aerodynamically shaped nozzles that produce uniform flows. The strain rate is then approximated by $(U_O + U_F)/L$, where U_O and U_F are the uniform velocities from the oxidizer and fuel nozzles respectively, and L is the separation distance between the nozzles. The analogous expressions for the porous sphere and cylinder cases are $(U_O + U_F)/2R$ and $(U_O + U_F)/R$, respectively. However, if we want to take into account of the different densities of the two streams, and the fact that the flows gradually change from being uniform at the nozzle exit to stagnating at the stagnation surface, then a can be approximated for the counterflow (Seshadri & Williams 1978) as

$$a_O = \frac{U_O}{L}\left(1 + \frac{U_F\sqrt{\rho_F}}{U_O\sqrt{\rho_O}}\right), \qquad a_F = \frac{U_F}{L}\left(1 + \frac{U_O\sqrt{\rho_O}}{U_F\sqrt{\rho_F}}\right), \tag{6.5.1}$$

for the oxidizer and fuel sides of the flow field, respectively.

The flow field, and thereby the strain rate, are modified from the above cold flow expressions by the presence of the flame. Thus an accurate specification of the flow would require either computational simulation or nonintrusive experimental mapping, and the subsequent determination of the local value of a upstream of the thermal layer constituting the flame. Since the strain rates on the oxidizer and fuel sides of the flame could be different, the appropriate strain rate to use is the one corresponding to the fuel or oxidizer side of the stagnation surface in which the flame is situated.

To demonstrate the reaction-sheet solution of the counterflow flame, we shall use the simple potential flow description, with $v = -ay$, as shown in Figure 6.5.1a. Then conservation of the coupling function β_i is

$$\rho a y \frac{d\beta_i}{dy} + (\lambda/c_p)\frac{d^2\beta_i}{dy^2} = 0, \tag{6.5.2}$$

which in nondimensional form is

$$\tilde{y}\frac{d\beta_i}{d\tilde{y}} + \frac{d^2\beta_i}{d\tilde{y}^2} = 0, \tag{6.5.3}$$

where $\tilde{y} = y/(\lambda/c_p\rho a)^{1/2}$, indicating that spatial quantities scale with $a^{1/2}$. Integrating Eq. (6.5.3) subject to the boundary conditions,

$$\begin{aligned} \tilde{y} \to -\infty:&\quad \tilde{Y}_F = \tilde{Y}_{F,\infty}, \quad \tilde{Y}_O = 0, \quad \tilde{T} = \tilde{T}_{-\infty} \\ \tilde{y} \to +\infty:&\quad \tilde{Y}_F = 0, \quad \tilde{Y}_O = \tilde{Y}_{O,\infty}, \quad \tilde{T} = \tilde{T}_{\infty}, \end{aligned} \tag{6.5.4}$$

we obtain the coupling functions,

$$\tilde{T} + \tilde{Y}_F = (\tilde{Y}_{F,-\infty} + \tilde{T}_{-\infty}) + [\tilde{T}_\infty - (\tilde{Y}_{F,-\infty} + \tilde{T}_{-\infty})]I(\tilde{y})/I(\infty) \tag{6.5.5}$$

$$\tilde{T} + \tilde{Y}_O = \tilde{T}_{-\infty} + [(\tilde{Y}_{O,\infty} + \tilde{T}_\infty) - \tilde{T}_{-\infty}]I(\tilde{y})/I(\infty), \tag{6.5.6}$$

where

$$I(\tilde{y}) = \int_{-\infty}^{\tilde{y}} e^{-\tilde{y}^2/2} d\tilde{y}.$$

Evaluating Eqs. (6.5.5) and (6.5.6) at the reaction sheet, we find that the reaction sheet location $\tilde{y}_f$ is implicitly given by

$$\frac{I(\tilde{y}_f)}{I(\infty)} = \Phi^*, \tag{6.5.7}$$

where $\phi^* = \tilde{Y}_{F,-\infty}/\tilde{Y}_{O,\infty}$ is the equivalence ratio, while the flame temperature is

$$\tilde{T}_f - \tilde{T}_{-\infty} = (\tilde{Y}_{O,\infty} + \tilde{T}_\infty - \tilde{T}_{-\infty})\Phi^*, \tag{6.5.8}$$

which are in the canonical forms identified before.

Systematic computational and experimental studies have been conducted for counterflow nonpremixed flames. Tsuji (1982) initiated systematic experimental studies by using the porous burner and with intrusive sampling. Recent investigations (Law et al. 1994) have favored the use of opposed nozzles because of the symmetry of the configuration and the ability to seed both sides of the flow with particles for LDV (laser Doppler velocimetry) or PIV (particle image velocimetry) mapping of the flow field. Other laser diagnostic techniques have also been employed to determine the temperature and concentration profiles. Furthermore, it is now a routine matter to computationally simulate quasi-one-dimensional variable-density counterflow flames.

The largest uncertainty in such studies is the quantification and specification of the strain rate of the flow. It has nevertheless been found that, by using the local strain rate determined just upstream of the thermal boundary of the flame, description of the flame structure becomes quite insensitive to that of the outer flow field. This point is demonstrated in Figure 6.5.2 for the computed temperature and velocity profiles of a symmetric counterflow flame, obtained by using potential flow, plug flow, and an arbitrary intermediate flow boundary condition (Sung, Liu & Law 1995). The comparison is based on collating the local axial velocity gradients immediately upstream of both sides of the thermal structure of the flame. It is seen that, despite the differences in the outer flows, the corresponding velocity and temperature profiles are surprisingly identical within the thermal mixing layer.

Figure 6.5.3 then compares the computed and measured temperature and major species profiles for the flame of Figure 6.5.2, with the experimental measurements conducted by using spontaneous Raman spectroscopy. The comparison is seen to be very favorable. At the present state of the art, this is the level of quantitative predictability for such a simple and well-characterized flow with reasonably well established chemistry, at least for the major species.

Figure 6.5.4 shows the computed and measured temperature profiles for different local strain rates. It is seen that the flame becomes thinner with increasing strain rate. This is reasonable because of the reduced residence time of the flow. Furthermore,

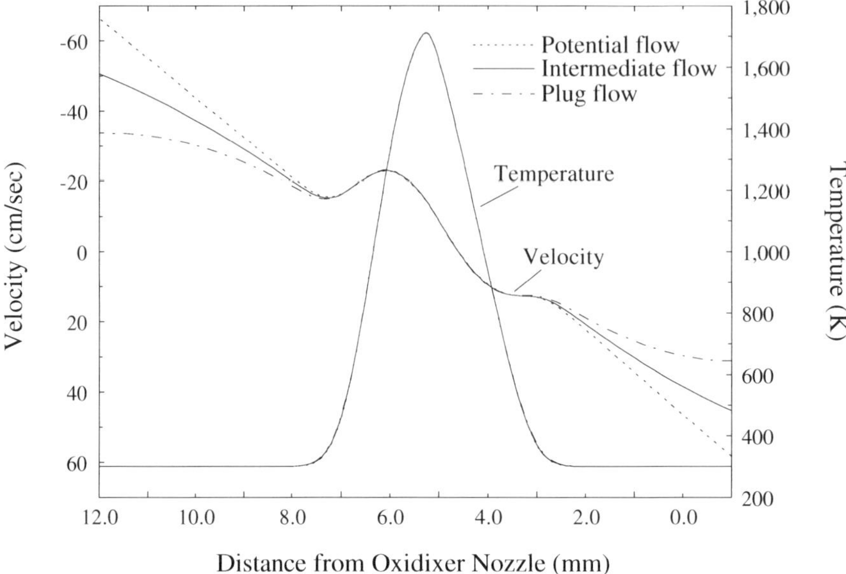

Figure 6.5.2. Calculated velocity and temperature profiles for a symmetric, nozzle-generated counterflow flame established by impinging a 23 percent methane-in-nitrogen stream against a 23 percent oxygen-in-nitrogen stream, for different descriptions of the flow (Sung, Liu & Law 1995).

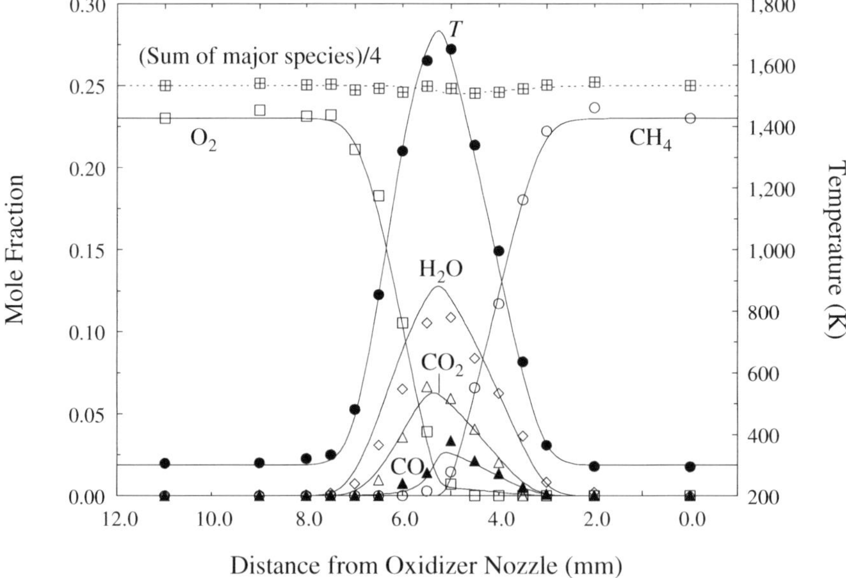

Figure 6.5.3. Calculated and experimental temperature and major species profiles for the flame of Figure 6.5.2, with $a = 56s^{-1}$.

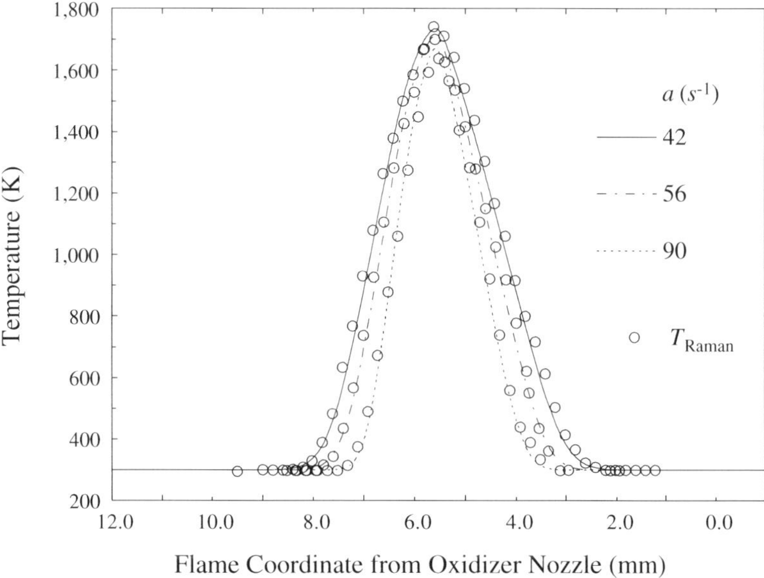

Figure 6.5.4. Calculated and experimental temperature profiles of the flame of Figure 6.5.3, with various strain rates.

from scaling considerations, we would expect that the flame thickness should scale inversely with $\sqrt{a}$. Thus if we characterize the flame thickness by its full width at half maximum (FWHM) value, then its product with $\sqrt{a}$ should be a constant with respect to a. Figure 6.5.5 shows that not only this is the case, but the same also holds if the

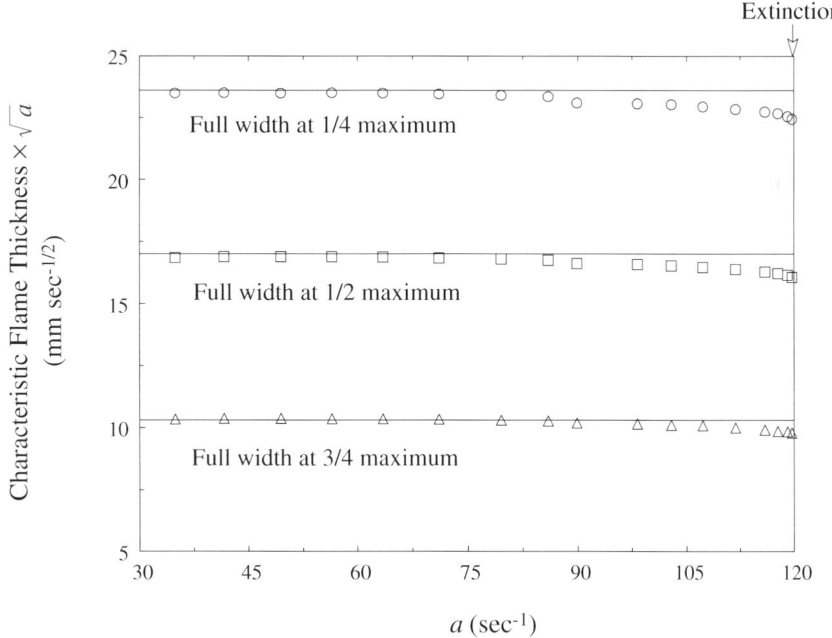

Figure 6.5.5. Results showing that the characteristic flame thickness scales inversely with $\sqrt{a}$.

flame thickness is alternately defined as the full width at 25 percent and 75 percent of the maximum temperature. It is also seen that there is a slight decrease in the "constant" value at higher strain rates, up to the state of extinction. This is due to the reduction in the maximum temperature with increasing strain rate (Figure 6.5.4). This in turn is a consequence of the reduced residence time within the flame, and, hence, an increased amount of the reactant leakage, indicating departure from the reaction-sheet behavior.

PROBLEMS

1. Rework the chambered flame problem of Figure 6.1.1 using the coupling function formulation, allowing for nonunity Lewis numbers for all species.

2. Rework the chambered flame problem of Figure 6.1.1, allowing for a constant mass flux $m = \rho u$ and constant values of $(\rho D_F)^-$, $(\rho D_O)^+$, $c_p^\pm$ and $\lambda^\pm$. Derive and discuss the solution for very small flow rate m.

3. Show that for the chambered flame problem of Figure 6.1.1 with constant mass flux $m = \rho u$, the final results should be identical to the $m = 0$ case if we identify a new spatial coordinate $\xi = (e^{\tilde{m}\tilde{x}} - 1)/(e^{\tilde{m}} - 1)$, which replaces $\tilde{x}$, where $\tilde{m} = (\rho u \ell)/(\rho D)$.

4. For a second-order ordinary differential equation with linear differential operator,

$$a_1(x)\frac{d^2 f}{dx^2} + a_2(x)\frac{df}{dx} = a_3(x, f)$$

and with boundaries at (x_1, x_2), show that the transformation of the independent variable to

$$\xi = \int_{x_1}^{x} F(x')dx' / \int_{x_1}^{x_2} F(x')dx'$$

eliminates the first-order differential such that the transformed equation is

$$\frac{d^2 f}{d\xi^2} = \left[\frac{1}{F(x)}\int_{x_1}^{x_2} F(x')dx'\right]^2 [a_3(x, f)/a_1(x)],$$

and the boundaries are now at (0,1). The function $F(x)$ is

$$F(x) = \exp\left[-\int_{x_1}^{x} \frac{a_2(x')}{a_1(x')}dx'\right].$$

Relate this result to Problem 3.

5. In the presence of Soret diffusion, with appropriate simplifications the conservation equations for the species i and enthalpy can be expressed as

$$\nabla \cdot \left[\rho \mathbf{v}\tilde{Y}_i - \rho D_i \nabla \tilde{Y}_i - \rho D_i \alpha_{T,i} \tilde{Y}_i \nabla \ln \tilde{T}\right] = -w$$
$$\nabla \cdot \left[\rho \mathbf{v}\tilde{T} - (\lambda/c_p)\nabla \tilde{T}\right] = w,$$

where $\alpha_{T,i} = D_{T,i}/(\rho D_i)$ and $D_{T,i}$ is the thermal diffusion coefficient. For the chambered flame problem of Figure 6.1.1, derive expressions for the flame

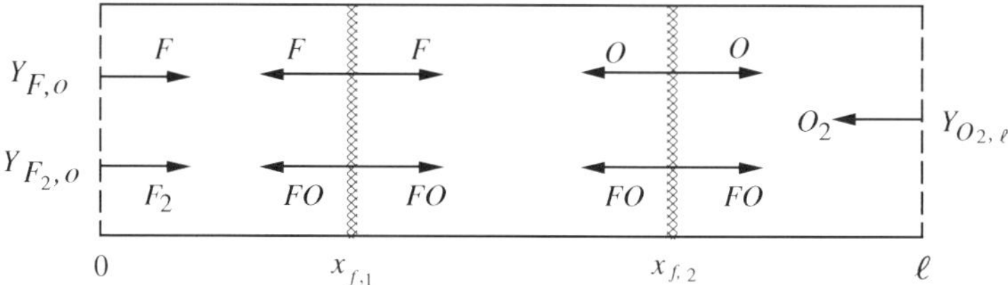

Figure 6.P.1 Flame structure for Problem 6.

temperature, flame location, and the fuel consumption rate in the limit of flame-sheet combustion. What can you say about the change in the burning characteristics in the presence of Soret diffusion when the oxidizer is air, and the fuel is: (a) hydrogen, and (b) heptane.

6. For the one-dimensional chambered flame problem, let the fuel and oxidizer gases be respectively represented by F_2 and O_2, which undergo a two-step reaction

$$F_2 + O \rightarrow FO + F$$
$$F + O_2 \rightarrow FO + O$$

occurring at two distinct infinitesimally thin flame fronts as shown in Figure 6.P.1. Consider initial dissociation of F_2 occurs so that both F_2 and F exist with known mass fractions at $x = 0$. Derive the fuel burning rate f_{F_2}, the flame locations $x_{f,1}$ and $x_{f,2}$, and the flame temperatures $T_{f,1}$ and $T_{f,2}$. Assume $Le = 1$.

7. For the Burke–Schumann flame: (a) Show that if $Le_O = Le_F = Le \neq 1$, the nonunity Lewis number effect on the species profiles and the flame shape is equivalent to modifying the Pe by a factor Le for the situation of $Le = 1$. (b) Show that as $Pe \rightarrow \infty$, the height of the Burke–Schumann flame scales with Pe, and that as $Pe \rightarrow 0$, the height becomes independent of Pe, as shown in Figure 6.2.1.

8. For the Burke–Schumann flame, the initial condition for the species concentrations, at $\tilde{y} = 0$, while correct, are not predictive because $Y_{i,o}$ is not the mass fraction of i, $Y_{i,-\infty}$, in the supply. With the porous plate defining the nozzle exit, the proper initial condition for the species concentrations should be

$$\rho v Y_i - (\lambda/c_p)\frac{\partial Y_i}{\partial y} = (\rho v)_i,$$

where $(\rho v)_i$ is the mass flux of i in its supply. Solve this problem in the same manner as that in the text.

The flame also loses heat to the porous plate. Derive an expression for the heat-loss rate.

9. Solve the Burke–Schumann flame in cylindrical coordinate. Use r and z to represent radial and streamwise coordinates, and r_{in} and r_{out} for the inner and outer tube radii.

10. Show that for $Le = 1$ systems the relation given by Eq. (6.3.16)

$$\frac{Y_{1,o} - Y_{1,\ell}}{1 - Y_{1,o}} = \frac{c_p(T_\ell - T_o)}{q_v}$$

is geometry independent and holds for the general vaporization problem described by

$$\nabla \cdot (\rho \mathbf{v} Y_i - \rho D \nabla Y_i) = 0$$
$$\nabla \cdot (\rho \mathbf{v} c_p T - \lambda \nabla T) = 0,$$

subject to the boundary conditions

$$\begin{aligned} \mathbf{x} = \mathbf{x}_o: \quad & Y_1 = Y_{1,o}, \quad T = T_o, \\ & \lambda \nabla T = f q_v, \quad f Y_1 - \rho D \nabla Y_1 = f; \\ \mathbf{x} = \mathbf{x}_\ell: \quad & Y_1 = Y_{1,\ell}, \quad T = T_\ell; \end{aligned}$$

where 1 is the vaporizing species and $\mathbf{x}_o$ the location of the vaporizing surface. From this problem we can also conclude that the wet-bulb temperature T_o determined in conjunction with the Clausius–Clapeyron relation is also a general result. (Hint: Review Section 5.5.2 on the general result for T_{ad}.)

11. Derive a relation that is analogous to that of Problem 10 but for the case of nonpremixed burning.

12. This is a variation of the Stefan flow problem treated in Section 6.3. Here the liquid in the container is a low-molecular weight, volatile, alcohol (say methanol) that tends to absorb water. Hygroscopic chips are placed in the alcohol to absorb the dissolved water such that the water vapor concentration at the alcohol surface is zero. A moist air with water vapor concentration $Y_{W,\ell}$ breezes over the container. Because of the high volatility of alcohol, it vaporizes with f_A and also maintains a low temperature T_o. The water vapor in the air, then, will tend to condense onto the alcohol surface with a rate f_W because of the low T_o and the presence of the chips. Determine $\tilde{f}_A$, $\tilde{f}_W$, and $\tilde{T}_o$, where $\tilde{f}_i = f_i/(\rho D/\ell)$. Discuss the conditions determining whether $\tilde{f} = \tilde{f}_A + \tilde{f}_W >$ or < 0. Assume $Le = 1$ for simplicity. This is an interesting heat transfer system in that part of the heat needed for alcohol vaporization comes from the condensation heat release of the water vapor.

13. Show that for slow rates of droplet vaporization:

(a) The mass vaporization rate m_v is almost independent of the specific heat c_p, provided T_s is given. Explain why this is so by considering the energy conservation equation.

(b) The effect of nonunity Lewis number (Le) on the liquid temperature T_o is equivalent to a modification of the specific heat c_p by a factor Le.

14. For a water droplet vaporizing in dry air with $T_\infty = 372$ K and $Y_{1,\infty} = 0$, what is the droplet temperature T_s and the nondimensional vaporization rate $\tilde{m}_v$ if the pressure is 1 atm? 5 atm? What is the boiling point at 5 atm? Take $q_v = 540$ cal/gm, $c_p = 0.3$ cal/gm-K, and $W_{\text{air}} = 29$.

15. A gasoline droplet is inducted at the beginning of the intake stroke into the cylinder of a four-stroke single-cylinder engine running at 1,000 rpm. The droplet subsequently vaporizes. Calculate the initial radius of the droplet that can just achieve complete vaporization at the end of the compression stroke. Assume for simplicity that the cylinder temperature and pressure remain constant at 1,500 K and 1 atm, respectively, and that the droplet temperature is at its boiling point of 370 K. Take $\rho = 10^{-3}$ gm/cm^3, $\rho_\ell = 0.8$ gm/cm^3, $D = 1$ cm^2/sec, $c_p = 0.3$ cal/gm-K, and $q_v = 70$ cal/gm.

16. Show that the droplet combustion problem can be reformulated from the very beginning by defining a scaled temperature $\hat{T} = \tilde{T} - (\tilde{T}_s - \tilde{q}_v)$ such that $\tilde{T}_s$ and $\tilde{q}_v$ do not appear explicitly in the governing equations and boundary conditions. Then state the entire problem, that is, the governing equations and boundary conditions, in terms of the new coupling function $\beta_i = \hat{T} + \tilde{Y}_i$.

17. Consider a fuel droplet burning in a reactive environment consisting of both an oxidizing gas and the fuel gas with concentrations $\tilde{Y}_{O,\infty}$ and $\tilde{Y}_{F,\infty}$, respectively. Assume $\tilde{Y}_{F,\infty} \ll \tilde{Y}_{O,\infty}$ and that the ambient temperature is very low such that reaction is still confined at a reaction sheet. Derive the burning rate and the flame size, and discuss their dependence with increasing $\tilde{Y}_{F,\infty}$.

18. When the reaction rate is not infinitely fast, a diffusion flame is broadened and both fuel and oxidizer will leak through it and remain unreacted. For the droplet problem this leakage will result in a finite oxidizer concentration at the droplet surface, $\tilde{Y}_{O,s}$, contrary to the boundary condition (6.4.27) used to derived the flame-sheet solution. With increasing leakage the reduction in the burning rate will eventually lead to flame extinction. Show that for small amount of leakage, $\tilde{Y}_{O,s} \ll 1$, the reduction in the burning rate is given by

$$\tilde{m}_c^o - \tilde{m}_c \approx \frac{\tilde{Y}_{O,s}}{\tilde{q}_v},$$

where $\tilde{m}_c^o$ is the flame-sheet solution.

19. For the stagnation flame studied, show that in the limit of $\tilde{Y}_{O,\infty} \ll 1$, the flame is located at

$$\tilde{y}_f \approx \sqrt{2 \ln \phi^*} \to \infty.$$

7 Laminar Premixed Flames

We now begin the study of premixed combustion. As we have learned from Chapter 6, a nonpremixed flame is supported by the stoichiometric, counterdiffusion of fuel and oxidizer. Thus, once ignited, a nonpremixed flame will situate itself somewhere between the fuel and oxidizer sources in order to satisfy this stoichiometry requirement. However, once ignition is achieved in a combustible fuel–oxidizer mixture, the resulting premixed flame tends to propagate into and consume the unburned mixture, if unrestrained through some aerodynamic means. Thus a premixed flame is a wave phenomenon.

In this chapter we shall study the simplest, idealized mode of wave propagation, namely the steady propagation of a one-dimensional, planar, adiabatic, wave relative to a stationary, combustible mixture in the doubly infinite domain. We shall call such a wave a standard wave or standard flame. In Section 7.1 we shall identify all such possible waves by constraining, through the conservation of mass, momentum, and energy, the states far upstream and downstream of the wave where the nonequilibrium processes of diffusion and reaction both vanish. Such an analysis yields the Rankine–Hugoniot relations, which show that two classes of waves can propagate in a combustible mixture, namely subsonic, deflagration waves and supersonic, detonation waves. These waves have distinctively different properties.

Since the wave structure is not described at the level of the Rankine–Hugoniot analysis, the problem is not closed in that the crucial parameter of the wave response, namely the wave propagation speed, needs to be given. It can be determined only by analyzing the wave structure. In this and several subsequent chapters, we shall first analyze the structure and propagation of deflagration waves, commonly called laminar premixed flames. Detonation waves will be studied under supersonic combustion in Chapter 14.

In Section 7.2 we shall present a phenomenological description of the standard premixed flame, resulting in an approximate derivation of its propagation speed, designated by s_u^o or s_L^o, with the superscript o indicating this particular mode of propagation. This parameter is usually called the laminar flame speed or the laminar burning velocity. It is a unique property of a mixture, indicating its reactivity and

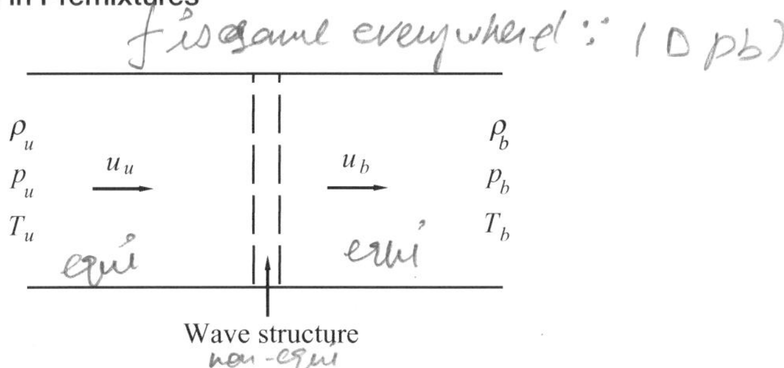

Figure 7.1.1. Schematic of the one-dimensional planar combustion wave in a premixture; the wave structure is governed by the nonequilibrium processes of diffusion and reaction.

exothermicity in a given diffusive medium. Furthermore, since it contains the physico-chemical information of the mixture, many premixed flame phenomena, such as extinction, flash back, blowoff, and turbulent flame propagation, can be characterized with s_u^o being a reference parameter.

The mathematical formulation of the standard premixed flame is stated in Section 7.3. This will be followed by approximate and then rigorous analyses of the flame structure in Sections 7.4 and 7.5. In Sections 7.6 and 7.7, we shall discuss the experimental and computational techniques used in determining s_u^o, and the dependence of s_u^o on the various system parameters. In Section 7.8, the detailed chemical structures of representative flames are studied.

We note in passing that while we shall use "laminar flame speed" and "laminar burning velocity" interchangeably throughout the text, whenever possible we shall use the former when referring to flame motion and the latter to flame property.

7.1. COMBUSTION WAVES IN PREMIXTURES

7.1.1. Rankine–Hugoniot Relations

Consider a steady, one-dimensional planar wave propagating into a quiescent combustible gaseous mixture with velocity u_u (Figure 7.1.1). Since diffusion and reaction vanish far upstream and downstream of the wave, in the wave-stationary frame we have the following conservation relations:

Mass:

$$\rho_u u_u = \rho_b u_b = f \tag{7.1.1}$$

Momentum:

$$\rho_u u_u^2 + p_u = \rho_b u_b^2 + p_b \tag{7.1.2}$$

Energy:

$$h_u + \frac{1}{2}u_u^2 = h_b + \frac{1}{2}u_b^2, \tag{7.1.3}$$

where f is the constant mass flux through the wave, and the subscripts u and b designate the unburned and burned equilibrium states of the gas far upstream and downstream of the wave, as defined previously in Chapter 5. We shall now examine all the possible solutions given by Eqs. (7.1.1)–(7.1.3), recognizing nevertheless that such a solution is only a possible one, in terms of its physical realization, in that it may also have to satisfy other auxiliary constraints such as stability, history and system effects, and entropy consideration.

7.1.1.1. Rayleigh Lines: From Eqs. (7.1.1) and (7.1.2) we have

$$(p_b - p_u) = -(\rho_u u_u)^2 \left(\frac{1}{\rho_b} - \frac{1}{\rho_u} \right) = -f^2(v_b - v_u), \tag{7.1.4}$$

where $v = 1/\rho$ is the specific volume.

If we define $\hat{p} = p_b/p_u$, $\hat{v} = v_b/v_u$, and flow Mach number $M = u/a$, where $a = \sqrt{\gamma p/\rho}$ is the speed of sound and γ the specific heat ratio, then Eq. (7.1.4) can be expressed in nondimensional form in terms of either M_u^2 or $M_b^2 = M_u^2 \hat{v}/\hat{p}$ as

$$M_u^2 = -\frac{\hat{p} - 1}{\gamma(\hat{v} - 1)}, \tag{7.1.5}$$

$$M_b^2 = -\frac{(\hat{p} - 1)\hat{v}}{\gamma(\hat{v} - 1)\hat{p}}, \tag{7.1.6}$$

where $\gamma M_u^2 = f^2/(p_u \rho_u)$ is a nondimensional expression of the square of the mass flux through the wave.

Curves satisfying Eq. (7.1.5) are called Rayleigh lines. They are straight lines passing through the $(\hat{p}, \hat{v}) = (1, 1)$ point with a negative slope of γM_u^2. Furthermore, since $\gamma M_u^2 > 0$, they exist only for values of $(\hat{p} > 1, \hat{v} < 1)$ or $(\hat{p} < 1, \hat{v} > 1)$. Therefore in crossing the wave either the pressure increases, specific volume decreases, and density increases, or the pressure decreases, specific volume increases, and density decreases. Waves with simultaneous increase or decrease in both the pressure and the specific volume are not possible. Figure 7.1.2 illustrates this situation.

7.1.1.2. Hugoniot Lines: From Eq. (7.1.3) we have

$$h_b - h_u = -\frac{1}{2}(\rho_u u_u)^2 \left(\frac{1}{\rho_b^2} - \frac{1}{\rho_u^2} \right) = -\frac{1}{2} f^2 (v_b - v_u)(v_b + v_u). \tag{7.1.7}$$

Using Eq. (7.1.4), Eq. (7.1.7) becomes

$$h_b - h_u = \frac{1}{2}(v_b + v_u)(p_b - p_u), \tag{7.1.8}$$

which is called the Hugoniot relation. Equation (7.1.8) is general in that it is independent of the equation of state and therefore the nature of the material within which the wave propagates.

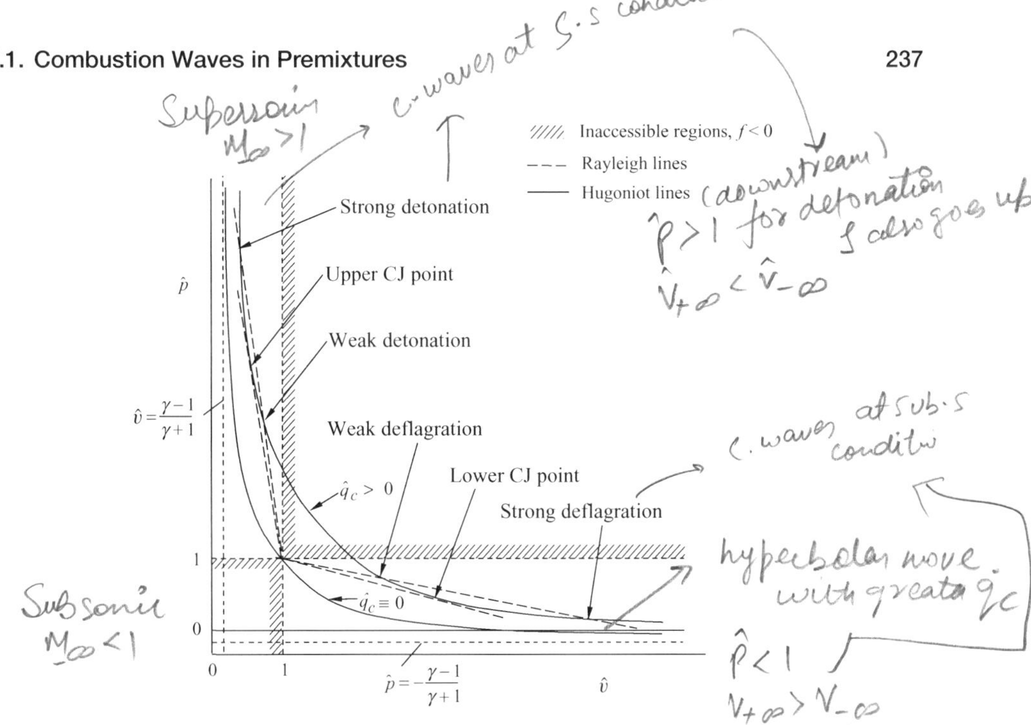

Figure 7.1.2. Schematic identifying the Rankine–Hugoniot solutions.

To be more specific, for q_c amount of heat release per unit mass flux and constant c_p, we have

$$h_b - h_u = -q_c + c_p(T_b - T_u). \tag{7.1.9}$$

Furthermore, using the equation of state $p = \rho RT$ for an ideal gas with constant molecular weight, and the relation $c_p/R = c_p/(c_p - c_v) = \gamma/(\gamma - 1)$, Eq. (7.1.9) becomes

$$h_b - h_u = -q_c + \frac{\gamma}{\gamma - 1}\left(\frac{p_b}{\rho_b} - \frac{p_u}{\rho_u}\right). \tag{7.1.10}$$

Equating (7.1.8) and (7.1.10), and rearranging, we obtain

$$\left(\hat{p} + \frac{\gamma - 1}{\gamma + 1}\right)\left(\hat{v} - \frac{\gamma - 1}{\gamma + 1}\right) = \frac{4\gamma}{(\gamma + 1)^2} + 2\hat{q}_c\left(\frac{\gamma - 1}{\gamma + 1}\right), \tag{7.1.11}$$

where $\hat{q}_c = (\rho_u/p_u)q_c$. Curves satisfying Eq. (7.1.11), shown in Figure 7.1.2, are called Hugoniot curves. These are hyperbolas with asymptotes of $\hat{p} \to -(\gamma - 1)/(\gamma + 1)$ and $\hat{v} \to (\gamma - 1)/(\gamma + 1)$ respectively. For $\hat{q}_c = 0$, the Hugoniot curve passes through the $(\hat{p}, \hat{v}) = (1, 1)$ point.

The problem is now reduced to the simultaneous solution of Eqs. (7.1.5) and (7.1.11), which are called the Rankine–Hugoniot relations. The solution and its implications are discussed next.

7.1.2. Detonation and Deflagration Waves

From Eqs. (7.1.5) and (7.1.11), after some algebraic manipulation it can be shown that

$$\hat{v}_{\pm} - 1 = \frac{(1 - M_u^2)}{(\gamma + 1)M_u^2}\left\{1 \pm \left[1 - \frac{2(\gamma^2 - 1)}{\gamma}\frac{M_u^2}{(1 - M_u^2)^2}\hat{q}_c\right]^{1/2}\right\}, \qquad (7.1.12)$$

$$\hat{p}_{\pm} - 1 = -\frac{(1 - M_u^2)\gamma}{(\gamma + 1)}\left\{1 \pm \left[1 - \frac{2(\gamma^2 - 1)}{\gamma}\frac{M_u^2}{(1 - M_u^2)^2}\hat{q}_c\right]^{1/2}\right\}. \qquad (7.1.13)$$

Equations (7.1.12) and (7.1.13) show that the system is characterized by three parameters, namely γ, $\hat{q}_c$, and M_u, which respectively represent the mixture's compressibility, exothermicity, and the wave propagation speed. The fact that M_u is a parameter that needs to be supplied, instead of determined, is because analysis at the present level does not include consideration of the wave structure, as noted earlier.

Thus for a given set of $(\gamma, \hat{q}_c, M_u)$, Eqs. (7.1.12) and (7.1.13) show that two solutions are possible. Furthermore, the characteristics of the solutions also change for $M_u \gtrless 1$. That is, since the term within the curly bracket is always positive, the signs of $(\hat{v}_{\pm} - 1)$ and $(\hat{p}_{\pm} - 1)$ are determined by the sign of $(1 - M_u^2)$. Thus for $M_u \gtrless 1$, we have $(\hat{p}_{\pm} - 1) \gtrless 0$ and $(\hat{v}_{\pm} - 1) \lessgtr 0$. We call solutions of the supersonic branch, $M_u > 1$, detonation waves, and those of the subsonic branch, $M_u < 1$, deflagration waves. These solutions are now separately discussed by using Figure 7.1.2, in which the Rayleigh line and Hugoniot curve are plotted. It is seen that for a given Hugoniot curve, two sets of Rayleigh lines can be drawn, respectively satisfying $(\hat{p} > 1, \hat{v} < 1)$ and $(\hat{p} < 1, \hat{v} > 1)$. The intersections of the Rayleigh line with the Hugoniot curve yield the solutions defined by Eqs. (7.1.12) and (7.1.13).

7.1.2.1. Detonation Waves: Detonation waves travel supersonically with $p_b > p_u$ and $v_b < v_u$, implying that in crossing such a wave the pressure and density increase while the velocity decreases. The possible increases in the pressure are bounded by $1 + (\gamma - 1)\hat{q}_c \leq \hat{p} \leq \infty$, with the lower limit corresponds to $\hat{v} = 1$ in Eq. (7.1.12). The possible reduction in the specific volume is bounded by $(\gamma - 1)/(\gamma + 1) \leq \hat{v} \leq 1$, with the lower limit corresponds to $\hat{p} = \infty$.

The solutions with the higher and lower values of the pressure jump are respectively known as strong detonation and weak detonation. There also exists a minimum Rayleigh line, tangent to the Hugoniot line, beyond which no solution exists. The point of tangency is called the upper Chapman–Jouguet (CJ) point and the corresponding wave is called a Chapman–Jouguet detonation. It is the "weakest" strong detonation that can exist. For $\hat{q}_c = 0$, which corresponds to the hydrodynamic shock wave, only the strong solution exists.

By considering the wave structure, Chapter 14 will show that weak detonations with exothermic reactions do not normally exist. Furthermore, under most experimental conditions detonations propagate at the Chapman–Jouguet wave speed.

7.1.2.2. Deflagration Waves: Deflagration waves travel subsonically with $p_b < p_u$ and $v_b > v_u$, implying that in crossing such a wave, both pressure and density decrease while the velocity increases. The possible reduction in pressure is bounded by $0 \leq \hat{p} \leq 1$, whereas the possible increase in specific volume is bounded by $1 + \hat{q}_c(\gamma - 1)/\gamma \leq \hat{v} \leq 2\hat{q}_c + (\gamma + 1)/(\gamma - 1)$, with the lower and upper limits respectively correspond to $\hat{p} = 1$ and $\hat{p} = 0$.

The solutions with higher and lower values of pressure reduction are respectively known as strong and weak deflagrations. There also exists a maximum Rayleigh line beyond which no solution exists. The point of tangency is called the lower CJ point and the corresponding wave is called a CJ deflagration.

It can be argued that strong deflagrations do not exist. That is, if we take the limit of a hydrodynamic discontinuity, with $q_c \equiv 0$, then Figure 7.1.2 shows that there is only one intersection between the Rayleigh and Hugoniot curves, yielding the strong deflagration solution. However, it can be readily shown (Liepmann & Roshko 1957) that entropy decreases in crossing such a wave, implying its nonexistence. Consequently, it is reasonable to expect that in the presence of heat release, with $q_c > 0$, strong deflagration also does not exist.

For weak deflagrations, $\hat{p} \stackrel{\sim}{<} 1$, which implies that the pressure change across the wave is very small. Therefore the wave structure can be treated as approximately isobaric in this case. For very slow flows ($M_u^2 \ll 1$) only the weak deflagration intersection exists because the strong deflagration will have $\hat{p} < 0$, which further rules out its existence.

7.1.3. Chapman–Jouguet Waves

Chapman–Jouguet waves have some special properties that provide insight into the general structure of detonation and deflagration waves. We first note that while M_u of a general detonation or deflagration wave is indeterminable from the Rankine–Hugoniot relations, the propagation velocities of the upper and lower Chapman–Jouguet waves are uniquely defined for given values of $(\gamma, \hat{q}_c)$ because of the additional tangency requirement. To determine the properties of the CJ waves, we first evaluate the slopes of the Rayleigh line and Hugoniot curves from Eqs. (7.1.6) and (7.1.11) respectively, yielding

$$\left(\frac{d\hat{p}}{d\hat{v}}\right)_{\text{Rayleigh}} = \frac{\hat{p} - 1}{\hat{v} - 1} \tag{7.1.14}$$

$$\left(\frac{d\hat{p}}{d\hat{v}}\right)_{\text{Hugoniot}} = -\frac{[(\gamma + 1)/(\gamma - 1)]\hat{p} + 1}{[(\gamma + 1)/(\gamma - 1)]\hat{v} - 1}. \tag{7.1.15}$$

It can then be readily shown that the relation

$$\left(\frac{d\hat{p}}{d\hat{v}}\right)_{\text{Hugoniot}} \gtreqless \left(\frac{d\hat{p}}{d\hat{v}}\right)_{\text{Rayleigh}} \tag{7.1.16}$$

is equivalent to

$$\frac{(\hat{p}-1)\hat{v}}{\gamma(1-\hat{v})\hat{p}} \lesseqgtr 1. \tag{7.1.17}$$

However, the LHS of (7.1.17) is simply M_b^2, given by Eq. (7.1.6). Thus the relation (7.1.16) is equivalent to

$$M_b^2 \gtreqless 1. \tag{7.1.18}$$

Equation (7.1.18) then implies that the downstream flow is sonic ($M_b = 1$) for the CJ wave. Furthermore, since, in general, the Hugoniot curve has a greater slope than the Rayleigh line for strong detonation and weak deflagration, as is evident from Figure 7.1.2, while the opposite holds for weak detonation and strong deflagration, we conclude that $M_b < 1$ for the former and $M_b > 1$ for the latter.

The propagation speed of the CJ wave can be readily determined by setting the radical term in Eq. (7.1.12) to zero, yielding

$$(M_{u,\mathrm{CJ}})_\pm^2 = 1 + \frac{(\gamma^2-1)\hat{q}_c}{\gamma}\left\{1 \pm \left[1 + \frac{2\gamma}{(\gamma^2-1)\hat{q}_c}\right]^{1/2}\right\}, \tag{7.1.19}$$

which shows that

$$(M_{u,\mathrm{CJ}})_+ > 1 \quad \text{and} \quad (M_{u,\mathrm{CJ}})_- < 1,$$

as they should be. Knowing $M_{u,\mathrm{CJ}}$, $\hat{v}_{\mathrm{CJ}}$ and $\hat{p}_{\mathrm{CJ}}$ are given by Eqs. (7.1.12) and (7.1.13) as

$$\hat{v}_{\mathrm{CJ},\pm} - 1 = \hat{q}_c\frac{(\gamma-1)}{\gamma}\left\{1 \mp \left[1 + \frac{2\gamma}{(\gamma^2-1)\hat{q}_c}\right]^{1/2}\right\} \tag{7.1.20}$$

$$\hat{p}_{\mathrm{CJ},\pm} - 1 = \hat{q}_c(\gamma-1)\left\{1 \pm \left[1 + \frac{2\gamma}{(\gamma^2-1)\hat{q}_c}\right]^{1/2}\right\}. \tag{7.1.21}$$

Note that the ($\pm$) signs in Eqs. (7.1.19) to (7.1.21) designate the upper and lower Chapman–Jouguet states, while those in Eqs. (7.1.12) and (7.1.13) represent the two possible solutions for a given M_u. Substituting Eqs. (7.1.20) and (7.1.21) into Eq. (7.1.6) again results in the sonic condition for $M_{b,\mathrm{CJ}}$. Furthermore, the fact that $M_u < 1$ and $M_b > 1$ for the strong deflagration indicates that the flow becomes more ordered in crossing the wave. This again supports the notion that such waves do not exist because of the decreased entropy.

7.1.4. Preliminary Discussion of Detonation Waves

Since we shall be concerned primarily with deflagration waves starting in Section 7.2, and shall return to detonations only in Chapter 14, it is informative to give a brief overview of the formation, propagation, and structure of detonation waves here.

A detonation wave can be formed through either direct or indirect initiation. In the direct mode of initiation, sufficient energy is rapidly deposited in a small volume of a mixture, resulting in an almost immediate emergence of the detonation wave without passing through an intermediate stage of a deflagration wave. The minimum amount of energy deposition should be at least equal to that needed to maintain a detonation wave structure, for a period at least as long as the chemical reaction time. However, detailed considerations of other effects such as the curvature of the blast wave that is formed from spherical or cylindrical sources will show that the actual energy required is substantially larger than this estimate.

In most of the indirect mode of initiation the ignition source is weaker, producing a deflagration wave that accelerates to a detonation wave. This mode of generation is called deflagration-to-detonation transition (DDT). The transition mechanism involves generation of compression waves by the hot combustion products and their subsequent coalescence to form shock waves, development of turbulence ahead of and within the flame, and interactions with solid boundaries that could further destabilize the flow.

A detonation wave can also be formed in a mixture that has, for example, a temperature or reactant concentration gradient such that successive autoignition along the gradient coincides with the arrival of the compression waves generated earlier. This synchronized mode of amplification can also be present during DDT, leading to rapid development of the detonation wave.

Gaseous detonations typically travel at $O(10^3)$ m/s, as compared to $O(1-10^2)$ cm/s for deflagration waves. The basic detonation wave structure consists of a leading shock that instantly heats and compresses the reactive gas to a state of high temperature and pressure. This is followed by an induction period and then a rapid state of reaction and, hence, heat release, generating compression waves that sustain the detonation propagation. Earlier studies were based on the one-dimensional interpretation of this model. However, it was subsequently realized that detonation waves are usually unstable, and that the structure is intrinsically three-dimensional, involving complex shock and rarefaction wave interactions. On the other hand, it has also been found that the propagation velocity of an established wave, well removed from limit situations, is very close to that of the Chapman–Jouguet wave. The reason being that since propagation of the CJ wave is defined without considering the wave structure, the detailed wave structure therefore should have minimum influence on its propagation. This is a remarkable result in that the detonation velocity can be calculated from thermodynamic and hydrodynamic considerations alone, without regard to the details of chemical kinetics and wave interactions.

7.2. PHENOMENOLOGICAL DESCRIPTION OF THE STANDARD FLAME

7.2.1. Flame Structure

We shall now restrict our study to the structure and propagation of the standard premixed flame. In the flame-stationary frame (Figure 7.2.1a) the upstream mixture approaches the flame with velocity $u_u = s_u^o$ and temperature T_u, and leaves the flame with velocity u_b^o and temperature T_b^o. If we assume that the mixture is sufficiently off-stoichiometric, such that the reaction is governed by the concentration Y of the

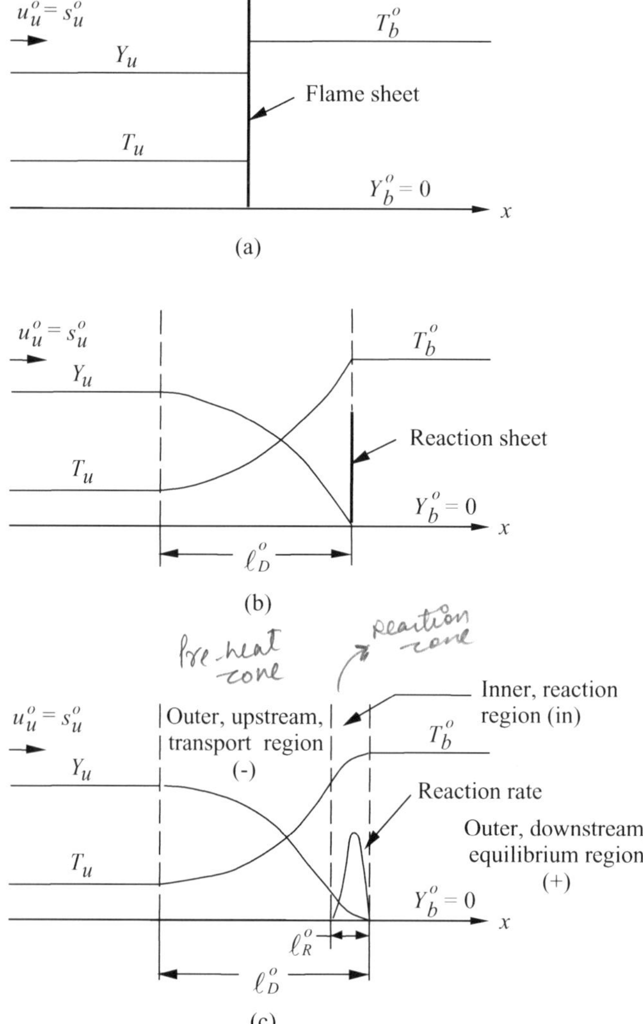

Figure 7.2.1. Schematic showing the premixed flame structure at successive levels of detail: (a) The hydrodynamic, flame-sheet level; (b) the transport, reaction-sheet level; and (c) detailed structure including the reaction zone.

deficient reactant, then a one-reactant reaction with

$$\text{Reactant} \rightarrow \text{Products} \tag{7.2.1}$$

can be used, with Y_u being the concentration of the fresh mixture and $Y_b^o \equiv 0$ indicating its complete consumption upon crossing the flame. Furthermore, since s_u^o is much smaller than the speed of sound, the combustion process can be assumed to be isobaric in accordance with the previous discussion in Section 5.2.4 on the properties of low-speed, subsonic flows, and in the previous section on the small pressure change across a weak deflagration wave.

The flame structure can be considered at three levels of detail. At the hydrodynamic level of the Rankine–Hugoniot relations (Figure 7.2.1a), the flame is simply an interface separating two fluid dynamical states of unburned and burned gases that are in thermodynamic equilibrium. They are related by the overall conservation of mass, species concentrations, and energy. At this flame sheet, the temperature and reactant concentration change discontinuously from T_u to T_b^o, and from Y_u to $Y_b^o = 0$, respectively.

At the next, more detailed, transport-dominated level of description, the flame sheet of Figure 7.2.1a is expanded to reveal a preheat zone, of characteristic thickness ℓ_D^o and governed by heat and mass diffusion processes, as shown in Figure 7.2.1b. Here, as the mixture approaches the flame, it is gradually heated up by the heat conducted forward from the chemical heat release region, resulting in a continuously increasing temperature profile until T_b^o is reached. The profile is not linear due to the presence of convective transport. The continuous heating of the mixture will eventually lead to its ignition and subsequent reaction. From large activation energy consideration, we expect that the reaction is activated only when the gas temperature is close to its maximum value. Furthermore, once reaction is initiated, it is completed rapidly as the deficient reactant is depleted. Thus at the transport level, the reaction zone can be considered to be concentrated at an interface—a reaction sheet, which serves as a source of heat and a sink for the reactant. At this surface, the temperature and reactant concentration are continuous and assume their respective burned values in the downstream. Their slopes, however, change discontinuously.

Vanishing of the reactant concentration at the reaction sheet establishes a concentration gradient in the preheat zone. Furthermore, for mixtures whose Lewis number is close to unity, the similar values of the heat and mass diffusivities imply that the rate of temperature increase should be correspondingly similar to the rate of concentration decrease. For a $Le = 1$ mixture, the two profiles, when properly normalized, are mirror images of each other (Figure 7.2.1b).

Description of the premixed flame at the diffusive transport level is similar to that of the reaction-sheet limit of a nonpremixed flame. However, while the global properties of a nonpremixed flame can be completely determined at this level through the requirement of stoichiometric reaction, this condition does not exist for a premixed flame. Thus a complete global characterization of a premixed flame cannot be achieved at this level and description of the reaction process is needed.

This then brings us to the third, and most detailed level of flame description. Here (Figure 7.2.1c) the reaction sheet is expanded to reveal the reaction rate profile, which has a characteristic thickness $\ell_R^o \ll \ell_D^o$ and is a highly peaked function, consisting of a rapidly increasing portion due to activation of the reaction, followed by a rapidly decreasing portion due to depletion of the reactant. This rapid rate of property change within a narrow region then implies that diffusive transport, which is described by a second-order differential, has a greater influence than convective transport, which is described by a first-order differential.

The flame structure can therefore be considered to consist of two distinct zones, namely the preheat zone in which convection and diffusion dominate and balance, and the reaction zone in which reaction and diffusion balance. Since $\ell_R^o \ll \ell_D^o$, the entire flame thickness representing the nonequilibrium processes of reaction and diffusion can be basically identified as ℓ_D^o. Across this flame, the overall conservation of mass and energy holds. Thus, from continuity, $d(\rho u)/dx = 0$, we have

$$f^o = \rho u = \rho_u u_u^o = \rho_b^o u_b^o, \tag{7.2.2}$$

where f^o is the constant mass flux and will be referred to as the laminar burning flux. Equation (7.2.2) demonstrates that the fundamental parameter characterizing the rate of flame propagation is the mass flux f^o instead of the propagation velocity u_u^o, or s_u^o, by itself.

From energy conservation across the flame, since all the deficient reactant is consumed, and if we further assume that there is no heat loss, then for constant c_p we have

$$c_p(T_b^o - T_u) = q_c Y_u. \tag{7.2.3}$$

Equation (7.2.3) simply states that all the chemical heat liberated is used to heat the incoming gas. Therefore the downstream temperature T_b^o is just the adiabatic flame temperature T_{ad}, given by

$$T_b^o = T_{\text{ad}} = T_u + q_c Y_u / c_p. \tag{7.2.4}$$

We have already identified this relation in Section 5.5.2.

In order to determine the laminar burning flux f^o and the characteristic flame thickness ℓ_D^o, we need to consider the nonequilibrium processes of diffusion and reaction within the flame structure. Derivations with increasing mathematical rigor will be presented in the following sections, although their dominant dependence on the diffusive and reactive nature of the flame structure can be readily assessed from a simple phenomenological analysis, to be discussed next.

7.2.2. Laminar Burning Flux and Flame Thickness

We first note that the characteristic temperature change across the reaction zone can be estimated by $[T_b^o - T(x_f^-)] \sim [w/(dw/dT)]_{T_b^o} = (T_b^o)^2/T_a$, using $w \sim \exp(-T_a/T)$. Then from continuity of the heat flux through the preheat and reaction zones, $(dT/dx)_{\ell_R^o} = (dT/dx)_{\ell_D^o}$, or $\ell_R^o/\ell_D^o \sim [T_b^o - T(x_f^-)]/(T_b^o - T_u)$, ℓ_R^o can be related to ℓ_D^o through

$$\frac{\ell_R^o}{\ell_D^o} \sim \frac{(T_b^o)^2}{(T_b^o - T_u)T_a} = Ze^{-1}. \tag{7.2.5}$$

Thus the reaction zone is thinner than the preheat zone by a factor $Ze^{-1} \ll 1$, which is to be expected.

We next note that since convection and diffusion balance in the preheat zone, we have

$$f^o \sim \frac{\lambda/c_p}{\ell_D^o}. \tag{7.2.6}$$

Furthermore, since all the reactant mass flux $Y_u f^o$ entering the flame is reacted at the same rate in passing through the reaction zone, we have the statement for overall mass conservation that

$$f^o \sim w_b^o \ell_R^o, \tag{7.2.7}$$

where $w_b^o = w(T_b^o)$ is the reaction rate evaluated at the temperature of the thin reaction zone, and we have set $Y_u = 1$ for simplicity in expression. Using Eqs. (7.2.5) to (7.2.7), the burning flux f^o and flame thickness ℓ_D^o can be solved, given by

$$(f^o)^2 \sim \frac{(\lambda/c_p)w_b^o}{Ze}, \tag{7.2.8}$$

$$(\ell_D^o)^2 \sim \frac{(\lambda/c_p)}{w_b^o} Ze. \tag{7.2.9}$$

Relations (7.2.6) and (7.2.7) can be alternately expressed as

$$f^o \ell_D^o \sim \lambda/c_p, \tag{7.2.10}$$

$$\frac{f^o}{\ell_D^o} \sim \frac{w_b^o}{Ze}, \tag{7.2.11}$$

which show that $f^o \ell_D^o$ and f^o/ℓ_D^o are respectively described by the transport and reactive aspects of the flame.

Relations (7.2.8) and (7.2.9) show that the laminar flame responses depend on the reaction kinetics through the characteristic reaction rate w_b^o, and on the transport processes through the density-weighted transport coefficient λ/c_p. Compared to the nonpremixed flame of Chapter 6 whose burning rate varies with λ/c_p, the burning flux for the premixed flame shows a weaker, square-root dependence. The results that f^o increases with increasing transport and reaction rates, while ℓ_R^o increases with the transport rate but decreases with the reaction rate, are physically reasonable.

The above discussion illustrates the fact that since laminar flame propagation is governed by two processes, its complete description also requires two representative parameters, whether they are λ/c_p and w_b^o, or f^o and ℓ_D^o, or $f^o \ell_D^o$ and f^o/ℓ_D^o. The fundamental importance of ℓ_D^o in characterizing laminar flame propagation, in addition to f^o, is to be noted. Thus when ℓ_D^o is evaluated through the frequently used relation $\ell_D^o \sim (\lambda/c_p)/f^o$, it does not imply that ℓ_D^o is inherently a derivable quantity. It is derivable in this case because λ/c_p has been selected as an independent parameter that relates ℓ_D^o to f^o. It is also of interest to note that the burning flux result given by (7.2.8) can be alternately interpreted as the geometric average of the rates of the two processes that control the phenomenon, that is, $f^o \sim \sqrt{(\lambda/c_p)w_b^o}$.

The dependence of f^o on the various chemical kinetic parameters can be demonstrated by noting that

$$w_b^o \sim p^n e^{-T_a/T_b^o}, \tag{7.2.12}$$

where n is a general overall reaction order. Thus, from (7.2.8), we have

$$f^o \sim [(\lambda/c_p)w_b^o]^{1/2} \sim [p^n(\lambda/c_p)_b e^{-T_a/T_b^o}]^{1/2}. \tag{7.2.13}$$

Furthermore, since $\rho_u \sim p$, we have $u_u \sim f^o/p$, or

$$s_u^o \sim p^{(\frac{n}{2}-1)}[(\lambda/c_p)_b e^{-T_a/T_b^o}]^{1/2}, \tag{7.2.14}$$

which shows that for a second-order reaction, with $n = 2$, s_u^o is independent of pressure. Physically, this result reflects the compensatory effect that while the reaction rate increases with pressure, the upstream gas also becomes denser for the flame to heat up and pass through.

Since the flame thickness ℓ_D^o varies inversely with f^o, then it should also vary inversely with pressure for $n > 0$, which is usually the case. This is physically reasonable because with increasing pressure, the rates of molecular collision and thereby reaction are facilitated, resulting in faster completion of the reaction as the mixture flows downstream. At the same time, the tendency for heat and mass diffusion to affect the flame is minimally influenced by changes in pressure because λ/c_p and ρD are insensitive to pressure variations, as shown in Chapter 4. The net effect is that pressure affects the flame thickness primarily through its influence on the reaction rate. This result forms the basis for conducting experimental flame structure studies under low pressures. That is, by stretching the flame thickness, the detailed flame structure can be mapped by using various experimental probes that have finite dimensions. The important point to note here is that flame thickening with decreasing pressure is due to reduced reaction rate instead of increased diffusive transport rates.

The upstream laminar flame speed s_u^o is related to the downstream laminar flame speed by the density ratio $\rho_u/\rho_b^o \approx T_b^o/T_u$, which is substantially greater than unity. Thus, care needs to be exercised in either defining or identifying the particular flame speed under study. Most reported experimental values are s_u^o. Typical values of s_u^o and flame thickness (ℓ_D^o) of hydrocarbon–air mixtures at atmospheric pressure are $O(1-10^2)$ cm/sec and $O(0.1-1)$ mm respectively.

7.3. MATHEMATICAL FORMULATION

7.3.1. Governing Equations

In the following, we shall formulate the mathematical problem for the standard premixed flame as governed by the situation of Figure 7.2.1. However, instead of directly using the nondimensional heat and mass conservation equations of the model system defined in Section 5.6, for illustrative purpose we shall start our derivation from the original dimensional equations (5.3.12) and (5.3.14),

$$f^o c_p \frac{dT}{dx} - \lambda \frac{d^2T}{dx^2} = q_c w, \tag{7.3.1}$$

$$f^o \frac{dY}{dx} - \rho D \frac{d^2Y}{dx^2} = -w, \tag{7.3.2}$$

where $w = B_c Y e^{-T_a/T}$, and B_C is given by Eq. (5.6.4). The dependence of w on the preexponential parameters and the concentration of the abundant species are absorbed in B_C, which can be considered to be a constant.

We next define $\tilde{T} = (c_p T)/(q_c Y_u)$ and $\tilde{Y} = Y/Y_u$ such that $\tilde{Y}_u = 1$, and scale x in units of the characteristic flame thickness $(\lambda/c_p)/f^o$ such that a nondimensional distance can be defined as

$$\tilde{x} = \frac{f^o}{\lambda/c_p} x. \tag{7.3.3}$$

Thus Eqs. (7.3.1) and (7.3.2) become, after linear combination of them to eliminate the reaction term in the latter,

$$\frac{d^2\tilde{T}}{d\tilde{x}^2} - \frac{d\tilde{T}}{d\tilde{x}} = -Da_C^o \tilde{Y} e^{-\tilde{T}_a/\tilde{T}}, \tag{7.3.4}$$

$$\frac{d^2\tilde{T}}{d\tilde{x}^2} + \frac{1}{Le}\frac{d^2\tilde{Y}}{d\tilde{x}^2} - \frac{d(\tilde{T}+\tilde{Y})}{d\tilde{x}} = 0, \tag{7.3.5}$$

where

$$Da_C^o = \frac{\lambda/c_p}{(f^o)^2} B_C \tag{7.3.6}$$

is the collision Damköhler number based on f^o. From the grouping of the parameters in Da_C^o, we anticipate that $f^o \sim (\lambda/c_p)^{1/2}$, as shown previously.

The boundary conditions for Eqs. (7.3.4) and (7.3.5) are

$$\tilde{x} = -\infty: \quad \tilde{T} = \tilde{T}_u, \quad \tilde{Y} = 1, \tag{7.3.7}$$

$$\tilde{x} = \infty: \quad \tilde{T} = \tilde{T}_b^o, \quad \tilde{Y} = 0. \tag{7.3.8}$$

Since $\tilde{T}$ and $\tilde{Y}$ attain constant and uniform values at $\tilde{x} = \pm\infty$, it is obvious that Eqs. (7.3.7) and (7.3.8) also imply

$$\tilde{x} = \pm\infty: \quad \frac{d\tilde{T}}{d\tilde{x}} = \frac{d\tilde{Y}}{d\tilde{x}} = 0. \tag{7.3.9}$$

Integrating Eq. (7.3.5) once and applying the boundary conditions at $\tilde{x} = -\infty$ yields

$$(\tilde{T}+\tilde{Y}) - \left(\frac{d\tilde{T}}{d\tilde{x}} + \frac{1}{Le}\frac{d\tilde{Y}}{d\tilde{x}}\right) = (\tilde{T}_u + 1). \tag{7.3.10}$$

By further evaluating Eq. (7.3.10) at $\tilde{x} = \infty$, we obtain

$$\tilde{T}_b^o = 1 + \tilde{T}_u, \tag{7.3.11}$$

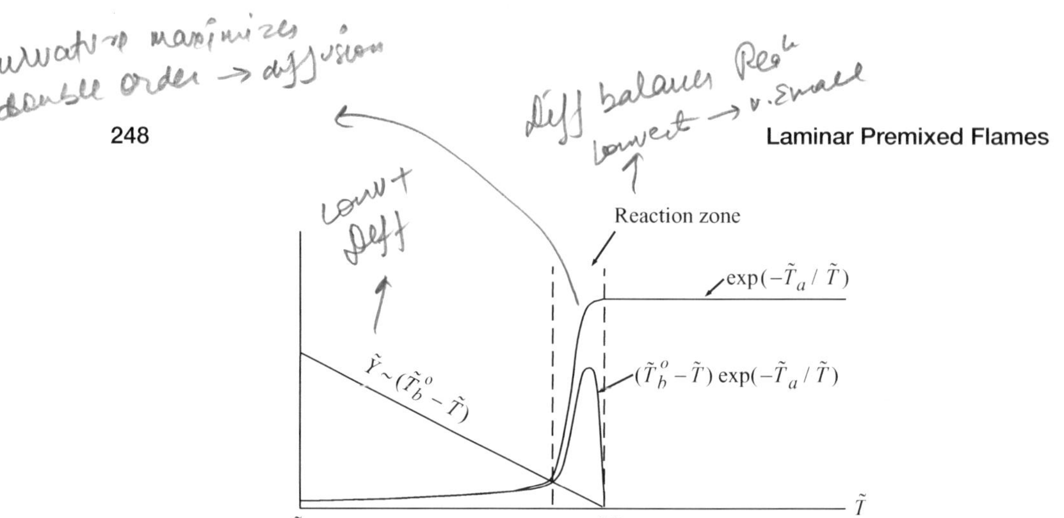

Figure 7.3.1. Schematic showing the dependence of the reaction rate on temperature.

which in dimensional form is Eq. (7.2.3) for the adiabatic flame temperature. Furthermore, according to the present nondimensionalization, $(\tilde{T}_b^o - \tilde{T}_u)$ assumes the value of unity in the expression for energy conservation, representing the temperature increase across the flame. It is also significant to note that for this particular flame, the flame temperature remains at $\tilde{T}_{\mathrm{ad}}$ even for $Le \neq 1$ because all the reactants are consumed for the same amount of the mixture flux.

For $Le = 1$, Eq. (7.3.10) can be further integrated with $(\tilde{T} + \tilde{Y})$ as a group to yield

$$\tilde{T} + \tilde{Y} = \tilde{T}_b^o + c_1 e^{\tilde{x}},$$

where c_1 is an integration constant. Since the solution has to be bounded as $\tilde{x} \to \infty$, c_1 must vanish, resulting

$$\tilde{T} + \tilde{Y} = \tilde{T}_b^o. \tag{7.3.12}$$

This allows $\tilde{Y}$ to be explicitly expressed in terms of $\tilde{T}$ as $\tilde{Y} = \tilde{T}_b^o - \tilde{T}$. Substituting $\tilde{Y}$ into Eq. (7.3.4), its dependence on $\tilde{Y}$ is decoupled, yielding

$$\frac{d^2\tilde{T}}{d\tilde{x}^2} - \frac{d\tilde{T}}{d\tilde{x}} = -Da_C^o \left(\tilde{T}_b^o - \tilde{T}\right) e^{-\tilde{T}_a/\tilde{T}}, \tag{7.3.13}$$

which is the equation to be solved subject to the boundary conditions for $\tilde{T}$ given by Eqs. (7.3.7) and (7.3.8).

Figure 7.3.1 shows the behavior of the reaction rate term $(\tilde{T}_b^o - \tilde{T})\exp(-\tilde{T}_a/\tilde{T})$ as a function of $\tilde{T}$, as the mixture is heated from $\tilde{T}_u$ to $\tilde{T}_b^o$. The factor $(\tilde{T}_b^o - \tilde{T})$ represents the reactant concentration $\tilde{Y}$ for $Le = 1$, which decreases linearly with $\tilde{T}$ and vanishes at $\tilde{T}_b^o$. The temperature-sensitive Arrhenius factor, $\exp(-\tilde{T}_a/\tilde{T})$, starts to rapidly increase only when $\tilde{T}$ is close to $\tilde{T}_b^o$. Thus the product of these two factors gives the highly peaked reaction rate profile spanning over a narrow temperature range neighboring $\tilde{T}_b^o$. This narrow temperature regime directly translates to a narrow spatial region in $\tilde{x}$.

Before solving either Eqs. (7.3.4) and (7.3.5) for $Le \neq 1$, or Eq. (7.3.13) for $Le = 1$, we shall first mention an important fundamental mathematical property of these equations.

7.3.2. The Cold Boundary Difficulty

Early attempts at solving either Eqs. (7.3.4) and (7.3.5), or their decoupled form Eq. (7.3.13), encountered what is known as the "cold boundary difficulty," which arises from the fact that the governing equations posed above are actually ill-defined mathematically at the cold boundary, $\tilde{x} = -\infty$. This point can be demonstrated by evaluating Eq. (7.3.4) at $\tilde{x} = -\infty$. Since

$$\left(\frac{d\tilde{T}}{d\tilde{x}}\right)_{-\infty} = \left(\frac{d^2\tilde{T}}{d\tilde{x}^2}\right)_{-\infty} \equiv 0,$$

the LHS of Eq. (7.3.4) is zero. However, substitution of $\tilde{Y} = 1$ and $\tilde{T} = \tilde{T}_u$ into its RHS yields a finite value, $-Da_C^o \exp(-\tilde{T}_a/\tilde{T}_u)$. Therefore the equation is not balanced at the cold boundary. A straightforward solution will yield meaningless results.

This cold boundary difficulty manifests the fact that since the one-step, irreversible reaction takes place at a finite temperature ($T_u \neq 0$) and hence a finite rate at $x = -\infty$, and since it also has an infinite time to proceed as it flows toward the flame, the reactant is completely reacted even at the freestream. Thus the concept of a steadily propagating flame of finite thickness becomes meaningless. The same difficulty does not exist for the hot boundary because $Y \to 0$ as $T \to T_b^o$ such that $w \to 0$ for the assumed one-step overall reaction.

The cold boundary difficulty can be removed by simply freezing the chemical reaction at a distance sufficiently far upstream of the flame. Such an approach can be interpreted on the basis of an ignition temperature, T_{ig}. That is, we simply set

$$w = 0 \quad \text{for} \quad T < T_{ig} \tag{7.3.14}$$

in Eq. (7.3.1) such that reaction at the cold boundary is automatically suppressed. This assumption is physically realistic because from large activation energy considerations the reaction rate is negligibly small except in the reaction region. Therefore, computationally, as long as the assumed T_{ig} is not close to either T_u or T_b^o, the solution obtained is insensitive to T_{ig}.

The cold boundary difficulty frequently arises in the steady-state formulation of phenomena in which a finite-rate process proceeds in one direction in the ambience. Another example is the combustion of a fuel droplet in an oxidizing environment consisting of a small amount of fuel vapor, as for the situation of fuel droplets in the interior of a spray.

In the next two sections, several solutions based on the concept of T_{ig} will be presented. The first solution (Chung & Law 1988) utilizes an integral approach commonly adopted in heat and mass transfer problems. We shall also use this opportunity to include the $Le \neq 1$ feature in the derivation. The second and third solutions are similar in their approaches, involving the separate analysis of the preheat, reaction, and equilibrium zones of the flame structure of Figure 7.2.1c, and matching the separate solutions at their respective common boundaries to assure continuity of their values and gradients. The second solution (Zel'dovich & Frank-Kamenetskii 1938) is physically more illuminating but mathematically less rigorous. The third solution

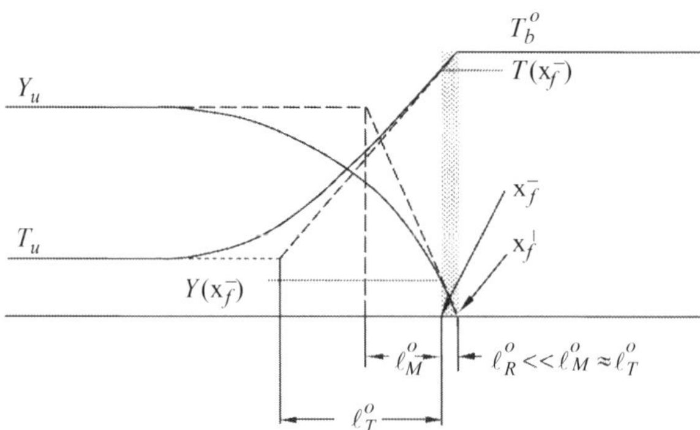

Figure 7.4.1. Schematic showing the definitions of the effective thicknesses of the heat and mass diffusion zones.

(Bush & Fendell 1970), given in Section 7.5, is based on activation energy asymptotic analysis and is mathematically more formal. The bulk response of many premixed flame phenomena can be analyzed by any of these three methods, with the results differing by at most some nonessential constant multiplicative factors. We shall have occasion to use all of them in subsequent studies. Further discussion on the analysis of the standard laminar flame can be found in Williams (1985).

7.4. APPROXIMATE ANALYSES

7.4.1. Integral Analysis

The configuration most suitable for visualizing this analysis is Figure 7.2.1b, which is reproduced in Figure 7.4.1, but with more precise definitions of the thicknesses of the various zones of the flame structure. The thin reaction zone is located at x_f when viewed from the broad diffusion and equilibrium zones, and the gradients of $\tilde{T}$ and $\tilde{Y}$ change discontinuously from finite values at x_f^- to zero at x_f^+.

We start with Eqs. (7.3.4) and (7.3.10). Integrating Eq. (7.3.4) from $\tilde{x} = -\infty$ to $\tilde{x}_f^-$, and recognizing that the reaction term is negligibly small in this region, that $(d\tilde{T}/d\tilde{x})_u = 0$, and that $\tilde{T}(\tilde{x}_f^-) \approx \tilde{T}_b^o$, we have

$$\left(\frac{d\tilde{T}}{d\tilde{x}}\right)_{\tilde{x}_f^-} = \tilde{T}_b^o - \tilde{T}_u = 1. \tag{7.4.1}$$

By also evaluating Eq. (7.3.10) at $\tilde{x}_f^-$ where $\tilde{Y} \approx 0$ and $\tilde{T} \approx \tilde{T}_b^o$, we obtain

$$\left(\frac{d\tilde{T}}{d\tilde{x}}\right)_{\tilde{x}_f^-} + \frac{1}{Le}\left(\frac{d\tilde{Y}}{d\tilde{x}}\right)_{\tilde{x}_f^-} = \tilde{T}_b^o - (\tilde{T}_u + 1) = 0. \tag{7.4.2}$$

We next define effective thicknesses of the thermal and mass diffusion zones, ℓ_T^o and ℓ_M^o, as

$$\ell_T^o = \frac{T_b^o - T_u}{(dT/dx)_{x_f^-}}, \qquad \ell_M^o = -\frac{Y_u}{(dY/dx)_{x_f^-}}. \tag{7.4.3}$$

The parameters ℓ_T^o and ℓ_M^o are simply alternate expressions for $(dT/dx)_{x_f^-}$ and $(dY/dx)_{x_f^-}$, and, hence, are properties of the flame to be determined from the analysis. No approximation is involved in writing these expressions.

Expressing ℓ_T^o and ℓ_M^o in nondimensional forms, we have

$$\tilde{\ell}_T^o = \frac{1}{(d\tilde{T}/d\tilde{x})_{\tilde{x}_f^-}}, \qquad \tilde{\ell}_M^o = -\frac{1}{(d\tilde{Y}/d\tilde{x})_{\tilde{x}_f^-}}. \tag{7.4.4}$$

Substituting (7.4.4) into Eqs. (7.4.1) and (7.4.2) yields

$$\tilde{\ell}_T^o = 1, \tag{7.4.5}$$

$$\frac{\tilde{\ell}_T^o}{\tilde{\ell}_M^o} = Le. \tag{7.4.6}$$

Equation (7.4.5) shows that the characteristic flame thickness $\ell_D^o = (\lambda/c_p)/f^o$ is simply the definition of ℓ_T^o in Eq. (7.4.3), representing one thermal thickness, and that $f^o\ell_T^o = \lambda/c_p$. Equation (7.4.6) shows that the ratio of the characteristic thicknesses for heat and mass diffusion is equal to Le, which is a reasonable outcome.

We next integrate Eq. (7.3.4) across the reaction zone, from $\tilde{x}_f^-$ to $\tilde{x}_f^+$, yielding

$$\left(\frac{d\tilde{T}}{d\tilde{x}}\right)_{\tilde{x}_f^+} - \left(\frac{d\tilde{T}}{d\tilde{x}}\right)_{\tilde{x}_f^-} - \left[\tilde{T}(\tilde{x}_f^+) - \tilde{T}(\tilde{x}_f^-)\right] = -Da_C^o \int_{\tilde{x}_f^-}^{\tilde{x}_f^+} \tilde{Y}e^{-\tilde{T}_a/\tilde{T}}d\tilde{x}. \tag{7.4.7}$$

Since $(d\tilde{T}/d\tilde{x})_{\tilde{x}_f^+} = 0$, $(d\tilde{T}/d\tilde{x})_{\tilde{x}_f^-} = 1$, and $\tilde{T} \approx \tilde{T}_b^o$ at $\tilde{x}_f^-$ and $\tilde{x}_f^+$, Eq. (7.4.7) becomes

$$1 = Da_C^o \int_{\tilde{x}_f^-}^{\tilde{x}_f^+} \tilde{Y}e^{-\tilde{T}_a/\tilde{T}}d\tilde{x}. \tag{7.4.8}$$

The cancelling of the convection terms in Eq. (7.4.7) indicates that convective transport is negligible in the reaction zone, resulting in a balance between diffusion and reaction. This is the consequence of the large activation energy nature of the reaction, which confines the reaction to a narrow, high temperature region within which the temperature does not change much. Equation (7.4.8) then clearly shows that the chemical energy generated is totally conducted upstream. Since this amount is directly transferred to the preheat zone, and since convection and diffusion balance there, the net effect of the chemical heat generation is to heat the fresh mixture from the unburned temperature to the burned temperature as it is convected from the freestream to the reaction zone. This point can also be readily demonstrated by integrating Eq. (7.3.4) from $\tilde{x} = -\infty$ to $\tilde{x}_f^+$. Since gradients at both boundaries vanish, we obtain a convective–reactive balance for the entire system, which turns out to be identical to Eq. (7.4.8).

It is worth noting that out of the three terms governing Eq. (7.3.4), only two independent relations can be obtained. Each of these relations represents a balance between two processes, whether it is convection and diffusion in the preheat zone, Eq. (7.4.1), or reaction and diffusion in the reaction zone, Eq. (7.4.8), or an overall convection and reaction conservation, which is again given by Eq. (7.4.8) through Eq. (7.4.1).

Using average values $\tilde{Y}_{\text{av}}$ and $\tilde{T}_{\text{av}}$ in Eq. (7.4.8), we have

$$Da_C^o \tilde{Y}_{\text{av}} e^{-\tilde{T}_a/\tilde{T}_{\text{av}}} \ell_R^o = 1, \tag{7.4.9}$$

where $\tilde{\ell}_R^o = \tilde{x}_f^+ - \tilde{x}_f^-$. To evaluate Eq. (7.4.9), we approximate $\tilde{T}_{\text{av}}$ by $\tilde{T}_b^o$ such that

$$e^{-\tilde{T}_a/\tilde{T}_{\text{av}}} \approx e^{-Ar}. \tag{7.4.10}$$

For $\tilde{Y}_{\text{av}}$, we note from Figure 7.4.1 that, based on simple geometrical considerations,

$$\tilde{Y}(\tilde{x}_f^-) \approx \frac{\tilde{\ell}_R^o}{\tilde{\ell}_M^o}, \tag{7.4.11}$$

which can be considered as a characteristic $\tilde{Y}_{\text{av}}$. Similarly, for $\tilde{\ell}_R^o/\tilde{\ell}_T^o$ we have, from Eq. (7.2.5),

$$\frac{\tilde{\ell}_R^o}{\tilde{\ell}_T^o} = \tilde{\ell}_R^o \approx \frac{1}{Ze}. \tag{7.4.12}$$

Substituting the above into Eq. (7.4.9), we have

$$\frac{Le\, Da_C^o e^{-Ar}}{Ze^2} = 1. \tag{7.4.13}$$

Writing out Da_C^o, we obtain the laminar burning flux as

$$(f^o)^2 = \frac{(\lambda/c_p) Le\, B_C e^{-Ar}}{Ze^2}, \tag{7.4.14}$$

which exhibits the same functional dependence as Eq. (7.2.8) derived through the phenomenological analysis, but with a general Lewis number and Ze^2 instead of Ze. The factor Le comes from the modification of the concentration in the reaction zone, given by Eq. (7.4.11) as $(\tilde{\ell}_R^o/\tilde{\ell}_M^o) = (\tilde{\ell}_R^o/\tilde{\ell}_T^o)(\tilde{\ell}_T^o/\tilde{\ell}_M^o) = Ze^{-1} Le$. Thus for fixed thermal and reaction zone thicknesses, a larger Le implies a smaller mass diffusivity, a shorter diffusion length ℓ_M^o, a steeper diffusion gradient, a higher concentration in the reaction zone, and therefore a higher burning flux. The opposite holds for a smaller Le.

The factor Ze^{-2} is simply $(\tilde{\ell}_R^o)^2$ as given by Eq. (7.4.12). Here one of the $\tilde{\ell}_R^o$s represents reduction in the reactant concentration in the reaction zone from the freestream value, as shown in Eq. (7.4.11). The second $\tilde{\ell}_R^o$ represents the fact that the reaction takes place in the reaction zone of thickness $\tilde{\ell}_R$, as shown in Eq. (7.4.9). This was captured in Eq. (7.2.8), although the first factor was missed.

Finally, we note that since

$$f^o(Le) = \sqrt{Le}\, f^o(Le = 1), \tag{7.4.15}$$

Eq. (7.4.14) shows that the dependence of the laminar burning flux on transport properties is actually

$$f^o \sim \frac{\lambda/c_p}{\sqrt{\rho D}}. \tag{7.4.16}$$

Thus f^o has a stronger dependence on thermal diffusion than mass diffusion.

7.4.2. Frank-Kamenetskii Solution

The analysis is again based on Eqs. (7.3.4) and (7.3.10). Referring to Figure 7.2.1c, we shall respectively designate the three zones constituting the flame structure by the superscript −, subscript in, and superscript +. We shall locate the origin, $\tilde{x} = 0$, at the boundary between the reaction zone and the downstream equilibrium zone. The boundary between the preheat and reaction zones is located at an ignition point, x_{ig}, at which $\tilde{T} = \tilde{T}_{ig}$. We shall now separately obtain the solution for each of these three zones, and then match them at their respective interfacial boundaries.

Thus in the preheat zone we have $w = 0$, and

$$\frac{d^2\tilde{T}^-}{d\tilde{x}^2} - \frac{d\tilde{T}^-}{d\tilde{x}} = 0, \tag{7.4.17}$$

subject to the boundary conditions

$$\tilde{T}^- = \tilde{T}_u, \quad \frac{d\tilde{T}^-}{d\tilde{x}} = 0 \quad \text{at} \quad \tilde{x} = -\infty. \tag{7.4.18}$$

Integrating Eq. (7.4.17) once and applying Eq. (7.4.18) yields

$$\frac{d\tilde{T}^-}{d\tilde{x}} = \tilde{T}^- - \tilde{T}_u. \tag{7.4.19}$$

In the downstream equilibrium zone, the solution is simply

$$\tilde{T}^+ = \tilde{T}_b^o, \quad \frac{d\tilde{T}^+}{d\tilde{x}} = 0, \tag{7.4.20}$$

because of complete uniformity here.

In the reaction zone, the convection term is negligible as demonstrated earlier. Equation (7.3.4) then becomes

$$\frac{d^2\tilde{T}_{in}}{d\tilde{x}^2} = -Da_C^o \tilde{Y}_{in} e^{-\tilde{T}_a/\tilde{T}_{in}}, \tag{7.4.21}$$

while Eq. (7.3.10), with $\tilde{T}_{in} \approx \tilde{T}_b^o$ and $\tilde{Y}_{in} \approx 0$, simplifies to

$$\frac{d\tilde{T}_{in}}{d\tilde{x}} + \frac{1}{Le}\frac{d\tilde{Y}_{in}}{d\tilde{x}} = 0. \tag{7.4.22}$$

Integrating Eq. (7.4.22) and applying the downstream boundary condition that $\tilde{T}_{in} = \tilde{T}_b^o$ at $\tilde{Y}_{in} = 0$, we have

$$\tilde{Y}_{in} = Le\left(\tilde{T}_b^o - \tilde{T}_{in}\right), \tag{7.4.23}$$

which shows the role of Le in modifying the reactant concentration in the reaction zone. Substituting $\tilde{Y}_{\text{in}}$ into Eq. (7.4.21), using the identity

$$\frac{d^2\tilde{T}_{\text{in}}}{d\tilde{x}^2} = \frac{d}{d\tilde{x}}\left(\frac{d\tilde{T}_{\text{in}}}{d\tilde{x}}\right) = \left(\frac{d\tilde{T}_{\text{in}}}{d\tilde{x}}\right)\frac{d}{d\tilde{T}_{\text{in}}}\left(\frac{d\tilde{T}_{\text{in}}}{d\tilde{x}}\right) = \frac{1}{2}\frac{d}{d\tilde{T}_{\text{in}}}\left(\frac{d\tilde{T}_{\text{in}}}{d\tilde{x}}\right)^2, \tag{7.4.24}$$

and integrating Eq. (7.4.21) once, from $\tilde{T}_b^o$ to $\tilde{T}_{\text{in}}$, we obtain

$$\left(\frac{d\tilde{T}_{\text{in}}}{d\tilde{x}}\right)^2 = -2\,Le\,Da_C^o\int_{\tilde{T}_b^o}^{\tilde{T}_{\text{in}}}(\tilde{T}_b^o - \tilde{T})e^{-\tilde{T}_a/\tilde{T}}d\tilde{T}, \tag{7.4.25}$$

where we have applied the matching condition

$$\left(\frac{d\tilde{T}_{\text{in}}}{d\tilde{x}}\right)_0 = \left(\frac{d\tilde{T}^+}{d\tilde{x}}\right)_0 = 0. \tag{7.4.26}$$

A straightforward integration of Eq. (7.4.25) is not possible because of the functional form of the Arrhenius factor. However, since in this region $\tilde{T}_{\text{ig}} \le \tilde{T} \le \tilde{T}_b^o$, and since $\tilde{T}_{\text{ig}}$ is only slightly smaller than $\tilde{T}_b^o$, the temperature variation in the reaction zone should be very small compared to the overall temperature variation, $(\tilde{T}_b^o - \tilde{T}_u) = 1$. Thus by defining

$$\tau = \frac{\tilde{T}_b^o - \tilde{T}_{\text{in}}}{\tilde{T}_b^o - \tilde{T}_u} = \tilde{T}_b^o - \tilde{T}_{\text{in}} \ll 1, \tag{7.4.27}$$

we have

$$\begin{aligned}\exp\left(-\frac{\tilde{T}_a}{\tilde{T}_{\text{in}}}\right) &= \exp\left(-\frac{\tilde{T}_a}{\tilde{T}_b^o - \tau}\right) = \exp\left(-\frac{Ar}{1-\tau/\tilde{T}_b^o}\right)\\ &\approx \exp\left[-Ar\left(1+\frac{\tau}{\tilde{T}_b^o}\right)\right] = e^{-Ar}e^{-Ze\tau}.\end{aligned} \tag{7.4.28}$$

Substituting τ and Eq. (7.4.28) into Eq. (7.4.25), we have

$$\left(\frac{d\tilde{T}_{\text{in}}}{d\tilde{x}}\right)^2 = 2\,Le\,Da^o\int_0^\tau \tau' e^{-Ze\tau'}d\tau', \tag{7.4.29}$$

where $Da^o = Da_C^o e^{-Ar}$ is the reaction Damköhler number, as identified in Section 5.6. Integrating Eq. (7.4.29) by parts yields

$$\left(\frac{d\tilde{T}_{\text{in}}}{d\tilde{x}}\right)^2 = \frac{2\,Le\,Da^o}{Ze^2}\left[1-(1+Ze\tau)e^{-Ze\tau}\right]. \tag{7.4.30}$$

If we evaluate Eq. (7.4.30) at the upstream boundary of the reaction zone, $\tilde{x} = \tilde{x}_{\text{ig}}$, then we should insist that the Arrhenius, exponential term representing chemical reactivity must vanish there because this is the boundary with the preheat zone within which reaction is negligible. That is, although Ze is very large and τ very small such that the product $Ze\tau$ is not obviously large as to make the exponential term vanish, the problem has been posed in such a way that it should vanish.

Equation (7.4.30) then becomes

$$\left[\left(\frac{d\tilde{T}_{\text{in}}}{d\tilde{x}}\right)_{\tilde{x}_{\text{ig}}}\right]^2 = \frac{2\,Le\,Da^o}{Ze^2}. \qquad (7.4.31)$$

We next evaluate the temperature gradient in the preheat zone, Eq. (7.4.19), at $\tilde{x}_{\text{ig}}$, yielding

$$\left(\frac{d\tilde{T}^-}{d\tilde{x}}\right)_{\tilde{x}_{\text{ig}}} = \tilde{T}_{\text{ig}} - \tilde{T}_u \approx \tilde{T}_b^o - \tilde{T}_u = 1. \qquad (7.4.32)$$

Since the temperature gradients in the preheat and reaction zones must match at $\tilde{x}_{\text{ig}}$ because of the continuity of the heat fluxes across their boundary, we have

$$\left(\frac{d\tilde{T}_{\text{in}}}{d\tilde{x}}\right)_{\tilde{x}_{\text{ig}}} = \left(\frac{d\tilde{T}^-}{d\tilde{x}}\right)_{\tilde{x}_{\text{ig}}} = 1. \qquad (7.4.33)$$

Thus Eq. (7.4.31) becomes

$$\frac{2\,Le\,Da^o}{Ze^2} = 1, \qquad (7.4.34)$$

from which the laminar burning flux f^o, defined through Da^o in Eq. (7.3.6), can be determined. Comparing Eq. (7.4.34) with Eq. (7.4.13) obtained from the integral analysis, it is seen that the two results differ by only a factor of 2.

7.5. ASYMPTOTIC ANALYSIS

In the Frank-Kamenetskii derivation we have to make the physically motivated assumption that the Arrhenius term vanishes at the upstream boundary of the reaction zone. We now present the activation energy asymptotic analysis through which such an assumption is a natural consequence of the analysis. Before doing so, it is necessary to first introduce the concept of distinguished limit in activation energy asymptotics.

7.5.1. Distinguished Limit

As discussed earlier, a particular feature of the laminar flame structure that leads to the Frank-Kamenetskii solution is the narrowness of the reaction zone relative to the preheat zone. For fixed amount of heat release, $c_p(T_b^o - T_u)$, the key parameter that controls this property is the activation energy—the larger the activation energy, the thinner is the reaction zone. To clearly demonstrate this property, let us consider the dependence of the reaction rate on temperature,

$$\tilde{w} \sim Da_C\tilde{Y}e^{-\tilde{T}_a/\tilde{T}},$$

and examine the effect of increasing $\tilde{T}_a$, with a fixed order of $\tilde{w}$, on the reaction rate profile. Since we are interested in the properties of the reaction zone, this reaction term must be of leading-order importance. Without loss of generality, we let $\tilde{w} \sim O(1)$ with the understanding that the following discussion can be conducted for $\tilde{w}$

having any fixed order of magnitude. Thus in order to maintain this $O(1)$ intensity, an increase in $\tilde{T}_a$, which leads to a decrease in the Arrhenius factor $\exp(-\tilde{T}_a/\tilde{T})$, must necessitate a corresponding increase in Da_C when interpreting the effect of varying $\tilde{T}_a$ on the reaction zone. The larger the $\tilde{T}_a$, the larger must also be Da_C. In the (distinguished) limit of $\tilde{T}_a \to \infty$, we also require $Da_C \to \infty$. Increasing $\tilde{T}_a$ without correspondingly increasing Da_C will rapidly suppress the reaction. The concentration term $\tilde{Y} \approx (\tilde{T}_b^o - \tilde{T})$ has minimal influence on this consideration because it varies algebraically, and therefore insensitively, with $\tilde{T}$.

To relate an algebraic increase in $\tilde{T}_a$ to an exponential increase in Da_C in order to maintain an $O(1)$ reaction rate, Da_C can be expressed as

$$Da_C \sim Da e^{\tilde{T}_a/\tilde{T}_b^o},$$

where T_b^o is the characteristic temperature in the reaction zone, and Da, the reaction Damköhler member, is an $O(1)$ proportionality constant and is actually the relevant Damköhler number in the reaction zone. Consequently, the reaction rate can now be expressed as

$$\tilde{w} \sim Da \exp\left[\tilde{T}_a\left(\frac{1}{\tilde{T}_b^o} - \frac{1}{\tilde{T}}\right)\right] \approx Da \exp\left[-Ze\left(\tilde{T}_b^o - \tilde{T}\right)\right],$$

where we have used the property that $\tilde{T}$ is very close to $\tilde{T}_b^o$ in the reaction zone.

The effect of increasing $\tilde{T}_a$, or rather Ze, on the flame thickness is now clear. Thus in order for $\tilde{w}$ to remain an $O(1)$ quantity, the exponent $Ze(\tilde{T}_b^o - \tilde{T})$ also has to remain as $O(1)$. Consequently an increase in Ze would demand a corresponding decrease in $(\tilde{T}_b^o - \tilde{T})$. In other words the temperature in the reaction zone becomes closer to the final flame temperature, which also implies that the region that can be identified as the reaction zone becomes narrower. In the limit of $Ze \to \infty$, the reaction zone collapses into a reaction sheet with a temperature $\tilde{T}_b^o$. Alternatively, for a fixed, large, value of Ze, reaction is significant only in the region within which $(\tilde{T}_b^o - \tilde{T}) = O(Ze^{-1})$. Reaction is frozen in neighboring regions in which $\tilde{T}$ is smaller than T_b^o by more than $O(Ze^{-1})$.

This discussion provides an alternate interpretation to that used in identifying the Zel'dovich number in Section 5.6.

7.5.2. Asymptotic Solution

The asymptotic analysis is based on the concept of large Zel'dovich number, Ze. For $Ze \to \infty$, we have the leading-order, structureless reaction-sheet solution of Figure 7.2.1b. For large but finite values of Ze, the reaction sheet is broadened, and the resulting reaction zone has a structure, as shown in Figure 7.2.1c. The structures of the neighboring upstream preheat zone and downstream equilibrium zones are correspondingly modified from those of the reaction-sheet limit by small amounts. All these changes are described as perturbations to the leading order solution. The asymptotic solution will therefore be sought in ascending powers of a small parameter $\epsilon(\ll 1)$, which for the time being is unspecified but will be systematically identified later as Ze^{-1}. Solutions will be separately obtained in the three zones constituting the flame structure, and then asymptotically matched. The reaction sheet is set at $\tilde{x}_f = 0$ because of the doubly infinite nature of the problem. It may also be noted that for simplicity in notation we have not attached the superscript o to ϵ and Ze in this deviation.

Further studies on the mathematical technique of asymptotic analysis and its application to combustion problems can be found in Buckmaster and Ludford (1982, 1983), and Williams (1985).

7.5.2.1. Upstream Preheat Zone: Since low temperature and high activation energy freeze the reaction to all orders, solution in this outer zone satisfies the chemistry-free form of Eq. (7.3.4),

$$\frac{d^2\tilde{T}^-_{\text{out}}}{d\tilde{x}^2} - \frac{d\tilde{T}^-_{\text{out}}}{d\tilde{x}} = 0, \quad \tilde{x} \le 0, \tag{7.5.1}$$

subject to the single boundary condition $\tilde{T}_{\text{out}}(-\infty) = \tilde{T}_u$. The second boundary condition, at the downstream boundary of this zone, is to be identified through matching with the inner solution. Let

$$\tilde{T}^-_{\text{out}}(\tilde{x}) = \tilde{T}^-_0(\tilde{x}) + \epsilon\tilde{T}^-_1(\tilde{x}) + O(\epsilon^2), \tag{7.5.2}$$

for which we expect that the maximum values of $\tilde{T}^-_0$ and $\tilde{T}^-_1$ are $O(1)$ quantities. Substituting Eq. (7.5.2) into Eq. (7.5.1) and the boundary condition yields the leading and first-order governing equations and boundary conditions,

$$O(1): \quad \frac{d^2\tilde{T}^-_0}{d\tilde{x}^2} - \frac{d\tilde{T}^-_0}{d\tilde{x}} = 0, \quad \tilde{T}^-_0(-\infty) = \tilde{T}_u, \tag{7.5.3}$$

$$O(\epsilon): \quad \frac{d^2\tilde{T}^-_1}{d\tilde{x}^2} - \frac{d\tilde{T}^-_1}{d\tilde{x}} = 0, \quad \tilde{T}^-_1(-\infty) = 0. \tag{7.5.4}$$

Solutions of (7.5.3) and (7.5.4) are

$$\tilde{T}^-_0(\tilde{x}) = \tilde{T}_u + c^-_0 e^{\tilde{x}}, \tag{7.5.5}$$

$$\tilde{T}^-_1(\tilde{x}) = c^-_1 e^{\tilde{x}}, \tag{7.5.6}$$

where c^-_0 and c^-_1 are the integration constants to be determined through matching with the inner solution.

We next expand the outer solution for $\tilde{Y}^-(\tilde{x})$ as

$$\tilde{Y}^-_{\text{out}}(\tilde{x}) = \tilde{Y}^-_0(\tilde{x}) + \epsilon\tilde{Y}^-_1(\tilde{x}) + O(\epsilon^2), \tag{7.5.7}$$

where it is again assumed that the maximum values of $\tilde{Y}^-_0$ and $\tilde{Y}^-_1$ are $O(1)$ quantities. Substituting $\tilde{Y}^-_{\text{out}}(\tilde{x})$ and $\tilde{T}^-_{\text{out}}(\tilde{x})$ into Eq. (7.3.10) and the boundary conditions $\tilde{Y}^-_{\text{out}}(-\infty) = 1$, and solving the resulting $O(1)$ and $O(\epsilon)$ equations with $\tilde{T}^-_0(\tilde{x})$ and $\tilde{T}^-_1(\tilde{x})$ given by Eqs. (7.5.5) and (7.5.6), yield

$$\tilde{Y}^-_0(\tilde{x}) = 1 + d^-_0 e^{Le\tilde{x}}, \tag{7.5.8}$$

$$\tilde{Y}^-_1(\tilde{x}) = d^-_1 e^{Le\tilde{x}}, \tag{7.5.9}$$

where d^-_0 and d^-_1 are the integration constants.

7.5.2.2. Downstream Equilibrium Zone: Since all reactants are consumed in crossing the reaction zone, solution in this outer zone again satisfies the chemistry-free form of Eq. (7.3.4),

$$\frac{d^2\tilde{T}^+_{\text{out}}}{d\tilde{x}^2} - \frac{d\tilde{T}^+_{\text{out}}}{d\tilde{x}} = 0, \quad \tilde{x} \geq 0, \tag{7.5.10}$$

subject to the single boundary condition $\tilde{T}^+_{\text{out}}(\infty) = \tilde{T}^o_b$. Substituting

$$\tilde{T}^+_{\text{out}}(\tilde{x}) = \tilde{T}^+_0(\tilde{x}) + \epsilon\tilde{T}^+_1(\tilde{x}) + O(\epsilon^2) \tag{7.5.11}$$

in Eq. (7.5.10) and the boundary condition, we have

$$O(1)\text{:} \quad \frac{d^2\tilde{T}^+_0}{d\tilde{x}^2} - \frac{d\tilde{T}^+_0}{d\tilde{x}} = 0, \quad \tilde{T}^+_0(\infty) = \tilde{T}^o_b, \tag{7.5.12}$$

$$O(\epsilon)\text{:} \quad \frac{d^2\tilde{T}^+_1}{d\tilde{x}^2} - \frac{d\tilde{T}^+_1}{d\tilde{x}} = 0, \quad \tilde{T}^+_1(\infty) = 0. \tag{7.5.13}$$

The solutions of (7.5.12) and (7.5.13) are

$$\tilde{T}^+_0(\tilde{x}) = \tilde{T}^o_b, \tag{7.5.14}$$

$$\tilde{T}^+_1(\tilde{x}) = 0, \tag{7.5.15}$$

because they must be bounded as $\tilde{x} \to \infty$.

Similarly, the outer solution

$$\tilde{Y}^+_{\text{out}}(\tilde{x}) = \tilde{Y}^+_0(\tilde{x}) + \epsilon\tilde{Y}^+_1(\tilde{x}) + O(\epsilon^2) \tag{7.5.16}$$

satisfying Eq. (7.3.10) and the boundary condition $\tilde{Y}^+_{\text{out}}(\infty) = 0$ is

$$\tilde{Y}^+_0(\tilde{x}) = 0, \tag{7.5.17}$$

$$\tilde{Y}^+_1(\tilde{x}) = 0. \tag{7.5.18}$$

7.5.2.3. Reaction Zone: Since the region here is exceedingly thin, in order to adequately resolve its structure the spatial coordinate is magnified, or "stretched," by defining a stretched inner variable

$$\chi = \tilde{x}/\epsilon, \tag{7.5.19}$$

such that $\chi = O(1)$. We let the inner solution assume the general forms

$$\tilde{T}_{\text{in}}(\chi) = \theta_0 - \epsilon\theta_1(\chi) + O(\epsilon^2), \tag{7.5.20}$$

$$\tilde{Y}_{\text{in}}(\chi) = \phi_0 + \epsilon\phi_1(\chi) + O(\epsilon^2), \tag{7.5.21}$$

where θ_0 and ϕ_0 are the leading-order solutions, while $\theta_1(\chi)$ and $\phi_1(\chi)$ are the $O(1)$ perturbation functions to be determined. Furthermore, since the inner zone is not in contact with either the upstream or the downstream boundaries, all boundary conditions for the inner solutions θ and ϕ are to be determined through matching with the outer solutions.

7.5.2.4. Matching: To match the inner and outer solutions at the downstream boundary of the reaction zone, where $\chi = \tilde{x}/\epsilon \to \infty$ for fixed $\tilde{x} > 0$ and $\epsilon \to 0$, we require

$$\lim_{\chi\to\infty} \tilde{T}_{\text{in}}(\chi) = \lim_{\chi\to\infty} \tilde{T}^{+}_{\text{out}}(\chi), \tag{7.5.22}$$

$$\lim_{\chi\to\infty} \tilde{Y}_{\text{in}}(\chi) = \lim_{\chi\to\infty} \tilde{Y}^{+}_{\text{out}}(\chi). \tag{7.5.23}$$

Evaluating Eqs. (7.5.22) and (7.5.23) by using Eqs. (7.5.11), (7.5.16), (7.5.20), and (7.5.21) for $\tilde{T}^{+}_{\text{out}}$, $\tilde{Y}^{+}_{\text{out}}$, $\tilde{T}_{\text{in}}$, and $\tilde{Y}_{\text{in}}$ respectively, we have

$$\lim_{\chi\to\infty} [\theta_0 - \epsilon\theta_1(\chi)] = \tilde{T}^{o}_{b}, \tag{7.5.24}$$

$$\lim_{\chi\to\infty} [\phi_0 + \epsilon\phi_1(\chi)] = 0. \tag{7.5.25}$$

Matching at the leading order yields the obvious solution $\theta_0 = \tilde{T}^{o}_{b}$ and $\phi_0 = 0$. Matching the next order solution and its derivatives yields

$$\theta_1(\infty) = 0, \tag{7.5.26}$$

$$\left(\frac{d\theta_1}{d\chi}\right)_{\infty} = 0, \tag{7.5.27}$$

$$\phi_1(\infty) = 0, \tag{7.5.28}$$

$$\left(\frac{d\phi_1}{d\chi}\right)_{\infty} = 0. \tag{7.5.29}$$

It may be noted that according to the principle of asymptotic matching, the limit of $\chi = \tilde{x}/\epsilon \to \infty$ in Eqs. (7.5.22) and (7.5.23) is effected by letting $\epsilon \to 0$ for finite $\tilde{x}$, such that the expansion is most accurate, instead of letting $\tilde{x} \to \infty$.

At the upstream side the matching is obtained by first expressing the outer solution in the inner variable for $\tilde{x} \to 0$, where the two solutions are supposed to match. Hence,

$$\begin{aligned}\lim_{\tilde{x}\to 0} \tilde{T}^{-}_{\text{out}} &= \tilde{T}_u + c_0^- e^{\tilde{x}} + \epsilon c_1^- e^{\tilde{x}} + \cdots, \\ &= \tilde{T}_u + c_0^-(1+\tilde{x}) + \epsilon c_1^-(1+\tilde{x}) + \cdots, \\ &= \tilde{T}_u + c_0^-(1+\epsilon\chi) + \epsilon c_1^-(1+\epsilon\chi) + \cdots, \\ &= (\tilde{T}_u + c_0^-) + \epsilon(c_0^-\chi + c_1^-) + O(\epsilon^2).\end{aligned} \tag{7.5.30}$$

Similarly,

$$\lim_{\tilde{x}\to 0} \tilde{Y}^{-}_{\text{out}} = (1 + d_0^-) + \epsilon(d_0^- Le\chi + d_1^-) + O(\epsilon^2). \tag{7.5.31}$$

To achieve matching, we let $\epsilon \to 0$, or $\chi \to -\infty$, for fixed $\tilde{x} < 0$; that is,

$$\lim_{\chi \to -\infty} \tilde{T}_{\text{in}}(\chi) = \lim_{\chi \to -\infty} \tilde{T}^{-}_{\text{out}}(\chi), \tag{7.5.32}$$

$$\lim_{\chi \to -\infty} \tilde{Y}_{\text{in}}(\chi) = \lim_{\chi \to -\infty} \tilde{Y}^{-}_{\text{out}}(\chi). \tag{7.5.33}$$

Evaluating Eqs. (7.5.32) and (7.5.33), we have

$$\lim_{\chi \to -\infty} \left[\tilde{T}^{o}_{b} - \epsilon\theta_1(\chi) + \cdots\right] = \lim_{\chi \to -\infty} [(\tilde{T}_u + c_0^-) + \epsilon(c_0^- \chi + c_1^-) + \cdots], \tag{7.5.34}$$

$$\lim_{\chi \to -\infty} \epsilon\phi_1(\chi) = \lim_{\chi \to -\infty} [(1 + d_0^-) + \epsilon(d_0^- Le\chi + d_1^-) + \cdots]. \tag{7.5.35}$$

Matching to the leading order for Eqs. (7.5.34) and (7.5.35) yields $c_0^- = \tilde{T}^{o}_{b} - \tilde{T}_u = 1$ and $d_0^- = -1$, which completes the solutions for $\tilde{T}_0^-$ and $\tilde{Y}_0^-$ of Eqs. (7.5.5) and (7.5.8) respectively. Matching to $O(\epsilon)$ yields

$$\lim_{\chi \to -\infty} (\theta_1 + \chi) = -c_1^-, \tag{7.5.36}$$

$$\left(\frac{d\theta_1}{d\chi}\right)_{-\infty} = -1, \tag{7.5.37}$$

$$\lim_{\chi \to -\infty} (\phi_1 + Le\chi) = d_1^-, \tag{7.5.38}$$

$$\left(\frac{d\phi_1}{d\chi}\right)_{-\infty} = -Le. \tag{7.5.39}$$

The above results also confirm that the maximum values of $\tilde{T}_0^-$ and $\tilde{Y}_0^-$, which are respectively $\tilde{T}^o_b = 1 + \tilde{T}_u \approx 1$ and 1, as well as $\theta_0 = \tilde{T}^o_b \approx 1$, are all $O(1)$ quantities. Consequently θ_1 is $O(1)$ because it has the same order as θ_0.

7.5.2.5. Structure Equation and Solution: We are now ready to solve for the inner solution $\theta_1(\chi)$ and $\phi_1(\chi)$. First, substituting χ, $\tilde{T}_{\text{in}}$ and $\tilde{Y}_{\text{in}}$, given respectively by Eqs. (7.5.19) to (7.5.21), into Eq. (7.3.10), we have to $O(\epsilon)$,

$$\frac{d\theta_1}{d\chi} - \frac{1}{Le}\frac{d\phi_1}{d\chi} = 0, \tag{7.5.40}$$

which identically satisfies the gradient boundary condition at $\chi \to \pm\infty$. Integrating Eq. (7.5.40) and applying the boundary conditions $\theta_1(\infty) = \phi_1(\infty) = 0$ yield

$$\theta_1 - Le^{-1}\phi_1 = 0, \tag{7.5.41}$$

which is the perturbed, local coupling function in the inner region. Equation (7.5.41) shows that ϕ_1 is of the same order as θ_1, being $O(1)$. Consequently $\tilde{Y}_1^-$ is $O(1)$.

We next substitute $\tilde{T}_{\text{in}}$, $\tilde{Y}_{\text{in}}$, and χ into Eq. (7.3.4), expand and use Eq. (7.5.41) to get

$$-\frac{d^2\theta_1}{d\chi^2} + \epsilon\frac{d\theta_1}{d\chi} = -(\epsilon^2 Le\, Da^o)\theta_1 e^{-\epsilon Ze\theta_1}. \tag{7.5.42}$$

The following observations can be made regarding Eq. (7.5.42). Since the diffusive term is the highest order derivative, it has to be important and is of order unity as shown. This then immediately implies that the convective term is $O(\epsilon)$ and therefore can be neglected. Physically, since properties change very rapidly within the thin reaction zone, a second order derivative will have a larger value than a first-order derivative. This result is equivalent to the reason in neglecting the convective term in developing the Frank-Kameneskii solution because a narrow reaction zone implies a small change in the temperature.

Next let us inspect the reaction term. Here, even though it has a multiplicative factor ϵ^2, we cannot say that it is $O(\epsilon^2)$ and again drop it. This is because we are now in the reaction zone and the reaction term has to be important. In particular, it must be $O(1)$. What this implies is that the rest of the factors in the reaction term will have to be $O(\epsilon^{-2})$.

Furthermore, we also cannot linearize the exponential factor for small ϵ. This factor is essential in describing the intrinsically nonlinear dependence of the reaction rate on temperature; an expansion in the form of $e^{-\epsilon x} \simeq 1 - \epsilon x \simeq 1$ will completely falsify the chemical kinetics. Therefore in order for this factor to be effective, we must require that the exponent be $O(1)$. Since θ_1 is $O(1)$, we must have $\epsilon Ze = O(1)$. Thus we have now identified our small parameter of expansion, ϵ, as

$$\epsilon = Ze^{-1}. \tag{7.5.43}$$

Having established that the reaction term is $O(1)$ and $\theta_1 e^{-\theta_1} = O(1)$, we also have the result that $Da^o = O(\epsilon^{-2})$.

We have therefore demonstrated that the inner zone is diffusive–reactive in nature, being governed by

$$\frac{d^2\theta_1}{d\chi^2} = \frac{\Delta^o}{2}\theta_1 e^{-\theta_1}, \tag{7.5.44}$$

where we have defined a laminar burning flux "eigenvalue"

$$\Delta^o = \frac{2\,Le\,Da^o}{Ze^2}. \tag{7.5.45}$$

The parameter Δ^o can be identified as the final, reduced Damköhler number based on the characteristic flow time in the reaction zone. Since diffusion and reaction are balanced in Eq. (7.5.44), we expect that Δ^o is an $O(1)$ quantity.

To solve Eq. (7.5.44), we first express it as

$$\frac{d}{d\theta_1}\left(\frac{d\theta_1}{d\chi}\right)^2 = \Delta^o\theta_1 e^{-\theta_1} \tag{7.5.46}$$

in accordance with Eq. (7.4.24). Integrating Eq. (7.5.46), we have

$$\left(\frac{d\theta_1}{d\chi}\right)^2 = \Delta^o \int \theta_1 e^{-\theta_1} d\theta_1 + c_{\text{in}} = -\Delta^o(1+\theta_1)e^{-\theta_1} + c_{\text{in}}. \tag{7.5.47}$$

Evaluating Eq. (7.5.47) at $\chi \to \infty$, where $\theta_1 = d\theta_1/d\chi = 0$ according to Eqs. (7.5.26) and (7.5.27), we have $c_{\text{in}} = \Delta^o$. Therefore Eq. (7.5.47) becomes

$$\left(\frac{d\theta_1}{d\chi}\right)^2 = \Delta^o\left[1-(1+\theta_1)e^{-\theta_1}\right]. \tag{7.5.48}$$

We next evaluate Eq. (7.5.48) at $\chi \to -\infty$ by using Eqs. (7.5.36) and (7.5.37). Noting that

$$\lim_{\chi\to-\infty}(1+\theta_1)e^{-\theta_1} = \lim_{\chi\to-\infty}\left[1-(c_1^- + \chi)\right]e^{(c_1^-+\chi)} = 0 \tag{7.5.49}$$

because of the dominance of the exponential term, $e^\chi \to 0$ as $\chi \to -\infty$, we have

$$\Delta^o = 1. \tag{7.5.50}$$

Thus Eq. (7.5.45) becomes

$$\frac{2\,Le\,Da^o}{Ze^2} = 1, \tag{7.5.51}$$

which is identical to Eq. (7.4.34) derived using the Frank-Kamenetskii solution. Note that had we not retained the nonlinear, exponential nature of the reaction rate term in the derivation between Eqs. (7.5.42) to (7.5.44), Eq. (7.5.48) would have become unbounded when taking the limit of $\chi \to -\infty$.

We have been able to determine Δ^o by integrating the inner structure equation only once, which allows a balance in the heat fluxes between the inner and outer zones. Since we are mainly interested in the laminar burning flux f^o or, equivalently, the laminar flame speed s_u^o, the problem can be considered to be solved at this stage. A second integration, which mathematically is slightly more complicated, is needed only if we wish to determine the temperature profile through the solution for θ_1. This fortunate convenience exists for many problems.

It is important to emphasize that the physical concepts underlying the Frank-Kamenetskii and asymptotic analyses are the same. Furthermore, the mathematical analysis for the reaction zone is also basically the same, although coordinate stretching is not used in the Frank-Kamenetskii solution. The only major difference between the two analyses is that the upstream outer solution is not perturbed in the Frank-Kamenetskii approach. This difference, however, turns out to be of no consequence as far as the determination of the laminar burning flux eigenvalue Δ^o is concerned. That is, since this determination involves matching of the perturbed temperature gradient of the inner zone with the leading-order temperature gradient in the preheat zone, as shown in Eqs. (7.5.34) and (7.5.37), the perturbed outer solution is not needed. Based on these considerations, it is therefore reasonable that the expression for Δ^o is identical for both analyses. It further implies that frequently only the leading-order outer solution is needed in determining the most important, bulk flame responses such as the burning flux.

7.5.3. Dependence of Burning Flux on Flame Temperature

The above analyses show that the laminar burning flux f^o varies with the flame temperature $T_b^o \equiv T_{\rm ad}$ through

$$(f^o)^2 \sim Ze^{-2} e^{-\tilde{T}_a/\tilde{T}_{\rm ad}} \sim T_{\rm ad}^4 e^{-\tilde{T}_a/\tilde{T}_{\rm ad}}. \tag{7.5.52}$$

As discussed earlier, the quadratic variation of f^o with the reaction rate is due to the diffusive-reactive nature of the problem, while the Ze^{-2} factor accounts for the diminished fuel concentration in the reaction zone ($\sim Ze^{-1}$) and the small reaction zone thickness ($\sim Ze^{-1}$).

The standard premixed flame is an idealized, conservative, system. In the presence of nonidealities, it is reasonable to expect that the relevant burning flux f could deviate from f^o, and that such a deviation would be most significant when the flame temperature is affected. Recognizing the fairly general manner in which Eq. (7.5.52) is identified, we anticipate that, for a nonconservative system with the flame temperature being T_f, the burning flux of the flame should vary with T_f in the same manner as Eq. (7.5.52), or

$$f^2 \sim T_f^4 e^{-T_a/T_f}. \tag{7.5.53}$$

Consequently, we have the relation

$$\tilde{f}^2 = \left(\frac{T_f}{T_{\rm ad}}\right)^4 \exp\left[-T_a\left(\frac{1}{T_f} - \frac{1}{T_{\rm ad}}\right)\right], \tag{7.5.54}$$

where $\tilde{f} = f/f^o$. Furthermore, we can also extend the result of, say, Eq. (7.2.6), $f^o \ell_T^o \sim \lambda/c_p$ to $f\ell_T \sim \lambda/c_p$, such that

$$\tilde{f}\tilde{\ell}_T = 1, \tag{7.5.55}$$

where $\tilde{\ell}_T = \ell_T/\ell_T^o$. We shall show in due course the usefulness of the above results.

7.6. DETERMINATION OF LAMINAR FLAME SPEEDS

Because of the fundamental significance of the laminar burning velocities of premixed flames, a considerable amount of effort has been expended toward their determination. A major difficulty in their determination is that a planar, stationary, and adiabatic flame rarely can be achieved. Frequently the upstream flow is nonuniform while the flame is also either propagating and/or curved. It is therefore useful to speak of an instantaneous, local flame speed, s_u, which, however, may not necessarily be s_u^o. Thus for an infinitesimal segment of the flame we can draw an instantaneous flow line as shown in Figure 7.6.1 for a Bunsen flame, without the flame structure. Here the upstream unburned mixture approaches the flame front with velocity u_u and at an angle α_u. After passing through the flame the flow is refracted and the

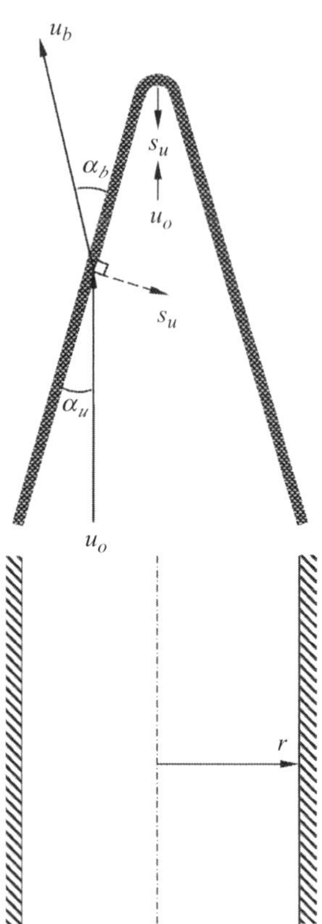

Figure 7.6.1. Definitions of the upstream and downstream laminar flame speeds of an instantaneous, quasi-planar flame segment as part of a Bunsen flame.

burned mixture leaves the flame with velocity u_b and at an angle α_b. Therefore if we assume that, in crossing the flame, continuity of mass flux holds in the normal direction whereas continuity of velocity holds in the tangential direction, then the laminar flame speed can be defined as the normal component of u_u, pointed in the direction away from the flame.

Another difficulty in flame speed measurement is the definition of the flame front and how it can be determined. Geometrically, since the flame itself has a finite thickness and structure, it becomes quite uncertain to define either the upstream boundary of the preheat zone for s_u or the downstream boundary of the reaction zone for s_b. If the flame is also curved, then there is the additional uncertainty of defining a local tangential plane for the evaluation of the flow velocities. Uncertainty also arises in the specific parameter selected to represent a flame surface. The most obvious ones are the constant temperature and density surfaces. For the latter the flame thickness and structure recorded photographically also depend on the optical method used, whether it is shadowgraph, schlieren, or interferometry, as discussed in, for example, Gaydon and Wolfhard (1970). Using laser diagnostics, surfaces of constant concentration of certain key radicals (e.g., CH and OH) have also been determined to represent flame surfaces of particular characteristics.

Measurements of flame speeds have employed either stationary burner flames held fixed by an upstream flow, or propagating flames in open and closed chambers. In the following we discuss several techniques that have proven to be quite versatile and accurate.

7.6.1. Bunsen Flame Method

In this method a premixture flows up a circular or two-dimensional tube and burns after it exits from the tube. If the tube is sufficiently long and its cross-sectional area constant, then the velocity profile at the exit is parabolic. This implies that the inclination of the flame surface changes along the flame, which therefore must be curved. The flame curvature and its thickness make it difficult to accurately determine the local inclination angle, and therefore the local burning velocity.

An averaging method has been used to determine s_u. Here it is assumed that s_u is constant over the flame surface whose total area is A_f. Therefore if the mass flow rate of the gas is m, then from mass conservation we have $m = fA_f = \rho_u s_u A_f$, or

$$s_u = \frac{m}{\rho_u A_f}. \tag{7.6.1}$$

The area of the photographed flame front can be easily determined graphically. This method is useful for rough estimations.

A more accurate determination of s_u can be achieved by using an aerodynamically contoured nozzle, which gives a uniform exit velocity profile. Then a nearly straight flame cone can be obtained over the shoulder region of the flame. Thus if the velocity at the nozzle exit is u_o and the half cone angle is $\alpha = \alpha_u$, the flame speed is given by

$$s_u = u_o \sin\alpha_u. \tag{7.6.2}$$

Local flame speeds along a flame segment can be determined by seeding the gas mixture with fine ceramic particles and measuring the particle velocity using either laser Doppler velocimetry, particle image velocimetry, or simply intermittent illumination, with the photographed particle tracks giving both the speed and direction of the streamlines. Figure 7.6.2 shows the flame speed of a flame of natural gas and air mixture determined by intermittent illumination (Lewis & von Elbe 1987). It is seen that s_u is a constant over most of the flame cone. For large radial distances the flame is close to the burner exit and s_u is reduced due to heat loss to the burner rim. Since the burner rim is invariably very cold relative to the flame, there is always a "dead" space between the flame and the rim. It is further seen that, for small radial distances, close to the flame tip, s_u increases. This increase is to be expected because, instead of being the apex of a sharp cone, the flame tip is rounded off. Consequently, $\alpha = \pi/2$ at the centerline, such that $s_u \equiv u_o$. The subtlety here is that, while we expect s_u to be only a function of the thermochemical properties of the mixture and therefore should be independent of the location over the flame surface as long as it is sufficiently far away from the burner rim, the behavior at the tip clearly shows that this is not the case. Mechanistically, while the flame segment at the shoulder region has the freedom to

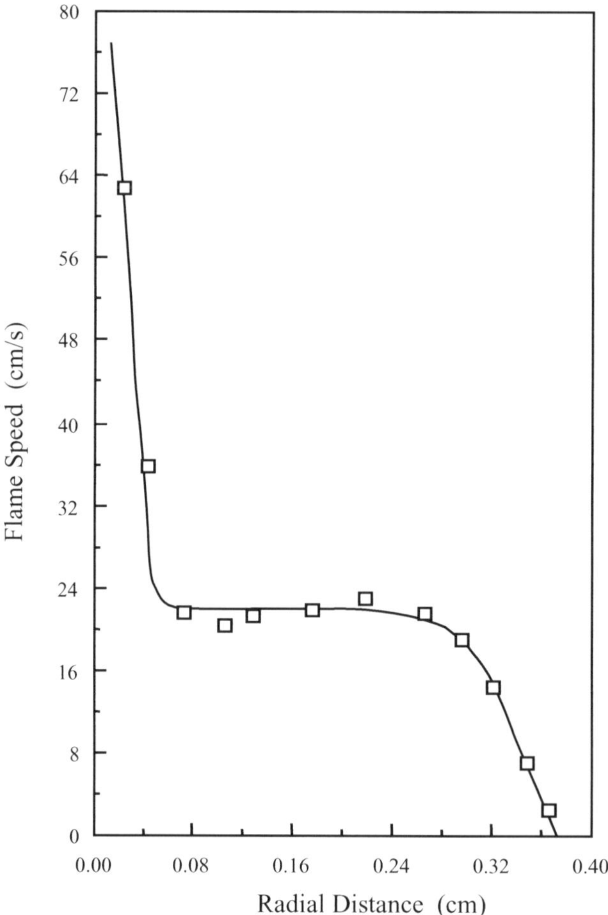

Figure 7.6.2. Measured flame speeds over the surface of a Bunsen flame (Lewis & von Elbe 1987).

adjust its inclination angle α_u in order to accommodate changes in u_o, for a given mixture of s_u^o, this flexibility is absent at the tip. Consequently, the flame structure there must qualitatively deviate from what we have learned about planar premixed flames. We shall show, in Chapter 10, that in this situation the flame curvature has a strong influence on the structure and thereby burning rate of the flame through what is collectively known as flame stretch effects.

It is noted that in the above discussion we have used s_u instead of s_u^o to designate the flame speed in order to distinguish the fact that the flame configuration here does not conform to that of the adiabatic, one-dimensional planar situation.

7.6.2. Flat and One-Dimensional Flame Methods

A major difficulty with the use of Bunsen flames is the identification of the flame surface, which is inclined to the freestream flow. This can be circumvented by using a flat-flame burner as shown in Figure 7.6.3. Here after ignition is achieved, the mixture flow rate is adjusted to produce a flat flame which is normal to the upstream flow direction. Environment effects can also be minimized by passing an inert shroud

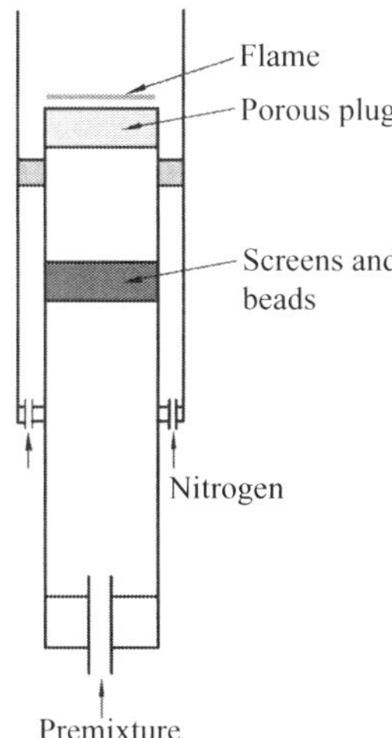

Figure 7.6.3. Schematic showing a typical design of the flat-flame burner.

gas around the burner. This gives a well-defined surface area of the flame, which when divided into the volumetric flow rate of the mixture yields the laminar burning velocity.

The primary limitation of this method is that because heat transfer to the burner is the mechanism through which the flame is stabilized over it, the flame is inherently nonadiabatic relative to the enthalpy of the freestream. The preheat zone starts immediately at the burner surface and consequently the flame has a finite temperature gradient at the burner surface, signifying the presence of heat transfer. The burning velocity determined by this method is therefore lower than s_u^o based on the freestream properties. Efforts to reduce heat loss by increasing the flow discharge rate could lead to severe distortion of the flame surface.

Botha and Spalding (1954) were able to manipulate the heat loss rate by cooling the porous plug. Thus by continuously varying the mixture flow rate and noting the corresponding cooling rate needed to obtain a flat flame, the burning velocity without heat loss can be estimated by extrapolating the cooling rate to zero, as shown in Figure 7.6.4. This method has been recently improved through manipulation and

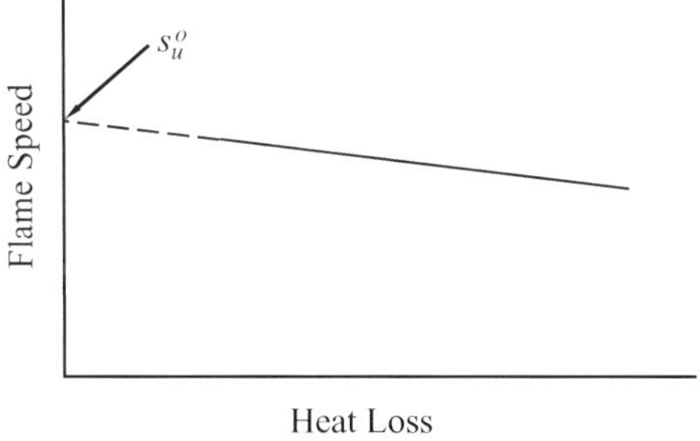

Figure 7.6.4. Determination of the laminar flame speed, s_u^o, through linear extrapolation to zero heat loss by using the flat-burner flame.

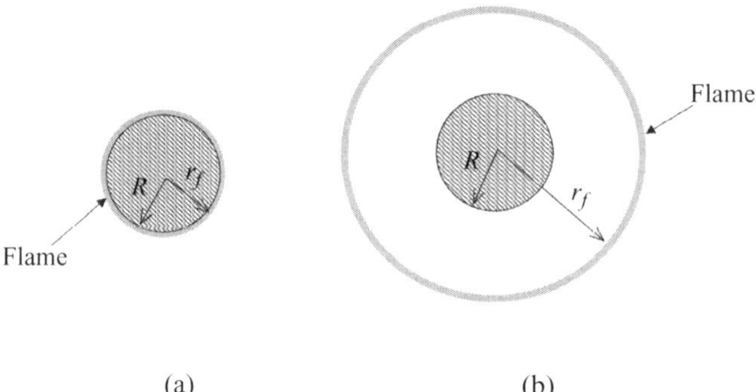

Figure 7.6.5. Schematic showing a cylindrical/spherical flame stabilized through: (a) heat loss to the burner surface ($r_f \approx R$), and (b) flow divergence ($r_f > R$).

more accurate measurement of the heat-loss rate to yield data of enhanced accuracy (De Goey, van Maaren & Quax 1993; Bosschaart & De Goey 2003).

The flat-flame burner method has been extended to that of the one-dimensional burner (Eng, Law & Zhu 1994), in which the combustible is ejected with a given mass flow rate m from either a porous tube or a sphere of radius R (Figure 7.6.5). For $m < m^o = f^o A_s$, where A_s is the surface area of the burner, the ejected mass flux is smaller than that of the adiabatic planar flame, and the flame is stabilized over the burner surface through heat loss to the burner. However, when $m > m^o$, the flame detaches from the burner surface and is subsequently stabilized by the divergent flow, without heat loss to the burner. Neglecting the influence of flame curvature on the flame burning intensity, the stabilization requirement of $m = f^o A_f$ yields the laminar burning velocity given by Eq. (7.6.1) for the measured flame radius r_f and, hence, flame surface area A_f. The method is rather straightforward. The main drawback is the requirement of either cylindrical or spherical symmetry for the flame, which requires that the experiments be conducted in buoyancy-free environments.

In Chapter 8, we shall study the influence of heat transfer at the burner surface on the burning intensity, stabilization of the burner-attached flame, and the essential nature with which it differs from the freely propagating adiabatic flame considered in this chapter.

7.6.3. Outwardly Propagating Spherical Flame Method

In this method a spherical chamber of radius R is filled with a combustible mixture and is centrally ignited by a spark. A spherical flame is developed and propagates outward with the laminar flame speed. As the amount of product in the chamber increases, the chamber pressure also uniformly increases while the unburned gas upstream of the flame is simultaneously heated through compression. If the data are taken when the flame size is not too large, then the chamber pressure and the temperature ahead of the flame can be considered to be those of the initial state. Otherwise they need to be measured separately.

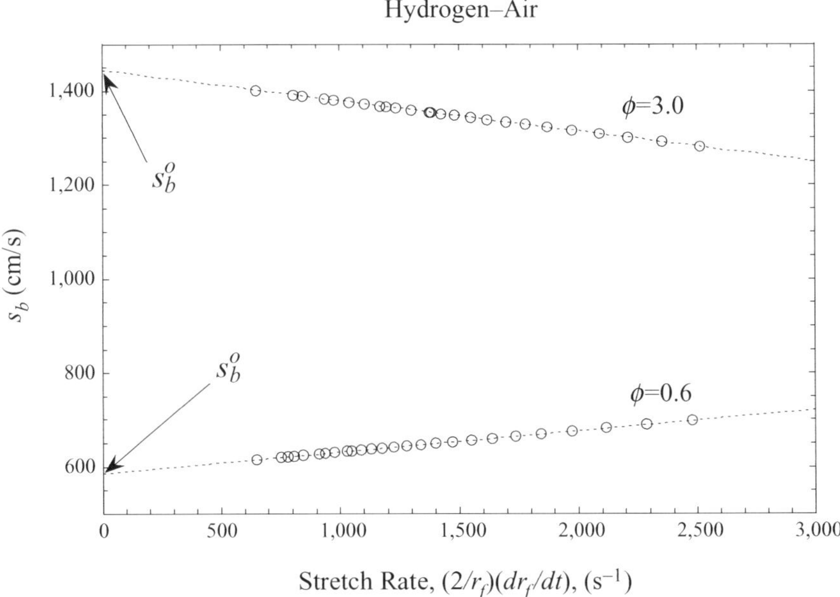

Figure 7.6.6. Numerically calculated downstream flame speeds of outwardly propagating lean and rich hydrogen–air flames at 1 atm pressure, demonstrating the determination of laminar flame speed through linear extrapolation to zero stretch rate.

Since the combustion product is stationary in the laboratory frame, the measured rate of increase of the flame radius, dr_f/dt, can be identified as the flame speed of the burned state, s_b. Thus from continuity and assuming that the flame is quasi-steady and quasi-planar, we have

$$s_u = s_b(\rho_b/\rho_u). \tag{7.6.3}$$

Since this method is static in operation, the design can be relatively simple and the amount of gas consumed is small. It is also well suited for the determination of flame speeds at moderately high pressures. The potential complications are the heat loss through the electrodes especially during the initial period of flame development, the distortion of the flame shape due to buoyancy especially for slowly burning flames, and the development of intrinsic pulsating and cellular instabilities over the flame surface, which will be studied in Chapter 10 on the aerodynamics of flames. Radiative heat loss from the large volume of the burned gas behind the flame could also reduce the flame temperature and hence the flame speed. Furthermore, the flame, being curved and nonstationary, does not conform to the one-dimensional steadily propagating planar flame used to define the laminar flame speed, and as such is subjected to flame stretch effects in the same manner as that of the Bunsen flame tip discussed earlier. These effects not only affect the flame speed, but the influence is also qualitatively dependent on the effective Lewis number of the mixture. For example, Figure 7.6.6 shows the numerically computed flame speeds of hydrogen–air mixtures with equivalence ratios of 0.6 and 3.0. The effective Le for these mixtures are smaller and larger than unity respectively. The intensity of stretch is represented by a stretch rate, defined as $(2/r_f)(dr_f/dt)$ and having the unit of s^{-1}. It is seen that the downstream flame speed can be significantly modified by stretch, that it varies approximately linearly with the stretch rate, and that the variations have opposite trends for the lean and rich mixtures. These are consequences of the deviation of Le from unity. The linear variation allows the extrapolation of these stretched flame speeds, s_b, to zero stretch rate, from which we obtain the unstretched flame speed for the burned state, s_b^o, and subsequently that of the unburned state through $s_u^o = (\rho_b^o/\rho_u)s_b^o$. A nonlinear extrapolation expression has also been derived (Dowdy et al. 1990; Kelley & Law 2009) that accounts for the slight nonlinearly in the stretch-affected flame speed.

A variation of the spherical-flame method is that of the soap bubble, in which the combustible mixture is introduced into a soap bubble and then ignited. The bubble subsequently expands freely as combustion proceeds, thus ensuring constant pressure when the experiment is conducted in an open atmosphere. The entire assembly can also be housed inside a sealed combustion chamber for reduced or elevated pressure experiments. The advantage of this method is that it only requires a small sample and is therefore particularly suitable for experimentation with gases that are toxic, highly explosive, or rare and expensive.

Ideally, the present problem is best studied by directly imaging the flame history, which yields the flame radius, $r_f(t)$, and, hence, the flame propagating rate. However, the requirement of windows for optical cinephotography can become rather demanding because they have to withstand the high postcombustion temperature and pressure, especially if the initial pressure is high. A dual-chamber design (Tse, Zhu & Law 2004) has however circumvented this difficulty, allowing optical access and an initial pressure that can be as high as 60 atm.

Chambers with optical windows are generally harder to design, therefore simpler designs of the experiment do not have windows. For such windowless chambers the flame traverse is recorded by sensing probes such as thermocouples and ionization probes. Because of spatial isobaricity, in principle it is only necessary to install two probes at closely spaced points to yield the flame speed for the chamber pressure at the instant of traverse. A particularly simple approach has also been developed (Lewis & von Elbe 1987) based on measuring the pressure history within the chamber, $p(t)$, after the flame has grown to a sufficiently large size such that the pressure variation is significant. Specifically, assuming that the unburned and burned states of the outwardly propagating spherical flame are spatially uniform, then at any instant of time, we have overall mass conservation

$$\frac{4\pi}{3}\left[\left(R^3 - r_f^3\right)\rho_u + r_f^3\rho_b\right] = \frac{4\pi}{3}R^3\rho_{u,o}. \tag{7.6.4}$$

If we further assume that these gases are compressed isentropically by the expanding flame sphere, then

$$\rho_u = \rho_{u,o}(p/p_o)^{1/\gamma}, \tag{7.6.5}$$

$$\rho_b = \rho_{b,o}(p/p_o)^{1/\gamma}, \tag{7.6.6}$$

where $p = p_u = p_b$. Using Eqs. (7.6.5) and (7.6.6) in Eq. (7.6.4), and defining $\tilde{r}_f = r_f/R$ and $\tilde{p} = p/p_o$, we have

$$\left[1 - \left(1 - \frac{\rho_{b,o}}{\rho_{u,o}}\right)\tilde{r}_f^3\right]\tilde{p}^{1/\gamma} = 1. \tag{7.6.7}$$

Differentiating Eq. (7.6.7) with respect to t, and letting $dr_f/dt = s_b$, it can be shown that

$$\frac{s_b}{R} = \left[3\gamma\left(1 - \frac{\rho_{b,o}}{\rho_{u,o}}\right)^{1/3}\tilde{p}^{(1+1/\gamma)}(1 - \tilde{p}^{-1/\gamma})^{2/3}\right]^{-1}\frac{d\tilde{p}}{dt}. \tag{7.6.8}$$

The factor $(\rho_{b,o}/\rho_{u,o})$ is given by energy conservation, $c_v(T_{b,o} - T_{u,o}) = q_c Y_o$, which yields

$$\frac{\rho_{u,o}}{\rho_{b,o}} = 1 + \tilde{q}, \tag{7.6.9}$$

where $\tilde{q} = (q_c Y_o)/(c_v T_{u,o})$ and Y_o is the initial reactant mass fraction.

Equation (7.6.8) shows that, for a given mixture characterized by γ, $\tilde{q}$ and p_o, the history of the downstream flame speed, $s_b(t)$, for propagation in a spherical vessel, can be determined by measuring the pressure history $p(t)$ alone.

Conceptually, this method is very attractive because pressure traces can be readily obtained in an experiment. Furthermore, in a single run, s_b can be mapped out not only as a function of the mixture strength, but also as functions of the instantaneous mixture temperature and pressure. The various constant property assumptions can also be removed through detailed computation. The major weakness of the method is that, since the flame is not imaged, there is no recourse in knowing whether the assumption of a spherically symmetric, smooth flame surface is violated due to buoyancy distortion and/or the development of cells over the flame surface.

We also note that the assumption of a spatially uniform downstream state is incorrect because during flame propagation the continuously increasing upstream temperature and pressure will lead to a corresponding increase in the downstream temperature. Thus there is a temperature gradient downstream of the flame. This feature can be integrated in the above formulation, at the expense of more involved algebraic manipulation.

7.6.4. Stagnation Flame Method

This method (Wu & Law 1985) involves first establishing a divergent stagnation flow field by impinging two identical, nozzle-generated combustible flows onto each other (Figure 7.6.7). Upon ignition two symmetrical flat flames are situated on the two sides of the stagnation surface. Figure 7.6.8 shows typical profiles of the normal velocity component v along the axis. It is seen that as the flow approaches the stagnation surface, but before reaching the main preheat region, the velocity decreases linearly, $v = ay$, in accordance with the characteristic of stagnation flow, where $a = dv/dy$ is the velocity gradient. However, as the flow enters the preheat region, intense heating and thereby thermal expansion reverse the decreasing trend and cause

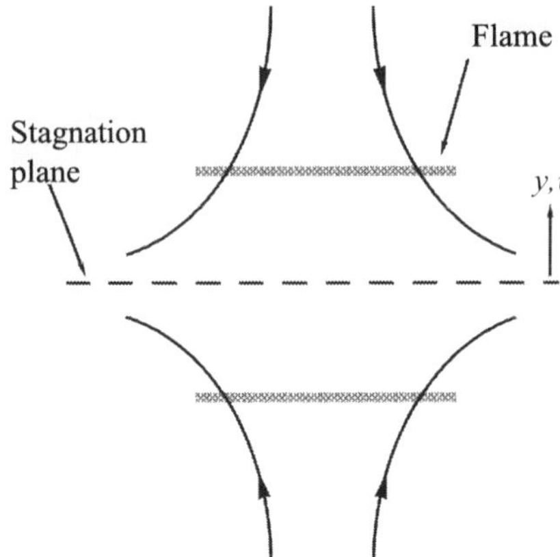

Figure 7.6.7. Schematic showing a typical counterflow, twin-flame configuration.

the velocity to increase. Eventually, upon almost complete heat release, the velocity decreases again as it approaches the stagnation surface.

From such a velocity profile we can determine the velocity gradient a, the minimum velocity point, v_{min}, which can be approximately identified as a reference flame speed ($s_{u,ref}$) at the upstream boundary of the preheat zone where the flame is stabilized, and the maximum velocity point, v_{max}, which can be approximately identified as a reference flame speed at the downstream boundary of the reaction zone. These values can also be considered to be obtained under adiabatic conditions because the upstream heat loss for the nozzle-generated flow is small while the downstream heat loss is also small due to symmetry. Small amount of radiative heat loss is of course always present.

Similar to the propagating spherical flame, the stagnation flame is also stretched, but now by the nonuniform flow whose stretch intensity is represented by the velocity gradient a. A plot of v_{min} versus a shows that the variation is approximately linear for small values of a, as shown in Figure 7.6.9. Thus, by extrapolating v_{min} to zero a, the intercept of v_{min} at $a = 0$ can be identified as s_u^o, evaluated at the upstream boundary, because both heat loss and flow nonuniformity effects are eliminated.

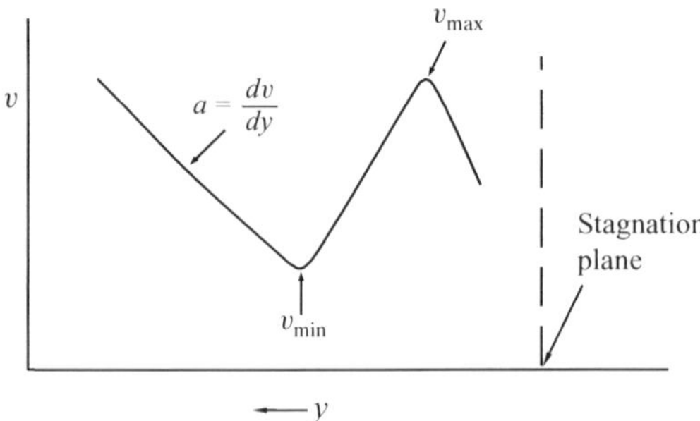

Figure 7.6.8. A typical axial velocity profile for one of the flames in the counterflow, twin-flame configuration.

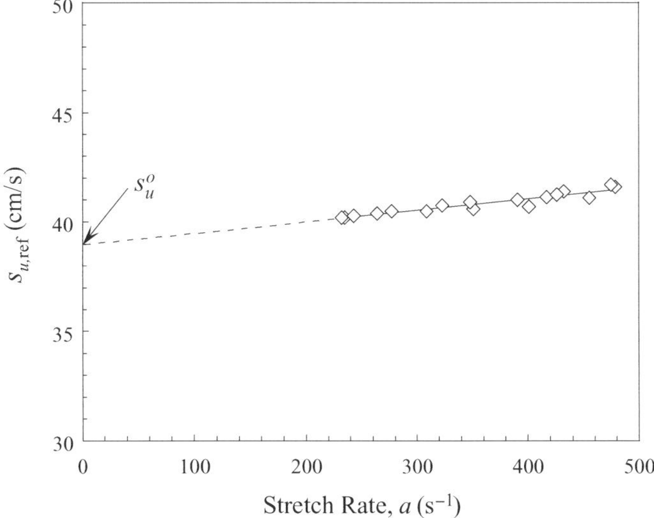

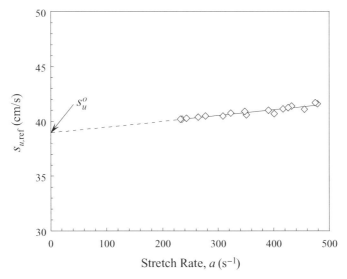

Figure 7.6.9. Determination of the laminar flame speed, s_u^o, through linear extrapolation to zero stretch by using the counterflow, twin-flame method.

Higher-order analysis has shown that the variation is slightly nonlinear for small a (Tien & Matalon 1991). This inaccuracy can be minimized by increasing the nozzle separating distance so that the flame can be better approximated as a surface. Alternatively, Vagelopoulos and Egolfopoulos (1998) achieved a flame segment of zero stretch by impinging a Bunsen flame onto a flat plate so that the resulting positively stretched stagnation flow neutralizes the negatively stretched flame tip.

7.6.5. Numerical Computation

Computational simulation is attractive for two reasons. First, the model of simulation truly conforms to the one in which the laminar burning velocity is defined. Second, such a simulation will also resolve the flame structure in terms of its temperature profile and concentration profiles of all the species including the radicals.

As is true with all computational simulations, a simulation is only as accurate and meaningful as the information supplied to the simulation program. For the present problem, there are three components constituting such a computational simulation program. The first is the numerical algorithm used to solve the conservation equations. This component is believed to be quite reliable as far as laminar flame calculation is concerned. The second is the specification of the transport properties and the thermochemical data, namely the specific heat, the heat of formation, and the entropy, of the various species, as discussed in Chapters 1 and 4. These values are considered to be quite well known for the major and stable species, although considerable uncertainty still exists for some of the unstable species. The third component is the data base for the chemical reaction schemes and the associated kinetic constants of the individual elementary reactions. This is the most uncertain, and definitely weakest part of the simulation. At present probably only the oxidation schemes of hydrogen, carbon monoxide, and methane are considered to be reasonably well established, as far as the laminar burning velocity calculation is concerned. In fact, the kinetic information on many reactant systems is so meager that the experimental burning velocity data have been used to help evaluate the kinetic constants through comparison with the computed results. It is clear that such a comparison and extraction of kinetic information are meaningful only if the experimental data are accurate.

7.6.6. Profile-Based Determination

The various methods discussed above all involve the direct determination of the laminar flame speed as a global response of the flame. Alternatively, the flame speed can also be indirectly determined through knowledge of the profiles of the velocity and reacting scalars across the flame. The basic concept can be readily appreciated by considering, say, Eq. (7.3.1), from which we can express the laminar burning flux as

$$f^o = \frac{q_c w + \lambda d^2 T/dx^2}{c_p dT/dx}, \tag{7.6.10}$$

where $w = w(Y, T)$. This relation holds for all x across the flame structure, requiring only the local profiles of $T(x)$ in order to evaluate the gradient terms, and $T(x)$ and $Y(x)$ to evaluate the reaction term $w(Y, T)$.

Another approach is to integrate Eq. (7.3.1) from $x = -\infty$ to ∞, resulting in

$$f^o = \frac{q_c}{c_p\left(T_b^o - T_u\right)} \int_{-\infty}^{\infty} w(Y, T)dx \tag{7.6.11}$$

from which f^o can be evaluated by determining the profiles of $T(x)$ and $Y(x)$. Flame speeds determined using relations analogous to Eqs. (7.6.10) and (7.6.11) are respectively called displacement speed and consumption speed.

Approach of this nature was originally used by Burgoyne and Weinberg (1954), who determined the complete thermal and dynamic structure of a flat-burner flame based on only the measured temperature profile across the flame. Recent advances in computation allows for more detailed specifications of the diffusion and reaction terms. The evaluation has also used the conservation of a particular species i, Eq. (7.3.2), to track the evolution of the chemical structure. For flames of complex geometry propagating unsteadily in an equally complex flow, as in the case of laminar flamelets embedded within a turbulent flame structure, relations similar to the above expressions can be defined, and evaluated, for the experimentally or computationally determined instantaneous mass flux entering some isovalue level surfaces within the structure of these laminar flamelets (Im & Chen 2000; Poinsot & Veynante 2005). For such situations evaluation of the differential and integral terms in Eqs. (7.6.1) and (7.6.2) is to be conducted in the direction normal to the flame front, while the flame speeds so determined are also affected by aerodynamic stretch. The primary interest in such studies is the understanding and quantification of the burning rates of these flamelets and their influence on the turbulent burning intensity, rather than the determination of the laminar burning fluxes.

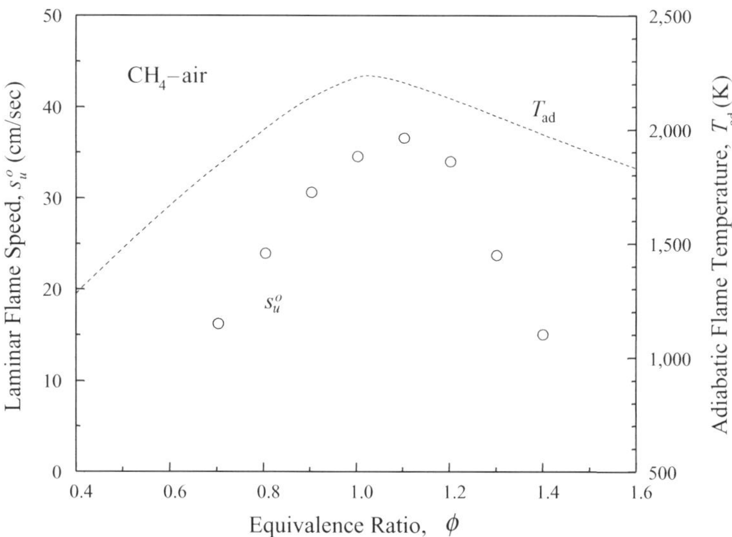

Figure 7.7.1. Calculated adiabatic flame temperatures and measured laminar flame speeds of atmospheric methane–air mixtures.

7.7. DEPENDENCE OF LAMINAR BURNING VELOCITIES

Out of the various methods of flame speed measurement presented above, the Bunsen burner method with contoured nozzle, the flat flame method with heat extraction, and the stagnation flame method have been extensively used for flame speed determination for atmospheric and near-atmospheric flames, while the spherical flame method has been used for both atmospheric and high pressure flames. Extensive investigations have been conducted on the dependence of the flame speed on the various physicochemical parameters of the mixture; the results are summarized in the following. In presenting these results, we shall reserve the symbol s_u^o only for data closely conforming to the requirements of the standard premixed flame.

7.7.1. Dependence on T_{ad} and Le

As expected, adiabatic flame temperature through the Arrhenius kinetics exerts a dominant influence on the laminar burning velocity. Since T_{ad} is directly controlled by the heat of combustion, fuels with larger heats of combustion tend to propagate a flame faster. Indeed, the clear evidence of such a strong dependence is the close correlation between s_u^o and T_{ad}, as shown for methane–air flames in Figure 7.7.1. It is seen that the two responses not only have the same shape, but they also peak on the rich side, at equivalence ratios that are close to each other. Furthermore, for the same T_{ad}, the laminar flame speeds of the lean mixtures and rich mixtures are quite close to each other.

The above correspondence is however sufficiently offset for hydrogen–air flames, for which T_{ad} peaks at $\phi = 1.07$ while s_u^o peaks at about $\phi = 1.75$, as shown in

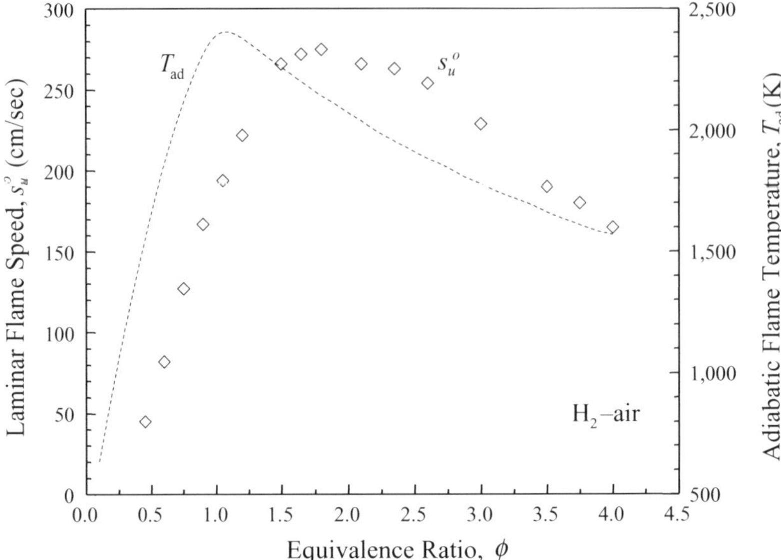

Figure 7.7.2. Calculated adiabatic flame temperatures and measured laminar flame speeds of atmospheric hydrogen–air mixtures.

Figure 7.7.2. This sufficiently off-stoichiometric rich peaking of s_u^o is a consequence of the highly diffusive nature of hydrogen. Specifically, since $s_u^o \sim \sqrt{Le}$, and since the freestream Le for sufficiently lean and rich hydrogen–air mixtures are 0.33 and 2.3 respectively, the effect of Le is to reduce s_u^o on the lean side but increase s_u^o on the rich side, leading to the observed rich shifting of the peaking.

Figures 7.7.3 and 7.7.4 show the measured s_u^o for n-alkanes (Davis & Law 1999). It is seen that the values of s_u^o for methane, ethane, and propane increase in that order, although the increase is smaller from ethane to propane than from methane

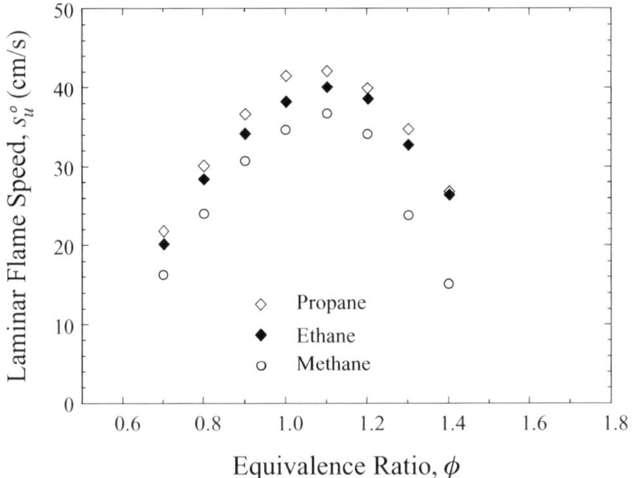

Figure 7.7.3. Measured laminar flame speeds of methane, ethane, and propane in air.

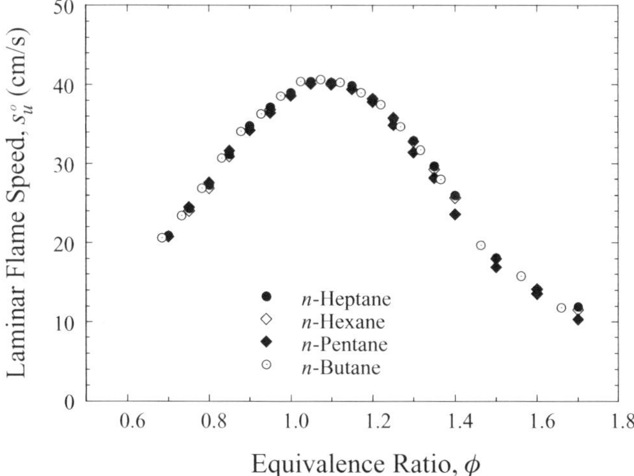

Figure 7.7.4. Measured laminar flame speeds of n-butane, n-pentane, n-hexane, and n-heptane in air.

to ethane. This diminishing trend subsequently leads to the result of Figure 7.7.4, showing that the differences between butane, pentane, hexane, and heptane are basically indistinguishable. These results are similar to those for T_{ad} shown in Figure 7.7.5, hence demonstrating again the dominant influence of T_{ad} on s_u^o.

7.7.2. Dependence on Molecular Structure

Next to the adiabatic flame temperature, the molecular structure of the fuel could also have a strong influence on the laminar flame speed. To assess the extent of such

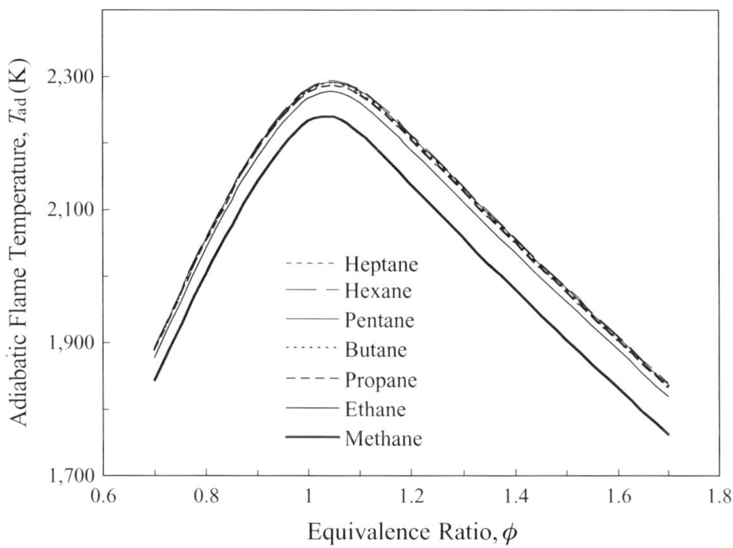

Figure 7.7.5. Calculated adiabatic flame temperatures for varies alkanes.

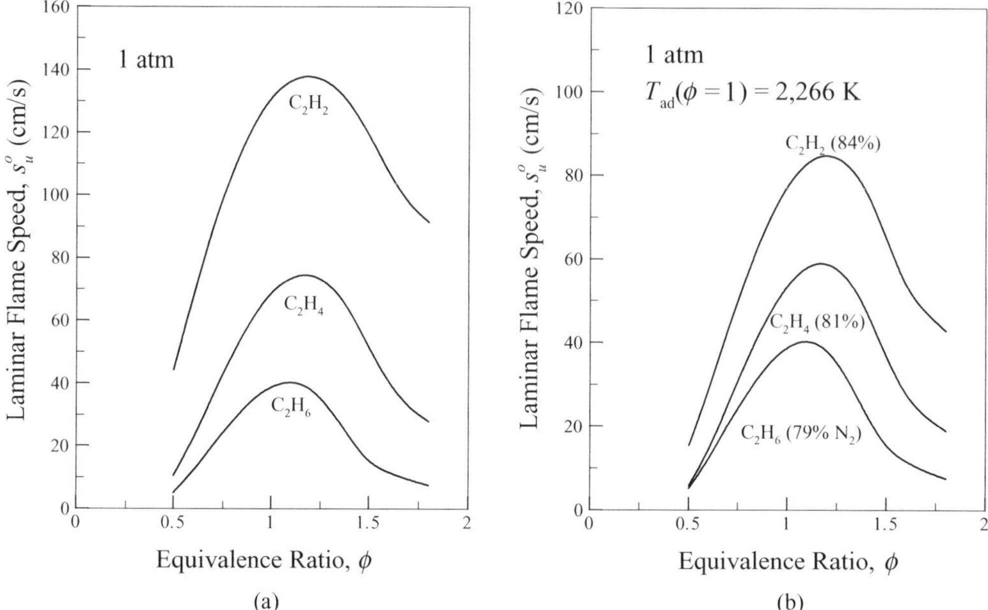

Figure 7.7.6. Calculated laminar flame speeds of ethane, ethylene, and acetylene with the oxidizer being: (a) air; (b) nitrogen-diluted air such that the adiabatic flame temperatures of the stoichiometric ethylene and acetylene mixtures match that of the ethane–air mixture.

an influence, we compare the computed laminar flame speeds of ethane, ethylene, and acetylene with air at atmospheric pressure, recognizing that the chemical reactivity increases from ethane to acetylene. Figure 7.7.6a shows that the laminar flame speeds indeed increase in this order.

Such a comparison, however, is not conclusive because of the corresponding increase in the adiabatic flame temperatures for these three fuels. Therefore for an unambiguous assessment, the laminar flame speeds of ethylene and acetylene are calculated with nitrogen dilution to such an extent that the adiabatic flame temperature of the respective stoichiometric mixture matches that of the ethane–air mixture. This eliminates the flame temperature effect for the stoichiometric mixture and minimizes it for the off-stoichiometric mixtures. Figure 7.7.6b then shows that the laminar flame speeds still increase substantially in the order of ethane, ethylene, and acetylene, hence demonstrating the influence of the molecular structure of the fuel.

7.7.3. Dependence on Pressure

Figure 7.7.7a, obtained for methane–air flames, shows that for a given ϕ, s_u^o decreases with increasing pressure. Since the eigenvalue for flame propagation is $f^o = \rho_u s_u^o$ instead of s_u^o alone, the data of Figure 7.7.7a are replotted in Figure 7.7.7b for f^o. It is then seen that f^o increases with pressure. Figure 7.7.8 shows the similar behavior for hydrogen–air flames. This is an important observation because it has been suggested in the literature that the decreasing trend of s_u^o with pressure, for a given ϕ, is a manifestation of the fundamental importance of the pressure-sensitive

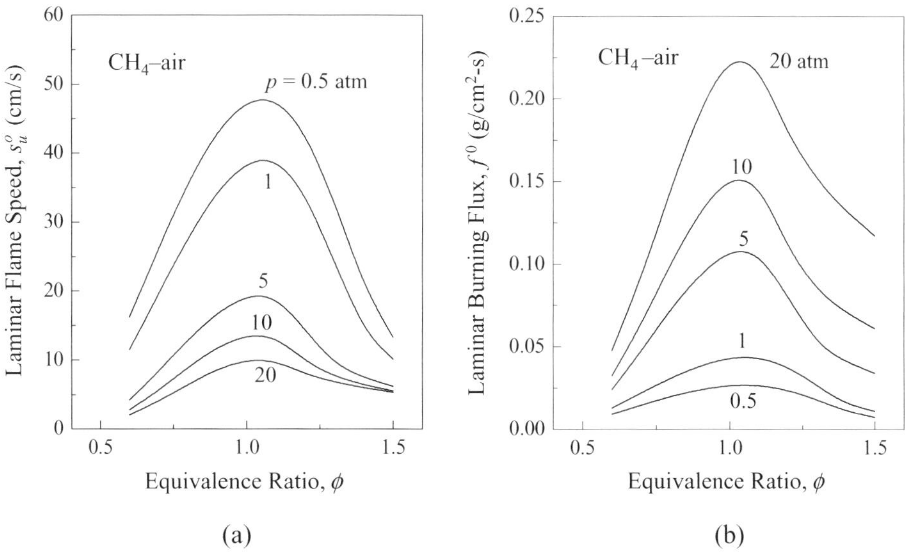

Figure 7.7.7. Computed (a) laminar flame speed, s_u^o, and (b) laminar burning flux, $f^o = \rho_u s_u^o$, for methane–air mixtures at various pressures, showing the trends of decreasing s_u^o but increasing f^o with increasing pressure.

chain mechanisms in the flame propagation process. Take the H–O_2 reactions as an example. As discussed previously, reaction (H1): $H + O_2 \rightarrow OH + O$ is a two-body, temperature-sensitive branching reaction, while reaction (H9): $H + O_2 + M \rightarrow HO_2 + M$ is a three-body, temperature-insensitive, inhibiting reaction. Hence by

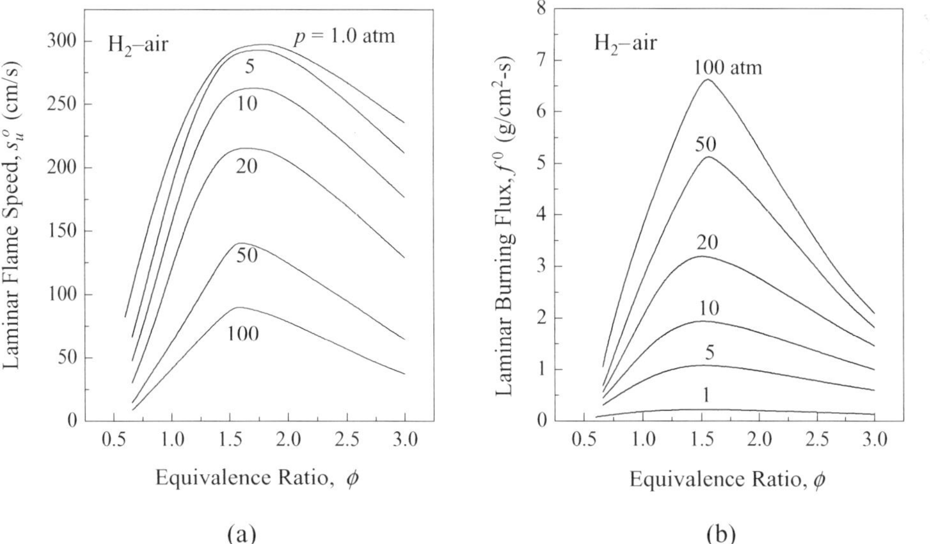

Figure 7.7.8. Computed (a) laminar flame speeds, s_u^o, and (b) laminar burning fluxes, f^o, of hydrogen–air mixtures at various pressures, showing the trend of decreasing s_u^o but increasing f^o with increasing pressure.

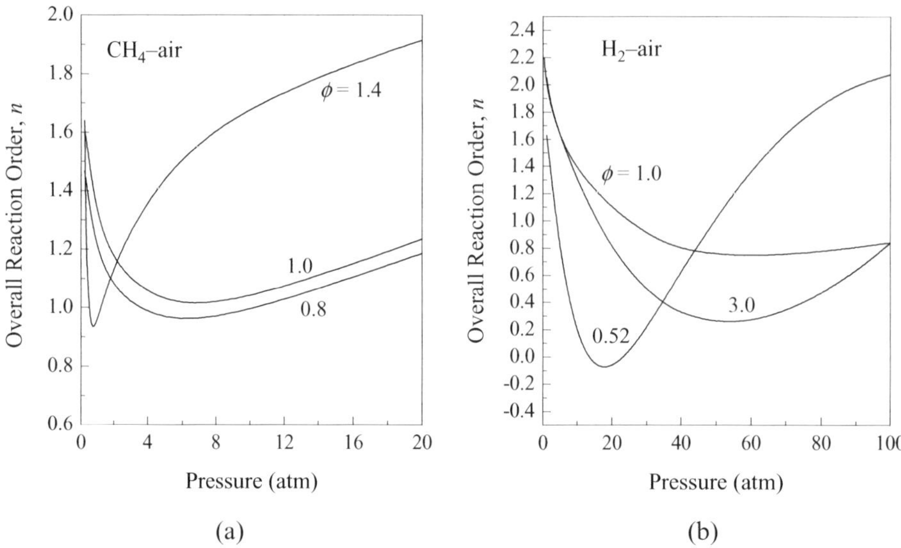

Figure 7.7.9. Computed overall reaction order, n, for (a) methane–air flames, and (b) hydrogen–air flames, showing decreasing and then increasing n with increasing pressure, as well as the existence of negative values.

fixing ϕ, T_{ad} and thereby the intensity of (H1) are approximately fixed, while by increasing pressure, (H9) is enhanced relative to (H1) because three-body reactions are favored over two-body reactions as pressure increases. A retarding effect is therefore imposed on the overall progress of the reaction with increasing pressure. What is shown in Figures 7.7.7b and 7.7.8b is that, although the chain-termination reactions indeed become more important with increasing pressure, the effect is not sufficiently large to be responsible for the observed decreasing trend of s_u^o, which is actually caused by the increasing density, ρ_u, with increasing pressure.

The pressure effect of the chain mechanism can be identified and quantified through the overall reaction order n, which can be locally defined according to the relation $f^o \sim p^{n/2}$, assuming negligible dependence of the thermodynamic and transport properties of f^o on p, such that

$$n = 2\left(\frac{\partial \ln f^o}{\partial \ln p}\right)_{T_{ad}}. \tag{7.7.1}$$

Figure 7.7.9a shows the calculated n as a function of p for methane–air flames of $\phi = 0.8$, 1.0, and 1.4; note that the $\phi = 0.8$ and 1.4 flames have about the same T_{ad} of 2,000 K, while $T_{ad} = 2230$ K for $\phi = 1.0$. It is seen that for all values of ϕ, n first decreases and then increases with increasing pressure. An analysis of the sensitivity of s_u^o to the individual reactions shows that the progressive importance of the termination reactions (H9) and $H + CH_3 + M \rightarrow CH_4 + M$, as compared to that of the main branching reaction (H1), causes the initial decrease in n. With further increase in pressure, the branching reaction $HO_2 + CH_3 \rightarrow OH + CH_3O$ becomes important. It reactivates HO_2 and, hence, contributes to the subsequent increase in

n by supplying the flame with the OH radical, which is further used by the chain-carrying step $CO + OH \to CO_2 + H$.

Figure 7.7.9b shows the overall reaction orders for hydrogen–air flames with $\phi = 0.52$, 1, and 3.0, such that the T_{ad} for the lean and rich flames have approximately the same value of 1,770 K. The T_{ad} for $\phi = 1$ is 2390 K. The behavior is again nonmonotonic, but now the minimum n takes place at higher pressures. Sensitivity analysis shows that the initial decrease of n with pressure is due to the competition between (H1) and (H9). However, as pressure further increases, the HO_2 reactions generate new radicals through $HO_2 + HO_2 \to H_2O_2 + O_2$, $H_2O_2 + M \to 2OH + M$, and $HO_2 + H \to 2OH$. This mechanism for the recovery of n with pressure is completely analogous to that of the explosion limits for H_2–O_2 mixtures.

There are three additional observations for the values of the overall reaction order. First, n seems to approach a value close to 2 with decreasing pressure, especially for the hydrogen–air flames. This is reasonable because as pressure decreases, two-body reactions dominate, not only when compared to three-body termination reactions but also as a consequence of the low-pressure behavior of the Lindemann mechanism. Second, based on the consideration of chain mechanism alone, n should be smaller than 2. This is because while the order of a termination reaction can be 3, its effect on the overall reaction order, which indicates the progress of reaction with increasing pressure, is negative. Third, for weakly burning flames, n can assume negative values, as in the case of the $\phi = 0.52$ H_2–air flames, providing evidence that the burning intensity of weak mixtures can actually decrease with increasing pressure. We have anticipated some aspects of the above observation in Section 2.1.7.

An overall activation energy, E_a, can also be determined through the relation $f^o \sim \exp(-E_a/2R^o T_{ad})$,

$$E_a = -2R^o \left[\frac{\partial \ln f^o}{\partial(1/T_{ad})} \right]_p . \tag{7.7.2}$$

Figures 7.7.10a and 7.7.10b show that E_a can vary substantially with pressure. This is again due to the competition between branching and termination reactions with increasing pressure.

The numerical calculation yields the temperature profile across the flame, from which a characteristic flame thickness ℓ_D^o can be defined. Figures 7.7.11a and 7.7.11b respectively show the calculated ℓ_D^o for the methane–air and hydrogen–air flames, with ℓ_D^o determined by the FWHM of the temperature gradient profile. It is seen that ℓ_D^o monotonically decreases with p for all methane–air flames, although the decrease is rather small at higher pressures. Furthermore, for the weakly burning hydrogen–air flames of $\phi = 0.52$, there is a range in p over which ℓ_D^o attains a slight local minimum. The result that the flame thickness becomes insensitive to pressure at higher pressures, and actually may not steadily decrease with increasing pressure, could be an important consideration in the study of flame phenomena under high pressures (Law 2006).

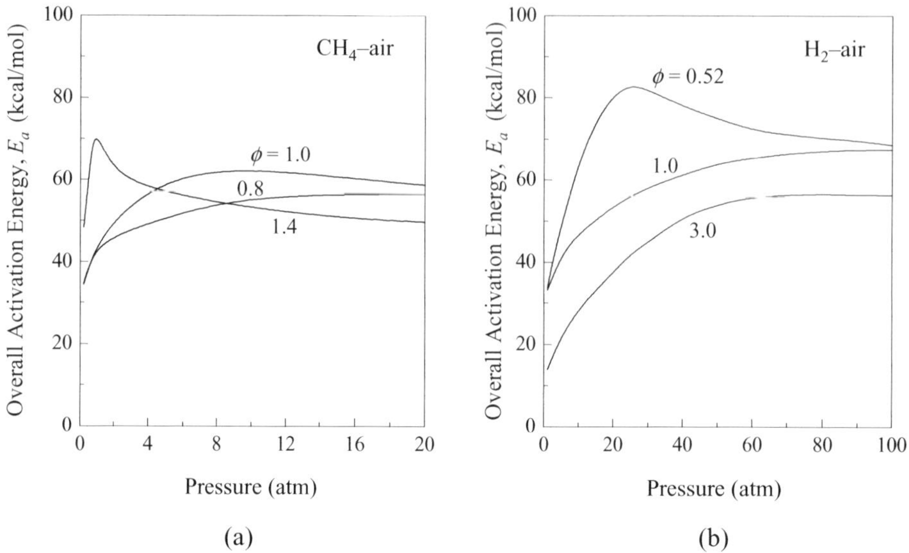

Figure 7.7.10. Computed overall activation energies, E_a, for the (a) methane–air flames, and (b) hydrogen–air flames of Figure 7.7.8.

7.7.4. Dependence on Freestream Temperature

The upstream temperature affects the flame speed in three ways. The first factor is through the adiabatic flame temperature $T_{ad} = T_u + (q_c/c_p)Y_u$, which influences the reaction rate. For low and small changes in the upstream temperature the influence is not expected to be strong because the chemical heat release, represented by q_c, is much larger than the thermal energy contained in the upstream flow. For larger values of T_u the dependence is more sensitive because of the Arrhenius factor.

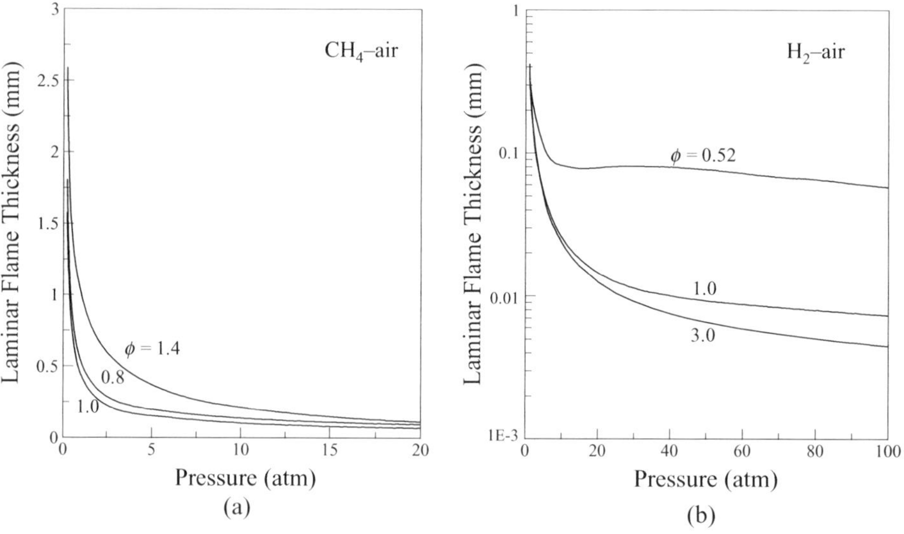

Figure 7.7.11. Computed effective flame thicknesses, ℓ_D^o, for the (a) methane–air flames, and (b) hydrogen–air flames of Figure 7.7.8.

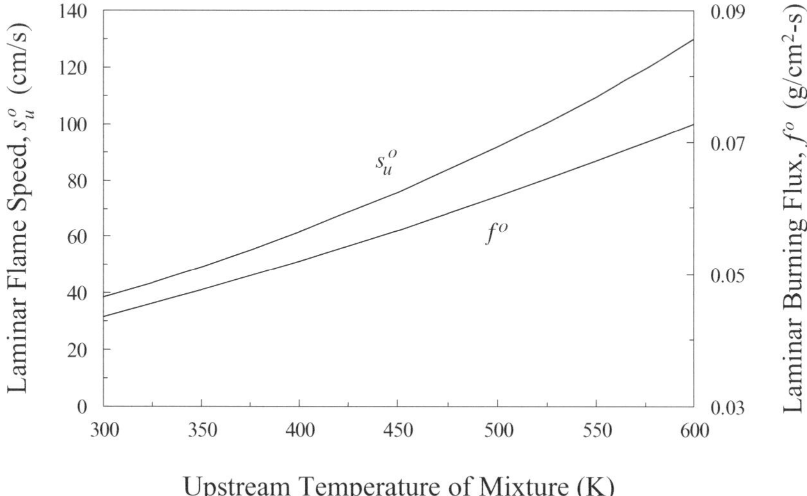

Figure 7.7.12. Computed laminar flame speed, s_u^o, and laminar burning flux, f^o, of stoichiometric methane–air flames with different mixture temperatures.

The second factor is due to the change in the transport properties. From the constant property derivation we have $f^o \sim (\lambda/c_p)^{1/2}$. Since $\lambda/c_p \sim T^\gamma$, with $\gamma < 1$, the temperature dependence through transport property variation is only mildly sensitive.

The third factor is through the change in density. That is, for a given mass flow rate $f^o = \rho_u s_u^o \sim s_u^o/T_u$, increasing T_u will lead to a corresponding increase in s_u^o. This effect is eliminated by considering f^o instead of s_u^o.

Figure 7.7.12 shows some calculated s_u^o and f^o for stoichiometric methane–air mixtures. The f^o curve shows that the combined effect of the Arrhenius factor and transport properties is somewhat linear. The curvature in s_u^o is then mostly due to the density variation.

7.7.5. Dependence on Transport Properties

The laminar flame speed s_u^o varies linearly with $(\lambda/c_p)^{1/2}$ and Arrheniusly with the flame temperature, which depends on the specific heat of the gas. These relations have been demonstrated by the measured and computed flame speeds of methane in different oxygen–inert mixtures, in which different inerts were used while the oxygen-to-inert molar ratio was fixed at 0.21/0.79. Figure 7.7.13a shows that the laminar flame speeds vary in the order of $(s_u^o)_{\mathrm{He}} > (s_u^o)_{\mathrm{Ar}} > (s_u^o)_{\mathrm{N_2}}$. The reason being that whereas He and Ar have the same specific heat (per mole), He is lighter and therefore has a higher λ/c_p because of the higher thermal conductivity coefficient; therefore $(s_u^o)_{\mathrm{He}} > (s_u^o)_{\mathrm{Ar}}$. Similarly, whereas Ar and N_2 have similar thermal conductivity coefficients because their molecular weights do not differ much, Ar, being monatomic, has a smaller specific heat and hence a higher flame temperature. Consequently $(s_u^o)_{\mathrm{Ar}} > (s_u^o)_{\mathrm{N_2}}$. Finally, in terms of the laminar burning flux, since the density of He is much smaller than that of Ar, Figure 7.7.13b shows that $(f^o)_{\mathrm{Ar}} > (f^o)_{\mathrm{He}} > (f^o)_{\mathrm{N_2}}$.

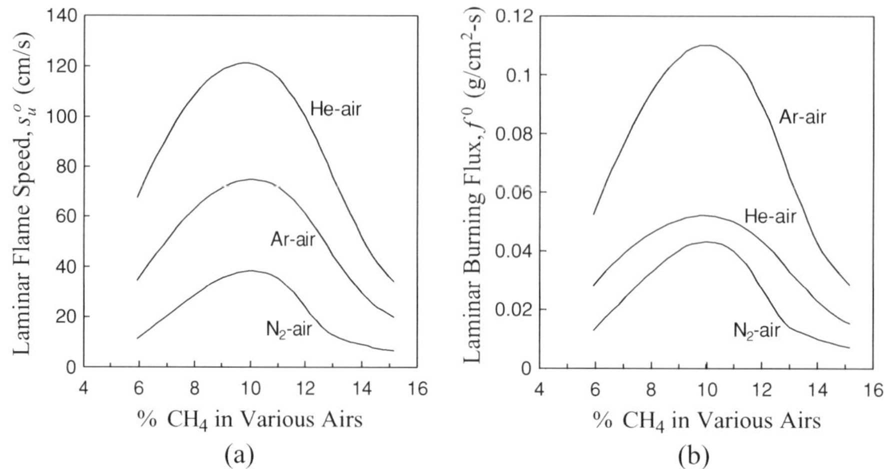

Figure 7.7.13. Computed (a) laminar flame speed, s_u^o, and (b) laminar burning flux, f^o, of stoichiometric methane–air flames in various airs.

7.8. CHEMICAL STRUCTURE OF FLAMES

It is instructive to recapitulate what we have studied so far on the standard premixed flame. We have focused on the theoretical derivation and experimental determination of the laminar burning velocity. The derivation is based on a flame structure that consists of a broad, diffusive-convective, preheat zone, followed by a much narrower diffusive-reactive zone within which a one-step, large-activation-energy reaction takes place, with the attendant heat release. Hence chemical activation and heat release occur in the same narrow region at the downstream end of the flame structure. We subsequently presented the experimentally and computationally determined laminar burning velocities (in the latter case detailed chemistry was used) and interpreted these data on the basis of the theoretically derived expression for the laminar burning velocity. It was emphasized that chain mechanisms, which are inherently multistepped, can profoundly affect the overall kinetic parameters used by the theory. Such an understanding not only points to the quantitative and for certain situations qualitative inadequacy of the one-step reaction used in the formulation, but it also argues against the physical realism of the associated two-zone flame structure, for three reasons. First, since chain mechanisms frequently consist of large-activation-energy branching reactions that are temperature sensitive, and terminating reactions that may not be temperature sensitive, as noted earlier, a chain mechanism may spread out over the entire flame structure, with the terminating reactions being effective also in the relatively low-temperature, preheat zone. Second, since some of these terminating reactions can be highly exothermic, the heat release profile within the flame structure is correspondingly affected. Third, chemical activation is purely thermal in nature for the one-step reaction, and occurs at the downstream end of the preheat zone where the temperature is close to that of the flame. However, from our studies of kinetic mechanisms we know that chemical activation is frequently initiated through the agency of radicals. These radicals can be produced in the high-temperature region of the flame structure and back diffuse to the colder region to initiate the reaction. This again implies the need for a more detailed description of the reaction mechanism and hence flame structure beyond that of the one-step.

In the following we shall first discuss the chemical structure using detailed mechanisms. Both the standard premixed flame and the counterflow nonpremixed flame will be studied and contrasted, after a brief discussion of the experimental aspects of studying the flame structure. We shall demonstrate that while homogeneous kinetics is preserved in the diffusive environment of flames, species transport plays an essential role in the flame structure. We shall then present results of the flame structure analysis based on reduced mechanisms, obtained by using asymptotics.

7.8.1. Experimental Methods

Although we shall conduct the following discussion on the chemical structure of flames based on calculated results, experimental studies have contributed essentially to our understanding of the fuel oxidation kinetics and the resulting flame structure. In particular, although studies of chemistry are best conducted in homogeneous systems such as the shock tube, flow reactor, and well-stirred reactor, the chemical structure of laminar flames provides additional information and scrutiny on the validity and completeness of the mechanism as it is subjected to environments of extensive variations in temperature and composition.

A flame structure is defined by its velocity, temperature, species, and reaction rate profiles. Experimental techniques developed for flame structure studies can be classified as either intrusive or nonintrusive. In terms of intrusive measurements, flow field, thermal structure, and species profiles can be respectively determined using hot wire velocimetry, thermocouple, and probe sampling (see, for example, Fristrom & Westenberg 1965; Gaydon & Wolfhard 1970). These techniques are relatively simple in design and usually inexpensive to acquire, although they have many limitations and their execution also requires skill and care. In general, the insertion of a measuring probe alters the flame structure and its properties at the location where the measurement is conducted, especially in view of the thinness of the flame and the finite dimension of the probe. Furthermore, hot wire velocimetry cannot be applied to flame studies because of the high temperature of the environment. For thermocouple measurements, there can be substantial uncertainty in the correction due to radiative and conductive heat loss. Furthermore, the heat release due to possible catalytic reactions at the thermocouple surface can also lead to higher measured temperatures. Although this catalytic effect can be mostly eliminated by coating the thermocouple, the coating increases its thickness and thereby disturbance to the flow field. For gas sampling of the species composition, and its subsequent analysis by using for example gas chromatography, it is essential that all chemical reactions are quenched, say through rapid expansion, once the sample is extracted by the probe.

Regarding nonintrusive, optical measurements, the most straightforward approach is imaging based on high-speed or multiple-exposure photography coupled to shadowgraph, schlieren, and interferometry. The possibility multiplies when laser is used as the light source, capitalizing on properties of light scattering, absorption, and emission, and based on the principles of linear and nonlinear optics. The probing can be spatially single point or multidimensional imaging, and temporally single pulse or continuous. As examples, the flow field can be quantified by laser Doppler velocimetry and particle image velocimetry, as mentioned earlier, while temperature can be measured via interferometry, Raman spectroscopy, coherent anti-Stokes Raman scattering (CARS), Rayleigh scattering, and laser induced fluorescence (LIF). Raman spectroscopy and CARS are also used to determine the major species concentrations, while LIF is suitable for the detection of minor species. These laser-based techniques provide fine spatial and temporal resolution without disturbing the flame structure, although their setup can be quite costly and involved. In addition, sophisticated post-processing procedure may be required. Finally, there are also physical processes affecting the fidelity of laser diagnostics that have to be taken into account, such as thermophoretic effects on the seeding particles in LDV and PIV, line broadening in Raman spectroscopy, and collisional quenching in LIF. Eckbreth (1996), Wolfrum (1998), Kohse-Höinghaus and Jeffries (2002), and Kohse-Höinghaus et al. (2005) provide comprehensive literature on the principles of laser diagnostics in combustion.

7.8.2. Detailed Structure

7.8.2.1. Premixed Flames: We first study the calculated structure of the standard premixed flame of a stoichiometric hydrogen–air mixture at one atmosphere pressure and room temperature, obtained by using the detailed mechanism of Kim, Yetter and Dryer (1994), listed in Table 3.1. Specifically, Figure 7.8.1a shows the profiles of the major species (H_2, O_2, and H_2O); Figure 7.8.1b the minor species (H, O, OH, HO_2, and H_2O_2); Figure 7.8.2a the total heat production rate, and the fractional amount of heat release; Figure 7.8.2b the molar production rates of H, H_2, O_2, and H_2O; Figure 7.8.3a the rates of key reactions involving the radicals H, O, and OH that account for the consumption of H_2 and O_2; and Figure 7.8.3b the heat production rates of the key exothermic and endothermic reactions. The temperature profile is superimposed in selected figures in order to provide a direct indication of the thermal environment experienced by the various flame properties.

Figure 7.8.1a shows that, due to the higher diffusivity of H_2 as compared to O_2 as well as heat, H_2 has a thicker diffusion zone than those of O_2 and T. This higher diffusivity actually causes a "bump" in the mole fraction of O_2 because of the rapid reduction in the concentration of H_2 as the mixture approaches the active reaction zone.

The active reaction zone, which approximately spans between 0.04 and 0.1 cm, can be considered to consist of two layers, namely a leading, H consumption layer,

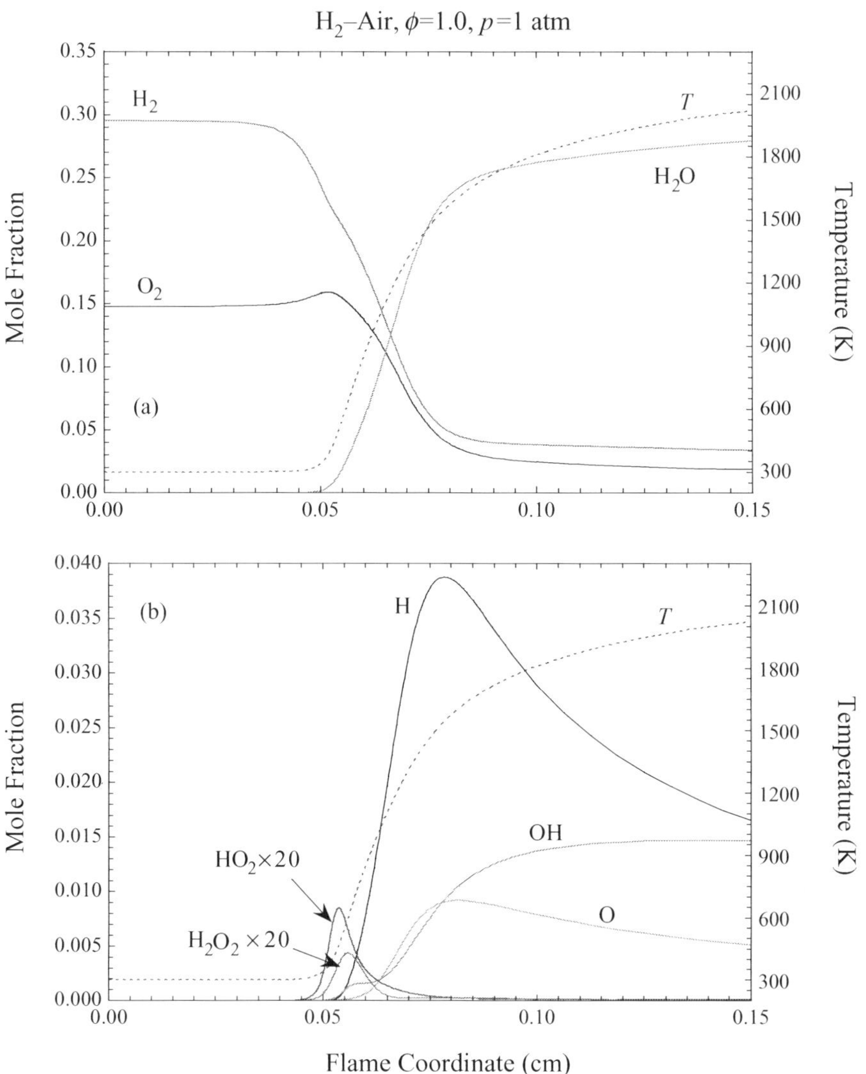

Figure 7.8.1. Chemical structure of the stoichiometric hydrogen–air premixed flame at standard conditions: (a) major species, and (b) minor species; temperature profile superimposed.

followed by an H production layer, as shown in Figure 7.8.2b. The H consumption layer coincides with the concentration profile of HO_2 shown in Figure 7.8.1b, which is the first noticeable radical as the freestream mixture enters the active reaction zone of the flame. However, while the emergence of HO_2 is also the indication of the initiation of the reaction zone for the ignition of a homogeneous H_2–air mixture, as discussed in Section 3.2, the production of HO_2 for the present flame case is not through reaction (−H10): $H_2 + O_2 \rightarrow HO_2 + H$. Rather, it is through reaction (H9): $H + O_2 + M \rightarrow HO_2 + M$, with H generated in the downstream, high-temperature layer through reactions (H2): $O + H_2 \rightarrow H + OH$ and (H3): $OH + H_2 \rightarrow H + H_2O$,

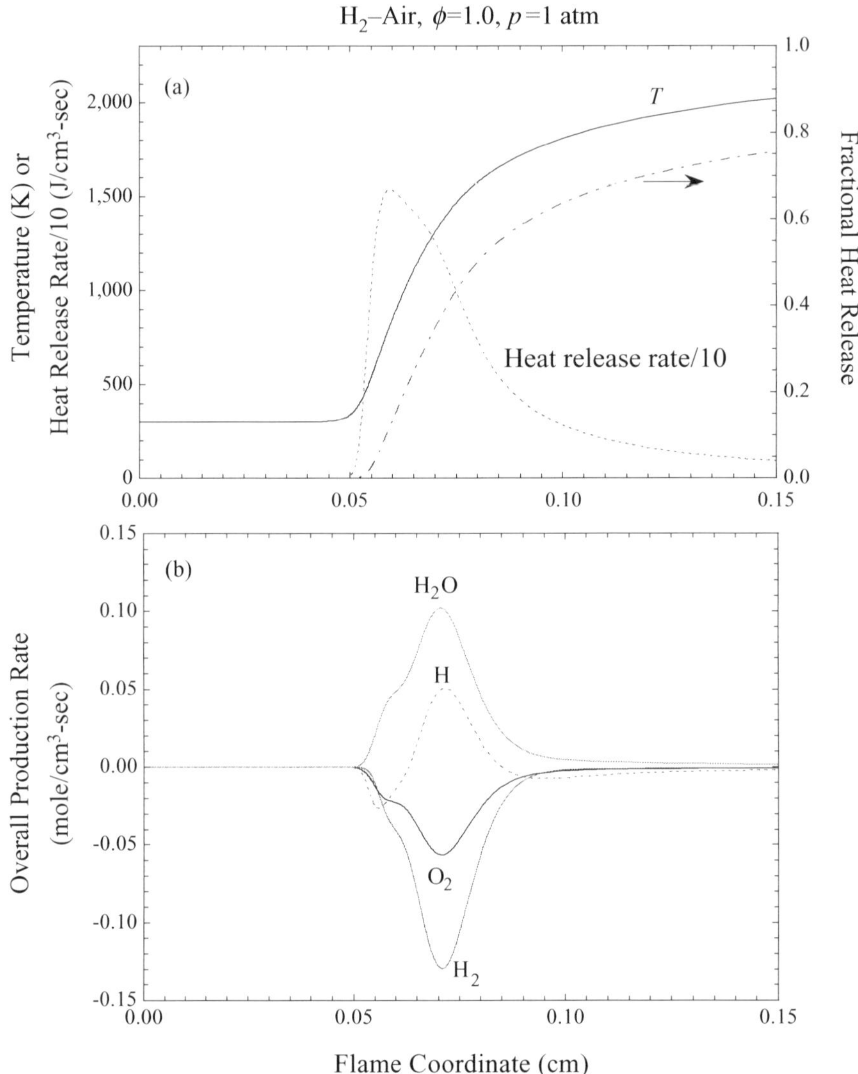

Figure 7.8.2. Chemical structure of the stoichiometric hydrogen–air premixed flame at standard conditions: (a) total heat release rate, fractional heat release, and temperature, and (b) production rates of the major species and H.

and back diffuses to react with O_2 in the incoming H_2–air mixture, as shown in Figure 7.8.2b. Consequently, the H production layer coincides with the reaction rate profiles of (H2) and (H3) in Figure 7.8.3a. Figure 7.8.1b also shows that the concentration profile of H_2O_2 slightly lags that of HO_2 because the production of H_2O_2 requires the presence of HO_2. Furthermore, comparing the reaction rate profile of (H9) with the temperature profile, it is seen that the temperature at which (H9) is activated is close to that of the freestream. This is reasonable because the activation energy for (H9) is zero.

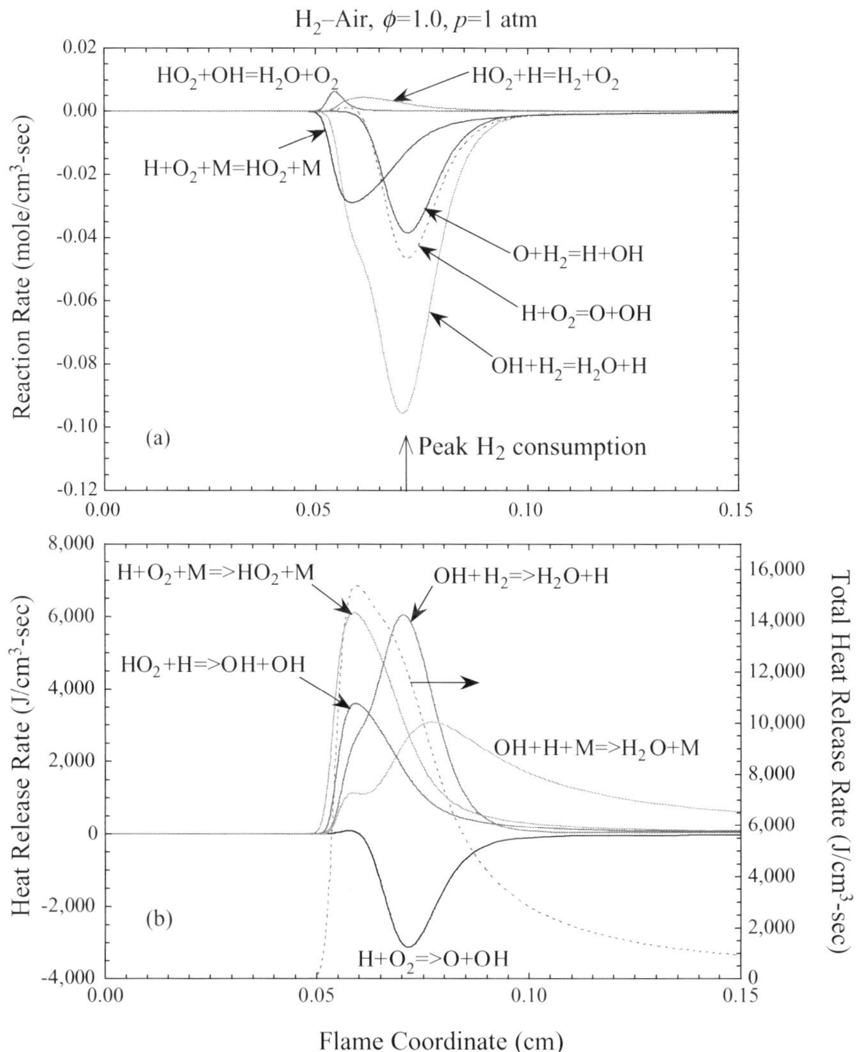

Figure 7.8.3. Chemical structure of the stoichiometric hydrogen–air premixed flame at standard conditions: (a) rates of key reactions involving H, O, and OH that account for the consumption of H_2 and O_2, and (b) heat production rates of key exothermic and endothermic reactions.

Next, we note from Figure 7.8.2b that the consumption of H_2 and O_2 and the generation of H_2O range over the entire reaction zone including both the H production and consumption layers, with the ratio of their maximum rates in approximate stoichiometric proportion. The major consumption of H_2 is through (H2) and (H3), while that of O_2 is dominated by (H1): $H + O_2 \rightarrow O + OH$, as shown in Figure 7.8.3a. The "bumps" in their respective leading segments are consequences of reactions involving HO_2 that produce H_2, O_2, and H_2O, namely (H13): $HO_2 + OH \rightarrow H_2O + O_2$ and (H10): $HO_2 + H \rightarrow H_2 + O_2$.

Figure 7.8.3b shows that the major exothermicity is contributed by (H9) and (H11): $HO_2 + H \to OH + OH$ in the H consumption layer, by (H3) in the H production layer, and by (H8): $H + OH + M \to H_2O + M$, which constitutes a long tail in the post active-reaction zone. The magnitudes of their maximum heat release rates are in the order of (H9) $\approx$ (H3) $>$ (H11) $\approx$ (H8). The major endothermic reaction is also the major chain-branching reaction (H1), as expected. Comparing these heat release rate profiles with those of the thermal parameters in Figure 7.8.2a, it is seen that the maximum heat release peaks at about 800 K, in the hydrogen radical consumption layer. Furthermore, about 30 percent of the total heat has already been released at the downstream boundary of this layer, where the temperature is about 1,000 K. Chemical activation, corresponding to the maximum production rate of H, occurs around 1,350–1,400 K, which is to be contrasted with the adiabatic flame temperature of 2,380 K.

The chemical structure that has emerged is therefore very different from that prescribed by the one-step reaction with large activation energy. It is shown that reactions of importance occur throughout the flame structure, that the H radical required by reaction (H9) at the leading edge of the flame is produced downstream and is made available to (H9) through back diffusion, that the maximum heat release occurs at the front of the active reaction zone instead of toward the back, and that substantial amount of chemical heat has already been released in the moderately low-temperature region of the flame.

7.8.2.2. Nonpremixed Flames: To contrast the difference in the chemical structure between premixed and nonpremixed flames, we now discuss the calculated chemical structure of a counterflow nonpremixed flame of hydrogen versus air at 1 atm pressure and with a strain rate of 300 s^{-1}. Specifically, Figure 7.8.4a shows the profiles of the major species (H_2, O_2, and H_2O); Figure 7.8.4b the minor species (H, O, OH, HO_2, and H_2O_2); Figure 7.8.5a the production rates of the important species (H, O, OH, H_2, O_2, and H_2O); Figures 7.8.5b and 7.8.6a the rates of the key reactions accounting for the consumption of H_2 and O_2, and Figure 7.8.6b the rates of the key exothermic and endothermic reactions. The temperature profile is again superimposed in selected figures to relate the relevant reactivity parameters to the thermal environment. Figure 7.8.4a then shows that the bulk temperature profile including its peak is located on the oxidizer side of the stagnation surface, indicated by an arrow at $\sim$ 0.5 cm. This is a result of the stoichiometry for the reaction between hydrogen and air. Elevating the oxygen concentration will shift the flame toward the fuel side.

Figure 7.8.4a further shows that while hydrogen and oxygen are initially separated, there is an interfacial region (from $\sim$ 0.7 to 0.8 cm) within which they coexist and the maximum heat release rate peaks (Figure 7.8.6b). Furthermore, the temperature and product H_2O peak slightly on the fuel side of this region. Comparing Figures 7.8.4a and 7.8.4b shows that the chemical structure of the flame, as represented by the span of the profiles of the radicals, is much broader (from 0.5 to 0.85 cm). In fact,

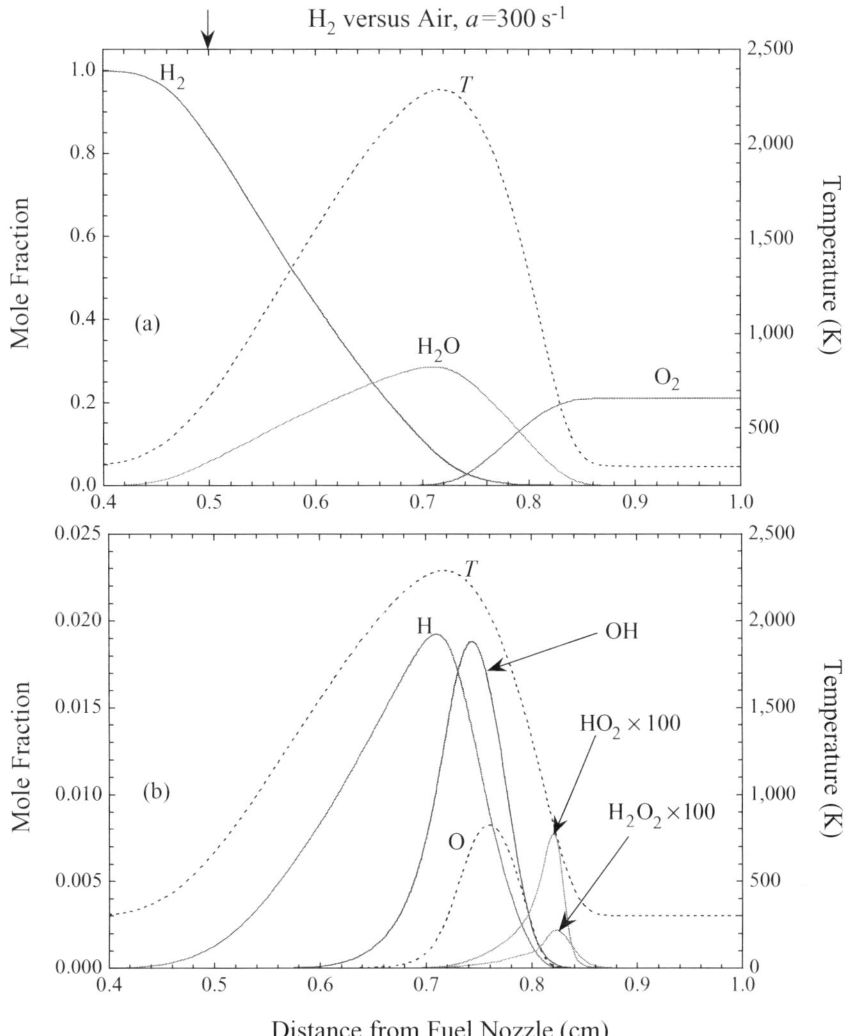

Figure 7.8.4. Chemical structure of a counterflow nonpremixed flame of hydrogen versus air at standard conditions: (a) major species, and (b) minor species; temperature profile superimposed. Vertical arrow indicates location of the stagnation surface.

the chemical structure basically overlaps with the thermal structure given by the temperature profile. On the other hand, the reaction rate profiles in Figures 7.8.5a, 7.8.5b, 7.8.6a, and 7.8.6b show that they mostly span a region slightly thicker than the interfacial region, between 0.65 and 0.85 cm, although the bulk of the profiles are located within this interfacial region. Consequently we can identify this region as the active reaction zone. Low-level reactivity, however, does occur throughout the flame structure, implying that substantial reaction takes place in the low-temperature region of the flame.

Let us now be more specific with the role of the individual reactions. Figure 7.8.5a shows that there is a slight shift in the locations between the maximum consumption

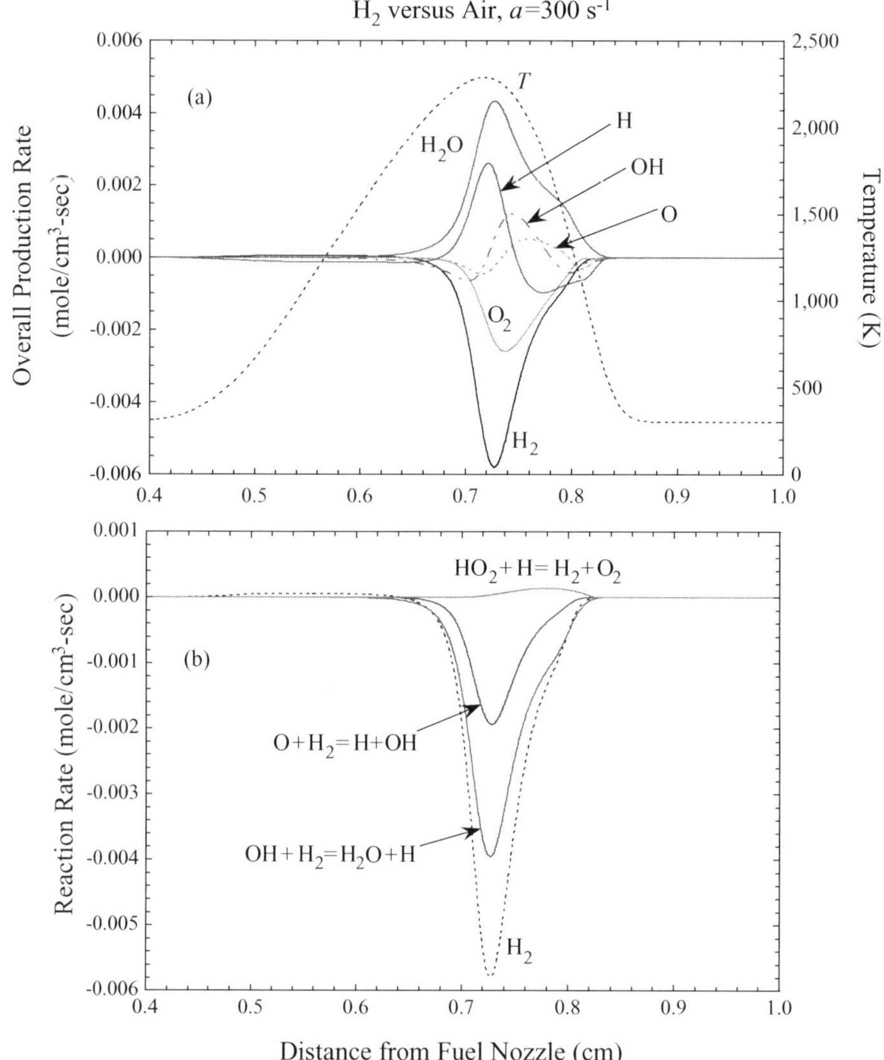

Figure 7.8.5. Chemical structure of a counterflow nonpremixed flame of hydrogen versus air at standard conditions: (a) production rates of the important species (H, O, OH, H_2, O_2, and H_2O), and (b) rates of some key reactions for the consumption of H_2.

rates of H_2 and O_2, which is to be contrasted with the case for the premixed flame for which their locations basically coincide. Their maximum values, however, are still in stoichiometric proportion. Furthermore, it is seen that H is produced on the H_2 side and consumed on the O_2 side of the flame structure, while the opposite holds for O and OH. These are expected according to reactions (H1) to (H3) in the production and consumption of these radicals.

Figure 7.8.5b shows that (H2) and (H3) are the dominant reactions leading to the consumption of H_2 and production of H, as expected, and that there is a slight production of H_2 via (H10) on the oxygen side where HO_2 peaks. Similarly, Figure 7.8.6a shows that (H1) and (H9) are the major reactions that consume O_2, producing

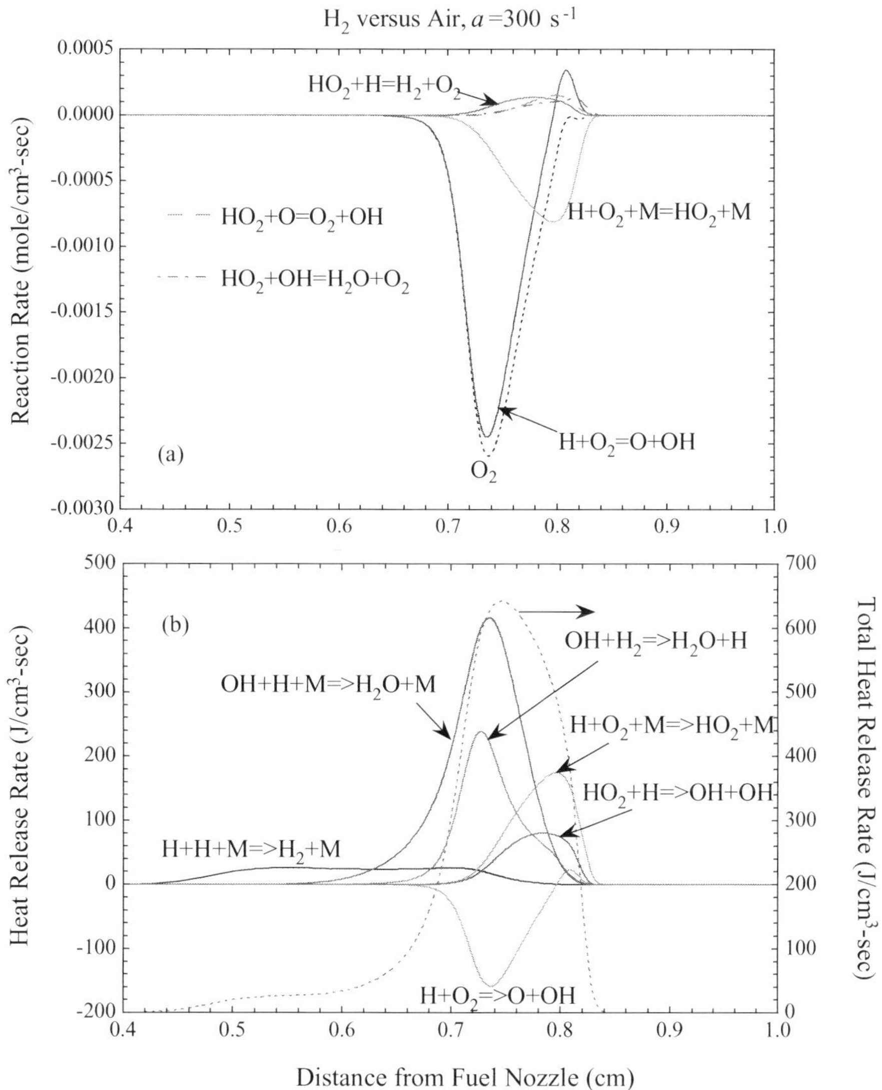

Figure 7.8.6. Chemical structure of a counterflow nonpremixed flame of hydrogen versus air at standard conditions: (a) rates of some key reactions accounting for the consumption of O_2, and (b) heat production rates of the key exothermic and endothermic reactions.

O and OH through the former and HO_2 through the latter. Furthermore, all reactions related to HO_2 are close to the oxygen side.

Figure 7.8.6b shows that the major heat release reactions are the same as those for the premixed flame, although the order of their relative magnitudes is now changed to (H8) > (H3) $\approx$ (H9) > (H11). Furthermore, (H8) and (H3) peak toward the hydrogen side of the flame, while (H9) and (H11) peak toward the oxygen side. This is reasonable in that (H3) and (H9) respectively require H_2 and O_2, while (H8) and (H11) respectively need the production of H on the hydrogen side and HO_2 on the oxygen side. The major endothermic reaction is still (H1), as in the premixed case.

It is also of interest to note the low-level, long-range presence of (−H5): $H + H + M \to H_2 + M$ into the upstream of the H_2 flow, even the temperature of the flow is fairly low and the reaction is that of three body, which requires high concentrations of H. The reasons are that the reaction is temperature insensitive, and that the high mobility of H enables it to diffuse rapidly upstream from the active reactive zone where it is formed, giving it the broad concentration profile shown in Figure 7.8.4b.

The structure of the nonpremixed flame is therefore again one in which reactions of significance take place throughout the flame, with substantial chemical activity including heat release occurring in the relatively cold part of the flame, and that transport is an essential mechanism in sustaining these reactions.

We close this section by emphasizing that we are only using this particular chemical system, namely reactions involving H_2 and O_2 of certain concentrations and under atmospheric pressure, to illustrate the complexity and richness of the flame structure that can emerge, and to demonstrate the rational manner through which such structures can be studied. The structure will change for different thermodynamic states of the reactants and for different fuels. For example, at higher pressures the reactions involving HO_2 and H_2O_2 will become more prominent, as discussed in Chapter 3, while for methane oxidation the reaction (M3): $CH_4 + H \to CH_3 + H_2$ could replace (H9) as the key chain breaking step. Furthermore, the premixed flame structure for lean and rich hydrocarbons could also be quite different. Thus care needs to be exercised when studying situations with different experimental conditions or fuel–oxidizer systems.

7.8.3. Asymptotic Structure with Reduced Mechanisms

The chemical structure of flames portrayed above demonstrates the inadequacy of the one-step, large-activation-energy reaction in the description of the flame structure. Since analytical study of the flame structure obviously is not amenable with detailed reaction mechanisms, which consist of large numbers of reactions and species, the concept of reduced mechanisms, discussed in Chapter 3, has been extensively exploited so as to capture the dominant influence of detailed mechanisms beyond those of the one-step chemistry. Furthermore, in order for the analysis to be tractable, the reduced mechanisms must necessarily be very small, consisting of only a few semiglobal steps. These are discussed in the following (Williams 2000).

7.8.3.1. Reduced Mechanisms: For H_2 oxidation, we have shown in Chapter 3 that the simplest multistep reaction mechanism beyond the one-step reaction is a two-step mechanism consisting of

$$3H_2 + O_2 \rightleftharpoons 2H_2O + 2H \tag{I}$$

$$H + H + M \rightleftharpoons H_2 + M, \tag{II}$$

which are basically reactions (II″) and (I′) identified in Section 3.11. Studies (Seshadri, Peters & Williams 1994) have shown that steps (I) and (II) primarily proceed at the rates of the elementary reactions (H1) and (H9) respectively. We further recall that these semiglobal reactions were derived from the elementary reactions of the detailed mechanism by assuming quasi-steadiness of O, OH, HO_2, and H_2O_2, and as such their reaction rates are dependent on those of the elementary reactions. The particular forms assumed by these semiglobal reactions also are not unique since any linear combination of them would be equally valid. For example, replacing (I) by $2H_2 + O_2 \rightleftharpoons 2H_2O$, obtained by adding (I) and (II), would result in an equally valid mechanism. Frequently a particular choice is based on the physical meaning that it conveys. Thus in the present representation step (I) can be considered to be chain branching in nature because two H atoms are produced, while step (II) is chain terminating because they are then eliminated.

For CO oxidation (Wang, Rogg & Williams 1993), since the presence of H is essential, its reduced chemistry consists of (I), (II), and the overall step

$$CO + H_2O \rightleftharpoons CO_2 + H_2, \tag{III}$$

which primarily proceeds at the rate of the elementary reaction (CO3): $CO + OH \rightleftharpoons CO_2 + H$.

For the oxidation of methane, which is the simplest hydrocarbon, a four-step mechanism has been developed (Peters & Williams 1987), consisting of (I) to (III) plus a fuel-consumption step

$$CH_4 + 2H + H_2O \rightleftharpoons CO + 4H_2, \tag{IV}$$

which proceeds at the rate of the elementary reaction (M3). Since C_2 species are not considered in (IV), this reduced mechanism is expected to be applicable only to lean to stoichiometric mixtures.

Step (IV) can be readily generalized to the consumption of other hydrocarbons. We can thus express the reaction for, say, alkanes, C_mH_{2m+2}, as

$$C_mH_{2m+2} + \gamma H + \left[\beta - \left(m + 1 + \frac{\gamma}{2}\right)\right] H_2O + \left\{m - \frac{\alpha}{2} - \frac{1}{2}\left[\beta - \left(m + 1 + \frac{\gamma}{2}\right)\right]\right\} O_2$$
$$\rightleftharpoons \alpha CO + \beta H_2 + (m - \alpha)CO_2 \tag{V}$$

where α, β, and γ respectively denote the number of CO and H_2 produced and H consumed.

The reduced mechanism for CO can be further simplified to a two-step mechanism by assuming partial equilibrium for the water–gas shift reaction (III), while the CH_4 mechanism can also be reduced to a two-step mechanism by further assuming H as a quasi-steady species, yielding

$$O_2 + 2H_2 \rightleftharpoons 2H_2O \tag{VI}$$

$$CH_4 + O_2 \rightleftharpoons CO + H_2 + H_2O. \tag{VII}$$

It has been shown (Peters & Williams 1987) that the rates of (VI) and (VII) are the same as those of (III) and (IV) respectively.

Extensive analyses have been performed for the structure of standard premixed flames and counterflow nonpremixed flames of H_2, CO, CH_4, alcohols, and a number of the higher hydrocarbons, using reduced mechanisms of two-, three-, and four-step mechanisms (Smooke 1991; Peters & Rogg 1993; Seshadri & Williams 1994; Seshadri 1996; Williams 2000). Such analyses typically involve identifying spatial regions of different thicknesses and controlling reactions, and linking these regions through appropriate matching conditions. However, unlike the asymptotic analysis for the one-step overall reaction in which the reaction is localized to a thin region by the large activation energy, the concept of large activation energy cannot be applied to all the elementary reactions in a multistep reaction mechanism because the activation energies for some of them are either zero or very small. It has therefore been found that the useful parameters for distinguishing reaction layers of different characteristics are the ratios of the rates of the various elementary reactions; this approach is referred to as rate-ratio asymptotics.

The flame structure analysis with reduced mechanisms is usually very involved. It is thus quite an accomplishment that useful analytical results such as the laminar burning velocities of premixed flames, the extinction strain rates of nonpremixed flames, and the formation of NO_x in flames (Williams 2000) have been derived for systems of such complexity. As an example of such analyses but without going into detail, we shall consider the premixed and nonpremixed methane flames with the four-step mechanism (I–IV) discussed above. Methane instead of hydrogen flame is selected for illustration because while the latter is well suited for demonstration of the salient features of the detailed structure because of the small number of reactions and species involved, the assumed asymptotic structure of methane is qualitatively similar and in addition is also simpler. Furthermore, the asymptotic structures of the higher hydrocarbons have also been built upon that of the methane flame.

7.8.3.2. Premixed Flames: The methane flame structure (Figure 7.8.7) is assumed to consist of three layers (Peters & Williams 1987), namely (1) the upstream chemically inert preheat zone of $O(1)$ thickness; (2) a thin, fuel consumption layer, also called the inner layer, of $O(\delta)$ thickness; and (3) another thin, oxidation layer of thickness $O(\epsilon)$; where $(\delta, \epsilon) \ll 1$. Downstream of the oxidation layer is the equilibrium, fully reacted state of the mixture. Thus the single reaction zone structure for the one-step overall reaction is now replaced by a two reaction zone structure, consisting of the fuel consumption layer and the oxidation layer. To be more specific, in the fuel consumption layer steps (I) and (IV) dominate, through which the primary fuel, CH_4, reacts with the H atom to form the secondary fuels, H_2 and CO, while in the oxidation layer these secondary fuels are converted to the products H_2O and CO_2 through steps (I) to (III). Since the characteristic reaction rate of step (IV) is much faster than those of (II) and (III), the fuel consumption layer is thinner than that of the oxidation layer, hence $\delta \ll \epsilon$. The H atoms needed for fuel consumption are supplied through back diffusion from the oxidation layer. It may also be noted that the major fuel consumption step, (M3), is in addition a chain-termination step. This

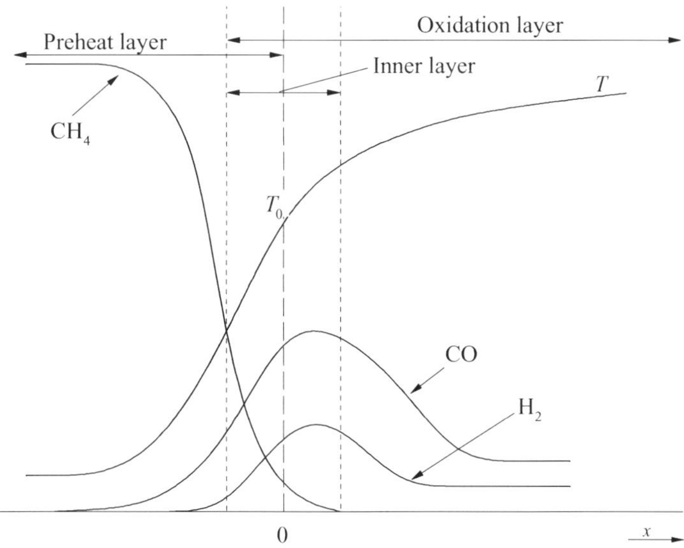

Figure 7.8.7. Schematic illustrating the asymptotic structure of the methane–air premixed flame (Williams 2000).

renders the analysis easier (Seshadri 1996) as compared to the hydrogen flame, for which the major fuel consumption steps, (H2) and (H3), are not chain breaking.

As a "reality check," Figures 7.8.8a and 7.8.8b respectively show the computed profiles of the production rates of the major and minor species for a stoichiometric CH_4–air flame at the standard condition, obtained by using a detailed mechanism. The profiles are qualitatively similar to those of the hydrogen flame, with the key radical H produced in the downstream region and consumed in the upstream region of the flame, while CH_3 is produced in the upstream region through (M3). Consumption of the fuel, CH_4, however, is not localized in the upstream region where CH_3 is produced, and neither is the production of H_2O localized to the downstream region. Instead, their reaction rate profiles span the entire flame structure.

Having been alerted to the differences between the real and assumed flame structures, we nevertheless continue with the discussion on the asymptotic structure of the methane flame. The inner layer, being very thin, has a characteristic temperature T_0, which assumes a central role in the asymptotic analysis. Together with the laminar burning velocity, they are the two most important parametric outcomes of the analysis. This temperature is basically the crossover temperature for the reduced mechanism, balancing the rates of branching and terminating reactions, and varies with the system pressure. It is important because a flame structure obviously cannot be established if the temperature in the inner layer is smaller than this value such that terminating reactions will dominate. As a consequence of this implication, it is then reasonable to anticipate that flame propagation ceases to be possible when its flame temperature, T_b, is less than T_0. In the absence of heat loss, T_b is just the adiabatic flame temperature, T_b^o. This therefore sets a limit on the lean concentration

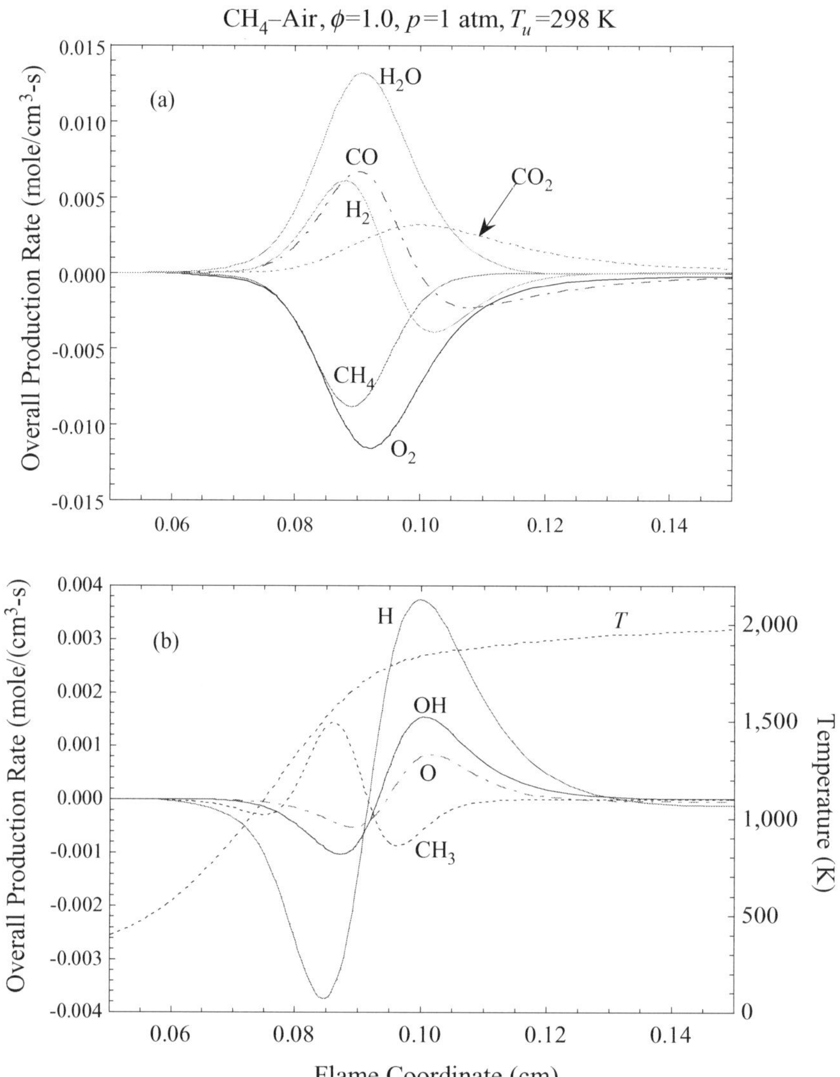

Figure 7.8.8. Chemical structure of the stoichiometric methane–air premixed flame at standard conditions: (a) major species, and (b) key radicals; temperature profile superimposed.

of a combustible mixture, which is known as the lean flammability limit of the given fuel–oxidizer system. We shall return to the concept and prediction of flammability limits in Section 8.5.

By further specializing to the two-step mechanism, Peters (1997) showed that T_0 is given by the relation

$$\frac{[k_{\mathrm{H1}}(T_0)]^2 R^o T_0}{k_{\mathrm{M3}}(T_0) k_{\mathrm{H9}}} = 1.5p, \tag{7.8.1}$$

while the laminar burning flux is

$$(f^o)^2 = Y_{F,u} A(T_0) \left(\frac{T_b^o - T_0}{T_b^o - T_u} \right)^4 p^2 \tag{7.8.2}$$

for unity Lewis numbers, where $A(T_0)$ is essentially a function of the thermodynamic and kinetic properties of the system through their dependence on T_0. Equation (7.8.1) shows the influence of the two pairs of chain-branching and chain-terminating reactions on the crossover temperature, namely those of (H1) and (M3), and of (H1) and (H9), as represented by the ratios of their respective rates, k_{H1}/k_{M3} and k_{H1}/k_{H9}. The unity exponent to p is due to the fact that the reaction rates of (H1), (M3), and (H9) are respectively proportional to p^2, p^2, and p^3. Thus their ratio through Eq. (7.8.1) is $p^{(4-2-3)} = p^{-1}$, which becomes p when moved to the RHS. Equation (7.8.1) also bears resemblance to that for the crossover temperature at the second explosion limit of the H_2–O_2 system, as shown in Eq. (3.2.7), and therefore exhibits the same trend of increasing T_0 with increasing pressure. This in turn implies that the flame temperature associated with the flammability limits will increase with increasing pressure, hence rendering the mixture less flammable.

Equation (7.8.2) shows that the laminar burning flux vanishes as $T_b^o \to T_0$. Realistically, the inevitable presence of loss mechanisms would have extinguished the flame before this state is reached, rendering the flame to assume a finite burning flux at the flammability limit. It also shows that f^o varies with pressure in the manner of $(T_b^o - T_0)^2 p$. Since the first factor decreases with increasing pressure, f^o is expected to first increase and then decrease with pressure.

It is further seen that f^o varies quadratically with $(T_b^o - T_0)$, which could in turn vary sensitively with T_b^o, especially when $T_b^o \to T_0$. To relate this behavior to the Arrhenius dependence of f^o for the one-step reaction formulation, an equivalent activation energy, E_a, can be determined for the present problem by evaluating Eq. (7.7.2) with f^o given by Eq. (7.8.2), yielding

$$\frac{E_a}{R^o T_b^o} \approx \frac{4 T_b^o}{T_b^o - T_0}, \tag{7.8.3}$$

obtained by recognizing that $T_u \ll T_0$. Using E_a, the corresponding equivalent Zel'dovich number is given by

$$Ze = \frac{E_a (T_b^o - T_u)}{R^o (T_b^o)^2} = \frac{4(T_b^o - T_u)}{T_b^o - T_0}. \tag{7.8.4}$$

Since T_b^o varies with the mixture composition while T_0 varies with the system pressure, E_a, and therefore Ze, cease to be constants.

7.8.3.3. Nonpremixed Flames: Figure 7.8.9 shows the asymptotic structure of a methane flame, with the mixture fraction Z as the independent variable. As discussed in Chapters 5 and 6, the structure of nonpremixed flames can be conveniently studied using Z because the results are independent of the flame configuration.

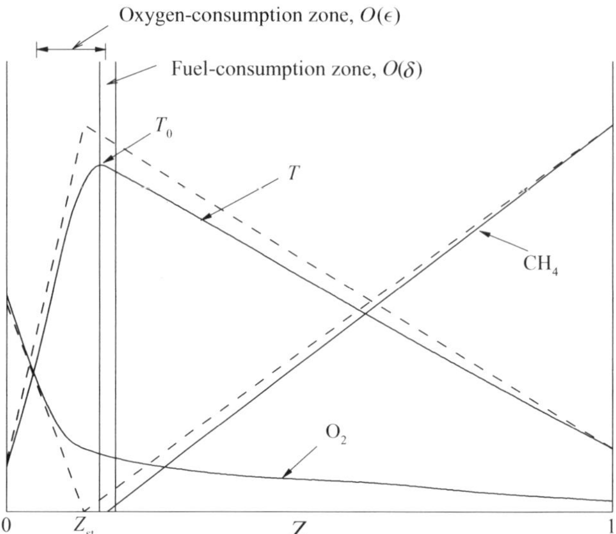

Figure 7.8.9. Schematic illustrating the asymptotic structure of a counterflow nonpremixed methane–air flame (Williams 2000).

The flame structure that emerges from rate-ratio asymptotics (Seshadri & Peters 1988; Seshadri & Ilincic 1995) is quite similar to that of the premixed flame. Specifically, the fuel consumption layer, controlled by step IV, is now on the fuel side of the flame, while the oxidation layer is now the oxygen consumption zone on the air side, where steps I to III occur. The structure analysis shows that the characteristic temperature, T_0, for the inner, fuel-consumption layer is again given by an expression similar to Eq. (7.8.3) for the crossover temperature.

A particularly interesting result of the multistep kinetics is that the strong demand of the radicals by step IV in the fuel consumption layer deprives the radicals needed to consume O_2 in the oxidation layer. Consequently, while there is no leakage of CH_4 through the fuel consumption layer, there is considerable leakage of O_2 through the flame, as shown schematically in Figure 7.8.9, especially for near-extinction situations when the flame is highly strained. This result, which is supported by experimental observations (Smooke, Puri & Seshadri 1986), is contrary to the prediction of one-step activation-energy asymptotics, which shows more fuel leakage than oxidizer leakage for methane–air flames (Liñán 1974).

Efforts have been made to quantitatively compare and correlate the asymptotic results, for both premixed and nonpremixed flames, with computed ones based on detailed mechanisms through adjustments of parameters, for example, those in the function $A(T_0)$. Extensive analyses have also been conducted using larger reduced mechanisms in order to incorporate additional chemical information. However, due to the complexity of the reaction mechanisms, reduced mechanisms at the level amenable for analysis are inherently incapable to be comprehensive in their description of the flame structure and response, at least quantitatively. On the other hand, the analytical results that they provide no doubt represent major improvements over those based on the one-step chemistry, removing some of the obviously erroneous features of the flame structure and consequently leading to more realistic descriptions of the flame response.

PROBLEMS

1. Derive Eqs. (7.1.14) and (7.1.15). Equating them at the CJ point results in a relation between $(\hat{p}, \hat{v})$,

$$\hat{p}_{\mathrm{CJ}} = \frac{\hat{v}_{\mathrm{CJ}}}{(\gamma+1)\hat{v}_{\mathrm{CJ}} - \gamma}.$$

Thus together with Eq. (7.1.11), $\hat{v}_{\mathrm{CJ}}$ and $\hat{p}_{\mathrm{CJ}}$ can be explicitly determined. Show that these are identical to Eqs. (7.1.20) and (7.1.21), as should be the case.

2. (a) From the first and second laws of thermodynamics, show that for a perfect gas with constant specific heat, its entropy is given by

$$s = c_p \ln T - R \ln p + \text{constant}.$$

(b) Show that the entropy change across a weak $(M_u^2 - 1 \ll 1)$ hydrodynamic shock $(q_c \equiv 0)$ is

$$\frac{s_b - s_u}{R} \approx \frac{2\gamma}{(\gamma+1)^2} \frac{(M_u^2 - 1)^3}{3}.$$

Since entropy always increases in an adiabatic flow, the above result suggests the nonexistence of strong deflagrations, for which $M_u < 1$.

3. Derive the entropy variation along the Hugoniot of the reaction products of a mixture of initial pressure p_u and temperature T_u. Plot your result in the entropy-specific volume plane. Show that the entropy is a minimum at the CJ detonation state and maximum at the CJ deflagration state.

4. For the laminar flame propagation problem treated, show that the temperature at which the maximum reaction rate occurs is smaller than the final, adiabatic flame temperature T_b^o by a fractional amount T_b^o/T_a. Generalize this to show that for an nth order reaction the amount is $n(T_b^o/T_a)$. This supports our assumption for the temperature distribution in the inner, reactive region.

5. If we define the reaction zone thickness as the distance between its downstream boundary and the inflection point on the upstream, ascending branch of the reaction rate profile, show that $\ell_R^o/\ell_D^o = O(\epsilon)$.

6. Derive asymptotically the laminar burning flux f^o for a two-reactant, stoichiometric mixture. Assume $Le = 1$ and that the reaction is first order with respect to both fuel and oxidizer.

7. Using the coupling function, derive the laminar burning flux f^o for a one reactant mixture with $Le = 1$ for the standard premixed flame propagation.

8. Derive an equation for the surface of a Bunsen flame supported by a parabolic flow issuing from a slot. Assume the burning velocity is fixed at s^o. Integrate over the surface to obtain an average s according to $s = \dot{V}/A$. Discuss the function s/s^o.

9. The possible existence of a stationary, spherical premixed flame, called a flame ball, in a quiescent combustible mixture of Y_u, T_u, and Le was first postulated by Zel'dovich. The existence was subsequently confirmed through microgravity experimentation.

(a) Derive the coupling function of this system.

(b) Derive and sketch the temperature and concentration profiles of the flame in the reaction-sheet limit, with a flame temperature T_f.

(c) Show that T_f and the reaction sheet location r_f are respectively given by

$$\tilde{T}_f = \tilde{T}_u + \frac{1}{Le}$$

$$r_f^2 = \frac{\lambda/c_p}{2Le^3\epsilon^2 B_C e^{-\tilde{T}_a/\tilde{T}_f}}.$$

8 Limit Phenomena

So far we have been concerned only with situations involving intense burning. In this chapter we shall study the transition between burning and nonburning states, namely phenomena involving ignition, extinction, flame stabilization and blowoff, and flammability.

There are many practical situations exhibiting ignition and extinction phenomena in our daily lives. For ignition, we can cite the striking of a match, turning on the gas stove by a pilot light or spark discharge, firing within an internal combustion engine through compressive heating or again by spark discharge, and the initiation of fires and explosions in mines, grain elevators, and upon the rupturing of fuel tanks by electric or frictional sparks. For extinction, we can cite firefighting through spraying of water and chemicals and the quenching of chemical reactions by the relatively cold wall of combustion chambers.

At the fundamental level, ignition can be achieved in one of two ways. One can either supply an amount of heat to part or all of a combustible mixture. The supply can be either momentary or continuous. The heated mixture responds Arrheniusly, reacts faster, and produces more heat. At the same time, however, being hotter it also tends to lose more heat to the walls and colder parts of the gas. Thus if the rate of heat generation exceeds that of cooling, then an accelerative, runaway process occurs that eventually leads to a state of intense burning. On the other hand if the cooling is sufficiently fast so that the heat input from the ignition stimulus and the subsequent chemical reactions can be drained away, then the ignition attempt is not successful.

In the second ignition mode a sufficient amount of chain-branching radicals are either supplied or produced. They will subsequently undergo branching, while at the same time they can also be destroyed through some deactivating reactions or collision with the wall. If chain branching dominates, a rapid rate of radical production is achieved, which will eventually lead to thermal runaway, and therefore ignition, through some exothermic reactions.

Similarly, flame extinction can be achieved by removing from the flame a certain amount of the chain-carrying or branching radicals, or a certain amount of heat. The former can be achieved by introducing chemical inhibitants into the reaction region or by placing the flame next to a surface that deactivates the radicals. The latter can be achieved by reducing the reaction rate by blowing it with cold gas, cooling it with a cold surface, or decreasing the reactant concentration or the system pressure, to mention just a few possibilities. There are, of course, inherent loss mechanisms associated with a flame, such as diffusive transfer of heat and radicals in the presence of steep temperature and concentration gradients at the flame, and radiative heat loss either from the high-temperature flame itself or the soot particles produced in the flame region.

In Section 8.1 we shall initiate our study on ignition and extinction by first discussing the practical phenomena of quenching distances and minimum ignition energies, using the laminar flame thickness as the critical parameter governing their response.

We shall then proceed to analyze ignition and extinction phenomena with more rigor, emphasizing their response to thermal influence described by the one-step overall reaction. Thus from large activation energy considerations, ignition and extinction phenomena are highly abrupt and transient events; the term "transient" is used here in the general sense of evolution, which may occur either in time or along a flow. We illustrate this characteristic by studying the classical thermal explosion theory that determines the transient ignition lag of a homogeneous mixture in which diffusive processes are absent. Both the adiabatic and nonadiabatic situations will be examined.

Ignition and extinction phenomena can also be studied from the steady-state viewpoint. That is, instead of analyzing the transient ignition and extinction processes, we ask whether steady-state solutions allowing for diffusive processes exist for either the vigorously or weakly burning situations. Thus the relevant information to be gained here is not how long it takes the system to ignite or extinguish, but rather whether the system is ignitable or extinguishable. This concept will be demonstrated through the S-curve analysis.

Thus by using the criticality test of ignitability and extinguishability, we shall first study in Section 8.2 the ignition of a premixture by a hot surface from the thermal point of view. This is the simplest situation possible involving premixtures and will therefore be studied in detail, using asymptotic analysis, in order to demonstrate the canonical nature of ignition induced by a one-step overall large activation energy reaction. In Section 8.3 we shall computationally study the ignition of a hydrogen jet by a counterflowing heated air jet, and through it show the intricate influence of chemical chain mechanisms. The corresponding asymptotic theory for the ignition of a nonpremixed system, say a cold fuel stream by a hot oxidizer stream, is conceptually similar to the premixed ignition, although mathematically more involved. It will be discussed in Chapter 9.

In Section 8.4 we study the classical problem of the extinction of the standard premixed flame by volumetric heat loss. Both the phenomenological and the Frank-Kamenetskii analyses will be presented. Other modes of premixed flame extinction, such as extinction due to aerodynamic stretch and insufficient residence time and the influence of mixture nonequidiffusion, will be covered in Chapter 10. Extinction of nonpremixed flames inherently involves reactant leakage and, hence, incomplete reaction. It will be discussed in Chapter 9.

In Section 8.5 the phenomena of flammability limits, which are the extreme concentration limits of lean and rich fuel–oxidizer mixtures beyond which steady combustion is absolutely not possible, will be discussed in terms of both empirical observations and fundamental concepts.

The disappearance of a flame does not necessarily imply the occurrence of some chemistry-limited extinction event. An intensely burning flame can still be "blown off" from a previously stabilized location as the flow velocity increases. In Section 8.6 we shall study the principles of flame stabilization, flashback, and blowoff.

Further exposition on the classical concepts and analyses of ignition and extinction, involving both one-step reactions and chain mechanisms, can be found in Semenov (1958, 1959), Frank-Kamenetskii (1969), Gray and Yang (1969), and Gray (1991), while extensive experimental results on flame quenching, extinction, and blowoff, and their interpretations, are presented in Lewis and von Elbe (1987). Recent theoretical developments based on asymptotic analysis, using both one-step reactions and reduced mechanisms, are discussed in Buckmaster and Ludford (1982), Williams (1985, 1992), and Buckmaster et al. (2005).

8.1. PHENOMENOLOGICAL CONSIDERATIONS OF IGNITION AND EXTINCTION

8.1.1. Quenching Distances and Minimum Ignition Energies

We now take the simplest viewpoint toward analyzing ignition and extinction phenomena. It is conceptually obvious that, because of heat loss, there exists a minimum separation distance, d_q, between two cold, flat plates beyond which a flame cannot pass. It is also reasonable to expect that this distance should be of the order of the laminar flame thickness ℓ_D^o because this is the characteristic distance through which the flame loses its heat to the plate through conduction. Thus d_q can be identified as the quenching distance for the flame, given by

$$d_q \sim \ell_D^o. \tag{8.1.1}$$

Consequently, flames cannot pass through screens with wire spacing less than the quenching distance. This is the principle underlying the miner's safety lamp.

By the same token, if we want to ignite a combustible by a pocket of hot gas, achieved through, say, spark discharge or laser irradiation, then the minimum amount of energy to achieve ignition, $E_{I,\min}$, should be proportional to the amount of energy needed to heat a spherical volume of the combustible from the unburned temperature to the burned temperature, with the radius of the sphere being of the order of ℓ_D^o. Thus we have

$$E_{I,\min} \sim \rho_u (\ell_D^o)^3 c_p (T_{\text{ad}} - T_u). \tag{8.1.2}$$

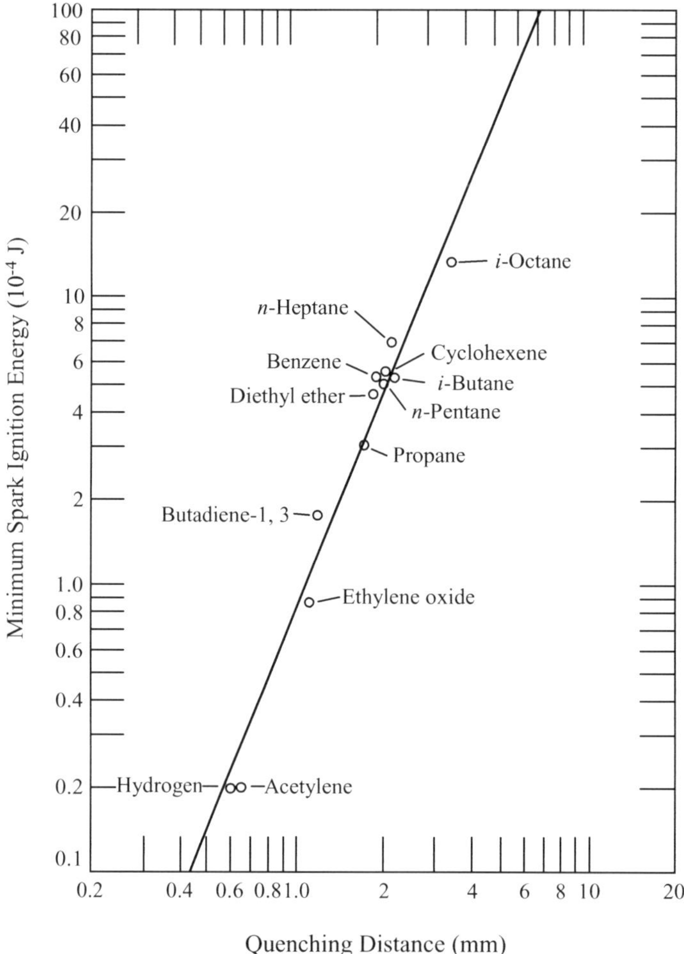

Figure 8.1.1. Correlation of the minimum spark ignition energy with quenching diameter (adapted from Calcote et al. 1952).

Two interesting observations can be made from these results. First, Eqs. (8.1.1) and (8.1.2) show that, for the same gas mixture, the quenching distance and minimum ignition energy are related through

$$E_{I,\min} \sim d_q^3. \tag{8.1.3}$$

Figure 8.1.1 (Calcote et al. 1952) plots the experimental values of $\ln(E_{I,\min})$ versus $\ln(d_q)$ for a variety of fuels, yielding a slope of about 2.5, which compares favorably with the value of 3 given by (8.1.3) and thereby demonstrates that this relation indeed holds approximately.

The second observation is concerned with the dependence of the minimum ignition energy on pressure. Since $\ell_D^o \sim (f^o)^{-1} \sim w^{-1/2} \sim p^{-n/2}$, and $\rho_u \sim p$, we have

$$E_{I,\min} \sim p^{1-3n/2}. \tag{8.1.4}$$

Thus for an overall reaction order of $n = 2$, we have $E_{I,\min} \sim p^{-2}$. This implies that $E_{I,\min}$ decreases quadratically with p, with the decrease being the net effect of the reduced flame thickness and, hence, reduced volume of the ignition kernel ($\sim p^{-3}$), and the increased density ($\sim p^{1}$).

Since n is always less than 2, as discussed in Section 7.7.3, the reduction of $E_{I,\min}$ with p is more moderate than quadratic. In particular, for sufficiently weak mixtures with smaller values of n, especially for $n < \frac{2}{3}$, $E_{I,\min}$ can actually increase with increasing pressure. In such cases the density effect dominates. Furthermore, since n is itself a function of p, either decreasing monotonically or reaching a minimum and then increasing again depending on whether the high-pressure branching cycle involving such radicals as HO_2 and H_2O_2 is activated, the possible responses of $E_{I,\min}$ with p can be rather rich. For example, $E_{I,\min}$ could first decrease and then increase with p, and thereby exhibit a minimum, if n itself continuously decreases with p. However, if high-pressure branching is activated before the minimum is activated, then $E_{I,\min}$ will just continuously decrease with p.

8.1.2. Adiabatic Thermal Explosion

We next study the classical thermal explosion problem. First consider the adiabatic situations. At time $t = 0$ a homogeneous body of gas of constant volume is instantaneously heated to a temperature T_o. If we assume that the rate of heat loss from all mechanisms is negligible compared to that of the chemical reaction, the mixture will ignite after a period of delay, which we aim to determine.

The governing equations are sufficiently simple such that we can directly write them down in dimensional form,

$$\rho_o c_v \frac{dT}{dt} = -Q_c \frac{dc_F}{dt} = BQ_c c_F \exp(-T_a/T), \tag{8.1.5}$$

where a one-step overall reaction is assumed and c_F is the molar concentration of the fuel with $c_{F,o}$ being its initial value. Furthermore, from continuity we have $d\rho/dt = 0$, or $\rho = \rho_o$.

With $\tilde{T} = (c_v \rho_o / Q_c c_{F,o})T = (c_v / q_c Y_{F,o})T$ and $\tilde{c}_F = c_F / c_{F,o}$, Eq. (8.1.5) becomes

$$\frac{d\tilde{T}}{dt} = -\frac{d\tilde{c}_F}{dt} = B\tilde{c}_F \exp(-\tilde{T}_a/\tilde{T}). \tag{8.1.6}$$

Using the coupling function approach, we have

$$\frac{d}{dt}(\tilde{T} + \tilde{c}_F) = 0, \tag{8.1.7}$$

which yields the solution relating $\tilde{c}_F$ and $\tilde{T}$,

$$\tilde{T} - \tilde{T}_o = 1 - \tilde{c}_F. \tag{8.1.8}$$

To investigate the behavior of the weakly reactive states leading to ignition, we assume that the temperature is perturbed from its initial value by a small amount,

$$\tilde{T} = \tilde{T}_o + \epsilon\theta(t) + O(\epsilon^2), \tag{8.1.9}$$

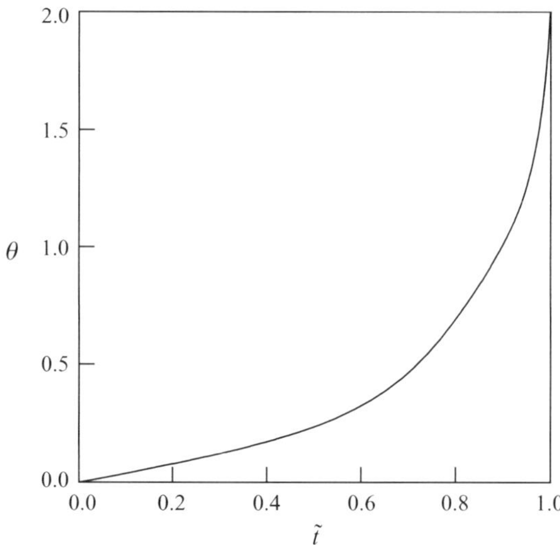

Figure 8.1.2. Increase in the mixture temperature with time for the adiabatic thermal explosion problem, indicating the nonlinear increase and the runaway, ignition behavior at $\tilde{t} = 1$.

where $\theta(0) = 0$ and $\epsilon = \tilde{T}_o^2/\tilde{T}_a \ll 1$ is identified following the same argument as that applied to the laminar flame analysis. Substituting Eq. (8.1.9) into Eq. (8.1.8) yields $\tilde{c}_F \approx 1 - \epsilon\theta$, which implies

$$\tilde{c}_F \approx 1 \tag{8.1.10}$$

to leading order. Equation (8.1.10) simply states that, if the initial concentration is not too low, then the reactant concentration is only slightly diminished during the induction period. Consequently the ignition process is primarily controlled by the temperature variation. If, however, the initial reactant concentration is exceedingly low ($Y_{F,o} \ll 1$) such that $\epsilon = O(1)$, then the approximation of Eq. (8.1.10) is not justified. Physically, reactant consumption as represented by the term $\epsilon\theta$ becomes significant during the induction period and will greatly affect the ignition lag (Kassoy & Liñán 1978).

Substituting Eqs. (8.1.9) and (8.1.10) into Eq. (8.1.6) yields

$$\frac{d\theta}{d\tilde{t}} = e^{\theta}, \tag{8.1.11}$$

where $\tilde{t} = t/\tau_c$ and $\tau_c = \epsilon/(Be^{-Ar})$ is a characteristic chemical time with $Ar = \tilde{T}_a/\tilde{T}_o$. Equation (8.1.11) is to be solved subject to the initial condition $\tilde{T}(0) = \tilde{T}_o$, or $\theta(0) = 0$. The solution is

$$\theta = -\ln(1 - \tilde{t}). \tag{8.1.12}$$

Figure 8.1.2 shows the function $\theta(\tilde{t})$. It is obvious that $\theta \to \infty$ as $\tilde{t} \to 1$. Thus if we identify this as the instant of ignition, then the ignition delay time is simply given by

$$\tilde{t}_I = 1, \tag{8.1.13}$$

which implies $t_I = \tau_c$. In dimensional form we have

$$t_I = \frac{c_v \left(T_o^2 / T_a\right)}{q_c Y_{F,o} B \exp(-T_a / T_o)}. \tag{8.1.14}$$

The meaning of t_I is now clear. That is, it decreases with the heat release q_c, the reactivity of the mixture, and the mass fraction of the fuel, and increases with the specific heat. The dependence on T_o and T_a is most sensitive through the Arrhenius factor in the reaction rate term. Interestingly it does not depend on the density of the gas. The reason is that, while an increase in density increases the reaction rate, there is also more gas to be heated. For the present first-order reaction the two effects cancel out. Otherwise, a density term will appear in t_I.

The final point to note is that since we have assumed $\theta(t)$ to be an $O(1)$ quantity in Eq. (8.1.9), the perturbation analysis breaks down as $\theta \to \infty$ at ignition. However, since θ assumes excessively large values only when $t \to t_I$, the approximation of t_I as the ignition time is quite adequate. A separate analysis (Kassoy 1975) can be conducted if one is interested to resolve the periods of $t \to t_I$ and the relaxation of the mixture to the equilibrium state. The reactant concentration is now very low and gradually decreases as $t \to t_I$.

8.1.3. Nonadiabatic Explosion and the Semenov Criterion

A combustible mixture undergoing a one-step exothermic reaction in the absence of heat loss will always attain the state of thermal explosion no matter how low is the initial temperature. When loss mechanisms are present, thermal explosion is not assured. To analyze these nonadiabatic situations, we envision that the thermal explosion process just discussed takes place in a vessel of volume V and surface area S. The temperature of its wall is maintained at a fixed value, say the initial gas temperature T_o for simplicity. As the gas temperature increases due to chemical heat release, the rates of reaction and conductive heat loss to the wall both increase; the former tends to promote the eventual thermal runaway while the latter tends to retard it. The final outcome then depends on the relative rates of these two competing processes.

To properly account for conductive heat loss, we must allow for the existence of temperature gradients in the vessel. This will significantly complicate the ensuing mathematical analysis. Fortunately, the essential physics can be captured by preserving spatial homogeneity through the use of a phenomenological heat transfer coefficient h such that the rate of heat loss is given by $-Sh(T - T_o)$. The energy conservation equation (8.1.5) can then be modified as

$$V \rho_o c_v \frac{dT}{dt} = V Q_c B c_F \exp(-T_a / T) - Sh(T - T_o). \tag{8.1.15}$$

By using the same perturbation variables as the previous problem, and neglecting reactant consumption during the induction period, the temperature perturbation θ

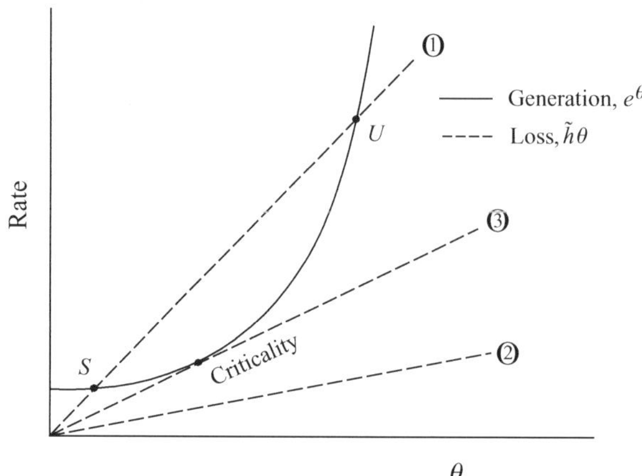

Figure 8.1.3. The Semenov criterion showing how the ignitability of a mixture depends on the competition between the nonlinear heat generation rate and linear heat loss rate.

is governed by

$$\frac{d\theta}{d\tilde{t}} = e^\theta - \tilde{h}\theta, \tag{8.1.16}$$

where $\tilde{h} = \tau_c/\tau_L$ and τ_L is a characteristic time for heat loss given by

$$\tau_L = \frac{c_v \rho_o}{(S/V)h}. \tag{8.1.17}$$

Thus heat loss becomes important with decreasing τ_L or increasing τ_c. Particularly, a vessel of larger surface-to-volume ratio, S/V, promotes heat loss and retards ignition. Consequently, for a functionally three-dimensional vessel, the effect of heat loss becomes progressively more serious as the vessel dimension is reduced. This is a major concern with the development of microengines.

Semenov (1958, 1959) discussed the behavior of Eq. (8.1.16) by comparing the relative magnitudes of the heat generation term e^θ and the heat loss term $\tilde{h}\theta$. Figure 8.1.3 shows a fixed heat generation rate and several heat-loss rates. It is seen that with a sufficiently large heat loss rate, given by the loss curve 1, the generation and loss curves intersect at two points, S and U, which respectively represent stable and unstable states. That is, by slightly increasing the mixture temperature from S, the rate of heat loss will exceed the rate of heat generation, causing a reduction in the mixture temperature and thereby bringing the system back to S. By the same token, a slight decrease in temperature will reduce the heat-loss rate and thereby increase the temperature back toward S. The mixture is therefore not conducive for ignition. A similar argument for U, however, shows a divergent behavior, indicating that ignition is favored.

For a system with a slower heat loss rate, as represented by state 2, the two curves do not intercept. Chemical heat generation is dominating and thermal runaway is therefore attainable.

Based on the above behavior, a state of criticality can be defined as one at which the heat loss curve, curve 3, is tangent to the heat generation curve. It satisfies the simultaneous requirements

$$e^{\theta_{\mathrm{ig}}} = \tilde{h}_{\mathrm{ig}}\theta_{\mathrm{ig}}, \tag{8.1.18}$$

$$e^{\theta_{\mathrm{ig}}} = \tilde{h}_{\mathrm{ig}}, \tag{8.1.19}$$

which yields $\theta_{\mathrm{ig}} = 1$ and $\tilde{h}_{\mathrm{ig}} = e$. The conditions of Eqs. (8.1.18) and (8.1.19) are known as the Semenov criterion.

8.1.4. The Well-Stirred Reactor Analogy

The Semenov criterion demonstrates the possibility of assessing the ignitability of a system by simply comparing the relative sensitivities of the heat generation and loss functions, without actually analyzing the transient process leading to ignition. We now extend this steady-state, criticality concept to include extinction by studying the operation limits of the well-stirred reactor, which is also known as the Longwell burner (Longwell & Weiss 1955), as first discussed in Section 2.5. The principle of this burner involves injecting a number of symmetrically located high-speed jets of a combustible mixture into a spherical or cylindrical vessel, within which combustion takes place. The high-intensity turbulence generated by these jets promotes almost instantaneous mixing and thereby creates a homogeneous combustion environment. The combustion "products" are then exhausted through a central port. The extent and intensity of the reaction within the reactor therefore depend on the reactivity of the mixture as well as the residence time of the flow.

An overall energy balance for a given volumetric flow rate $\dot{V}$ in steady operation can be written as

$$\dot{V}\rho_o c_p(T_f - T_o) = V Q_c B c_F e^{-T_a/T_f}, \tag{8.1.20}$$

where V is the reactor volume, T_o the temperature of the initial mixture, and T_f and c_F the reaction temperature and reactant concentration in the burner; c_p instead of c_v is used because the phenomenon involves a flow. Using Eq. (8.1.8) for c_F, Eq. (8.1.20) can be expressed in nondimensional form as

$$\tilde{T}_f - \tilde{T}_o = Da_C(\tilde{T}_{\mathrm{ad}} - \tilde{T}_f)e^{-\tilde{T}_a/\tilde{T}_f}, \tag{8.1.21}$$

where $\tilde{T} = (c_p/q_c Y_{F,o})T$, $\tilde{T}_{\mathrm{ad}} - \tilde{T}_f \equiv \tilde{c}_F$, with $\tilde{T}_{\mathrm{ad}} = 1 + \tilde{T}_o$ being the adiabatic flame temperature, and

$$Da_C = \frac{B}{\dot{V}/V} = \frac{\text{Characteristic flow time}}{\text{Characteristic collision time}} \tag{8.1.22}$$

is the collision Damköhler number. It is important to note that we have now allowed for the effect of reactant consumption by using Eq. (8.1.8) instead of assuming $\tilde{c}_F \approx 1$, which only holds for ignition situations. We therefore expect that Eq. (8.1.21) can describe both ignition and extinction.

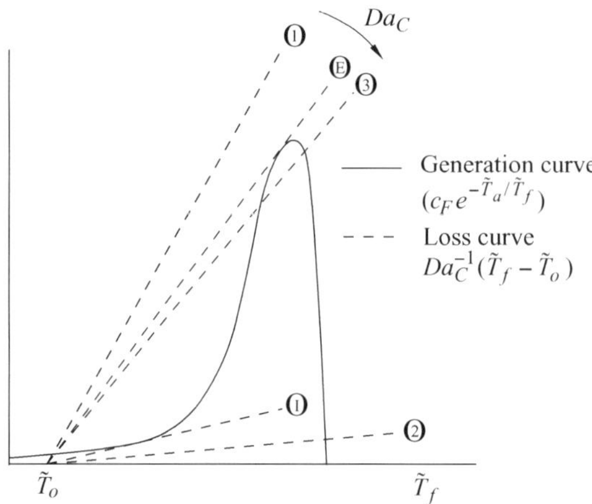

Figure 8.1.4. Extension of the Semenov ignition criterion to include the extinction criterion, through consideration of multiplicity and stability of solutions.

Equation (8.1.21) represents the balance between chemical heat release and convective heat transport. Reaction is promoted with increasing Da_C and retarded otherwise. The characteristics of its solution can be studied by locating the intersection between the transport term $Da_C^{-1}(\tilde{T}_f - \tilde{T}_o)$ and the reaction term $\tilde{c}_F \exp(-\tilde{T}_a/\tilde{T}_f)$, as shown in Figure 8.1.4, in which we hold the reaction term fixed and vary the convection intensity by changing Da_C^{-1}. The shape of the reaction rate curve represents the initial increase of the Arrhenius factor with $\tilde{T}_f$ and its eventual decay due to reactant depletion, as $c_F \to 0$.

Line 1 in Figure 8.1.4 indicates a situation of fast flow rate and hence small Da_C. The intersection point thus represents a state of low reaction temperature, weak chemical reaction, and very little reactant consumption. At the other extreme, line 2 indicates a situation of slow flow rate and, hence, large Da_C. The intersection point thus represents a state of high reaction temperature, intense chemical reaction, and almost complete reaction.

For an intermediate Da_C, say line 3, there are three intersection points. It can be easily argued, in the same manner as that for Figure 8.1.3, that the middle point is unstable while the other two points respectively represent stable states of weak and strong reactions. Consequently we can identify an ignition collision Damköhler number, $Da_{C,I}$, as given by the tangency line I, and an extinction collision Damköhler number, $Da_{C,E}$, as given by the tangency line E. The points of tangency then respectively represent the states of ignition and extinction: ignition is not possible for $Da_C < Da_{C,E}$ while extinction is not possible for $Da_C > Da_{C,I}$. We have therefore demonstrated that distinct ignition and extinction states can be identified through steady-state analysis. It is also important to note from Figure 8.1.4 that state 1 has to go through the extinction state E in order to attain the ignition state I. Similarly state 2 has to go through the ignition state I in order to attain the extinction state E. This indicates the hysteresis nature of the ignition–extinction processes.

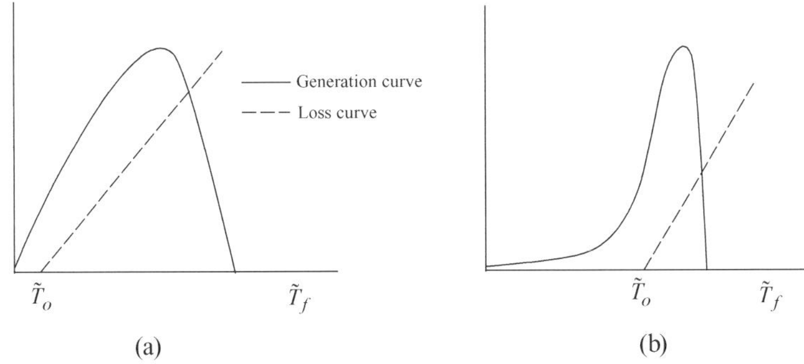

Figure 8.1.5. Situations in which multiplicity of solutions may not exist: (a) low activation energy reactions; (b) high mixture temperatures.

It is clear that the basic concept of ignition–extinction criticality is the same as the Semenov ignition criterion, except now reactant consumption is allowed, which enables us to identify the extinction state on the decreasing segment of the reaction rate curve.

There also exist situations in which the triple intersection does not occur for any loss rates. The first situation involves reactions of low activation energies. Since reaction is now readily activated at low temperatures, the reaction rate curve would be parabola-like, as shown in Figure 8.1.5a. This can be easily demonstrated by noting that the ascending portion of the curve, representing the Arrhenius factor, would concave upward for large activation energy, but concave downward for small activation energy because $\exp(-\tilde{T}_a/\tilde{T}_f) \approx 1 - \tilde{T}_a/\tilde{T}_f$ for $\tilde{T}_a \ll \tilde{T}_f$. For such a profile there is only one intersection point between the heat production and loss curves for all positive values of the heat loss rate.

The second situation involves reactions of high activation energy but also high initial temperature. Figure 8.1.5b shows that a single intersection point again results.

For both situations discussed above, distinct, abrupt ignition and extinction states or events are not expected. Chemical reaction becomes progressively more complete with increasing Da_C.

Extension of the above well stirred reactor analogy to assess the criticality of diffusive–convective systems is straightforward. Here the well-stirred reactor can be identified as the reaction zone, in an average sense. Transport of heat away from the thin reaction zone is now effected by diffusion, with a rate that is proportional to the difference between the flame temperature and a boundary temperature. The precise definition of the Damköhler number is not essential, as long as it represents a ratio of some characteristic transport time of the system to the characteristic reaction time within the flame.

8.1.5. The *S*-Curve Concept

The concepts of Figures 8.1.4 and 8.1.5 can be presented in a more illuminating manner by plotting either the reaction temperature or the burning rate versus a

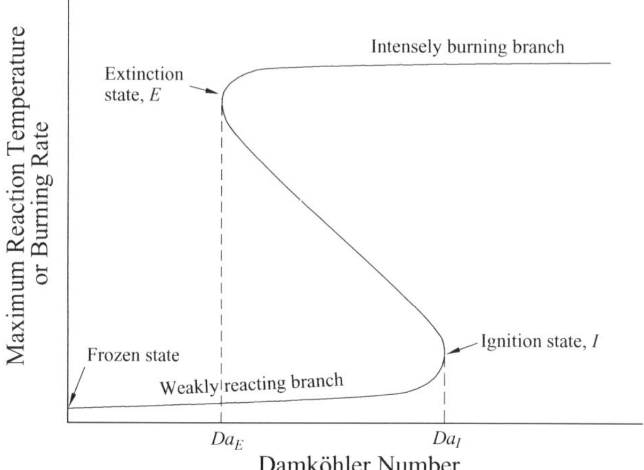

Figure 8.1.6. Representative folded S-curve with multiple solutions and distinct ignition and extinction states.

system Damköhler number Da. This results in either a regular, folded, S-shaped curve (Figure 8.1.6) or a stretched S-shaped curve (Figure 8.1.7), as first studied by Fendell (1965) and Liñán (1974). The folded S-curve, corresponding to the situations of Figure 8.1.4, consists of three branches. For $Da = 0$ we have the chemically frozen flow limit representing, for example, the pure droplet vaporization case of Chapter 6. By increasing Da along the lower branch that exists for low Damköhler numbers and thereby slow reaction rates, we cover all possible weakly reactive, nearly frozen, states the system can have. However, beyond Da_I no such solution exists. At Da_I the change in the reaction intensity, T_f, with respect of the change in the system's reactivity, Da, becomes infinite. The system consequently "jumps" to the upper branch, which exists for higher Damköhler numbers and faster reaction rates, and therefore represents

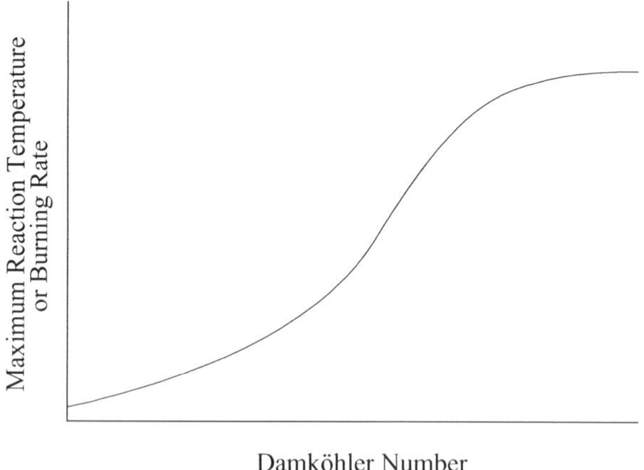

Figure 8.1.7. Representative stretched S-curve with unique solution and no distinct ignition and extinction states.

all the intensely burning states the system can have. It is then reasonable to identify point I as the state of ignition and Da_I the ignition Damköhler number.

If we move along the upper branch with increasing Da, we will attain the reaction-sheet limit as $Da \to \infty$. Examples are the reaction-sheet nonpremixed combustion cases of Chapter 6. If we next move along the upper branch with decreasing Da, then the system again jumps at point E back to the nearly frozen, lower branch. Therefore point E represents the extinction state and Da_E is the corresponding extinction Damköhler number.

Physically, the existence of turning points implies that there exist states for which the chemical reaction rate cannot balance the heat transport rate in steady state. Thus for the lower branch, beyond Da_I, the chemical heat is generated so fast in the reaction zone that it cannot be transported away in a steady manner. Similarly, for the upper branch, the effect of a finite Da and thereby a finite reaction rate is that not all of the available chemical energy can be released during the finite residence time in the reaction region. Thus extinction occurs when the heat loss from the flame becomes excessive to sustain steady burning.

It is also seen that while the solution is unique for $Da > Da_I$ and $Da < Da_E$, there exist three possible solutions for a given Da when $Da_E < Da < Da_I$. Within this regime, the solutions on the middle branch show decreasing reaction rate with increasing Da, and is therefore unstable as discussed previously and can be considered to be physically unrealistic. Furthermore, since $Da_E < Da_I$, the system exhibits hysteresis as mentioned earlier. The distinction between solutions on the lower and upper branches as the relevant solution can only be made by knowing the initial conditions.

The critical states of ignition and extinction can be readily determined for the well-stirred reactor model, with Da_C being the representative Damköhler number. These critical, turning-point states are defined by their vertical tangents on the S-curve, which imply

$$\left(\frac{d \ln Da_C}{d\tilde{T}_f}\right)_{\text{cr}} = 0, \tag{8.1.23}$$

where the subscript cr designates such a state. Applying Eq. (8.1.23) to (8.1.21), we get

$$\frac{1}{\tilde{T}_{f,\text{cr}} - \tilde{T}_o} + \frac{1}{\tilde{T}_{\text{ad}} - \tilde{T}_{f,\text{cr}}} = \frac{\tilde{T}_a}{\tilde{T}^2_{f,\text{cr}}}, \tag{8.1.24}$$

in which the first and second terms on the LHS respectively designate effects due to heat transfer and variation in the reactant concentration. Equation (8.1.24) readily shows the characteristics of the ignition and extinction states. For ignition, since $\tilde{T}_{f,\text{cr}} = \tilde{T}_{f,I}$ is expected to be close to $\tilde{T}_o$, the first term dominates over the second term, yielding

$$\tilde{T}_{f,I} \approx \tilde{T}_o + \frac{\tilde{T}_o^2}{\tilde{T}_a}. \tag{8.1.25}$$

Substituting $\tilde{T}_{f,I}$ into Eq. (8.1.21), expanding for $\tilde{T}_o/\tilde{T}_a \ll 1$, and noting that $\tilde{T}_{\text{ad}} = 1 + \tilde{T}_o$, we obtain the ignition Damköhler number,

$$Da_{C,I} = e^{-1}\left(\tilde{T}_o^2/\tilde{T}_a\right)e^{\tilde{T}_a/\tilde{T}_o}. \tag{8.1.26}$$

Similarly, for extinction we expect that $\tilde{T}_{f,\text{cr}} = \tilde{T}_{f,E}$ is close $\tilde{T}_{\text{ad}}$ such that the second term in Eq. (8.1.24) dominates, yielding

$$\tilde{T}_{f,E} = \tilde{T}_{\text{ad}} - \frac{\tilde{T}_{\text{ad}}^2}{\tilde{T}_a} \tag{8.1.27}$$

$$Da_{C,E} = e\left(\tilde{T}_a/\tilde{T}_{\text{ad}}^2\right)e^{\tilde{T}_a/\tilde{T}_{\text{ad}}}. \tag{8.1.28}$$

Referring to Eq. (8.1.21), the above results imply that ignition is directly affected by heat loss, while extinction is affected by the extent of reduced reaction rate and consequently the flame temperature.

A more general expression for these critical states can be obtained by solving for $\tilde{T}_{f,\text{cr}}$ from Eq. (8.1.24), which yields a quadratic equation in $\tilde{T}_{f,\text{cr}}$,

$$\tilde{T}_{f,\text{cr}} = \frac{(\tilde{T}_{\text{ad}} + \tilde{T}_o) \pm \{1 - 4(\tilde{T}_o\tilde{T}_{\text{ad}}/\tilde{T}_a)\}^{1/2}}{2(1 + 1/\tilde{T}_a)}. \tag{8.1.29}$$

Expanding Eq. (8.1.29) for small $1/\tilde{T}_a$, we readily obtain $\tilde{T}_{f,I}$ and $\tilde{T}_{f,E}$ from the lower and upper roots, respectively.

Equation (8.1.29) also shows that distinct ignition and extinction states, as manifested by folded S-curves, exist only when the term under the radical sign is positive. There is no solution when it is negative, implying the existence of stretched S-curves. Thus the criterion for folded S-curve to exist is $1 - 4(\tilde{T}_o\tilde{T}_{\text{ad}}/\tilde{T}_a) > 0$, or

$$\tilde{T}_a > 4\tilde{T}_o(1 + \tilde{T}_o). \tag{8.1.30}$$

Figure 8.1.8 plots Eq. (8.1.30), showing that stretched S-curves are favored for large $\tilde{T}_o$ and small $\tilde{T}_a$, as shown in Figure 8.1.5.

Although numerical efforts are usually needed to generate the entire S-curve for more complex situations, separate analysis can be performed for the lower and upper branches including the respective turning point regimes. Thus by knowing the ignition or extinction Damköhler numbers, Da_I or Da_E, the ignitability or extinguishability of a system can be assessed by comparing its Da with either Da_I or Da_E. The S-curve is a useful concept in the study of ignition and extinction phenomena, as will be demonstrated subsequently. The penalties of using this criticality test are: (a) the loss of influence of the initial conditions; (b) the loss of information regarding the ignition/extinction time lag because, for example, a mixture predicted to be ignitable may take so long to ignite that for all practical purposes they can be considered to be unignitable; (c) the existence of multiple solutions whose uniqueness and physical existence may have to be resolved through transient analysis with given initial conditions; and (d) there are processes that do not have steady states even for the chemically inert flow; for them a transient analysis is essential. Furthermore,

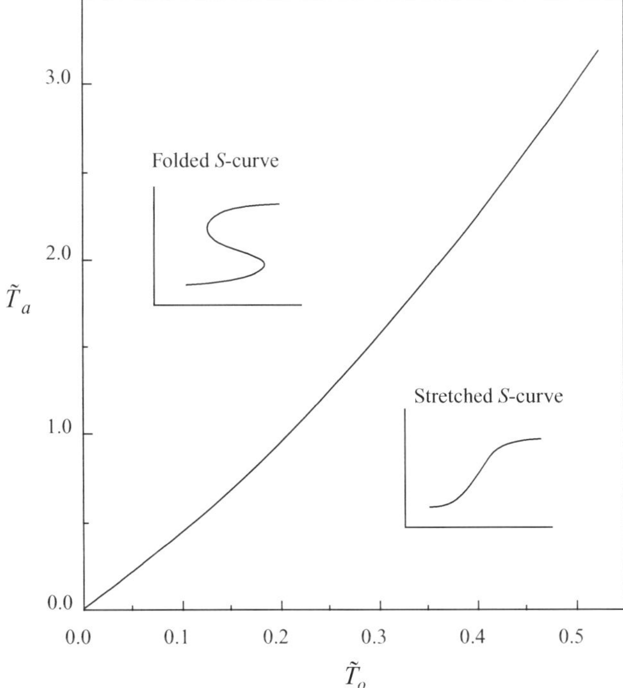

Figure 8.1.8. Regimes of folded and stretched S-curves according to whether $\tilde{T}_a$ is greater or less than $4\tilde{T}_o(1 + \tilde{T}_o)$.

as the flame becomes weakened and is sensitive to disturbances near the extinction point, it can lose temporal or spatial stability and begin to oscillate or wrinkle before it reaches the S-curve extinction state. In such cases the extinguishability of the flame should be assessed by allowing for pulsations and wrinklings.

The final point to note is that the folded S-curve demonstrated herein is just one manifestation of the existence of multiple solutions, for the simplified one-step reaction. For systems with more complex reaction schemes, response curves with multiple "wiggles" have been observed, implying the possible existence of multiple steady-state flame solutions with different burning intensities. It has also been found that for systems that are strongly influenced by chain mechanisms, it may be more beneficial to use the (maximum) concentration of some crucial radicals as the flame response in generating the S-curve. We shall encounter such situations in Section 8.3.

8.2. IGNITION BY A HOT SURFACE

In this section, we shall be more specific in the description of the transport process in inducing ignition. We shall use the ignition of a combustible by a hot surface as an example to illustrate the various physical concepts and mathematical techniques involved in ignition analysis.

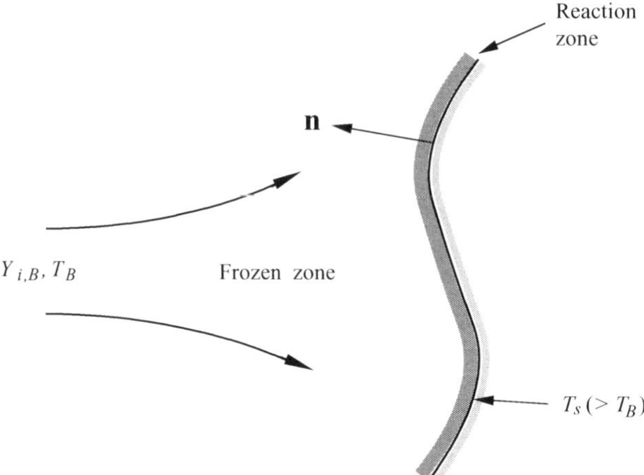

Figure 8.2.1. Schematic for the generalized reaction zone analysis for the hot surface ignition of a combustible.

8.2.1. Asymptotic Analysis of the Reaction Zone

In many practical situations, ignition is caused by a hot surface situated in a cold premixture. Thus from large activation energy considerations we expect that the chemical reaction responsible for ignition is confined to a thin layer next to the hot surface, with the rest of the flow field being chemically frozen. From what we have learned in Chapter 7, problems of this nature can be systematically tackled through analysis involving matching of the inner and outer solutions.

We shall also use this class of problem to introduce an important concept regarding the structure of the reaction zone. As we have seen in the analysis of the standard premixed flame, a flame structure frequently consists of an inner, reaction zone and outer, nonreactive zones away from it. Since the outer zones are convective–diffusive in nature, their structures are directly influenced by the particulars of the flow system, such as flow nonuniformity, flame curvature, and heat loss. However, they are not affected by the nature of the chemical reaction. On the other hand, the thin reaction zone is reactive–diffusive in nature, with convection being of higher order. Consequently, its structure may not be directly affected by the particulars of the system. Indirectly, of course, processes in the inner and outer zones interact through the boundary conditions they provide for each other when their solutions are matched.

In view of the above properties, we then recognize the possibility of performing a generalized analysis of the reaction zone structure, with the solution expressed in terms of some general boundary conditions that are to be supplied by the outer solutions for individual flow situations. We shall now demonstrate this approach for the hot surface ignition problem (Semenov 1958, 1959; Law 1978a).

Figure 8.2.1 shows a hot body of arbitrary shape and constant temperature T_s situated in an environment of temperature T_B and reactant mass fraction $Y_{i,B}$. The environment is sufficiently cold such that reaction is frozen there during the characteristic time of interest. As such, reaction is initiated only in a thin layer next to the hot surface.

To simplify matter, we shall assume that the surface is noncatalytic and nonpermeable such that in the frozen limit the reactant concentration is uniform, at $Y_{i,B}$. If we further assume that the extent of reactant consumption is small in effecting ignition, then the reactant concentration can be simply approximated by $Y_{i,B}$. This allows us to dispense with analyzing the species conservation equation.

It must be cautioned, however, that such an assumption cannot always be made. For example, if the surface is permeable such that the reactant concentration is maintained to be zero there, then the reactant concentration next to the surface is spatially varying and is also very small even in the frozen limit. As such, the extent of reactant consumption can be of the same order as the amount originally present and therefore should be properly accounted.

We again adopt the simplified diffusion-controlled system of Eq. (5.6.6) for illustration, with energy conservation given by

$$\frac{\partial}{\partial \tilde{t}}(\tilde{\rho}\tilde{T}) + \tilde{\nabla}\cdot(\tilde{\rho}\tilde{\mathbf{v}}\tilde{T}) - \tilde{\nabla}^2\tilde{T} = Da_C\tilde{Y}_O\tilde{Y}_F e^{-\tilde{T}_a/\tilde{T}}, \tag{8.2.1}$$

where $Da_C = (\ell^2 B_C)/(\lambda/c_p)$ and ℓ is a characteristic length of the system. Equation (8.2.1) is to be solved subject to an initial condition and known boundary temperatures of T_B and T_s.

8.2.1.1. Inner Expansion: If we assume that the reaction zone thickness is much smaller than the radius of curvature of the surface, and the characteristic times of any transient processes are much longer than those of diffusion and reaction in the reaction zone, then the reaction zone can be treated as quasi-steady and quasi-planar. The inner, diffusive–reactive, solution $\tilde{T}_{\text{in}}$ is then governed by

$$\frac{d^2\tilde{T}_{\text{in}}}{d\tilde{n}^2} = -Da_C\tilde{Y}_{O,B}e^{-\tilde{T}_a/\tilde{T}_{\text{in}}}, \tag{8.2.2}$$

subject to the boundary conditions at the surface,

$$\tilde{n} = 0: \quad \tilde{T}_{\text{in}} = \tilde{T}_s, \tag{8.2.3}$$

where $\tilde{n}$ is the coordinate normal to the surface. In writing the reaction rate term in Eq. (8.2.2), we note that $\tilde{Y}_O \approx \tilde{Y}_{O,B}$ and $\tilde{Y}_F \approx \tilde{Y}_{F,B} = 1$ because of negligible reactant consumption prior to ignition.

Using the inner expansion

$$\tilde{T}_{\text{in}} = \tilde{T}_s + \epsilon\theta(\chi) + O(\epsilon^2), \tag{8.2.4}$$

where $\epsilon = \tilde{T}_s^2/\tilde{T}_a = Ze^{-1}$ and $\chi = \beta\tilde{n}/\epsilon$, Eqs. (8.2.2) and (8.2.3) become

$$\frac{d^2\theta}{d\chi^2} = -\frac{\Delta}{2}e^{\theta} \tag{8.2.5}$$

$$\theta(0) = 0, \tag{8.2.6}$$

where

$$\Delta = \frac{2\epsilon Da_C \tilde{Y}_{O,B} e^{-\tilde{T}_a/\tilde{T}_s}}{\beta^2}. \tag{8.2.7}$$

A parameter $\beta = -(\partial \tilde{T}_0/\partial \tilde{n})_0$ is introduced into the definition of χ so as to simplify subsequent expressions, where $\tilde{T}_0$ is the leading order outer solution. It is the negative of the temperature gradient at the hot surface in the chemically frozen limit, and is a characteristic heat loss rate to the cold ambience from the reaction zone, which is next to the hot surface. Since $(\partial \tilde{T}_0/\partial \tilde{n})_0 < 0$, β is positive.

8.2.1.2. Outer Expansion: We express the outer expansion as

$$\tilde{T}_{\text{out}}(\tilde{\mathbf{x}}, \tilde{t}) = \tilde{T}_0(\tilde{\mathbf{x}}, \tilde{t}) + \epsilon \tilde{T}_1(\tilde{\mathbf{x}}, \tilde{t}) + O(\epsilon^2). \tag{8.2.8}$$

Since the outer solution is to match with the inner solution around $\tilde{\mathbf{x}} = 0$, we Taylor expand $\tilde{T}_{\text{out}}(\tilde{\mathbf{x}})$ around $\tilde{\mathbf{x}} = 0$ in the direction of $\tilde{n}$, and then express $\tilde{n}$ in terms of χ, as

$$\begin{aligned}\tilde{T}_0(\tilde{n}, \tilde{t}) &= \tilde{T}_0(0, \tilde{t}) + \tilde{n}\left(\frac{\partial \tilde{T}_0}{\partial \tilde{n}}\right)_0 + O(\tilde{n}^2)\\ &= \tilde{T}_s - \epsilon\chi + O(\epsilon^2).\end{aligned}$$

The outer expansion thus becomes

$$\tilde{T}_{\text{out}}(\chi; \tilde{t}) = \tilde{T}_s + \epsilon[\tilde{T}_1(0; \tilde{t}) - \chi] + O(\epsilon^2), \tag{8.2.9}$$

where $\tilde{t}$ is now treated as a parameter.

8.2.1.3. Matching: Matching the inner and outer expansions, we have

$$(\theta + \chi)_{\chi\to\infty} = (\tilde{T}_1)_{\tilde{n}\to 0} \tag{8.2.10}$$

$$\left(\frac{d\theta}{d\chi}\right)_{\chi\to\infty} = -1. \tag{8.2.11}$$

8.2.1.4. Inner Solution: Integrating Eq. (8.2.5) once, using the boundary conditions at $\chi \to \infty$, and noting that $e^{\theta} \sim e^{-\chi} \to 0$ in this limit, we obtain

$$\left(\frac{d\theta}{d\chi}\right) = \mp(1 - \Delta e^{\theta})^{1/2}. \tag{8.2.12}$$

Integrating Eq. (8.2.12), with the boundary condition $\theta(0) = 0$, the final solution is given by

$$\theta(\chi) = \ln\left\{\frac{1}{\Delta}\left[1 - \left(\frac{\Delta' e^{\pm\chi} - 1}{\Delta' e^{\pm\chi} + 1}\right)^2\right]\right\}, \tag{8.2.13}$$

where

$$\Delta' = \frac{1 + (1 - \Delta)^{1/2}}{1 - (1 - \Delta)^{1/2}}.$$

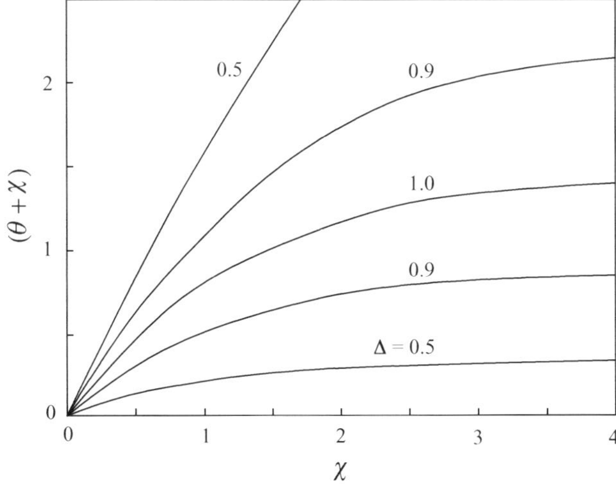

Figure 8.2.2. Profiles of the temperature perturbation as function of the Damköhler number for the hot surface ignition of a combustible.

Evaluating Eq. (8.2.13) at $\chi \to \infty$, we obtain

$$(\theta + \chi)_{\chi\to\infty} = \ln\left\{\frac{4}{\Delta}\left[\frac{1 \mp (1-\Delta)^{1/2}}{1 \pm (1-\Delta)^{1/2}}\right]\right\}, \tag{8.2.14}$$

which gives $(\tilde{T}_1)_{\tilde{n}\to 0}$ according to the matching relation (8.2.10).

8.2.1.5. Ignition Characteristics and Criterion: The ignition characteristics can be adequately assessed from the inner solution. Figure 8.2.2 shows $(\theta + \chi)$ as a function of χ, for various Δ. It is seen that this function increases from zero at $\chi = 0$ to a constant value $(\theta + \chi)_\infty = \tilde{T}_1(0)$ as $\chi \to \infty$, indicating the attainment of the maximum temperature increase as the boundary with the outer region is approached. Furthermore, two solutions exist for each $\Delta < 1$, as can be seen clearly in Eqs. (8.2.13) or (8.2.14). The two solutions merge at $\Delta = 1$. Thus by plotting $(\theta + \chi)_\infty$ versus Δ, we obtain the lower and part of the middle branch of the characteristic S-curve, as shown in Figure 8.2.3. The ignition turning point occurs at an ignition Damköhler number of $\Delta_I = 1$, with a corresponding $(\theta + \chi)_{\infty,I} = \ln 4$. Thus ignition is expected to be possible for

$$\Delta \geq \Delta_I = 1, \tag{8.2.15}$$

or

$$\frac{2\epsilon\, Da\, \tilde{Y}_{O,B}}{\beta^2} \geq 1, \tag{8.2.16}$$

where $Da = Da_C e^{-\tilde{T}_a/\tilde{T}_s}$. Explicit evaluation of Δ_I requires solution of the outer flow in order to specify $\beta = -(\partial\tilde{T}_0/\partial\tilde{n})_0$, and therefore depends on the specific nature of the problem.

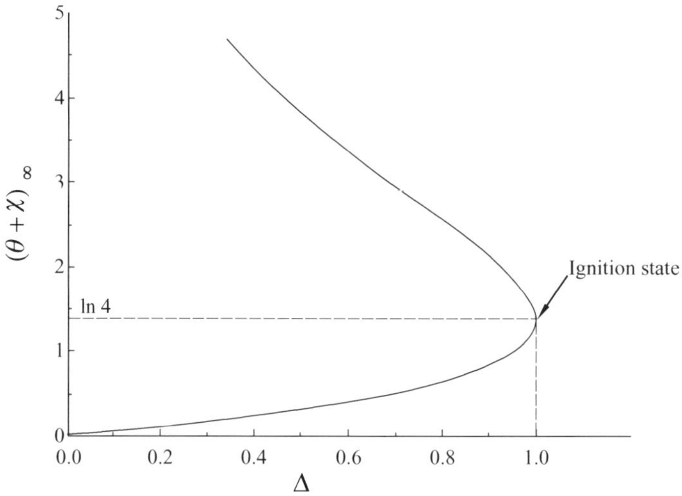

Figure 8.2.3. Plot of maximum temperature increase parameter, $(\theta + \chi)_\infty$, versus Δ, for the hot surface ignition of a combustible, showing the lower and middle branches of the S-curve with its ignition turning point occurring at $\Delta = 1$.

8.2.1.6. Heat Transfer Characteristics: Heat transfer at the wall is given by

$$\left(\frac{\partial \tilde{T}_{\text{in}}}{\partial \tilde{n}}\right)_0 = -\beta \left(\frac{d\theta}{d\chi}\right)_0. \tag{8.2.17}$$

Evaluating $(d\theta/d\chi)$ of Eq. (8.2.12) at $\theta(0) = 0$, Eq. (8.2.17) becomes

$$\left(\frac{\partial \tilde{T}_{\text{in}}}{\partial \tilde{n}}\right)_0 = -\beta(1-\Delta)^{1/2}, \tag{8.2.18}$$

where we have taken the negative root that represents the physically realistic lower branch of the S-curve. Equation (8.2.18) shows that before ignition is achievable, there is always heat transfer from the hot wall to the cold gas. However, at the first instant of ignitability, when $\Delta = 1$, the heat transfer ceases. Physically, there is now so much heat generation in the gas that it does not need to receive heat from the wall in order to sustain the reaction, implying ignition becomes possible. Indeed, this adiabaticity criterion has been extensively used in the past to assess ignitability. We have thus shown here that the criticality, turning-point criterion also implies the adiabaticity criterion.

8.2.2. Ignition of a Confined Mixture by a Flat Plate

To demonstrate the utility of the generalized analysis of the reaction zone, let us consider the simplest system possible, namely the ignitability of a stagnant combustible confined by a hot plate situated at $x = 0$ and a cold plate situated at $x = \ell$ (Figure 8.2.4). Thus the hot plate is the ignition source while the cold plate is the heat sink. The hot plate is nonpermeable, in conformity with the concentration boundary condition adopted in the general analysis, while the cold plate is permeable such

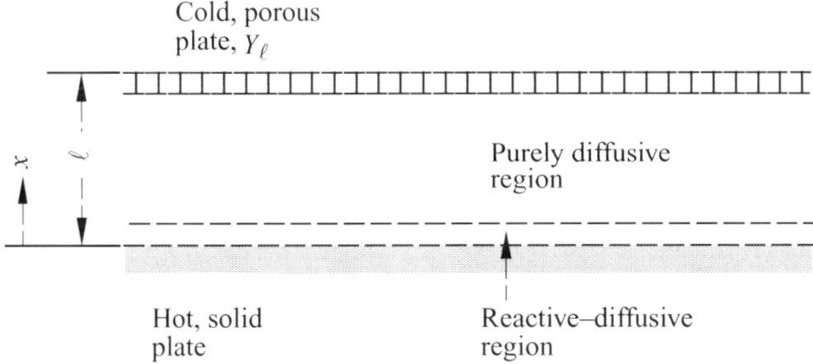

Figure 8.2.4. Ignition of a combustible by a hot surface, with simultaneous heat loss to an oppositely located cold surface.

that the reactant concentrations at its surface can be maintained at the constant value $Y_{i,\ell}$.

The outer solution is governed by $d^2\tilde{T}_{\text{out}}/d\tilde{x}^2 = 0$ subject to $\tilde{T}(1) = \tilde{T}_\ell$, where $\tilde{x} = x/\ell$. Using the expansion Eq. (8.2.8) and matching it with the inner solution Eq. (8.2.4), the leading order solution is described by $d^2\tilde{T}_0/d\tilde{x}^2 = 0$, $\tilde{T}_0(0) = \tilde{T}_s$, $\tilde{T}_0(1) = \tilde{T}_\ell$, and is given by

$$\tilde{T}_0(\tilde{x}) = \tilde{T}_s + (\tilde{T}_\ell - \tilde{T}_s)\tilde{x}, \tag{8.2.19}$$

which yields

$$\beta = -\left(\frac{d\tilde{T}_0}{d\tilde{x}}\right)_0 = \tilde{T}_s - \tilde{T}_\ell. \tag{8.2.20}$$

Knowing β, the ignition criterion (8.2.16) is completely specified as

$$\frac{2\epsilon\, Da\, \tilde{Y}_{O,\ell}}{(\tilde{T}_s - \tilde{T}_\ell)^2} \geq 1. \tag{8.2.21}$$

As an auxiliary result, we note that the $O(\epsilon)$ outer solution, $\tilde{T}_1(\tilde{x})$, can also be determined from $d^2\tilde{T}_1/d\tilde{x}^2 = 0$ and the boundary conditions $\tilde{T}_1(1) = 0$ and $\tilde{T}_1(0) = (\theta + \chi)_\infty$ given by Eq. (8.2.10) determined through matching. Thus,

$$\tilde{T}_1(\tilde{x}) = (\theta + \chi)_\infty(1 - \tilde{x}), \tag{8.2.22}$$

with $(\theta + \chi)_\infty$ given by Eq. (8.2.14).

Knowing both $\tilde{T}_{\text{in}}(\chi)$ and $\tilde{T}_{\text{out}}(\tilde{x})$, the composite solution, which is valid in both the inner and outer regions, is given by

$$\tilde{T} = \tilde{T}_{\text{in}} + \tilde{T}_{\text{out}} - \text{common part}. \tag{8.2.23}$$

The common part is determined by taking either the inner limit of the outer solution or the outer limit of the inner solution. Thus if we use the former, then

$$\lim_{\tilde{x}\to 0} \tilde{T}_{\text{out}} = \tilde{T}_s + \epsilon\tilde{T}_1(0) = \tilde{T}_s + \epsilon(\theta + \chi)_\infty. \tag{8.2.24}$$

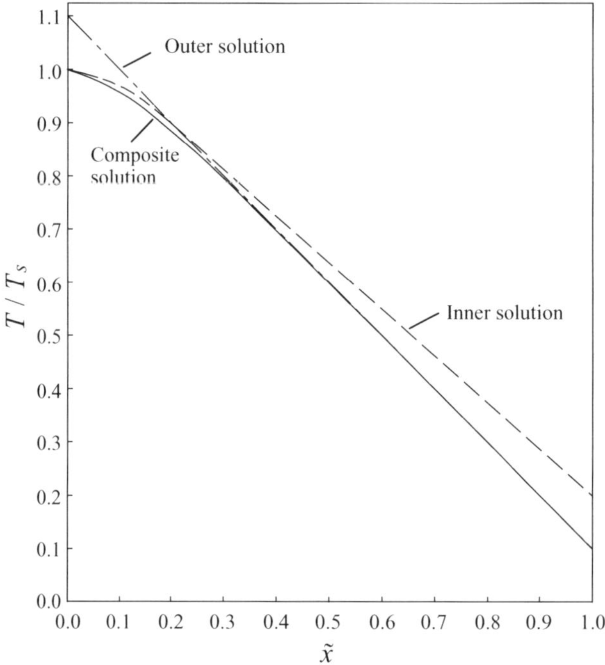

Figure 8.2.5. Inner, outer, and composite solutions for the ignition problem of Figure 8.2.4 ($\epsilon = 0.1$, $\alpha = 0.1$, $\tilde{T}_s = 1$, $\Delta = 0.95$).

Substituting Eqs. (8.2.4) and (8.2.8) for $\tilde{T}_{\text{in}}$ and $\tilde{T}_{\text{out}}$ respectively, we have

$$\tilde{T}(\tilde{x}) = \tilde{T}_s - \beta\tilde{x} + \epsilon\left\{\ln\left\{\frac{1}{\Delta}\left[1 - \left(\frac{\Delta' e^{\pm\chi} - 1}{\Delta' e^{\pm\chi} + 1}\right)^2\right]\right\} - (\theta + \chi)_\infty \tilde{x}\right\}, \qquad (8.2.25)$$

which can be rearranged to give

$$\frac{\tilde{T}}{\tilde{T}_s} = 1 - (1 - \alpha)\tilde{x} + \frac{\epsilon}{\tilde{T}_s}\left\{\ln\left\{\frac{1}{\Delta}\left[1 - \left(\frac{\Delta' e^{\pm(1-\alpha)\tilde{x}/\epsilon} - 1}{\Delta' e^{\pm(1-\alpha)\tilde{x}/\epsilon} + 1}\right)^2\right]\right\} - (\theta + \chi)_\infty \tilde{x}\right\}. \qquad (8.2.26)$$

Figure 8.2.5 shows typical temperature profiles for the inner, outer, and composite solutions, for the meaningful upper root. Here we see that the composite solution agrees well with the inner solution for $\tilde{x}$ near 0, where the outer solution breaks down, and with the outer solution for $\tilde{x}$ near 1, where the inner solution breaks down, as should be the case.

8.2.3. Ignition of an Unconfined Mixture by a Flat Plate

The ignition criterion for the confined mixture becomes meaningless for an unconfined mixture, with $\ell \to \infty$ and thereby $Da_C \to \infty$. In this situation the problem is intrinsically unsteady in that, even in the absence of any chemical reaction, the gas temperature distribution will continuously evolve with time. Thus the relevant question to ask here is how long it will take the mixture to ignite rather than whether the mixture is ignitable.

If we assume that the thin reaction zone is quasi-steady, then the unsteadiness it experiences is imposed on it by that of the outer flow, through its leading order solution governed by the classical heat conduction equation,

$$\frac{\partial \tilde{T}_0}{\partial \tilde{t}} - \frac{\partial^2 \tilde{T}_0}{\partial \tilde{x}^2} = 0, \tag{8.2.27}$$

where $\tilde{x} = x/\ell_o$ and $\tilde{t} = [(\lambda/c_p\rho)/\ell_o^2]t$. Since the nonreactive system has no characteristic length, ℓ_o has only an interim role here for nondimensionalization, and will be cancelled out in the final solution. In writing Eq. (8.2.27) we have also assumed $\rho =$ constant. This assumption is not essential and can be removed by transforming $\tilde{x}$ to a density-weighted coordinate, $\int \rho d\tilde{x}$.

Equation (8.2.27) is to be solved subject to the initial and boundary conditions

$$\tilde{T}_0(\tilde{x}, \tilde{t} = 0) = \tilde{T}_\infty, \quad \tilde{T}_0(\tilde{x} = 0, \tilde{t}) = \tilde{T}_s, \quad \tilde{T}_0(x = \infty, \tilde{t}) = \tilde{T}_\infty. \tag{8.2.28}$$

Since the problem has no characteristic dimension, we seek a similarity solution with the similarity variable $\eta = \tilde{x}/\sqrt{\pi \tilde{t}}$. Noting that

$$\frac{\partial}{\partial \tilde{t}} \equiv -\frac{\eta}{2\tilde{t}}\frac{d}{d\eta}, \qquad \frac{\partial^2}{\partial \tilde{x}^2} \equiv \frac{1}{\pi \tilde{t}}\frac{d^2}{d\eta^2},$$

with $\tilde{t} = 0$ and $\tilde{x} = \infty$ correspond to $\eta = \infty$, and $\tilde{x} = 0$ corresponds to $\eta = 0$, Eq. (8.2.27) becomes

$$\frac{d^2 \tilde{T}_0}{d\eta^2} + \frac{\pi}{2}\eta \frac{d\tilde{T}_0}{d\eta} = 0, \tag{8.2.29}$$

with the boundary conditions $\tilde{T}_0(0) = \tilde{T}_s$ and $\tilde{T}_0(\infty) = \tilde{T}_\infty$. The solution is

$$\tilde{T}_0(\eta) = \tilde{T}_s - (\tilde{T}_s - \tilde{T}_\infty)\int_0^\eta e^{-(\pi/4)\eta^2} d\eta. \tag{8.2.30}$$

The heat transfer parameter can now be readily evaluated as

$$\beta = -\left(\frac{\partial \tilde{T}_0}{\partial \tilde{x}}\right)_0 = -\frac{\partial \eta}{\partial \tilde{x}}\left(\frac{\partial \tilde{T}}{\partial \eta}\right)_0 = \frac{\tilde{T}_s - \tilde{T}_\infty}{\sqrt{\pi \tilde{t}}}.$$

Substituting β into the definition of Δ given by Eq. (8.2.7), we see that, by redefining $Da_C = (\ell^2 B_C)/(\lambda/c_p)$ as

$$Da_{C,t} = \pi(B_C/\rho)t, \tag{8.2.31}$$

we again obtain the ignition criterion (8.2.21) identified for the confined mixture. The present collision Damköhler number, however, is a transient one, being linearly dependent on time, t. Thus given sufficiently long time, ignition is always possible.

A more direct approach to identify $Da_{C,t}$ is to start with the original governing equation

$$\frac{\partial \tilde{T}}{\partial \tilde{t}} - \frac{\partial^2 \tilde{T}}{\partial \tilde{x}^2} = Da_C \tilde{Y}_{O,\infty} e^{-\tilde{T}_a/\tilde{T}}. \tag{8.2.32}$$

By applying the similarity transformation, we have

$$\frac{d^2 \tilde{T}}{d\eta^2} + \frac{\pi}{2}\eta \frac{d\tilde{T}}{d\eta} = -Da_{C,t} \tilde{Y}_{O,\infty} e^{-\tilde{T}_a/\tilde{T}}. \tag{8.2.33}$$

The problem is now completely recast as a one-dimensional steady flow, with a collision Damköhler number $Da_{C,t}$, which, together with $\beta = -(\partial \tilde{T}_0/\partial \eta)_{\eta=0} = \tilde{T}_s - \tilde{T}_\infty$, defines the ignition criterion.

It is perhaps somewhat remarkable that the ignition of these two apparently very different problems, involving confined and unconfined combustibles, can share the same solution. Physically, as discussed earlier, the basic ignition process, which involves thermal runaway in the thin layer next to the hot surface, is the same for both cases. Thus by properly identifying the relevant Damköhler number for each case, the state of thermal runaway can be rendered the same. For example, for the confined case Da_C compares the characteristic flow time, $\ell^2/(\lambda/c_p\rho)$, with the collision time, ρ/B_C, whereas for the unconfined case, which lacks any characteristic flow time, Da_C simply measures the real time t in units of the collision time.

The issue of similarity in chemically reacting flows will be discussed in detail in Chapter 12 on boundary-layer flows.

8.2.4. Nusselt Number Correlation

The evaluation of the ignition criterion requires a knowledge of $(\partial \tilde{T}_0/\partial \tilde{n})_0$. Since these are obtained from the leading order, chemically frozen, outer solutions, they can be determined independent of the chemical aspects of the problem, as we have seen for the confined and unconfined combustible cases. Indeed, their determination is a major focus in the field of nonreactive heat and mass transfer. In such studies the nature and extent of the transfer is frequently expressed in terms of a Nusselt number, Nu, defined as

$$Nu = \frac{\ell_o(\partial \tilde{T}_0/\partial \tilde{n})_0}{\tilde{T}_B - \tilde{T}_s}. \tag{8.2.34}$$

The ignition criterion (8.2.16) is then generalized to

$$\frac{2\epsilon \ell_o^2 Da \tilde{Y}_{O,B}}{Nu^2(\tilde{T}_s - \tilde{T}_B)^2} \geq 1, \tag{8.2.35}$$

with a properly defined Da_C for use in Da.

8.2.5. Convection-Free Formulation

The problem of the confined mixture is a purely one-dimensional diffusive–reactive system such that the transport operator is simply given by $d^2(\ )/d\tilde{x}^2$. For the unconfined system an additional convection-like term, $(\pi/2)\eta(d\tilde{T}/d\eta)$, is introduced upon transformation to the similarity coordinate, as given by Eq. (8.2.33). It is also clear that the presence of convection in other one-dimensional, or self-similar two-dimensional flows will introduces a first-order, convection term into the transport operator.

It is, however, well known that in a second-order linear ordinary differential equation, the first-order differential can be transformed away. For example, by defining

$$\xi = \int_0^\eta e^{-(\pi/4)\eta^2} d\eta \tag{8.2.36}$$

such that $\xi = [0, 1]$ for $\eta = [0, \infty]$, Eq. (8.2.33) becomes

$$\frac{d^2\tilde{T}}{d\xi^2} = e^{(\pi/2)\eta^2} Da_C \tilde{Y}_{O,B} e^{-\tilde{T}_a/\tilde{T}}, \tag{8.2.37}$$

subject to the boundary conditions $\tilde{T}(\xi = 0) = \tilde{T}_s$ and $\tilde{T}(\xi = 1) = \tilde{T}_\infty$. In addition to eliminating the first-order convection term, the transformation also places the domain of interest to $\xi = [0, 1]$, with the reaction region next to $\xi = 0$. The cost of achieving this simplicity is the introduction into the reaction term an additional factor, which is a function of the original independent variable and is $f(\eta) = e^{(\pi/2)\eta^2}$ in this case. However, since this factor is to be evaluated where the reaction is concentrated, it becomes a constant in the reaction term. In this problem, the reaction zone is next to $\eta = 0$, and we therefore have $f(\eta) \approx 1$.

Comparing the above result with that of the purely diffusive system in Section 8.2.2, we see that through this transformation we have rendered the two systems canonically identical. Thus all results derived for the purely diffusion case can be readily used for the present situation.

It is therefore clear that problems of this nature can all be rendered diffusive–reactive in nature. Furthermore, the reaction-independent quantities such as the coupling function and the frozen solution will vary linearly in ξ-space. This implies that the existence and severity of chemical reactions can be qualitatively assessed by examining the deviation from linearity of a given quantity when plotted in ξ-space. Therefore this transformation frequently simplifies mathematical manipulation, facilitates physical interpretation, and brings about a degree of similarity between problems with different geometries and nature of convection. The approach is most convenient for $Le_i = 1$ mixtures. When $Le_i \neq 1$, the problem can sometimes still be brought to a simplified form with some additional manipulation.

8.3. IGNITION OF HYDROGEN BY HEATED AIR

Our studies in this chapter have so far been based on one-step Arrhenius kinetics, which introduces the needed nonlinearity for ignition and extinction through its dependence on temperature. Consequently, at ignition (or extinction) a small temperature perturbation to the frozen (equilibrium) branch of the S-curve provided

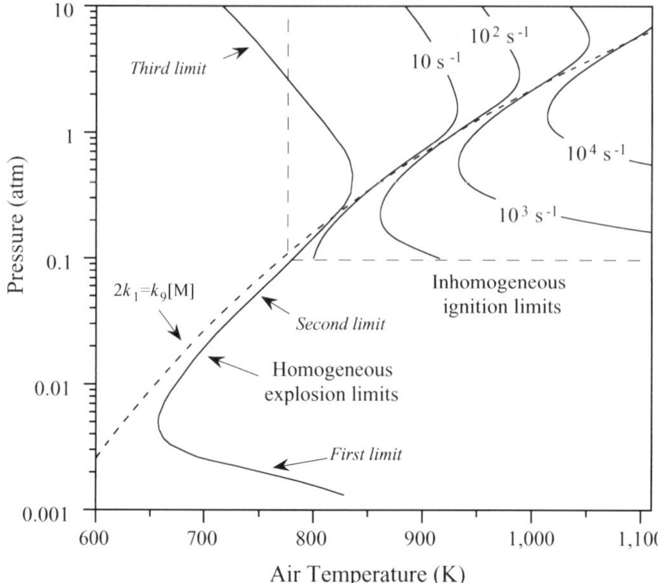

Figure 8.3.1. Ignition temperatures of counterflowing heated air against a cold mixture of 60 percent H_2 in N_2 as functions of pressure and density-weighted strain rate. The homogeneous explosion limits of the hydrogen–air mixture are shown as reference.

the necessary thermal feedback to cause an ignition (extinction) turning point in the solution. However, in Chapter 3 we also discussed the role of chemical kinetic mechanisms, involving chain-branching and -termination reactions, in effecting a nonlinear feedback in the concentrations of certain radicals and consequently system runaway. A notable example is the nonmonotonic, explosion limits of the hydrogen–oxygen system in response to temperature and pressure variations. Clearly, a satisfactory description of such phenomena must consider both thermal and kinetic influences. To demonstrate the intricacies of these two influences, in the following we shall present a unified chain–thermal interpretation of the ignition of a cold hydrogen jet by a heated air jet in the counterflow (Kreutz & Law 1996, 1998). Ignition in such situations can be brought about by either slowly increasing the air temperature, with the flow strain rate, a, fixed, or slowly decreasing a, with the air temperature fixed. Experimentally, the instant of ignition can be readily identified by the sudden appearance of a flame. Computationally, ignition can be defined as the narrow range of states over which some thermodynamic quantities, such as the temperature or the concentration of the hydrogen radical, increase rapidly.

8.3.1. Global Response to Strain Rate Variations

Figure 8.3.1 compares the homogeneous explosion limits of hydrogen–air mixtures, discussed in Chapter 3 (Figure 3.2.1), with the calculated ignition limits of a 60 percent hydrogen-in-nitrogen jet against a heated air jet, for various density-weighted strain rates $\tilde{a} = (\rho_{ox}/\rho_{ox,ref})a$, where ρ_{ox} is the density of the oxidizer jet and $\rho_{ox,ref}$ is ρ_{ox}

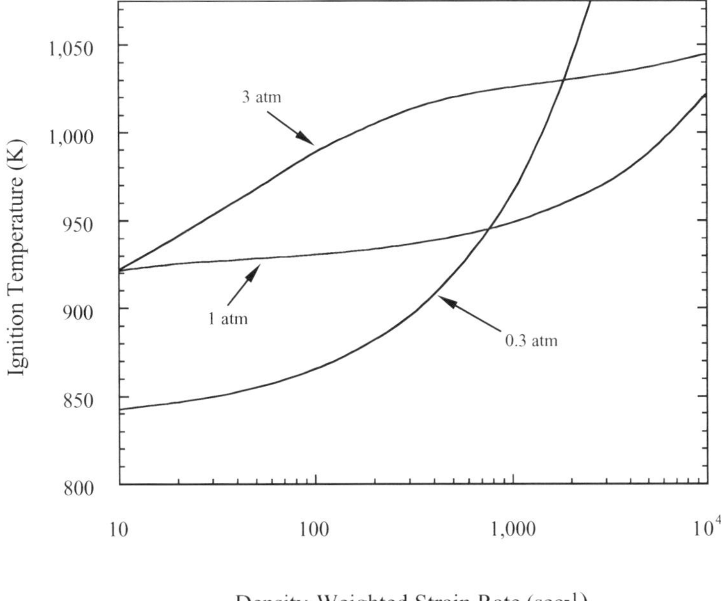

Figure 8.3.2. Ignition temperature as a function of $\tilde{a}$ for three different pressures, showing the insensitivity of the ignition temperature to strain rate variations near the second limit. Same condition as that of Figure 8.3.1.

at a reference pressure, say the normal pressure. Because of the dependence of the mass flux on the system pressure, the strain rate needs to be density weighted when its effects at different pressures are compared. This is the same reason that the laminar burning flux f^o instead of the laminar flame speed s_u^o is the more appropriate quantity in designating the burning rate of a flame.

Figure 8.3.1 shows that the ignition limits of the present diffusive, nonpremixed system basically track those of the homogeneous explosion limits, shifting along the second limit to higher temperatures and pressures with increasing strain rate. Consequently, the regime of nonignition is widened, which is reasonable because of the reduced residence time. Furthermore, one also expects that the kinetics governing the three ignition limits should be the same as those for the three homogeneous explosion limits.

An interesting feature of Figure 8.3.1 is that the ignition curves of different strain rates "bunch up" along the second ignition limits, which roughly correspond to the crossover temperature curve. Consequently, by plotting the ignition temperature as a function of $\tilde{a}$ (Figure 8.3.2), we see that, for each pressure, there exists a wide range of strain rates ($\approx 10^1$ to 10^3/s at 1 atm) over which the ignition temperatures are close to the corresponding crossover temperature and therefore vary only minimally with the strain rate. This behavior, which has been completely substantiated experimentally, implies that the reaction mechanism under such situations fundamentally cannot be approximated by a one-step overall reaction. This is because such an approximation

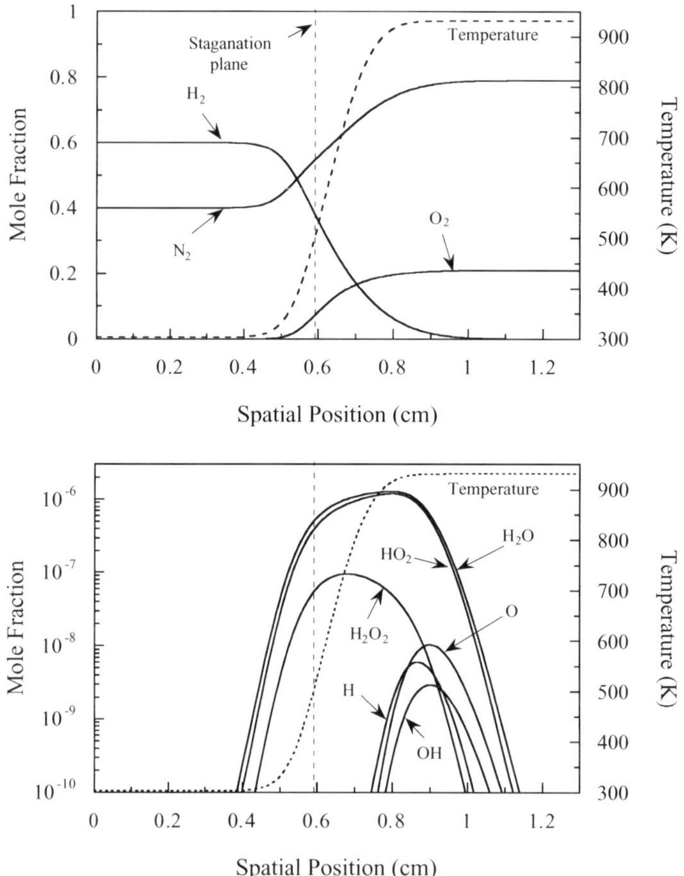

Figure 8.3.3. Spatially resolved axial temperature and species concentration profiles in the second ignition limit ($p = 1$ atm, $\tilde{a} = 100\,s^{-1}$, 60 percent H_2) at the ignition point, $T_{air} = 930.7$ K.

would require a balance between diffusion and reaction in the reaction zone, whose thickness varies with the strain rate. Consequently, an increase in the strain rate must necessarily bring about a corresponding increase in the ignition temperature which, however, is practically a constant in the present situation. This lack of sensitivity also implies that the dominant reactions occur at rates much faster than that of diffusion such that ignition can be considered to be kinetically controlled.

The above behavior can be understood by examining the structure of the reaction zone for the three limits of ignition, to be discussed next.

8.3.2. Second Ignition Limit

Figures 8.3.3 and 8.3.4 show representative second-limit spatial profiles of temperature, the major and minor species concentrations, the creation and destruction rates of H through various reactions, and the convection, diffusion, and net production rates of H, at 1 atm pressure. It is seen that the concentration profiles of the minor species can be grouped into two categories, namely "steady-state" and "sink" species,

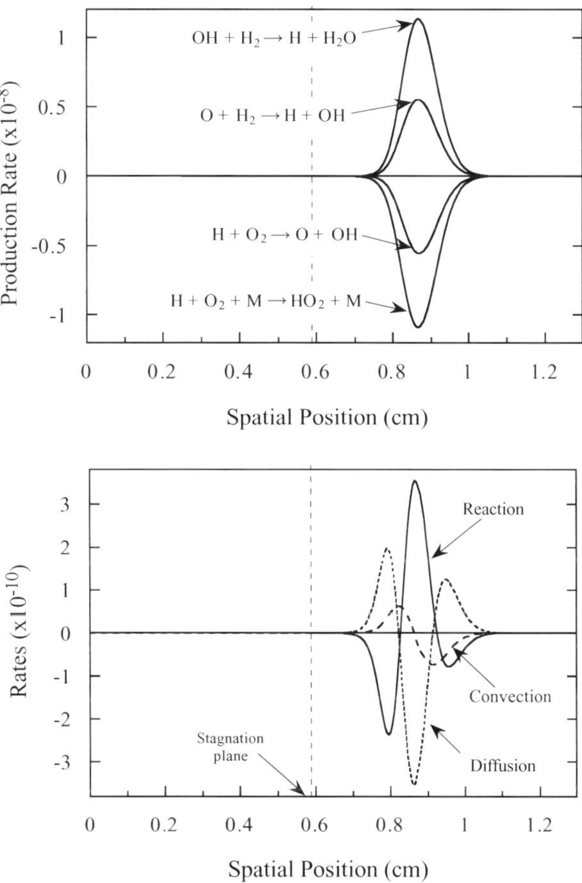

Figure 8.3.4. Spatially resolved rates for hydrogen radical production, destruction, and transport (moles/cm^3-s) in the second ignition limit. Same condition as that of Figure 8.3.3.

in analogy with the situation in homogeneous explosions. The steady-state species, O, H, and OH, exhibit relatively narrow peaks that are located near the hot oxidizer boundary, reflecting the high activation energy and temperature sensitivity of their formation reactions. The region within which these radicals peak can therefore be identified as the ignition kernel. The sink species, H_2O, HO_2, and H_2O_2, are either stable or metastable, and can therefore diffuse over long distances without reacting, resulting in the broad profiles shown in Figure 8.3.3. They are subsequently swept away upon crossing the stagnation surface, analogous to the diffusive loss to the vessel wall in homogeneous explosions.

Figure 8.3.4 shows that the individual reaction rates leading to the creation and consumption of H are at least two orders larger than its transport and net production rates. Similar results hold for O and OH. As such, these individual reaction rates can be decoupled from the conservation equations for their concentrations, thereby justifying the steady-state assumption for these radicals. The picture that emerges is that the ignition kernel is simply a chemical reactor within which the hydrogen–oxygen

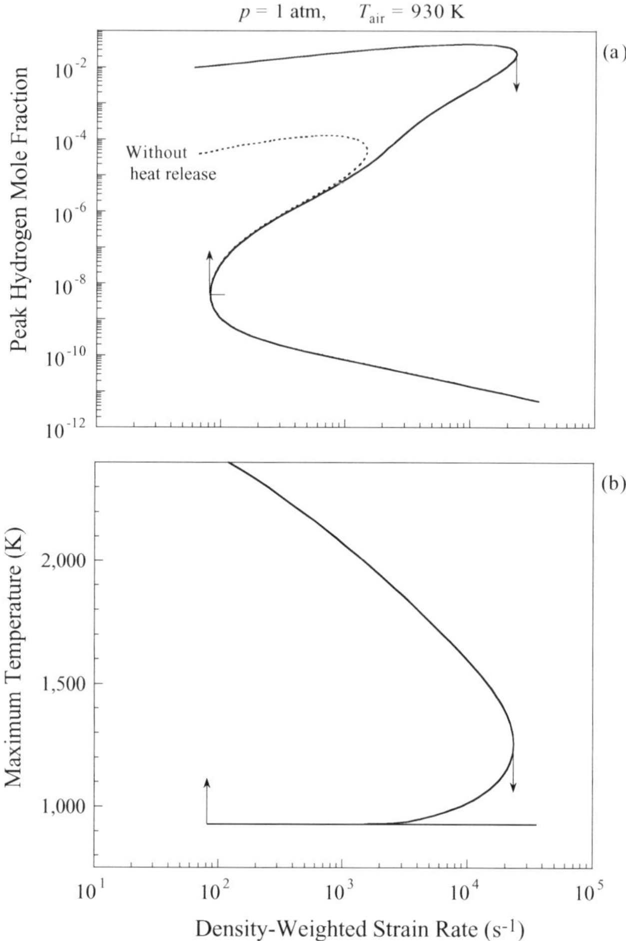

Figure 8.3.5. Solution S-curves for the second ignition limit as functions of $\tilde{a}$. Upward (downward)-pointing vertical arrows denote the ignition (extinction) point. Same condition as that of Figure 8.3.3, with $T_{air} = 930$ K.

reactions proceed rapidly in the manner similar to that of the homogeneous second limit explosion. Furthermore, because of the small concentrations of these radicals, the temperature change associated with the radical growth is expected to be negligible, thereby substantiating the observed temperature insensitivity shown in Figure 8.3.2.

Figure 8.3.5 shows the S-curve responses for the maximum H concentration and temperature as functions of the strain rate, for a representative second limit behavior. The lower, weakly reactive branch of the maximum temperature curve exhibits a cuspy response, emphasizing the fact that there is practically no change in the temperature due to chemical reaction within the ignition kernel, which is basically situated at the boundary of the hot air jet.

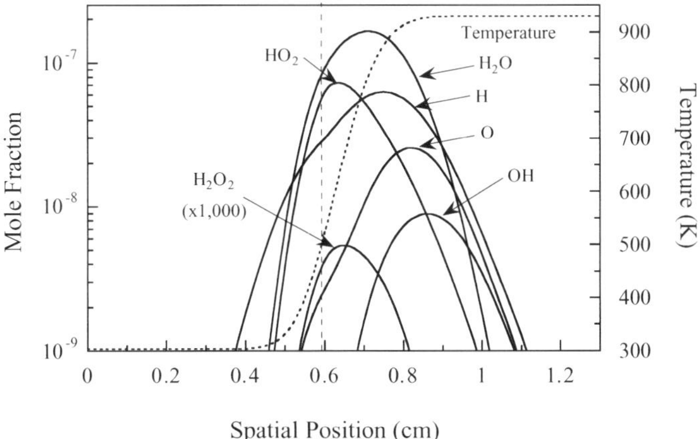

Figure 8.3.6. Spatially resolved minor species concentration profiles in the first ignition limit ($p = 0.1$ atm, $\tilde{a} = 100\ s^{-1}$, 60 percent H_2) at the ignition point, $T_{\text{air}} = 915.8$ K.

8.3.3. First and Third Ignition Limits

The first ignition limit is the low-pressure regime where the ignition temperature rises with decreasing pressure. In contrast to the second limit, diffusive transport of the active radicals plays an important role here. This is demonstrated by the broad profiles and hence diffusive loss of the active radicals O, H, and OH to the fuel stream, shown in Figure 8.3.6. Furthermore, results for the transport rate and the reaction rates for the individual radical species such as H show that they are of equal order. The *S*-curve for this limit (Figure 8.3.7) is again cuspy, demonstrating the minimal evolvement of heat during ignition. It is also of interest to note that the particular *S*-curve here contains four turning points and five simultaneous solutions over a range of the strain rate. Realistically, it is reasonable to expect that the primary ignition and extinction points are the physically observable ones because the secondary points, designated by α and β, are not accessible as one traverses the response curve either from the lowest, weakly reactive branch or the uppermost, strongly burning branch. Computationally, care needs to be exercised in not identifying β as the ignition point because the cuspy nature of the true ignition point can be easily missed.

The third limit occurs at high pressures where the ignition temperature decreases with increasing pressure. HO_2 now actively participates in the chain mechanism, generating OH and H. An ignition kernel consisting of H, O, OH, and HO_2 can be identified (Figure 8.3.8), although HO_2, being less reactive, diffuses to a larger distance. H_2O_2 is also an active species, with the maximum of its concentration located in the neighborhood of the ignition kernel. This is to be contrasted with the behavior of the second limit at which its peak is offset from those of H, O, and OH. Furthermore, unlike H, O, OH, and HO_2, its reactivity is weaker, rendering it to migrate across the stagnation surface and subsequently convected away. As such, the third limit is affected by transport.

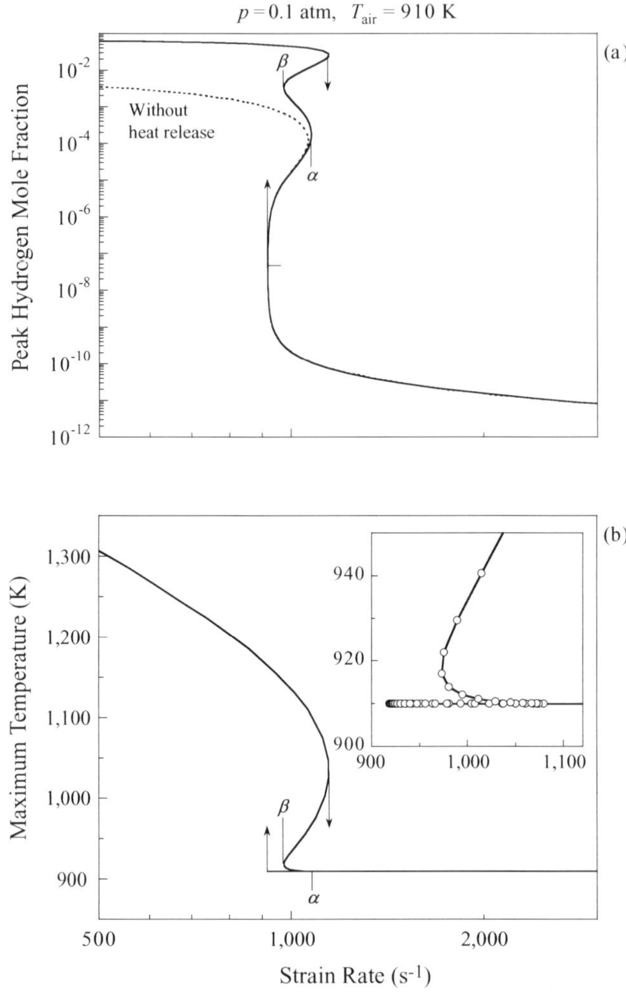

Figure 8.3.7. Solution S-curves for the first ignition limit as functions of $\tilde{a}$. Same condition as that of Figure 8.3.6, with $T_{air} = 910$ K.

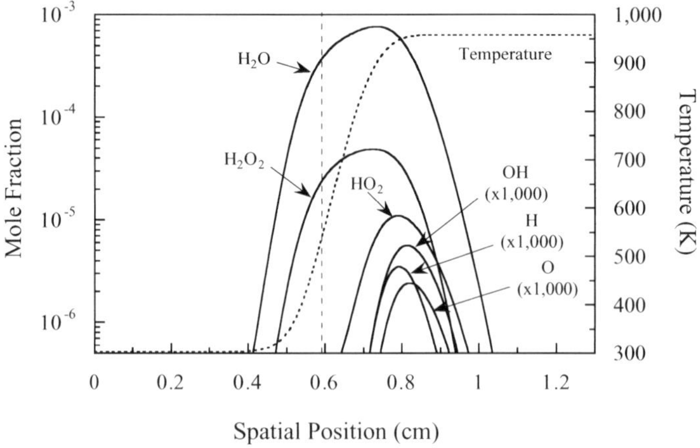

Figure 8.3.8. Spatially resolved minor species concentration profiles in the third ignition limit ($p = 10$ atm, $\tilde{a} = 100\ s^{-1}$, 60 percent H_2) at the ignition point, $T_{air} = 954.3$ K.

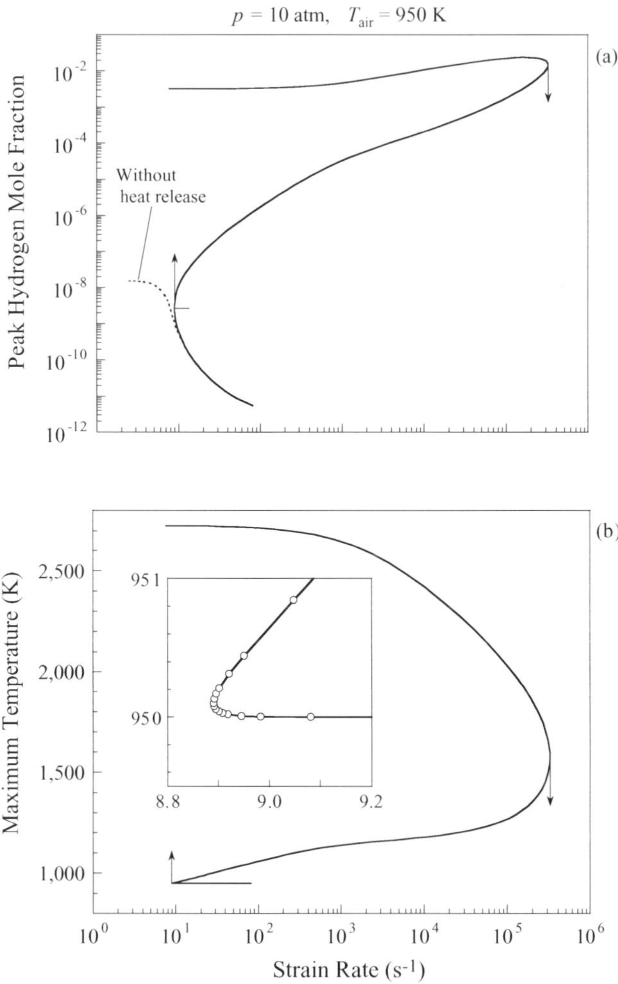

Figure 8.3.9. Solution S-curves for the third ignition limit as functions of $\tilde{a}$. Same condition as that of Figure 8.3.8, with $T_{air} = 950$ K.

Chemical heat generation still remains unimportant in the lower pressure range of the third limit, resulting in cuspy ignition turning points found for the first and second limits (Figures 8.3.5 and 8.3.7). However, at higher pressures the turning is more like a corner than a cusp, as shown in Figure 8.3.9, indicating the emerging importance of heat release and thermal feedback in ignition.

8.3.4. Decoupled Environment and Kinetic versus Thermal Feedback

The above results seem to indicate that, because of the small size of the radical pool required for system runaway, the chemical heat release at the first and second limits is so small that there is essentially no temperature perturbation, or thermal feedback, responsible for ignition. To be conclusive of the absence of thermal feedback, Figures 8.3.5, 8.3.7, and 8.3.9 show calculated results obtained by suppressing the

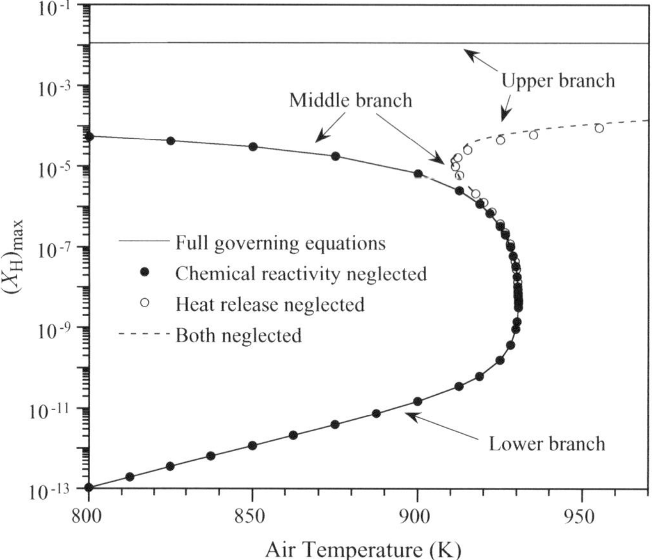

Figure 8.3.10. A comparison between system response S-curves calculated using the full governing equations versus those using various decoupled ignition environment approximations. Same condition as that of Figure 8.3.3, with T_{air} varying.

chemical enthalpy term in the energy conservation equation. These results basically reproduce the lower branch and the ignition turning point segment of the complete solution for the first and second limits. For the third limit, however, the ignition turning point fails to be reproduced without chemical heat release. Consequently, ignition at the first and second limits is controlled by kinetic instead of thermal feedback, while both feedbacks are required for the third limit.

The result that the mixture temperature is basically unaffected by ignition in the first and second limits suggests that the concentrations of the major species may also be effectively unaffected by the chemical reactivity. This possibility is substantiated by repeating the calculation without the chemical reactivity terms in the governing equations for the major species, H_2, O_2, and N_2. Figure 8.3.10 shows that the lower branch and the ignition turning point segment are indeed reproduced with this assumption.

The viability of this assumption leads to a new view of the ignition process that simplifies the system both conceptually and practically. That is, in simulating systems of this nature, the ignition environment can be first predetermined by simply solving the nonreactive conservation equations for the reactants and enthalpy. The response of the minor species can then be calculated by using profiles of this "decoupled environment."

We finally address the reaction mechanism governing the ignition turning point for the second limit. Obviously, the basic hydrogen–oxygen chain mechanism must be included, namely (H1): $H + O_2 \rightarrow O + OH$; (H2): $O + H_2 \rightarrow H + OH$; (H3): $OH + H_2 \rightarrow H + H_2O$; and (H9): $H + O_2 + M \rightarrow HO_2 + M$. Furthermore, the initiation reaction (–H10): $H_2 + O_2 \rightarrow HO_2 + H$ is also needed because of the low level of

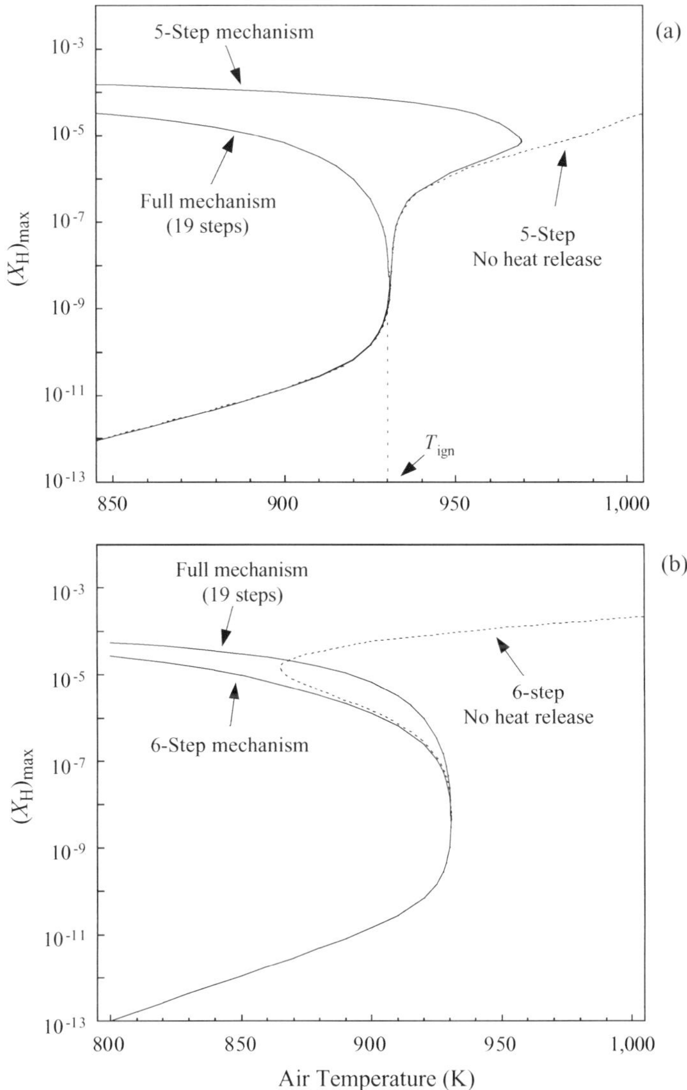

Figure 8.3.11. A comparison between system response S-curves calculated using the full reaction mechanism, (a) the 5-step mechanism, and (b) the 6-step mechanism. Same condition as that of Figure 8.3.3, with T_{air} varying.

H radicals formed in the ignition kernel that can back diffuse and initiate the chain reactions through (H1).

A simulation based on the above five reactions, however, fails to reproduce the turning point, as shown in Figure 8.3.11 in which the air temperature is used as the independent variable. The missing reaction turns out to be (H11): $HO_2 + H \rightarrow 2OH$. With the addition of (H11), Figure 8.3.11 shows that the ignition response is now well described.

To clarify the role of (H11), we first note that the lower branches of the 5- and 6-step mechanisms are very close to each other until the state of turning attained by the 6-step mechanism. Around this state the 6-step solution turns while the 5-step

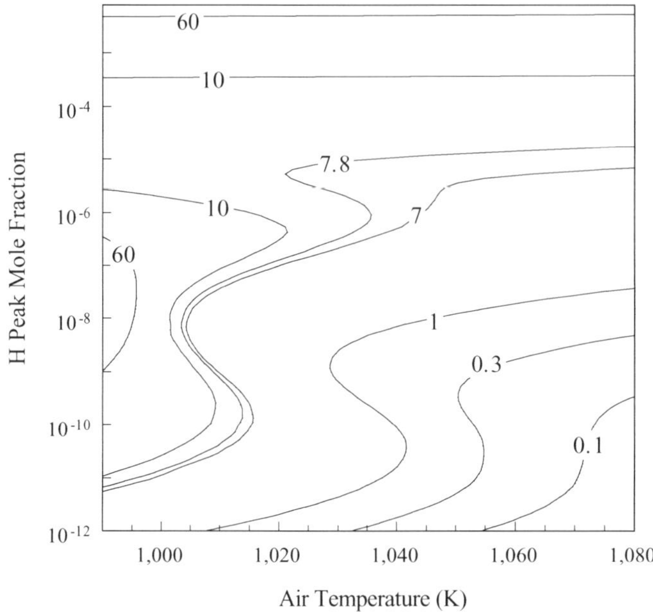

Figure 8.3.12. Solution S-curves for $p = 4$ atm, $\tilde{a} = 300\,s^{-1}$, and various hydrogen-in-nitrogen volumetric concentrations.

solution exhibits an inflection point and fails to turn. Thus there is sufficient radical generation in both cases, and (H11) simply effects the turning. The reason that (H11) is needed for turning is the following. Our earlier discussion on the S-curve shows that turning points are possible only for systems that possess a sensitive nonlinear source function as well as a sink function that can be either linear or nonlinear. In the thermal-based discussion of Section 8.1.5, temperature is the system response, the Arrhenius factor with the large activation energy constitutes the nonlinear source function, and conductive heat loss being the sink function. For the present problem we can use any radical species as the system response, with the chain-branching mechanism as the nonlinear source function. However, there is no loss mechanism for all the active radicals, H, O, and OH, in the 5-step mechanism. As such, while the system solution can runaway, it cannot turn. On the other hand, by bringing in reaction (H11) in the 6-step scheme, HO_2 is included as an actively participating radical. Since it is also a sink species (see Figure 8.3.3), its presence introduces both branching and loss to the system, and as such effects the turning.

8.3.5. Multiple Criticality and Staged Ignition

Figure 8.3.7 shows the existence of S-curves that could have two ignition turning points, two extinction turning points, and five steady-state solutions. While we have dismissed the significance of the secondary turning points for Figure 8.3.7, multiple criticality does exist. This is demonstrated in Figure 8.3.12 for hydrogen with different amounts of nitrogen dilution.

It is seen that the regular, triple-solution S-curve is obtained for 60 percent hydrogen, and ignition is kinetically controlled as discussed earlier. However, as the hydrogen concentration is reduced to, say, 7.8 percent, the response curve has two ignition and two extinction points. This curve is also qualitatively different from that of Figure 8.3.7 in that the lower, primary ignition point occurs at a lower air temperature than that of the secondary ignition point. Similarly, the upper, primary extinction point occurs at a higher temperature than that of the secondary extinction point. What this means is that, by gradually increasing the air temperature from a low value, in an attempt to effect ignition, the system will first ignite when encountering the primary ignition turning point. The system then jumps to a steady burning state corresponding to the lower branch of the secondary ignition point. With further increase of the air temperature, a second ignition will be effected at the secondary ignition point. The system then transitions to the uppermost, intense burning branch. As such, the transition of the system from a weakly reactive, near-frozen state to a rigorously burning, near-equilibrium state occurs through a two-staged ignition process. A similar discussion can be conducted to show that extinction also occurs in a two-staged manner.

Further computation shows that the primary ignition event is kinetically controlled, while the secondary ignition additionally requires thermal feedback. The existence of the staged ignition, as well as three steady states respectively representing near-frozen flow, weak burning, and strong burning have been experimentally verified (Fotache et al. 1998).

By further reducing the hydrogen concentration to, say, 7 percent, the upper turning point disappears, having changed to a stretched S-curve behavior. This is reasonable because the upper turning point requires thermal feedback, which however becomes excessively weak for low hydrogen concentrations. The system can therefore only sustain a weakly "burning" state, with minimal amount of heat generation.

Finally, with excessive amount of dilution (e.g., 0.1 percent), even the lower turning point transitions to the stretched S-curve behavior. All system nonlinearities, both kinetic and thermal, are now eliminated.

8.4. PREMIXED FLAME EXTINCTION THROUGH VOLUMETRIC HEAT LOSS

The various possible phenomena involved in, and the associated mechanisms responsible for, premixed flame extinction are very rich and subtle. To appreciate the role of loss in premixed flame extinction, we note that the expression for the laminar flame speed s_u^o for the propagation of the adiabatic, one-dimensional planar flame, say Eq. (7.2.14), shows that, even for very small values of the freestream reactant concentration, the flame will be able to propagate with $\tilde{T}_{\text{ad}}$ and a finite flame speed. Thus the standard flame, which is an adiabatic system, does not possess an extinguished state.

As discussed in Section 8.1, an obvious mechanism to extinguish this flame is heat loss through the concept of quenching distance. A further example of flame

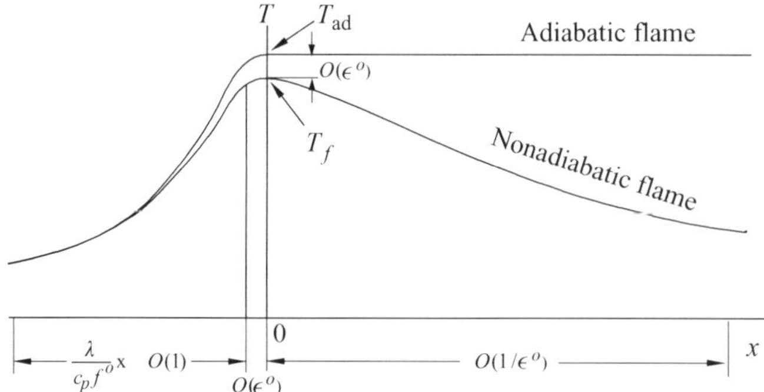

Figure 8.4.1. Temperature profiles of the adiabatic and nonadiabatic freely propagating planar flames.

quenching is to seed its upstream with cold, inert particles. In the case of preventing mine explosions, this is achieved by purposely spreading dust particles on the mine floor, through the practice of "inerting." Then in the event of an accidental ignition, the disturbance produced either in the ignition region or upstream of the propagating flame will lift both the coal and dust particles into the air. While the coal particles tend to promote the development and sustenance of a flame, the dust particles will serve as a heat sink, tending to retard these events.

In this section, we shall study the classical problem of the extinction of the standard premixed flame due to volumetric heat loss, as first formulated by Spalding (1957). The sources of such losses could be, for example, radiative heat transfer from the flame to the cold ambience or the heating of the cold inert particles present in the mixture.

Figure 8.4.1 schematically compares the temperature profiles for the adiabatic and nonadiabatic cases. Here we see that, in the adiabatic case, the temperature increases from the ambient value to the adiabatic flame temperature at the flame location, and remains at that value in the downstream region. However, when there is volumetric heat loss in the bulk gas, the temperature in the upstream, preheat region is slightly reduced, leading to a corresponding reduction in the flame temperature from T_{ad}. In the downstream cooling region, the continuous action of the heat loss causes a steady reduction in the temperature until the ambient temperature is eventually reached. This downstream heat loss induces a further reduction in the flame temperature, leading to its steady-state value T_f. Intuitively, one expects that extinction can be brought about by an $O(\epsilon^o)$ reduction in the flame temperature, which in term results in an $O(1)$ reduction in the burning flux and hence the flame speed because of Arrhenius kinetics, where $\epsilon^o = \tilde{T}^2_{ad}/\tilde{T}_a$ is the value of $\epsilon = \tilde{T}^2_f/\tilde{T}_a$ in the adiabatic limit. This substantial reduction in the flame speed then indicates the possibility of extinction. An $O(1)$ reduction in the flame temperature will lead to an exponential reduction in the flame speed, which is unrealistically large.

For small rates of heat loss, Figure 8.4.1 shows that the thicknesses of the preheat and reaction zones respectively remain at $O(1)$ and $O(\epsilon^o)$ as in the adiabatic case.

The downstream cooling zone, which is infinitely thick in the adiabatic limit, now has a finite, though still large, characteristic thickness because the cooling will eventually bring the temperature down to the ambient value. The excessive thickness implies that diffusive transport is now much slower than convective transport and the cooling region is therefore convective-loss in nature. A balance between convection and loss, with loss being an $O(\epsilon^o)$ quantity, readily shows that the thickness of the cooling zone is $O(1/\epsilon^o)$. We note, however, that this excessively thick downstream cooling region is only characteristic of the unbounded nature of this flame. A shorter cooling region would prevail if it is rendered finite by, say, placing a cold plate normal to the flow motion.

8.4.1. Phenomenological Derivation

We shall first demonstrate phenomenologically that volumetric heat loss can indeed lead to the nonlinear phenomenon of flame extinction (Zel'dovich et al. 1985). Let $L(T)$ be the volumetric heat loss rate, which is a general function of temperature with the only requirement that $L(T) \to 0$ as $T \to T_u$. The heat loss flux q can be separately estimated for each of the upstream preheat zone, the reaction zone, and the downstream cooling zone. Furthermore, the heat loss function $L(T)$ should also be distinguished for each of these zones because, for example, the nature of the radiating gas can be quite different before and after the reaction.

For the preheat zone, since it has a characteristic thickness ℓ_D, we have

$$q^- = \int_0^{\ell_D} L^- dx \approx \ell_D L^- \approx \frac{\lambda/c_p}{f} L^-, \tag{8.4.1}$$

where we have assumed an average L^- when summing over all the loss, and have used the relation $f\ell_D \sim \lambda/c_p$ given in Eq. (7.5.55). Equation (8.4.1) shows that an increase in L^- not only causes a direct increase in q^-, but it also reduces the flame burning flux f, which in turn increases the flame thickness ℓ_D. A larger ℓ_D implies a larger volume from which loss occurs, and therefore further increases q^-. This indirect effect of L^- through changes in the flame thickness is essential because the final result shows that it is this nonlinear feedback mechanism in the flame response that causes the abrupt extinction behavior.

A similar estimate for the reaction zone shows that its loss rate is only $O(\epsilon^o)$ of q^- because the reaction zone thickness is only $O(\epsilon^o)$ of ℓ_D. Therefore heat loss in the reaction zone can be neglected due to its small volume.

For the cooling zone, we can write a balance between heat loss and convection as

$$L^+ = f c_p \frac{dT}{dx}, \tag{8.4.2}$$

from which the heat loss flux from the reaction zone to the cooling zone is

$$q^+ = \lambda \left(\frac{dT}{dx}\right)_{0^+} = \frac{\lambda/c_p}{f} L^+. \tag{8.4.3}$$

Here L^+ again affects q^+ both directly and indirectly. The indirect effect, through the presence of f in the denominator of q^+, accounts for the fact that the temperature gradient, and hence the rate of heat loss, increases with decreasing convection intensity. This constitutes the second nonlinear feedback mechanism responsible for the abrupt extinction.

It is significant to note that although q^- and q^+ basically have the same expression, they are derived on entirely different physical reasoning. That is, the volumetric heat loss from the preheat zone affects the reaction zone in a direct manner in that it reduces the amount of enthalpy entering the reaction zone. Consequently all the heat loss from the preheat zone needs to be taken into account. The volumetric heat loss from the cooling zone, however, affects the reaction zone in an indirect manner, through modification of its temperature gradient at the downstream boundary of the reaction zone. The amount of heat loss from the reaction zone through this mechanism is obviously much smaller than the total amount of heat loss from the cooling zone. The loss from the cooling zone is of little relevance to the reaction zone structure, and consequently flame extinction, because it takes place subsequent to the completion of reaction. Indeed, if all of the downstream heat loss were subtracted from the enthalpy of the flow in the reaction zone, in addition to the amount already subtracted due to loss from the preheat zone, and since the cooling zone temperature must approach that of the unburnt mixture, the flame temperature must necessarily be that of the ambience due to overall energy conservation, which is an obviously unacceptable outcome.

Thus accounting for the losses from both zones, we can write the overall energy conservation, from the upstream ambience to the downstream boundary of the reaction zone, for a mixture with an initial enthalpy flux of $fc_p(T_{\rm ad} - T_u)$ as

$$fc_p(T_{\rm ad} - T_u) = fc_p(T_f - T_u) + \frac{\lambda/c_p}{f}(L^- + L^+). \tag{8.4.4}$$

This yields an expression for the flame temperature, which in nondimensional form, can be expressed as

$$\tilde{T}_f = \tilde{T}_{\rm ad} - \epsilon^o\theta, \tag{8.4.5}$$

where $\tilde{T} = (c_pT)/(q_cY_u)$, $\theta = \tilde{L}_v/\tilde{f}^2$, $\tilde{L}_v = \tilde{L}^- + \tilde{L}^+$, $\tilde{f} = f/f^o$, and where we have defined a nondimensional heat loss parameter

$$\tilde{L}^{\mp} = \frac{\lambda/c_p}{\epsilon^o(f^o)^2q_cY_u}L^{\mp}. \tag{8.4.6}$$

Note that in writing Eq. (8.4.5) we have deliberately inserted the factor ϵ^o to indicate that the temperature perturbation is expected to be $O(\epsilon^o)$ such that θ is $O(1)$.

From Eq. (7.5.53), we have

$$f^2 \sim T_f^4 e^{-\tilde{T}_a/\tilde{T}_f}. \tag{8.4.7}$$

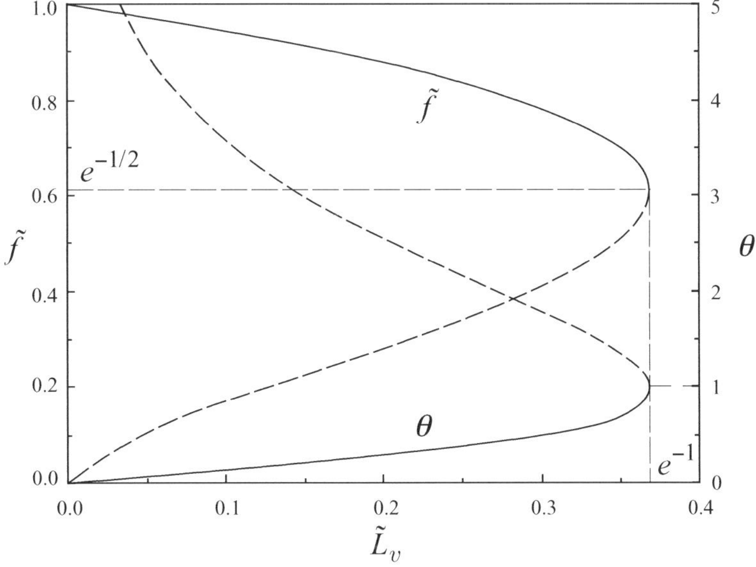

Figure 8.4.2. Dependence of the burning flux and flame temperature reduction on the heat loss rate for the freely propagating planar flame. Solid segment is the stable branch and dashed segment the unstable branch.

Substituting T_f given by Eq. (8.4.5) into (8.4.7), and expanding for small heat loss and hence small reduction in the flame temperature from T_{ad}, we have

$$f^2 \sim \left(T_{ad}^4 e^{-\tilde{T}_a/\tilde{T}_{ad}}\right) e^{-\tilde{L}_v/\tilde{f}^2}. \tag{8.4.8}$$

Note that the nonlinear term, $e^{-\tilde{L}_v/\tilde{f}^2}$, represents the indirect effects of flame thickening due to upstream heat loss and gradient steepening due to downstream heat loss.

Referencing (8.4.8) to its adiabatic limit $(f^o)^2 \sim T_{ad}^4 e^{-\tilde{T}_a/\tilde{T}_{ad}}$, we have $\tilde{f}^2 = e^{-\tilde{L}_v/\tilde{f}^2}$, or

$$\tilde{f}^2 \ln \tilde{f}^2 = -\tilde{L}_v. \tag{8.4.9}$$

Substituting $\tilde{f}^2 = \tilde{L}_v/\theta$ into Eq. (8.4.9) yields

$$\theta e^{-\theta} = \tilde{L}_v. \tag{8.4.10}$$

Equation (8.4.9) is an important result in flame extinction studies and has more generality than implied by the above phenomenological derivation. By plotting $\tilde{f}$ and θ versus $\tilde{L}_v$, Figure 8.4.2 shows that the characteristic extinction turning point is exhibited. It is seen that, starting from the adiabatic limit of $\tilde{f} = 1$ with $\tilde{L}_v = 0$, an increase in $\tilde{L}_v$ reduces $\tilde{f}$ and increases θ as is reasonable. However, there is a maximum heat loss rate, $\tilde{L}_{v,E} = e^{-1}$, beyond which flame propagation is not possible. This can therefore be identified as the extinction limit of the system. At this state the burning flux has been reduced to its minimum value of $\tilde{f}_E = e^{-1/2} \approx 0.61$ while the flame temperature has also been reduced from T_{ad} by an ϵ^o amount.

8.4.2. Frank-Kamenetskii Solution

We shall now conduct a more rigorous analysis based on the Frank-Kamenetskii approach. The governing equations are

$$\tilde{f}\frac{d\tilde{T}}{d\tilde{x}} - \frac{d^2\tilde{T}}{d\tilde{x}^2} = -\tilde{w}_F - \epsilon^o \tilde{L}(\tilde{T}) \tag{8.4.11}$$

$$\tilde{f}\frac{d\tilde{Y}}{d\tilde{x}} - \frac{d^2\tilde{Y}}{d\tilde{x}^2} = \tilde{w}_F, \tag{8.4.12}$$

where $\tilde{x} = [f^o/(\lambda/c_p)]x$. The boundary conditions are

$$\tilde{x} \to -\infty: \quad \tilde{T} = \tilde{T}_u, \quad \tilde{Y} = 1 \tag{8.4.13}$$

$$\tilde{x} \to \infty: \quad \tilde{T} = \tilde{T}_u, \quad \tilde{Y} = 0. \tag{8.4.14}$$

In the preheat zone $\tilde{w}_F$ is neglected such that energy conservation becomes

$$\tilde{f}\frac{d\tilde{T}^-}{d\tilde{x}} - \frac{d^2\tilde{T}^-}{d\tilde{x}^2} = -\epsilon^o \tilde{L}^-. \tag{8.4.15}$$

Integrating Eq. (8.4.15) once and applying the boundary conditions at $\tilde{T}^-(-\infty) = \tilde{T}_u$ and $(d\tilde{T}^-/d\tilde{x})_{-\infty} = 0$, we have

$$\frac{d\tilde{T}^-}{d\tilde{x}} = \tilde{f}(\tilde{T}^- - \tilde{T}_u) + \epsilon^o \int_{-\infty}^{\tilde{x}} \tilde{L}^- d\tilde{x}. \tag{8.4.16}$$

Similarly, in the cooling region,

$$\tilde{f}\frac{d\tilde{T}^+}{d\tilde{x}} - \frac{d^2\tilde{T}^+}{d\tilde{x}^2} = -\epsilon^o \tilde{L}^+. \tag{8.4.17}$$

In this region the heat loss term must be important because if it were neglected, then we would get a constant $\tilde{T}^+$. Furthermore, convection must also be present because it gives the flame propagation rate. A balance between these two terms shows that the relevant spatial variable is a compressed variable $\eta = \epsilon^o \tilde{x} = O(1)$. This then indicates that the thickness of the cooling zone is $O(1/\epsilon^o)$, and diffusion is of higher order, as noted earlier. Thus we have

$$\tilde{f}\frac{d\tilde{T}^+}{d\tilde{x}} = -\epsilon^o \tilde{L}^+, \quad \text{or} \quad \tilde{f}\frac{d\tilde{T}^+}{d\eta} = -\tilde{L}^+. \tag{8.4.18}$$

The flame temperature $\tilde{T}_f$ can be evaluated through overall energy balance by adding Eqs. (8.4.11) and (8.4.12) to eliminate the reaction term, integrating the resulting expression from $\tilde{x} = -\infty$ to the downstream boundary of the reaction zone $\tilde{x} = 0^+$, using Eq. (8.4.18) to evaluate $(d\tilde{T}^+/d\tilde{x})_{0^+}$, and by noting that $\tilde{Y}(0^+) = (d\tilde{Y}/d\tilde{x})_{0^+} = 0$. This leads to

$$\tilde{T}_f = \tilde{T}_{\mathrm{ad}} - \epsilon^o \left[\frac{1}{\tilde{f}} \int_{-\infty}^{0} \tilde{L}^- d\tilde{x} + \frac{\tilde{L}^+(\tilde{T}_f)}{\tilde{f}^2}\right] \approx \tilde{T}_{\mathrm{ad}} - \epsilon^o \frac{\tilde{L}_v}{\tilde{f}^2}, \tag{8.4.19}$$

where we have defined

$$\tilde{L}_v = \int_{\tilde{T}_u}^{\tilde{T}_{\rm ad}} \frac{\tilde{L}^-(\tilde{T})}{\tilde{T} - \tilde{T}_u} d\tilde{T} + \tilde{L}^+(\tilde{T}_{\rm ad}). \tag{8.4.20}$$

In deriving Eq. (8.4.19), we have approximated $\tilde{L}^+(\tilde{T}_f)$ by $\tilde{L}^+(\tilde{T}_{\rm ad})$, and have used the leading-order expression of Eq. (8.4.16) to change the integration variable from $d\tilde{x}$ to $d\tilde{T}$ because $\tilde{L}$ is a function of $\tilde{T}$. We have therefore derived $\tilde{L}_v$, which was identified phenomenologically earlier, especially the different functional forms for the loss terms due to $\tilde{L}^-$ and $\tilde{L}^+$.

The reaction zone is governed by

$$\frac{d^2\tilde{T}_{\rm in}}{d\tilde{x}^2} = -Da_C^o \tilde{Y}_{\rm in} e^{-\tilde{T}_a/\tilde{T}_{\rm in}}. \tag{8.4.21}$$

To relate $\tilde{Y}_{\rm in}$ to $\tilde{T}_{\rm in}$, we note that in an adiabatic system $\tilde{T}_{\rm in} + \tilde{Y}_{\rm in} = \tilde{T}_{\rm ad}$. For the present problem the flame temperature, which represents the total enthalpy of the flow, has been reduced to $\tilde{T}_f$ from $\tilde{T}_{\rm ad}$. Therefore we expect the total enthalpy in the reaction zone to be

$$\tilde{T}_{\rm in} + \tilde{Y}_{\rm in} = \tilde{T}_f, \tag{8.4.22}$$

which expresses $\tilde{Y}_{\rm in}$ in terms of $\tilde{T}_{\rm in}$.

If we next substitute $(\tilde{T}_f - \tilde{T}_{\rm in}) = O(\epsilon^o)$ into the RHS of Eq. (8.4.21) and expand for small ϵ^o, perform the integration from the upstream to the downstream boundaries of the reaction zone in the same manner as that in Section 7.4.2, and observe that $\tilde{T}_{\rm in} \to \tilde{T}_f$ at the downstream boundary so that $\tilde{Y}_{\rm in} \to 0$, while $e^{-(\tilde{T}_f - \tilde{T}_{\rm in})} \to 0$ at the upstream boundary so that all chemical reactivities vanish there, we obtain

$$\left[\left(\frac{d\tilde{T}_{\rm in}}{d\tilde{x}}\right)_{0^+}\right]^2 - \left[\left(\frac{d\tilde{T}_{\rm in}}{d\tilde{x}}\right)_{0^-}\right]^2 = -2Da_C^o(\epsilon^o)^2 e^{-\tilde{T}_a/\tilde{T}_f}. \tag{8.4.23}$$

To evaluate Eq. (8.4.23), we substitute $(d\tilde{T}^+/d\tilde{x})_{0^+}$ from Eq. (8.4.18) for $(d\tilde{T}_{\rm in}/d\tilde{x})_{0^+}$, $(d\tilde{T}^-/d\tilde{x})_{0^-}$ from Eq. (8.4.16) for $(d\tilde{T}_{\rm in}/d\tilde{x})_{0^-}$, and Eq. (8.4.19) for $\tilde{T}_f$. The resulting leading-order expression is Eq. (8.4.9), where we have observed that in the adiabatic limit $2Da_C^o(\epsilon^o)^2 \exp(-\tilde{T}_a/\tilde{T}_{\rm ad}) = 1$. We have therefore rederived the nonlinear burning rate response for this nonadiabatic flame propagation problem, with $\tilde{L}_v$ now given by a more generalized expression, Eq. (8.4.20).

To be more specific with the loss function $L^{\pm}$, we note that an obvious volumetric loss mechanism from the gaseous medium is radiation, with $L = h(T^4 - T_u^4)$, where h is an appropriate dimensional heat transfer coefficient. Another expression that is commonly used in heat transfer analysis is the linear function $L = h(T - T_u)$, which relates conductive heat loss from a one-dimensional flow to the tube wall. By using this simpler function for demonstration, we have

$$\tilde{L}^+(\tilde{T}_{\rm ad}) = \tilde{h}^+, \qquad \int_{\tilde{T}_u}^{\tilde{T}_{\rm ad}} \frac{\tilde{L}^-(\tilde{T})}{(\tilde{T} - \tilde{T}_u)} d\tilde{T} = \tilde{h}^-,$$

because $\tilde{T}_{ad} - \tilde{T}_u = 1$. Thus, from Eq. (8.4.20),

$$\tilde{L}_v = \tilde{h}^+ + \tilde{h}^-. \tag{8.4.24}$$

For $\tilde{h}^+ = \tilde{h}^-$, Eq. (8.4.24) shows that the upstream and downstream contributions in heat loss are equal. Similar, but somewhat more lengthy expressions can be derived when radiation is the loss mechanism.

8.5. FLAMMABILITY LIMITS

8.5.1. Empirical Limits

It is empirically known that sufficient dilution of a combustible mixture with excess oxidizer or fuel will make it nonflammable in that, even after every effort has been made to reduce heat loss from the system, a flame would still fail to propagate after it can be somehow initially started through, say, a sufficiently intense ignition source. The critical composition delineating combustible and noncombustible mixtures is called the flammability limit. Typically a lean, or lower, flammability limit (LFL) and a rich, or upper, flammability limit (UFL) can be identified for a system of given temperature and pressure, and consisting of fuel, oxidizer, and inert, which respectively represent the excessively fuel-lean and fuel-rich situations. A quantitative knowledge of these limits is of importance for assessment of fire hazards. Furthermore, the interest in lean combustion, and thereby improved combustor performance in terms of emissions and efficiency characteristics, also imply that flammability considerations could ultimately impose limitations on the system performance. Since air usually constitutes the oxidizer–inert components in practical situations, mentioning of a flammability limit of a certain fuel implicitly implies that the limit is for the fuel in air. From safety considerations the LFL is more relevant because a rich nonflammable gas can be easily rendered flammable upon dilution by air.

The former U.S. Bureau of Mines (Zabetakis 1965) has standardized the determination of flammability limits. In this procedure the test mixture is placed in a vertically oriented tube of 51 mm diameter and 1.5 m length. The tube is closed at the upper end and open to the atmosphere at the lower end. Ignition by either spark discharge or a small pilot flame at the lower end is then attempted. A mixture is said to be flammable if the resulting flame can propagate all the way to the top of the tube, and nonflammable otherwise. The flammability limit is then the boundary mixture composition separating these two states. Table 8.1 lists the lean and rich flammability limits of some common fuels.

The specific dimension of the tube was chosen because further increases minimally affect the limits. Furthermore, by flipping the tube upside down and igniting the mixture from the upper end, downward propagation limits have also been determined. The flammable range is always wider for the upward propagation limits because flame propagation is now assisted by the buoyantly rising hot combustion products below the flame front. Thus from safety considerations the upward limits are more conservative.

Table 8.1. Flammability limits of some common fuel–air mixtures at 1 atm pressure, in mole % and (ϕ)[†]

Fuel	Lean limit	Rich limit
Hydrogen	4.00 (0.10)	75.0 (7.14)
Carbon monoxide	12.5 (0.34)	74.0 (6.8)
Ammonia	15.0 (0.63)	28.0 (1.4)
Methane	5.00 (0.50)	14.9 (1.67)
Ethane	3.0 (0.52)	12.4 (2.4)
Propane	2.1 (0.56)	9.5 (2.7)
Butane	1.8 (0.57)	8.4 (2.8)
Ethylene	2.7 (0.40)	36.0 (8.0)
Acetylene	2.5 (0.31)	100.0 (∞)
Benzene	1.3 (0.56)	7.9 (3.7)
Methanol	6.7 (0.51)	36.0 (4.0)
Ethanol	3.3 (0.41)	19.0 (2.8)

[†] Data compiled from M. G. Zabetakis, 1965. Flammability characteristics of combustible gases and vapors, *U.S. Department of Mines Bulletin* **627**.

There are some additional empirical observations of interest. First, increasing the mixture temperature always widens the flammability limits because of the corresponding increase of the flame temperature. Increasing pressure slightly narrows the lean flammability limits of hydrogen as well as the common hydrocarbons fuels, hence rendering the mixtures less flammable. However, for rich mixtures, the hydrogen limit is narrowed while the hydrocarbon limits are extended.

An interesting empirical result on flammability limits is the limit temperature concept of Burgess and Wheeler (1911), which states that, for many hydrocarbon–air mixtures, the product of the volume percent of the fuel at its lean limit, $(LFL)_i$, and the heating value of the fuel, $q_{c,i}$, is approximately a constant, say C, that is,

$$(LFL)_i q_{c,i} = C. \tag{8.5.1}$$

This constant can be estimated by, say, taking the lean flammability limit of methane–air as 5.0 percent and its heat of combustion as 192 kcal/mol, yielding a value of $C = 9.6$ kcal/mole. The existence of such a constant then implies that the adiabatic flame temperatures of these hydrocarbons are about the same at the lean limit. This temperature range has been found to be around 1,450 K.

The Burgess–Wheeler law also allows the estimation of the LFL of a hydrocarbon mixture by noting that since all the hydrocarbon components of the mixture obey Eq. (8.5.1), we can treat the fuel mixture as a single fuel such that

$$(LFL)_{\text{mix}} q_{c,\text{mix}} = C. \tag{8.5.2}$$

But since

$$q_{c,\text{mix}} = \Sigma_i X_i q_{c,i} = \Sigma_i X_i \frac{C}{(LFL)_i}, \tag{8.5.3}$$

where X_i is the mole fraction of the ith component of the fuel mixture, we have

$$(LFL)_{\text{mix}} = \left[\Sigma_i \frac{X_i}{(LFL)_i} \right]^{-1}. \tag{8.5.4}$$

Equation (8.5.4) is known as Le Châtelier's rule.

This rule has also been used to estimate the effect of inert dilution. Thus if $(LFL)_F$ is the limit without dilution, X_F the volume fraction of the fuel in the mixture of fuel and inert, and $X_I = 1 - X_F$ the corresponding inert fraction, then treating the inert as a fuel species but with $(LFL)_I \to \infty$, we have from Le Châtelier's rule,

$$(LFL)_{\text{dilution}} = \left[\frac{X_F}{(LFL)_F} + \frac{X_I}{(LFL)_I} \right]^{-1} = \frac{(LFL)_F}{X_F}. \tag{8.5.5}$$

The LFL for a diluted fuel mixture can thus be assessed by first determining the $(LFL)_{\text{mixture}}$ assuming no dilution, and then calculating $(LFL)_{\text{dilution}}$ by treating all the fuel components as a single species.

Clearly one would not expect that Le Châtelier's rule, which is based on simple molar mixing considerations, would have universal applicability. It has been found to work well only for fuel mixtures containing hydrogen, carbon monoxide, and the common hydrocarbons. It is not applicable in the presence of chemical inhibitants or such unusual compounds as carbon disulfide.

The above empirical results have proven very useful for safety estimates. The existence of Le Châtelier's rule and the limit temperature also point to the possibility of a unifying concept in the flammability limit phenomena. Such a possibility is presented next.

8.5.2. Fundamental Limits

In spite of its practical importance, "flammability limit" has remained an empirical observation without an established fundamental understanding. Since steady propagation is possible for the standard premixed flame even as the reactant concentration approaches zero, mechanisms and processes in addition to those responsible to propagate the standard flame must exist to bring about these limits, if they in fact exist.

In order to identify these limits, we recognize that if "flammability limit" is indeed a useful fundamental concept, then the lean and rich flammability limits of a mixture, and the associated limit mechanisms, must be unique physicochemical properties of this mixture. These are to be contrasted with the extinction limits, which can exist at any mixture concentration, bounded by the two flammability limits, as long as the externally imposed, system-dependent loss intensities are sufficiently strong. Thus identification of the flammability limits should require the elimination of all external, system-dependent loss mechanisms such as conductive and convective heat-loss, aerodynamic straining, and buoyancy-related phenomena as manifested by the differences between the limits for the upward- and downward-propagating flames.

With the above requirements, the configuration based on which flammability limit can be best defined would be the state at which steady propagation of the planar premixed flame in the one-dimensional doubly infinite domain fails to be possible. This state should also be completely specifiable by the mixture properties. There are two mechanisms that can weaken flame propagation and are also properties of the mixture. The first is radiative heat loss from the flame to the cold environment. The extent of loss depends on the radiative properties of the mixture, the flame temperature and thickness, and the ambient, unburned gas temperature, which can all be specified based on properties of the unburned mixture. The viability of this extinction mechanism depends on the relative sensitivities of the Arrhenius chemical reactivity and the radiative heat loss to temperature variations. That is, since the flame temperature decreases as the mixture concentration approaches the flammability limit, both the reaction rate and heat-loss rate will decrease. If the heat loss rate is more sensitive to the temperature reduction, then the loss process will be "quenched" first, rendering the flame to be essentially adiabatic. In this case radiative heat loss may not be a viable mechanism for flammability. However, if chemical reaction is more temperature sensitive and, hence, is quenched first in the presence of radiative heat loss, then radiation is a viable flammability mechanism.

Indeed, the analysis of Spalding (1957) for flame extinction with volumetric heat loss discussed in the previous section is meant to be a study of flammability limits. By using large activation energies, more recent and rigorous analysis (Buckmaster & Ludford 1982) implicitly assumes that chemical reaction is more sensitive to temperature variations than heat loss. The extinction turning point derived can therefore be identified as a flammability limit. The quantitative realism of the limit, however, is inadequate because of the grossly simplified descriptions of chemical kinetics as well as the radiative and diffusive transport properties.

The second mechanism that could contribute to the flammability limit is chemical kinetics. Here as the flame temperature is decreased, the overall reactivity is reduced. Furthermore, while this temperature reduction has a significant influence on the temperature-sensitive two-body branching reaction, its influence on the temperature-insensitive three-body termination reaction is minimal. However, since the system is still a conservative one even when considering a detailed reaction mechanism, kinetic retardation alone is not sufficient to bring about flammability. On the other hand, the much reduced reaction rate due to chain termination is expected to have so severely weakened the flame that it can be readily extinguished at a certain concentration due to the unavoidable loss disturbances present, such as radiative heat loss. This concentration can thus be identified as the flammability limit.

The separate concepts of radiative heat loss and kinetic termination can be thus interpreted from a unified viewpoint (Law & Egolfopoulos 1992; Miller 1996). To demonstrate this, we first numerically calculate the flame structure and propagation speed, with detailed descriptions of chemical kinetics as well as diffusive and radiative transport. Figure 8.5.1 shows calculated flame speed and maximum flame temperature as functions of fuel equivalence ratio ϕ for lean methane–air mixtures. It is found that in the absence of heat loss, s_u^o decreases monotonically with decreasing ϕ. However, when radiative heat loss is included, the extinction turning point is exhibited.

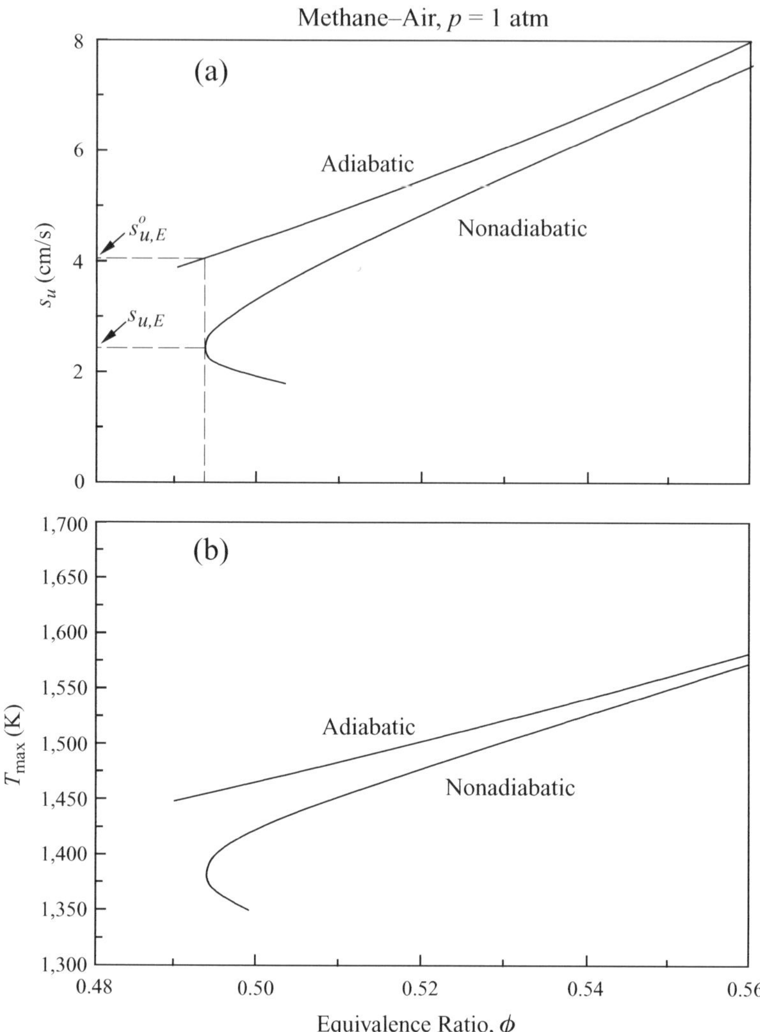

Figure 8.5.1. Computed (a) laminar flame speed and (b) maximum flame temperature as functions of equivalence ratio for methane–air flames at 1 atm pressure, with and without radiative heat loss.

Thus the fuel concentration at this state can be identified as the lean flammability limit. The calculated limit, $\phi = 0.493$, agrees well with the empirical value of 0.48. It is also significant to note that at the extinction state $s_u/s_u^o = f/f^o \approx 0.6$, which is very close to the value of $e^{-1/2} \approx 0.61$ predicted by the simplified theory of last section, and that the flame temperatures are about 1,380 K and 1,450 K for propagations with and without radiative heat loss respectively. The adiabatic flame temperature agrees with the adiabatic limit temperature of about 1,450 K determined for the empirically observed extinction concentrations. The exact numerical agreement is, of course, fortuitous.

From the calculated results one can also use sensitivity analysis to evaluate the dominant branching (B) and termination (T) reactions as the flammability limit is

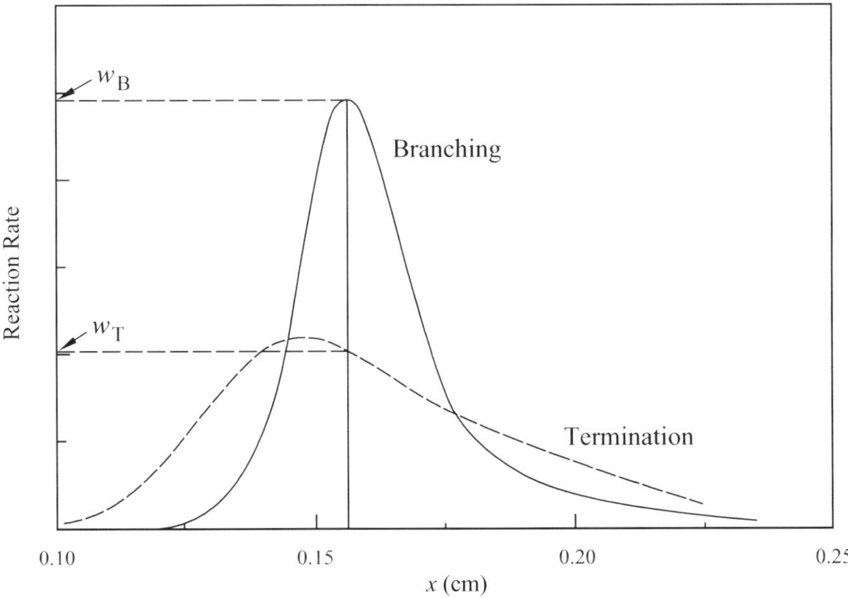

Figure 8.5.2. Profiles of the dominant branching and termination reactions needed for the evaluation of the flammability exponent.

approached. For lean methane–air mixtures, the reaction pair is found to be

$$H + O_2 \rightarrow OH + O \tag{B}$$

$$H + O_2 + M \rightarrow HO_2 + M, \tag{T}$$

as expected, where (B) and (T) are respectively reactions (H1) and (H9).

The relative sensitivity of the termination and branching reactions to concentration variations can be assessed by evaluating a sensitivity exponent defined as

$$\alpha(\phi) = \frac{\partial \ln w_T / \partial \ln \phi}{\partial \ln w_B / \partial \ln \phi} = \frac{\partial(\ln w_T)}{\partial(\ln w_B)}, \tag{8.5.6}$$

where w_B is a representative value of (B), say its maximum value in the reaction zone, and w_T is the value of (T) at the location where w_B is evaluated, as shown in Figure 8.5.2. Since Eq. (8.5.6) is equivalent to the relation

$$w_T \sim w_B^{\alpha}, \tag{8.5.7}$$

the flammability exponent has the significance that w_T responds to changes in w_B in either a gradual or decelerative manner for $\alpha < 1$, and in an accelerative manner for $\alpha > 1$. Thus $\alpha = 1$ can be considered to be the state at which the terminating reaction becomes dominating.

Figure 8.5.3 shows the variation of α with ϕ, for lean methane–air flames with and without radiative heat loss. For the nonadiabatic flame, it is seen that shortly after the state of $\alpha = 1$ is reached, at $\phi = 0.501$, the turning-point state is also attained, at $\phi = 0.493$. Thus the state of flammability limit is characterized by thermal extinction

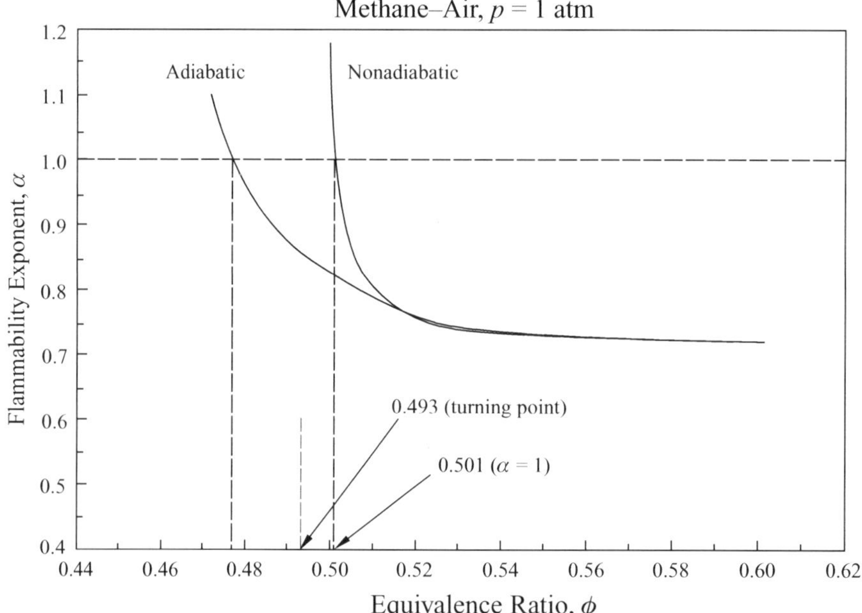

Figure 8.5.3. The flammability exponents for the flames of Figure 8.5.1 needed to assess fundamental flammability limits.

as well as by chain termination overwhelming chain branching in the kinetic mechanism. This is physically reasonable because as soon as chain termination becomes dominating, the overall chemical reaction rate slows down, consequently leading to thermal extinction.

It is also seen that in the absence of radiative heat loss, the attainment of $\alpha = 1$ is only slightly delayed, now occurring at $\phi = 0.477$. Thus the flammability limit can be determined with reasonable accuracy using the kinetic termination criterion ($\alpha = 1$) alone, without considering radiative loss.

It has also been shown that reactions (H1) and (H9) are the dominant branching and termination reactions for many lean hydrocarbon–air mixtures as well as lean and rich hydrogen–air mixtures for the pressure range over which the H–O_2 kinetics dominates. For rich hydrocarbon–air mixtures, the dominant termination/retardation reaction is more fuel specific, but is frequently a two-body reaction. For example, for rich methane–air flames, it is

$$CH_4 + H \rightarrow CH_3 + H_2, \qquad \text{(M3)}$$

through which the reactive H radical is lost and a less reactive CH_3 radical generated.

The nature of the termination reactions explains the experimentally observed pressure dependence of the flammability limits. That is, since (H9) is a three-body reaction that becomes more efficient with increasing pressure, relative to the two-body reaction (H1), the flammability limits are mostly narrowed for those mixtures that are sensitive to (H9). On the other hand, since (M3) is a two-body reaction and has no advantage over (H1) with increasing pressure, the flammability limit is extended.

Calculated flame structure also shows that as the flammability limit is approached, the termination reaction rate profile merges with the branching reaction rate profile, indicating that scavenging of the radicals becomes the most efficient. Thus with further weakening of the flame and radical production, the overall reaction rate decreases precipitously. Indeed, in systems in which several retarding reactions are found to have the same extent of influence on flame propagation, the flammability limit is determined by the reaction whose profile merges the closest with that of the branching reaction.

8.6. FLAME STABILIZATION AND BLOWOFF

We have established in Chapter 7 that a given fuel–air mixture possesses a definite propagation speed s_u^o, which is defined by the thermochemical properties of the mixture. Thus once ignited, the flame will propagate into the fresh mixture with a velocity close to s_u^o. However, from the practical point of view we frequently desire to keep this flame stationary in space such that a continuous mode of operation can be achieved within a combustor, such as the gas turbine engine and the industrial furnace. In order to achieve this, it appears that the oncoming fresh mixture must have precisely the same speed to balance the flame speed so as to render the flame stationary. This not only is basically impossible to accomplish but, even if it can be done, is also too restrictive in the operational range of the combustor. The principle of flame stabilization is to provide some mechanism through which the flame burning intensity can be automatically modified so that the flame is afforded sufficient flexibility to adjust its location, orientation, and configuration in a nonuniform, temporally varying flow field. Thus not only static equilibrium can be attained between the flow velocity and the flame speed in a localized region, there is also sufficient flexibility for the flame to accommodate changes in the operation conditions so as to attain dynamic equilibrium.

In the following we shall discuss how such a dynamic equilibrium can be achieved to effect flame stabilization, and situations under which stabilization is not possible, leading to flame blowoff.

8.6.1. The Flat-Burner Flame

The flat-burner flame, introduced in Section 7.6.2, has been extensively used in combustion studies, especially those related to the chemical structure of flames. In this configuration (Figure 8.6.1) a combustible mixture issues from a porous plate that may or may not be internally cooled. Heat is conducted from the preheat zone to the plate, and is the basic mechanism through which the flame can remain situated above the burner at a given standoff distance. Since heat transfer to the plate is through conduction, the plate must be in contact with the preheat zone. Furthermore, since

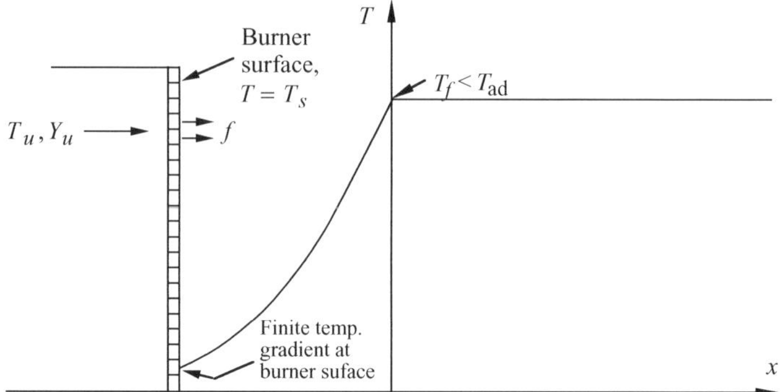

Figure 8.6.1. Schematic and the temperature profile for the flat-burner flame.

there is always a finite amount of heat transfer at the burner surface, the temperature in the flow field is overall reduced from the adiabatic limit based on the freestream mixture, while the mass burning flux of the flame, f, which is simply the flow discharge flux from the burner, is correspondingly smaller than the adiabatic freely propagating flame. Only in the limit of vanishing heat loss does the flame degenerate to the adiabatic situation.

There has been considerable confusion regarding the role of heat loss in the flat-burner flame as compared to the volumetric heat loss in the freely propagating flame of Section 8.4. In fact, after formulating the theory of Section 8.4, Spalding and Yumlu (1959) attempted to verify this theory by using the flat-burner flame. The experimental result on the burning flux $\tilde{f}$ as a function of the heat loss rate $\tilde{L}_s$ to the burner not only yielded the upper, strongly burning branch for large $\tilde{f}$, but also a small segment of the lower, slower-burning branch extending from the turning point, as shown in Figure 8.6.2a. This seemed to suggest that there could exist two flame speeds for a given heat loss rate. Since stability analysis of the nonadiabatic

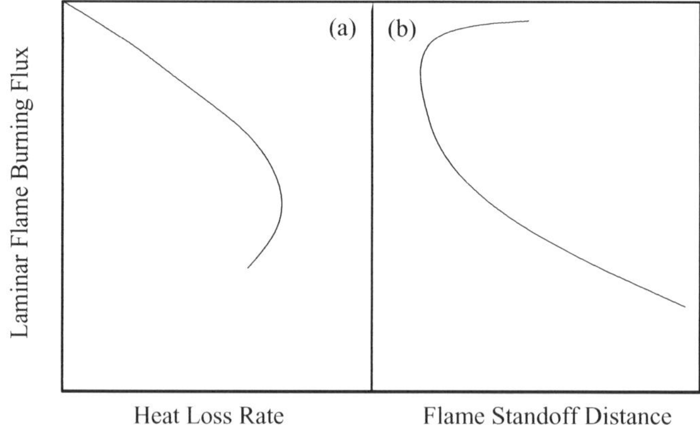

Figure 8.6.2. Schematic showing the experimentally observed dual flame speed phenomenon: (a) for a given heat loss rate to the burner surface; (b) for a given flames standoff distance.

freely propagating flame showed that the slow-burning branch solution is unstable, the meaning of the observed slow branch result has been unclear. This has become the classical dual flame speed anomaly. The existence of the experimental turning point, however, has been cited frequently as the verification of Spalding's theory, especially on the identification of the turning point as the state of extinction or even flammability limit.

Further studies by Ferguson and Keck (1979) revealed an alternate manifestation of the dual flame speed phenomena, namely the existence of two flame speeds for the same flame standoff distance, as shown in Figure 8.6.2b. It was further suggested that the distance of closest approach be identified as the quenching distance of the flame. Let us now solve and examine this problem in detail (Eng, Zhu & Law 1995).

A combustible mixture of temperature T_u and reactant concentration Y_u is discharged with a mass flux f from a porous plate that has a surface temperature T_s. Because of heat loss to the plate, the resulting flame temperature T_f is lower than the adiabatic flame temperature $T_{\rm ad}$ based on T_u and Y_u. There is, however, no heat loss from the flame structure downstream of the burner surface. In nondimensional form, total energy conservation is

$$\tilde{f}\frac{d(\tilde{T}+\tilde{Y})}{d\tilde{x}} - \frac{d^2(\tilde{T}+\tilde{Y})}{d\tilde{x}^2} = 0, \tag{8.6.1}$$

where $\tilde{x} = (f^o/\rho D)x$, $\tilde{Y} = Y/Y_u$, and $\tilde{T} = (c_p T)/(q_c Y_u)$. Integrating Eq. (8.6.1) from the burner surface ($\tilde{x} = 0^+$) to the downstream ($\tilde{x} = \infty$), and by using the boundary conditions

$$\tilde{x} = 0^+: \quad \tilde{T} = \tilde{T}_s, \quad \tilde{f}\tilde{Y} - \frac{d\tilde{Y}}{d\tilde{x}} = \tilde{f}, \quad \left(\frac{d\tilde{T}}{d\tilde{x}}\right)_{0^+} = \left(\frac{d\tilde{T}}{d\tilde{x}}\right)_{0^-} = \tilde{L}_s \tag{8.6.2}$$

$$\tilde{x} = \infty: \quad \tilde{T} = \tilde{T}_f, \quad \tilde{Y} = 0, \quad \frac{d\tilde{T}}{d\tilde{x}} = \frac{d\tilde{Y}}{d\tilde{x}} = 0, \tag{8.6.3}$$

where $\tilde{L}_s$ is a surface heat loss rate parameter, we obtain

$$\tilde{f}[\tilde{T}_{\rm ad} + (\tilde{T}_s - \tilde{T}_u)] - \tilde{L}_s = \tilde{f}\tilde{T}_f. \tag{8.6.4}$$

Equation (8.6.4) states that, as the freestream of total energy $\tilde{f}\tilde{T}_{\rm ad}$ passes through the porous plate, it picks up an additional sensible heat $\tilde{f}(\tilde{T}_s - \tilde{T}_u)$ as it is heated to $\tilde{T}_s$, while at the same time it loses $\tilde{L}_s$ amount of energy due to cooling. The remaining energy is then simply used to heat the mixture to $\tilde{T}_f$. Also note that $Y_s \neq Y_u$.

The amount of heat loss can be easily determined by solving $\tilde{f}\tilde{T}' - \tilde{T}'' = 0$ for the temperature distribution in the preheat zone, which yields

$$\tilde{T} = \tilde{T}_s + (\tilde{T}_f - \tilde{T}_s)\left(\frac{e^{\tilde{f}\tilde{x}} - 1}{e^{\tilde{f}\tilde{x}_f} - 1}\right), \qquad \tilde{x} \leq \tilde{x}_f, \tag{8.6.5}$$

from which

$$\tilde{L}_s = \left(\frac{d\tilde{T}}{d\tilde{x}}\right)_{0^+} = \frac{\tilde{f}(\tilde{T}_f - \tilde{T}_s)}{e^{\tilde{f}\tilde{x}_f} - 1}. \tag{8.6.6}$$

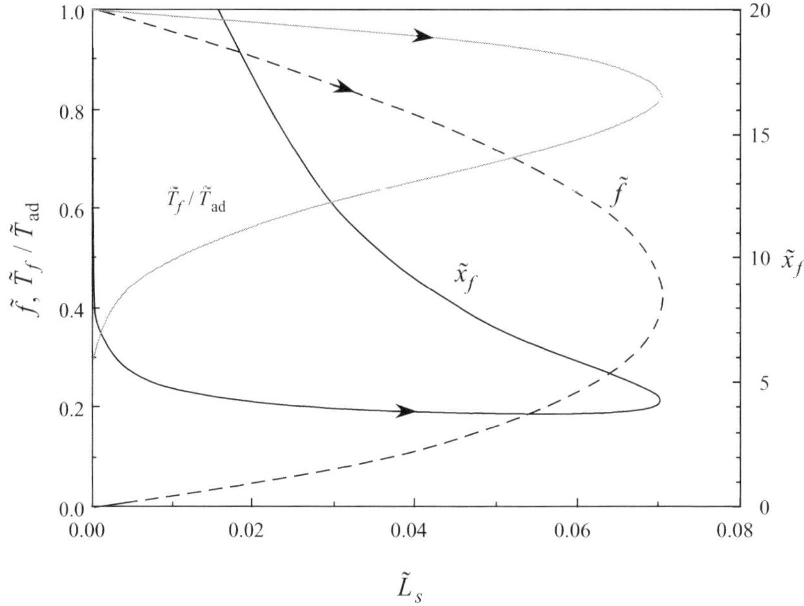

Figure 8.6.3. Responses of the flat-burner flame from analytical solution ($\tilde{T}_u = \tilde{T}_s = 0.12$) with heat-loss rate as the independent parameter. Arrows indicate the corresponding branches of solution for different responses.

Furthermore, from Eq. (7.5.53) we have $f^2 \sim T_f^4 \exp(-\tilde{T}_a/\tilde{T}_f)$, which, when referenced to the adiabatic state, yields Eq. (7.5.54),

$$\tilde{f}^2 = \left(\frac{\tilde{T}_f}{\tilde{T}_{ad}}\right)^4 \exp\left[-\tilde{T}_a\left(\frac{1}{\tilde{T}_f} - \frac{1}{\tilde{T}_{ad}}\right)\right], \qquad (8.6.7)$$

where $\tilde{T}_f$ is given by Eq. (8.6.4).

The problem is completely solved in that, for a given flow discharge rate and, hence, mass burning flux $\tilde{f}$, the flame temperature $\tilde{T}_f$, the heat loss rate $\tilde{L}_s$, and the flame location $\tilde{x}_f$ are given by Eqs. (8.6.7), (8.6.4), and (8.6.6). Note that unlike the problem of the standard premixed flame with volumetric heat loss, in which $\tilde{T}_f$ is only slightly reduced from $\tilde{T}_{ad}$, the reduction can be substantial for the present problem, depending on the extent of heat transfer at the burner surface.

Figures 8.6.3 and 8.6.4 respectively show typical plots of the flame response according to Eqs. (8.6.4), (8.6.6), and (8.6.7), with either $\tilde{L}_s$ or $\tilde{f}$ as the independent variable. Specifically, the proper limiting behaviors are exhibited in that $\tilde{T}_f \to \tilde{T}_{ad}$, $\tilde{x}_f \to \infty$ and $\tilde{L}_s \to 0$ in the adiabatic limit of $\tilde{f} \to 1$, while $\tilde{T}_f \to \tilde{T}_u$, $\tilde{x}_f \to \infty$ and $\tilde{L}_s \to 0$ in the nonreactive limit of $\tilde{f} \to 0$. Furthermore, there are two solutions for $\tilde{f}$ for a given heat loss rate $\tilde{L}_s$, and two $\tilde{f}$ for a given flame standoff distance. Consequently, there exists a maximum heat loss rate, $\tilde{L}_{s,\max}$, and a minimum flame standoff distance, $\tilde{x}_{f,\min}$, beyond which there is no solution. These dual solution behaviors are in accord with the experimental observations discussed above.

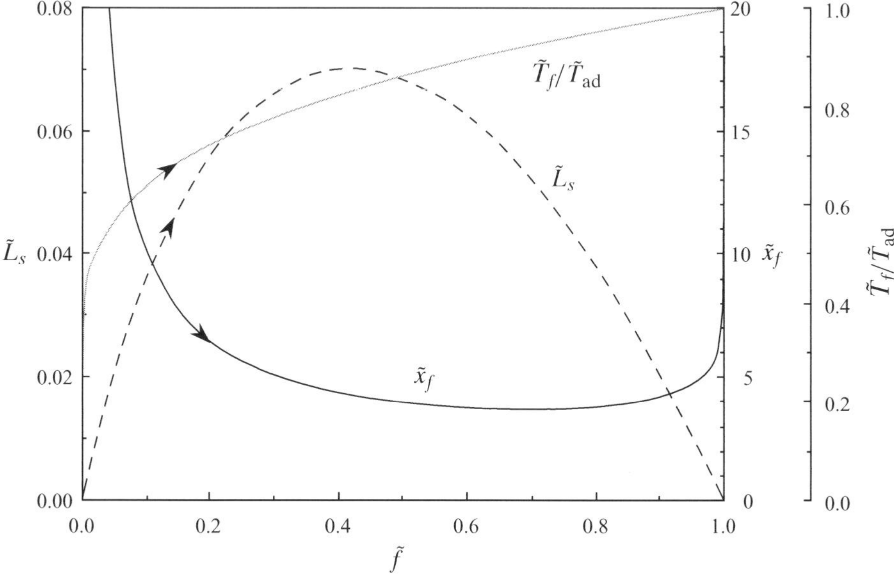

Figure 8.6.4. Same result as those of Figure 8.6.3, but with mass flow rate as the independent parameter.

It is, however, important to recognize the fundamental differences between the nonadiabatic, freely propagating flame with volumetric heat loss and the present burner-stabilized flame with heat transfer to the upstream boundary, even though they both seem to exhibit the dual flame speed, turning-point behavior when the burning flux $\tilde{f}$ is plotted versus the heat loss parameter, as in Figures 8.4.2 and 8.6.3. First, from an operation point of view, the heat loss rate can be considered to be independently specifiable for the freely propagating flame. Thus by continuously increasing the heat loss rate by changing, say, the heat transfer coefficient $\tilde{h}$, the flame can eventually be brought to the state of extinction, at $\tilde{L}_{v,E}$. For the burner-stabilized flame, with a fixed surface temperature, however, it is operationally more difficult to control the heat loss rate because the flame has the flexibility to adjust its location and thereby the extent of conductive heat transfer. Rather, it is the flow discharge rate $\tilde{f}$ that can be independently controlled. Thus if we take $\tilde{f}$ as the independent variable, then Figure 8.6.4 shows that for a given $\tilde{f}$ there is a unique heat loss rate $\tilde{L}_s$ and a unique flame location $\tilde{x}_f$. The system response is therefore single valued. Furthermore, since it is difficult to control the heat loss rate, the burner-stabilized flame cannot be easily extinguished through increasing heat loss. In fact, the most viable mechanism to "eliminate" the flame is to continuously increase the discharge rate until it exceeds the adiabatic, freely propagating value. Since flame elimination is caused by the approaching flow having a velocity exceeding that of the adiabatic flame, the flame in this case is actually blown off rather than extinguished.

Consequently, the turning points of the response curves for the flat-burner flame can be simply interpreted as manifestations of the nonmonotonic flame responses

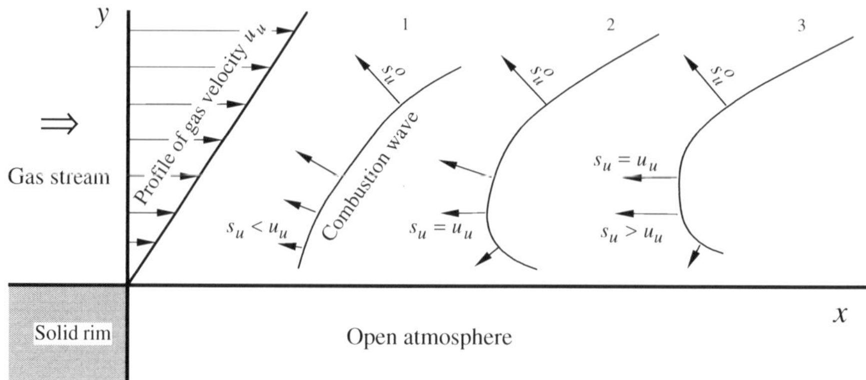

Figure 8.6.5. Stabilization mechanism of Bunsen flames (adapted from Lewis & von Elbe 1987).

to variations in the flow discharge rate. Thus neither is the turning point of the $\tilde{f}$-versus-$\tilde{L}_s$ curve an extinction state, nor does the turning point of $\tilde{x}_f$-versus-$\tilde{f}$ curve designate the quenching distance.

The classical "dual-flame speed" anomaly is therefore an artificial one in that the nonadiabatic freely propagating flame and the flat-burner flame are two fundamentally different systems. The anomaly arises when one tries to explain the experimental results of the latter based on the theory of the former. The anomaly disappears when the explanation is based on the theory for the flat-burner flame. Perhaps also equally important to note is that the presence of a turning point does not automatically imply the existence of an extinction state, and the presence of heat loss also does not imply that the flame will be extinguished at some state.

The viability of the above concept has been further substantiated by additional experiments as well as computation with detailed chemistry and transport (Eng, Zhu & Law 1995), with close agreement between them.

8.6.2. Stabilization of Premixed Flame at Burner Rim

We next discuss flame stabilization through heat loss and flow nonuniformity. As an example let us consider the stabilization of the Bunsen flame, which, as we know, can stay over the burner exit for a considerable range of the flow velocity as well as the mixture concentration. The local region within which the flame is stabilized is that of the burner rim; the stabilization process is shown in Figure 8.6.5. We mention in passing that equivalent terminologies for flame stabilization in this manner are flame holding, flame anchoring, and flame attachment.

Following Lewis and von Elbe (1987), we consider the flow in the burner tube to be well developed such that it has a parabolic velocity profile at the exit plane. Furthermore, since the region of interest is only confined to a narrow one adjacent to the rim, the velocity variation there can be approximated to be linear. In this localized region, the speed of the flame segment, s_u, can be significantly modified from s_u^o by two factors. First, cooling from the rim causes $s_u < s_u^o$. Second, the concentration of the mixture is modified by the ambient air through diffusion and entrainment. If the mixture is fuel lean, then it will become more so as a result of mixing, causing the flame speed to further decrease. However, if the mixture is fuel rich, then it will become more stoichiometric upon mixing with air and thereby burn more strongly. As we move away from the rim into the center, the flame tends to resume its laminar flame speed s_u^o based on the upstream composition of the flame segment, and adjusts its configuration in accordance with the local velocity profile.

Now consider the three possible flame configurations shown in Figure 8.6.5, assuming a fuel-lean mixture. In configuration 1, the flame is too close to the rim such that excessive heat loss causes s_u to be less than the approach flow velocity everywhere over this flame segment. The flame is pushed back. In the equilibrium configuration 2, the flame is further away from the rim, and suffers less heat loss. Consequently, s_u increases except at the outermost fringe where it further decreases because dilution due to air entrainment now becomes important. Therefore at one point on the flame segment the flame speed exactly balances the flow velocity, providing anchoring of the entire flame. Finally, in configuration 3 the flame is too far from the rim such that the laminar flame speed exceeds the flow velocity somewhere, causing the flame to move forward toward the equilibrium position.

It is clear that this flame anchoring mechanism allows considerable flexibility for the flame segment to respond to changes in the mixture velocity and concentration by adjusting its location relative to the rim, such that a local velocity balance can be achieved. When such a balance cannot be achieved, flashback or blowoff occurs, to be discussed next.

By continuously decreasing the mixture velocity or increasing its reactivity, the flame is situated closer to the rim. When this velocity has been reduced to a certain level, characterized by its velocity gradient, the gas velocity at some point in the flow becomes smaller than the flame speed. The flame would then propagate against the flow into the tube, resulting in flashback. The condition for flashback can be quantified in Figure 8.6.6, in which the linear flow velocity profile and the flame speed variation next to the burner wall are compared. It is seen that if the flow velocity is higher than the flame speed (curve 3), then the flame will be expelled out of the tube. Flashback occurs for the opposite situation (curve 1). Thus curve 2 gives the incipient state at which flashback is possible.

For blowoff, we refer to Figure 8.6.7, again schematized for the situation of a fuel-lean mixture. Thus with increasing flow velocity the flame is lifted higher and higher above the rim. The flame speed at the base of the flame also increases because of the reduced heat loss to the rim. There is, however, an upper limit beyond which the flame speed cannot be further increased. Therefore with an excessively high flow velocity, represented by curve 4, the gas velocity exceeds the flame speed everywhere and the flame blows off. Thus curve 3 gives the incipient state of blowoff.

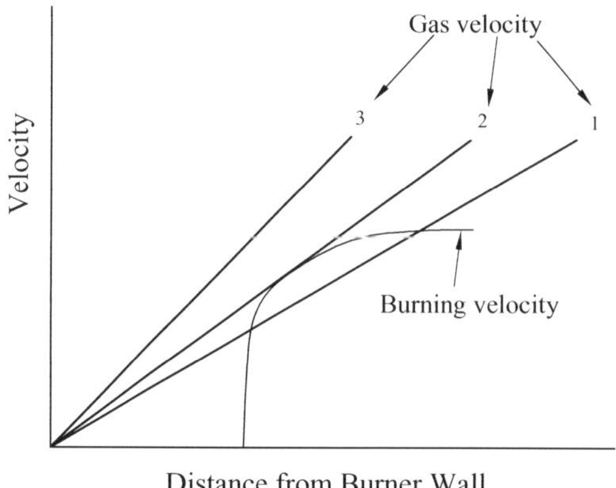

Figure 8.6.6. Flashback mechanism of Bunsen flames (adapted from Lewis & von Elbe 1987).

If we let y be the distance of the flame base above the burner rim, and $v(y)$ the local flow velocity, stabilization is possible when the conditions

$$s_u = v_u, \qquad \left(\frac{\partial s_u}{\partial y}\right) > \left(\frac{\partial v}{\partial y}\right)_u \tag{8.6.8}$$

are simultaneously satisfied. The state of incipient blowoff is reached when the gradients of s_u and v become equal.

It is important to recognize that flame blowoff and flame extinction are two entirely different phenomena. Blowoff is caused by a lack of dynamic balance between the flame speed and the flow velocity. When blowoff is effected by increasing gas velocity, the burning intensity of the flame at the stabilization point attains its local maximum at incipient blowoff, and the flame retains its structure when blown off. On the other hand, extinction is caused by the precipitous drop in the chemical reactivity, which completely annihilates the flame. The burning intensity of the flame

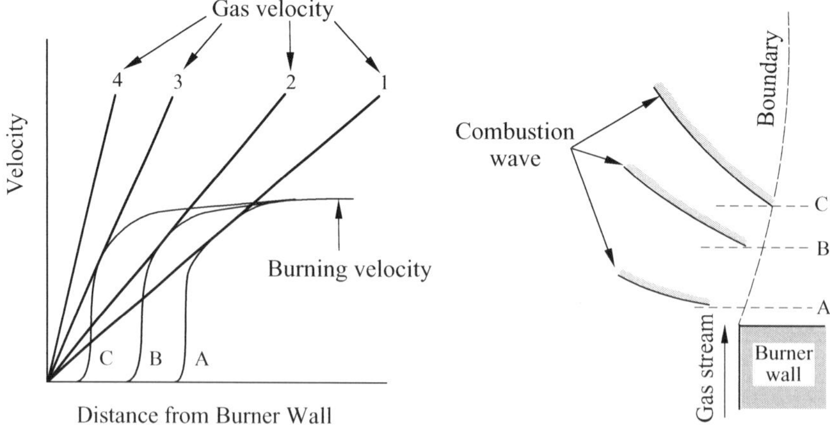

Figure 8.6.7. Blowoff mechanism of Bunsen flames (adapted from Lewis & von Elbe 1987).

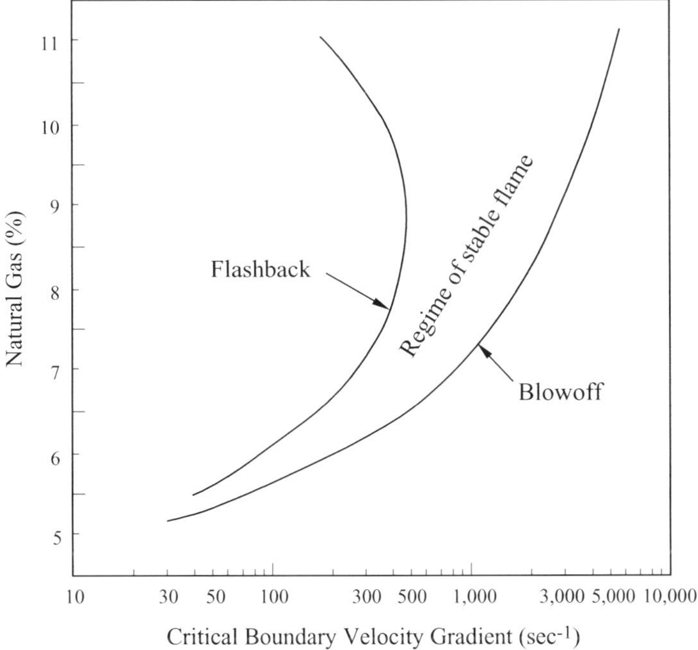

Figure 8.6.8. Blowoff, stable, and flashback regimes of the laminar Bunsen flame (adapted from Lewis & von Elbe 1987).

therefore attains its local minimum at incipient extinction in response to the imposed extinction mechanism.

Figure 8.6.8 shows the boundaries of flashback and blowoff as functions of the fuel concentration and flow velocity gradient for mixtures of natural gas and air. Conditions for the stable operation of a Bunsen flame are bounded by these flashback and blowoff curves. It is seen that flashback can be suppressed for either fuel-lean or fuel-rich mixtures, as is reasonable to expect. For blowoff such symmetry does not exist in that blowoff becomes progressively more difficult with increasing fuel concentration and thereby increasing burning intensity of the flame.

In Chapter 10 we shall show that in addition to heat loss, the flame speed can also be modified by aerodynamic straining, flame curvature, and mixture nonequidiffusion. The concept of flame stabilization is therefore significantly modified and enriched.

8.6.3. Stabilization of Nonpremixed Flame at Burner Rim

When studying the configuration of the Burke–Schumann flame in Chapter 6, a reaction sheet was assumed to exist at the burner rim and the issue of flame anchoring was not addressed. In realistic situations it is clear that there must exist a dead space at the burner rim because of its low temperature, and that with sufficiently high flow velocity the flame will be blown off.

There is increasing evidence that the bulk nonpremixed flame is stabilized via a premixed flame segment adjacent to the burner rim (Figure 8.6.9). That is, as the

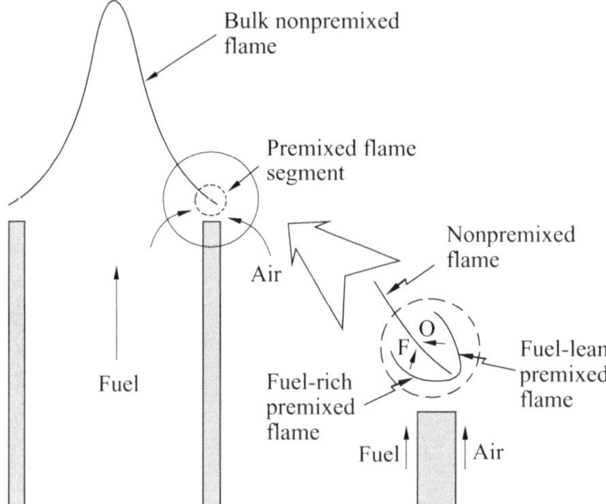

Figure 8.6.9. Schematic showing the stabilization of a nonpremixed Bunsen flame by the premixed flame segment at the flame base: (a) overall flame configuration; (b) structure of flame base.

fuel gas issues from the burner tube, it first mixes with the ambient air in the burner rim region where reaction cannot take place because of the low temperature. Thus when reaction does occur, the segment of the flame at the burner rim assumes the character of a premixed flame. The bulk of the flame, away from the burner rim, is of course still a nonpremixed flame. Stabilization of the entire flame, however, is controlled by the stabilization of the premixed flame segment in the same manner as just discussed.

The premixed flame segment needed for stabilization actually has a rather interesting flame structure, shown as the inset in Figure 8.6.9. Here fuel and oxidizer mix immediately downstream of the burner rim, creating a mixture with a stratified composition ranging from fuel lean to fuel rich. The resulting premixed flame therefore also has a stratified burning intensity over its surface, yielding combustion products that consist only of oxygen and fuel from the fuel lean to the fuel rich flame segments respectively. These oxygen- and fuel-based combustion products in turn form a nonpremixed flame of increasing strength that eventually becomes the bulk nonpremixed flame shown in the overall flame configuration. Such a flame structure is called a triple flame or a tribrachial flame.

We shall revisit the problem of nonpremixed flame stabilization, via the triple flame structure, in Chapter 12.

8.6.4. Stabilization of Lifted Flames

Figure 8.6.8 shows that, with continuous increase in the fuel concentration, flame blowoff becomes increasingly more difficult. Figure 8.6.10 shows a complete stabilization mapping covering the entire range of fuel concentration in air (Wohl, Kapp & Gazley 1949); the blowoff limit when the environment is inert is also indicated. The first point to note is that when the mixture is sufficiently fuel rich, the entire flame

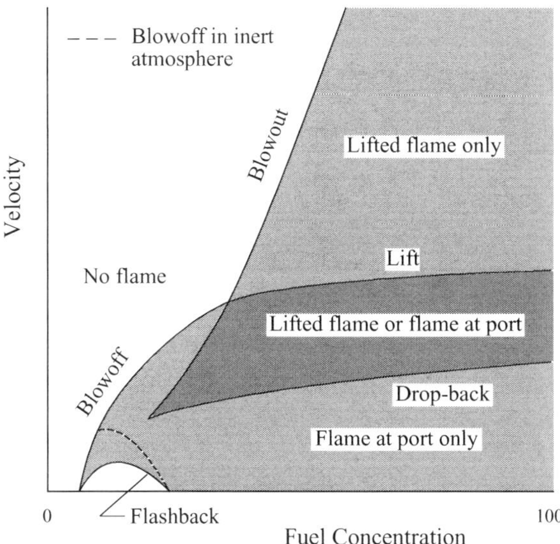

Figure 8.6.10. Complete stabilization map of the Bunsen flame, recognizing that the flame is turbulent at high flow velocities (adapted from Wohl, Kapp & Gazley 1949).

behaves like a nonpremixed flame in the air environment. Thus for fuel concentrations beyond this limit it is reasonable to treat the system as being nonpremixed. For such mixtures, blowoff cannot be readily achieved. Instead, with excessive blowing the flame will be lifted from the burner rim and suspend itself somewhere above the burner. The flow can become turbulent under these situations because of the high flow rate. There is considerable amount of premixing between the fuel-rich jet and the ambient air before the flow reaches the flame. The resulting flame thus has the characteristics of both premixed and nonpremixed flames. With further increase in velocity, the lifted flame will eventually be blown out.

If we start with a lifted flame and gradually decrease the flow velocity, then eventually the lifted flame will drop back to the burner rim. There is, however, a considerable range of hysteresis between lift off and drop back. The hysteresis is due to the change in the flow pattern and the rim temperature, which is influenced by whether a flame is in its vicinity.

When the environment is inert, then the propensity to be blown off is similar for lean and rich mixtures. The blowoff velocity therefore peaks around stoichiometric as shown in Figure 8.6.10. Furthermore, since flashback is not affected by entrainment of the ambient gas, it also exhibits the somewhat symmetric sequence as shown.

Two mechanisms have been proposed for the stabilization and blowout of a lifted flame. The first assumes that because of sufficient mixing, the base of the lifted flame is primarily a premixed flame, which may or may not be turbulent. Thus stabilization and blowout can be assessed on the dynamic balance between the flow velocity and the premixed flame speed at the flame base. The second mechanism assumes inadequate mixing such that the flame base is primarily a nonpremixed flame. Stabilizaton and blowout in this case are based on the sustenance and extinction of either the bulk laminar nonpremixed flame or the laminar nonpremixed flamelets constituting the bulk turbulent nonpremixed flame base.

PROBLEMS

1. Show that the vertical tangent criterion in determining the ignition and extinction states of the S-curve is identical to the Semenov criterion on the tangency of the heat generation and loss curves.

2. An idealized structure for the one-dimensional planar detonation wave, referred to as the ZND (Zel'dovich–von Neumann–Döring) structure, is one consisting of a leading hydrodynamic shock of infinitesimal thickness and Mach number M_u, followed by a period of induction before the shocked mixture ignites. The flow velocity is usually so high that diffusion can be considered negligible throughout the detonation structure. Derive an expression for the ignition distance behind the shock for an unburned mixture of temperature T_u and density ρ_u in the limit of a strong shock ($M_u \gg 1$.)

3. Derive the ignition criterion of a combustible by a hot sphere of radius r_s and constant temperature T_s.

4. Consider the problem of Section 8.2.2 for the hot plate ignition of a stagnant confined mixture, except now the hot plate is porous such that the reactant concentration there is $Y_{i,s} \equiv 0$. Let $Le_i = 1$ for simplicity. Just derive the structure equation for the reaction zone without solving it.

5. Derive the ignition criterion for a heated, isothermal, catalytic plate of fixed temperature T_s in the stagnation flow. First derive the coupling function and make sure that consumption of the reactants at the surface is allowed. Why is it more difficult to ignite for the present problem, in the presence of a catalyst?

6. A metal particle of radius r_s and fixed temperature T_s undergoes surface burning with a burning rate $w = k_s \rho_s Y_{O,s}$ in a quiescent environment of oxidizer concentration $Y_{O,\infty}$, where k_s is the reaction rate constant for the surface reaction, and ρ_s and $Y_{O,s}$ the density and oxidizer mass fraction at the surface, respectively.

(a) Show that (Glassman & Law 1991) the nondimensional mass burning rate of the particle is

$$\tilde{m}_F = \frac{1}{\sigma_O} \ln\left(\frac{1 - \sigma_O \tilde{m}_F / \tilde{k}_s}{1 - Y_{O,\infty}}\right)$$

where $\tilde{m}_F = m_F/(4\pi\rho D r_s)$, $\tilde{k}_s = (k_s \sigma_O r_s \rho_s)/(\rho D)$, and σ_O is the stoichiometric oxidizer-to-fuel mass ratio.

(b) Show that the burning intensity is very sensitive to $Y_{O,\infty}$ for a highly reactive surface, with $\tilde{k}_s \gg 1$. This is the principle of deliberately doping pure oxygen in storage with a trace amount of inert gas so as to reduce its explosion hazard.

7. For the flame ball problem (Problem 9 in Chapter 7), let us now take into account the unavoidable radiation heat loss. If we assume for simplicity that the loss is only significant in the upstream region, and that $Le = 1$, derive expressions for the modified flame temperature and flame location, as well as the critical flame radius beyond which the steady-state flame cannot be established. You can use either the phenomenological or the Frank-Kamenetskii approach.

9 Asymptotic Structure of Flames

In 1974 Liñán published the first complete analysis of the asymptotic structure of nonpremixed flames based on the one-step second-order reaction. Although the analysis was conducted for the steady counterflow flame with unity Lewis number, it was subsequently recognized that the analysis is of such a canonical nature that the results are applicable to many laminar flame situations. Specifically, since one of the flame regimes identified has the characteristics of a premixed flame, the analysis is actually applicable to both nonpremixed and premixed flames. Furthermore, since the flame response is ultimately determined by that of the reaction zone, many subsequent analyses of other flame configurations (e.g. Law 1975) invariably end up with the structure equations and, hence, solutions first derived by Liñán. We shall study these structures in this chapter.

We shall first present a separate analysis of the reaction zone structure of premixed flames, with the solution expressed in terms of some general boundary conditions that are to be supplied by the outer solutions for individual situations. From our study of the ignition of a combustible by a hot surface in Section 8.2, we anticipate that such a generalized formulation is possible for situations involving thin reaction zones embedded within flames. The concept of delta-function closure will also be introduced.

For nonpremixed flames, we have studied their combustion characteristics in the reaction-sheet limit in Chapter 6. It was demonstrated that much can be learned about the bulk combustion and flame behavior without any specification of the nature of the chemical reaction, except for the implicit assumption that they occur infinitely fast relative to diffusion. However, it was also clear that without considering the finite nature of the chemical reaction rate, as well as the thermal and chemical structure of the reaction zone, we are incapable of predicting such intrinsically chemistry-based phenomena as ignition and extinction. In this chapter, we shall therefore analyze the structure of these reaction zones and thereby derive appropriate criteria for ignition and extinction of nonpremixed systems.

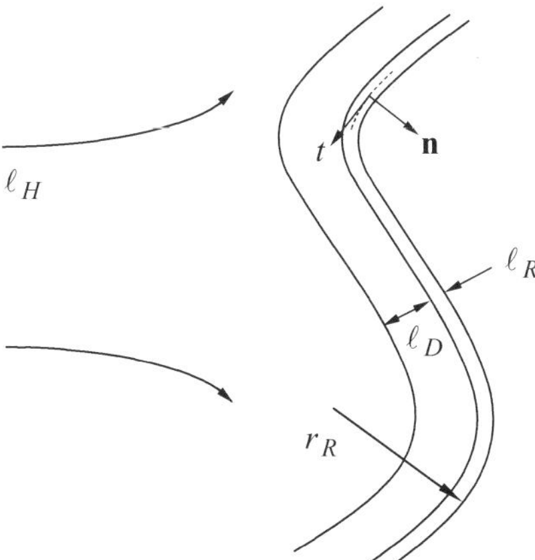

Figure 9.1.1. Schematic showing the characteristic length scales of a wrinkled flame in a nonuniform flow field.

9.1. STRUCTURE OF PREMIXED FLAMES

We consider the simplified diffusion-controlled system of Eqs. (5.6.6) and (5.6.7),

$$\frac{\partial}{\partial \tilde{t}}(\tilde{\rho}\tilde{T}) + \tilde{\nabla}\cdot(\tilde{\rho}\tilde{\mathbf{v}}\tilde{T}) - \tilde{\nabla}^2\tilde{T} = -\tilde{w} - \tilde{L}, \tag{9.1.1}$$

$$\frac{\partial}{\partial \tilde{t}}(\tilde{\rho}\tilde{Y}) + \tilde{\nabla}\cdot(\tilde{\rho}\tilde{\mathbf{v}}\tilde{Y}) - \frac{1}{Le}\tilde{\nabla}^2\tilde{Y} = \tilde{w}, \tag{9.1.2}$$

with a one-reactant, one-step reaction rate $\tilde{w} = -Da_C\tilde{Y}\exp(-\tilde{T}_a/\tilde{T})$, and a suitably nondimensionalized volumetric energy loss function $\tilde{L}$, such as that of radiation.

Using the approach similar to the one we adopted in Section 8.2 for the analysis of a combustible by a hot surface, we consider a general flame segment (Figure 9.1.1), with the sole restriction that the reaction zone thickness ℓ_R is much smaller than its radius of curvature r_R, the diffusion zone thickness ℓ_D, and the characteristic scale of the flow nonuniformity ℓ_H, such that $\ell_R \ll (r_R, \ell_D, \ell_H)$, where the subscript H designates the hydrodynamic zone. Note that we have not made any statement regarding the relative dimensions of r_R, ℓ_D and ℓ_H. Thus the entire flame does not need to be thin at the hydrodynamic scale.

Since the reaction zone is so thin, we shall treat it as a quasi-planar reaction sheet as viewed from the diffusion zone, and consider variations in the direction **n** that is locally normal to it. By further assuming that in the reaction zone the temporal and convective variations are small, and that volumetric heat loss is negligible as compared to chemical heat generation, the reaction zone is described by transverse diffusion and reaction such that Eqs. (9.1.1) and (9.1.2) become

$$\frac{d^2\tilde{T}_{\text{in}}}{d\tilde{n}^2} = \tilde{w}, \tag{9.1.3}$$

$$\frac{d^2}{d\tilde{n}^2}\left(\tilde{T}_{\text{in}} + \frac{\tilde{Y}_{\text{in}}}{Le}\right) = 0, \tag{9.1.4}$$

where $Da_C = (\lambda/c_p)B_C/f_n^2$ with f_n being the n-component of the mass burning flux f, and we have also applied the coupling function approach to obtain Eq. (9.1.4). We shall now derive the structure equation governing the inner reaction zone as well as the jump conditions relating the outer solutions across the reaction sheet.

9.1.1. Structure Equation

Using the inner expansions,

$$\tilde{T}_{\text{in}}(\chi) = \tilde{T}_f - \epsilon\theta(\chi) + O(\epsilon^2), \tag{9.1.5}$$

$$\tilde{Y}_{\text{in}}(\chi) = \epsilon\phi(\chi) + O(\epsilon^2), \tag{9.1.6}$$

where $\chi = (\tilde{n} - \tilde{n}_f)/\epsilon$, $\epsilon = \tilde{T}_f^2/\tilde{T}_a$, and $\tilde{n}_f$ and $\tilde{T}_f$ the reaction-sheet location and temperature respectively, Eqs. (9.1.3) and (9.1.4) become

$$\frac{d^2\theta}{d\chi^2} = \frac{\Delta}{2}\phi e^{-\theta}, \tag{9.1.7}$$

$$\frac{d^2}{d\chi^2}\left(\theta - \frac{\phi}{Le}\right) = 0, \tag{9.1.8}$$

where $\Delta = 2\epsilon^2 Da_C \exp(-\tilde{T}_a/\tilde{T}_f)$.

To demonstrate matching, we first consider the energy equation. We expand the outer solution in the general form as

$$\tilde{T}_{\text{out}}^{\pm}(\tilde{\mathbf{x}}, t) = \tilde{T}_0^{\pm}(\tilde{\mathbf{x}}, t) + \epsilon\tilde{T}_1^{\pm}(\tilde{\mathbf{x}}, t) + O(\epsilon^2), \tag{9.1.9}$$

where the superscripts $\pm$ respectively denote the burned and unburned sides of the reaction sheet. Since the outer solutions are to match with the inner solution at the reaction sheet, we Taylor-expand $\tilde{T}_{\text{out}}^{\pm}$ around $\tilde{n} = \tilde{n}_f$ in the direction of $\tilde{\mathbf{n}}$. For example,

$$\begin{aligned}\tilde{T}_0^{\pm}(\tilde{n}, \tilde{t}) &= \left(\tilde{T}_0^{\pm}\right)_{\tilde{n}_f} + \left(\frac{\partial\tilde{T}_0^{\pm}}{\partial\tilde{n}}\right)_{\tilde{n}_f}(\tilde{n} - \tilde{n}_f) + \cdots,\\ &= \left(\tilde{T}_0^{\pm}\right)_{\tilde{n}_f} + \left(\frac{\partial\tilde{T}_0^{\pm}}{\partial n}\right)_{\tilde{n}_f}(\epsilon\chi) + O(\epsilon^2).\end{aligned} \tag{9.1.10}$$

Thus we have, in terms of the inner variable χ,

$$\tilde{T}_{\text{out}}^{\pm}(\chi) = \left(\tilde{T}_0^{\pm}\right)_{\tilde{n}_f} + \epsilon\left[\left(\tilde{T}_1^{\pm}\right)_{\tilde{n}_f} + \left(\frac{\partial\tilde{T}_0^{\pm}}{\partial\tilde{n}}\right)_{\tilde{n}_f}\chi\right] + O(\epsilon^2). \tag{9.1.11}$$

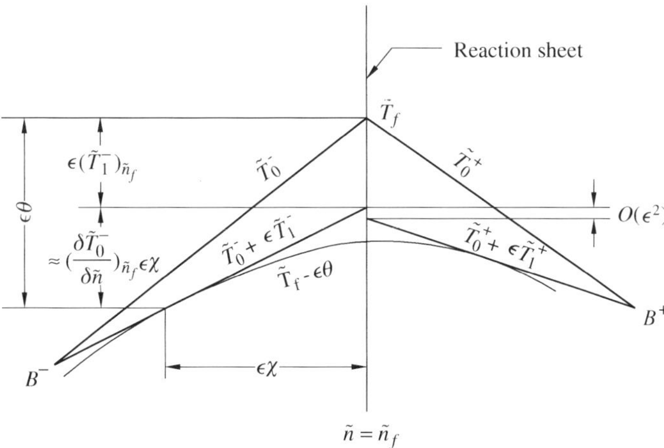

Figure 9.1.2. Schematic showing the asymptotic matching between the inner and outer solutions.

Matching the $O(1)$ and $O(\epsilon)$ values and slopes of $\tilde{T}_{\text{in}}$ with $\tilde{T}^{\pm}_{\text{out}}$, in the limit $\chi \to \pm\infty$, we obtain

$$\left(\tilde{T}_0^{\pm}\right)_{\tilde{n}_f} = \tilde{T}_f, \tag{9.1.12}$$

$$\left[\theta + \left(\frac{\partial \tilde{T}_0^{\pm}}{\partial \tilde{n}}\right)_{\tilde{n}_f} \chi\right]_{\chi \to \pm\infty} = -\left(\tilde{T}_1^{\pm}\right)_{\tilde{n}_f}, \tag{9.1.13}$$

$$\left(\frac{d\theta}{d\chi}\right)_{\chi \to \pm\infty} = -\left(\frac{\partial \tilde{T}_0^{\pm}}{\partial \tilde{n}}\right)_{\tilde{n}_f}. \tag{9.1.14}$$

Equation (9.1.12) simply states that the leading-order temperatures $\tilde{T}_0^{\pm}$ are continuous at the reaction sheet, and are equal to $\tilde{T}_f$. The matching relations (9.1.13) and (9.1.14) can be interpreted by considering the temperature profiles of the various expansions, as shown in Figure 9.1.2. For simplicity and clarity the profiles for the outer solutions are shown as linear, and those in the inner and matching regions are disproportionately magnified.

First we see that for finite ϵ, the perturbed temperatures, $\tilde{T}_0^{\pm} + \epsilon\tilde{T}_1^{\pm}$, are reduced from $\tilde{T}_0^{\pm}$ and coincide with $\tilde{T}_0^{\pm}$ at the respective freestream boundaries, $B^{\pm}$, where the temperatures are fixed. At the reaction sheet, the perturbed temperatures do not meet, indicating the fact that so far we do not know the relation between $(\tilde{T}_1^-)_{\tilde{n}_f}$ and $(\tilde{T}_1^+)_{\tilde{n}_f}$, and therefore we cannot assume that the temperature is continuous to this order at the reaction sheet. Thus our previous assumption that values are continuous in deriving the jump relations across a reaction sheet in Section 5.5.1 holds rigorously only for the leading-order solution.

Now consider the negative side of the temperature profile. Here the inner solution, $\tilde{T}_f - \epsilon\theta$, "merges" with the perturbed outer solution at a tangency point located at

$\tilde{n} = \tilde{n}_f + \epsilon\chi$ as $\chi \to -\infty$. According to the inner solution, this tangency point is located below $\tilde{T}_f$ by an amount $\epsilon\theta$, while according to the outer solution, this point is below $\tilde{T}_f$ by an amount equal to $\epsilon(\tilde{T}_1^-)_{\tilde{n}_f}$ plus

$$\left(\frac{\partial \tilde{T}_0^-}{\partial \tilde{n}} + \epsilon \frac{\partial \tilde{T}_1^-}{\partial \tilde{n}}\right)_{\tilde{n}_f} \epsilon\chi = \left(\frac{\partial \tilde{T}_0^-}{\partial \tilde{n}}\right)_{\tilde{n}_f} \epsilon\chi + O(\epsilon^2). \tag{9.1.15}$$

Their equality gives the matching condition (9.1.13). Differentiating (9.1.13) then yields the matching condition (9.1.14) for the gradients at the tangency point.

The matching conditions for the species can be similarly derived, resulting in

$$\left(\tilde{Y}_0^{\pm}\right)_{\tilde{n}_f} = 0, \tag{9.1.16}$$

$$\left[\phi - \left(\frac{\partial \tilde{Y}_0^{\pm}}{\partial \tilde{n}}\right)_{\tilde{n}_f} \chi\right]_{\chi\to\pm\infty} = \left(\tilde{Y}_1^{\pm}\right)_{\tilde{n}_f}, \tag{9.1.17}$$

$$\left(\frac{d\phi}{d\chi}\right)_{\chi\to\pm\infty} = \left(\frac{\partial \tilde{Y}_0^{\pm}}{\partial \tilde{n}}\right)_{\tilde{n}_f}. \tag{9.1.18}$$

Note that since $(\tilde{Y}_0^+)_{\tilde{n}_f} = 0$ and, hence, $\tilde{Y}_0^+(\tilde{n}) = 0$, we have $(\partial \tilde{Y}_0^+/\partial \tilde{n})_{\tilde{n}_f} = 0$.

We can now solve Eq. (9.1.8) for the local total enthalpy. Integrating twice leads to $(\theta - \phi/Le) = c_1 + c_2\chi$, which, when evaluated at, say, $\chi \to \infty$ by using the matching relations (9.1.13), (9.1.14), (9.1.17), and (9.1.18), yields c_1 and c_2. Thus we can express ϕ in terms of θ such that the inner, structure equation (9.1.7) becomes

$$\frac{d^2\theta}{d\chi^2} = \frac{Le\Delta}{2}\left[\theta + \left(\frac{\partial \tilde{T}_0^+}{\partial \tilde{n}}\right)_{\tilde{n}_f} \chi + \left(\tilde{T}_1^+ + \frac{\tilde{Y}_1^+}{Le}\right)_{\tilde{n}_f}\right] e^{-\theta}. \tag{9.1.19}$$

Equation (9.1.19) is to be solved subject to the two boundary conditions of Eq. (9.1.14). The solution, however, would contain seven unknowns, namely $(\tilde{T}_1^{\pm})_{\tilde{n}_f}$, $(\tilde{Y}_1^{\pm})_{\tilde{n}_f}$, $\tilde{T}_f, \tilde{n}_f$, and the burning flux eigenvalue Δ. These are to be determined by the two additional boundary conditions given by Eq. (9.1.13), and five jump relations for the outer solutions to be determined next.

9.1.2. Delta Function Closure and Jump Relations

To the outer solution the reaction term is simply a source of heat whose strength and nature of singularity must depend on the inner structure of the reaction zone. Thus if we were to solve, say, Eqs. (9.1.1) and (9.1.2) for $\tilde{T}_{\text{out}}$, then $\tilde{w}$ would have to be represented by appropriate source terms. The problem is somewhat analogous to the description of an electric field generated by a charge distribution. That is, when viewed from a large distance, the electric field produced by a uniformly charged sphere behaves as a point source, or a monopole, which can be described by $q\delta(r)$, where q is the total charge of the sphere, $\delta(r)$ a delta function, and r the radial distance. Now if the charge distribution is not uniform, but is concentrated and oppositely located with different amounts at the two poles of the sphere, then the point source would behave like the sum of a monopole and a dipole when viewed from the far field, with the dipole having a different nature of the source strength and a higher-order singularity representation from those of the monopole. Thus the source strength of a body of arbitrary charge distribution needs to be described by the summation of singularities of all orders.

Recognizing that the charge distribution simply corresponds to the reaction rate function of the present problem, and that this function is far from being spatially uniform, it is therefore clear that in order to properly represent the reaction rate function as a concentrated source term, for the outer solutions, it needs to be expanded into a complete set of orthogonal functions of concentrated nature. One such set is the Hermite polynomial, which consists of the delta function and all of its derivatives. We thus expand the reaction rate function $\tilde{w}(\tilde{n})$ around the reaction-sheet location $\tilde{n}_f$ in terms of this basis set as

$$-\tilde{w}(\tilde{n}) = \sum_{i=0}^{\infty} M_i \frac{d^i}{d\tilde{n}^i} \delta(\tilde{n} - \tilde{n}_f). \tag{9.1.20}$$

The coefficients M_i are the strengths of the singularities at the various orders, and are given by successively integrating (9.1.20) across the reaction sheet to yield

$$M_0 = -\int_{\tilde{n}_f^-}^{\tilde{n}_f^+} \tilde{w} d\tilde{n}, \tag{9.1.21}$$

$$M_1 = -\int_{\tilde{n}_f^-}^{\tilde{n}_f^+} (\tilde{n} - \tilde{n}_f) \tilde{w} d\tilde{n}, \tag{9.1.22}$$

and higher order terms.

In asymptotic studies frequently only the leading order, monopole term (M_0) is kept for $\tilde{w}$, resulting in the delta function representation. However, Eq. (9.1.22) shows that only under situations of complete symmetry, around $\tilde{n}_f$ in the reaction zone, does the contribution from the first moment, dipole term (M_1) vanish such that the field strength can be approximated by the monopole, delta function term with an $O(\epsilon)$ accuracy. Since the reaction rate function is not spatially symmetric no matter where $\tilde{n}_f$ is located, the dipole term does not vanish and the use of the delta function term is only accurate to $O(1)$. Furthermore, since asymptotic studies require at least $O(\epsilon)$ analysis, the delta function representation is therefore not consistent with the overall accuracy.

The problem, however, can be made to be accurate to $O(\epsilon)$ by setting $M_1 \equiv 0$. From Eq. (9.1.22) it is seen that this is equivalent to locating the reaction sheet at

$$\tilde{n}_f = \frac{\int_{\tilde{n}_f^-}^{\tilde{n}_f^+} \tilde{n} \tilde{w} d\tilde{n}}{\int_{\tilde{n}_f^-}^{\tilde{n}_f^+} \tilde{w} d\tilde{n}} \tag{9.1.23}$$

within the reaction zone. Drawing analogy with the definition of "center of mass," for which the weighting function is the mass density, Eq. (9.1.23) shows that the reaction sheet is located at the "center of reaction," for which the weighting function is the reaction rate.

The fact that the location of the reaction sheet needs to be independently specified is reasonable. Since the reaction sheet is strictly a mathematical creation instead of a physical quantity, determination of its properties including the location must naturally evolve from the mathematical formulation of the asymptotic analysis. Furthermore, we also anticipate that since the flame properties in the outer zone are continuous to $O(1)$ at the reaction sheet, and since the $O(\epsilon)$ singularity is suppressed by locating $\tilde{n}_f$ at the center of reaction, these flame properties should also be continuous to $O(\epsilon)$. To demonstrate this point, we substitute $\tilde{T}_{\text{in}}$ of Eq. (9.1.5) in Eq. (9.1.3) to get

$$\frac{d^2\theta}{d\chi^2} = -\epsilon\tilde{w}, \tag{9.1.24}$$

which, when used in Eq. (9.1.23), leads to the requirement

$$\int_{-\infty}^{\infty} \chi \frac{d^2\theta}{d\chi^2} d\chi = 0. \tag{9.1.25}$$

Integrating Eq. (9.1.25) by parts, and by using the matching relations (9.1.14) and then (9.1.13), we have

$$\begin{aligned} \left[\chi\frac{d\theta}{d\chi}\right]_{-\infty}^{\infty} - [\theta]_{-\infty}^{\infty} &= -\left[\theta + \left(\frac{\partial \tilde{T}_0^+}{\partial \tilde{n}}\right)_{\tilde{n}_f}\chi\right]_{\infty} + \left[\theta + \left(\frac{\partial \tilde{T}_0^-}{\partial \tilde{n}}\right)_{\tilde{n}_f}\chi\right]_{-\infty} \\ &= (\tilde{T}_1^+)_{\tilde{n}_f} - (\tilde{T}_1^-)_{\tilde{n}_f} = 0, \end{aligned}$$

which implies

$$(\tilde{T}_1^-)_{\tilde{n}_f} = (\tilde{T}_1^+)_{\tilde{n}_f}. \tag{9.1.26}$$

Using the local coupling function $\theta - \phi/Le = c_1 + c_2\chi$, we see that Eq. (9.1.25) also holds for ϕ. A similar manipulation then yields

$$(\tilde{Y}_1^-)_{\tilde{n}_f} = (\tilde{Y}_1^+)_{\tilde{n}_f}. \tag{9.1.27}$$

We have therefore demonstrated that the outer flame properties are continuous to $O(\epsilon)$ when the reaction rate is represented by the delta function and the reaction sheet located at the center of reaction. This procedure (Law, Chao & Umemura 1992) can be referred to as the delta function closure.

It is, however, important to mention that the delta function closure for the problem is only one of several possible closure schemes that locate the reaction sheet somewhere within the reaction zone and can adequately describe the leading-order flame responses such as the burning rate and flame location. For example, the reaction sheet can be independently located at the downstream boundary of the reaction zone, or at the maximum reaction rate location. Alternatively, Liñán solved for the flame location by suppressing all perturbations upstream of the flame. The use of these alternate closure schemes implicitly assumes that the singularity of the reaction sheet is higher than that of the delta function such that $M_1 \neq 0$. Consequently the outer solution is also discontinuous at $O(\epsilon)$.

With $\tilde{w}$ represented by Eq. (9.1.20), Eqs. (9.1.1) can be readily integrated across the reaction sheet for $\tilde{T}_{\text{out}} = \tilde{T}_0 + \epsilon \tilde{T}_1$ to yield the strength of the reaction source

$$M_0 = \left(\frac{\partial \tilde{T}_0^-}{\partial \tilde{n}}\right)_{\tilde{n}_f} - \left(\frac{\partial \tilde{T}_0^+}{\partial \tilde{n}}\right)_{\tilde{n}_f}. \tag{9.1.28}$$

Equation (9.1.28) shows that the leading order strength of the delta function, M_0, is the sum of the heat conducted away to both sides of the reaction sheet, and is therefore the heat release rate at the sheet. A similar integration of the enthalpy equation (9.1.2), and by using Eqs. (9.1.26) and (9.1.27), lead to

$$\left[\frac{\partial}{\partial \tilde{n}}\left(\tilde{T}_0^- + \frac{\tilde{Y}_0^-}{Le}\right)\right]_{\tilde{n}_f} = \left(\frac{\partial \tilde{T}_0^+}{\partial \tilde{n}}\right)_{\tilde{n}_f}, \tag{9.1.29}$$

$$\left[\frac{\partial}{\partial \tilde{n}}\left(\tilde{T}_1^- + \frac{\tilde{Y}_1^-}{Le}\right)\right]_{\tilde{n}_f} = \left[\frac{\partial}{\partial \tilde{n}}\left(\tilde{T}_1^+ + \frac{\tilde{Y}_1^+}{Le}\right)\right]_{\tilde{n}_f}. \tag{9.1.30}$$

Equations (9.1.26) to (9.1.30) are the five jump relations needed to completely solve the problem, with Δ related to M_0.

9.1.3. Reduction to Canonical Form

The structure equation (9.1.19) and the boundary conditions (9.1.14) can be reduced to a canonical form characterized by only a single parameter. This is achieved by defining

$$\tilde{\theta} = \theta - \mu\eta + \ln\left(\frac{M_0^2}{Le\Delta}\right), \tag{9.1.31}$$

$$\eta = M_0\chi - \frac{p}{\mu}, \tag{9.1.32}$$

where

$$\mu = -\frac{1}{M_0}\left(\frac{\partial \tilde{T}_0^+}{\partial \tilde{n}}\right)_{\tilde{n}_f}, \tag{9.1.33}$$

$$p = \left(\tilde{T}_1^+ + \frac{\tilde{Y}_1^+}{Le}\right)_{\tilde{n}_f} - \ln\left(\frac{M_0^2}{Le\Delta}\right). \tag{9.1.34}$$

Equation (9.1.19) is then reduced to

$$2\frac{d^2\tilde{\theta}}{d\eta^2} = \tilde{\theta}e^{-(\tilde{\theta}+\mu\eta)}, \tag{9.1.35}$$

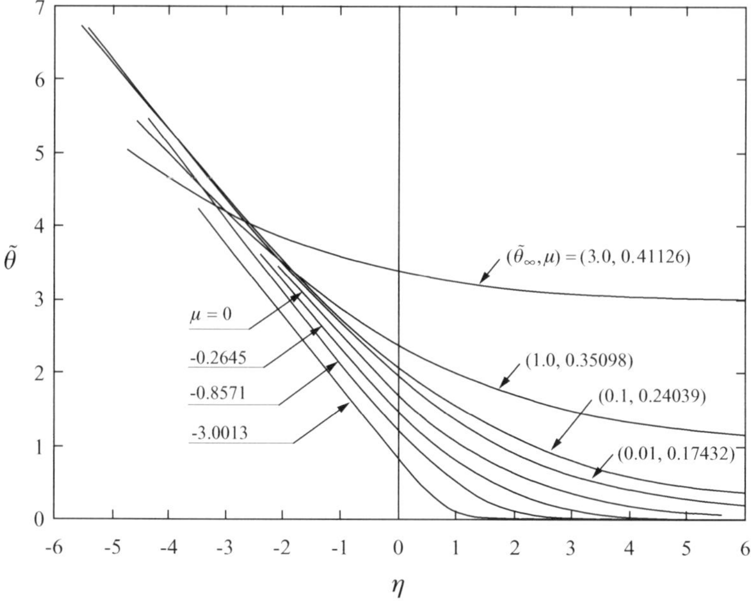

Figure 9.1.3. Inner structure of the premixed flame (Liñán 1974).

subject to the boundary conditions

$$\left(\frac{d\tilde{\theta}}{d\eta}\right)_{-\infty} = -1, \tag{9.1.36}$$

$$\left(\frac{d\tilde{\theta}}{d\eta}\right)_{\infty} = 0, \tag{9.1.37}$$

showing that this system depends only on the parameter μ. The matching conditions of Eq. (9.1.13) are then reduced to

$$(\tilde{\theta} + \eta)_{-\infty} = \frac{1}{Le}\left(\tilde{Y}_1^+\right)_{\tilde{n}_f} - \frac{p}{\mu}, \tag{9.1.38}$$

$$(\tilde{\theta})_\infty = \frac{1}{Le}\left(\tilde{Y}_1^+\right)_{\tilde{n}_f}. \tag{9.1.39}$$

Equation (9.1.33) shows that the characterizing parameter μ is the ratio of the heat conducted downstream from the reaction sheet to the total amount of heat generated at the sheet. Thus the factor $\exp(-\mu\eta)$ in Eq. (9.1.35) represents the effect on the reaction rate due to heat loss ($\mu > 0$) or gain ($\mu < 0$) by the reaction sheet through heat transfer with the downstream. For $\mu = 0$, the downstream is adiabatic. Furthermore, $(\tilde{\theta})_\infty$ given by Eq. (9.1.39) represents the amount of fuel leakage through the reaction zone and hence the extent of incomplete reaction.

Figure 9.1.3 shows the numerical solution of Eqs. (9.1.35) to (9.1.37) for the reduced temperature profile $\tilde{\theta}(\eta)$ for both positive and negative values of μ. The extent of fuel leakage is also indicated. From the behavior of these solutions and further analysis of Eqs. (9.1.35)–(9.1.37), the following system behavior is obtained. First, for the nearly adiabatic situation of small μ, $(\tilde{\theta}+\eta)_{-\infty} = 1.344$, while the reactant leakage vanishes, with $(\tilde{\theta})_{\infty} = 0$. Consequently, there is no fuel leakage for the super-adiabatic situations of $\mu < 0$, which is reasonable. Furthermore, both $-(\tilde{\theta}+\eta)_{-\infty}$ and $(\tilde{\theta})_{\infty}$ approach infinity as $\mu \to 1/2$, and there is no solution for $\mu \geq 1/2$. Since $\mu = 1/2$ represents equal rates of heat transfer to the upstream and the downstream, this result shows that the reaction sheet always transfers more heat to the upstream than to the downstream. As $\mu \to 1/2$, leakage of the reactant through the reaction zone becomes excessively large because of the significant and rapid reduction in the temperature downstream of the reaction zone.

The matching values $(\tilde{\theta}+\eta)_{-\infty}$ and $(\tilde{\theta})_{\infty}$ have also been numerically fitted as functions of μ,

$$\begin{aligned}\mu(\tilde{\theta}+\eta)_{-\infty} &= -\ln(1-1.344\mu+0.6307\mu^2), \quad \mu \leq 0\\ &= 1.344\mu - \frac{4\mu^2(1-\mu)}{(1-2\mu)} + 1.2\mu^2 - \ln(1-4\mu^2), \quad \mu > 0 \end{aligned} \tag{9.1.40}$$

$$\begin{aligned}(\tilde{\theta})_{\infty} &= \frac{0.000246}{(0.5-\mu)^4} - \frac{0.01001}{(0.5-\mu)^3} + \frac{0.15450}{(0.5-\mu)^2}\\ &\quad - \frac{0.62026}{(0.5-\mu)} + 0.72168 \qquad \text{for} \quad 0.15 < \mu < 0.5\\ &= 0, \qquad \text{for} \quad \mu < 0.15. \end{aligned} \tag{9.1.41}$$

From the definitions of p in Eqs. (9.1.34) and (9.1.38), the system Damköhler number can be related to the particulars of the flame,

$$M_0^2 = Le\Delta \exp\left\{\left(\tilde{T}_1^+ + \frac{\tilde{Y}_1^+}{Le}\right)_{\tilde{n}_f} + \mu\left[(\tilde{\theta}+\eta)_{-\infty} - (\tilde{\theta})_{\infty}\right]\right\}. \tag{9.1.42}$$

The burning rate of the flame, $\tilde{f}_n$, is either given by Eq. (9.1.42) through the definition of Δ and, hence, Da_C, or is related to $\tilde{n}_f$ through the fluid mechanical aspects of the problem. The problem is now completely solved.

As mentioned earlier, the above results are quite general and hence applicable to a large number of premixed flame phenomena, including the description of flame extinction through the extinction turning point behavior. For the standard laminar flame studied in Chapter 7, since the downstream state is uniform and its temperature is simply the adiabatic flame temperature, all downstream perturbations vanish such that $\tilde{T}_1^+(\tilde{x}) = \tilde{Y}_1^+(\tilde{x}) = 0$. Consequently, $(\tilde{T}_1^{\pm})_{n_f} = (\tilde{Y}_1^{\pm})_{n_f} = (\tilde{\theta})_{\infty} = \mu(\tilde{\theta}+\eta)_{-\infty} \equiv 0$. Furthermore, with $\tilde{T}_0^-(\tilde{x}) = \tilde{T}_u + e^{\tilde{n}-\tilde{n}_f}$, obtained by solving $d^2\tilde{T}_0^-/d\tilde{n}^2 - d\tilde{T}_0^-/d\tilde{n} = 0$, with $\tilde{T}_0^- = \tilde{T}_u$, at $\tilde{n} \to -\infty$ and $\tilde{T}_0 = \tilde{T}_f$ at $\tilde{n} = \tilde{n}_f$, we have $M_0 = (d\tilde{T}_0^-/d\tilde{n})_{\tilde{n}_f} = 1$ from Eq. (9.1.28). Substituting these results into Eq. (9.1.42) and evaluating the various terms, we have

$$Le\Delta = 1, \tag{9.1.43}$$

which is equivalent to Eq. (7.5.50), noting that the factor Le is absorbed in the definition for Δ^o in Eq. (7.5.45).

9.2. STRUCTURE OF NONPREMIXED FLAMES: CLASSIFICATION

Although we can also perform a generalized analysis for the reaction zone structures for the nonpremixed flames, for conceptual simplicity we shall demonstrate the basic methodology and flame structure by using the simplest flame configuration possible, namely the one-dimensional chambered flame studied in Chapter 6 for the reaction-sheet limit of nonpremixed flames. Since the final structure equations are expressed in canonical form anyway, use of this simple configuration facilitates discussion.

Here we have a chamber of length ℓ and constant cross-sectional area (Figure 6.1.1). Two porous walls, at $x = 0$ and ℓ, maintain constant concentrations and temperatures according to

$$\begin{aligned} x = 0:&\quad Y_O = Y_{O,o},\quad Y_F = 0,\quad Y_P = 0,\quad T = T_o,\\ x = \ell:&\quad Y_O = 0,\quad Y_F = Y_{F,\ell},\quad Y_P = 0,\quad T = T_\ell. \end{aligned} \tag{9.2.1}$$

Note that we have reversed the locations of fuel and oxidizer from those of Section 6.1. This is done strictly for conceptual convenience in that the relatively hot reactant, which is frequently the oxidizer (e.g., air), is located at the origin as in Liñán's canonical formulation.

Assuming unity Lewis number for simplicity, the nondimensional governing equations for this purely diffusive system are

$$\frac{d^2\tilde{T}}{d\tilde{x}^2} = -Da_C\tilde{Y}_O\tilde{Y}_F e^{-\tilde{T}_a/\tilde{T}}, \tag{9.2.2}$$

$$\frac{d^2(\tilde{T}+\tilde{Y}_i)}{d\tilde{x}^2} = 0, \quad i = O, F, \tag{9.2.3}$$

where $\tilde{x} = x/\ell$ and $Da_C = (\ell^2 B_C)/(\lambda/c_p)$. Equation (9.2.3) can be readily solved subject to the boundary conditions in (9.2.1), yielding

$$\tilde{T} + \tilde{Y}_O = \tilde{T}_\ell + (\beta + \tilde{Y}_{O,o})(1 - \tilde{x}), \tag{9.2.4}$$

$$\tilde{T} + \tilde{Y}_F = \tilde{T}_o + (1 - \beta)\tilde{x}, \tag{9.2.5}$$

where

$$\beta = \tilde{T}_o - \tilde{T}_\ell \tag{9.2.6}$$

is a heat transfer parameter. For the sake of discussion we shall assume that $\tilde{T}_o > \tilde{T}_\ell$ such that $\beta > 0$. A completely analogous discussion can be conducted for the inverse situation of $\beta < 0$.

The problem is now reduced to finding the asymptotic solution of Eq. (9.2.2), subject to the boundary conditions

$$\tilde{T}(0) = \tilde{T}_o, \tag{9.2.7}$$

$$\tilde{T}(1) = \tilde{T}_\ell, \tag{9.2.8}$$

with $\tilde{Y}_{O,o}$, β and $\tilde{T}_o$ of order unity while Da_C can assume values ranging from zero to infinity.

We also recall that, as shown in Chapters 6 and 8, the convection term in a general quasi-one-dimensional flow can be suppressed through a suitable transformation, while the system boundaries can also be normalized to be located at (0, 1). Thus flows of this nature are described by the same set of equations as that defined above, with the reaction term modified by a multiplicative function that depends on $\tilde{x}$.

9.2.1. Classification of Flow Types

We first recognize that there are three situations under which a region of the flow field can be rendered chemically nonreactive, with $\tilde{w} \sim Da_C \tilde{Y}_O \tilde{Y}_F \equiv 0$. This is achieved for $Da_C \equiv 0$, $\tilde{Y}_O \equiv 0$, or $\tilde{Y}_F \equiv 0$. Their respective solutions are given in the following.

Frozen Flow $(Da_C \equiv 0)$*:* In this limit, the solution satisfying $d^2\tilde{T}/d\tilde{x}^2 = 0$ for Eq. (9.2.2) is simply

$$\tilde{T} = c_1 + c_2\tilde{x}, \tag{9.2.9}$$

where c_1 and c_2 are to be determined from boundary conditions appropriate for the particular region in the flow field.

Equilibrium Flow with Vanishing Oxidizer Concentration $(\tilde{Y}_O \equiv 0)$*:* In this region the temperature profile is described by setting $\tilde{Y}_O = 0$ in Eq. (9.2.4), yielding

$$\tilde{T} = (\tilde{T}_o + \tilde{Y}_{O,o}) - (\beta + \tilde{Y}_{O,o})\tilde{x}. \tag{9.2.10}$$

Equilibrium Flow with Vanishing Fuel Concentration $(\tilde{Y}_F \equiv 0)$*:* In this region the temperature profile is described by setting $\tilde{Y}_F = 0$ in Eq. (9.2.5), yielding

$$\tilde{T} = \tilde{T}_o + (1 - \beta)\tilde{x}. \tag{9.2.11}$$

It is emphasized that while the reaction term in Eq. (9.2.2) is made to vanish in all three types of flows such that $d^2\tilde{T}/d\tilde{x}^2 = 0$ and the temperature profiles are linear, the cause for the nonreactiveness of each flow is different.

9.2.2. Classification of Flame Regimes

Based on the above possible flow regions, the following four classes of flame structure can be identified (Figure 9.2.1).

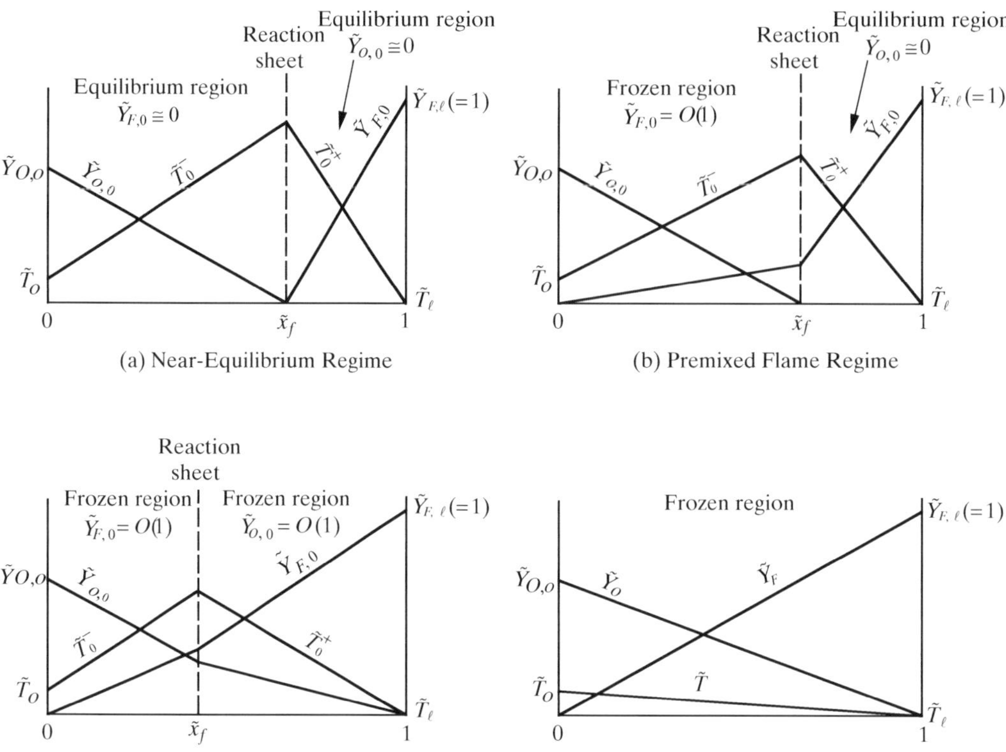

Figure 9.2.1. Leading-order temperature and species profiles showing the characteristics of different flame regimes.

9.2.2.1. Near-Equilibrium (N.E.) Regime: We start with the situation in which the burning is most intense. Thus to the leading order fuel and oxidizer are totally consumed in crossing the reaction sheet such that equilibrium exists on both sides of the reaction sheet located at $\tilde{x}_f$. The temperature and species profiles (Figure 9.2.1a) are then given by

$$\tilde{x} < \tilde{x}_f: \quad \tilde{Y}_{F,0}^- = 0, \quad \tilde{T}_0^- = \tilde{T}_o + (1-\beta)\tilde{x} \tag{9.2.12}$$

$$\tilde{x} > \tilde{x}_f: \quad \tilde{Y}_{O,0}^+ = 0, \quad \tilde{T}_0^+ = \tilde{T}_\ell + (\beta + \tilde{Y}_{O,o})(1-\tilde{x}), \tag{9.2.13}$$

where the subscript 0 denotes the reaction sheet solution. Equating (9.2.12) and (9.2.13) at the reaction sheet, we obtain the flame temperature and reaction-sheet location as

$$\tilde{T}_f = \tilde{T}_o + (1-\beta)\frac{\tilde{Y}_{O,o}}{1+\tilde{Y}_{O,o}} = \tilde{T}_\ell + \left(\frac{\beta + \tilde{Y}_{O,o}}{1+\tilde{Y}_{O,o}}\right), \tag{9.2.14}$$

$$\tilde{x}_f = \frac{\tilde{Y}_{O,o}}{1+\tilde{Y}_{O,o}}. \tag{9.2.15}$$

Furthermore, $\tilde{Y}_{O,0}^-$ and $\tilde{Y}_{F,0}^+$ can be obtained by substituting $\tilde{T}_0^-$ in Eq. (9.2.4) and $\tilde{T}_0^+$ in Eq. (9.2.5) respectively. The above results are simply the reaction-sheet solution of nonpremixed flames studied in Chapter 6. All the reaction-sheet properties are completely determined at this level of the analysis.

With gradual reduction in the reaction intensity from the reaction-sheet limit, say by decreasing Da_C, both fuel and oxidizer will leak through the reaction zone by small amounts and will eventually lead to extinction.

9.2.2.2. Premixed Flame (P.F.) Regime: In this regime substantial leakage of one reactant is possible before the onset of extinction, and the reactant becomes frozen after leaking through the flame. Thus to the leading order the reaction sheet separates a frozen flow region from an equilibrium flow region. Let us consider the situation in which substantial, $O(1)$, leakage of the fuel occurs such that the oxidizer side is frozen while the fuel side is in equilibrium, as shown in Figure 9.2.1b. Completely parallel discussion can be conducted when the oxidizer is the excessively leaked reactant. It will be shown in Section 9.2.3 that these two situations correspond to $\tilde{Y}_{O,o} + 2\beta < 1$ and $\tilde{Y}_{O,o} + 2\beta > 1$ respectively.

For the present situation, we have

$$\tilde{x} < \tilde{x}_f: \quad \tilde{T}_0^- = \tilde{T}_o + \left(\tilde{T}_f - \tilde{T}_o\right)\frac{\tilde{x}}{\tilde{x}_f}, \tag{9.2.16}$$

$$\tilde{x} > \tilde{x}_f: \quad \tilde{Y}_{O,0}^+ = 0, \quad \tilde{T}_0^+ = \tilde{T}_\ell + (\beta + \tilde{Y}_{O,o})(1 - \tilde{x}), \tag{9.2.17}$$

where Eq. (9.2.16) is obtained by evaluating the frozen profile of Eq. (9.2.9) with $\tilde{T}(0) = \tilde{T}_0$ and $\tilde{T}(\tilde{x}_f) = \tilde{T}_f$, while $\tilde{Y}_{F,0}^+$ is given by substituting $\tilde{T}_0^+$ in Eq. (9.2.5).

Evaluating Eq. (9.2.17) at the reaction sheet yields a relation between $\tilde{T}_f$ and $\tilde{x}_f$,

$$\tilde{T}_f = \tilde{T}_\ell + (\beta + \tilde{Y}_{O,o})(1 - \tilde{x}_f). \tag{9.2.18}$$

Unlike the near-equilibrium regime for which both $\tilde{T}_f$ and $\tilde{x}_f$, and indeed the complete reaction-sheet solution, are determined at this level of analysis, the reaction-sheet solution for the premixed flame regime is determined to within one unknown, say $\tilde{x}_f$. The additional relation is obtained by analyzing the reaction zone structure at the next order, because the problem depends on the rate of chemical reaction in the reaction region.

At this point it must have become obvious as to why this regime is called the premixed flame regime. In fact, if we identify the deficient reactant in the previous premixed flame analysis as the oxidizer in the present case, the upstream preheat zone as the present frozen flow, and the downstream burned zone as the present equilibrium flow, then complete analogy exists with the previous analysis (Figure 9.2.2). The concentrations of the previous abundant species (from the upstream) and the present fuel (from the fuel side) decrease by small amounts in crossing the reaction zone, and therefore are $O(1)$ quantities in the reaction zone. As such, these abundant reactants affect the reaction rate only in a passive manner. The structure of the reaction zone is instead controlled by the $O(\epsilon)$ concentration of the deficient reactant for the premixed flame and oxidizer for the present premixed flame regime.

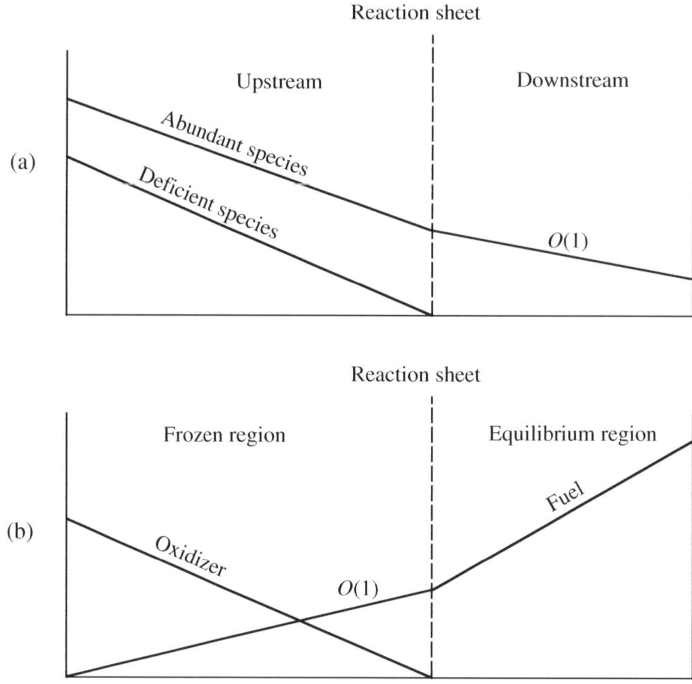

Figure 9.2.2. Schematic demonstrating the basic structural similarity between (a) the premixed flame and (b) the premixed flame regime of nonpremixed flames.

Flame structure analysis for this regime therefore completely parallels that of the premixed flame in Section 9.1, resulting in the same extinction criterion when leakage of the fuel becomes excessive, as we shall demonstrate shortly.

9.2.2.3. Partial Burning (P.B.) Regime: With further reduction in the reaction intensity, excessive leakage of both fuel and oxidizer occurs. A very weak reaction sheet now separates two frozen regions, and the concentrations of both reactants are only slightly affected (Figure 9.2.1c) when crossing the flame. The temperature profiles are now given by evaluating $\tilde{T} = c_1 + c_2\tilde{x}$ in the two frozen regions,

$$\tilde{x} < \tilde{x}_f: \ \tilde{T}_0^- = \tilde{T}_o + (\tilde{T}_f - \tilde{T}_o)\frac{\tilde{x}}{\tilde{x}_f}, \tag{9.2.19}$$

$$\tilde{x} > \tilde{x}_f: \ \tilde{T}_0^+ = \tilde{T}_\ell + (\tilde{T}_f - \tilde{T}_\ell)\frac{1-\tilde{x}}{1-\tilde{x}_f}, \tag{9.2.20}$$

for given values of flame temperature $\tilde{T}_f$ and reaction-sheet location $\tilde{x}_f$.

The reaction-sheet solution in this regime is given in terms of two unknowns, namely $\tilde{T}_f$ and $\tilde{x}_f$, both of which need to be determined by analyzing the flame structure. It will be shown in Section 9.3.2 that to the leading order the reaction sheet conducts heat to the fuel and oxidizer sides at an equal rate. Thus by equating the magnitude of the slopes of $\tilde{T}_0^-$ and $\tilde{T}_0^+$, we obtain a relation between $\tilde{x}_f$ and $\tilde{T}_f$,

$$\tilde{x}_f = \frac{(\tilde{T}_f - \tilde{T}_o)}{(\tilde{T}_f - \tilde{T}_o) + (\tilde{T}_f - \tilde{T}_\ell)}. \tag{9.2.21}$$

Equation (9.2.21) simply states that the location of the flame is proportional to the heat transfer to the oxidizer side.

9.2.2.4. Nearly Frozen (N.F.) Regime: In this regime chemical reaction is frozen throughout the flow field to the leading order. The temperature profile (Figure 9.2.1d) is simply given by

$$\tilde{T}_0 = \tilde{T}_o - \beta\tilde{x}. \tag{9.2.22}$$

The fuel and oxidizer concentrations are then given by

$$\tilde{Y}_{F,0} = \tilde{x}, \tag{9.2.23}$$

$$\tilde{Y}_{O,0} = (1 - \tilde{x})\tilde{Y}_{O,o}. \tag{9.2.24}$$

By allowing the reaction to proceed at a slow, but finite rate, chemical heat is released at a finite rate. Ignition is expected to occur when the heat release rate exceeds the heat dissipation rate through diffusion. The ignition analysis will be conducted subsequently.

9.2.3. Parametric Boundaries of Flame Regimes

For the partial burning regime, the fuel and oxidizer concentrations in the reaction zone are $O(1)$ quantities. Applying the requirements $\tilde{Y}_{O,f} > 0$ and $\tilde{Y}_{F,f} > 0$ to the coupling functions $(\tilde{T} + \tilde{Y}_O)$ and $(\tilde{T} + \tilde{Y}_F)$ given by Eqs. (9.2.4) and (9.2.5) respectively, and by using Eq. (9.2.21), it can be shown that

$$\tilde{Y}_{O,f} > 0 \quad \Rightarrow \quad \tilde{Y}_{O,o} > 2(\tilde{T}_f - \tilde{T}_o), \tag{9.2.25}$$

$$\tilde{Y}_{F,f} > 0 \quad \Rightarrow \quad (1 - 2\beta) > 2(\tilde{T}_f - \tilde{T}_o). \tag{9.2.26}$$

Since heat is conducted to the fuel and oxidizer sides at an equal rate, $\tilde{T}_f$ has the maximum value in the flow field such that $(\tilde{T}_f - \tilde{T}_o) > 0$. Equation (9.2.26) then implies that the partial burning regime exists only for $\beta < \frac{1}{2}$.

The relations in (9.2.25) and (9.2.26) further indicate the dependence of the premixed flame regime on $\tilde{Y}_{O,o}$ and β. Since in this regime the reaction zone separates a frozen region from an equilibrium region with vanishing concentration of one of the reactants, then at the reaction sheet the concentration of one reactant vanishes while that of the other is $O(1)$. Thus if oxidizer is the vanishing species and fuel the

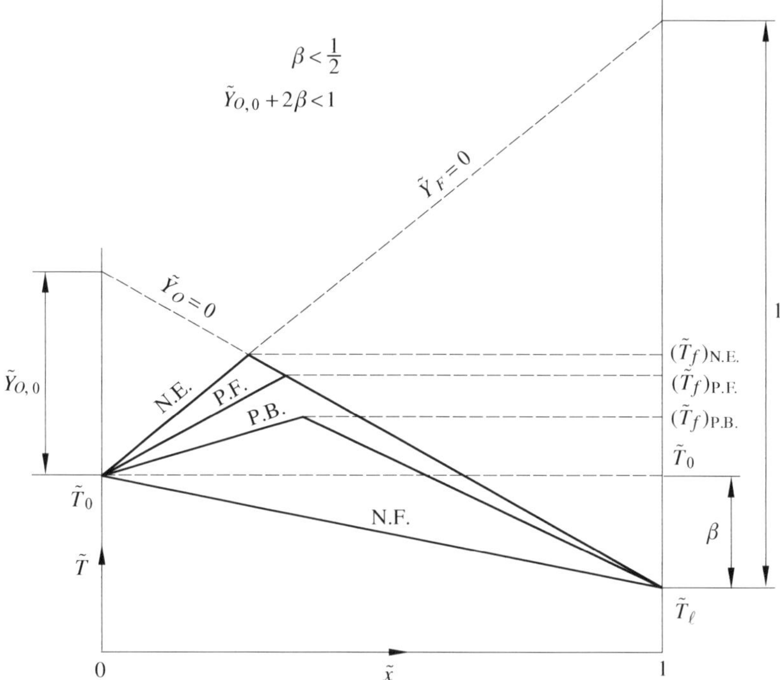

Figure 9.2.3. Classification of different flame regimes according to energy levels, for $\beta < \frac{1}{2}$ and $\tilde{Y}_{O,o} + 2\beta < 1$ (Liñán 1974).

excessively leaked species, as is the case represented by Eqs. (9.2.16) and (9.2.17), then clearly $\tilde{Y}_{F,f} > \tilde{Y}_{O,f}$ and we must require $(1 - 2\beta) > \tilde{Y}_{O,o}$ according to (9.2.25) and (9.2.26). By the same reasoning, the premixed flame regime with oxidizer being the excessively leaked species, such that equilibrium and frozen flows respectively prevail for $\tilde{x} < \tilde{x}_f$ and $\tilde{x} > \tilde{x}_f$, is constrained by $1 - 2\beta < \tilde{Y}_{O,o}$.

It is instructive to superimpose and compare the leading-order temperature profiles of these four regimes, as shown in Figure 9.2.3 for the situation with $\beta < \frac{1}{2}$ and oxidizer being the vanishing species such that $\tilde{Y}_{O,o} + 2\beta < 1$. Here the temperature profile for the frozen flow (N.F.) is simply the straight line joining $\tilde{T}_o$ and $\tilde{T}_\ell$, which has a slope of $-\beta$ given by Eq. (9.2.22). The near-equilibrium flame (N.E.) profile is given by the two lines respectively representing $\tilde{Y}_F = 0$, which has a slope of $(1 - \beta)$ according to Eq. (9.2.12), and $\tilde{Y}_O = 0$, which has a slope of $-(\tilde{Y}_{O,o} + \beta)$ according to Eq. (9.2.13). Since $\beta < \frac{1}{2}$ in Figure 9.2.3, $(\tilde{T}_f)_{\text{N.E.}} > \tilde{T}_o$ as shown in Eq. (9.2.14), and the reaction sheet "loses" heat to both sides. Furthermore, since the near-equilibrium flame is the strongest burning mode among the four flame regimes, its flame temperature is also the highest.

For the premixed flame (P.F.) regime with oxidizer being the completely depleted species to leading order, the temperature profile is composed of the equilibrium flow line of $\tilde{Y}_O = 0$ and a frozen flow line whose slope, $(\tilde{T}_f - \tilde{T}_o)/\tilde{x}_f$ given by Eq. (9.2.16),

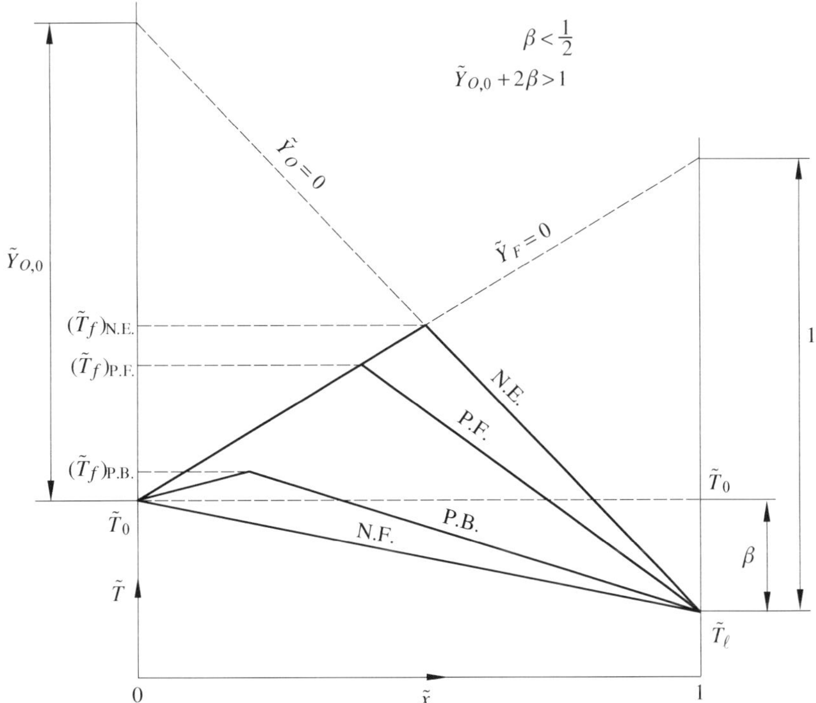

Figure 9.2.4. Classification of different flame regimes according to energy levels, for $\beta < \frac{1}{2}$ and $\tilde{Y}_{O,o} + 2\beta > 1$ (Liñán 1974).

must necessarily be smaller than that of the equilibrium flow line with $\tilde{Y}_F = 0$. The intersection of these two lines gives the premixed flame temperature $(\tilde{T}_f)_{\text{P.F.}}$, which is smaller than $(\tilde{T}_f)_{\text{N.E.}}$. The extent of deviation depends on the extent of leakage of the fuel, and hence the reaction rate in the reaction zone.

For the partial burning (P.B.) regime, leakage of the oxidizer also becomes important. The temperature profile on the fuel side now deviates from the $\tilde{Y}_O = 0$ line and its slope assumes the frozen value of $(\tilde{T}_f - \tilde{T}_\ell)/(1 - \tilde{x}_f)$ given by Eq. (9.2.20). The flame temperature, $(T_f)_{\text{P.B.}}$, is smaller than the premixed flame temperature, $(T_f)_{\text{P.F.}}$, and follows the locus defined by Eq. (9.2.21), namely the equality of the slopes on the two sides of the reaction sheet. The upper limit of this locus is the point where it intersects the $\tilde{Y}_O = 0$ line, indicating that $(\tilde{Y}_{O,f})_{\text{P.B.}} = 0$ and premixed flame regime prevails for higher values of the flame temperature. The lower limit of the locus is the hot boundary, $(\tilde{T}_f)_{\text{P.B.}} = \tilde{T}_o$, and the system degenerates to the frozen flow regime. Analysis of the partial burning regime breaks down as these limits are approached.

Figure 9.2.4 shows a similar plot for $\beta < \frac{1}{2}$ and $\tilde{Y}_{O,o} + 2\beta > 1$. The primary difference with Figure 9.2.3 is that now the fuel side is the frozen flow while the oxidizer side is the equilibrium flow for the premixed flame regime. Consequently, substantial amount of the oxidizer leaks into the fuel side in this regime.

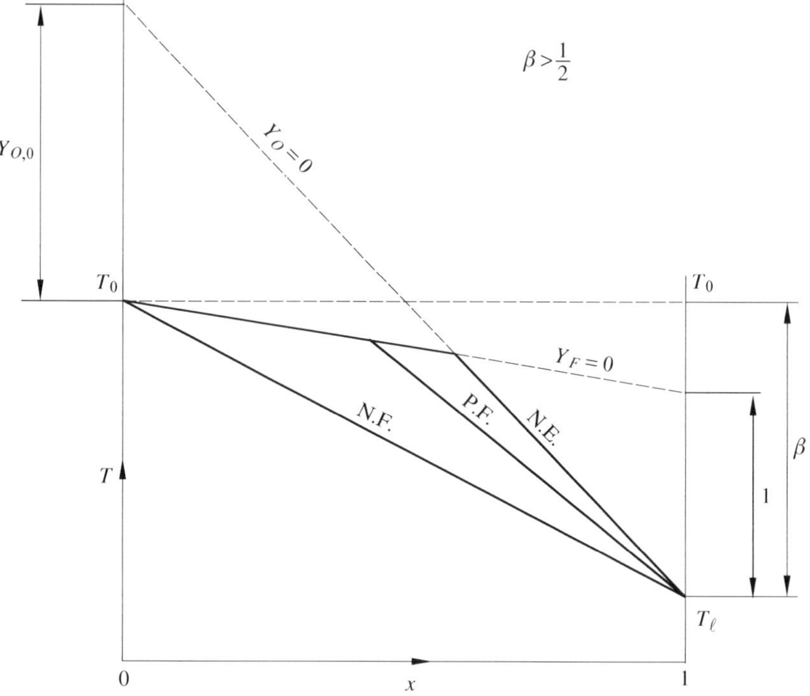

Figure 9.2.5. Classification of different flame regimes according to energy levels, for $\beta > \frac{1}{2}$ (Liñán 1974).

Finally, Figure 9.2.5 shows the situation of $\beta > \frac{1}{2}$, which of course also implies that $\tilde{Y}_{O,o} + 2\beta > 1$ such that only complete fuel consumption in the premixed flame regime is possible. Here $\tilde{T}_o > (\tilde{T}_f)_{\text{N.E.}}$ such that the hot boundary temperature is the highest in all flame regimes. Thus all flames receive heat from the oxidizer side and lose heat to the fuel side. The partial burning regime obviously also cannot exist here.

Based on the evolution of the various flame regimes as a function of the flame burning intensity shown in Figures 9.2.3–9.2.5, these regimes can also be related to the S-response curve obtained by plotting, say, the flame temperature versus a suitably defined Damköhler number. As shown in Figure 9.2.6, the complete upper branch and a small part of the middle branch around the turning point are expected to be described by the near-equilibrium regime; a good part of the middle branch and a small part of the upper branch by the premixed flame regime; the complete middle branch and small parts of the upper and lower branches by the partial burning regime; and the complete lower branch with a small part of the middle branch by the nearly frozen, ignition regime. As such, depending on the various system parameters, the extinction states are expected to be described by analyses of the near-equilibrium, premixed flame, and partial burning regimes, while the ignition state is described by the partial burning and nearly frozen regimes.

We shall now study the flame structure for each of these regimes.

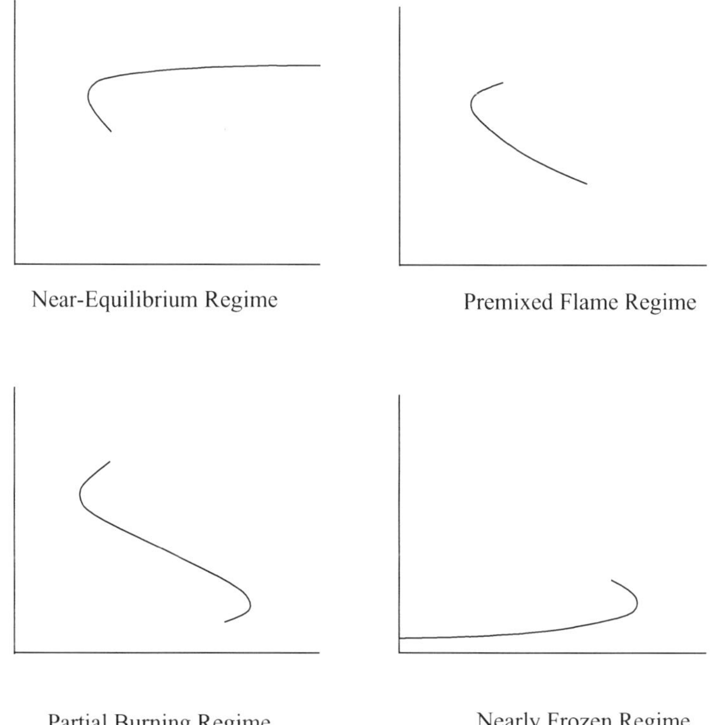

Figure 9.2.6. Segments of the response curve described by different flame regimes.

9.3. STRUCTURE OF NONPREMIXED FLAMES: ANALYSIS

9.3.1. Nearly Frozen Regime

In this regime, to the leading order the flow is frozen and the temperature and species concentrations are given by Eqs. (9.2.22)–(9.2.24). For large but finite values of T_a, the flow becomes weakly reactive, which may eventually lead to ignition for sufficiently large values of the Damköhler number.

When $\tilde{T}_o$ is much larger than $\tilde{T}_\ell$ such that β is $O(1)$, ignition is expected to be initiated near the hot boundary, with $\tilde{x} = O(\epsilon)$ and $\epsilon = \tilde{T}_o^2/\tilde{T}_a$. Thus letting

$$\begin{aligned}\tilde{T}_{\text{in}}(\chi) &= \tilde{T}_0(\tilde{x}) + \epsilon\theta(\chi) + O(\epsilon^2)\\ &= \tilde{T}_o + \epsilon(\theta - \beta\chi) + O(\epsilon^2),\end{aligned} \tag{9.3.1}$$

where $\chi = \tilde{x}/\epsilon$ and $\tilde{T}_0$ is given by Eq. (9.2.22), and substituting $\tilde{T}_{\text{in}}$ into Eqs. (9.2.2), (9.2.4), (9.2.5), and (9.2.7), we obtain $\tilde{Y}_{O,\text{in}} = \tilde{Y}_{O,o} + O(\epsilon)$, $\tilde{Y}_{F,\text{in}} = \epsilon(\chi - \theta) + O(\epsilon^2)$, and

$$\frac{d^2\theta}{d\chi^2} = -\Delta(\chi - \theta)e^{(\theta - \beta\chi)}, \tag{9.3.2}$$

with the boundary condition

$$\theta(0) = 0, \tag{9.3.3}$$

where $\Delta = \epsilon^2 Da_C \tilde{Y}_{O,o} \exp(-\tilde{T}_a/\tilde{T}_o)$ is the reduced Damköhler number. In Eq. (9.3.2) the preexponential factor $(\chi - \theta)$ represents the $O(\epsilon)$ fuel concentration and its variation in the reaction zone. Since the oxidizer concentration is $O(1)$ here and therefore not affected to the leading order by the reaction, $\tilde{Y}_O \approx \tilde{Y}_{O,o}$ and is absorbed in Δ.

The outer solution can be expressed as

$$\begin{aligned}\tilde{T}_{\text{out}}(\tilde{x}) &= \tilde{T}_0(\tilde{x}) + \epsilon \tilde{T}_1(\tilde{x}) + O(\epsilon^2)\\ &= \tilde{T}_o - \beta\tilde{x} + \epsilon c_1(1-\tilde{x}) + O(\epsilon^2),\end{aligned} \tag{9.3.4}$$

which satisfies $d^2\tilde{T}_{\text{out}}/d\tilde{x}^2 = 0$ and the boundary condition (9.2.8), $\tilde{T}(1) = \tilde{T}_\ell$. Matching $\tilde{T}_{\text{in}}$ with $\tilde{T}_{\text{out}}$ in the limit $\chi \to \infty$ yields

$$\theta(\infty) = c_1, \tag{9.3.5}$$

$$\left(\frac{d\theta}{d\chi}\right)_{\chi\to\infty} = 0. \tag{9.3.6}$$

Equation (9.3.6) provides the additional boundary condition for Eq. (9.3.2), while Eq. (9.3.5) determines c_1.

Unlike the flat plate case for which analytic solution exists, the present equation needs to be solved numerically because of the presence of the concentration term $(\chi - \theta)$. For a given $\beta < 1$, a plot of $\theta(\chi)$ on constant Δ yields a dual solution behavior similar to Figure 8.2.2, while a subsequent plot of $\theta(\infty)$ versus Δ reproduces the lower branch and part of the middle branch of the S-curve, similar to Figure 8.2.3. These curves yield a universal relation between the ignition Damköhler number Δ_I and the heat transfer parameter β. A semiempirical correlation (Makino, 1991) results in an explicit expression

$$\Delta_I(\beta) = \frac{\beta^2/2}{1 + 2.17(1-\beta)/\beta}, \tag{9.3.7}$$

which can be readily used to assess ignitability of a system. The analysis breaks down for $\beta \geq 1$, for which the hot boundary temperature is higher than the nonpremixed flame temperature of the mixture, implying that distinct ignition cannot be defined.

9.3.2. Partial Burning Regime

In this regime, the reaction zone separates two frozen regions. The reaction is sufficiently weak such that the reactants are only slightly depleted in crossing the flame, thereby resulting in substantial leakage.

This regime is the most difficult to analyze because the reaction-sheet solution is determined to within two unknown flame properties, namely the flame temperature $\tilde{T}_f$ and location $\tilde{x}_f$, as shown earlier. Thus it is necessary to carry the analysis to $O(\epsilon^2)$.

We shall not present this analysis because the mathematical details are quite involved. Furthermore, since this regime mainly corresponds to the middle branch of the S-curve, the solutions may not be stable anyway. We shall, however, demonstrate the feature that the flame conducts heat to the fuel and oxidizer sides at equal rates because this result was previously used in classifying the flame regimes.

For $T_{\text{in}}(\chi) = T_f + \epsilon\theta(\chi) + O(\epsilon^2)$ and $O(1)$ leakages of both fuel and oxidizer across the flame, the structure equation for the reaction zone is $d^2\theta/d\chi^2 = -\Delta e^\theta$, which can be readily integrated to yield

$$\theta(\chi) = \theta_m - \ln\left\{\cosh^2\left[\left(\frac{\Delta}{2}e^{\theta_m}\right)^{1/2}(\chi - \chi_m)\right]\right\}, \tag{9.3.8}$$

where $\Delta = \epsilon\, Da_C \tilde{Y}_{O,f} \tilde{Y}_{F,f} \exp(-\tilde{T}_a/\tilde{T}_f)$ and the state of the maxima,

$$\chi = \chi_m: \quad \theta = \theta_m, \quad \left(\frac{d\theta}{d\chi}\right)_{\chi_m} = 0, \tag{9.3.9}$$

has been used to evaluate the integration constants. It is seen that θ is an even function in $(\chi - \chi_m)$, and hence is symmetrical about χ_m. Furthermore, we have

$$\lim_{\chi\to\pm\infty} \theta(\chi) = \theta_m \mp (2\Delta e^{\theta_m})^{1/2} \lim_{\chi\to\pm\infty} (\chi - \chi_m) + \ln 4, \tag{9.3.10}$$

$$\left(\frac{d\theta}{d\chi}\right)_{\pm\infty} = \mp\left(2\Delta e^{\theta_m}\right)^{1/2}. \tag{9.3.11}$$

Equation (9.3.11) clearly shows that $(d\theta/d\chi)$ decreases at equal rates as $\chi \to \pm\infty$. Since these values are to be matched with the leading-order outer temperature gradients, these gradients must also have the same magnitude. We have therefore demonstrated that the reaction sheet loses heat to the fuel and oxidizer sides at equal rates. Furthermore, by applying the closure relation Eq. (9.1.26) and using Eq. (9.3.10), it can be shown that the reaction sheet is at the location of the maximum temperature.

9.3.3. Premixed Flame Regime

In this regime, the reaction zone separates a frozen region from an equilibrium region. We shall analyze the case of $\tilde{Y}_{O,o} + 2\beta < 1$, for which the oxidizer is mostly consumed in crossing the reaction zone while substantial amount of fuel leakage occurs. Therefore the oxidizer side is frozen and the fuel side is in equilibrium. The case of $\tilde{Y}_{O,o} + 2\beta > 1$ can be similarly analyzed, resulting in an identical structure equation in appropriately defined variables.

The inner and outer expansions can be written as

$$\tilde{T}_{\text{in}}(\chi) = \tilde{T}_f - \epsilon\theta(\chi) + O(\epsilon^2), \tag{9.3.12}$$

$$\begin{aligned}\tilde{T}^-_{\text{out}}(\tilde{x}) &= \tilde{T}^-_0(\tilde{x}) - \epsilon\tilde{T}^-_1(\tilde{x}) + O(\epsilon^2)\\ &= \tilde{T}_o + (\tilde{T}_f - \tilde{T}_o)\frac{\tilde{x}}{\tilde{x}_f} - \epsilon c^-_1\tilde{x} + O(\epsilon^2),\end{aligned} \tag{9.3.13}$$

$$\begin{aligned}\tilde{T}^+_{\text{out}}(\tilde{x}) &= \tilde{T}^+_0(\tilde{x}) - \epsilon\tilde{T}^+_1(\tilde{x}) + O(\epsilon^2)\\ &= (\tilde{Y}_{O,o} + \tilde{T}_o) - (\tilde{Y}_{O,o} + \beta)\tilde{x} - \epsilon c^+_1(1 - \tilde{x}) + O(\epsilon^2),\end{aligned} \tag{9.3.14}$$

where $\chi = (\tilde{x} - \tilde{x}_f)/\epsilon$, $\epsilon = \tilde{T}_f^2/\tilde{T}_a$, and $\tilde{T}_0^{\mp}$ are respectively given by Eqs. (9.2.16) and (9.2.17).

From Eqs. (9.3.13) and (9.3.14), we readily identify

$$\left(\frac{d\tilde{T}_0^-}{d\tilde{x}}\right)_{\tilde{x}_f} = \frac{\tilde{T}_f - \tilde{T}_o}{\tilde{x}_f}, \qquad \left(\frac{d\tilde{T}_0^+}{d\tilde{x}}\right)_{\tilde{x}_f} = -(\tilde{Y}_{O,o} + \beta).$$

Furthermore, Eqs. (9.2.4) and (9.2.17) show that $(\tilde{Y}_{O,1}^+ + \tilde{T}_1^+)_{\tilde{x}_f} \equiv 0$. Thus following the definition of Section 9.1.3 for the canonical formulation, we have

$$\tilde{\theta} = \theta - \mu\eta - p, \qquad \eta = M_0\chi - \frac{p}{\mu},$$

$$\frac{1}{\mu} = 1 + \frac{\tilde{T}_f - \tilde{T}_o}{\tilde{x}_f(\tilde{Y}_{O,o} + \beta)}, \qquad M_0 = \frac{\tilde{Y}_{O,o} + \beta}{\mu},$$

$$p = \ln\left[\Delta\left(\frac{\mu}{\tilde{Y}_{O,o} + \beta}\right)^2\right].$$

Using these relations, the canonical structure equations (9.1.35)–(9.1.37) are retrieved. The problem is therefore completely solved.

9.3.4. Near-Equilibrium Regime

In this regime the reaction zone separates two near-equilibrium regions, with fuel and oxidizer leaking through it in $O(\epsilon)$ amounts. The reaction-sheet location $\tilde{x}_f$ and flame temperature $\tilde{T}_f$ are determined at the leading order through stoichiometry considerations. Thus the primary purpose of the reaction zone analysis is to determine the state of extinction.

We write the inner and outer expansions as

$$\tilde{T}_{\text{in}}(\chi) = \tilde{T}_f - \epsilon\theta(\chi) + O(\epsilon^2), \tag{9.3.15}$$

$$\begin{aligned}\tilde{T}_{\text{out}}^-(\tilde{x}) &= \tilde{T}_0^-(\tilde{x}) - \epsilon\tilde{T}_1^-(\tilde{x}) + O(\epsilon^2)\\ &= \tilde{T}_o + (1-\beta)\tilde{x} - \epsilon c_1^-\tilde{x} + O(\epsilon^2),\end{aligned} \tag{9.3.16}$$

$$\begin{aligned}\tilde{T}_{\text{out}}^+(\tilde{x}) &= \tilde{T}_0^+(\tilde{x}) - \epsilon\tilde{T}_1^+(\tilde{x}) + O(\epsilon^2)\\ &= (\tilde{Y}_{O,o} + \tilde{T}_o) - (\tilde{Y}_{O,o} + \beta)\tilde{x} - \epsilon c_1^+(1-\tilde{x}) + O(\epsilon^2),\end{aligned} \tag{9.3.17}$$

where $\chi = (\tilde{x} - \tilde{x}_f)/\epsilon$, $\epsilon = \tilde{T}_f^2/\tilde{T}_a$, and $\tilde{T}_0^{\mp}$ are respectively given by Eqs. (9.2.12) and (9.2.13). Substituting $\tilde{T}_{\text{in}}$ into Eqs. (9.2.2), (9.2.4), and (9.2.5) results in

$$\tilde{Y}_{O,\text{in}} = \epsilon[\theta - (\tilde{Y}_{O,o} + \beta)\chi] + O(\epsilon^2),$$

$$\tilde{Y}_{F,\text{in}} = \epsilon[\theta + (1-\beta)\chi] + O(\epsilon^2),$$

$$\frac{d^2\theta}{d\chi^2} = \Delta[\theta - (\tilde{Y}_{O,o} + \beta)\chi][\theta + (1-\beta)\chi]e^{-\theta}, \tag{9.3.18}$$

where $\Delta = \epsilon^3 Da_C e^{-\tilde{T}_a/\tilde{T}_f}$. It is seen that $\tilde{Y}_{F,\text{in}}$ and $\tilde{Y}_{O,\text{in}}$ are both $O(\epsilon)$ quantities in the reaction zone.

Matching between the inner and outer solutions yields

$$\lim_{\chi \to -\infty} [\theta + (1-\beta)\chi] = c_1^- \tilde{x}_f, \tag{9.3.19}$$

$$\left(\frac{d\theta}{d\chi}\right)_{-\infty} = -(1-\beta), \tag{9.3.20}$$

$$\lim_{\chi \to \infty} [\theta - (\tilde{Y}_{O,o} + \beta)\chi] = c_1^+ (1 - \tilde{x}_f), \tag{9.3.21}$$

$$\left(\frac{d\theta}{d\chi}\right)_{\infty} = \tilde{Y}_{O,o} + \beta. \tag{9.3.22}$$

By defining

$$\theta = \tilde{\theta} + \gamma\eta, \qquad \chi = \frac{2}{1 + \tilde{Y}_{O,o}}\eta,$$

$$\delta = \frac{4\Delta}{1 + \tilde{Y}_{O,o}^2}, \qquad \gamma = 1 - \frac{2(1-\beta)}{1 + \tilde{Y}_{O,o}} = \frac{2(\tilde{Y}_{O,o} + \beta)}{1 + \tilde{Y}_{O,o}} - 1,$$

the above system of equations can be cast into the canonical form

$$\frac{d^2\tilde{\theta}}{d\eta^2} = \delta(\tilde{\theta} - \eta)(\tilde{\theta} + \eta)e^{-(\tilde{\theta} + \gamma\eta)}, \tag{9.3.23}$$

$$\left(\frac{d\tilde{\theta}}{d\eta}\right)_{-\infty} = -1, \tag{9.3.24}$$

$$\left(\frac{d\tilde{\theta}}{d\eta}\right)_{\infty} = 1, \tag{9.3.25}$$

with the additional matching conditions given by

$$\lim_{\eta \to -\infty} (\tilde{\theta} + \eta) = c_1^- \tilde{x}_f, \tag{9.3.26}$$

$$\lim_{\eta \to \infty} (\tilde{\theta} - \eta) = c_1^+ (1 - \tilde{x}_f), \tag{9.3.27}$$

which allow the determination of c_1^+ and c_1^-.

Since $(\tilde{\theta} - \eta)$ and $(\tilde{\theta} + \eta)$ respectively represent the oxidizer and fuel concentrations in the reaction zone, the quantities $(\tilde{\theta} - \eta)_\infty$ and $(\tilde{\theta} + \eta)_{-\infty}$ then respectively indicate the extent of oxidizer and fuel leakages into the fuel and oxidizer sides of the reaction zone. The consequence of the oxidizer leakage is lowering of the outer flame temperature on the fuel side by an amount proportional to $c_1^+(1 - \tilde{x}_f)$, as given by the matching condition Eq. (9.3.27). Similarly, fuel leakage leads to lowering of the oxidizer side of the flame temperature by an amount proportional to $c_1^- \tilde{x}_f$ through Eq. (9.3.26).

The parameter which governs the relative amounts of fuel and oxidizer leakages is γ, which has the following significance. Since the heat loss rates to the oxidizer and

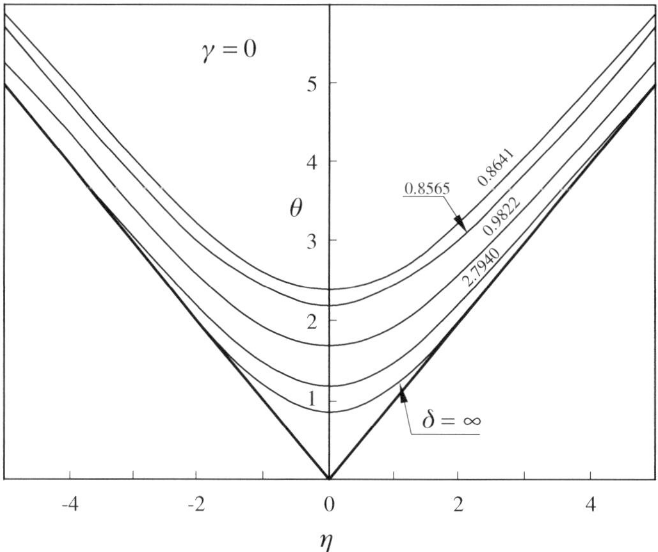

Figure 9.3.1. Inner structure of the near-equilibrium flame, for $\gamma = 0$ (Liñán 1974).

fuel sides of the flame are respectively proportional to the temperature gradients, $(1 - \beta)$ and $(\tilde{Y}_{O,o} + \beta)$, therefore $(1 + \tilde{Y}_{O,o})$ in the definition of γ is simply the total heat loss rate from the flame, which of course is also the total chemical heat release rate. Thus the ratios $(1 - \beta)/(1 + \tilde{Y}_{O,o})$ and $(\tilde{Y}_{O,o} + \beta)/(1 + \tilde{Y}_{O,o})$ respectively represent the heat loss rate to the oxidizer and fuel sides of the flame as fractions of the total heat loss rate. Consequently, when the heat loss rates are equal, these ratios are $\frac{1}{2}$ and, hence, $\gamma = 0$. When $\gamma > 0$ there is more heat loss toward the fuel side while $\gamma < 0$ implies more heat loss toward the oxidizer side.

For given γ and δ, $\tilde{\theta}(\eta)$ can be numerically solved from Eqs. (9.3.23) to (9.3.25). Figure 9.3.1 shows $\tilde{\theta}(\eta)$ on constant δ for $\gamma = 0$. Because of the equal heat loss rates to both sides of the flame, it is seen that $\tilde{\theta}(\eta)$ is symmetrical. The results also show that there are two solutions for $\delta > 0.8564$, and no solution for smaller values. The characteristic dual solution, extinction turning point behavior is therefore exhibited, yielding an extinction Damköhler number, δ_E.

Figure 9.3.2 shows the case for $\gamma = 0.5$. Since the flame now loses more heat to the fuel side, burning is expected to be stronger on the oxidizer side. This is demonstrated by having the minimum values of $\tilde{\theta}$, which implies the maximum flame temperatures, occurring on the oxidizer side of the flame. Furthermore, because of the relatively strong burning intensity there, near-equilibrium burning is favored, leading to the decreased sensitivity on the reduced Damköhler δ and hence merging of the various δ curves.

When $\gamma = 1$, Figure 9.3.3 shows that the solution becomes single valued and extinction is not possible. This corresponds to the situation of $\beta = 1$ and hence $\tilde{T}_f = \tilde{T}_o$, implying that there is no heat loss to the oxidizer boundary. The Arrhenius

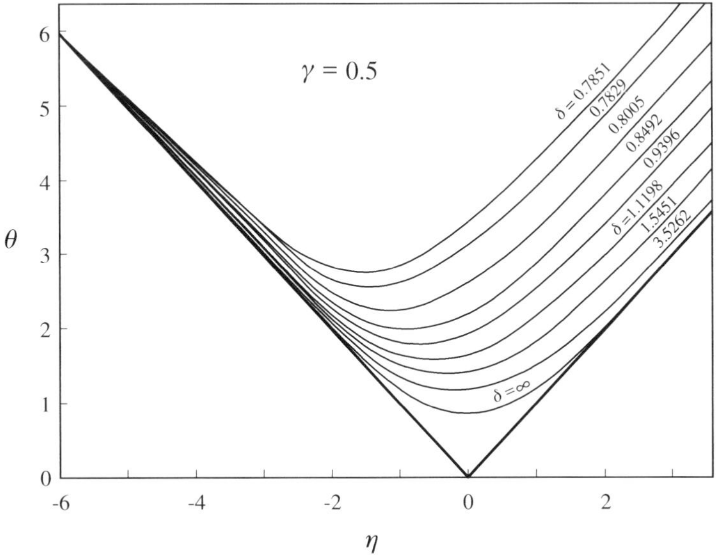

Figure 9.3.2. Inner structure of the near-equilibrium flame, for $\gamma = 0.5$ (Liñán 1974).

factor $\exp\{-(\tilde{\theta} + \gamma\eta)\}$ in the structure equation (9.3.23) becomes $\exp\{-(\tilde{\theta} + \eta)\} \to \exp(-c_1^- \tilde{x}_f) \to$ constant as $\eta \to -\infty$. Then the only mechanism to suppress the reaction rate as $\eta \to -\infty$ is for the leakage term $(\tilde{\theta} + \eta)_{-\infty} \to 0$ in an exponential manner. This implies complete fuel consumption in the reaction region, resulting in vanishing fuel leakage. The structure of the reaction zone now resembles that of the premixed flame.

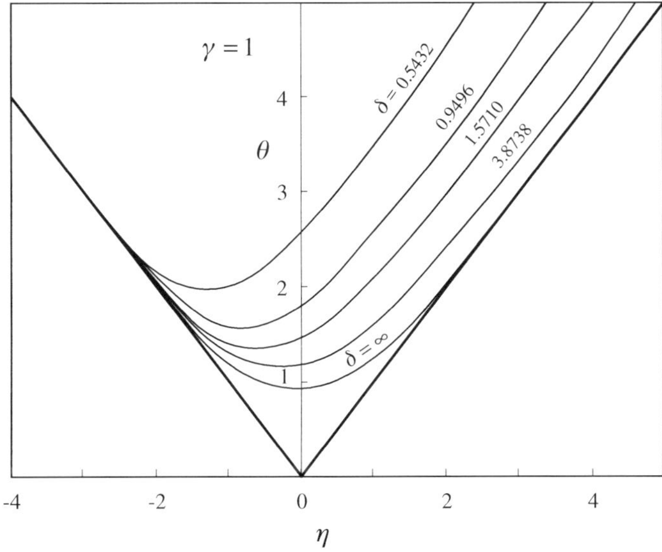

Figure 9.3.3. Inner structure of the near-equilibrium flame, for $\gamma = 1$ (Liñán 1974).

So far the discussion has been for $\gamma > 0$. The problem, however, is completely symmetrical in γ, in that for $\gamma < 0$, the character of the equation system (9.3.23) to (9.3.27) remains unchanged by substituting $\gamma \to -\gamma$ and $\eta \to -\eta$.

From the reduced extinction Damköhler number, δ_E, numerically determined for various values of γ, an explicit correlation has been obtained as

$$\delta_E = \{(1-|\gamma|) - (1-|\gamma|)^2 + 0.26(1-|\gamma|)^3 + 0.055(1-|\gamma|)^4\}e. \quad (9.3.28)$$

The Damköhler number for extinction, $Da_{C,E}$, can thus be determined through the definitions of δ and Δ.

Finally, the leakage factor at extinction has also been numerically determined and correlated as

$$\left\{\lim_{\eta\to\infty}(\tilde{\theta}-\eta)\right\}_E = -0.27 + \frac{4.19}{(1-\gamma)} - \frac{0.894}{(1-\gamma)^2} + \frac{0.117}{(1-\gamma)^3} - \frac{0.00545}{(1-\gamma)^4}, \quad \gamma \geq -0.3. \quad (9.3.29)$$

The leakage factor $\{(\tilde{\theta}+\eta)_{-\infty}\}_E$ is given by substituting γ by $-\gamma$ in Eq. (9.3.29). The fitting breaks down for $\gamma \leq -0.3$ in that the leakage becomes negative. However, since the leakage is extremely small under these situations, $(\tilde{\theta}-\eta)_\infty$ can be simply set to zero.

We close this section by noting that the reduced Damköhler numbers, Δ, for the partial burning regime, the premixed flame regime, and the near-equilibrium regime are respectively proportional to ϵ, ϵ^2, and ϵ^3. Here one of the ϵs comes from the conversion between $d^2\tilde{T}_{in}/d\tilde{x}^2$ and $d^2\theta/d\chi^2$ due to the need to stretch the thickness of the reaction region. The remaining ϵs represent the influence of the variation of the reactant concentration. Specifically, since both reactants are $O(1)$ quantitative in the reaction zone for the partial burning regime, concentration variations are not important and therefore their influence is $O(1)$. By the same reasoning, the influence is $O(\epsilon)$ for the premixed flame regime because the concentration of one of the reactants is $O(\epsilon)$, while it is $O(\epsilon^2)$ for the near-equilibrium regime because the concentrations of both fuel and air are $O(\epsilon)$ quantities.

9.4. MIXTURE FRACTION FORMULATION FOR NEAR-EQUILIBRIUM REGIME

In Section 5.4.5 we introduced the concept that, by assuming equidiffusivity, analysis of nonpremixed flames can be usefully conducted in the phase space of the mixture fraction Z such that the results obtained are independent of the physical configurations (Williams 1985; Peters 2000). Such a formulation therefore provides a unifying interpretation of the flame structure in terms of the variations of its thermochemical properties such as the temperature, species concentrations, and reaction rates. This formulation was applied to the reaction-sheet limit in Section 6.1.3. We now extend the analysis to the reaction zone structure.

To demonstrate this approach, we consider the energy equation in the Z-coordinate, Eq. (5.4.53),

$$\rho\frac{\partial\tilde{T}}{\partial t}+\rho\mathbf{v}_t\cdot\nabla_t\tilde{T}=-w+\rho D|\nabla Z|^2\frac{\partial^2\tilde{T}}{\partial Z^2}+\nabla_t\cdot(\rho D\nabla_t\tilde{T}), \tag{9.4.1}$$

where the subscript t designates the tangential component, and we have written $\tilde{T}=\tilde{h}^s$ for simplicity. Furthermore, the coupling functions are

$$\tilde{T}+\tilde{Y}_O=(\tilde{T}_o+\tilde{Y}_{O,o})-(\beta+\tilde{Y}_{O,o})Z, \tag{9.4.2}$$

$$\tilde{T}+\tilde{Y}_F=\tilde{T}_o+(1-\beta)Z, \tag{9.4.3}$$

where we have exchanged the locations of fuel and oxidizer from those in Section 5.4, and we also note that $\tilde{Y}_{F,B^-}\equiv 1$. The mixture fraction Z, being a conserved scalar, is governed by

$$\rho\frac{\partial Z}{\partial t}+\rho\mathbf{v}\cdot\nabla Z-\nabla\cdot(\rho D\nabla Z)=0. \tag{9.4.4}$$

In the reaction-sheet limit, the species and temperature distributions are given by

$$Z<Z_f:\quad \tilde{Y}_{O,0}=\tilde{Y}_{O,o}-(1+\tilde{Y}_{O,o})Z, \tag{9.4.5}$$

$$\tilde{T}_0^-=\tilde{T}_o+(1-\beta)Z, \tag{9.4.6}$$

$$\tilde{Y}_{F,0}=0, \tag{9.4.7}$$

$$Z>Z_f:\quad \tilde{Y}_{O,0}=0, \tag{9.4.8}$$

$$\tilde{T}_0^+=(\tilde{T}_o+\tilde{Y}_{O,o})-(\beta+\tilde{Y}_{O,o})Z, \tag{9.4.9}$$

$$\tilde{Y}_{F,0}=-\tilde{Y}_{O,o}+(1+\tilde{Y}_{O,o})Z, \tag{9.4.10}$$

where

$$Z_f=\frac{\tilde{Y}_{O,o}}{1+\tilde{Y}_{O,o}} \tag{9.4.11}$$

is the value of Z at the reaction sheet, and we also have the flame temperature

$$\tilde{T}_f=\tilde{T}_o+(1-\beta)Z_f. \tag{9.4.12}$$

Comparing Eqs. (9.4.11) with (9.2.15), it is seen that Z_f is the flame location in Z-space. Since Z_f is the state where the fuel and oxidizer are stoichiometrically consumed, it is conventionally designated by Z_{st}.

To analyze the flame structure, we note that since flame properties vary most significantly in the direction normal to the flame surface defined by Z_f, the reaction zone is described by a balance between reaction and "diffusion" in the normal direction. Thus Eq. (9.4.1) simplifies to

$$\rho\frac{\partial\tilde{T}}{\partial t}-\frac{\rho\chi}{2}\frac{\partial^2\tilde{T}}{\partial Z^2}\approx -w, \tag{9.4.13}$$

where

$$\chi(t, Z) = 2D\left(\frac{\partial Z}{\partial n}\right)^2 \qquad (9.4.14)$$

is called the scalar dissipation rate. It has the dimension of s^{-1} and its inverse, χ^{-1}, can be interpreted as the characteristic diffusion time across the flame. An inspection of Eq. (9.4.13), and recognizing that ρD instead of D is the proper parameter indicating the efficiency of diffusion, a natural representation of the characteristic diffusion time should therefore be through a density-weighted scalar dissipation rate

$$X \equiv \rho\chi = 2\rho D\left(\frac{\partial Z}{\partial n}\right)^2. \qquad (9.4.15)$$

We have therefore encountered again the importance of attaching the density to a flow-dependent quantity in variable density flows, such as the density-weighted laminar flame speed $f^o = \rho s_u^o$ and density-weighted strain rate ρa for the flame of Section 6.5. We also note that Eq. (9.4.13) holds rigorously for one-dimensional or quasi-one-dimensional flows, such as the droplet and counterflow problems respectively, for which the tangential components vanish and the first-order spatial differentials can be transformed away.

Let us now solve the chambered flame problem using the mixture fraction formulation. Equations (9.4.4) and (9.4.13) now respectively simplify to

$$\frac{d^2 Z}{d\tilde{x}^2} = 0, \qquad (9.4.16)$$

$$\frac{X}{2}\frac{d^2\tilde{T}}{dZ^2} = w. \qquad (9.4.17)$$

The solution of Eq. (9.4.16) subject to $Z(0) = 0$ and $Z(1) = 1$ is simply $Z = \tilde{x}$, which, when substituted into Eq. (9.4.15), yields $X = 2\rho D/\ell^2$. Using the above, it is immediately clear that Eq. (9.4.17) is identical to Eq. (9.2.2). Thus the problem to be solved in Z-space is the same as that in $\tilde{x}$-space, because $Z = \tilde{x}$ for this simple problem. For more complicated problems, we follow the same asymptotic analysis procedure, using Z as the independent variable, stretching around the reaction zone through an inner variable $(Z - Z_f)/\epsilon$, and performing the matchings between the inner and outer solutions.

PROBLEMS

1. Show that the generalized premixed flame result of Section 9.1 specializes to that of the nonadiabatic premixed flame with volumetric loss of Section 8.4.
2. Show that the generalized premixed flame result of Section 9.1 specializes to that of the flat-burner flame of Section 8.6.1.
3. In Liñán's original asymptotic formulation, closure was achieved by suppressing all perturbations on the fuel side. Reformulate the premixed flame problem by

using this closure scheme, and show that the final result can be brought to the same form as the present one.

4. Perform the analysis for the near-equilibrium regime of the chambered flame, for $Le_i \neq 1$.

(a) Write down the governing equations.

(b) Show that the inner structure equation and its boundary conditions are given by

$$\frac{d^2\theta}{d\chi^2} = Da\, Le_F\, Le_O \left(\theta + h_F + \frac{d\tilde{T}_0^-(\tilde{x}_f)}{d\tilde{x}}\chi\right)\left(\theta + h_O + \frac{d\tilde{T}_0^+(\tilde{x}_f)}{d\tilde{x}}\chi\right)e^{-\theta} \tag{9.P.1}$$

$$\left(\frac{d\theta}{d\chi}\right)_{\chi\to-\infty} = \frac{d\tilde{T}_0^-(\tilde{x}_f)}{d\tilde{x}}, \qquad \left(\frac{d\theta}{d\chi}\right)_{\chi\to\infty} = \frac{d\tilde{T}_0^+(\tilde{x}_f)}{d\tilde{x}}, \tag{9.P.2}$$

where $Da = \varepsilon^3 Da_C \exp(-\tilde{T}_a/\tilde{T}_f)$,

$$h_F \equiv \tilde{T}_1^-(\tilde{x}_f) + \frac{1}{Le_F}\tilde{Y}_{F,1}^-(\tilde{x}_f), \qquad h_O \equiv \tilde{T}_1^+(\tilde{x}_f) + \frac{1}{Le_O}\tilde{Y}_{O,1}^+(\tilde{x}_f)$$

and θ is the $O(\varepsilon)$ expansion term of the temperature in the inner reaction zone, namely $\tilde{T}_{\text{in}}(\chi) = \tilde{T}_f - \varepsilon\theta(\chi) + O(\varepsilon^2)$.

(c) Transform the above structure equation into Liñán's canonical form, Eqs. (9.3.23) to (9.3.25), by defining

$$\delta = 4 Le_F\, Le_O\, Da \left(\frac{d\tilde{T}_0^+(\tilde{x}_f)}{d\tilde{x}} - \frac{d\tilde{T}_0^-(\tilde{x}_f)}{d\tilde{x}}\right)^{-2} \exp\left(\frac{1-\gamma}{2}h_O + \frac{1+\gamma}{2}h_F\right)$$

$$\gamma = \left(\frac{d\tilde{T}_0^+(\tilde{x}_f)}{d\tilde{x}} + \frac{d\tilde{T}_0^-(\tilde{x}_f)}{d\tilde{x}}\right) \Bigg/ \left(\frac{d\tilde{T}_0^+(\tilde{x}_f)}{d\tilde{x}} - \frac{d\tilde{T}_0^-(\tilde{x}_f)}{d\tilde{x}}\right).$$

10 Aerodynamics of Laminar Flames

10.1. GENERAL CONCEPTS

In Chapter 7 we studied the simplest possible mode of laminar flame propagation, namely the steady propagation of an adiabatic, one-dimensional, planar flame into an unburned combustible medium of temperature T_u and reactant concentrations $Y_{i,u}$, as shown in Figure 7.2.1 in the flame-stationary frame. This flame is characterized by its final, burned state of equilibrium temperature T_b^o, which is simply the adiabatic flame temperature T_{ad} of the mixture, and by its mass burning flux f^o, which is a function of the exothermicity, diffusivity, and reactivity of the mixture. We have also used the superscript o to designate properties pertaining to this specific, standard premixed flame propagation mode. Thus if we symbolize the dependence on exothermicity by the specific heat release q_c, diffusivity by the Lewis numbers $Le_{i,j} = \lambda/(c_p \rho D_{i,j})$, and reactivity by the reaction rate w_k of reaction k, then

$$f^o = f^o(q_c, Le_{i,j}, w_k), \qquad T_b^o = T_b^o(q_c).$$

Furthermore, if the characteristic chemical times of the major heat release reactions are much shorter than those of heat and mass diffusion, then the flame structure can be considered to consist of a thin reaction zone preceded by a broader transport-dominated, diffusive–convective zone, which are respectively characterized by thicknesses ℓ_R^o and ℓ_D^o satisfying $\ell_R^o \approx \ell_D^o/Ze \ll \ell_D^o$. From $f^o = \rho u$ we can then define an unburned, upstream laminar flame speed $s_u^o = f^o/\rho_u$ at the upstream boundary of the transport zone, and a burned, downstream laminar flame speed $s_b^o = f^o/\rho_b^o$ for the entire reaction zone because velocity does not change much owing to the thinness of this zone.

Practical flames seldom behave in this idealized manner. For example, among the various techniques used to determine the laminar flame speed, the Bunsen flame cone possesses a curvature that is maximized at its tip, an outwardly propagating flame from an ignition kernel is unsteady, and in most situations the flow field ahead of the flame can also be quite nonuniform. Furthermore, small amount of heat loss

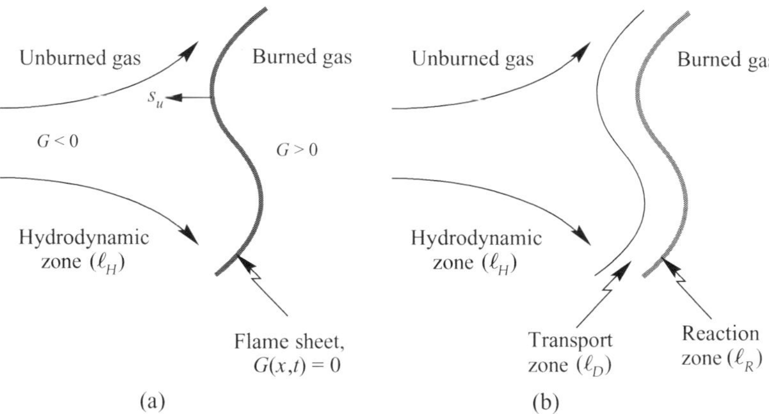

Figure 10.1.1. Structure of a wrinkled flame at: (a) hydrodynamic, flame-sheet level, and (b) transport, reaction-sheet levels.

from the flame is usually unavoidable. It is therefore reasonable to expect that these influences can cause departure of the flame behavior from that in the idealized limit.

Figure 10.1.1 shows the general situation of a wrinkled flame in a nonuniform flow, with Figure 10.1.1a depicting the flame-sheet limit and Figure 10.1.1b exposing the flame structure. It is seen that the presence of flow nonuniformity and flame curvature introduces into the problem an additional, hydrodynamic, length scale, ℓ_H. If we now symbolize influences due to such aerodynamic factors as flow nonuniformity, flame curvature, and flame unsteadiness by a Karlovitz number Ka, which will be defined later, and system nonadiabaticity by a "loss" parameter L, then the flame behavior is expected to be described in general by

$$f = f(q_c, Le_{i,j}, w_k; Ka, L), \qquad T_b = T_b(q_c, Le_{i,j}, w_k; Ka, L),$$

which should deviate from f^o and T_b^o of the idealized situation. It may be noted that although in f and T_b we have purposely separated out the functional dependence into the two groups of parameters $(q_c, Le_{i,j}, w_k)$ and (Ka, L), which, respectively, represent parameters pertaining to the standard premixed flame and, hence, are properties of the mixture, and those that can affect its burning characteristics, strong coupling between the individual effects is expected. Furthermore, while T_b^o depends only on q_c, the presence of hydrodynamic and/or loss mechanisms can cause T_b to also depend on the diffusive, reactive, as well as hydrodynamic aspects of the problem.

In this chapter, we shall study the effects of aerodynamics on the response of laminar flames. Emphasis will be on premixed flames because of the richness of the responses. The general phenomena of interest are the influences of the nonuniformity and unsteadiness of the flow on the response of propagating, wrinkled flames. A flame subjected to these aerodynamic influences is called a stretched flame. The behavior of nonpremixed flames will be discussed whenever appropriate.

For the standard premixed flame of Chapter 7, effects due to the unequal diffusivities of heat and the various species are minor: the flame temperature is not affected and the burning flux f^o is only modified by a constant factor, which is $\sqrt{Le}$ for one-reactant mixtures. For the stretched flame, however, it will become clear that the flame temperature, and through it the entire flame response, can be affected in an essential manner by the mixture property of unequal diffusivities.

The influence of stretch on the flame response can be discussed based on the length scales as well as the tangential and normal components of the flow field at the flame. Let us first consider stretch at the hydrodynamic scale (Figure 10.1.1a). In this limit, diffusion and reaction zones are not resolved and the entire flame collapses into a flame sheet, with $\ell_D = 0$. The flame propagates into the fresh mixture with a local flame speed s_u, which is normal to the flame surface and can also differ from s_u^o. The presence of stretch, through the tangential velocity gradient at the flame, changes the flame surface area A and consequently the mass burning rate $\int f_u dA$. Depending on whether the local tangential velocity increases or decreases with distance along the surface, the local mass burning rate can also increase or decrease with stretch. In stationary situations the normal velocity gradient allows adjustment of the flame location in the normal direction so that the flame situates where the local flame speed, s_u, dynamically balances the local normal velocity, u_u. In nonstationary situations the difference between s_u and u_u yields the net propagation velocity of the flame segment in the laboratory frame. Thus the combined effects of the tangential and normal components of the velocity are the displacement of the flame surface, distortion of its geometry, and modification of the mass burning rate. We shall refer to this stretch as hydrodynamic stretch.

Resolving the transport and reaction zones (Figure 10.1.1b), the tangential velocity variation in the transport zone directly affects the normal mass flux f_b entering the reaction zone. Furthermore, through interaction with heat and mass diffusion, it can also modify the temperature and concentration profiles in the transport zone and consequently the burning intensity, T_b and f_b, in the reaction zone, as will be shown in Section 10.4. We shall refer to this stretch as flame stretch. The normal velocity variation also affects the residence time within the reaction zone and consequently T_b and the completeness of reaction. However, it is also important to note the flexibility with which a premixed flame can adjust its location to accommodate changes in the normal velocity gradient and to achieve complete reaction. Thus a change in the stretch rate does not necessarily lead to a change of corresponding extent in the residence time. We shall call a flame with total freedom of adjustment either a freely propagating or a freely standing flame, depending on whether the flame is in motion in the frame of reference under consideration.

The hydrodynamic stretch and flame stretch are strongly coupled in that the former imposes the stretch intensity within the flame, constituting the flame stretch, while the latter not only yields the propagating speed of the hydrodynamic flame surface, but it also affects such critical phenomena as extinction. The direct influence of stretch in the reaction zone is expected to be small because of the secondary importance of convective transport in this very thin zone.

In the next section, the dynamics of structureless stretched flame surfaces in hydrodynamic flow fields are discussed. The phenomena of corner formation and burning rate augmentation through flame wrinkling are analyzed. In Sections 10.3–10.6 the influences of stretch on the flame structure and response are described and compared with experimental and computational results. The implications of flame stretch on combustion studies are also discussed. In Section 10.7 we integrate results from the flame structure analysis to the dynamics of flame surfaces through the model problems of corner broadening and the configurations of Bunsen flames. In Section 10.8 the flame response in oscillating flow fields is studied, and in Section 10.9 the phenomena of flamefront instability are discussed.

Further exposition on flame dynamics can be found in Williams (1985), Clavin (1985, 2000), Law (1989), Dixon-Lewis (1990), Peters (2000), Law and Sung (2000), Williams (2000), and Candel (2002).

10.2. HYDRODYNAMIC STRETCH

10.2.1. The *G*-Equation

When a flame is much thinner than the hydrodynamic length scale of a flow, it can be treated as a surface propagating in the flow. Let the geometry of the surface be described by $G(\mathbf{x}, t) =$ constant (Figure 10.1.1). In particular, we shall set

$$G(\mathbf{x}, t) = 0 \tag{10.2.1}$$

for convenience such that $G < 0$ and > 0 for the unburned and burned states, respectively. This surface is assumed to be smooth and continuous so that its unit normal vector,

$$\mathbf{n} = \frac{-\nabla G}{|\nabla G|}, \tag{10.2.2}$$

is uniquely defined everywhere. The negative sign in Eq. (10.2.2) indicates the convention that $\mathbf{n}$ is defined to be positive when pointed in the upstream direction of the flame. Consequently, a flame segment that is convex toward the unburned mixture has a positive curvature and as such is considered to be positively stretched.

On the surface $G(\mathbf{x}, t)$, the relation

$$\frac{dG}{dt} = \frac{\partial G}{\partial t} + \mathbf{V}_f \cdot \nabla G = 0 \tag{10.2.3}$$

holds, where $\mathbf{V}_f = d\mathbf{x}/dt$ is the local propagation velocity of the surface. Furthermore, the local flame speed, s_u, is by definition

$$s_u = (\mathbf{V}_f - \mathbf{v}\,|_{G=0^-}) \cdot \mathbf{n}, \tag{10.2.4}$$

where $\mathbf{v}$ is the flow velocity. Substituting Eq. (10.2.4) into Eq. (10.2.3), and using Eq. (10.2.2), we obtain the G-equation (Kerstein, Ashurst & Williams 1988; Peters 2000),

$$\frac{\partial \hat{G}}{\partial \hat{t}} + \tilde{\mathbf{v}}\,|_{\hat{G}=0^-} \cdot \hat{\nabla}\hat{G} = \tilde{s}_u\,|\hat{\nabla}\hat{G}|, \tag{10.2.5}$$

with its density-weighted version being

$$\rho_u \frac{\partial \hat{G}}{\partial \hat{t}} + (\rho \tilde{\mathbf{v}})_{\hat{G}=0^-} \cdot \hat{\nabla} \hat{G} = \rho_u \tilde{s}_u \, |\hat{\nabla} \hat{G}|, \tag{10.2.6}$$

where $\tilde{s}_u = s_u/s_u^o$, $\tilde{f}_u = f_u/f^o$, and we have also nondimensionalized $\mathbf{v}$ by s_u^o, all space variables by the hydrodynamic scale ℓ_H, and t by ℓ_H/s_u^o. For consistency all quantities referenced to the hydrodynamic scale and the flame properties are superscripted by ˆ and ˜ respectively.

We note that the LHS of Eq. (10.2.5) is simply the substantial derivative of G while the RHS represents the source term that causes the flame surface to propagate normal to itself with the flame speed s_u relative to the motion of the unburned mixture. The G-equation is coupled to the governing equations in the hydrodynamic zone through the term $\mathbf{v}\,|_{G=0^-}$. The coupling is quite complicated, representing the interaction between the flame front and the outer, hydrodynamic flow: the outer flow convects the front while the front affects the outer flow through thermal expansion. The problem, however, can be decoupled and, hence, significantly simplified by making the constant density assumption such that the surface is a passive scalar being convected and distorted by $\mathbf{v}\,|_{G=0^-}$, which can be considered to be prescribed. The constant density assumption is equivalent to the statement that there is negligible heat release in crossing the flame (Matkowsky & Sivashinsky 1979). Although this assumption is obviously a gross violation of what constitutes a conventional flame, its use does facilitate the description of flame dynamics. Furthermore, there are also liquid systems in which a (liquid) flame can spread with small heat release and consequently minimal density change (Shy, Jang & Ronney 1996).

Another major assumption in the above derivation is that the flame is much thinner than the hydrodynamic length scale, which does not always hold. For such "thick flame" situations, it is more appropriate to treat the much thinner reaction zone as the surface of interest (Peters 2000). Identical derivation then yields

$$\frac{\partial \hat{G}}{\partial \hat{t}} + \tilde{\mathbf{v}}\,|_{\hat{G}=0^-} \cdot \hat{\nabla} \hat{G} = \tilde{s}_b \, |\hat{\nabla} \hat{G}|, \tag{10.2.7}$$

and its density-weighted version

$$\rho_b \frac{\partial \hat{G}}{\partial \hat{t}} + (\rho \tilde{\mathbf{v}})_{\hat{G}=0^-} \cdot \hat{\nabla} \hat{G} = \rho_b \tilde{s}_b \, |\hat{\nabla} \hat{G}|, \tag{10.2.8}$$

for the evolution of the geometry and dynamics of the reaction sheet, where s_b and f_b are respectively the downstream flame speed and flux.

10.2.2. Corner Formation in Landau Propagation

Solutions of Eq. (10.2.5) require a knowledge of s_u, which is sensitively affected by stretch effects, as will be shown in Section 10.4. In the following analysis, however, we shall assume that s_u is not affected by stretch such that $s_u = s_u^o$ over the entire

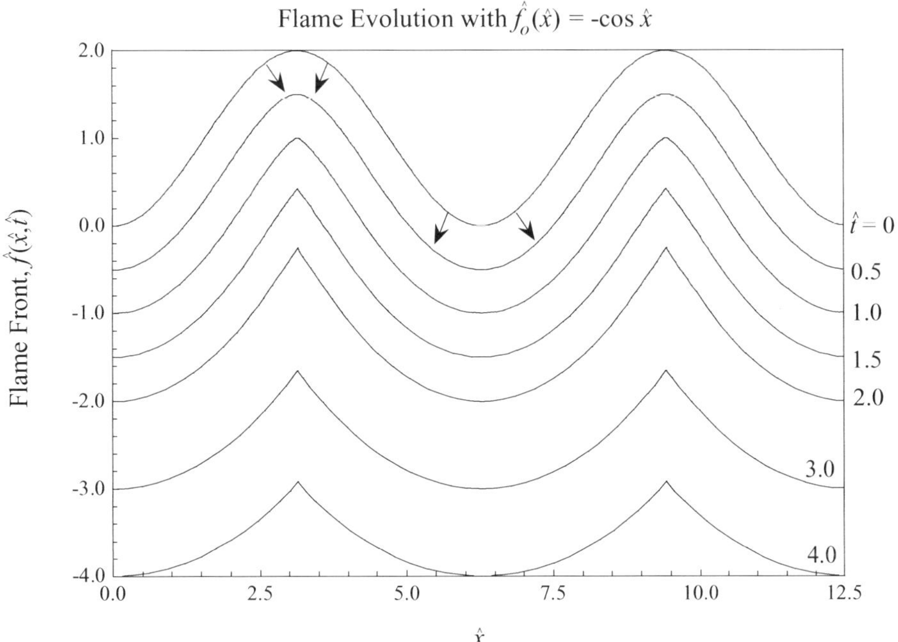

Figure 10.2.1. Evolution and propagation of an initially sinusoidal flame surface in a quiescent, constant-density, medium, showing the formation of corners.

flame surface. We shall call this mode of flame propagation Landau propagation because Landau was among the first in using this limit to describe flame dynamics. The G-equation then becomes

$$\frac{\partial \hat{G}}{\partial \hat{t}} + \tilde{\mathbf{v}}\,|_{\hat{G}=0^-} \cdot \hat{\nabla}\hat{G} = |\hat{\nabla}\hat{G}|. \tag{10.2.9}$$

A characteristic of Landau propagation is the propensity of corner formation over the flame surface. This phenomenon can be readily appreciated by following the evolution of a flame surface with an initial sinusoidal profile in a quiescent environment, as shown in Figure 10.2.1 (Law & Sung 2000). It is seen that, since the flame surface propagates normal to itself, in the same manner as Huygens' principle of ray optics, the surface tends to spread out in the crest region but collide in the trough region. Thus depending on the initial flame shape, corners can develop in the trough region.

To study the problem in a more general manner, we consider the evolution of a flame surface in a quiescent flow field. Equation (10.2.9) then becomes

$$\frac{\partial \hat{G}}{\partial \hat{t}} = |\hat{\nabla}\hat{G}|. \tag{10.2.10}$$

If we assume that the flame is not folded or multiply connected, then the flame surface in two dimensions can be described by $\hat{G}(\hat{x}, \hat{y}, \hat{t}) = \hat{y} - \hat{f}(\hat{x}, \hat{t})$, where $\hat{f}(\hat{x}, \hat{t})$ is the flame shape function with $\hat{f}(\hat{x}, 0) = \hat{f}_o(\hat{x})$ being its initial shape; the use of the symbol

$\hat{f}$ here is not to be confused with the mass burning flux f. Equation (10.2.10) can thus be expressed as

$$\frac{\partial \hat{f}}{\partial \hat{t}} = -\left[1 + \left(\frac{\partial \hat{f}}{\partial \hat{x}}\right)^2\right]^{1/2}. \tag{10.2.11}$$

Taking the partial derivative of Eq. (10.2.11) with respect to $\hat{x}$ and letting $\hat{g} = \partial \hat{f}/\partial \hat{x}$, Eq. (10.2.11) becomes

$$\frac{\partial \hat{g}}{\partial \hat{t}} + \frac{\hat{g}}{(1+\hat{g}^2)^{1/2}} \frac{\partial \hat{g}}{\partial \hat{x}} = 0, \tag{10.2.12}$$

with the initial condition $\hat{g}(\hat{x}, 0) = \hat{g}_o(x) = d\hat{f}_o(\hat{x})/d\hat{x}$.

Equation (10.2.12) is a quasi-linear wave equation whose solution can be obtained in the same manner as that for shock formation in supersonic flows by using the method of characteristics, which will be studied in Chapter 14. Thus using

$$\frac{d\hat{g}}{d\hat{t}} = \frac{\partial \hat{g}}{\partial \hat{t}} + \frac{d\hat{x}}{d\hat{t}} \frac{\partial \hat{g}}{\partial \hat{x}},$$

and by comparing it with Eq. (10.2.12), we see that Eq. (10.2.12) is equivalent to the two equations

$$\frac{d\hat{x}}{d\hat{t}} = \frac{\hat{g}}{(1+\hat{g}^2)^{1/2}}, \tag{10.2.13}$$

$$\frac{d\hat{g}}{d\hat{t}} = 0. \tag{10.2.14}$$

Equations (10.2.13) and (10.2.14) show that $\hat{g}$ is constant along the characteristic defined by Eq. (10.2.13). Its slope, $\hat{g}/(1+\hat{g}^2)^{1/2}$, is therefore also a constant. The slope is different for different characteristics, which at $\hat{t} = 0$ start out at different $\hat{x}$. Thus at any $\hat{x}_o$, we have

$$\hat{g}(\hat{x}, \hat{t}) = \hat{g}(\hat{x}_o, 0) = \hat{g}_o(\hat{x}_o). \tag{10.2.15}$$

Furthermore, since

$$\hat{x} = \frac{\hat{g}}{(1+\hat{g}^2)^{1/2}} \hat{t} + \hat{x}_o, \tag{10.2.16}$$

obtained by integrating Eq. (10.2.13), $\hat{g}(\hat{x}, \hat{t})$ is given by $\hat{g}_o\,[\cdot]$ as

$$\hat{g}(\hat{x}, \hat{t}) = \hat{g}_o\left[\hat{x} - \frac{\hat{g}}{(1+\hat{g}^2)^{1/2}} \hat{t}\right]. \tag{10.2.17}$$

The characteristics can either diverge or converge, depending on $\hat{x}_o$ and the functional form of the slope. In the latter situation multiple solutions result as the characteristics collide, which physically implies the formation of discontinuities. Mathematically these discontinuities are called corners.

To investigate the instant at which the corner is first formed, we substitute Eq. (10.2.17) back to Eq. (10.2.12), yielding

$$\left[\frac{\partial \hat{g}}{\partial \hat{t}} + \frac{\hat{g}}{(1+\hat{g}^2)^{1/2}}\frac{\partial \hat{g}}{\partial \hat{x}}\right]\left[1 + \frac{(d\hat{g}_o/d\hat{x})\hat{t}}{(1+\hat{g}^2)^{3/2}}\right] = 0. \tag{10.2.18}$$

Equation (10.2.18) shows that as long as its second term is finite, Eq. (10.2.12) is retrieved and the flame shape is uniquely determined. However, as the flame propagates, at a certain time $\hat{t}^*$ the system could reach a state at which the second term vanishes. This then implies that the first term does not need to vanish and indeed can assume any value. In other words, the flame shape now becomes multiple valued, as manifested by the formation of corners that are characterized by the discontinuity in slope at a given point (or line) on the flame surface. Thus

$$\hat{t}^* = \min\left\{\frac{-(1+\hat{g}_o^2)^{3/2}}{d\hat{g}_o/d\hat{x}}\right\} \tag{10.2.19}$$

can be identified as the minimum time for the corner to form.

It is also of interest to determine the propagation velocity $\tilde{v}^*$ of the corner once it is formed. Transforming Eq. (10.2.12) to a new frame of reference attached to the corner, with $\hat{\xi} = \hat{x} - \int^{\hat{t}} \tilde{v}^*(\hat{t}')d\hat{t}'$ and $\hat{\tau} = \hat{t}$, we have

$$\frac{\partial \hat{g}}{\partial \hat{\tau}} - \tilde{v}^* \frac{\partial \hat{g}}{\partial \hat{\xi}} + \frac{\partial (1+\hat{g}^2)^{1/2}}{\partial \hat{\xi}} = 0. \tag{10.2.20}$$

Integrating Eq. (10.2.20) across the corner and neglecting the contribution from the $\partial \hat{g}/\partial \hat{\tau}$ term, we obtain

$$\tilde{v}^* = \frac{(1+\hat{g}_+^2)^{1/2} - (1+\hat{g}_-^2)^{1/2}}{\hat{g}_+ - \hat{g}_-} \tag{10.2.21}$$

where $\hat{g}_+$ and $\hat{g}_-$ are values of $\hat{g}$ across the corner.

For the example given in Figure 10.2.1, the initial flame shape is described by $\hat{f}_o(\hat{x}) = -\cos\hat{x}$, such that $\hat{g}_o(\hat{x}) = \sin\hat{x}$. Equation (10.2.19) then readily shows that the corner is formed at $\hat{t}^* = 1$, as shown in the figure. Furthermore, $\tilde{v}^* = 0$ due to the symmetry of $(1+\hat{g}^2)$.

10.2.3. Burning Rate Increase through Flame Wrinkling

Conceptually, it can be readily accepted that, by wrinkling a flame through hydrodynamic stretch, the total flame surface area is increased. Thus if the local flame speed at the flame surface is not affected by stretch, then the burning rate of a combustible through which the wrinkled flame propagates will increase with increasing wrinkling simply due to the increase in the flame surface area. Indeed, this is the primary mechanism through which the propagation speed of a turbulent flame is increased over that of the laminar flame.

To demonstrate such an increase in the flame speed, Figure 10.2.2 shows an arbitrarily wrinkled flame situated in a flow field of uniform velocity $\tilde{s}_T\mathbf{j}$ in the y-direction, where the subscript T designates turbulent flame in anticipation of our latter studies

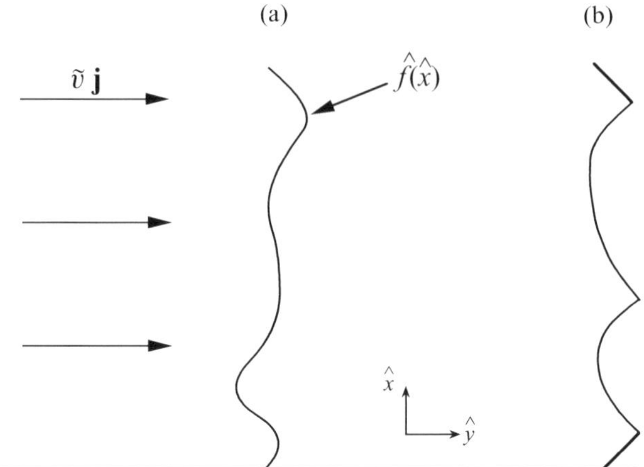

Figure 10.2.2. Schematic demonstrating the increase of bulk flame propagation rate due to flame wrinkling: (a) before corner formation, (b) after corner formation.

on this subject. Assuming $\partial\hat{G}/\partial\hat{t} = 0$, use of $\tilde{s}_T\mathbf{j}$ for $\tilde{\mathbf{v}}$ in Eq. (10.2.9) readily yields

$$\tilde{s}_T = \left[1 + \frac{\left(\partial\hat{G}/\partial\hat{x}\right)^2}{\left(\partial\hat{G}/\partial\hat{y}\right)^2}\right]^{1/2}, \tag{10.2.22}$$

which shows that $\tilde{s}_T = s_T/s_u^o > 1$, hence demonstrating that the flame speed of a wrinkled flame is higher than that of the laminar flame.

If the flame is also not folded, then $\hat{G}(\hat{x}, \hat{y}) = \hat{y} - \hat{f}(\hat{x})$. Substituting $\hat{G}$ into Eq. (10.2.22) yields

$$\tilde{s}_T = \left[1 + \left(\frac{d\hat{f}}{d\hat{x}}\right)^2\right]^{1/2}. \tag{10.2.23}$$

Extending Eq. (10.2.23) to an arbitrary three-dimensional flame surface, we have

$$\tilde{s}_T = \left(1 + |\hat{\nabla}_t \hat{f}\,|^2\right)^{1/2}, \tag{10.2.24}$$

where $\hat{\nabla}_t$ is the tangential gradient operator over the flame surface.

Since the flame surface continuously evolves with time, as shown in Figure 10.2.1, it is clear that for a quiescent flow in the Landau limit, the convex segment continuously grows, while the concave segment continuously shrinks until it is totally eliminated when the corner is formed. Subsequently the increase in the surface area and, hence, the burning rate are solely contributed by the convex flame segments (Figure 10.2.2b). This then implies that flame wrinkles tend to be dominated by those with positive curvatures. Furthermore, the area of such a convex segment will continuously decrease as the opposite sides of the corner collide and annihilate each other, thereby reducing the extent of wrinkling of the flame surface. Thus sustenance or even amplification of flame wrinkling requires either external forcing through flow nonuniformity or development of inherent flamefront cellular instabilities.

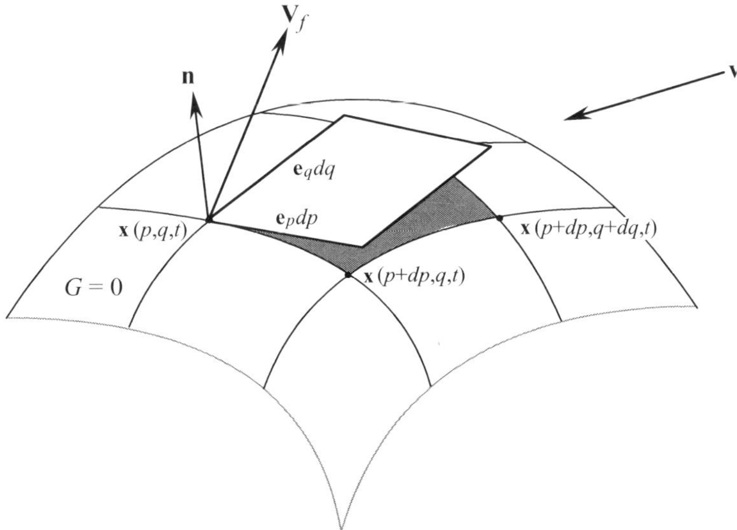

Figure 10.2.3. Schematic of a surface element with velocity $\mathbf{V}_f$ and unit normal vector **n** in a flow field of velocity **v**.

10.2.4. The Stretch Rate

The G-equation describes the dynamics and geometry of the flame surface through the knowledge of the flow velocity at the flame surface, $\mathbf{v}|_{G=0^-}$, the flame speed s_u, and the geometry of the flame through its unit normal vector **n**. However, for an observer stationed on the flame surface, the individual influences of the flow and flame motion cannot be distinguished. Rather, the observer simply perceives an unsteady and nonuniform flow approaching it with some effective velocity. Its influence on the flame response, either the flame surface area in the hydrodynamic limit or the flame speed s_u and thereby the burning intensity, is through the extent of the unsteadiness of the flow and the nonuniform tangential velocity over the flame surface. Consequently it is reasonable to expect that the various influences due to flow nonuniformity, flame curvature, and flame/flow unsteadiness can be collectively described by a single parameter – the stretch rate, to be defined next.

Refer to the general flame surface $G = 0$ in Figure 10.2.3. This surface has a velocity $\mathbf{V}_f$ while the flow has a velocity **v**. A general definition of stretch rate at any point on this surface is the Lagrangian time derivative of the logarithm of the area A of an infinitesimal element of the surface (Williams 1975),

$$\kappa = \frac{1}{A}\frac{dA}{dt}, \tag{10.2.25}$$

with the boundary of this surface element moving tangentially along the surface at the local tangential component of the fluid velocity. The stretch rate κ has the unit of s^{-1}.

The deceptively simple expression of Eq. (10.2.25) actually contains the various factors that contribute to the influence of stretch. To show this (Matalon 1983; Chung

& Law 1984b) we now express Eq. (10.2.25) in terms of the dynamics of a general surface $\mathbf{x}(p, q, t)$ as shown in Figure 10.2.3, where (p, q) are the two curvilinear coordinates on it. The instantaneous velocity of the surface is therefore $\mathbf{V}_f(p, q, t) = \partial \mathbf{x}(p, q, t)/\partial t$. Since $d\mathbf{x}(t) = \mathbf{e}_p dp + \mathbf{e}_q dq$, where $\mathbf{e}_p$ and $\mathbf{e}_q$ are the unit vectors in the directions of p and q, an elemental area $\mathbf{A}(p, q, t)$ of the surface at time t is simply

$$\mathbf{A}(p, q, t) = (\mathbf{e}_p dp) \times (\mathbf{e}_q dq) = (dpdq)\mathbf{n}, \tag{10.2.26}$$

where $\mathbf{n} = \mathbf{e}_p \times \mathbf{e}_q$ is the unit vector of the elemental surface pointed in the direction in which the surface is propagating, as defined.

At a later time $t + \delta t$, we have

$$\mathbf{x}(p, q, t + \delta t) = \mathbf{x}(p, q, t) + \frac{\partial \mathbf{x}}{\partial t}\delta t = \mathbf{x}(p, q, t) + \mathbf{V}_f \delta t$$

$$\begin{aligned} d\mathbf{x}(p, q, t + \delta t) &= d\mathbf{x}(p, q, t) + \left(\frac{\partial \mathbf{V}_f}{\partial p}dp + \frac{\partial \mathbf{V}_f}{\partial q}dq\right)\delta t \\ &= \left(\mathbf{e}_p + \frac{\partial \mathbf{V}_f}{\partial p}\delta t\right)dp + \left(\mathbf{e}_q + \frac{\partial \mathbf{V}_f}{\partial q}\delta t\right)dq. \end{aligned}$$

Using the p and q components of $d\mathbf{x}(p, q, t + \delta t)$, we have

$$\mathbf{A}(p, q, t + \delta t) = \left(\mathbf{e}_p + \frac{\partial \mathbf{V}_f}{\partial p}\delta t\right) \times \left(\mathbf{e}_q + \frac{\partial \mathbf{V}_f}{\partial q}\delta t\right) dpdq. \tag{10.2.27}$$

Therefore if we now write $A = \mathbf{A} \cdot \mathbf{n}$, and use Eqs. (10.2.26) and (10.2.27), then

$$\begin{aligned} \kappa &= \frac{1}{A(t)} \lim_{\delta t \to 0} \frac{A(t + \delta t) - A(t)}{\delta t} \\ &= \left(\mathbf{e}_p \times \frac{\partial \mathbf{V}_f}{\partial q} + \frac{\partial \mathbf{V}_f}{\partial p} \times \mathbf{e}_q\right) \cdot \mathbf{n} = \left(\mathbf{e}_p \cdot \frac{\partial \mathbf{V}_f}{\partial p} + \mathbf{e}_q \cdot \frac{\partial \mathbf{V}_f}{\partial q}\right), \end{aligned} \tag{10.2.28}$$

where we have used the cyclic law of triple scalar products, which states that for any vectors $\mathbf{a}$, $\mathbf{b}$, and $\mathbf{c}$, we have $(\mathbf{a} \times \mathbf{b}) \cdot \mathbf{c} = (\mathbf{b} \times \mathbf{c}) \cdot \mathbf{a} = (\mathbf{c} \times \mathbf{a}) \cdot \mathbf{b}$.

The stretch rate κ can be further developed as follows. First, by using the relations

$$\begin{aligned} \nabla_t \cdot \mathbf{V}_f &= \frac{\partial(\mathbf{e}_p \cdot \mathbf{V}_f)}{\partial p} + \frac{\partial(\mathbf{e}_q \cdot \mathbf{V}_f)}{\partial q} \\ &= \left(\mathbf{e}_p \cdot \frac{\partial \mathbf{V}_f}{\partial p} + \mathbf{e}_q \cdot \frac{\partial \mathbf{V}_f}{\partial q}\right) + \mathbf{V}_f \cdot \left(\frac{\partial \mathbf{e}_p}{\partial p} + \frac{\partial \mathbf{e}_q}{\partial q}\right) \end{aligned}$$

and

$$\frac{\partial \mathbf{e}_p}{\partial p} + \frac{\partial \mathbf{e}_q}{\partial q} = -(\nabla \cdot \mathbf{n})\mathbf{n},$$

we have

$$\kappa = \nabla_t \cdot \mathbf{V}_f + (\mathbf{V}_f \cdot \mathbf{n})(\nabla \cdot \mathbf{n}). \tag{10.2.29}$$

If we next decompose $\mathbf{V}_f$ into its tangential and normal components as $\mathbf{V}_f = \mathbf{V}_{f,t} + (\mathbf{V}_f \cdot \mathbf{n})\mathbf{n}$, where $\mathbf{V}_{f,t}$ is the tangential velocity of the surface, and assume that $\mathbf{V}_{f,t}$ is equal to the tangential component of the flow velocity $\mathbf{v}_s$ at the flame,

$$\mathbf{V}_{f,t} = \mathbf{v}_{s,t}, \tag{10.2.30}$$

Eq. (10.2.29) becomes

$$\kappa = \nabla_t \cdot \mathbf{v}_{s,t} + (\mathbf{V}_f \cdot \mathbf{n})(\nabla \cdot \mathbf{n}). \tag{10.2.31}$$

Equation (10.2.31) shows the two sources of stretch a flame can be subjected to. The first term represents the influence of flow nonuniformity along the flame surface. Since

$$\mathbf{v}_{s,t} = \mathbf{n} \times (\mathbf{v}_s \times \mathbf{n}), \tag{10.2.32}$$

this term embodies the effects due to flow nonuniformity through $\mathbf{v}_s$ and flame curvature through the variation in $\mathbf{n}$. Furthermore, it exists only if the flow is oblique to the surface such that $\mathbf{v}_s \times \mathbf{n} \neq 0$. The second term in Eq. (10.2.31) represents stretch experienced by a nonstationary flame through $\mathbf{V}_f$, although the flame also has to be curved because $\nabla \cdot \mathbf{n}$ vanishes otherwise. These three stretch-induced effects can be separately referred to as those due to aerodynamic straining, flame curvature, and flame motion. We further note that since diffusion is in the direction of $\mathbf{n}$, the nonorthogonality requirement of $\mathbf{v}_s \times \mathbf{n}$ leads us to anticipate the importance of heat and mass diffusive transport in the dynamics of stretched flames, even though the discussion so far has been kinematic in nature.

Although the use of the tangential gradient operator at the surface, ∇_t, provides a clear physical interpretation of stretch, mathematical specification of ∇_t can be somewhat cumbersome, especially for curved flames. However, writing $\nabla = \nabla_t + \nabla_n$, where ∇_n is the normal component of the gradient operator on the surface, and noting that $\nabla_n \cdot \mathbf{v}_{s,t} \equiv 0$, Eq. (10.2.31) can be expressed in alternate forms as

$$\kappa = \nabla \cdot \mathbf{v}_{s,t} + (\mathbf{V}_f \cdot \mathbf{n})(\nabla \cdot \mathbf{n}), \tag{10.2.33}$$

$$\begin{aligned} \kappa &= \nabla \cdot [\mathbf{n} \times (\mathbf{v}_s \times \mathbf{n})] + (\mathbf{V}_f \cdot \mathbf{n})(\nabla \cdot \mathbf{n}) \\ &= -\mathbf{n} \cdot \nabla \times (\mathbf{v}_s \times \mathbf{n}) + (\mathbf{V}_f \cdot \mathbf{n})(\nabla \cdot \mathbf{n}). \end{aligned} \tag{10.2.34}$$

As examples, let us compute the stretch rate κ for some common flame configurations shown in Figure 10.2.4. The flames are infinitely thin so that it constitutes the stretched surface.

Stationary Planar Flame in Stagnation Flow: Figure 10.2.4a shows a planar flame situated in a divergent stagnation flow. Assuming potential flow, the velocity

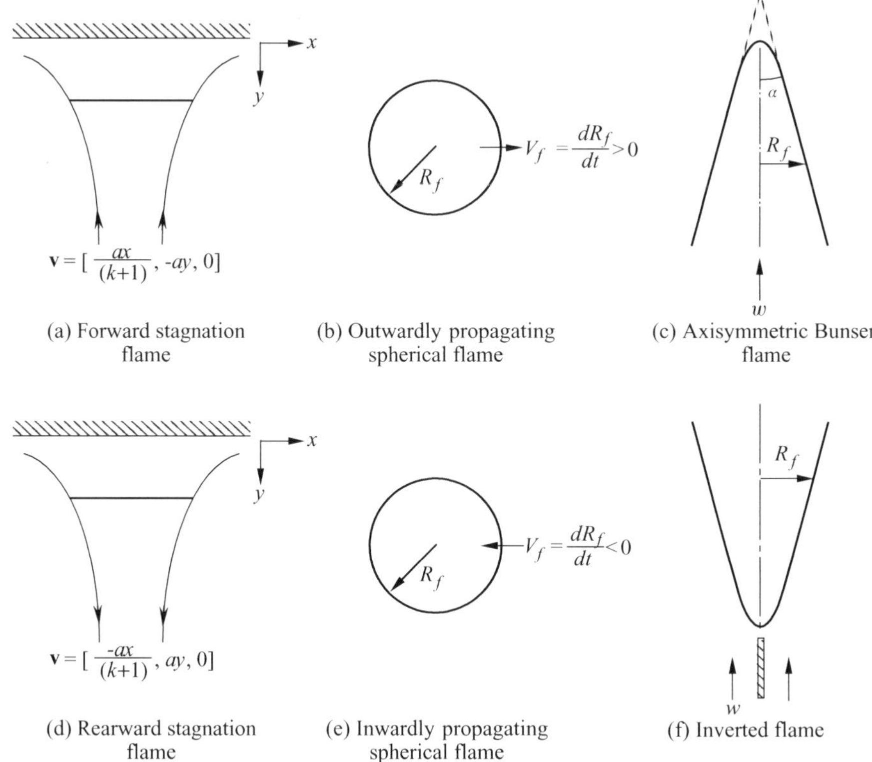

Figure 10.2.4. Configurations of various model stretched flames.

vector is

$$\mathbf{v} = \left\{ \frac{a}{(k+1)} x, -ay, 0 \right\}, \tag{10.2.35}$$

where a is the strain rate of the flow, $k = 0, 1$ for cartesian and cylindrical coordinates respectively, and the x- and y-velocities in the cylindrical coordinates are those in the radial and axial directions respectively. Using Eq. (10.2.31), since $\mathbf{V}_f \cdot \mathbf{n} = 0$ and $\nabla_t \cdot \mathbf{v}_{s,t} = a$, we have

$$\kappa = a. \tag{10.2.36}$$

Nonstationary Spherical Flame: We again use Eq. (10.2.31) for evaluation. Here the flame propagates normal to its surface, implying that $\mathbf{v}_{s,t} = 0$ and $\mathbf{V}_f \cdot \mathbf{n} = V_f = dR_f/dt$, where R_f is the instantaneous flame radius. Furthermore,

$$\nabla \cdot \mathbf{n} = \pm \left(\frac{1}{R_1} + \frac{1}{R_2} \right), \tag{10.2.37}$$

where $\pm$ respectively refer to outwardly and inwardly propagating flames because by definition $\partial \mathbf{e}_p / \partial p$ and $\partial \mathbf{e}_q / \partial q$ are the two principal radii of curvature, R_1 and R_2, pointed away from the flame surface toward the center of curvature, and $\mathbf{n}$

respectively points outward and inward for these flames. Note that as the flame propagates, the stretch effect becomes weaker for an expanding flame and stronger for an inwardly propagating flame.

If the flame is spherically symmetric (Figures 10.2.4b and 10.2.4e), $R_1 = R_2 = R_f$, we have

$$\kappa = \pm \frac{2}{R_f} \frac{dR_f}{dt}. \tag{10.2.38}$$

Axisymmetric Flame: This configuration includes the Bunsen flame (Figure 10.2.4c). Adopting the cylindrical (r, θ, z) coordinate, and using Eq. (10.2.34) in which $\mathbf{V}_f \cdot \mathbf{n} = 0$, $\mathbf{v} = (0, 0, -w)$, and $\mathbf{n} = (-\cos\alpha, 0, \sin\alpha)$, we have

$$\kappa = -\frac{\sin\alpha}{r} \frac{\partial}{\partial r}(rw\cos\alpha) - \cos\alpha \frac{\partial}{\partial z}(w\cos\alpha) \tag{10.2.39}$$

evaluated at the flame surface.

The above result is general in that w and α can be general functions of r and z. If we further assume that the flame surface is a circular cone with a sharp apex such that w and α are constants, then

$$\kappa = -\frac{w \sin 2\alpha}{2R_f}. \tag{10.2.40}$$

Stretch in this case is derived from the three-dimensional nature of the curved surface. Note that while the stagnation and expanding spherical flames are positively stretched, the axisymmetric flame is negatively stretched. This indicates that the Bunsen flame actually suffers compression, which tends to reduce the flame surface area in the direction of the flow. The intensity of compression increases with decreasing R_f as the flow moves upward. The expression breaks down around the apex of the cone where $R_f \to 0$.

Similar to the outwardly and inwardly propagating flames, the counterpart of the (forward) stagnation flame is the rearward stagnation flame (Figure 10.2.4d) whose stretch rate is negative, while that of the Bunsen flame is the inverted flame (Figure 10.2.4f), whose stretch rate is positive based on curvature alone. Note that a rearward stagnation flame in the adiabatic limit is inherently unstable because a small displacement of the flame either toward or away from the stagnation surface will render the flame to be situated in a region of either lower or higher flow velocity, leading to a continuous motion in the direction of the displacement. One mechanism to stabilize such a flame is through heat loss to the stagnation surface. Indeed, we have invoked heat loss in our previous consideration of stabilization of the Bunsen flame, in Section 8.6.2.

There are also stretchless flames. Examples are the stationary and nonstationary one-dimensional planar flames, and the stationary cylindrical and spherical flames respectively sustained by line and point sources.

In preparation for discussions on flame stretch in the next section, we shall define a nondimensional stretch rate, the Karlovitz number, which is a measure of the flame time in terms of the aerodynamic time, as

$$Ka = \frac{\ell_T}{s_u}\kappa = \frac{\ell_T}{f_u}(\rho_u \kappa)$$
$$\sim \frac{\lambda/c_p}{f_u^2}(\rho_u \kappa) \sim \frac{\rho_u \kappa}{w_f}, \qquad (10.2.41)$$

where we have used the relations $f_u = \rho_u s_u$, $f_u \ell_T \sim \lambda/c_p$, and $f_u^2 \sim (\lambda/c_p)w_f$ from Chapter 7. The above scaling evolves naturally in subsequent derivations. Since Ka increases with increasing stretch rate κ, the above expressions then imply that the effect of stretch is expected to be stronger for either a more diffusive mixture or a weaker-burning flame. This is reasonable because of the central role played by diffusion in modifying flame stretch. We also see that the proper parameter to measure the effect of stretch is the density-weighted stretch rate, $\rho_u \kappa$, instead of κ alone.

10.3. FLAME STRETCH: PHENOMENOLOGY

We now study the effects of stretch on the flame structure and response, particularly the flame speed s_u used in the G-equation. The response has been found to be particularly strong for mixtures with unequal species and thermal diffusivities because the flame temperature is directly affected. For mixtures with equal diffusivities, the influence has also been found to be quite subtle, as we shall show in the following.

To demonstrate the influence of nonequidiffusion on the response of stretched flames, we first note that there are at least three diffusivities of interest for an inert-abundant mixture, namely those associated with heat (D_T), the deficient reactant (D_i), and the excess reactant (D_j). From these three diffusivities two interpretations for the effects of different diffusivities can be conducted, based on comparing D_i with D_T, and D_i with D_j, for sufficiently off- and near-stoichiometric situations respectively. These two interpretations can be respectively termed nonunity Lewis number ($Le = D_T/D_i$) effect and preferential diffusion (D_i/D_j) effect. The general phenomena related to unequal diffusion rates will be referred to as nonequidiffusion effects.

10.3.1. Effects of Flow Straining: The Stagnation Flame

We consider the flame response in a stagnation flow, which imposes a well-defined strain rate on the flame (Figure 10.3.1a), and draw a control volume enclosing the transport zone and the divergent streamlines as shown. By assuming that the stagnation surface is adiabatic, we can study the coupled effects of flow straining and mixture nonequidiffusion without complications from system heat loss, flame curvature, and flame unsteadiness.

Recognizing that diffusive transport is normal to the reaction surface, then with the nonunity Lewis number interpretation, shown in the left portion of the figure, the control volume loses thermal energy to the external streamlines while it gains chemical energy from them due to an increase of the deficient reactant concentration.

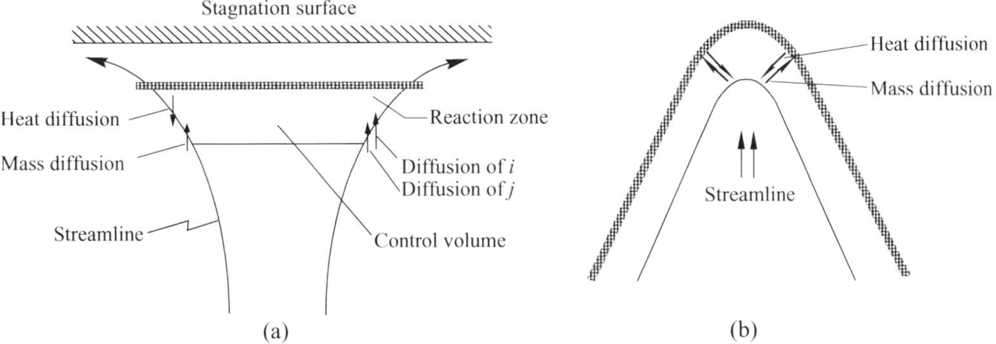

Figure 10.3.1. Conceptual demonstration of the effects of stretch on the flame response in the presence of nonequidiffusion: (a) stagnation flame, (b) tip of Bunsen flame.

Thus the flame behavior, especially the flame temperature, depends on the relative rates of heat and mass diffusion. If the diffusivities are equal such that $Le = 1$, then total energy is conserved and the flame temperature is the adiabatic flame temperature. However, if $Le > 1$, heat loss exceeds the gain in the concentration of the deficient species and we expect $T_b < T_{ad}$. Conversely, $T_b > T_{ad}$ if $Le < 1$.

We next take the preferential diffusion interpretation, shown in the right portion of the figure. Then if the leaner reactant is also the more diffusive one, the reactant composition in the reaction zone will become more stoichiometric such that the flame temperature is higher and the burning more intense. The converse holds if the leaner reactant is less diffusive.

In the following we shall adopt the nonunity Lewis number interpretation because off-stoichiometric burning is more prevalent and because heat diffusion is a crucial mechanism in flame propagation.

If we now increase the stretch rate, the flame will be pushed closer to the stagnation surface in order to maintain kinematic balance between the local flame speed and the normal flow velocity immediately upstream of the flame surface. The movement of the flame is unrestrained and the lean reactant is completely depleted in crossing the flame. The flame, however, suffers stronger flame stretch and thereby nonequidiffusion effects. Consequently the reaction rate and the flame temperature will either decrease or increase, depending on whether $Le > 1$ or < 1, as shown in Figures 10.3.2 and 10.3.3. Thus it is clear that for $Le > 1$, there exists a critical stretch rate at which T_b will be reduced to such an extent that burning is not possible. Extinction occurs when the flame, being unrestrained, is located at a finite distance away from the surface, as shown in Figure 10.3.4a. On the other hand, since increasing stretch elevates T_b for the $Le < 1$ flame, extinction cannot occur until the downstream boundary of the reaction zone is pushed onto the stagnation surface and the flame movement becomes restrained. With further stretching, reaction cannot be completed because of the reduced residence time. Only then will the reaction rate and, hence, the flame temperature start to decrease (Figures 10.3.2 and 10.3.3),

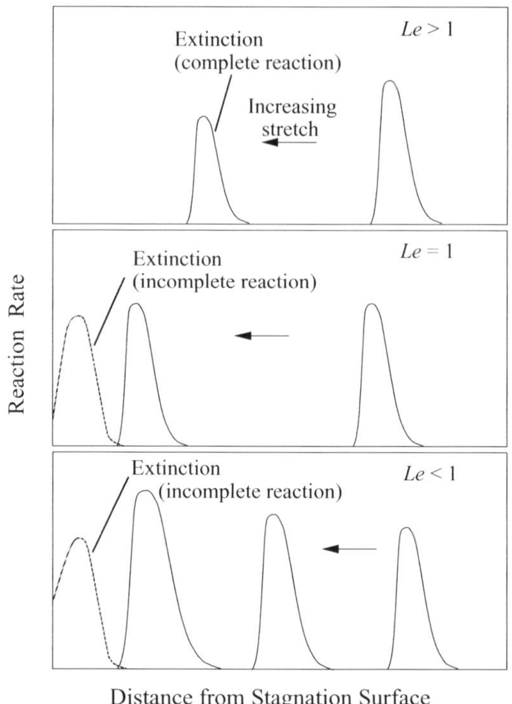

Figure 10.3.2. Effects of nonequidiffusion on the reaction rate with increasing stretch, especially the completeness of reactant consumption at extinction.

leading eventually to extinction, with the flame in direct contact with the stagnation surface, as shown in Figure 10.3.4a.

For the diffusionally neutral case of $Le = 1$, the flame temperature and reaction rate remain unchanged until the reaction zone reaches the surface. With further increase in the stretch rate, incomplete reaction and eventually extinction occurs. Thus incomplete reaction is essential in causing flame extinction for $Le \leq 1$ for an adiabatic and impermeable surface.

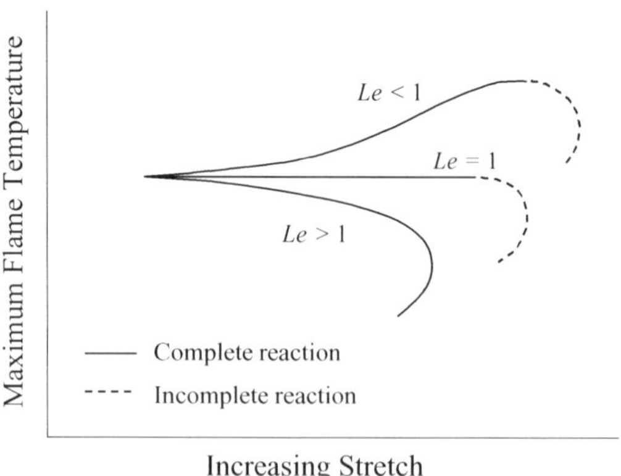

Figure 10.3.3. Effects of stretch, nonequidiffusion, and completeness of reaction, on the extinction turning point behavior.

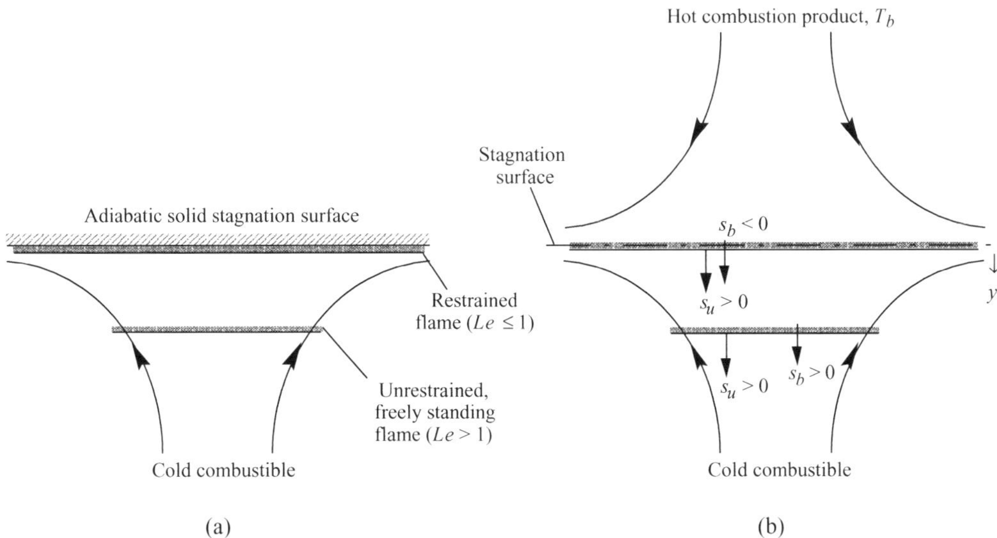

Figure 10.3.4. Schematic showing different extinction mechanisms and the associated flame locations for (a) the stagnation flame, and (b) the counterflow flame.

However, if the stagnation surface is permeable but still adiabatic, as in the case of impinging the combustible mixture against a hot product gas stream of temperature T_b (Figure 10.3.4b), then the reaction zone can actually migrate across the stagnation surface and situate in the product side of the counterflow. Since the propagation speed of the flame is conventionally defined to be positive in the direction opposite to that of the flow in which it is embedded, the downstream flame speed therefore assumes a negative value in this case. The fuel consumption and heat release rates are of course always positive. Combustion is now supported by diffusion of the reactant across the stagnation surface against convection from the product stream. Since there is no loss mechanism involved in such a situation, extinction is not possible. The existence of negative flame speeds has been experimentally observed for superadiabatic flames whose downstream temperature is higher than T_b (Sohrab, Ye & Law 1985). Negative flame speeds for adiabatic flames have yet to be observed.

10.3.2. Effects of Flame Curvature: The Bunsen Flame

The second situation illustrating flame stretch effects, due to curvature, involves the burning intensity over the curved surface of an axisymmetric Bunsen cone (Figure 10.3.1b). Here if we assume for simplicity that the flow is uniform, then flame stretch is manifested through curvature effects over the flame surface, especially in the tip region, which has the largest curvature. Thus for a closed tip, its concave curvature towards the fresh mixture focuses the heat ahead of the flame. Consequently, for a given streamtube that traverses a certain segment of the reaction sheet, it is diffusively heated by a larger segment of the reaction sheet. Thus the flame temperature tends to be raised to a value above T_{ad}. On the other hand, this curvature has a defocusing effect on the concentration of the deficient reactant approaching the flame, which tends to reduce the flame temperature. Thus the temperature of the flame again depends on the relative rates of heat and mass diffusion, with the extent of deviation increasing progressively from the flame base toward the flame tip because of the corresponding increase in the stretch intensity. Specifically, for $Le > 1$, the tip will burn more intensely relative to the shoulder region of the flame, while for $Le < 1$ the burning is less intense and can lead to local extinction, commonly known as the tip opening phenomenon. The response to Lewis number variations is therefore completely opposite to that of the positively stretched stagnation flame as a result of the negatively stretched, compressive nature of the flame curvature.

It is important to recognize that this effect is eliminated when the streamlines are normal to the flame surface, as would be the situation involving stationary cylindrical and spherical flames. We also note that the tip segment of the Bunsen flame is actually restrained in that while the flame segment of the shoulder region can freely adjust its orientation to achieve dynamic balance with the freestream flow, the orientation and, hence, flame speed at the tip is fixed, as noted earlier in Section 7.6.1. Since the tip segment is "held" by the flame shoulders, while the orientation of the shoulder changes with increasing flow rate, there could exist situations under which the tip segment would lose its flexibility to adjust its curvature. This could lead to incomplete reaction and, hence, extinction of the tip regardless of the mixture Lewis number.

10.3.3. Effects of Flame Motion: The Unsteady Spherical Flame

We now study the effects of stretch due to flame motion via the outwardly and inwardly propagating spherical flames. Consider first the positively stretched, outwardly propagating flame whose radius R_f is much larger than its thickness (Figure 10.3.5). In an interval δt, the flame radius grows by an amount $\delta R_f \ll (R_f, R_T, R_M)$, where R_T and R_M are respectively the radii for the thermal and limiting reactant layers. Then the volume for the thermal energy will be increased by an approximate amount $\frac{4}{3}\pi(R_T + \delta R_f)^3 - \frac{4}{3}\pi R_T^3 \approx 4\pi R_T^2 \delta R_f$, while that for the reactant concentration by $4\pi R_M^2 \delta R_f$. The increase in the thermal energy in the flame structure represents an increased extent of heat transfer away from the reaction region, while an increase in the reactant concentration represents an increased amount of reactant supply to the reaction region. Consequently, if $R_T > R_M$, that is, $Le > 1$, then the flame temperature is expected to be reduced from T_{ad}, while the opposite holds for $Le < 1$.

The same argument can be applied to the negatively stretched, inwardly propagating flame, recognizing that with a reduction in the flame radius there is a greater reduction of heat loss than mass gain, which implies that $T_f > T_{ad}$ for $Le > 1$ and $T_f < T_{ad}$ for $Le < 1$.

There are some additional points to note. First, the stretch intensity of an expanding flame is the largest at the initial stage of the flame development, and steadily decreases

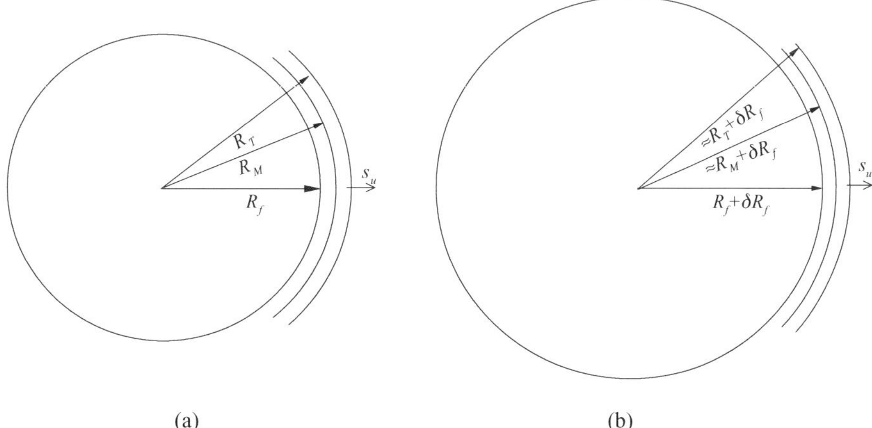

Figure 10.3.5. Conceptual demonstration of the effects of stretch on the flame response in the presence of nonequidiffusion for an expanding spherical flame at (a) $t = 0$, (b) $t = \delta t$.

as the flame expands. The flame degenerates to the stretchless, planar case as $R_f \to \infty$. Since the flame temperature is temporally varying with the stretch rate, the burned state downstream of the flame is not uniform, decreasing toward the center for $Le > 1$, and increasing otherwise. Furthermore, if a spherical flame can be first established, and if there is no external loss such as that through radiation, then stretch-induced extinction is not possible and the flame expands to approach the planar limit, with T_f continuously increasing and decreasing for $Le > 1$ and < 1 mixtures.

For a shrinking flame, the intensity of stretch continuously increases. Since the burning intensity is reduced for a $Le < 1$ flame, extinction occurs when the flame has decreased to a finite size, with a finite amount of the fresh reactant upstream of the flame left unreacted. However, for a $Le > 1$ flame the burning intensity continuously increases until the upstream boundary of the flame reaches the center such that $\ell_T^o / R_f = O(1)$. The reactant concentration at the center then rapidly decreases, being diffusively drawn away toward the reaction zone. The burning intensity therefore also decreases, rapidly leading to flame extinction. Thus extinction in this case is caused by reactant depletion, with almost complete reactant consumption at extinction.

10.3.4. Effects of Heat Loss

Heat loss, both volumetric and through surface conduction, is of course still frequently the major cause of extinction. For volumetric heat loss in the presence of stretch, the relevant factors that need to be considered are the influence of stretch on the flame temperature and thickness, both of which affect the extent of loss. For surface heat loss, it is necessary to consider the influence of stretch on the proximity of the flame to the surface. In the case of the stagnation flow, increasing stretch pushes the flame closer to the surface and, hence, increases the heat loss rate. This downstream heat loss will induce flame extinction before the flame can approach the surface and become restrained.

We have therefore identified several factors responsible for the extinction of stretched flames, namely nonequidiffusion, incomplete reaction due to flame restraining, upstream reactant depletion, and heat loss. Except for upstream reactant depletion, extinction in each of the other three cases is caused by a loss of enthalpy. The enthalpy loss can be in the form of the unburned reactant in the diffusion zone due to diffusional imbalance or in the reaction region due to incomplete reaction, or the volumetric and surface heat loss from either the upstream or downstream regions of the flame.

10.4. FLAME STRETCH: ANALYSES

Asymptotic analyses have been performed for the structure and response of a curved premixed flame situated in a general nonuniform flow field (Sivashinsky 1976; Pelce & Clavin 1982; Clavin & Williams 1982; Matalon & Matkowsky 1982). We shall, however, use the simpler, conceptually more transparent integral analysis for solution (Chung & Law 1988). Furthermore, we shall break the analysis into two parts. In Section 10.4.1 we shall capture the effects of stretch on the flame response through the model problem of the freely standing planar flame situated in a strained flow field. In Section 10.4.2 we shall study the unstretched, stationary spherical flame to show that there is a pure curvature effect that can also modify the flame speed. In the linearized limit of small stretch rate and flame curvature, results from these two separate analyses can be added to yield the final expressions for the flame response, as shown in Section 10.4.3. Extinction is described by retaining the Arrhenius nonlinearity, and is controlled by the reduction of the flame temperature due to stretch. These approximate analyses are fairly straightforward, providing clear quantification of, and insight into, the various phenomena described above. Finally, in Section 10.4.4 an asymptotic analysis is applied to solve the problem of Figure 10.3.4b.

10.4.1. Effects of Flame Stretch

The stagnation flame problem analyzed is shown schematically in Figure 10.4.1a. Here we shall treat the flow motion in traversing the flame structure as quasi-one-dimensional, with a varying streamtube area $A(x)$. The flame boundaries are planar. The quasi-one-dimensional governing equations are the following:

Continuity:

$$\frac{d(fA)}{dx} = 0 \tag{10.4.1}$$

Energy Conservation:

$$\frac{d}{dx}\left[A\left(fc_pT - \lambda\frac{dT}{dx}\right)\right] = A_b q_c B_C Y e^{-T_a/T} \tag{10.4.2}$$

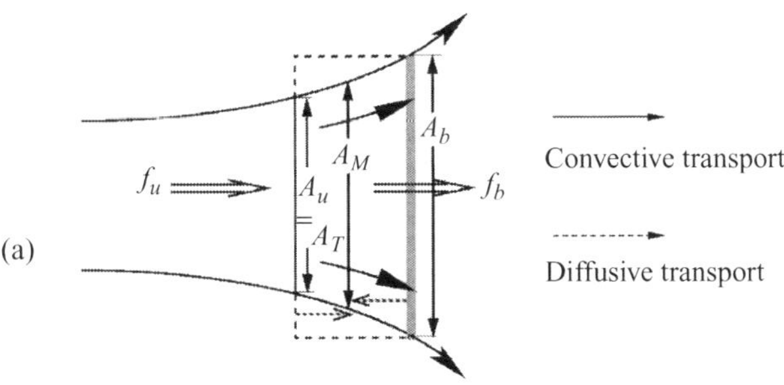

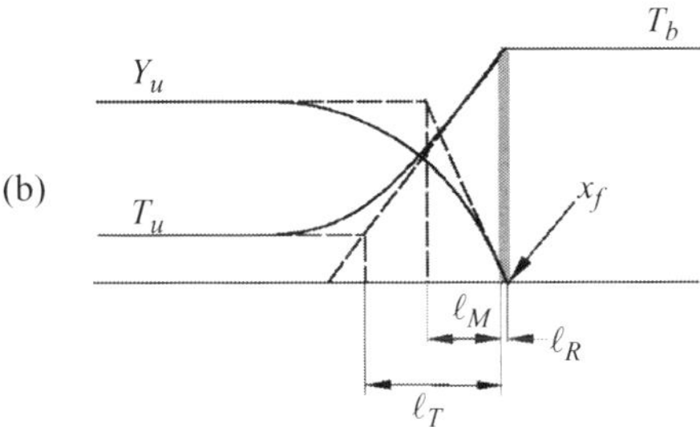

Figure 10.4.1. Schematic for control volume analysis of the planar stretched flame: (a) Definitions of various transport zones and directions; and (b) thicknesses of the transport and reaction zones.

Species Conservation:

$$\frac{d}{dx}\left[A\left(fY - \rho D\frac{dY}{dx}\right)\right] = -A_b B_C Y e^{-T_a/T}, \tag{10.4.3}$$

where A_b is the area of the thin reaction zone, and $f = \rho u$ the local mass flux.

Equation (10.4.1) readily yields the constant flow rate, and, hence, mass burning rate m in the streamtube,

$$m = f_u A_u = f_b A_b. \tag{10.4.4}$$

We next integrate Eq. (10.4.2) over the preheat zone, from the unburned state to x_f^-, where $A = A_u$ and A_b respectively. The integration, however, is performed by recognizing (Figure 10.4.1a) that while convective transport follows the streamline over the entire streamtube such that A varies from A_u to A_b and $\dot{m}$ is fixed, diffusive transport occurs only in the direction normal to the reaction "sheet" such that only the diffusive heat flux from the projection of A_u on A_b is utilized in heating the unburned mixture. Thus diffusive heat transport within the streamtube takes place only over an area A_u. Consequently, we have

$$mc_p(T_b - T_u) - \lambda\left(\frac{dT}{dx}\right)_{x_f^-} A_u = 0. \tag{10.4.5}$$

We then integrate Eq. (10.4.2) across the reaction zone, which has a constant area A_b and uniform downstream temperature T_b, to obtain

$$\lambda\left(\frac{dT}{dx}\right)_{x_f^-} = q_c B_C \int_{x_f^-}^{x_f^+} Y e^{-T_a/T} dx. \tag{10.4.6}$$

Multiplying Eq. (10.4.6) by A_b, and adding the resulting expression to Eq. (10.4.5), we obtain an expression for overall energy transport across the entire flame,

$$mc_p(T_b - T_u) + \lambda\left(\frac{dT}{dx}\right)_{x_f^-}(A_b - A_u) = A_b q_c B_C \int_{x_f^-}^{x_f^+} Y e^{-T_a/T} dx. \tag{10.4.7}$$

Equation (10.4.7) clearly shows the nonconservative nature of the thermal energy transport. That is, if thermal energy were conserved, then all the chemical heat release is used to heat the mixture from the freestream and the diffusion term should identically vanish, as for the one-dimensional planar flame in which $A_u \equiv A_b$. For the present problem, however, a finite amount of the thermal energy is lost from the control volume because of the change in the streamtube area and the fact that diffusive transport occurs normal to the reaction zone.

A similar manipulation for the reactant concentration yields the corresponding relations for the mass diffusion zone, the reaction zone, and the overall nonconservative reactant transport across the flame,

$$mY_u + \rho D\left(\frac{dY}{dx}\right)_{x_f^-} A_M = 0 \tag{10.4.8}$$

$$\rho D\left(\frac{dY}{dx}\right)_{x_f^-} = -B_C \int_{x_f^-}^{x_f^+} Y e^{-T_a/T} dx \tag{10.4.9}$$

$$mY_u - \rho D\left(\frac{dY}{dx}\right)_{x_f^-}(A_b - A_M) = A_b B_C \int_{x_f^-}^{x_f^+} Y e^{-T_a/T} dx, \tag{10.4.10}$$

where A_M is the area of the upstream boundary of the mass diffusion zone.

If we multiply Eq. (10.4.10) by q_c and subtract the resulting expression from Eq. (10.4.7), and further assume equal diffusivity ($Le = 1$) such that $A_T = A_u = A_M$, then we obtain

$$(A_b - A_u)\left[c_p\left(\frac{dT}{dx}\right)_{x_f^-} + q_c\left(\frac{dY}{dx}\right)_{x_f^-}\right] = 0. \tag{10.4.11}$$

Since $(A_b - A_u)$ is arbitrary, the only solution of Eq. (10.4.11) is for the remaining term to vanish, which in turn implies that the loss in thermal energy is balanced by the gain in chemical energy. The system is therefore rendered conservative again, even when it is stretched. This obviously does not hold for $Le \neq 1$, which we shall show next.

Following Section 7.4.1 for the integral analysis of the one-dimensional unstretched flame, we define effective thicknesses for the thermal and mass diffusion zones as (Figure 10.4.1b)

$$\ell_T = \frac{T_b - T_u}{(dT/dx)_{x_f^-}}, \qquad \ell_M = -\frac{Y_u}{(dY/dx)_{x_f^-}}.$$

Equations (10.4.5), (10.4.6), (10.4.8), and (10.4.9) then respectively become

$$f_u \ell_T = \lambda/c_p, \tag{10.4.12}$$

$$\lambda \frac{T_b - T_u}{\ell_T} = q_c B_C \int_{x_f^-}^{x_f^+} Y e^{-T_a/T} dx, \tag{10.4.13}$$

$$f_u \ell_M = \rho D \frac{A_M}{A_u}, \tag{10.4.14}$$

$$\rho D \frac{Y_u}{\ell_M} = B_C \int_{x_f^-}^{x_f^+} Y e^{-T_a/T} dx, \tag{10.4.15}$$

which provide four relations to solve for the four flame responses f_u, T_b, ℓ_T, and ℓ_M in terms of the area ratio A_M/A_u.

If we next assume that the change in the streamtube area is gradual (see Figure 10.4.1b), then from geometrical consideration we have

$$\frac{A_b - A_u}{\ell_T} \approx \frac{A_b - A_M}{\ell_M}, \tag{10.4.16}$$

which relates A_M to $A_T = A_u$. Furthermore, recognizing that the Karlovitz number is simply a nondimensional measure of the extent of flow nonuniformity across the flame, it can be represented by the fractional area change along the streamtube. We shall therefore define Ka as

$$Ka = \frac{\Delta A}{A_b} = \frac{A_b - A_u}{A_b} = 1 - \frac{A_u}{A_b}. \tag{10.4.17}$$

The final point to note is that Ka itself, given by Eq. (10.2.41), depends on the burning flux f_u through the flame time and, hence, the flame thickness. This dependence needs to be accounted for so as not to suppress the feedback between them. Consequently, we write

$$Ka = \frac{(\lambda/c_p)_u}{f_u^2}(\rho_u \kappa) = \frac{Ka^o}{\tilde{f}_u^2}, \tag{10.4.18}$$

where $Ka^o = \left[(\lambda/c_p)_u/(f^o)^2\right](\rho_u \kappa)$.

By further approximating the reaction integral term $\int Y\exp(-T_a/T)dx$ in Eqs. (10.4.13) and (10.4.15) by $(Y_u/Ze^o)\ell_R\exp(-T_a/T_b)$, Eqs. (10.4.12) to (10.4.16) can be solved, after some algebraic manipulation, to yield $A_u/A_M \approx 1 + S^o$,

$$\frac{\tilde{\ell}_T}{\tilde{\ell}_M} \approx 1 + \frac{S^o}{\tilde{f}_u^2} \approx 1 + S^o \tag{10.4.19}$$

$$\tilde{T}_b - \tilde{T}_b^o \approx \frac{S^o}{\tilde{f}_u^2} \approx S^o \tag{10.4.20}$$

$$\tilde{f}_u^2 = (\tilde{\ell}_T)^{-2} = \exp\left(\frac{2\sigma^o}{\tilde{f}_u^2}\right), \tag{10.4.21}$$

where $\tilde{f}_u = f_u/f^o$, $\tilde{\ell}_T = \ell_T/\ell_T^o$, $\tilde{\ell}_M = \ell_M/\ell_M^o$, and $\tilde{T}_b^o = 1 + \tilde{T}_u$. Furthermore, through the above derivation, we have identified two fundamental parameters for stretched flames, namely

$$S^o = \left(\frac{1}{Le} - 1\right) Ka^o, \tag{10.4.22}$$

which exhibits the intrinsic coupling between stretch and nonequidiffusion, through Ka^o and Le respectively, and

$$\sigma^o = \frac{Ze^o}{2} S^o = \frac{Ze^o}{2}\left(\frac{1}{Le} - 1\right) Ka^o, \tag{10.4.23}$$

which represents the chemical weighting of S^o, through Ze^o.

From Eq. (10.4.23), a Markstein number,

$$Ma^o = \frac{Ze^o}{2}\left(\frac{1}{Le} - 1\right), \tag{10.4.24}$$

which is a property of the mixture, can also be defined. Consequently, σ^o can be alternately expressed as

$$\sigma^o = Ma^o Ka^o, \tag{10.4.25}$$

which properly identifies the influences of the mixture's chemical-diffusive properties (Ma^o) and the system dynamics (Ka^o).

We have therefore determined all the flame responses of interest. These results degenerate to those of the standard premixed flame for $Ka^o = 0$, as should be the case. Furthermore, although the above derivation was performed only for the aerodynamically stretched planar flame, the general nature with which κ and, hence, the Karlovitz number Ka are defined implies that the above results are applicable to flames subjected to various forms of stretch as imposed by flow nonuniformity, flame curvature, and flame motion. We do note, however, that for simplicity effects of thermal expansion in crossing the flame were not accounted for in the derivation. Results incorporating these effects, obtained by using the integral analysis, are given in Sun et al. (1999).

Several important observations can be made regarding the above results. First, expressing Eq. (10.4.20) in dimensional form, we have

$$c_p(T_b - T_u) = (1 + S^o)Y_u q_c, \tag{10.4.26}$$

which shows clearly that the influence of stretch on the flame temperature is equivalent to modifying the freestream concentration of the reactant, Y_u, by a factor $(1 + S^o)$. Second, the parameter S^o directly affects the flame temperature through energy conservation due to the combined effects of nonequidiffusion and stretch. Furthermore, since deviation of the flame temperature from the adiabatic flame temperature can occur only in the simultaneous presence of stretch ($Ka^o \neq 0$) and nonequidiffusion ($Le \neq 1$) such that $S^o \neq 0$, and since Ka^o can be positive and negative, while Le can be greater or less than unity, we expect

$$\begin{aligned} T_b > T_b^o = T_{\text{ad}} \quad &\text{for} \quad \{Ka^o > 0, Le < 1\} \quad \text{or} \quad \{Ka^o < 0, Le > 1\} \\ T_b < T_b^o = T_{\text{ad}} \quad &\text{for} \quad \{Ka^o > 0, Le > 1\} \quad \text{or} \quad \{Ka^o < 0, Le < 1\}. \end{aligned} \tag{10.4.27}$$

Thus the flame behavior for a gas of given Le is expected to be completely opposite for the positively stretched stagnation flame and outwardly propagating flame as compared to the negatively stretched Bunsen flame and inwardly propagating flame, as anticipated previously based on phenomenological considerations.

The modification of T_b through S^o leads to corresponding modifications of the burning flux and flame thickness through the factor σ^o. Expressing Eq. (10.4.21) as

$$\tilde{f}_u^2 \ln \tilde{f}_u^2 = 2\sigma^o, \tag{10.4.28}$$

we readily recognize that it has the same functional form as that for the nonadiabatic planar flame propagation, Eq. (8.4.9), studied in Section 8.4. Thus $2\sigma^o$ assumes the same role as the heat loss parameter, $-\tilde{L}_v$, for the nonadiabatic flame. As such, for $\sigma^o < 0$ (and $S^o < 0$) the response of $\tilde{f}_u$ with σ^o is described by the extinction curves of Figure 8.4.2, with extinction occurring at $-2\sigma_E^o = e^{-1}$ and $\tilde{f}_E = e^{-1/2}$. This is reasonable in that $S^o < 0$ flames are subadiabatic, causing the flame temperature to be reduced from T_{ad}. Thus the role of diffusive loss of enthalpy due to nonequidiffusion is entirely analogous to that of heat loss through, say, radiation.

For the downstream mass flux, f_b, from continuity, $f_u A_u = f_b A_b$, and using Eq. (10.4.17), we have

$$\tilde{f}_b = \tilde{f}_u(1 - Ka), \tag{10.4.29}$$

where $\tilde{f}_b = f_b/f^o$.

When the exponent of Eq. (10.4.21) is of order $1/Ze^o$, the Arrhenius factor can be linearized, yielding the weakly stretched burning flux

$$\tilde{f}_u \approx 1 + \sigma^o \tag{10.4.30}$$

$$\tilde{f}_b \approx 1 + \sigma^o - Ka^o. \tag{10.4.31}$$

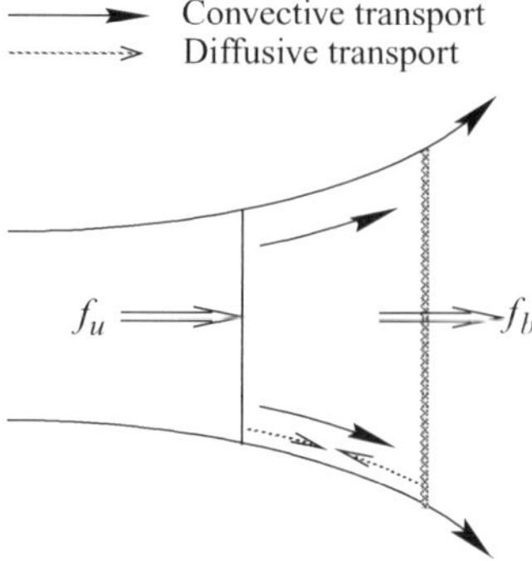

Figure 10.4.2. Incorrect control volume analysis, with diffusion occurring along the streamline direction.

It is of particular significance to recognize that, for an equidiffusive mixture, the effects of stretch vanish identically ($S^o \equiv 0$), regardless of the intensity of stretch (Ka^o), such that $\tilde{T}_b = \tilde{T}_b^o$, $\tilde{f}_u = 1$, $\tilde{f}_b = 1 - Ka^o$, and $\tilde{\ell}_T = 1$. Thus for an adiabatic, equidiffusive, freely propagating stretched planar flame, the flame temperature, the upstream burning flux, and the flame thickness are not affected by flow straining, being identical to their unstretched values. The downstream burning flux is modified from the unstretched value f^o, by a factor $(1 - Ka^o)$, due to flow divergence. As corollaries, we then expect that the thermal and concentration structures of the flame, in the direction normal to its surface, should be insensitive to strain rate variations. Furthermore, since the burning intensity in the reaction zone, as indicated by the flame temperature $\tilde{T}_b$, is not affected by stretch, we should also expect that this flame cannot be extinguished by stretch alone, contrary to early suggestions (Klimov 1963).

The final point to note is that in writing the energy balance of Eq. (10.4.5) for the diffusion zone, it is essential that the diffusive flux is normal to the reaction zone as shown in Figure 10.4.1. If, however, the diffusive flux is along the streamlines (Figure 10.4.2), then this amount is conserved. A similar analysis yields

$$\tilde{f}_u = \frac{A_b}{A_u}, \tag{10.4.32}$$

$$\tilde{f}_b = 1, \tag{10.4.33}$$

which shows that, with increasing (positive) straining, f_b remains unchanged from f^o, while f_u is increased. These results are completely contrary to what we have just shown, and are incorrect. Thus it is necessary to preserve the orthogonality relation between the reaction surface and the diffusive fluxes when performing a "quasi-one-dimensional" analysis.

10.4.2. Effects of Pure Curvature

Markstein (1964) first showed that the flame speed can be affected by curvature alone. To demonstrate this effect, we consider the situation of Figure 10.4.3 in which

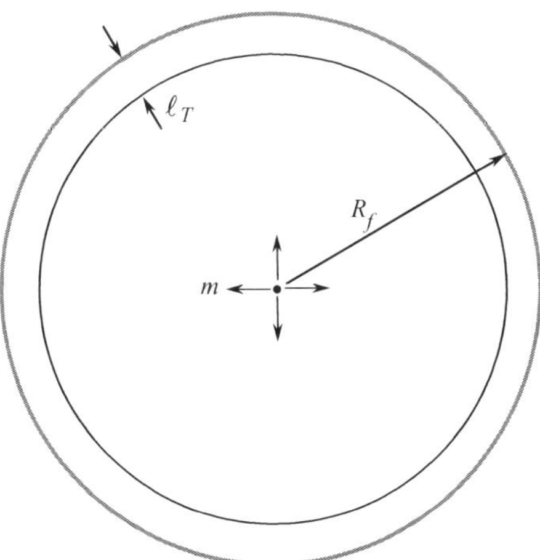

Figure 10.4.3. Control volume analysis of the stationary spherical flame supported by a point source.

a stationary spherical flame ($V_f = 0$) is supported by a point source of constant mass flow rate $m = fA$. Since diffusion now occurs in the same direction as convection, the flame is not stretched ($\mathbf{v}_s \times \mathbf{n} \equiv 0$) and is basically the spherical analog of the one-dimensional planar flame.

A balance of thermal energy and species transport through a radial streamtube across the entire flame respectively yields

$$mc_p(T_b - T_u) = A_b q_c B_C \int_{R_f^-}^{R_f^+} Ye^{-T_a/T} dr \tag{10.4.34}$$

$$mY_u = A_b B_C \int_{R_f^-}^{R_f^+} Ye^{-T_a/T} dr. \tag{10.4.35}$$

Comparing Eqs. (10.4.7) and (10.4.34), we see that there is no diffusive loss in the present case and, hence, thermal energy is conserved across the flame. This is due to the fact that because diffusion now takes place along the streamline, the total amount of diffusive transport at A_u and A_b must be the same. A similar observation can be made for species conservation. Thus subtracting Eq. (10.4.34) from Eq. (10.4.35) results in $c_p(T_b - T_u) = q_c Y_u$, which shows that $T_b = T_b^o = T_{\text{ad}}$.

Analysis for the rest of the flame responses follows the same procedure as that for the stretched flame. In particular, we have

$$\tilde{\ell}_T = 1, \qquad \frac{\tilde{\ell}_T}{\tilde{\ell}_M} = Le \tag{10.4.36}$$

$$\tilde{f}_u = \frac{A_b}{A_u} = \left(\frac{R_f}{R_f - \ell_T}\right)^2 = \frac{1}{(1 - \ell_T/R_f)^2} \tag{10.4.37}$$

$$\tilde{f}_b = \frac{A_u}{A_b}\tilde{f}_u = 1. \tag{10.4.38}$$

Recognizing that $-2/R_f$ is simply the curvature term $\nabla \cdot \mathbf{n}$ for the spherical flame, $\tilde{f}_u$ can be generalized to an arbitrary curvature,

$$\tilde{f}_u = \left(1 + \frac{1}{2}\tilde{\nabla} \cdot \mathbf{n}\right)^{-2}, \qquad (10.4.39)$$

where $\tilde{\nabla} = \ell_T^o \nabla$. For $\tilde{\nabla} \cdot \mathbf{n} \ll 1$, Eq. (10.4.39) can be linearized to

$$\tilde{f}_u \approx 1 - \tilde{\nabla} \cdot \mathbf{n}. \qquad (10.4.40)$$

We have therefore shown that for a purely curved flame without stretch effect, the downstream burning flux remains the same as that of the one-dimensional planar flame, while the upstream burning flux is increased by an amount that is dependent on its curvature. Thus f_u is increased for a flame with negative curvature, and decreased otherwise. This is the mechanism that allows the flame speed s_u at the tip of a $Le = 1$ Bunsen flame to exceed that at the shoulder (Poinsot, Echekki & Mungal 1992). Furthermore, since the flame is adiabatic and its temperature not affected by curvature, pure curvature cannot cause flame extinction. It is also significant to note that, with $Le = 1$, f_u is unaffected by straining for the stretched planar flame but is affected by curvature for the purely curved flame. Consequently, through flow divergence f_b is affected for the stretched planar flame but unaffected for the purely curved flame.

10.4.3. Combined Solution

It is reasonable to expect that a generalized linear solution would be the sum of the linear solutions for the stretched planar flame and the unstretched curved flame (Chung & Law 1988). Since $\tilde{T}_b$, $\tilde{\ell}_T$, and $\tilde{\ell}_T/\tilde{\ell}_M$ are unaffected by curvature for the purely curved flame, their respective generalized responses are the same as those for the stretched flame. For $\tilde{f}_u$, such a combination yields

$$\tilde{f}_u \approx 1 - \tilde{\nabla} \cdot \mathbf{n} + \sigma^o, \qquad (10.4.41)$$

while $\tilde{f}_b$ is still given by Eq. (10.4.31).

Generalized nonlinear analyses (Sun & Law 2000) again yield the extinction turning point behavior identified for the general stretched flame presented earlier. This point can be readily appreciated by recognizing that reaction, and hence extinction, take place near the downstream boundary, and because both $\tilde{f}_b$ and the flame temperature are not affected by the pure curvature effect.

10.4.4. Asymptotic Analysis of the Counterflow Flame

It is instructive to present an asymptotic analysis of the stretched flame phenomena. We choose to analyze the problem of Figure 10.3.4b, in which a stagnation flame is established in the counterflow of a combustible of temperature T_u and reactant concentration Y_u against a hot inert of T_b such that the flame has the freedom to migrate across the stagnation surface in order to achieve complete reaction. For simplicity we shall assume that the flow is potential such that $\rho =$ constant and its x- and y-velocities are respectively $ax/(k+1)$ and $(-ay)$, as discussed in relation to Figure 10.2.4a. The stagnation surface is located at $y = 0$. Energy and reactant conservations in the

y-direction are respectively described in nondimensional form as

$$\frac{d^2\tilde{T}}{d\tilde{y}^2} + \tilde{y}\frac{d\tilde{T}}{d\tilde{y}} = -Da_C\tilde{Y}e^{-\tilde{T}_a/\tilde{T}} \tag{10.4.42}$$

$$\frac{1}{Le}\frac{d^2\tilde{Y}}{d\tilde{y}^2} + \tilde{y}\frac{d\tilde{Y}}{d\tilde{y}} = Da_C\tilde{Y}e^{-\tilde{T}_a/\tilde{T}}, \tag{10.4.43}$$

where $\tilde{y} = y/(\lambda/c_p\rho a)^{1/2}$ and $Da_C = B_C/\rho a$ is the collision Damköhler number, which measures the characteristic flow time ($\sim 1/a$) to the characteristic collision time ($\sim 1/B_C$).

The boundary conditions are

$$\tilde{y} = \infty: \quad \tilde{T} = \tilde{T}_u, \quad \tilde{Y} = 1 \tag{10.4.44}$$

$$\tilde{y} = -\infty: \quad \tilde{T} = \tilde{T}_f = \tilde{T}_b, \quad \tilde{Y} = 0. \tag{10.4.45}$$

It is important to note that in writing the boundary condition $\tilde{T}_f = \tilde{T}_b$, we have assumed downstream adiabaticity and, hence, eliminated possible influences on the flame response due to external heat loss. Thus all predicted flame behavior should have their origin in flame stretch and nonequidiffusion.

In the reaction-sheet limit in which the flame temperature and location are respectively $\tilde{T}_f$ and $\tilde{y}_f$, the temperature and concentration distributions are respectively given by

$$\tilde{y} \geq \tilde{y}_f: \quad \tilde{T}_0^+(\tilde{y}) = \tilde{T}_u + (\tilde{T}_f - \tilde{T}_u)I(\tilde{y};1)/I(\tilde{y}_f;1) \tag{10.4.46}$$

$$\tilde{Y}_0^+(\tilde{y}) = 1 - I(\tilde{y};Le)/I(\tilde{y}_f;Le) \tag{10.4.47}$$

$$\tilde{y} \leq \tilde{y}_f: \quad \tilde{T}_0^-(\tilde{y}) = \tilde{T}_f \tag{10.4.48}$$

$$\tilde{Y}_0^-(\tilde{y}) = 0, \tag{10.4.49}$$

where

$$I(\tilde{y};Le) = \int_\infty^{\tilde{y}} e^{-(Le)s^2/2}ds.$$

For large but finite values of $\tilde{T}_a$, in the inner region $\tilde{T}$ and $\tilde{Y}$ assume the forms

$$\tilde{T}_{\text{in}} = \tilde{T}_f - \epsilon\theta(\chi) + O(\epsilon^2), \qquad \tilde{Y}_{\text{in}} = \epsilon\phi(\chi), \tag{10.4.50}$$

where $\chi = (\tilde{y} - \tilde{y}_f)/\epsilon, \epsilon = \tilde{T}_f^2/\tilde{T}_a$, and we write $\epsilon^o = \epsilon$ for simplicity. Substituting $\tilde{T}_{\text{in}}$ and $\tilde{Y}_{\text{in}}$ into Eqs. (10.4.42) and (10.4.43) respectively, and solving for the local coupling function with the downstream boundary conditions that $\theta(-\infty) = \phi(-\infty) = 0$ and $(d\theta/d\chi)_{-\infty} = (d\phi/d\chi)_{-\infty} = 0$, we obtain

$$\frac{d^2\theta}{d\chi^2} = \frac{\Delta}{2}\phi e^{-\theta} \tag{10.4.51}$$

$$\frac{d}{d\chi}\left(\theta - \frac{\phi}{Le}\right) = 0, \tag{10.4.52}$$

$$\theta - \frac{\phi}{Le} = 0, \tag{10.4.53}$$

where $\Delta = 2\epsilon^2 Da_C \exp(-\tilde{T}_a/\tilde{T}_f)$.

In the outer region, the perturbed temperature and concentration distributions that satisfy the reaction-free forms of Eqs. (10.4.42) and (10.4.43), as well as the upstream boundary conditions, are

$$\tilde{T}^+_{\text{out}} = \tilde{T}^+_0(\tilde{y}) - \epsilon c_1 I(\tilde{y}; 1)/I(\tilde{y}_f; 1) \tag{10.4.54}$$

$$\tilde{Y}^+_{\text{out}} = \tilde{Y}^+_0(\tilde{y}) + \epsilon c_2 I(\tilde{y}; Le)/I(\tilde{y}_f; Le), \tag{10.4.55}$$

where c_1 and c_2 are integration constants to be determined through matching.

To perform the matching as $\chi \to \infty$, we need to expand the outer solution around $\tilde{y}_f$. We first note that

$$\begin{aligned} I(\tilde{y}; Le) &= \int_\infty^{\tilde{y}} e^{-(Le)s^2/2} ds = \int_\infty^{\tilde{y}_f + \epsilon\chi} e^{-(Le)s^2/2} ds \\ &= \int_\infty^{\tilde{y}_f} e^{-(Le)s^2/2} ds + \int_{\tilde{y}_f}^{\tilde{y}_f + \epsilon\chi} e^{-(Le)s^2/2} ds \\ &= I(\tilde{y}_f; Le) + \int_0^{\epsilon\chi} e^{-Le(\tilde{y}_f + s)^2/2} ds \\ &= I(\tilde{y}_f; Le) + e^{-Le\tilde{y}_f^2/2} \int_0^{\epsilon\chi} e^{-Le(\tilde{y}_f s + s^2/2)} ds. \end{aligned}$$

The integrand in the last expression is approximately unity because s is a small number in the integration interval $0 < s < \epsilon\chi$. Thus we have

$$I(\tilde{y}; Le) \simeq I(\tilde{y}_f; Le) + \epsilon e^{-(Le)\tilde{y}_f^2/2} \chi \quad \text{as} \quad \chi \to \infty. \tag{10.4.56}$$

Using Eq. (10.4.56), matching between the inner and outer solutions yields

$$\theta(\infty) = c_1 - \frac{(\tilde{T}_f - \tilde{T}_u)}{I(\tilde{y}_f; 1)} e^{-\tilde{y}_f^2/2} \lim_{\chi \to \infty} \chi \tag{10.4.57}$$

$$\left(\frac{d\theta}{d\chi}\right)_\infty = -\frac{(\tilde{T}_f - \tilde{T}_u)}{I(\tilde{y}_f; 1)} e^{-\tilde{y}_f^2/2}, \tag{10.4.58}$$

$$\phi(\infty) = c_2 - \frac{1}{I(\tilde{y}_f; Le)} e^{-(Le)\tilde{y}_f^2/2} \lim_{\chi \to \infty} \chi \tag{10.4.59}$$

$$\left(\frac{d\phi}{d\chi}\right)_\infty = -\frac{1}{I(\tilde{y}_f; Le)} e^{-(Le)\tilde{y}_f^2/2}. \tag{10.4.60}$$

Putting Eqs. (10.4.58) and (10.4.60) into Eq. (10.4.52), we have

$$\tilde{T}_f - \tilde{T}_u = \frac{e^{(1-Le)\tilde{y}_f^2/2}}{Le} \frac{I(\tilde{y}_f; 1)}{I(\tilde{y}_f; Le)}. \tag{10.4.61}$$

The final step is to solve the inner equation (10.4.51), with $\phi = Le\theta$ and the boundary conditions $\theta(-\infty) = (d\theta/d\chi)_{-\infty} = 0$. The solution yields

$$\left[\frac{(\tilde{T}_f - \tilde{T}_u)}{I(\tilde{y}_f; 1)} e^{-\tilde{y}_f^2/2}\right]^2 = Le\Delta. \tag{10.4.62}$$

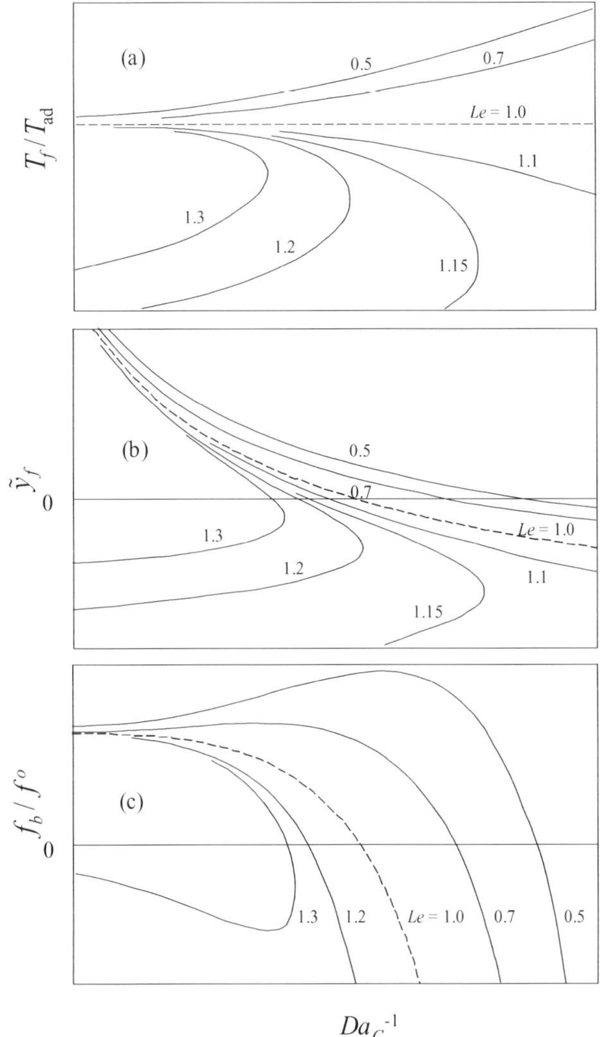

Figure 10.4.4. (a) Flame temperature, (b) flame location, and (c) downstream burning flux for the counterflow configuration of combustible versus product.

Equations (10.4.61) and (10.4.62) are the final solutions of the problem, from which the flame temperature $\tilde{T}_f$ and reaction zone location $\tilde{y}_f = \tilde{y}_b$ can be determined. Knowing $\tilde{y}_f$, the downstream burning flux is simply given by

$$f_b = (\rho v)_f = \rho a y_f. \tag{10.4.63}$$

Figure 10.4.4 shows typical solutions of T_f / T_{ad}, $\tilde{y}_f$, and f_b / f^o with increasing stretch ($\sim Da_C^{-1}$), obtained from Eqs. (10.4.61) to (10.4.63). It is clear that the characteristic turning point behavior, representing the occurrence of extinction, is captured from the nonlinear analysis. The results for $Le > 1$ situations are easy to interpret: increasing stretch reduces the flame temperature and causes flame extinction. For Le

sufficiently greater than unity, extinction occurs on the reactant side of the stagnation surface ($\tilde{y}_f > 0$). However, for smaller values of *Le*, the reaction sheet migrates across the stagnation surface and extinguishes on the product side of the flow. Flames in this region possess negative downstream flame speeds as mentioned earlier. For $Le < 1$, the flame temperature continuously increases with stretch. The burning flux first increases and then decreases. The increase is due to the increase in the flame temperature while the decrease is due to flow divergence. Since there is no chemical limitation in the flame behavior in that complete reaction is assumed, extinction does not occur for these situations. This is supported by the lack of turning points for these curves.

The above result therefore substantiates our previous anticipation that extinction cannot be achieved by stretch alone. In order to attain extinction, the additional factor of nonequidiffusion is needed for $Le > 1$ mixtures and incomplete reaction is needed for $Le \le 1$ mixtures.

It is also of interest to derive the linearized results for weak stretch. Under such a situation the flame is located close to the freestream, with $\tilde{y}_f \to \infty$. Using the expansion

$$\int_{\infty}^{\tilde{y}_f} e^{-(Le)s^2/2} ds \simeq -\frac{e^{-(Le)\tilde{y}_f^2/2}}{Le\tilde{y}_f}\left[1 - \frac{1}{Le\tilde{y}_f^2} + \frac{3}{(Le\tilde{y}_f^2)^2} + \cdots\right], \quad \text{for } \tilde{y}_f \gg 1$$

Eq. (10.4.61) becomes

$$\tilde{T}_f - \tilde{T}_u \simeq 1 + \left(\frac{1}{Le} - 1\right)\frac{1}{\tilde{y}_f^2}. \tag{10.4.64}$$

Equation (10.4.64) again shows that $\tilde{T}_f > \tilde{T}_{\text{ad}}$ for $Le < 1$ and $\tilde{T}_f < \tilde{T}_{\text{ad}}$ for $Le > 1$.

If we next expand Eq. (10.4.62) for large $\tilde{y}_f$, using the definition of Δ, it can be shown that the following expression results:

$$Ka^o\tilde{y}_f^4 - \tilde{y}_f^2 + \left[\frac{2}{Le} - \left(\frac{1}{Le} - 1\right)\frac{1}{\epsilon}\right] = 0, \tag{10.4.65}$$

where we have identified $Ka^o = (\lambda/c_p)(\rho a)/(f^o)^2$. Furthermore, since $\tilde{y}_f = y_f/(\lambda/c_p\rho a)^{1/2}$, and $f_b = \rho a y_f$, f_b^2 can be solved from Eq. (10.4.65) for small a, yielding $\tilde{f}_b = f_b/f^o = (1 + \sigma^o - Ka^o)$, which is Eq. (10.4.31). Further noting that since $f_u = (\rho v)_u = \rho a y_u = \rho a(y_b + \ell_T)$, while $\ell_T = \ell_T^o + O(\epsilon) = \lambda/(c_p f^o)$, we have $\tilde{f}_u = 1 + \sigma^o$, which is simply Eq. (10.4.30). Finally, an expansion for Eq. (10.4.64) also leads the linearized result for the flame temperature, Eq. (10.4.20).

10.5. EXPERIMENTAL AND COMPUTATIONAL RESULTS

10.5.1. Equidiffusive Flames

Perhaps one of the most interesting properties predicted for stretched flames is that, for an equidiffusive, freely standing or freely propagating planar stretched flame, its flame temperature and thickness, as well as the upstream burning flux and velocity, are independent of the magnitude of stretch. This then further implies that the flame structure in the direction normal to the flame surface should also be insensitive to strain rate variations. In order to verify this prediction, the temperature and major species profiles across a basically adiabatic, equidiffusive flame in a symmetrical counterflow have been experimentally determined by using laser Raman spectroscopy (Law et al. 1994). Figure 10.5.1a shows the temperature profiles for four strain rates, with the highest strain rate being close to the extinction state. It is seen that, with increasing straining, the flame recedes toward the stagnation surface located at the origin. If we now superimpose these temperature profiles by shifting their spatial locations such that the locations of their maximum temperature gradient coincide, then Figure 10.5.1b shows that, in this "flame coordinate," the temperature profiles basically overlap. To provide an even more stringent comparison, the temperature gradients were computed and their profiles again overlap. Similar results were obtained for the major species profiles. Furthermore, these experimental results also quantitatively agree well with the computed ones using detailed chemistry and transport, with an agreement similar to that of the nonpremixed counterflow flame of Figure 6.5.3. As such, it is reasonable to conclude that the structure of equidiffusive, planar flames is insensitive to strain rate variations.

Figure 10.5.1 also tabulates the computed Karlovitz numbers, $Ka^o = (\ell_T^o/s_u^o)\kappa$, for the four flames, with $s_u^o = 17.0$ cm/s independently calculated. The flame thickness is found to be 0.785 mm based on the definition $\ell_T^o = (T_{\rm ad} - T_u)/(dT/dx)_{\rm max}$. Using this value, Figure 10.5.1 shows that the estimated Karlovitz numbers including that for the near extinction state are either smaller than, or of the order of, unity.

10.5.2. Nonequidiffusive Flames

For nonequidiffusive, stretched flames, theoretical results show that the flame response exhibits opposite behavior when the stretch changes from positive to negative, and when the mixture's effective Lewis number is greater or smaller than a critical value, which is unity for the flame temperature. These completely opposite trends should provide definitive verification of the concept of flame stretch with nonequidiffusion.

Two groups of mixtures are especially suitable for the study of nonequidiffusion effects (Table 10.1). The first group consists of lean hydrogen–air, lean methane–air

Table 10.1. Mixtures for the study of nonequidiffusion effects on stretched flames

Mixture for Simulation	$Le \neq 1$ Interpretation	$D_i \neq D_j$ Interpretation
Lean hydrogen–air	$Le_{H_2} < 1$	
Lean methane–air	$Le_{CH_4} < 1$	
Rich propane–air	$Le_{O_2} < 1$	$D_{H_2} > D_{CH_4} > D_{O_2} > D_{C_3H_8}$
Rich hydrogen–air	$Le_{O_2} > 1$	
Rich methane–air	$Le_{O_2} > 1$	
Lean propane–air	$Le_{C_3H_8} > 1$	

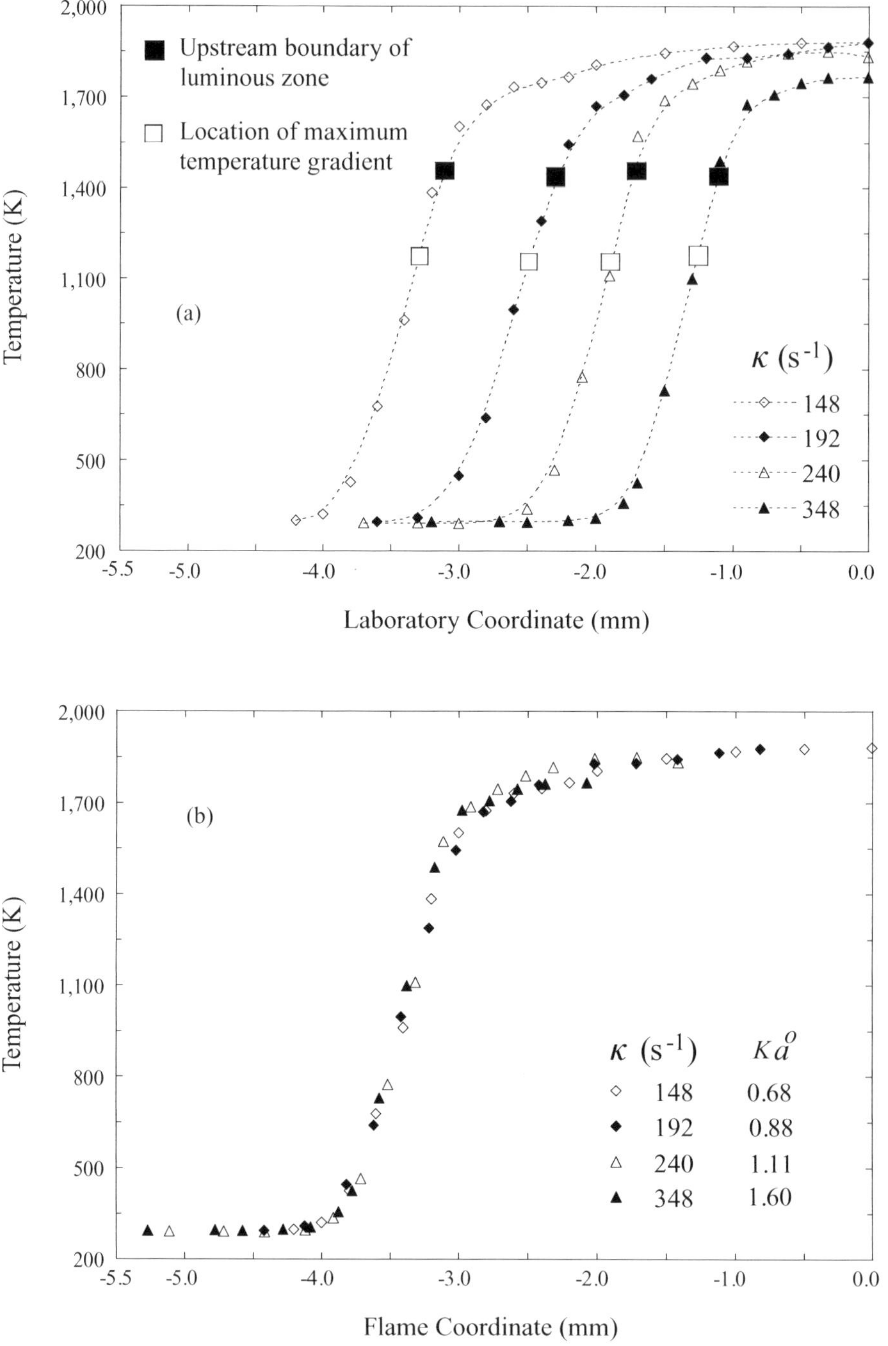

Figure 10.5.1. Experimentally determined temperature profiles of equidiffusive counterflow premixed flames with different strain rates, in (a) laboratory coordinate, and (b) flame coordinate, demonstrating the insensitivity of the flame structure to strain rate variations. (Fuel: methane, $\phi = 0.95$, $N_2/O_2 = 5$).

and rich propane–air mixtures. Estimates show that their effective Lewis numbers, based on the deficient species, are smaller than unity. Thus positive (negative) stretch is expected to increase (decrease) the flame temperature of such mixtures from T_{ad}. This argument still holds even if we just consider the relative mass diffusivities of the fuel and oxidizer species. That is, based on molecular weight considerations, we expect the diffusivities of the various reactants relative to nitrogen should increase in the order of propane, oxygen, and methane. Thus positive (negative) stretch will increase (decrease) the methane concentration of a lean methane–air mixture but decrease (increase) the propane concentration of a rich propane–air mixture at the flame. Both mixtures are consequently rendered more (less) stoichiometric at the flame, leading to enhanced (reduced) flame temperature.

The second group consists of rich hydrogen–air, rich methane–air, and lean propane–air mixtures, whose effective Lewis numbers are greater than unity while positive (negative) stretch also renders the mixture less (more) stoichiometric. Thus the responses of these two groups of mixtures to stretch are expected to be qualitatively opposite.

Several flame configurations have been found useful to demonstrate the effects of positive or negative stretch, either experimentally and/or computationally. For positive stretch, the symmetric counterflow flame (CFF) and the outwardly propagating flame (OPF) can be readily investigated through both experiment and computation. For negative stretch, the inwardly propagating flame (IPF) can be computationally simulated but not easily experimentally established. The tip of the Bunsen flame is also negatively stretched, and can provide qualitative nonequidiffusive indications for such flame response as the flame temperature. However, accurate quantification of, say the local flame speed, is hindered by the difficulty in defining the local flame curvature.

Figure 10.5.2a (Sun et al. 1999) shows the computationally determined s_b as a function of the stretch rate for lean and rich hydrogen–air counterflow flames, with s_b defined as the axial flow velocity at the location of the maximum heat release rate. The use of s_b avoids the ambiguity from choosing the spatial location at which s_u is defined (Tien & Matalon 1991). It is seen that while s_b increases with κ for lean flames, it decreases for rich flames. This is in agreement with the anticipated behavior of the positively stretched flames with Le smaller and greater than unity, respectively. The increasing trend for the lean flame due to nonequidiffusion is particularly significant because, as shown in Eq. (10.4.31), pure stretch alone would cause s_b to decrease because of flow divergence.

To further demonstrate the importance of nonequidiffusion, Figure 10.5.3a shows the corresponding plot for the lean and rich propane–air flames. Since the Le behavior for lean and rich mixtures are switched for hydrogen–air and propane–air flames, it is seen that s_b now exhibits completely opposite behavior, decreasing for lean mixtures while increasing for rich mixtures.

Figures 10.5.2b and 10.5.3b show the s_b for the outwardly propagating hydrogen and propane flames; s_b is chosen because, in addition to being well-defined, the

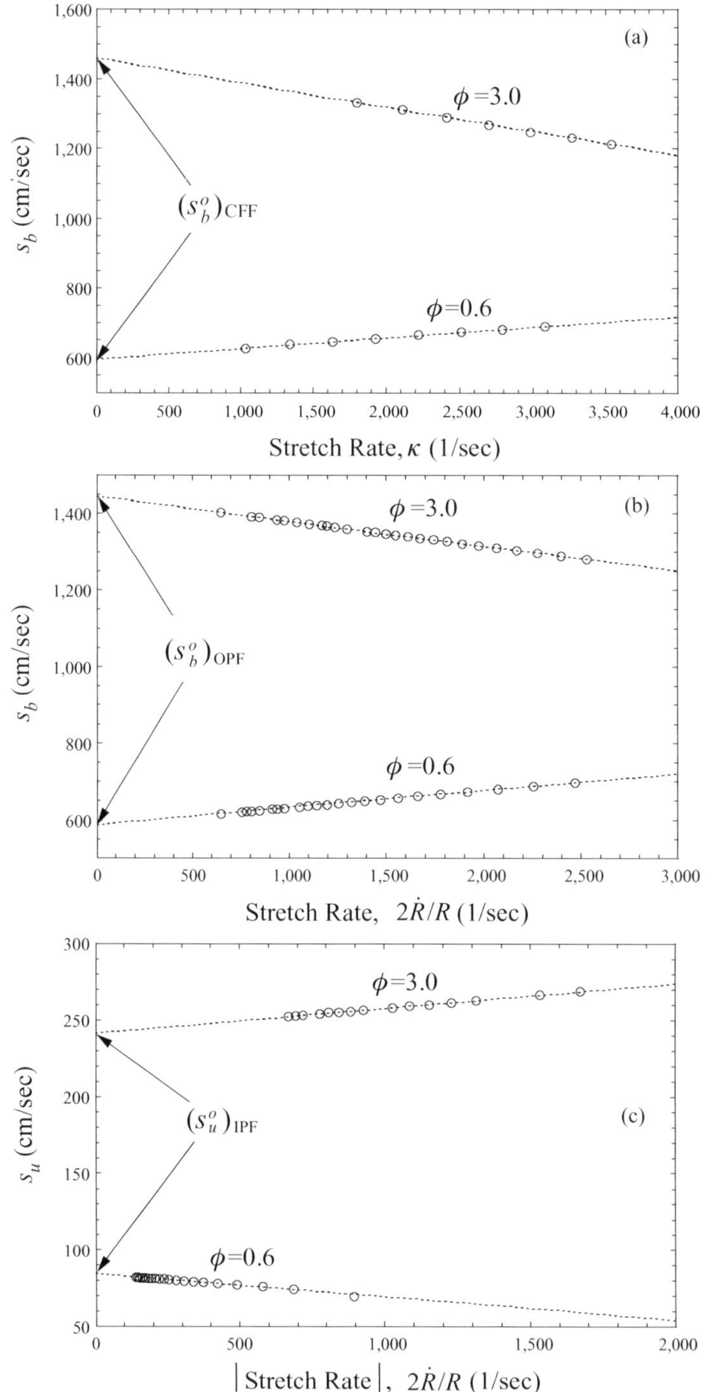

Figure 10.5.2. Calculated flame speeds of 1 atm lean and rich hydrogen–air mixtures as functions of stretch rate for: (a) counterflow flame, (b) outwardly propagating flame, and (c) inwardly propagating flame.

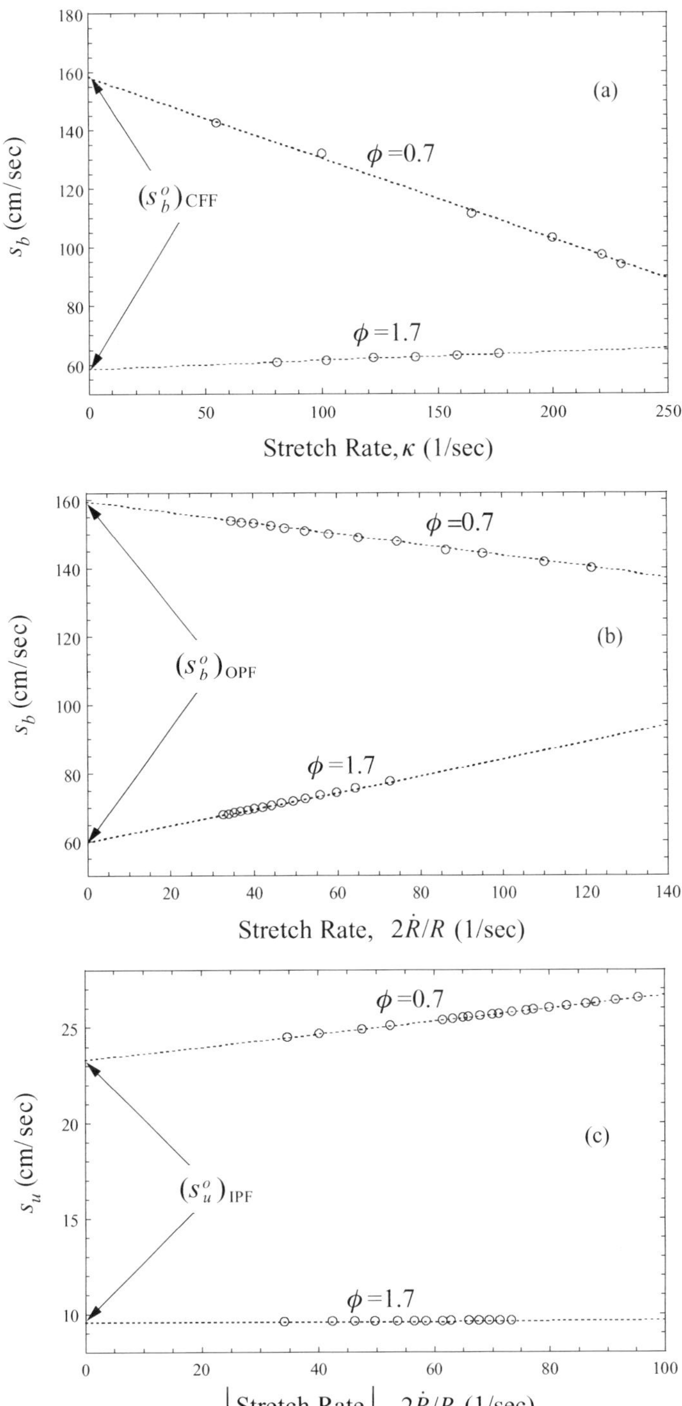

Figure 10.5.3. Calculated flame speeds of 1 atm lean and rich propane–air mixtures as functions of stretch rate for: (a) counterflow flame, (b) outwardly propagating flame, and (c) inwardly propagating flame.

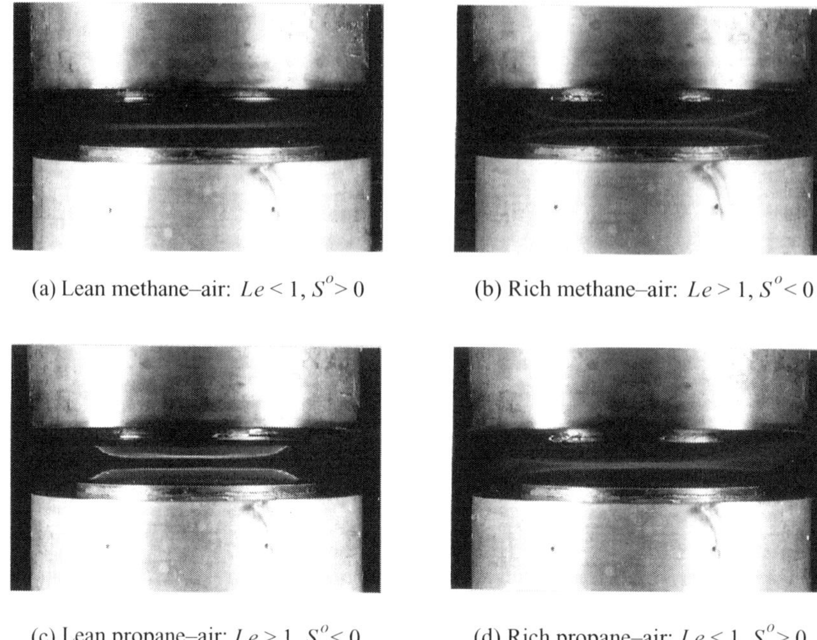

(a) Lean methane–air: $Le < 1, S^o > 0$

(b) Rich methane–air: $Le > 1, S^o < 0$

(c) Lean propane–air: $Le > 1, S^o < 0$

(d) Rich propane–air: $Le < 1, S^o > 0$

Figure 10.5.4. Counterflow ($\kappa > 0$) twin flame images just prior to the state of extinction, demonstrating the influence of Le and, hence, the stretch parameter $S^o \sim (Le^{-1} - 1)\kappa$.

downstream state is stationary for the OPF in the laboratory frame, and as such can be readily determined experimentally, as mentioned in Section 7.6.3. It is seen that since OPF is also positively stretched, the flame responses are qualitatively similar to those for the CFF. However, unlike the CFF whose s_b is affected by both nonequidiffusion and flow divergence, the s_b for the OPF is affected only by nonequidiffusion. Thus the opposite behavior for lean and rich flames here is a clear indication of the influence of nonequidiffusion alone.

Figures 10.5.2c and 10.5.3c show the s_u for the negatively stretched, inwardly propagating flames. Since the upstream state of an IPF is stationary in the laboratory frame, s_u is now a more logical choice than s_b in determining the flame speed. It is seen that, since the nature of the stretch is now reversed as compared to those of the CFF and OPF, the flame response is also reversed, with s_u decreasing and increasing with increasing stretch rate for the hydrogen–air flames, and increasing and decreasing for the propane–air flames.

The variations shown in Figures 10.5.2 and 10.5.3 are all linear, indicating that the flames computed are all weakly stretched and, hence, can be described by the linear results presented earlier.

We next examine the influence of nonlinear stretch on the flame response. Figure 10.5.4 shows the photographic images of the binary flame configuration for lean and rich methane–air and propane–air mixtures at the state just prior to extinction if stretch is further increased by increasing the freestream flow velocities

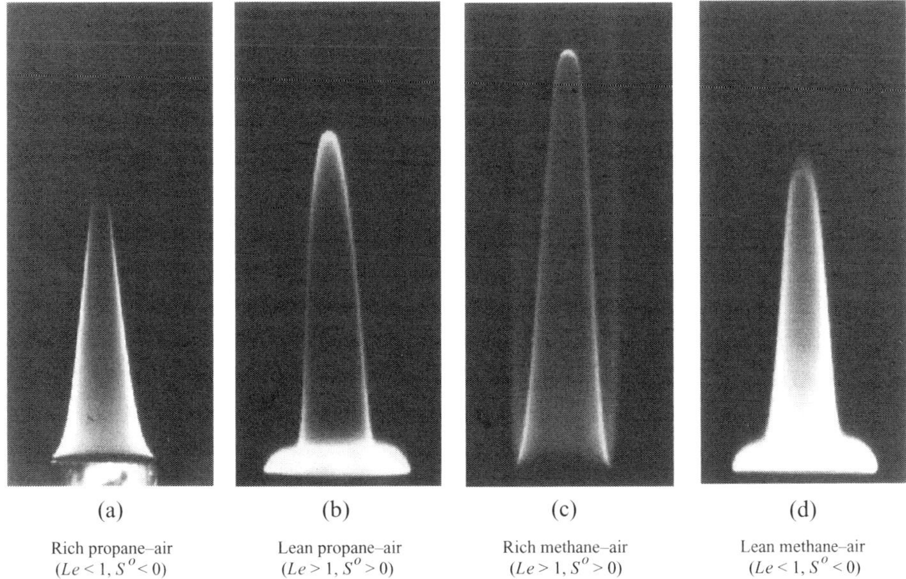

Figure 10.5.5. Images of Bunsen flames ($\kappa < 0$) with open and closed tips, demonstrating the influence of Le and, hence, the stretch parameter $S^o \sim (Le^{-1} - 1)\kappa$.

(Ishizuka & Law 1983; Tsuji & Yamaoka 1983). It is seen that while the lean propane–air and rich methane–air flames are quite separated at extinction, implying that the flames are located away from the stagnation surface, the lean methane–air and rich propane–air flames merge at extinction. These results, which have also been substantiated computationally, agree with our previous discussion that $Le > 1$ flames extinguish when situated away from the stagnation surface, and $Le < 1$ flames extinguish at the stagnation surface due to incomplete reaction. The slight separation for the rich propane–air flames is likely due to the presence of soot and, hence, radiative heat loss downstream of the flames, causing them to extinguish before reaching the stagnation surface.

Next we examine the flame temperature response to negative stretch, provided by the increasing curvature along the surface of a Bunsen cone. Figure 10.5.5 shows the photographic images of the flame configurations of lean and rich propane–air and methane–air mixtures. It is clear that with increasing curvature, and thereby increasing negative stretch along the surface as the flame tip is approached from the flame base, the burning intensity increases for the lean propane–air and rich methane–air mixtures, but decreases for rich propane–air and lean methane–air mixtures. The reduction in the flame temperature can be so severe that extinction occurs at the flame tip, which suffers the largest stretch, hence exhibiting the tip opening phenomenon (Law, Ishizuka & Cho 1982).

To quantify the above observation, Figure 10.5.6 shows the measured maximum temperatures along the flame surface (Mizomoto et al. 1985). Excluding the segment

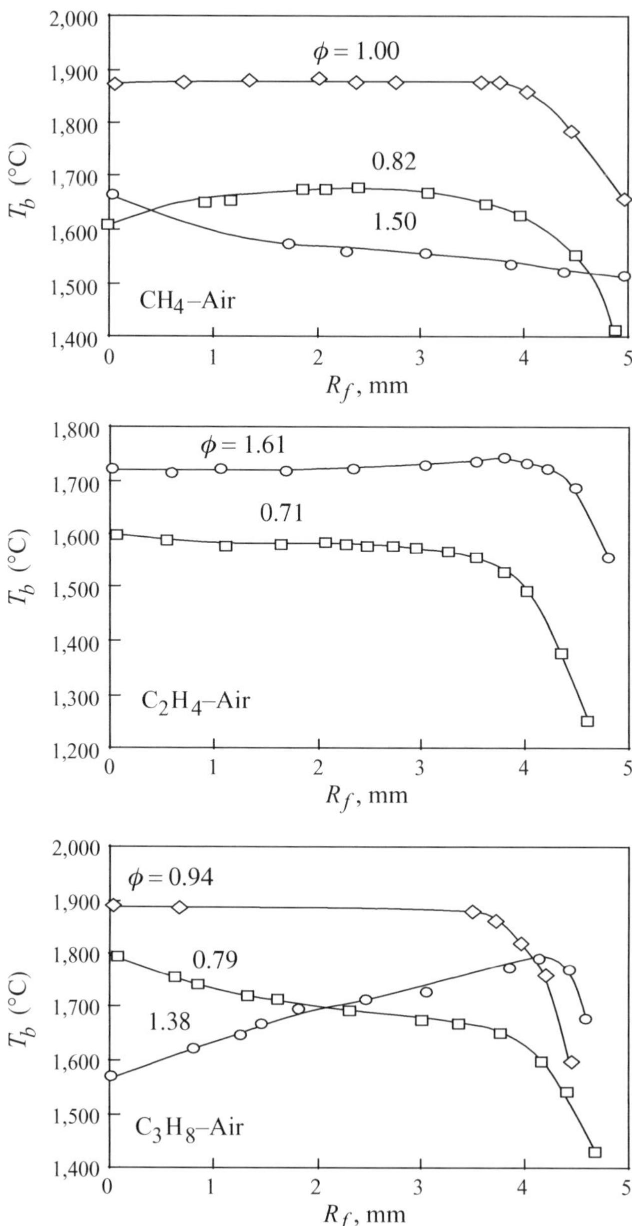

Figure 10.5.6. Measured flame temperature distribution over the flame surface for lean, near-stoichiometric, and rich methane–air, ethylene–air, and propane–air flames.

near the flame base where burning is weak due to heat loss to the burner rim, it is seen that as we move toward the flame tip, the flame temperature increases for the rich methane–air and lean propane–air flames, but decreases for the lean methane–air and rich propane–air flames. The neutral compositions are found to be approximately $\phi = 1.00$ and 0.94 for methane–air and propane–air flames respectively.

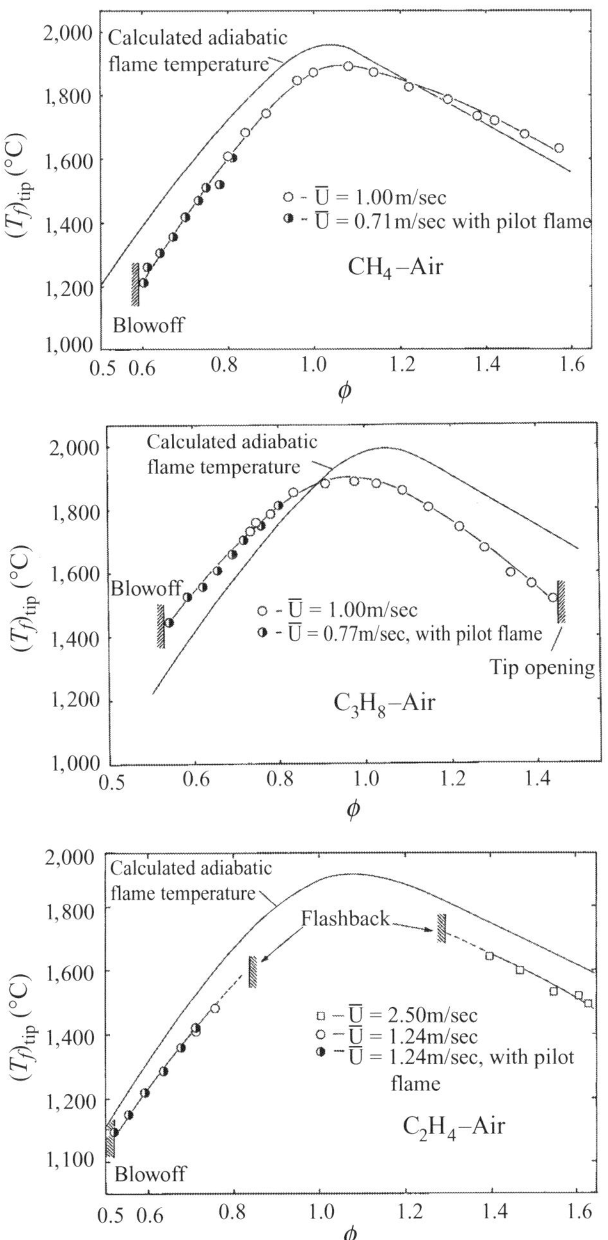

Figure 10.5.7. Comparison of measured flame tip temperatures and calculated adiabatic flame temperatures of methane–air, ethylene–air and propane–air Bunsen flames with various equivalence ratios.

Recognizing that the flame responds in opposite trends for methane–air and propane–air mixtures, and that the molecular weight of ethylene is between those of methane and propane, Figure 10.5.6 shows that for the ethylene–air flame, the flame temperature varies very slightly not only with flame curvature but also for rich and

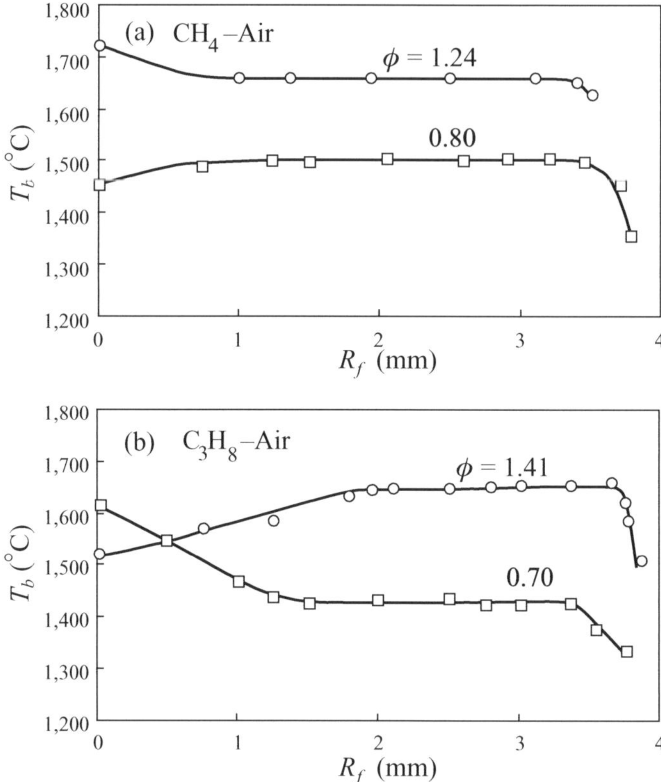

Figure 10.5.8. Temperature profiles of the two-dimensional Bunsen flame surface for methane–air and propane–air flames, showing the absence of stretch effects over the unstretched flame shoulder region.

lean mixtures. This is in agreement with the theoretical result that for an equidiffusive mixture, stretch has no effect on the flame temperature.

The theory not only indicates the increasing/decreasing trends of a given mixture, but it also specifies that such deviations are relative to the adiabatic flame temperature. In Figure 10.5.7 the measured flame temperatures at the tip, $(T_f)_{\text{tip}}$, are compared with the calculated T_{ad}. It is seen that $T_{\text{ad}} > (T_f)_{\text{tip}}$ for lean methane–air and rich propane–air flames and $T_{\text{ad}} < (T_f)_{\text{tip}}$ for rich methane–air and lean propane–air mixtures. The fact that the crossover points do not occur at $\phi \simeq 1$ is due to radiative loss from the thermocouple. If such losses were compensated for in analyzing the data, then the data would be shifted upward and the crossover point would be closer to $\phi = 1$. For ethylene–air flames there is no crossover point; T_{ad} exceeds the uncorrected $(T_f)_{\text{tip}}$ over the entire range of ϕ.

We further note that since the flame temperature can deviate from the adiabatic flame temperature only in the simultaneous presence of stretch and nonequidiffusion, the deviation should also be suppressed for a nonequidiffusive mixture if the flame is not stretched. By using the shoulder region of a two-dimensional Bunsen flame to simulate such an unstretched flame, Figure 10.5.8 shows that the flame temperature is indeed minimally affected for the lean and rich methane–air and propane–air flames (Law et al. 1988).

It is therefore clear that all experimental and computational results are in agreement with the concept of flame stretch in the presence of nonequidiffusion.

10.6. FURTHER IMPLICATIONS OF STRETCHED FLAME PHENOMENA

Except for some specially created laboratory-scale flames, such as those stabilized over flat-flame burners or two-dimensional slot burners, flames are invariably under the influence of stretch. Since stretch coupled with nonequidiffusion can significantly affect the flame behavior, failure to account for these effects can lead to quantitative as well as qualitative errors in the study of fundamental flame phenomena and the modeling of complex combustion processes. Some examples are given in this section.

10.6.1. Determination of Laminar Flame Parameters

With understanding gained herein, it is now clear that practically all the flame configurations adopted in the determination of the laminar burning velocity s_u^o suffer stretch. Notable examples are the widely adopted, negatively stretched Bunsen flame and the positively stretched, outwardly propagating spherical flame. These stretch effects have to be subtracted out when determining s_u^o from stretched flames, as previously discussed in Section 7.6.

The need for an accurate knowledge of s_u^o can be illustrated by the practice of extracting and/or partially validating chemical kinetic schemes and constants by numerically modeling the standard premixed flame and comparing the predicted laminar burning velocities with the measured ones, as briefly mentioned in Chapter 3. Implicit in such a comparison is the assumption that the experimental configurations are truly representative of the chemical system under simulation. However, as we have seen, experimentally determined burning velocities can be contaminated by stretch effects. Furthermore, since stretch can induce opposite effects as a hydrogen–air or hydrocarbon–air mixture changes from fuel lean to fuel rich, lean and rich kinetics of the mixture can also be misinterpreted accordingly.

Results of the previous section show that, for a given stretch rate κ, the global responses of a stretched flame depend only on the four global properties of the corresponding unstretched flame, namely f^o, ℓ_T^o, $Ze^o \sim E_a$, and Le, recognizing that ℓ_T^o is needed in the evaluation of Ka^o and $\tilde{\nabla}$. In Sections 7.6 and 7.7, we discussed the determination of f^o, ℓ_T^o and E_a. We shall now consider the evaluation of Le.

Conventionally, Le is evaluated based on the freestream properties of the mixture, with the mass diffusivity being that of the deficient reactant and the abundant inert. This evaluation therefore embodies two assumptions, namely Le is only a diffusive property of the flame, and it is only applicable to sufficiently off-stoichiometric mixtures. The potential inadequacy of the first assumption can be appreciated by recognizing that since the flame is a diffusive–reactive system, Le should be a global diffusive–reactive property of the flame in the same manner as E_a. Specifically, although E_a is meant to describe the global response of the detailed reaction chemistry, the progress of the individual reactions obviously depends on the availability and concentrations of the intermediates, which in turn depend on the transport aspects of the mixture and the flame structure. By the same token, although Le is superficially a parameter representing the transport of the freestream reactants, there are many intermediates, with different diffusivities, that could affect the entire reaction progress and manifest their effects through some nonequidiffusive phenomena. As such, E_a and Le are fundamentally flame-dependent properties.

The procedure used to extract E_a from f^o, as discussed in Section 7.7.3, satisfies the above consideration. The extraction of Le is however best conducted based on the response of stretched flames, which is sensitive to Le variations (Sun et al. 1999). As an example, consider the CFF of Figure 10.5.2a. Since the burning velocity varies linearly with the stretch rate κ for small κ, we can express the result as

$$s_{b,\mathrm{CFF}} = s_b^o + L_{b,\mathrm{CFF}}\kappa, \tag{10.6.1}$$

where L_b is the slope of the linear variation and is called the Markstein length (based on the downstream burning velocity) of the flame, measuring its response to stretch rate variation. However, $s_{b,\mathrm{CFF}}$ is also given by Eq. (10.4.31) as

$$s_{b,\mathrm{CFF}} \approx s_b^o + \left[\frac{Ze^o}{2}\left(\frac{1}{Le} - 1\right) - 1\right]\left(\frac{\ell_T^o}{\alpha^o}\right)\kappa, \tag{10.6.2}$$

where $\alpha^o = \rho_b^o/\rho_u$ and use has also been made of Eq. (10.4.23). Thus by equating the slopes of Eqs. (10.6.1) and (10.6.2), we have

$$L_{b,\mathrm{CFF}} = \left[\frac{Ze^o}{2}\left(\frac{1}{Le} - 1\right) - 1\right]\left(\frac{\ell_T^o}{\alpha^o}\right), \tag{10.6.3}$$

from which Le can be determined.

Figure 10.6.1 shows the global Le for hydrogen–air and propane–air mixtures, computationally determined in this manner for various flame configurations as a function of normalized equivalence ratio Φ. These values were also found to be only weakly dependent on the flame configuration, and, hence, the upstream or downstream location at which they are evaluated. Additional results show that they are also insensitive to pressure variations.

Then by using f^o, ℓ_T^o, E_a, and the extracted values of Le, determined for the entire range of stoichiometry, the response of stretched flames can be predicted while the various nondimensional parameters, S^o, Ma^o, Ka^o, and σ^o can also be evaluated. This procedure is conceptually complete, with the various aspects of the phenomena well described.

A simpler, though less fundamental approach is to use the Markstein length, L_u or L_b, as the empirical parameter, and express the rest of the stretched flame responses in terms of it. The dependence on Le and E_a are then embedded within L_u or L_b. Evaluation of the stretched flame speed, say s_u, then just depends on s_u^o, ℓ_T^o, and L_u.

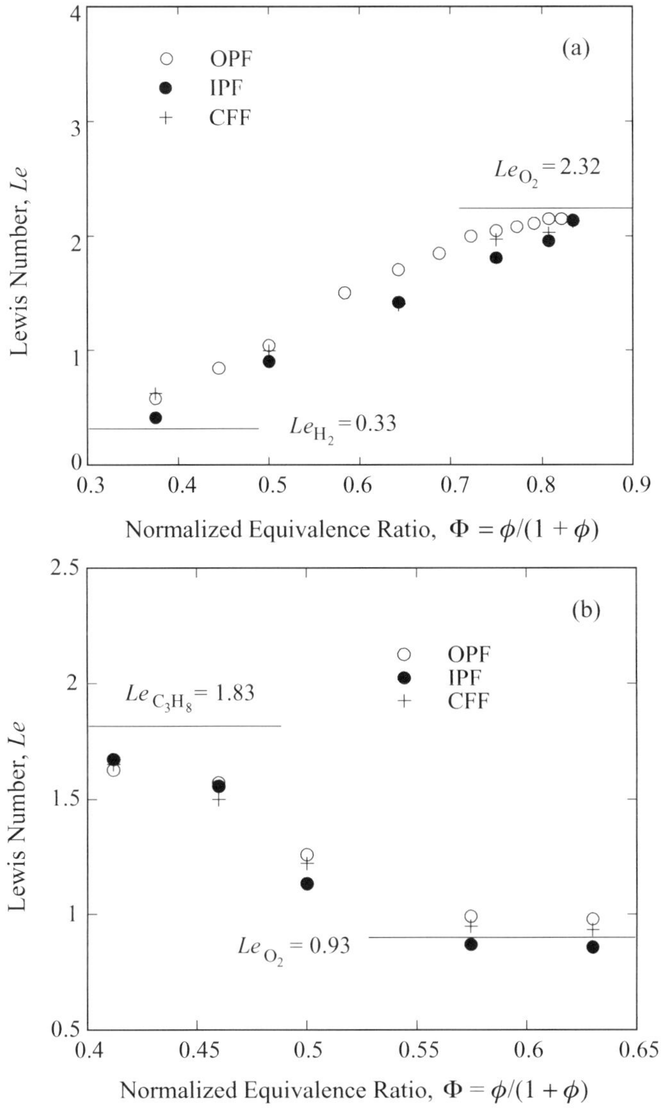

Figure 10.6.1. Extracted *Le* of stretched flames of various flame configurations for: (a) hydrogen–air, and (b) propane–air mixtures.

In terms of utility it is simpler to use the upstream flame speed expression

$$s_u = s_u^o + L_u \kappa \qquad (10.6.4)$$

instead of the downstream expression, Eq. (10.6.1). In evaluating L_u, it is important to recognize that since the flame is treated as being infinitesimally thin in the general formulation, the upstream flame speed s_u should correspond to the local flow velocity obtained by extrapolating the upstream flow velocity from the location of the upstream boundary to that of the thin reaction zone. Failure to do so could result in different values of L_u for different flame configurations.

Further discussion on the dependence of the Markstein length on the stoichiometry of several fuel–air mixtures can be found in Bechtold and Matalon (2001).

10.6.2. Dual Extinction States and Extended Flammability Limits

In Section 8.4, we studied the extinction of the planar flame in the doubly infinite domain due to volumetric heat loss, and through it defined the fundamental flammability limits of fuel–oxidizer systems in terms of its equivalence ratio or the system pressure. In this chapter, we have shown that the burning intensity of a flame can also be modified by stretch. Thus it is reasonable to expect that the flammability limits of a combustible mixture can be either extended or contracted when the flame is subjected to a stretch which tends to modify its burning intensity through the parameter S^o, which can be either greater or smaller than zero. This concept can be demonstrated by combining the effects of volumetric heat loss and stretch modification of the flame temperature, which are respectively given by Eqs. (8.4.19) and (10.4.20), as

$$\tilde{T}_b = \tilde{T}_b^o + \frac{S^o}{\tilde{f}_u^2} - \frac{1}{Ze^o}\frac{\tilde{L}_v}{\tilde{f}_u^2}. \qquad (10.6.5)$$

Using the modified $\tilde{T}_b$, it can be shown that the nonlinear flux relation Eq. (10.4.21) becomes

$$\tilde{f}_u^2 \ln \tilde{f}_u^2 = 2\sigma^o - \tilde{L}_v. \qquad (10.6.6)$$

Recognizing that $\sigma^o \sim (Le^{-1} - 1)\kappa$, it is clear that, for a positively stretched flame, extinction is promoted and the flammability limit narrowed for $Le > 1$, while extinction is delayed and the flammability limit extended for $Le < 1$. The opposite holds for a negatively stretched flame. Specifically, the extinction criterion is given by the state of the turning point of Eq. (10.6.6),

$$\tilde{L}_v - 2\sigma^o \geq e^{-1}. \qquad (10.6.7)$$

The extinction turning point behavior is shown in Figure 10.6.2.

The simultaneous presence of heat loss and stretch can also lead to a rather interesting, dual extinction state phenomenon. Take the positively stretched, counterflow flame as an example (Maruta et al. 1996; Sung & Law 1996). Figure 10.6.3 shows the computed results for the maximum temperature T_f as a function of the strain rate a, for lean methane–air and propane–air flames whose Lewis numbers are respectively smaller and greater than unity. It is seen that extinction occurs for both flames with increasing strain rate, as would be expected. However, with decreasing strain rate, the methane flame exhibits another extinction turning point while T_f monotonically increases for the propane flame. The reason for the difference is that since the burning rate of a freely standing methane flame decreases with decreasing a, its flame becomes thicker and, hence, loses more heat because of the larger volume. This further reduces the flame temperature, and, hence, the flame speed, which in turn increases the flame thickness and consequently leads to even more heat loss;

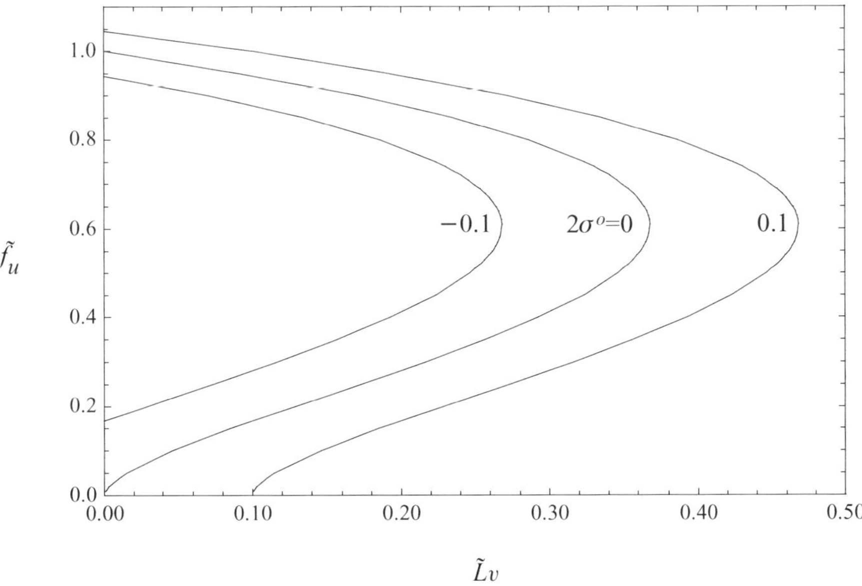

Figure 10.6.2. Extinction turning point response for stretched flames with radiative heat loss.

this feedback process eventually ends in flame extinction. Consequently for such a flame there is an upper, stretch-induced extinction state and a lower, loss-induced extinction state, characterized by their respective strain rates, $a_{E,\text{upper}}$ and $a_{E,\text{lower}}$. The temperature reduction is small for the former but is quite substantial for the latter. Figure 10.6.3a further shows that the flame response is a closed curve, called an isola. Consequently, with continuous reduction of the burning intensity of the flame, say through decreasing the fuel concentration or increasing the heat loss tendency, the extent of the combustible regime shrinks until a critical value is reached beyond which combustion is not possible. This behavior can be represented by plotting the extinction strain rates as a function of, say, the fuel equivalence ratio ϕ, as shown in Figure 10.6.4a. Here we see that with continuous decrease in ϕ, the dual extinction state phenomenon is exhibited, with $a_{E,\text{upper}}$ decreasing with decreasing ϕ, which is characteristic of stretch-induced extinction, and $a_{E,\text{lower}}$ increasing with decreasing ϕ, which is contrary to stretch-induced extinction but is characteristic of loss-induced extinction. We further recognize that since the loss-induced extinction state must degenerate to that of the fundamental flammability limit as the strain rate vanishes, mixtures leaner than that of the fundamental flammability limit can sustain burning, demonstrating the potential of extended flammability through aerodynamic stretching. The flames in this extended flammability regime are very weak, as shown by their propensity to be extinguished by exceedingly small strain rates.

For the $Le > 1$, propane flame, the burning rate increases with decreasing straining. Consequently the flame becomes thinner, heat loss is reduced, and extinction is not possible. Thus the propane flame exhibits only the upper, stretch-induced extinction state, as shown in Figures 10.6.3b and 10.6.4b. Combining the results for the methane

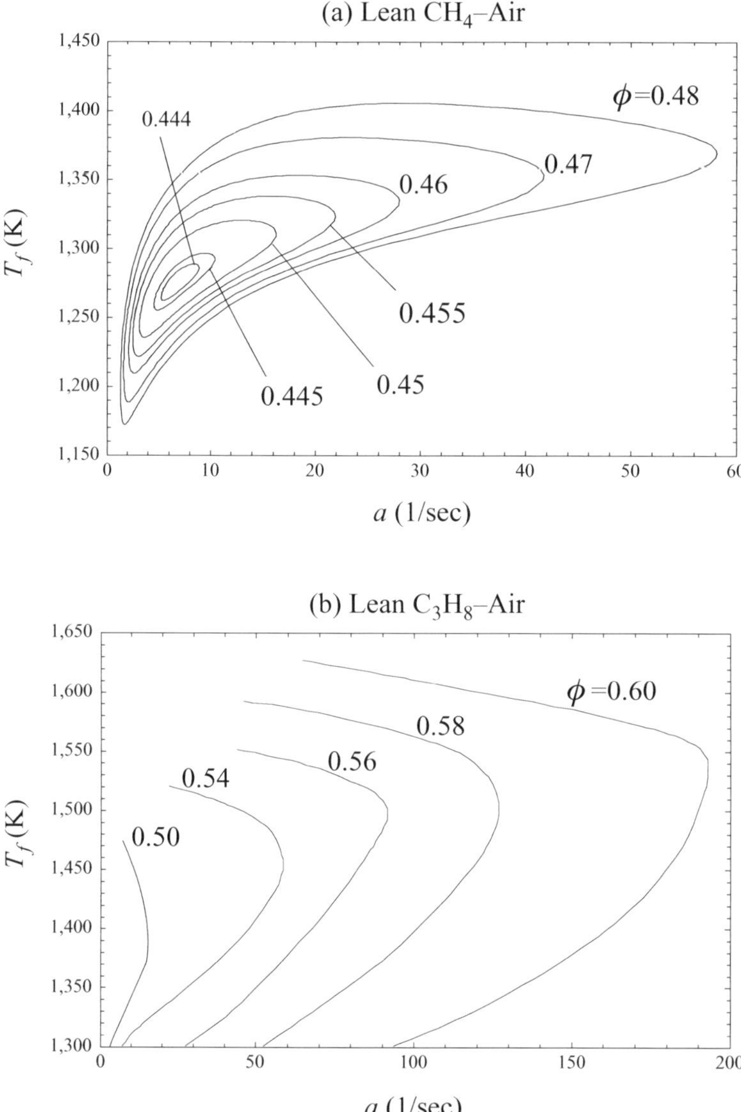

Figure 10.6.3. Calculated maximum temperature for: (a) lean methane–air, and (b) lean propane–air counterflow flames with radiative heat loss, showing the phenomenon of dual turning points for the former.

and propane flames, it can then be stated that dual extinction states and, hence, extension of the fundamental flammability limits are possible for $S^o > 0$ flames. For $S^o < 0$ flames, stretch-induced extinction is the only extinction mode and the flammability limit is contracted in the presence of stretch.

The responses of Figures 10.6.4a and 10.6.4b have been confirmed through microgravity experiments (Maruta et al. 1996) in which counterflows with very small strain rates, uncomplicated by buoyancy, can be produced.

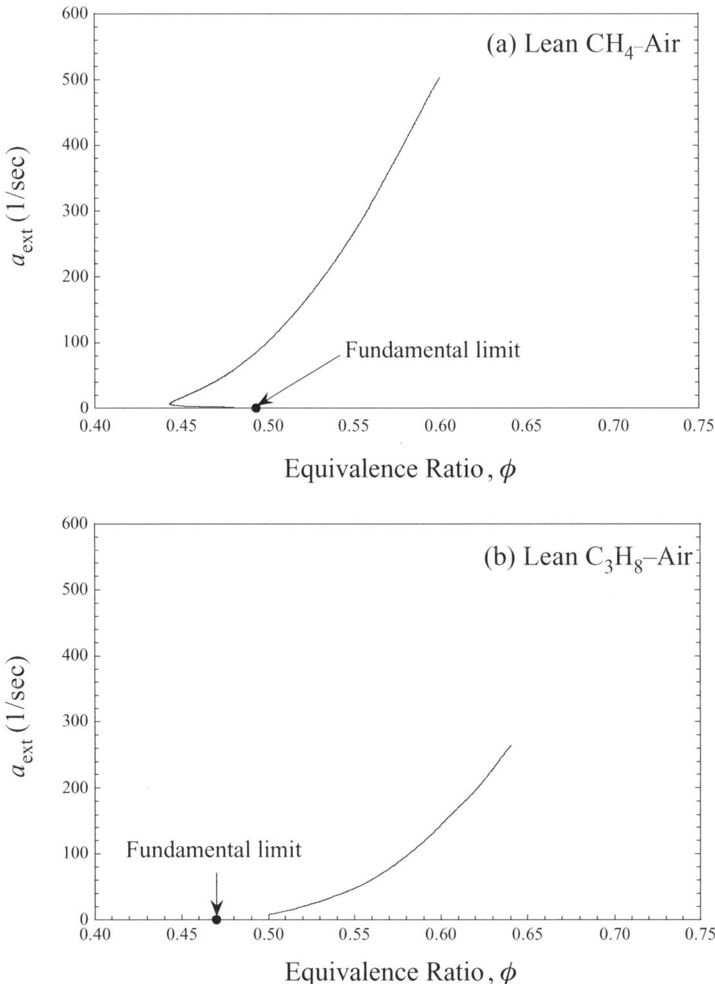

Figure 10.6.4. Calculated extinction strain rates for: (a) lean methane–air, and (b) lean propane–air counterflow flames, with radiative heat loss, showing the extended lean flammability limit for the former.

The extended flammability phenomenon also explains the existence of the so-called "self-extinguishing flames" (Ronney & Wachman 1985; Ronney 1985). Specifically, it has been observed in microgravity experiments that while sublimit, $Le < 1$ mixtures can support the initial outward propagation of a spark-ignited spherical flame, the flame will suddenly extinguish when it reaches a certain size. As recognized in Section 10.3.3, the stretch effect is stronger when the flame radius is smaller for such positively stretched, outwardly propagating flames. Flame propagation beyond the fundamental flammability limit therefore can be initially supported by stretch when the flame is small. The effect, however, diminishes as the flame grows. At the same time, heat loss increases because of the continuously increasing volume of the flame sphere. The enhancing effect of stretch is eventually eliminated when the flame attains a certain size, leading to abrupt extinction. The practical implication of this phenomenon in the firing of spark-ignition engines is obvious.

The dual extinction turning point behavior has also been identified for other flame systems. Of particular interest is the response of nonpremixed flames (Chao, Law & Tien 1991) whose thickness varies with the stretch rate through κ, as shown in Section 6.5. Thus while extinction occurs with increasing κ, it can also be induced with decreasing κ. Interestingly though not unexpectedly, the final structure equation obtained from asymptotic analysis is that of Liñán's near-equilibrium regime and is found to describe both turning points, indicating that extinction is ultimately caused by reactant leakage.

The phenomenon of dual extinction turning points is expected to be fairly prevalent, with the loss-induced extinction being particularly relevant for weakly strained or weakly burning flames of extended dimension, typically found in limit situations.

10.6.3. Other Phenomena

10.6.3.1. Concentration and Temperature Modifications in Flame Chemistry: The study of fundamental chemical kinetics is sometimes conducted by employing a particular flame as a "chemical reactor" fed by the freestream mixture, which is assumed to react either at the adiabatic flame temperature or in a temperature environment prescribed by, say, the one-dimensional laminar flame. Recognizing that some of these flames suffer stretch, which can cause modifications in both the mixture composition as well as the flame temperature, failure to account for these modifications can lead to inaccuracies in the deduced chemical information. In fact, the need for an accurate determination of s_u^o, as just discussed, stems from such a concern. Another example of the concentration modification is soot formation in flames, as evidenced by the formation of soot streak from the tip of the Bunsen flame. Studies of polyhedral flames, to be discussed later, have also shown substantial composition variations from the crests to the troughs. Because of the nonlinear nature of chemical kinetics, these effects obviously do not average out. A study on the counterflow ignition of a hydrogen–air mixture by a heated inert jet (Zheng et al. 2002) has shown that, because of the large mobility of hydrogen, the concentration of an ultra-lean hydrogen–air jet can be modified to be very rich as it reaches the hot ignition boundary (Figure 10.6.5). This causes the system to exhibit the counterintuitive result that ignition becomes more difficult with increasing hydrogen concentration in the freestream.

10.6.3.2. Flame Stabilization and Blowoff: It was discussed in Section 8.6.2 that the mechanism through which a Bunsen flame is stabilized at, and subsequently blown off from, the burner rim is heat loss from the flame base to the burner rim. This heat loss allows the flame base to have the flexibility to adjust its local flame speed so that a dynamic balance with the local flow velocity can be continuously maintained.

Recognizing now that in addition to heat loss, the flame speed can also be modified by stretch as well as pure curvature effects, the possible mechanisms which can lead to flame stabilization and blowoff are significantly enriched. Consider, for example, the

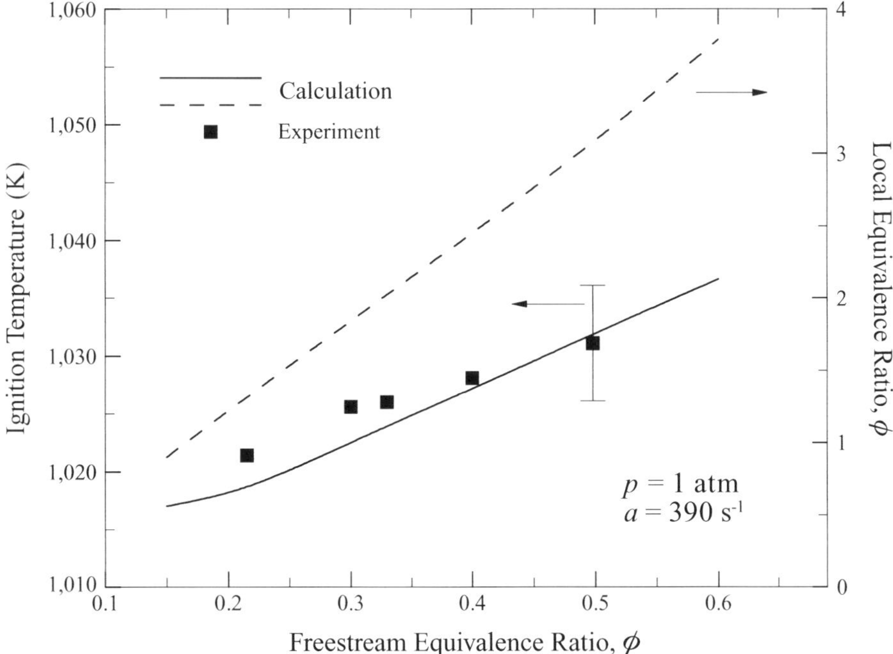

Figure 10.6.5. Variation of the fuel equivalence ratio at the ignition kernel and the ignition temperature of a lean hydrogen–air jet counterflowing against a heated inert jet, showing the enrichment of the hydrogen–air jet due to the preferential diffusion of hydrogen.

possible modification of the flame speed at the flame base of an equidiffusive mixture through pure curvature effects. Then, since the flame base is convex toward the approach flow, Eq. (10.4.39) shows that the flame speed is reduced from s_u^o. Therefore, this convex curvature could assume the same role as heat loss in maintaining flame stabilization through its continuous adjustment. By the same reasoning, the flame can also be blown off in the absence of heat loss.

The possibility of adiabatic stabilization and blowoff has been experimentally investigated by measuring the temperature and heat flux at the burner rim (Sun, Sung & Law 1994). Figure 10.6.6 shows the measured rim temperature and the standoff distance of the flame base for different average flow velocities issuing from the Bunsen tube. It is seen that, with increasing flow velocity, the flame base is continuously lifted higher until blowoff, which is indicated by the last datum. The temperature at the rim initially decreases but then attains a constant value until blowoff occurs. Furthermore, this constant value is 320 K, which is only slightly higher than the room temperature. Additional interferometric studies have shown that there is practically no temperature gradient near the burner rim. These results then substantiate the possibility that flames can be adiabatically stabilized and blown off. In general, the effects of stretch and pure curvature should be taken into consideration when considering flame stabilization and blowoff processes.

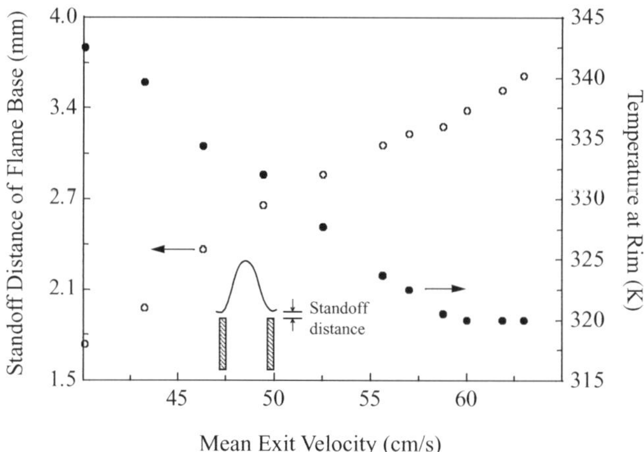

Figure 10.6.6. Variation of the flame base standoff distance and rim temperature with mean exit velocity for a methane–air premixed flame, demonstrating the possibility of adiabatic flame stabilization.

10.6.3.3. Flamefront Instability and Modeling of Turbulent Flames: When a smooth flamefront is subjected to spatial or temporal disturbances, the resulting unsteadiness or wrinkles can either grow or decay. This is a stretched flame phenomenon because the susceptibility to instability depends on the flame response to the development of the flame unsteadiness or curvature over its surface. We shall study the flamefront instability phenomena in Section 10.9.

Stretch is also of relevance to the modeling of turbulent flames. In a turbulent flow the fluctuating nonuniform local velocities impose aerodynamic stretching on the highly convoluted flame surface. The results of stretched laminar flames can be applied to the structure of turbulent premixed flames when the laminar flame thickness is smaller than the smallest turbulent eddy sizes. The subject of turbulent combustion will be studied in the next chapter.

10.7. SIMULTANEOUS CONSIDERATION OF HYDRODYNAMIC AND FLAME STRETCH

Our discussion has demonstrated that while the hydrodynamic stretch affects the geometry of the flame surface, the flame stretch affects the flame structure. These effects, while manifested at different scales, are intimately coupled. One approach through which such a coupling can be effected is the use of the G-equation, Eq. (10.2.5), with $\tilde{s}_u$ given by results of Section 10.4. In the following we shall study two problems that clearly show the coupling and influence of these two stretch effects, namely the smoothing of flame corners identified in Section 10.2.2, and the possible configurations of Bunsen flames (Sung, Yu & Law 1994).

10.7.1. Curvature-Induced Corner Broadening

In Section 10.2.2 we have shown the propensity of corner formation for wrinkled flame surfaces. The flame itself, however, possesses an intrinsic response that tends

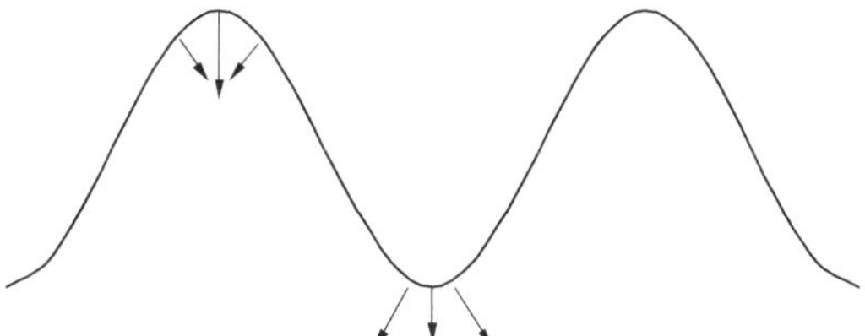

Figure 10.7.1. Schematic showing the smoothing effect of curvature on wrinkled flames.

to inhibit corner formation. As shown in Section 10.4.2, the upstream flame speed of an equidiffusive mixture in the presence of curvature is modified from $\tilde{f}_u = 1$ for the one-dimensional flame to

$$\tilde{s}_u = 1 - \ell_T^o \nabla \cdot \mathbf{n}. \tag{10.7.1}$$

Using this curvature-affected flame speed expression in the G-equation, Eq. (10.2.5), with $\mathbf{v} = 0$ for the quiescent flow example studied in Section 10.2.2, Eq. (10.2.12) is modified to

$$\frac{\partial \hat{g}}{\partial \hat{t}} + \frac{\hat{g}}{(1+\hat{g}^2)^{1/2}} \frac{\partial \hat{g}}{\partial \hat{x}} = \frac{\partial}{\partial \hat{x}} \left[\frac{\hat{\ell}_T^o}{(1+\hat{g}^2)} \frac{\partial \hat{g}}{\partial \hat{x}} \right], \tag{10.7.2}$$

which governs the evolution of the flame surface. Comparing Eq. (10.7.2) with Eq. (10.2.12), we see that the additional, second-order term assumes the role of diffusional broadening that tends to smoothen the corner. This "diffusion" term has a corresponding "diffusion coefficient," given by $\hat{\ell}_T^o/(1+\hat{g}^2)$. Physically, as shown in Eq. (10.7.1), since the negative flame curvature associated with the receding, trough region of the flame (Figure 10.7.1) enhances the burning velocity ($\tilde{s}_u > 1$), while the positive curvature in the protruding, crest region tends to reduce it ($\tilde{s}_u < 1$), the aggravating tendency for the flame segment in the trough region to collide is moderated.

For weakly wrinkled flames ($\hat{g} \ll 1$), Eq. (10.7.2) simplifies to

$$\frac{\partial \hat{g}}{\partial \hat{t}} + \hat{g} \frac{\partial \hat{g}}{\partial \hat{x}} = \hat{\ell}_T^o \frac{\partial^2 \hat{g}}{\partial \hat{x}^2}, \tag{10.7.3}$$

which is Burger's equation. Analytical solutions are available for this well-known equation.

For nonequidiffusive mixtures, the flame speed will be further affected by the stretch term σ^o, which can be either positive or negative depending on the nature of stretch and nonequidiffusion. Thus following similar reasoning for the burning intensity of Bunsen flames (Figure 10.3.1b), for the present wrinkled flame we expect that the tendency to form sharp segments is respectively moderated and aggravated for $Le > 1$ and < 1 mixtures. Furthermore, when the burning intensity in the trough region is reduced by flame stretch, local extinction may also occur, leading to the formation of "holes" over the flame surface. We shall now demonstrate this phenomenon by considering the opening of the Bunsen flame tip.

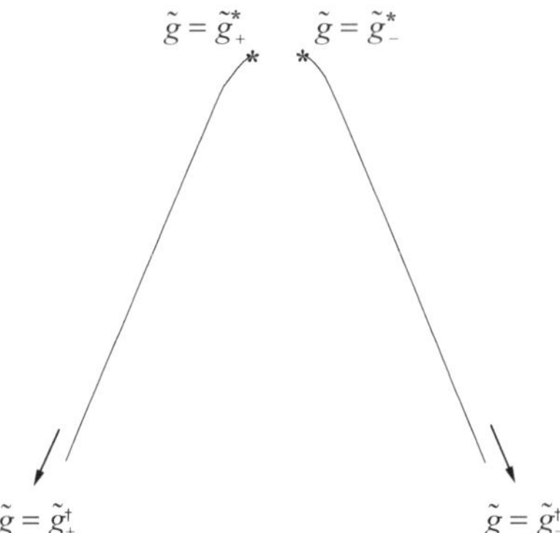

Figure 10.7.2. Analytical description of the configuration of an open-tipped Bunsen flame.

10.7.2. Inversion and Tip Opening of Bunsen Flames

We consider the steady-state configuration of a two-dimensional Bunsen flame situated in a uniform flow of velocity $\mathbf{v} = (0, v)$. With $\mathbf{V}_f = 0$, space variables nondimensionalized by ℓ_T^o, $\tilde{G}(\tilde{x}, \tilde{y}) = \tilde{y} - \tilde{f}(\tilde{x})$, and using the stretch-affected flame speed given by Eq. (10.4.41), the G-equation becomes

$$\frac{d\tilde{g}}{d\tilde{x}} = \frac{(1+\tilde{g}^2)^{3/2}[(1+\tilde{g}^2)^{1/2} - \tilde{v}]}{(1+\tilde{g}^2)^{1/2} - (Ma^o)\tilde{v}}, \tag{10.7.4}$$

which describes the flame configuration for given flow velocity and Markstein number, Ma^o. For this problem it is more convenient to reference quantities to the flame scale instead of the hydrodynamic scale because the hydrodynamic scale here is the curvature of the flame, which however is a response of the analysis. Furthermore, the present reference facilitates analysis of the flame structure and extinction.

Although Eq. (10.7.4) can be integrated to yield an analytical solution, the characteristic of the flame configuration can be more clearly illuminated by using the critical-point analysis. Thus setting the numerator and denominator of Eq. (10.7.4) to zero, we obtain the critical points that respectively correspond to the states at which the slopes of the flame are constant or become discontinuous,

$$\tilde{g}^\dagger_\pm = \pm(\tilde{v}^2 - 1)^{1/2}, \tag{10.7.5}$$

$$\tilde{g}^*_\pm = \pm[(Ma^o\tilde{v})^2 - 1]^{1/2}. \tag{10.7.6}$$

The particular value of $\tilde{g}_\pm = \tilde{g}^\dagger_\pm$ corresponds to the Landau limit, which describes the constant slope of the flame shoulder as shown in Figure 10.7.2 for an open tipped Bunsen flame. It is also clear that $\tilde{g}^\dagger_- < \tilde{g} < \tilde{g}^\dagger_+$. Further setting $\tilde{g}^*_\pm = 0$ and $\tilde{g}^*_\pm = \tilde{g}^\dagger_\pm$ respectively yields the following two critical Markstein numbers:

$$Ma_1^o = \frac{1}{\tilde{v}} \quad \text{and} \quad Ma_2^o = 1, \tag{10.7.7}$$

where $0 < Ma_1^o < Ma_2^o = 1$.

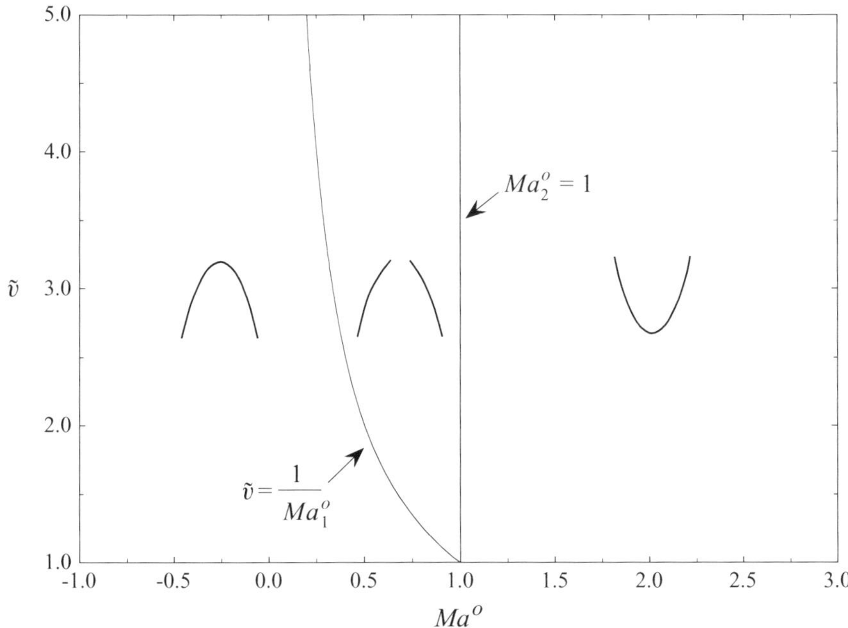

Figure 10.7.3. Regimes with different Bunsen-type flame configurations as functions of flow velocity and Markstein number.

Based on the above critical Markstein numbers, the following observations can be made regarding the possible configurations of Bunsen flames. First,

$$\frac{d\tilde{g}}{d\tilde{x}} > 0 \quad \text{for} \quad Ma^o > Ma_2^o = 1 \tag{10.7.8a}$$

$$\frac{d\tilde{g}}{d\tilde{x}} < 0 \quad \text{for} \quad Ma^o < Ma_1^o. \tag{10.7.8b}$$

Recognizing that $d\tilde{g}/d\tilde{x}$ is the second derivative of the flame shape function, (10.7.8a) and (10.7.8b) then respectively define regimes for the existence of inverted and normal Bunsen flames, as shown in Figure 10.7.3. The prediction of inverted Bunsen flames is of particular interest because they have not been experimentally observed. This is not surprising because special stabilization mechanisms are probably needed to hold such freely propagating flames at the burner rim.

In the regime $Ma_1^o < Ma^o < 1$, the behavior is more complex:

$$\begin{aligned} \frac{d\tilde{g}}{d\tilde{x}} &< 0 \quad \text{when} \quad \tilde{g}_-^\dagger < \tilde{g} < \tilde{g}_-^* \\ \frac{d\tilde{g}}{d\tilde{x}} &> 0 \quad \text{when} \quad \tilde{g}_-^* < \tilde{g} < \tilde{g}_+^* \\ \frac{d\tilde{g}}{d\tilde{x}} &< 0 \quad \text{when} \quad \tilde{g}_+^* < \tilde{g} < \tilde{g}_+^\dagger. \end{aligned} \tag{10.7.9}$$

In this intermediate regime of the Markstein number, the differential equation is singular at $\tilde{g} = \tilde{g}^*_{\pm}$ where $|\tilde{g}^*_{\pm}| < |\tilde{g}^{\dagger}_{\pm}|$. Consequently, an integration starting from $\tilde{g} = \tilde{g}^{\dagger}_{\pm}$ at the shoulder will stop at $\tilde{g}^*_{\pm}$. The flame takes the shape of an open tip Bunsen flame, which is concave toward the unburned mixture (Figures 10.7.2 and 10.7.3). Moreover, since increasing either Ma^o or $\tilde{v}$ causes $|\tilde{g}^*_{\pm}|$ to increase, the opening becomes wider with increasing flow velocity and decreasing mixture diffusivity. The above characteristics of tip opening are in agreement with experimental observations.

10.8. UNSTEADY DYNAMICS

The stretch effects considered so far are quasi-steady in nature in that the flame structure and location can respond almost instantaneously to any change in the externally imposed stretch rate. Conceptually this is expected to be the situation when the characteristic time associated with the flow unsteadiness is much longer than the flame time. When the flow time is much shorter, however, a lag in response is expected and the instantaneous flame structure would correspond to the stretch rate at an earlier time. In the particular case of an oscillating stretch rate, the responsiveness of the flame can be substantially weakened by the periodic reversal of the perturbation stretch rate, to be demonstrated in the following.

We first study the response of nonpremixed flames (Sung & Law 2000). Figure 10.8.1a shows the computed temporal variations of the FWHM flame thickness of a basically equidiffusive methane–air counterflow flame in response to symmetrical oscillations from both the fuel and oxidizer streams, for different frequencies. The flow has a steady strain rate of 134/s and 20 percent oscillation amplitude, A, and the flame has a steady-state extinction strain rate of 172/s. It is seen that the amplitude of oscillation of the flame thickness steadily decreases with the increasing oscillation frequency, indicating the weakening in the flame responsivity. Since the characteristic (inverse) flow time here is of the order of the steady strain rate, $O(100/\text{s})$, we expect that this inertia effect should become important for oscillation frequencies that are $O(100/\text{s})$ and higher. To demonstrate this concept, Figure 10.8.1b shows the scaled flame thickness, FWHM$\sqrt{a}$, for the results of Figure 10.8.1a, with a being the instantaneous, local strain rate evaluated on the oxidizer side of the flame. Recalling that this quantity is basically a constant for the steady-state flames, as shown in Figure 6.5.5, it is seen that it indeed remains fairly constant for $f = 10$ Hz, demonstrating the quasi-steady nature of the response for low-frequency oscillations. For high-frequency oscillations, the gradual loss of response of the flame structure then causes the quantity FWHM$\sqrt{a}$ to behave as $\sqrt{a}$, hence the observed oscillation. These results therefore show that the nonpremixed flame structure scales with $\sqrt{a}$ for low-frequency oscillations, but is minimally affected for high-frequency oscillations.

Figure 10.8.2 shows the temporal variation of the maximum flame temperature for the same flame except the oscillation amplitude is increased to 60 percent such that the maximum strain rate of 218/s exceeds the steady-state extinction strain rate. It

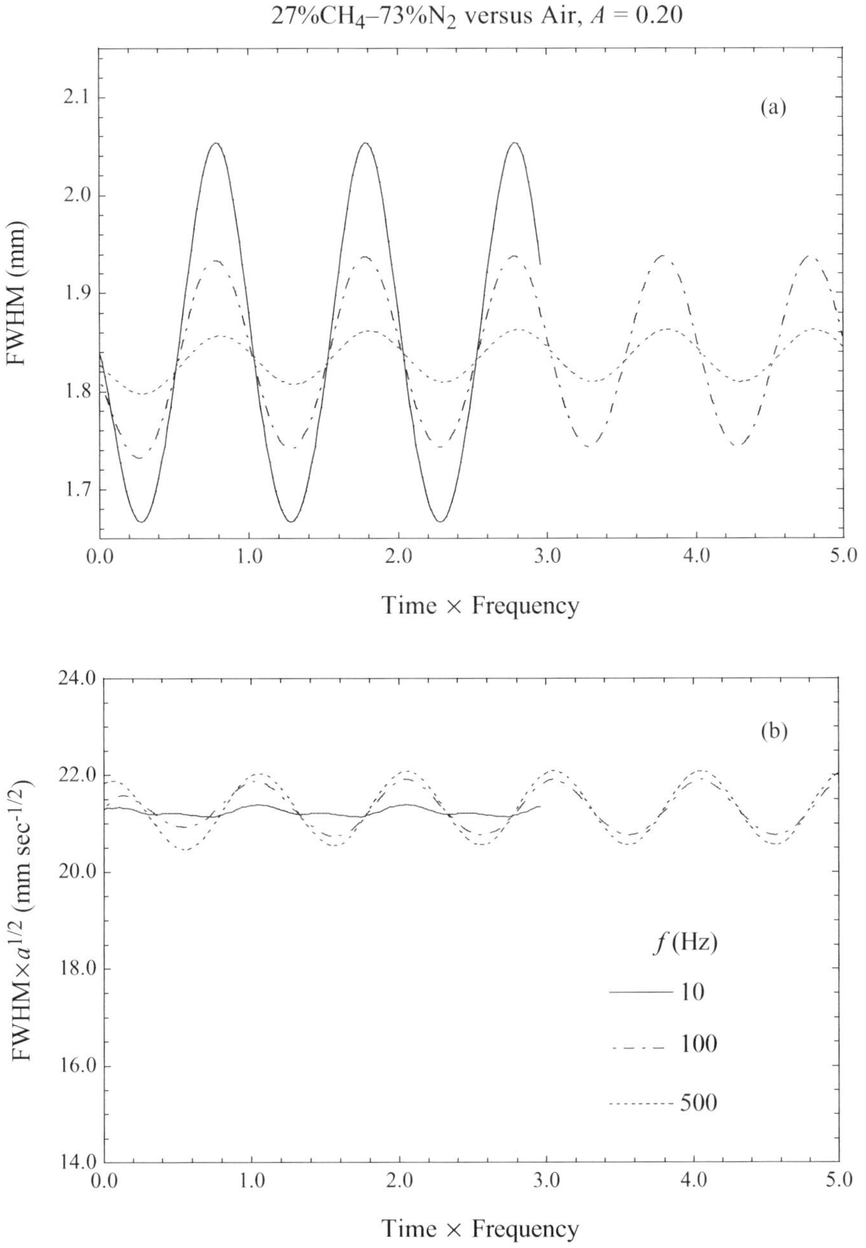

Figure 10.8.1. Variations of: (a) the full-width-at-half-maximum (FWHM) thickness, and (b) scaled FWHM of a methane-versus-air nonpremixed flame when subjected to flow oscillations of different frequencies.

is seen that, for the relatively low frequency of 50 Hz, extinction occurs within the first cycle. However, extinction is delayed with increasing frequency (500 Hz) and is totally inhibited for the high-frequency situation of 1,000 Hz. Thus high-frequency oscillation not only delays but can also inhibit flame extinction. The physical reasoning being that before the flame can respond to the high strain rate phase of the oscillation

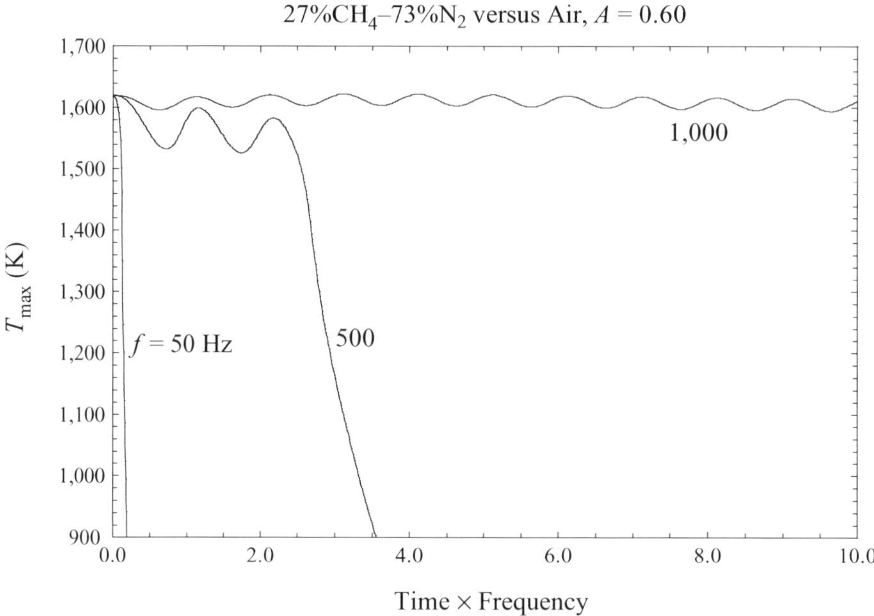

Figure 10.8.2. Variation of the flame temperature of a methane-versus-air nonpremixed flame when subjected to flow oscillations of different frequencies, demonstrating delayed extinction with increasing frequency.

that induces extinction, the flow has already reversed to the low strain rate phase that favors burning.

We now examine the response of premixed flames (Egolfopoulos & Campbell 1996; Sung & Law 2000). Figure 10.8.3a shows the temporal variation of the location of a near-equidiffusive, counterflow, methane–air flame with $\phi = 1$, when it is subjected to an oscillation amplitude that is approximately half of the steady strain rate. Since a premixed flame can freely adjust its location in response to changes in the flow so as to achieve dynamic balance, provided there is enough time to achieve the relocation, Figure 10.8.3a shows that for low frequencies the flame indeed translates readily and, hence, exhibits large movements. However, for high-frequency oscillations, the flame does not have enough time to adjust to changes in the flow field and its movement is considerably diminished.

Figure 10.8.3b shows the corresponding variation of the flame thickness in terms of the FWHM value. It is seen that the flame thickness is not sensitive to the oscillation frequency for all frequencies. This interesting behavior is due to the fact that, for a near-equidiffusive mixture, the flame thickness is basically invariant to strain rate variations in the steady, and hence low-frequency, limit, as discussed earlier, while it is also insensitive to high-frequency oscillations because of the reduced response time.

The effects of nonequidiffusion are illustrated in Figure 10.8.4 for two mixtures whose Lewis numbers are respectively larger and smaller than unity. It is seen that while the maximum flame temperature and heat release rate are out of phase with

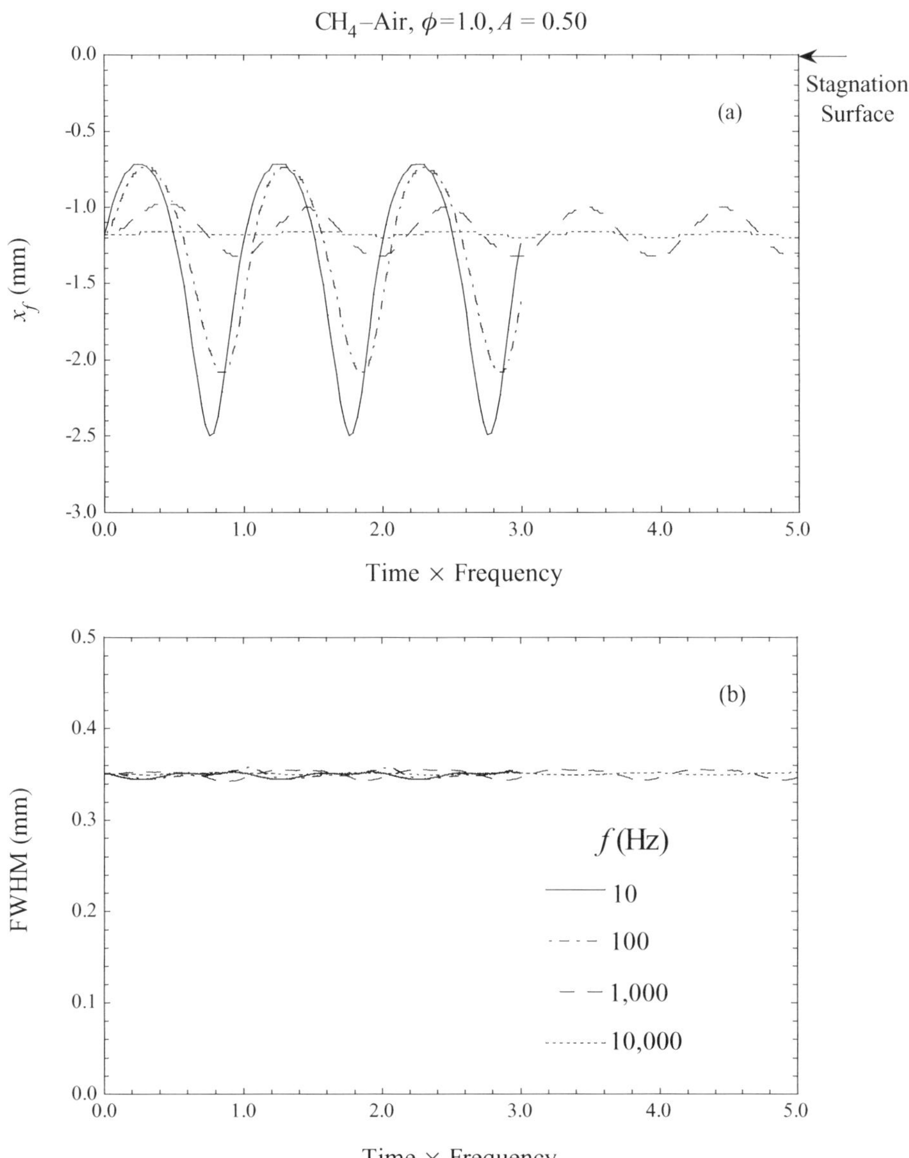

Figure 10.8.3. Variations of: (a) flame location, and (b) flame thickness of a stoichiometric methane–air premixed flame when subjected to flow oscillations of different frequencies, demonstrating the insensitivity of the flame structure to oscillation frequencies.

the imposed oscillatory strain for the $Le > 1$ flame, the behavior is reversed for the $Le < 1$ flame. This result is in agreement with the understanding from the steady-state response that shows that, for mixtures with Le greater or smaller than unity, the burning intensity respectively decreases and increases with increasing strain rate. The practical implication of this result in combustion instability within combustion chambers can be quite significant.

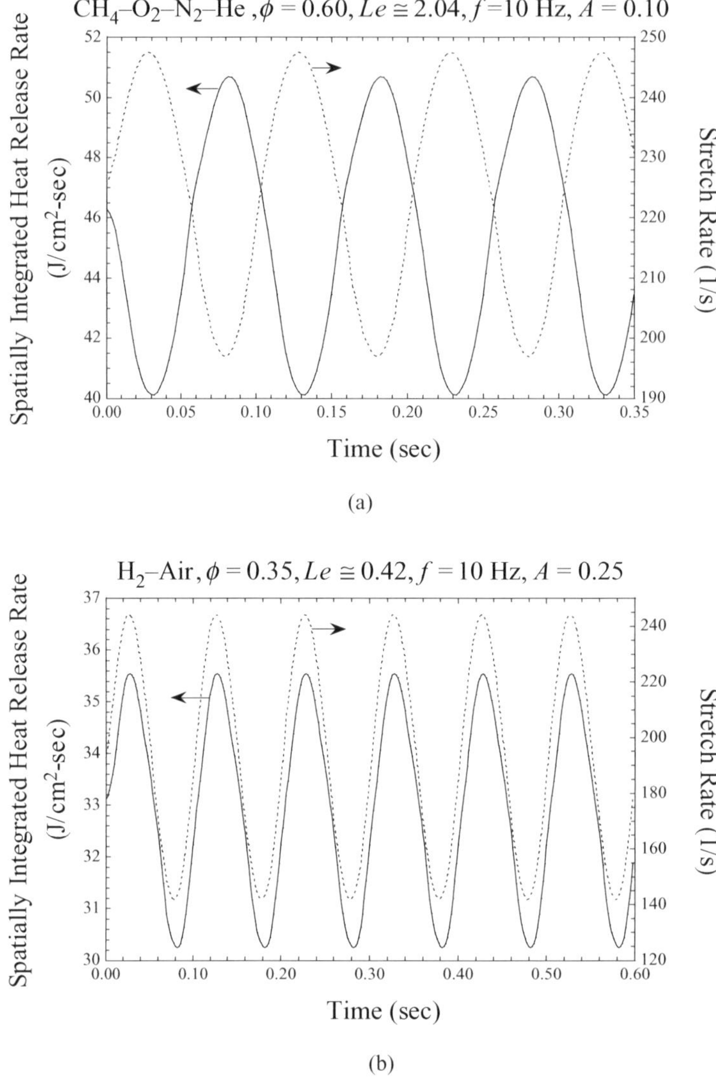

Figure 10.8.4. Phase relations between stretch rate, maximum flame temperature, and integrated heat release rate for oscillated premixed flames with: (a) $Le > 1$, and (b) $Le < 1$.

Finally, similar to the result of Figure 10.8.2 for diffusion flame extinction, high-frequency oscillation has also been found to inhibit extinction of premixed flames, as is reasonable to expect.

10.9. FLAMEFRONT INSTABILITIES

10.9.1. Mechanisms of Cellular Instabilities

Perhaps one of the most beautiful and fascinating phenomena in flame dynamics is the presence of instabilities in the forms of cells and ridges of characteristic sizes over the flame surface (Smith & Pickering 1928; Markstein 1964). These nonplanar flame

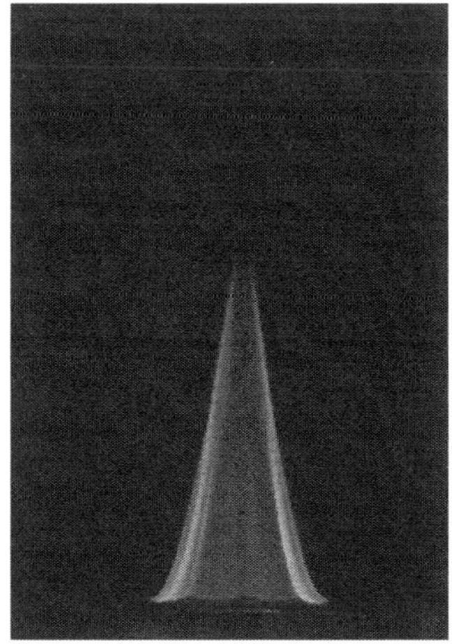

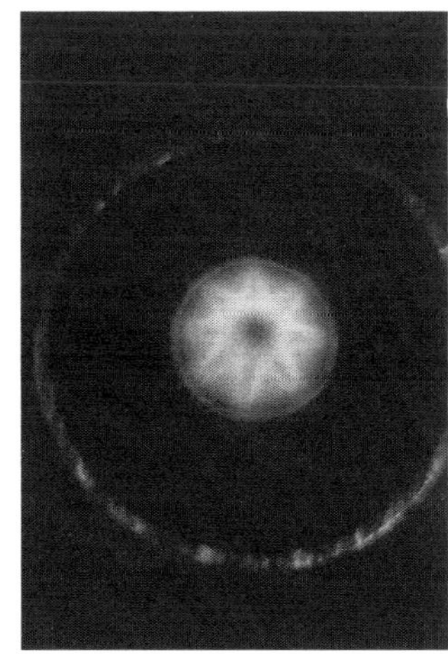

Front View Top View

Figure 10.9.1. Front and planar photographic images of a polyhedral rich propane–air flame.

patterns, which represent alternating regions of intensified and weakened burning, can be either stationary or nonstationary; in the latter case they can also be either steadily or chaotically evolving.

The earliest observation of flamefront instability is that of the polyhedral flame, manifested by the presence of regularly spaced ridges over a Bunsen flame in which the deficient reactant is also the lighter one. Figure 10.9.1 shows the frontal and top views of such a flame, observed for a rich propane–air flame. The flame pattern consists of petals of strongly burning flame surfaces separated by weakly burning regions of ridges, with the petals being convex toward the unburned mixture. For a given tube diameter, the number of ridges has been found to vary with the mixture concentration and the flow velocity (Sohrab & Law 1985). Furthermore, the polyhedral pattern can spin about its vertical axis, with a spinning rate that sometimes can even exceed the laminar flame speed of the mixture. The spinning, however, does not appear to have any preference for either clockwise or counterclockwise direction. One may also note that burning at the tip is particularly weak for the present $Le < 1$, negatively stretched flame in accordance with our previous discussion.

There are three modes of intrinsic cellular instability, namely diffusional-thermal instability (Markstein 1964; Sivashinsky 1977), hydrodynamic instability (Darrieus 1938; Landau 1945), and buoyancy-driven instability. We shall now discuss the mechanisms and characteristics of these instabilities.

The flame images of Figure 10.9.1 clearly demonstrate the importance of nonequidiffusion on the generation of the flame wrinkles. Indeed, if we perturb an initially planar flame into one consisting of alternating convex and concave segments toward the unburned mixture (Figure 10.9.2), then the subsequent evolution of these flame segments can be considered in the same manner as that for the intensification or weakening of the Bunsen flame tip (Figure 10.3.1b). Specifically, for a $Le > 1$ flame,

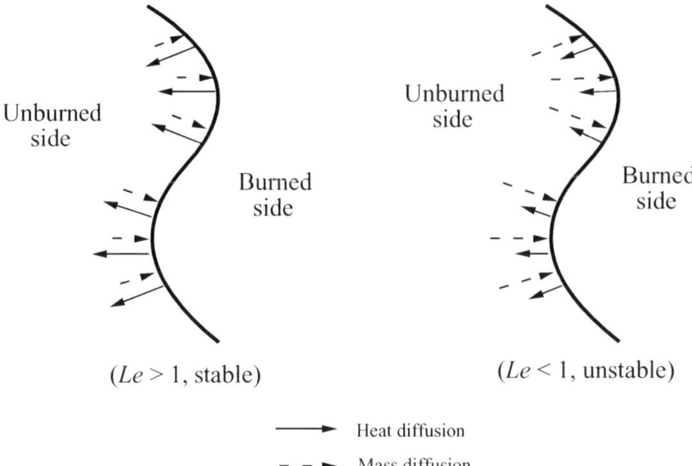

Figure 10.9.2. Schematic showing the mechanism of diffusional-thermal cellular instability.

the burning is intensified at the concave segment and weakened at the convex segment, leading to smoothing of the wrinkles. Consequently such a flame is cellularly stable. Conversely, by the same reasoning a $Le < 1$ flame is cellularly unstable. This phenomenon can of course also be interpreted on the basis of the different diffusivities of the deficient and abundant species, leading to the conclusion that a flame is diffusively unstable if it is deficient in the more mobile reactant. Since the instability is caused by the active modification of the diffusional structure of the flame, the cell size is expected to be of the order of the flame thickness. We shall refer to this mode of instability as nonequidiffusional instability.

In addition to nonequidiffusion, we have shown that the upstream flame speed can also be modified by pure curvature effects. Thus if we again consider Figure 10.9.2, but for an equidiffusive mixture, it is then apparent that since the flame speed is reduced for the convex segment and increased for the concave segment, as schematically shown in Figure 10.7.1, curvature tends to stabilize the flame. This is expected to shift the stability boundary based on nonequidiffusion considerations away from $Le = 1$ to a smaller value of Le. We shall refer to the combined nonequidiffusive and pure curvature instabilities as diffusional-thermal instability.

Hydrodynamic instability, also known as the Landau–Darrieus instability, is caused by the density jump across the flame. The mechanism (Figure 10.9.3) considers the flame being infinitely thin, separating the upstream region of constant density ρ_u from the downstream region of constant density ρ_b. The flame surface propagates in the Landau mode, with a constant flame speed that can be taken to be the laminar flame speed, s_u^o. By again perturbing the flame surface, and recognizing that the areas of the streamtube should remain the same both far upstream and downstream of the flame because of the lack of disturbance there, that because the normal component of the downstream flow velocity at the flame surface is larger than that of the upstream

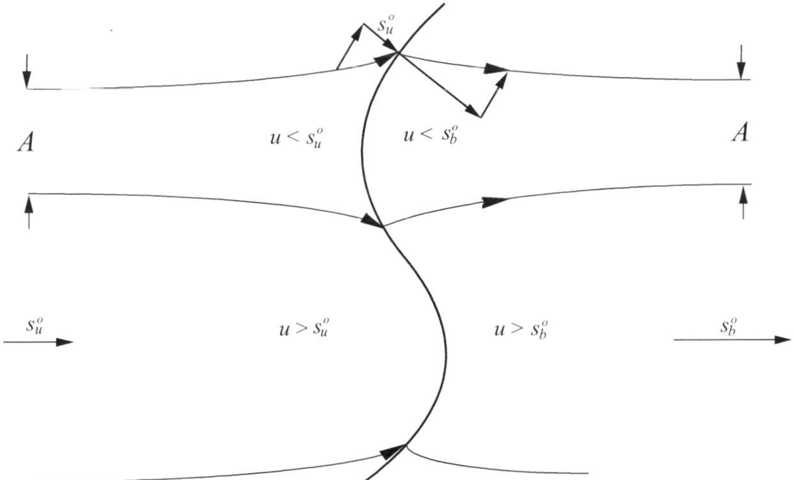

Figure 10.9.3. Schematic showing the mechanism of hydrodynamic instability (Williams 1985).

velocity due to thermal expansion, and that the tangential components of the upstream and downstream velocities should be continuous, the streamlines must respectively diverge and converge in approaching the convex and concave segments of the flame. Thus for the convex segment of the flame, the widening of the streamtube causes the flow to slow down. However, since the flame speed remains unaffected as specified, the local velocities of the approach flow and the flame can no longer balance each other, hence resulting in further advancement of this flame segment into the unburned mixture. A similar argument for the concave segment shows that it will further recede into the burned mixture. Thus this hydrodynamic mode of instability is absolutely unstable. Furthermore, because this discussion does not involve any length scales, the flame is unstable to perturbations of all wavelengths.

The buoyancy-driven instability, commonly known as the Rayleigh–Taylor instability, occurs for fluids that have negative density stratification in the direction of a body force such as gravity. Thus an upwardly propagating flame is buoyantly unstable because the denser, unburned mixture is over the lighter, burned product, while the converse holds for a downwardly propagating flame. Furthermore, since an accelerating flame experiences a body force directed from the unburned to the burned mixtures, it is also subjected to this mode of body force instability.

The discussion on the hydrodynamic and buoyancy-driven instabilities is based on the flame being infinitely thin. In the presence of finite flame thickness, the curvature-stabilization mechanism just considered for the diffusional-thermal instability must also be operative. This renders a characteristic dimension to the cells, and consequently a stabilization mechanism for disturbances with the dimension of the flame thickness. Thus while the dimensions of the diffusional-thermal cells are expected to be of the order of the flame thickness, those of the hydrodynamic and buoyancy-driven cells are larger, but with the smallest sizes still being that of the flame thickness.

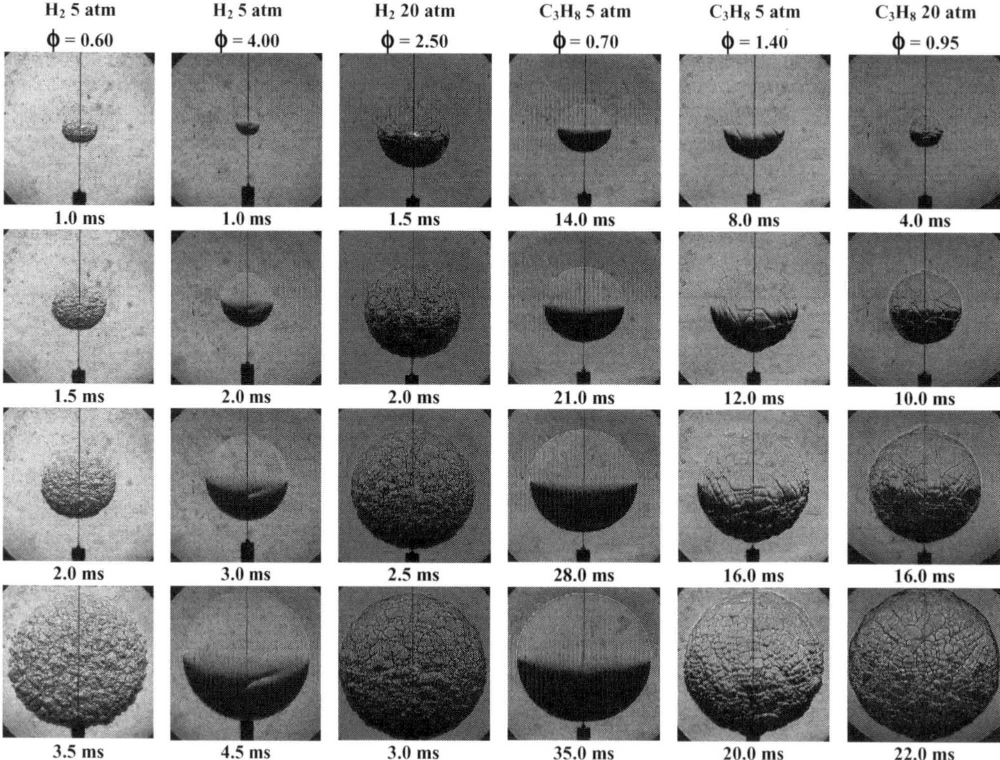

Figure 10.9.4. Photographic images of spherically expanding flames of hydrogen–air and propane–air mixtures, showing the presence of diffusional-thermal instability for the lean hydrogen–air and rich propane–air flames, and the presence of hydrodynamic instability for the rich hydrogen–air and lean propane–air flames at high pressures because of the reduced flame thickness. The horizontal physical dimension of each frame is 5.4 cm.

Consequently, development of the hydrodynamic and buoyancy-driven cells is promoted with decreasing flame thickness.

Figure 10.9.4 shows the time-resolved Schlieren images of spark-ignited expanding spherical flames of lean and rich hydrogen–air and propane–air mixtures, at pressures of 5 and 20 atm (Law 2006). It is seen that, at 5 atm, the rich hydrogen flame remains fairly smooth except for some large wrinkles as it propagates outward, while its lean counterpart rapidly develops profuse small-scale wrinkles after ignition. The behavior of the propane flames, however, is completely opposite in that its lean flame is remarkably smooth while large and then small wrinkles are developed for the rich flame even from its early stage of propagation. These results are in agreement with the theoretical prediction in that the *Le* for lean and rich hydrogen–air mixtures are respectively smaller and larger than unity, while the opposite holds for propane–air mixtures.

Figure 10.9.4 further shows that, although the rich hydrogen–air and lean propane–air flames are stable at 5 atm, they become wrinkled at 20 atm. Since these flames are diffusionally stable, the wrinkling must be caused by the hydrodynamic instability. Thus the stability of the 5-atm flames is due to the stabilizing, $Le > 1$, diffusional-thermal effect as well as the fact that the flame is sufficiently thick that the curvature effect stabilizes the omnipresent hydrodynamic instability. However, with increasing pressure, the flame becomes thinner and stabilization by curvature becomes less effective. Consequently, when the pressure is sufficiently high, the flame is so thin that the hydrodynamic instability dominates over the stabilizing diffusional-thermal and curvature effects, causing the flame to wrinkle.

The fact that the hydrodynamic instability is promoted with increasing pressure could have important implications on analyzing and understanding the flame morphology and dynamics in internal combustion engines. That is, in addition to turbulence, flame-front instability could also contribute substantially to flame wrinkling and thereby the burning rate of turbulent flames.

10.9.2. Analysis of Cellular Instabilities

Extensive asymptotic analyses have been performed on flamefront instabilities (Sivashinsky 1977, 1983; Pelce & Clavin 1982; Matalon & Matkowsky 1982; Clavin 1985, 2000; Bychkov & Liberman 2000). We shall, however, adopt an approximate, though analytically and conceptually apparent, analysis which incorporates the three modes of instabilities mentioned above (Law & Sung 2000). The approach involves analyzing the stability of flame surfaces in the manner of Landau and Darrieus, but allowing the flame speed to be affected by the generalized flame stretch effects. In this manner both the large-scale hydrodynamic and body-force instabilities, and the small-scale diffusional-thermal instabilities, are captured. The analysis is that of linear stability, relevant for the initial growth of the disturbance and, hence, small departures of the flame surface configuration from the planar one. Such an analysis yields the stability boundaries and the dispersion relations of a given system. For simplicity we shall also restrict the analysis to two-dimensional disturbances.

As discussed in Section 10.2, in the hydrodynamic limit the slightly-perturbed, unfolded flame surface can be represented by the function $\hat{G}(\hat{x}, \hat{y}, \hat{t}) = \hat{y} - \hat{f}(\hat{x}, \hat{t}) = 0$, whose geometry and dynamics are described by the G-Equation (10.2.5). We aim to determine whether this disturbance will cause the instantaneous flame surface $f(x, t)$, along with other quantities, to grow or decay.

On the unburned and burned sides of the flame, respectively designated by the subscripts $(-)$ and $(+)$, the densities are assumed to be uniform, given by

$$\hat{\rho}_- = \hat{\rho}_u = 1 \quad \text{and} \quad \hat{\rho}_+ = \hat{\rho}_b = \alpha^o < 1, \tag{10.9.1}$$

where $\alpha^o = 1/(1 + \tilde{q}_c)$ is the density ratio. The velocity and pressure variations for the inviscid and incompressible flows are governed by continuity and the Euler equation

$$\hat{\nabla} \cdot \tilde{\mathbf{v}}_\mp = 0 \tag{10.9.2}$$

$$\hat{\rho}_\mp \left(\frac{\partial \tilde{\mathbf{v}}_\mp}{\partial \hat{t}} + \tilde{\mathbf{v}}_\mp \cdot \hat{\nabla} \tilde{\mathbf{v}}_\mp \right) = -\hat{\nabla} \tilde{p}_\mp + \hat{\rho}_\mp \hat{g} \mathbf{e}_g \tag{10.9.3}$$

where $\mathbf{e}_g$ is the unit vector in the direction of gravity. In the above, length, velocity, time, ρ, p, g and the heat of reaction, q_c, are respectively nondimensionalized by the physical quantities ℓ_H, s_u^o, ℓ_H/s_u^o, ρ_u, $\rho_u(s_u^o)^2$, $(s_u^o)^2/\ell_H$, and $c_p T_u/Y_u$. We further require that the upstream and downstream solutions must be bounded as $y \to \pm\infty$, and that they are related to each other across the flame sheet through the conservations of mass and the normal and tangential momenta, which are simply the jump relations of Eqs. (5.1.27) and (5.1.29):

$$[\hat{\rho}(\tilde{\mathbf{v}} - \tilde{\mathbf{V}}_f) \cdot \mathbf{n}]_-^+ = 0 \tag{10.9.4}$$

$$[\hat{\rho}(\tilde{\mathbf{v}} - \tilde{\mathbf{V}}_f) \cdot \mathbf{n}(\tilde{\mathbf{v}} \cdot \mathbf{n}) + \tilde{p} - \hat{\rho}\hat{g}(\hat{f} - \hat{y})]_-^+ = 0 \tag{10.9.5}$$

$$[\tilde{\mathbf{v}} \times \mathbf{n}]_-^+ = 0, \tag{10.9.6}$$

where $[\phi]_-^+ = \phi(\hat{y} = \hat{f}_+) - \phi(\hat{y} = \hat{f}_-)$ for the quantity ϕ. The jump relation of (10.9.6) states that the tangential components of the velocities are continuous across the flame.

The problem is completed by specifying the upstream flame speed $\tilde{s}_u$ needed for the G-equation (10.2.5). We shall use the linear expression for the stretched flame, (10.4.40). When expressed in the hydrodynamic scale ℓ_H, this expression can be written as

$$\tilde{s}_u = 1 + \hat{\ell}_T^o\left(-\hat{\nabla} \cdot \mathbf{n} + Ma^o \hat{K}a^o\right), \tag{10.9.7}$$

where the Karlovitz number $\hat{K}a^o$ is now measured in units of ℓ_H and therefore differs from Ka^o given by Eq. (10.2.41) by the factor $\hat{\ell}_T^o = \ell_T^o/\ell_H$, which compares the flame thickness to the hydrodynamic scale.

The steady one-dimensional solution of the above problem, corresponding to a planar flame front with the velocity field $\tilde{\mathbf{v}} = \tilde{\mathbf{v}}_o = (0, \tilde{v}_o)$, is given by

$$\tilde{v}_o = \begin{cases} 1 \\ 1 + \tilde{q}_c \end{cases}, \quad \tilde{p}_o = \begin{cases} -\hat{g}\hat{y} & \hat{y} < 0 \\ -\tilde{q}_c - \hat{g}\hat{y}/(1 + \tilde{q}_c) & \hat{y} > 0. \end{cases} \tag{10.9.8}$$

Here the subscript o designates this basic solution, $\hat{g} > 0$ corresponds to a downward propagating flame, and we have also located the planar flame front at $\hat{y} = 0$.

To perform a linear stability analysis in order to determine the response of this solution to small arbitrary disturbances, we represent the disturbed quantities as $\tilde{\mathbf{v}} = \tilde{\mathbf{v}}_o + \tilde{\mathbf{v}}'$, $\tilde{p} = \tilde{p}_o + \tilde{p}'$, and $\hat{f} \equiv \hat{f}$, where $\tilde{\mathbf{v}} = (\tilde{u}', \tilde{v}_o + \tilde{v}')$, $\tilde{\mathbf{v}}' = (u', v')$, and the perturbations are assumed to be small compared to the basic steady-state solution. Substituting these expressions into Eqs. (10.9.2) and (10.9.3), and linearizing about the basic steady state, we obtain

$$\hat{\nabla} \cdot \tilde{\mathbf{v}}'_\mp = \frac{\partial \tilde{u}'_\mp}{\partial \hat{x}} + \frac{\partial \tilde{v}'_\mp}{\partial \hat{y}} = 0 \tag{10.9.9}$$

$$\hat{\rho}_\mp\left(\frac{\partial \tilde{\mathbf{v}}'_\mp}{\partial \hat{t}} + \tilde{\mathbf{v}}_{o,\mp} \cdot \hat{\nabla}\tilde{\mathbf{v}}'_\mp + \tilde{\mathbf{v}}'_\mp \cdot \hat{\nabla}\tilde{\mathbf{v}}_{o,\mp}\right) = -\hat{\nabla}\tilde{p}'_\mp \tag{10.9.10}$$

whose $\hat{x}$- and $\hat{y}$-components are

$$\hat{\rho}_{\mp}\frac{\partial \tilde{u}'_{\mp}}{\partial \hat{t}} + \frac{\partial \tilde{u}'_{\mp}}{\partial \hat{y}} = -\frac{\partial \tilde{p}'_{\mp}}{\partial \hat{x}} \tag{10.9.11}$$

$$\hat{\rho}_{\mp}\frac{\partial \tilde{v}'_{\mp}}{\partial \hat{t}} + \frac{\partial \tilde{v}'_{\mp}}{\partial \hat{y}} = -\frac{\partial \tilde{p}'_{\mp}}{\partial \hat{y}}. \tag{10.9.12}$$

Note that Eq. (10.9.12) can be replaced by

$$\frac{\partial^2 \tilde{p}'_{\mp}}{\partial \hat{x}^2} + \frac{\partial^2 \tilde{p}'_{\mp}}{\partial \hat{y}^2} = 0, \tag{10.9.13}$$

obtained by adding the differentials of Eqs. (10.9.11) and (10.9.12) by $\hat{x}$ and $\hat{y}$ respectively. Similar linearization for the flame speed expression (10.9.7), the G-equation (10.2.5), and the jump relations (10.9.4), (10.9.5), and (10.9.6) then respectively yield

$$\tilde{s}_u = 1 - \hat{\ell}^o_T\frac{\partial^2 \hat{f}}{\partial \hat{x}^2} + \hat{\ell}^o_T Ma^o\frac{\partial^2 \hat{f}}{\partial \hat{x}^2} = 1-(1-Ma^o)\hat{\ell}^o_T\frac{\partial^2 \hat{f}}{\partial \hat{x}^2} \tag{10.9.14}$$

$$\tilde{v}'_-(0) = \frac{\partial \hat{f}}{\partial \hat{t}} - (1-Ma^o)\hat{\ell}^o_T\frac{\partial^2 \hat{f}}{\partial \hat{x}^2} \tag{10.9.15}$$

$$[\tilde{v}']^+_- = -\tilde{q}_c(1-Ma^o)\hat{\ell}^o_T\frac{\partial^2 \hat{f}}{\partial \hat{x}^2} \tag{10.9.16}$$

$$[\tilde{p}']^+_- = 2\tilde{q}_c(1-Ma^o)\hat{\ell}^o_T\frac{\partial^2 \hat{f}}{\partial \hat{x}^2} - \frac{\hat{g}\tilde{q}_c\hat{f}}{1+\tilde{q}_c} \tag{10.9.17}$$

$$\tilde{q}_c\frac{\partial \hat{f}}{\partial \hat{x}} + [\tilde{u}']^+_- = 0. \tag{10.9.18}$$

In obtaining these results we have used the following relations:

$$\mathbf{n} = -\frac{\hat{\nabla}\hat{G}}{|\hat{\nabla}\hat{G}|} = \frac{(\partial\hat{f}/\partial\hat{x}, -1)}{\{1+(\partial\hat{f}/\partial\hat{x})^2\}^{1/2}} \approx \left(\frac{\partial \hat{f}}{\partial \hat{x}}, -1\right)$$

$$\tilde{\mathbf{V}}_f\cdot\mathbf{n} = \frac{1}{|\hat{\nabla}\hat{G}|}\frac{\partial \hat{G}}{\partial \hat{t}} = \frac{-\partial\hat{f}/\partial\hat{t}}{\{1+(\partial\hat{f}/\partial\hat{x})^2\}^{1/2}} \approx -\frac{\partial \hat{f}}{\partial \hat{t}}.$$

Note that in the Landau–Darrieus model $[\tilde{u}']^+_- = 0$ and $[\tilde{p}']^+_- = 0$.

To solve Eqs. (10.9.9), (10.9.11), and (10.9.13) subject to Eqs. (10.9.15)–(10.9.18), we look for solutions of the form

$$\begin{aligned} \tilde{u}'_{\mp} &= \bar{\tilde{u}}_{\mp}(\hat{y})\exp(\hat{\omega}\hat{t}+i\hat{k}\hat{x}), \quad \tilde{v}'_{\mp} = \bar{\tilde{v}}_{\mp}(\hat{y})\exp(\hat{\omega}\hat{t}+i\hat{k}\hat{x}) \\ \tilde{p}'_{\mp} &= \bar{\tilde{p}}_{\mp}(\hat{y})\exp(\hat{\omega}\hat{t}+i\hat{k}\hat{x}), \quad \hat{f} = \hat{A}\exp(\hat{\omega}\hat{t}+i\hat{k}\hat{x}). \end{aligned} \tag{10.9.19}$$

Substituting (10.9.19) into Eqs. (10.9.9)–(10.9.18), we obtain

$$\frac{d\bar{\bar{v}}_\mp}{d\hat{y}} = -i\hat{k}\bar{\bar{u}}_\mp \tag{10.9.20}$$

$$\frac{d\bar{\bar{v}}_\mp}{d\hat{y}} + \hat{\rho}_\mp\hat{\omega}\bar{\bar{v}}_\mp = -\frac{d\bar{\bar{p}}_\mp}{d\hat{y}} \tag{10.9.21}$$

$$\frac{d^2\bar{\bar{p}}_\mp}{d\hat{y}^2} = \hat{k}^2\bar{\bar{p}}_\mp \tag{10.9.22}$$

$$\bar{\bar{v}}_-(0) = \{\hat{\omega} + \hat{k}^2(1 - Ma^o)\hat{\ell}_T^o\}\hat{A} \tag{10.9.23}$$

$$[\bar{\bar{v}}]_-^+ = \bar{\bar{v}}_+(0) - \bar{\bar{v}}_-(0) = \hat{k}^2\tilde{q}_c(1 - Ma^o)\hat{\ell}_T^o\hat{A} \tag{10.9.24}$$

$$[\bar{\bar{p}}]_-^+ = \bar{\bar{p}}_+(0) - \bar{\bar{p}}_-(0) = \left\{-2\hat{k}^2\tilde{q}_c(1 - Ma^o)\hat{\ell}_T^o - \frac{\hat{g}\tilde{q}_c}{(1+\tilde{q}_c)}\right\}\hat{A} \tag{10.9.25}$$

$$[\bar{\bar{u}}]_-^+ = \bar{\bar{u}}_+(0) - \bar{\bar{u}}_-(0) = -i\hat{k}\tilde{q}_c\hat{A}. \tag{10.9.26}$$

The solution of Eqs. (10.9.20)–(10.9.22), subject to Eqs. (10.9.23)–(10.9.26), and the requirement of boundedness as $\hat{y} \to \pm\infty$, is straightforward. It can thus be shown that in order for a nontrivial solution to exist, $\hat{\omega}$ must satisfy the following dispersion relation:

$$(2+\tilde{q}_c)\hat{\omega}^2 + 2(1+\tilde{q}_c)\hat{k}\hat{\omega} - \tilde{q}_c(1+\tilde{q}_c)\hat{k}\left\{\hat{k} - \frac{2(1-Ma^o)\hat{\ell}_T^o(1+\tilde{q}_c)}{\tilde{q}_c}\hat{k}^2 - \frac{\hat{g}}{(1+\tilde{q}_c)}\right\} = 0. \tag{10.9.27}$$

Equation (10.9.27) possesses two roots. One root has a negative real part, and therefore it does not predict instability. The other root is given by

$$\hat{\omega} = \left(\frac{1+\tilde{q}_c}{2+\tilde{q}_c}\right) \times \left\{-\hat{k} + \sqrt{\left[1 + \frac{\tilde{q}_c(2+\tilde{q}_c)}{(1+\tilde{q}_c)}\right]\hat{k}^2 - 2(2+\tilde{q}_c)(1-Ma^o)\hat{\ell}_T^o\hat{k}^3 - \frac{(2+\tilde{q}_c)\tilde{q}_c}{(1+\tilde{q}_c)^2}\hat{g}\hat{k}}\right\}. \tag{10.9.28}$$

We now study the implications of the dispersion relation (10.9.28) for real values of $\hat{\omega}$, in particular the relative importance of the three terms under the radical sign. These three terms respectively represent effects due to thermal expansion, diffusional-thermal instability, and body-force instability. Furthermore, they are respectively proportional to $\hat{k}^2$, $\hat{k}^3$, and $\hat{k}$, implying that the diffusional-thermal cells are the smallest while the body-force cells the largest.

We first examine the Landau limit, corresponding to the Landau–Darrieus instability. Setting $\hat{\ell}_T^o \equiv 0$ and $\hat{g} \equiv 0$, Eq. (10.9.28) becomes

$$\hat{\omega} = \left(\frac{1+\tilde{q}_c}{2+\tilde{q}_c}\right)\left\{-1 + \sqrt{1 + \frac{\tilde{q}_c(2+\tilde{q}_c)}{(1+\tilde{q}_c)}}\right\}\hat{k}. \tag{10.9.29}$$

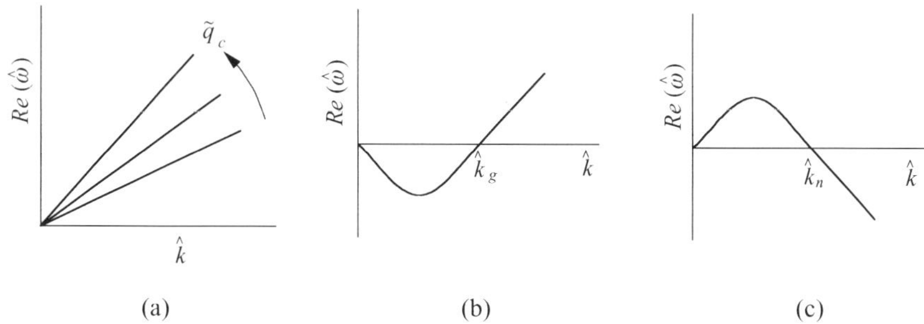

Figure 10.9.5. Stability diagrams for: (a) hydrodynamic instability, (b) hydrodynamic instability in the presence of a stabilizing body force, and (c) stabilizing influence of pure curvature on the hydrodynamic instability.

Since $\hat{\omega} > 0$ for all $\tilde{q}_c > 0$, Eq. (10.9.29) shows that the flame is unconditionally unstable to disturbances of all wavelengths. Because this mode of hydrodynamic instability is caused by the thermal expansion of the gas upon crossing the flame, it grows faster with increasing α^o and, hence, increasing heat release, $\tilde{q}_c$, as shown in Figure 10.9.5a.

We next consider body-force instability. Specifically, for an upwardly propagating flame in the presence of gravity, $\hat{g} < 0$ and we again have $\hat{\omega} > 0$ from Eq. (10.9.28) for $\hat{\ell}^o_T \equiv 0$. Such flames are exposed to both hydrodynamic and body-force instabilities. However, for a downwardly propagating flame, $\hat{g} > 0$, the above root possesses both real and imaginary parts. The real part passes through zero at the critical value $\hat{k}_g = \hat{g}/(1 + \tilde{q}_c)$. Since in the radical term of Eq. (10.9.28) the gravity term varies with the wavenumber $\hat{k}$ while the thermal expansion term varies with $\hat{k}^2$, the gravity term dominates for small $\hat{k}$. The flame is therefore unstable to short wavelength disturbances with $\hat{k} > \hat{k}_g$, and stable to long wavelength disturbances with $\hat{k} < \hat{k}_g$, as shown in Figure 10.9.5b. Thus buoyancy can stabilize long wavelength disturbances for downwardly propagating flames.

We next study the pure flame curvature effect by setting $Ma^o = 0$ (and $\hat{g} = 0$) in Eq. (10.9.28). It is seen that the term representing its influence in the radical term is always negative, and it therefore tends to moderate the destabilizing effect of thermal expansion. Furthermore, because this curvature term varies with $\hat{k}^3$ as compared to the $\hat{k}^2$ variation of the thermal expansion term, we expect that the flame is rendered stable by curvature for short wave disturbances with $\hat{k} > \hat{k}_n = \tilde{q}_c/[2\hat{\ell}^o_T(1 + \tilde{q}_c)]$, as shown in Figure 10.9.5c. This is in agreement with our earlier anticipations and explains the fact that smooth flames are routinely observed in the laboratory even though they are absolutely unstable based on hydrodynamic considerations.

We finally study the nonequidiffusive instability, as determined by the term in Eq. (10.9.28) with the factor $Ma^o = (Le^{-1} - 1)Ze^o/2$. Since this term is positive for $Ma^o > 0$ and, hence, $Le < 1$, the flame is rendered unstable for $Le < 1$ mixtures. The converse holds for $Le > 1$ mixtures. Thus stability is promoted for short wavelength

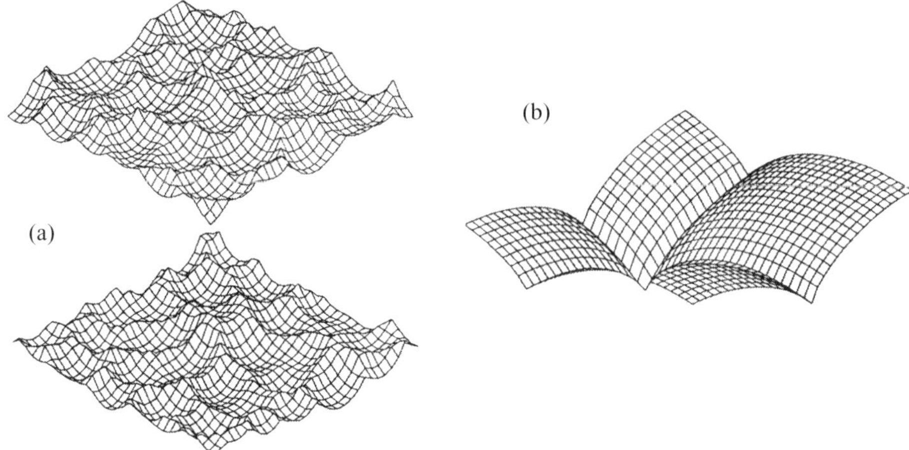

Figure 10.9.6. Numerical simulation of the cellular flame structure exhibiting: (a) diffusional-thermal instability, and (b) hydrodynamic instability. Note the chaotic nature of the former and the regular folds of the latter (Sivashinsky 1983).

disturbances in the same manner as the pure curvature effect. All these results are again in agreement with our earlier discussion.

By combining the pure curvature and nonequidiffusive effects as represented by the factor $(1 - Ma^o)$ for the diffusional-thermal instability, it is clear that their influence is destabilizing for $Ma^o > 1$, or $Le < 1/(1 + 2/Ze^o) \approx (1 - 2/Ze^o)$, and stabilizing otherwise. Thus the pure curvature effect extends the regime of stabilizing Lewis number by $2/Ze^o$ in that without considering it the flame will lose stability for $Le < 1$.

The linear stability analysis discussed above only describes the initial response of the flame. Nonlinear analyses, frequently aided by numerical solutions, are needed to trace through the development of the instabilities until the formation of the cellular flame pattern. Such numerical simulation has found that diffusive-thermal instability generates cells of a chaotic nature (Figure 10.9.6a), while hydrodynamic instability generates steady cells of regular sizes and shapes (Figure 10.9.6b). It has thus been suggested that diffusional-thermal instability could lead to self-turbulization of flames (Sivashinsky 1983).

10.9.3. Mechanisms of Pulsating Instabilities

In addition to cellular instability, a flame can also propagate in a pulsating or spinning mode due to temporal instability. The controlling factor in inducing the pulsating instability is again diffusional-thermal in nature, and the mechanism is depicted in Figure 10.9.7, for a planar flame. Specifically, consider a disturbance momentarily applied to the reaction zone, causing it to move forward. Because of the relatively larger inertia of the heat and mass diffusion zones, the flame structure cannot instantaneously adjust itself to accommodate such a disturbance. Consequently the diffusion zones become thinner, and the corresponding temperature and concentration

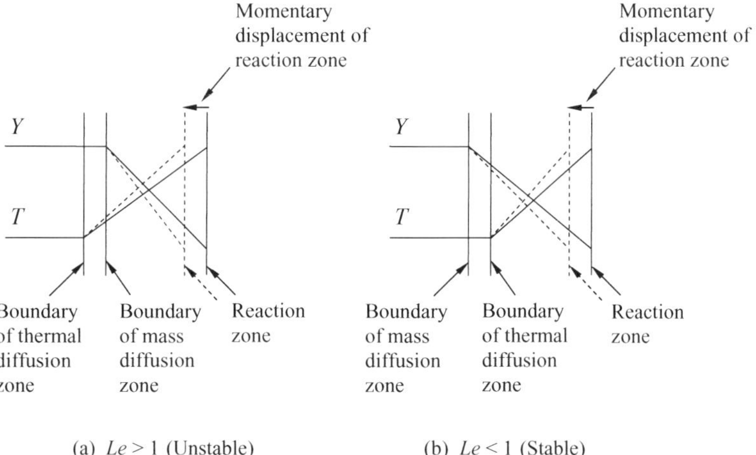

Figure 10.9.7. Schematic showing the mechanism of diffusional-thermal pulsating instability: (a) $Le > 1$, unstable; (b) $Le < 1$, stable.

gradients also steepen. For a $Le > 1$ flame, since the thermal diffusion zone is initially thicker than the mass diffusion zone, a reduction of their respective thicknesses by the same amount implies that the thermal gradient suffers less steepening than the concentration gradient. Thus the reaction zone will burn stronger because it now gains more reactant from the freestream than loses heat to it, and consequently will propagate forward with a higher velocity. By the same token, if the reaction zone is displaced backward, the burning will be weakened, causing it to further lag behind. Thus $Le > 1$ flames can be pulsatingly unstable while $Le < 1$ flames are pulsatingly stable. The dependence on Le for the pulsating instability is therefore completely opposite to that of diffusional-thermal cellular instability in that cellular instability is promoted for $Le < 1$ flames and suppressed for $Le > 1$ flames.

Asymptotic analysis (Sivashinsky 1977) shows that the standard, planar flame is pulsatingly unstable for $Ze^o(Le - 1) > 4(1 + \sqrt{3}) \approx 10.9$. Since Ze^o for many practical flames are actually not too large, typically having values around six or seven, and since Le for gaseous mixtures are close to unity, the tendency for gaseous flames to exhibit pulsating instability is actually not strong. Thus studies of pulsating instability have been focused on solid flames, which are of interest to materials synthesis (Merzhanov 1990, 1994; Makino 2001), for which Le is very large. Indeed, pulsating or spinning modes of propagation are frequently observed in the solid-phase synthesis of materials, resulting in undesirable laminated products.

While pulsating instability is not expected for strongly burning flames, the global activation energy and, hence, Ze^o are expected to be quite large for weakly burning flames. Thus the above instability criterion can possibly be satisfied for such flames. Take, for example, the computed results for the planar freely propagating rich hydrogen–air flames with radiative heat loss (Christiansen, Sung & Law 1998). A steady-state calculation shows that the flame extinguishes at $\phi = 10.4$ through the

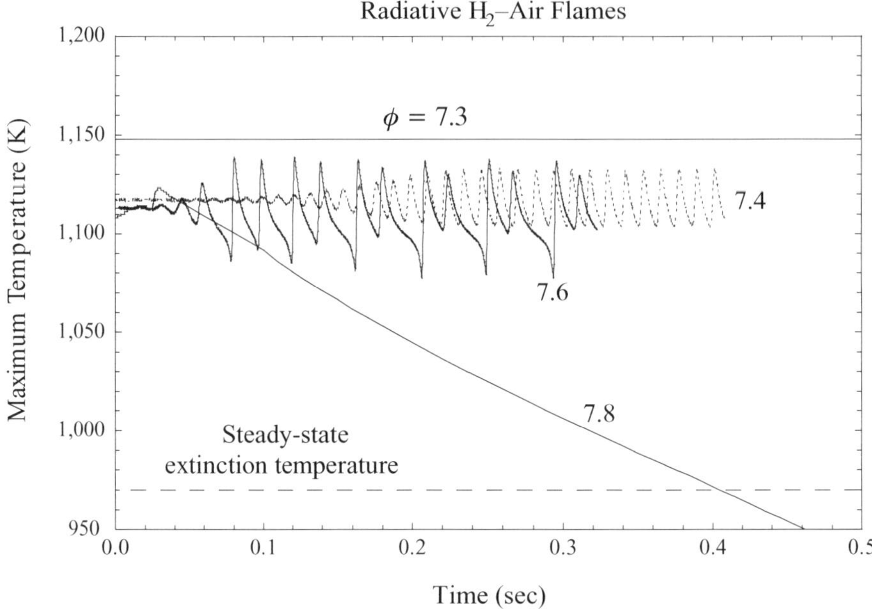

Figure 10.9.8. Calculated temporal variations of the maximum temperature of radiative, freely propagating planar rich hydrogen–air flames, showing the gradual transition from steady propagation, to pulsating propagation, to extinction, with increasing ϕ.

turning point criterion. However, when unsteadiness is allowed in the calculation, a rich variety of propagation modes is captured, as shown in Figure 10.9.8. It is seen that while the flame is still stable at $\phi = 7.3$, it loses stability at $\phi = 7.4$ and propagates in an oscillatory mode, with a single frequency. The overall propagation rate, however, is slower, as shown in Figure 10.9.9, because the flame spends more time in the negative phase of the oscillation due to the slower flame speed. At $\phi = 7.6$ the propagation mode transitions to that of period doubling, with two frequencies. Finally, the flame fails to propagate at $\phi = 7.8$. These results clearly show that the flame extinguishes in the pulsating mode, and that the extinction limit is narrowed due to pulsation. Thus pulsation promotes extinction. The reason is that although the positive phase of the oscillation enhances the intensity of the flame, which however is already burning anyway, the negative phase can reduce the burning intensity to a state of temporary extinction, which is unrecoverable. Indeed, further studies have shown that permanent extinction occurs when the instantaneous, oscillating flame temperature dips below the steady-state extinction flame temperature, indicating the quasi-steady nature of the flame pulsation process. This is reasonable in that, compared to the phenomena of forced oscillation for which the imposed frequency can be arbitrarily large and the flame can lose its sensitivity in response, the intrinsic pulsating instability is self-generated such that the frequency and magnitude of the oscillation are controlled by the diffusive characteristics of the flame.

It is also reasonable to expect that the instability boundary can be sensitively affected by the kinetic mechanism through the global activation energy. For the

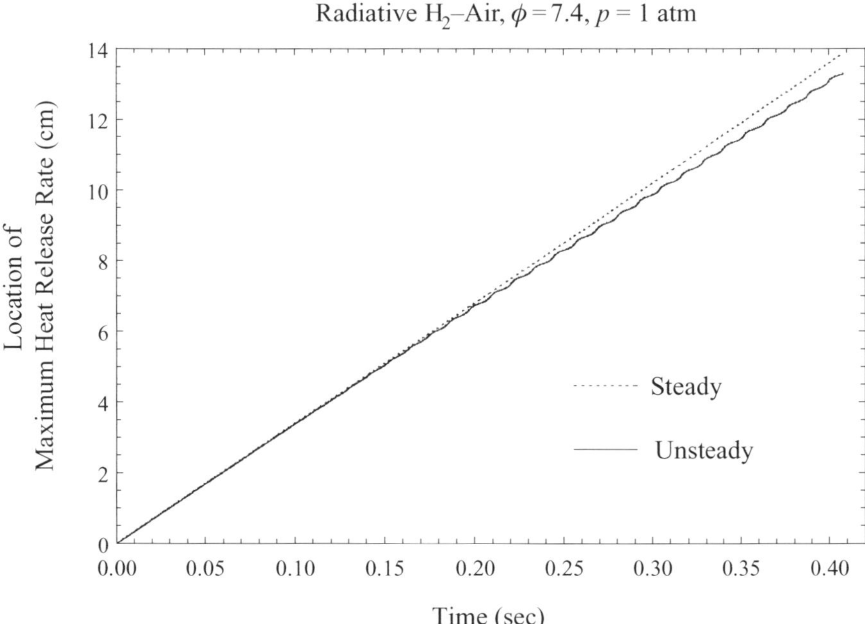

Figure 10.9.9. Trajectory of steady and pulsating flames, showing that the propagation rate is smaller for the latter.

hydrogen–oxygen system, since the global reaction rate will first increase and then decrease with pressure, one would expect that the propensity to pulsate will first increase and then decrease. Figure 10.9.10 shows the calculated stability diagram in terms of ϕ and pressure, indicating the respective limits for the onset of pulsation, extinction in the steady mode, and extinction in the pulsating mode. It is seen that while pulsation is indeed promoted with increasing pressure for the lower pressure range, an island of stability is identified at higher pressures.

10.9.4. Effects of Heat Loss and Aerodynamic Straining

Results from computation and stability analysis (Joulin & Clavin 1979) for the doubly infinite flame with volumetric heat loss show that heat loss has a destabilizing effect and therefore narrows the regime of stability. This is conceptually reasonable because heat loss tends to weaken the flame burning intensity and, hence, increase the flame time, which favors the onset of diffusional-thermal cellular and pulsating instabilities.

Bulk aerodynamic stretching has been found to be stabilizing for cellular instability if it is positive (Sivashinsky, Law & Joulin 1982), and destabilizing otherwise. Conceptually, a positively stretched flame such as the counterflow flame and the outwardly propagating flame tends to continuously "stretch out" and "carry away" any wrinkles which may develop over the flame surface. Figure 10.9.11 shows the flame configurations in a stagnation flow with increasing stretch rate. It is seen that for low rates of stretch, the flame exhibits the same cellular structure as observed

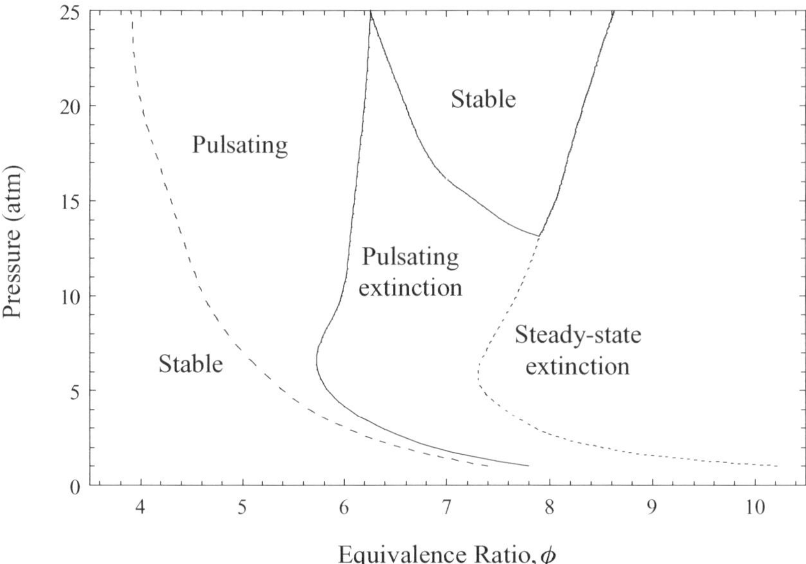

Figure 10.9.10. Calculated pulsating and extinction boundaries of rich hydrogen–air flames, with radiative heat loss, from 1 to 25 atm (Christiansen, Law & Sung 2001).

for unstretched flames. However, with increasing stretch, the instability in the radial direction is first suppressed, resulting in ridges emanating in the radial direction. With strong enough stretching, even these radial ridges are suppressed. By the same reasoning, flamefront cellular instability tends to be aggravated in a compressive flow whose stretch is negative, such as the formation of the polyhedral Bunsen flames.

The tendency to form hydrodynamic and diffusional-thermal cells can also be delayed for the outwardly expanding flames (Bechtold & Matalon 1987), as shown

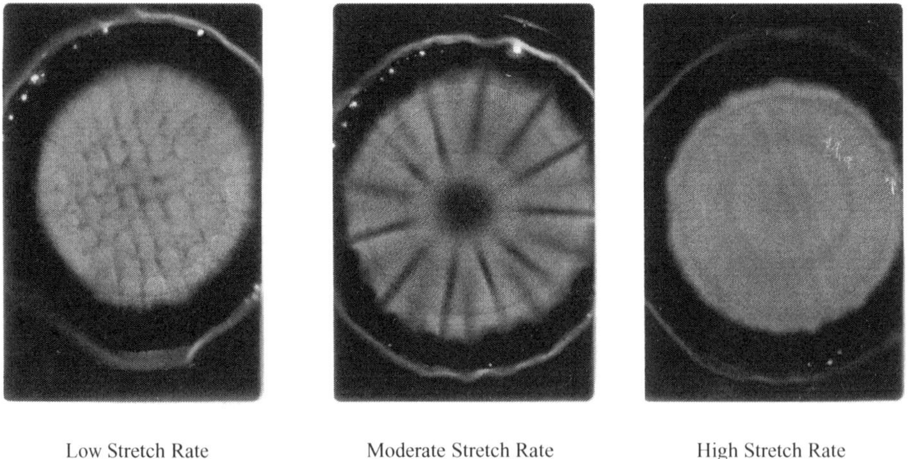

Low Stretch Rate Moderate Stretch Rate High Stretch Rate

Figure 10.9.11. Photographic images showing the stabilizing effect of positive stretch on diffusional-thermal instability (Courtesy: S. H. Sohrab).

in Figure 10.9.4. This is because upon ignition, the flame kernel is small and therefore the positive stretch can be sufficiently strong to suppress the development of the cells. The intensity of stretch subsequently decreases as the flame propagates outward and cells will then develop quite abruptly at a state when the stretch is substantially weakened.

For pulsating instability, the effect of bulk aerodynamic stretching is completely opposite in that positive stretch tends to be destabilizing and negative stretch stabilizing (Sung, Makino & Law 2002). This is reasonable because pulsating and cellular instabilities are both affected by the combined effects of stretch and nonequidiffusion. Since these processes are governed by the combined parameter $S^o \sim (Le^{-1} - 1)\kappa$, and since these two instability modes have opposite Le dependence, it is then reasonable that their dependence on κ is also reversed. Mechanistically, since positive (negative) stretch tends to reduce (increase) the thickness of the diffusion zone (Figure 10.9.7) during transient, effects due to the disparity in the heat and mass diffusion rates are aggravated (relaxed).

PROBLEMS

1. For the stationary saw-tooth flame shown in Figure 10.P.1, derive the effective flame speed s_T in the Landau limit.

2. (a) For a stationary flame with a sinusoidal profile $\hat{f} = A \sin \hat{x}$, derive s_T in the Landau limit. Show that in the limit of weak wrinkling,
$$s_T \approx s_u^o \left(1 + \frac{A^2}{4}\right).$$
(b) Repeat (a) but allowing for $Le = 1$, stretch effects. Compare and comment on results from (a) and (b).

3. (a) For a two-dimensional slot Bunsen flame of cone angle α and a parabolic exit velocity $u = u_o[1 - (x/a)^2]$, show that its stretch factor is
$$\kappa = -\left(\frac{x_f}{a^2}\right) u_o \sin \alpha,$$
where a is the half width of the slot.
(b) Repeat (a) for a tube with radius a. Show that the stretch rate vanishes at
$$r_f = \frac{a}{\sqrt{3}}.$$

4. For a uniform slot flow of velocity u_o, assume the tip of a Bunsen flame is given by the arc of a circle, as shown in Figure 10.P.2. For $Le = 1$, show that the height of the flame is
$$h = \frac{a}{\tan \alpha} - \frac{s_u^o \ell_T^o}{(u_o - s_u^o)} \frac{(1 - \sin \alpha)}{\sin \alpha}.$$

5. As discussed in Section 7.6.3, an experimental method used to determine the laminar flame speed is the outwardly propagating flame. Measuring the rate of change of the radius of the expanding flame, $r(t)$, yields the downstream flame

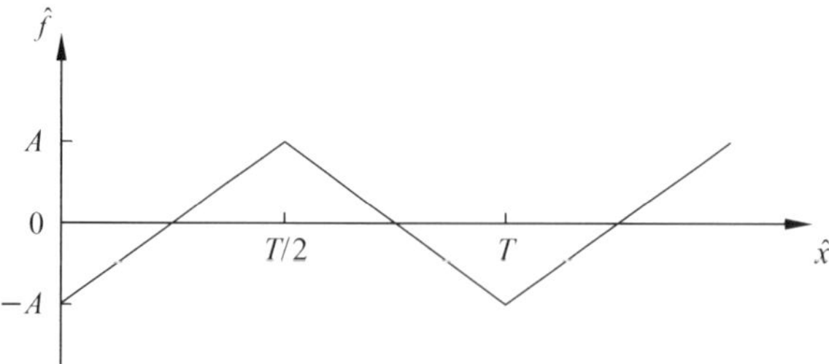

Figure 10.P.1. Schematic of the saw-tooth flame.

speed, s_b, as a function of the instantaneous strain rate κ that the flame experiences. Assume the variation is linear, with the proportionality constant being the Markstein length L_b, show that

$$r(t) + 2L_b \ln r(t) = s_b^o t + \text{constant}.$$

Thus by fitting $r(t)$ as a function of t according the above relation, we can simultaneously determine L_b and the downstream laminar flame speed s_b^o and, hence, the upstream flame speed through $s_u^o = s_b^o(\rho_b^o/\rho_u)$.

6. In Section 10.4.4 we solved the stagnation flame problem for the situation of a combustible mixture impinging upon an inert at the flame temperature. Consequently the flame can migrate across the stagnation surface. Let us now solve the problem when the stagnation surface is an adiabatic, noncatalytic wall so that the flame movement is restrained, as shown in Figure 10.3.4a.

(a) For a general Le and assuming that the flame is located away from the surface, determine the flame temperature and location for a given strain rate κ, as well as the state of extinction characterized by the extinction strain rate. Show

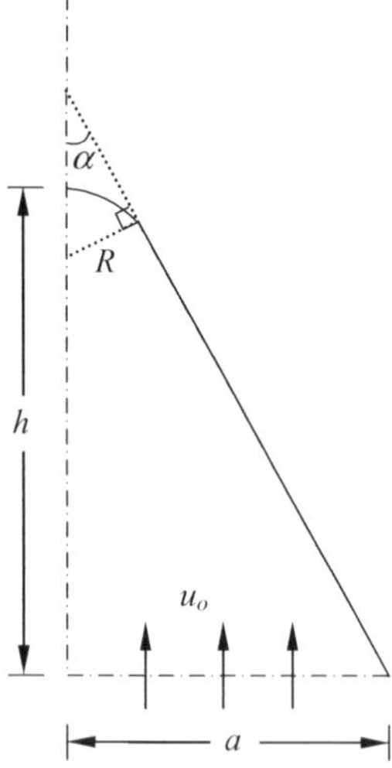

Figure 10.P.2. Schematic of the Bunsen flame with a circular tip.

that for this flame location extinction through increasing stretch is possible only for $Le > 1$.

(b) For $Le < 1$, the flame is extinguished after it is in contact with the wall and a separate analysis is needed. This is accomplished by assuming that to the leading order the flame is located at the stagnation surface, and that reaction is not complete at the downstream boundary of the reaction zone, which is the stagnation surface. Derive the extinction criterion.

11 Combustion in Turbulent Flows

11.1. GENERAL CONCEPTS

Most flows in practical combustion devices are turbulent, characterized by the presence of rapid, random fluctuations of the flow velocity and scalar properties at a given point in space. These fluctuations spread out in a manner similar to molecular diffusion as the flow evolves in time and/or proceeds downstream. Figure 11.1.1 illustrates various canonical flow configurations that are often encountered in practical combustion systems: unconfined flows such as jets and mixing layers, semiconfined flows over solid surfaces, confined flows in ducts, reverse flows in wakes, and buoyant flows.

Turbulence remains one of the most challenging and unsolved problems in physics. The complexity further increases when chemical reactions are also present. Because of these difficulties, studies on turbulent combustion have been mostly empirical until the late 1970s. Advances since then have identified fruitful paths for rational investigation. In this chapter we present a brief account of the current state of understanding.

In the next two sections the general concepts and solution techniques of turbulent flows, mostly nonreacting, are presented. These are followed by separate discussions on turbulent premixed and nonpremixed combustion. For a more detailed exposition, the reader is referred to Monin and Yaglom (1965), Tennekes and Lumley (1972), Launder and Spalding (1972), Hinze (1975), Schlichting et al. (1999), and Pope (2000) for nonreacting turbulent flows, and to Libby and Williams (1980, 1994), Williams (1985), Peters (2000), and Poinsot and Veynante (2005) for reacting turbulent flows.

11.1.1. Origin and Structure

To appreciate the possibility that laminar flows can become unstable when subjected to disturbances of infinitesimal intensity, consider the simple example of two parallel, uniform streams of an inviscid fluid with different initial velocities, as shown in Figure 11.1.2a. This flow satisfies the equations of motion and is therefore a possible

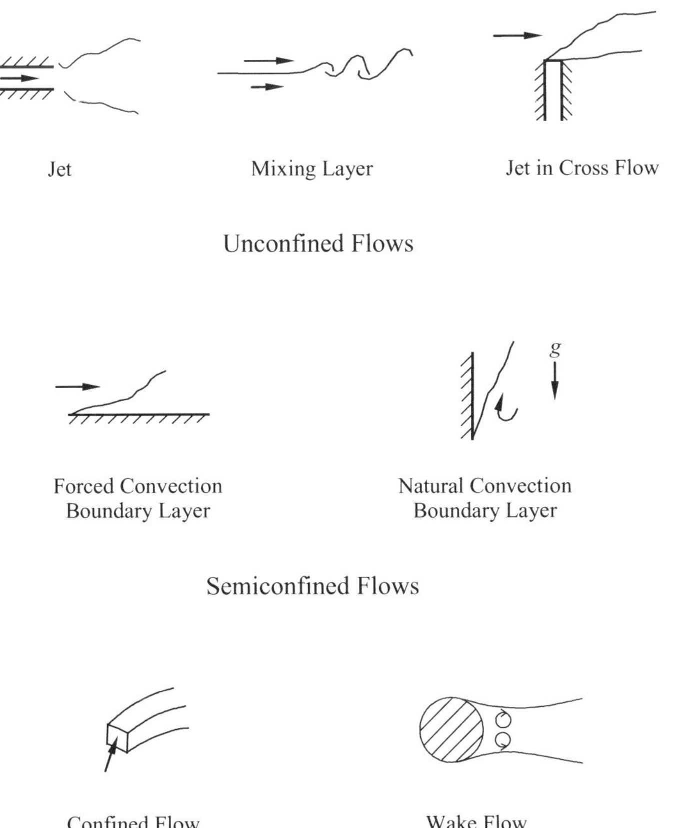

Figure 11.1.1. Schematic showing various classes of turbulent flows.

flow configuration. Let a disturbance be introduced such that one of the streamlines is distorted, as shown in Figure 11.1.2b. At the bulge of the disturbance, the increase in area of the streamtube causes a decrease in the flow velocity and hence an increase in pressure. By the same reasoning, the pressure in the region of constriction is reduced. This imbalance in pressure further distorts the streamlines (Figure 11.1.2c), implying that the intensity of the original disturbance tends to increase through this feedback mechanism, causing the flow to be unstable.

In order to sustain the growth of the instability, the depleted mass from the original lower streamtube must be replenished from that of the upper streamtube. Furthermore, in order to close the flow circuit, in a suitable coordinate the upper streamline must be flowing in the opposite direction from that of the lower streamline, hence forming a vortex. Thus turbulent flows are characterized by the presence of vortexes, or eddies, of various strengths and dimensions.

The growth of the disturbance can be limited by the finite viscosity of the fluid, which tends to damp the disturbance and thereby produce a stabilizing effect on the flow. If the viscous force is sufficiently large as compared to the inertial force, the damping is strong enough to render the flow stable and, hence, laminar. Furthermore,

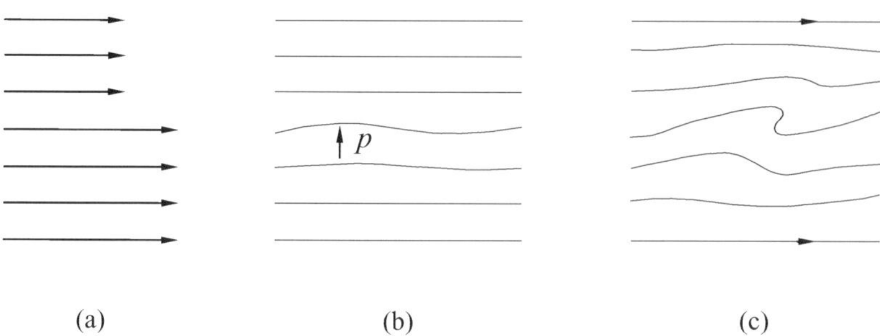

Figure 11.1.2. Destabilization mechanism in an inviscid flow.

an initially destabilized flow may not necessarily develop into a turbulent flow. It can instead evolve into a laminar, multidimensional flow of organized pattern, for example, the Benard cells in a thermally stratified flow subjected to the gravitational field in the direction of decreasing density. The flow becomes turbulent only if it exhibits a chaotic behavior subsequent to the initial destabilization.

Recognizing the importance of inertia and viscosity on flow stability, the relevant nondimensional parameter characterizing the tendency for a flow to become unstable and subsequently turbulent is the Reynolds number, $Re = \rho UL/\mu$, which is the ratio of the inertial force to the viscous force in the flow, where U and L are respectively the characteristic velocity and dimension of the flow. Thus, flows with small Re tend to be laminar, and those with large Re tend to be turbulent. Another reason that large Re promotes the existence of turbulent flows is that a large amount of kinetic energy is needed to sustain the generation of turbulent eddies, which are eventually dissipated through viscous action. Consequently, the general structure of a turbulent flow is one that consists of large eddies created by abstracting energy from the mean flow motion. These eddies then continuously break up into smaller ones until a certain size range is reached over which viscous dissipation becomes effective. This is the cascade concept of turbulent flows.

A complete description of a turbulent flow must also include the presence of intermittency. That is, if we were to measure the turbulence properties in the transition region of a turbulent flow, say close to the edge of a turbulent boundary layer, then the measurements will record periods of total randomness separated by periods of laminar-like flows. Thus processing and interpretation of measurements need to be conditioned for this intermittency behavior. A prominent class of turbulent flows exhibiting intermittency is that of coherent structures (Brown & Roshko 1974), which are large, rather organized parcels of turbulent flows with relatively long lifetimes and their own dynamic behavior. They may also contain a substantial amount of the turbulent kinetic energy of a turbulent flow. Figure 11.1.3 shows the spark-shadowgraph of the mixing layer in a two-dimensional shear flow, illustrating the characteristics of such coherent structures. Contrary to the cascading mechanism of energy dissipation described above, coherent structures may actually grow in size in the downstream

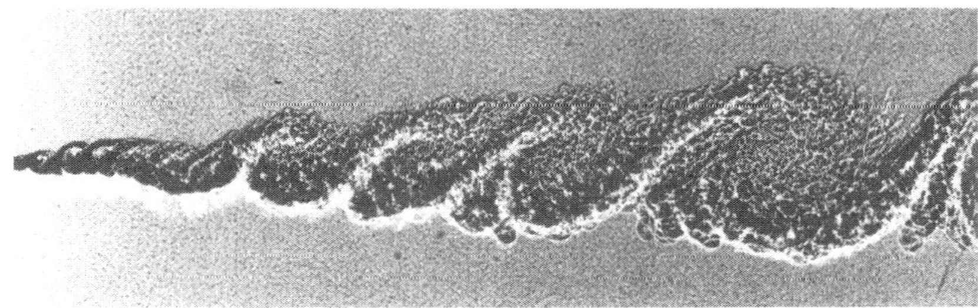

Figure 11.1.3. Spark-shadowgraph of the mixing layer in a two-dimensional shear flow, showing the characteristics of coherent structures (Brown & Roshko 1974).

direction through pairing of two structures into one. Thus a turbulent flow could consist of the organized nature of coherent structures and the chaotic nature of turbulence.

In the presence of chemical reactions, the complexity of turbulent flows takes on a new dimension. Mechanistically, turbulence is expected to generally increase the burning intensity by enhancing the mixing of reactants and the transport of heat, and by wrinkling and, hence, increasing the total flame surface area available for reaction to take place. Excessively intense turbulent eddies, however, can also cause local extinction in a turbulent flow, thereby adversely affecting the combustion efficiency.

While turbulence affects the intensity and extent of reactions in the flow, the turbulence intensity and structure are in turn affected by the large exothermicity that is characteristic of combustion reactions. Specifically, although chemical heat release is an energy source in sustaining the structure of a turbulent flow, it also has a laminarizing effect on the flow because for gases $Re \sim \rho/\mu \sim T^{-(1+\alpha)}$, with $\alpha > 0$. Consequently Re is reduced in the presence of heat release because of the simultaneous reduction in density and increase in viscosity.

Counteracting the laminarizing effect of heat release, vorticity and, hence, turbulence are generated in the flow through the baroclinic torque, $\nabla(1/\rho) \times \nabla(p)$ developed on the flame surface. This is caused by the coupling between the density jump across the flame and the pressure variation over it when it is wrinkled.

The burning intensity of a turbulent flame can also be enhanced by the development of flamefront instabilities. Clearly, the potential of generating hydrodynamic cells due to the density jump across flame surfaces is always present. Furthermore, cells generated by the diffusional-thermal instability have been found to be chaotic, as mentioned in Section 10.9, and could constitute a source of flame-generated turbulence. The presence of these cells is expected to increase the local flame surface area and, hence, the total burning rate.

11.1.2. Probabilistic Description

Although the Navier–Stokes equation and the conservation equations for energy and species are deterministic in that unique solutions should exist for properly specified boundary and initial conditions, for large Reynolds number flows the solutions are highly sensitive to conditions at points sufficiently away from the boundaries and/or at sufficiently early times. In other words, minute changes in these conditions can lead to huge changes in the solutions. As such, even the most powerful computer attainable in the foreseeable future cannot adequately resolve the solutions in space and time. It is then logical to describe the flow from the statistical viewpoint, yielding solutions that are probabilistic in nature.

For a probabilistic description of a variable u, we introduce the probability density function (*pdf*), $P(u;\mathbf{x},t)$, such that $P(u;\mathbf{x},t)du$ is the probability of observing u within a small range du about u, at fixed values of $\mathbf{x}$ and t. P must satisfy the realization requirement that

$$\int_{-\infty}^{\infty} P(u;\mathbf{x},t)du = 1. \tag{11.1.1}$$

Once P is known for the stochastic variable u, then its nth moment can be defined as

$$\overline{u(\mathbf{x},t)^n} = \int_{-\infty}^{\infty} u^n P(u;\mathbf{x},t)du, \tag{11.1.2}$$

where the overbar indicates the mean, or expectation, value of u^n. The first moment $(n=1)$ then gives the mean value of $u(\mathbf{x},t)$, namely $\bar{u}(\mathbf{x},t)$. Similarly, the mean value for a function $g(u)$ is given by

$$\bar{g}(\mathbf{x},t) = \int_{-\infty}^{\infty} g(u;\mathbf{x},t)P(u;\mathbf{x},t)du. \tag{11.1.3}$$

We can also define the nth central moment by

$$\overline{[u(\mathbf{x},t)-\bar{u}(\mathbf{x},\mathbf{t})]^n} = \int_{-\infty}^{\infty} (u-\bar{u})^n P(u;\mathbf{x},t)du. \tag{11.1.4}$$

The second central moment, $n=2$, is the variance. By splitting u into its mean $(\bar{u})$ and random fluctuation (u'),

$$u(\mathbf{x},t) = \bar{u}(\mathbf{x},t) + u'(\mathbf{x},t), \tag{11.1.5}$$

such that $\overline{u'} \equiv 0$, the variance is given in terms of the first and second moments as

$$\overline{u'^2} = \overline{(u-\overline{u})^2} = \overline{u^2} - 2\bar{u}^2 + \bar{u}^2 = \overline{u^2} - \bar{u}^2. \tag{11.1.6}$$

Thus while $\overline{u'} \equiv 0$, $\overline{u'^2} \neq 0$.

Since in a given flow, fluctuation of all of its properties $(u, v, w, T, \rho, \ldots.)$ can be related to each other, we are frequently interested to know the joint probability of observing some or all of them about their respective values. A joint probability density function can thus be similarly defined for these stochastic variables. For example, for the variables u and v, the joint *pdf*, $P(u, v;\mathbf{x},t)$, gives the probability

$Pdudv$ of observing u and v within the range du and dv about u and v, at fixed values of $\mathbf{x}$ and t. Furthermore,

$$P(u;\mathbf{x},t) = \int_{-\infty}^{\infty} P(u,v;\mathbf{x},t)dv. \tag{11.1.7}$$

From stochastic theory it is shown that the joint *pdf* of two independent variables can be expressed as the product of the *pdf* of one variable conditioned on the *pdf* of the other,

$$P(u,v;\mathbf{x},t) = P(u \mid v;\mathbf{x},t)P(v;\mathbf{x},t), \tag{11.1.8}$$

where $P(u \mid v;\mathbf{x},t)$ is the probability density of u conditioned at a fixed value of v. If u and v are statistically independent, that is, not correlated, then

$$P(u,v;\mathbf{x},t) = P(u;\mathbf{x},t)P(v;\mathbf{x},t). \tag{11.1.9}$$

In this case it can be easily shown that $\overline{u'v'} \equiv 0$. In turbulent flows $\overline{u'v'} \neq 0$ in general.

A random field may also exhibit some special statistical properties. Specifically, it is called statistically stationary, statistically homogeneous, and statistically isotropic if all statistics are, respectively, invariant under a shift in time, a shift in position, and a rotation and reflection of the coordinate system.

Statistical treatment based on (11.1.5) is called Reynolds averaging, and is used for constant density flows. For high-speed or chemically reacting flows, large changes in density take place. As we have shown several times before, the relevant flow variable representing convective transport for such flows is $\rho\mathbf{u}$ instead of $\mathbf{u}$ alone. It is therefore appropriate to introduce a density-weighted average $\tilde{u}(\mathbf{x},t)$, called the Favre average, such that all fluid mechanical quantities except the pressure are density weighted. Thus by splitting $u(\mathbf{x},t)$ into a Favre-averaged quantity $\tilde{u}(\mathbf{x},t)$ and a corresponding fluctuating component $u''(\mathbf{x},t)$,

$$u(\mathbf{x},t) = \tilde{u}(\mathbf{x},t) + u''(\mathbf{x},t), \tag{11.1.10}$$

we define the Favre average as

$$\tilde{u} \equiv \frac{\overline{\rho u}}{\bar{\rho}}. \tag{11.1.11}$$

Substituting Eq. (11.1.10) into Eq. (11.1.11), it is readily seen that

$$\overline{\rho u''} \equiv 0. \tag{11.1.12}$$

In general, in terms of the probability density function, Favre-averaged quantities are defined as

$$\bar{\rho}\tilde{g}(\mathbf{x},t) = \overline{\rho g}(\mathbf{x},t) = \int_{0}^{\infty}\int_{-\infty}^{\infty} \rho g(u;\mathbf{x},t)P(\rho,u;\mathbf{x},t)dud\rho, \tag{11.1.13}$$

where

$$\bar{\rho} = \int_0^\infty \int_{-\infty}^\infty \rho P(\rho, u; \mathbf{x}, t) du d\rho.$$

To illustrate the usefulness of Favre averaging, consider a typical convection term, ρuv, in the Navier–Stokes equation. By respectively using the Reynolds and Favre averagings, we have

$$\overline{\rho uv} = \bar{\rho}\bar{u}\bar{v} + \bar{\rho}\overline{u'v'} + \bar{u}\overline{\rho' v'} + \bar{v}\overline{\rho' u'} + \overline{\rho' u' v'}, \tag{11.1.14}$$

$$\overline{\rho uv} = \bar{\rho}\tilde{u}\tilde{v} + \overline{\rho u'' v''} = \bar{\rho}\tilde{u}\tilde{v} + \bar{\rho}\widetilde{u'' v''}. \tag{11.1.15}$$

Comparing Eqs. (11.1.14) and (11.1.15), it is clear that their respective first terms correspond to the mean quantities, and their respective second terms also resemble each other. The three additional terms involving density fluctuations in Reynolds averaging, however, do not appear in Favre averaging, which automatically incorporates the influence of these various modes of momentum exchange into a smaller number of terms, with better physical clarity. Indeed, Eq. (11.1.15) has the similar expressions as those of a constant density flow for which $\overline{uv} = \bar{u}\bar{v} + \overline{u'v'}$.

11.1.3. Turbulence Scales

A turbulent flow is often characterized by a spectrum of eddies. An eddy is a canonical structure represented by a vortical flow unit riding on the mean flow, for which the average rotational velocity and diameter characterize the relevant velocity and length scales. The magnitude of the vortical velocity of an eddy is a measure of the intensity of the turbulent fluctuation. To estimate the length scales of the eddies, a normalized space correlation can be defined based on the velocities at two adjacent points in the flow, $\mathbf{x}$ and $\mathbf{x} + \mathbf{r}$, where $\mathbf{r}$ is a spatial distance emanating from $\mathbf{x}$. In the three-dimensional Cartesian coordinate, the normalized correlation functions are written as

$$R_{11}(\mathbf{x}, \mathbf{r}, t) = \overline{u'(\mathbf{x}, t) u'(\mathbf{x} + \mathbf{r}, t)} / \overline{u'^2(\mathbf{x}, t)}, \tag{11.1.16a}$$

$$R_{22}(\mathbf{x}, \mathbf{r}, t) = \overline{v'(\mathbf{x}, t) v'(\mathbf{x} + \mathbf{r}, t)} / \overline{v'^2(\mathbf{x}, t)}, \tag{11.1.16b}$$

etc., where R_{11} and R_{22} are the longitudinal and transverse correlations, respectively.

For homogeneous isotropic turbulence, the correlations are independent of $\mathbf{x}$ and are also rotationally invariant. Figure 11.1.4 schematically shows its variation with r, indicating the progressive decrease in the intensity of interaction between two points as their separation distance increases.

A characteristic length, called the integral scale, can thus be defined based on these correlations, such as

$$\ell_o(\mathbf{x}, t) = \int_0^\infty R_{11}(\mathbf{x}, r, t) dr. \tag{11.1.17}$$

Eddies at the integral scale are called energy-containing eddies to indicate the fact that the largest concentration of turbulent kinetic energy occurs in the neighborhood

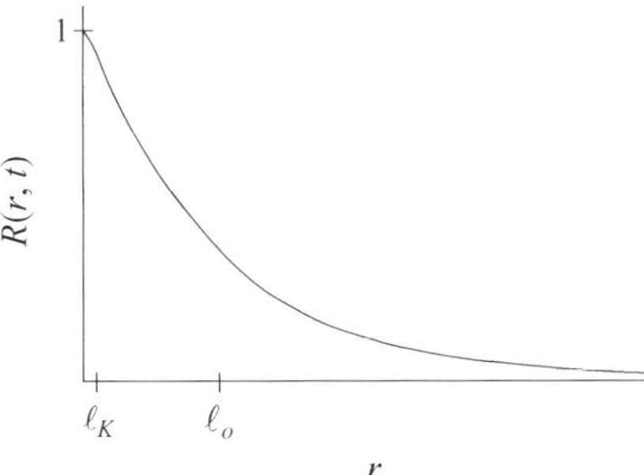

Figure 11.1.4. The normalized two-point velocity correlation for homogeneous isotropic turbulence as a function of the distance r between the two points (Peters 2000).

of ℓ_o. Integral scale eddies are associated with the large scale instabilities of the flow during its transition to turbulence, and the scale is typically set by the dimension of the device or phenomenon, or confinement of the flow field.

By further identifying the characteristic velocity fluctuation at the integral scale as $u'_o = (\overline{u'^2})^{1/2}$, the turbulent kinetic energy of the flow, k, can be expressed as

$$k \approx \frac{3u'^2_o}{2}, \tag{11.1.18}$$

allowing for the three components of the fluctuation and assuming isotropy. Based on the integral scale ℓ_o and the turbulent velocity fluctuation u'_o, a turbulent Reynolds number can be defined as

$$Re_o = \frac{u'_o \ell_o}{\nu}, \tag{11.1.19}$$

where $\nu = \mu/\rho$ is a characteristic kinematic viscosity of the fluid. The turbulent Reynolds number is a measure of the overall turbulence intensity.

Eddies of different sizes in a turbulent flow have different amounts of kinetic energy, determined by the intensity of the velocity fluctuations at each scale of the eddies. The kinetic energy of eddies at one scale is transferred to another through various modes. This energy cascade concept, first established through the Kolmogorov's postulate, states that for turbulent flows at sufficiently high Reynolds numbers, there exists a range of scales through which the energy transfer rate is independent of the molecular viscosity. This universal range of scales through which the energy cascade occurs is called the inertial subrange. Dimensional consideration yields the rate of energy transfer in this range as

$$\epsilon \approx \frac{u'^3_o}{\ell_o} \approx \frac{k^{3/2}}{\ell_o}. \tag{11.1.20}$$

The period of cascade, which can be identified as either the turbulent time or the turnover time of the integral scale eddies, is then

$$\tau_o \approx \frac{\ell_o}{u'_o} \approx \frac{k}{\epsilon}. \tag{11.1.21}$$

Consequently, the turbulent kinetic energy contained in eddies of the integral scale is continuously transferred at a rate ϵ to eddies of smaller sizes, until they eventually reach the smallest size at which viscous dissipation dominates. This minimum scale of eddies is the Kolmogorov scale, ℓ_K. From energy conservation, the rate of energy dissipation in the Kolmogorov eddy must be equal to the rate of energy transfer in the inertial subrange, ϵ. Thus by using ϵ and ν, dimensional analysis leads to the determination of the Kolmogorov time, length, and velocity as

$$\tau_K \approx \left(\frac{\nu}{\epsilon}\right)^{1/2}, \qquad \ell_K \approx \left(\frac{\nu^3}{\epsilon}\right)^{1/4}, \qquad u'_K \approx (\nu\epsilon)^{1/4}. \tag{11.1.22}$$

Using the definition of ϵ, given by (11.1.20), the above relations then show that the Kolmogorov length is related to the integral length through

$$\frac{\ell_o}{\ell_K} \approx Re_o^{3/4}, \tag{11.1.23}$$

which implies that the disparity between the integral and Kolmogorov scales becomes larger as the turbulent Reynolds number increases. Similar manipulation shows that the Kolmogorov time is related to the integral time through

$$\frac{\tau_o}{\tau_K} \approx Re_o^{1/2}, \tag{11.1.24}$$

where we have used (11.1.21).

Since the rate of energy transfer ϵ is constant throughout the entire inertial subrange, we also have

$$\epsilon \approx \frac{u_o'^3}{\ell_o} \approx \frac{u_o'^2}{\tau_o} \approx \frac{u_K'^3}{\ell_K} \approx \frac{u_K'^2}{\tau_K}. \tag{11.1.25}$$

The dependence of the turbulent kinetic energy of a flow on its eddy size can be represented by the turbulent kinetic energy spectrum, $E(K) \sim dk/dK$, where K is the wave number, which is inversely related to the length scale of the eddies. In the inertial subrange, we have

$$E(K) = \frac{dk}{dK} \sim \frac{k}{K} \approx \frac{(\epsilon\ell_o)^{2/3}}{K} \approx \frac{\epsilon^{2/3}}{K^{5/3}}, \tag{11.1.26}$$

which is known as the "$-5/3$ law," indicating the decay rate of the turbulent kinetic energy in the inertial subrange.

A typical energy spectrum of a large Reynolds number flow is shown in Figure 11.1.5. In general, the energy spectrum peaks at the integral length scale, followed by the inertial subrange of the $-5/3$ decay, and then the viscous subrange where the kinetic energy is dissipated by molecular viscosity. As (11.1.23) suggests, the inertial subrange spans a wider range of wave number as the turbulent Reynolds number increases.

We note in passing that in addition to the integral and Kolmogorov scales, studies on nonreactive flows have used an intermediate scale, namely the Taylor scale, ℓ_λ,

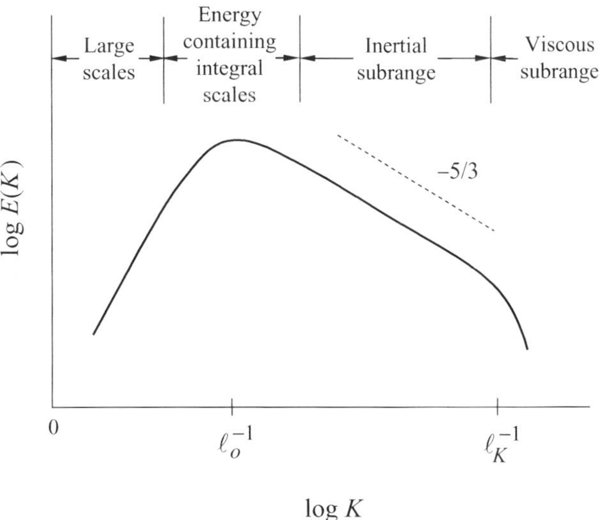

Figure 11.1.5. Schematic representation of the turbulent kinetic energy spectrum as a function of the wavenumber K (Peters 2000).

that represents the distance a Kolmogorov eddy travels during its turnover time when it is convected by an integral-scale eddy. Thus, we have

$$\ell_\lambda \approx u_o' \tau_K \approx \frac{\ell_o}{\tau_o}\tau_K \approx \frac{\ell_o}{Re_o^{1/2}}, \qquad (11.1.27)$$

which shows that for $Re_o \gg 1$, $\ell_K \ll \ell_\lambda \ll \ell_o$. This scale has not been shown to be of importance to the description of turbulent combustion.

In the following, we shall discuss various computational simulation and modeling approaches for turbulent reacting flows. Starting from the Navier–Stokes and scalar transport equations, Favre-averaged equations will be derived and terms representing the averages of the products of fluctuating quantities are generated. The determination of these quantities requires additional independent input, which constitutes the closure problem of turbulence modeling. Various modeling approaches to provide closure of the problem will be discussed. Since the purpose of the present discussion is to illustrate the salient features of turbulence modeling, assumptions will be made on the nonessential aspects of the formulation, leaving the reader to consult advanced literature on turbulence and turbulent combustion, such as the texts by Pope (2000) and Peters (2000) respectively, for more detailed formulations, and review articles by Bradley (1992), Ashurst (1994), Bray (1996), Kerstein (2002), Lipatnikov and Chomiak (2002), and Veynante and Vervisch (2002).

11.2. SIMULATION AND MODELING

We first start from the general conservation equations for reacting flows, described in Chapter 5. For compactness of notation and ease of referencing to the literature, we shall use the Cartesian tensor notation here, with subscripts i and j to indicate quantities associated with the coordinate direction, and the repeated subscripts α and β to indicate summation as in, for example,

$$u_\alpha \frac{\partial u_i}{\partial x_\alpha} = u_1 \frac{\partial u_i}{\partial x_1} + u_2 \frac{\partial u_i}{\partial x_2} + u_3 \frac{\partial u_i}{\partial x_3}$$
$$= u \frac{\partial u_i}{\partial x} + v \frac{\partial u_i}{\partial y} + w \frac{\partial u_i}{\partial z},$$

for $u_i = (u_1, u_2, u_3) = (u, v, w)$. For ease of comprehension, we shall also write out the tensor expressions in terms of the individual Cartesian components for some complicated expressions.

Thus for a mixture of constant c_p, the conservation equations in Cartesian tensor notation are:

Continuity:

$$\frac{\partial \rho}{\partial t} + \frac{\partial(\rho u_\alpha)}{\partial x_\alpha} = 0 \tag{11.2.1}$$

Momentum:

$$\frac{Du_i}{Dt} = -\frac{\partial p}{\partial x_i} + \frac{\partial \sigma_{i\alpha}}{\partial x_\alpha} + \rho g_i \tag{11.2.2}$$

Energy:

$$\frac{Dh^s}{Dt} = \frac{\partial p}{\partial t} + \frac{\partial}{\partial x_\alpha}\left(\frac{\lambda}{c_p}\frac{\partial h^s}{\partial x_\alpha}\right) - q_c w \tag{11.2.3}$$

Species:

$$\frac{DY}{Dt} = \frac{\partial}{\partial x_\alpha}\left(\rho D \frac{\partial Y}{\partial x_\alpha}\right) + w, \tag{11.2.4}$$

where

$$\frac{D}{Dt} \equiv \rho \frac{\partial}{\partial t} + \rho u_\alpha \frac{\partial}{\partial x_\alpha} = \frac{\partial \rho}{\partial t} + \frac{\partial(\rho u_\alpha)}{\partial x_\alpha}$$

is the material derivative,

$$\sigma_{ij} = \mu \left(\frac{\partial u_i}{\partial x_j} + \frac{\partial u_j}{\partial x_i} - \frac{2}{3}\frac{\partial u_\alpha}{\partial x_\alpha}\delta_{ij}\right) \tag{11.2.5}$$

the viscous stress tensor, δ_{ij} the Kronecker delta, and we have not attached the subscripts i to Y to indicate the ith species in order to avoid confusion with the subscript i for the directional coordinate. The formulation is complete with the equation of state, $p = \rho RT$.

11.2.1. Direct Numerical Simulation

Allowing for the restrictions and assumptions imposed in deriving Eqs. (11.2.1) to (11.2.4), the numerical solutions to these equations are correct to the accuracy of the embedded numerical methods. This approach, namely direct numerical simulation (DNS), was started in the 1980s for nonreacting flows and in the 1990s for reacting flows (Poinsot, Candel & Trouvé 1995; Poinsot 1996). It yields first-principle solutions free from turbulence modeling errors such that, with careful implementation of robust high-fidelity numerical schemes, it serves as a valuable tool for the investigation of all the fine-scale physics with utmost realism. To ensure spatially and temporally accurate solutions, the main challenge of DNS is the development of stable, accurate, and efficient numerical methods. It includes high-order spatial discretization and time integration schemes, accurate implicit time integration for the terms with large stiffness, and adaptive mesh refinement to resolve small scales efficiently. Moreover, because the computational domain size is limited, accurate physical boundary conditions that minimally affect the interior solutions are crucial in achieving the fidelity of simulation.

The superior accuracy of DNS, however, comes at the expense of enormous computational demand. Relations (11.1.23) and (11.1.24) put in a nutshell the almost impossible task of computationally resolving a high Re_o turbulent flow field without resorting to statistical methods. Specifically, since $\ell_o/\ell_K \approx Re_o^{3/4}$, then in a flow of D dimension the minimum number of grid points, N, needed to spatially resolve the flow structure down to the Kolmogorov scale is

$$N = (\ell_o/\ell_K)^D = Re_o^{3D/4}.$$

Furthermore, the number of time steps, M, needed to resolve the temporal variation down to the turnover time of a Kolmogorov eddy is

$$M = \tau_o/\tau_K = Re_o^{1/2}.$$

Consequently, the total computational demand would be

$$NM = Re_o^{(3D/4)+(1/2)},$$

which is $NM = Re_o^{11/4} (\approx Re_o^3)$ for a three-dimensional flow ($D = 3$). The steep rise in NM with increasing Re_o is quite evident. For example, if we take $Re_o = 10^4$, then $NM = 10^{11}$, which is a huge number.

The computational demand is further multiplied in the presence of chemical reactions, for it is obvious that there are laminar flames whose thicknesses are much smaller than the Kolmogorov scale, and there are reactions whose rates are much faster than the turnover time of a Kolmogorov eddy. For such situations, resolutions finer than those just mentioned are required. More detailed discussion of various combustion regimes based on the length scales of flames and turbulent flows will be given in later sections. Suffice to note herein that the relevant spatial and temporal scales are respectively given by $(\ell_o/\ell_L) \approx Re_o^{3/4} Ka_L^{-1/2}$ and $(\tau_o/\tau_L) \approx Re_o^{1/2} Ka_L^{-1}$, where ℓ_L and τ_L are respectively the characteristic thickness and time of the laminar flame, and Ka_L is a Karlovitz number based on the Kolmogorov eddy. Thus for $Ka_L < 1$ the demands on the number of grid points and time steps are further increased from the nonreactive flow, with the computational burden being

$$NM = Re_o^{(3D/4)+(1/2)} Ka_L^{-(D/2)-1},$$

which becomes $NM = Re_o^{11/4} Ka_L^{-5/2}$ for $D = 3$.

While it is expected that the role of DNS will continue to increase with advances in high-speed computing hardware, due to its extreme cost, such an approach is not applicable to most practical engineering simulations. Some compromises in resolution must be made and the fine-scale details have to be modeled. In the next three sections we shall present three major classes of approaches that have been developed toward modeling turbulent flows. The first approach is to solve only the statistical mean and variance of the solution variables in Eqs. (11.2.1)–(11.2.5) by averaging the equations, as was done in the last section. This approach is called Reynolds-averaged Navier–Stokes (RANS) models, for which additional modeling approximations must be provided for the unclosed Reynolds stress and flux terms, $\bar{\rho}\widetilde{u''v''}$ and $\bar{\rho}\widetilde{u''\phi''}$, respectively, where ϕ is a reaction scalar. This approach is further subdivided into two levels of modeling approximation. The simpler level, called either the gradient transport model or the turbulent viscosity model, involves approximating the Reynolds transport terms as diffusively transported quantities, with the associated turbulent diffusivity modeled either algebraically or through differential equations. This approach will be discussed in detail in the next section. The second level is called the Reynolds stress model, which pushes the closure assumption to higher moments.

The second approach is called large eddy simulation (LES), which has attracted significant interest in recent years (Piomelli 1999; Janicka & Sadiki 2005). LES is introduced to overcome limitations of the conventional RANS approach by exactly solving the transient and three-dimensional flow field up to a certain length scale that is amenable to the computational capacity, and then modeling the scale that is not resolved. This will involve derivation of spatially filtered equations instead of the statistically averaged ones. The similarities and differences between LES and RANS approaches will be discussed.

The third approach is based on the probability density function (PDF) technique in which modeled transport equations are developed to solve for the *pdf*s.

11.2.2. Reynolds-Averaged Navier–Stokes Models

Applying Favre averaging discussed in Section 11.1.2 and Eqs. (11.2.1)–(11.2.5), we obtain the first moment equations,

$$\frac{\partial \bar{\rho}}{\partial t} + \frac{\partial (\bar{\rho}\tilde{u}_\alpha)}{\partial x_\alpha} = 0 \tag{11.2.6}$$

$$\frac{\tilde{D}\tilde{u}_i}{\tilde{D}t} = -\frac{\partial \bar{p}}{\partial x_i} + \frac{\partial}{\partial x_\alpha}(\bar{\sigma}_{i\alpha} - \bar{\rho}\widetilde{u''_\alpha u''_i}) + \bar{\rho}g_i \tag{11.2.7}$$

$$\frac{\tilde{D}\tilde{h}^s}{\tilde{D}t} = \frac{\partial \bar{p}}{\partial t} + \frac{\partial}{\partial x_\alpha}\left(\overline{\frac{\lambda}{c_p}\frac{\partial h^s}{\partial x_\alpha}} - \bar{\rho}\widetilde{u''_\alpha h^{s\prime\prime}}\right) - q_c\bar{w} \tag{11.2.8}$$

$$\frac{\tilde{D}\tilde{Y}}{\tilde{D}t} = \frac{\partial}{\partial x_\alpha}\left(\overline{\rho D\frac{\partial Y}{\partial x_\alpha}} - \bar{\rho}\widetilde{u''_\alpha Y''}\right) + \bar{w} \tag{11.2.9}$$

where

$$\frac{\tilde{D}}{\tilde{D}t} \equiv \bar{\rho}\frac{\partial}{\partial t} + \bar{\rho}\tilde{u}_\alpha\frac{\partial}{\partial x_\alpha}$$

$$\bar{\sigma}_{ij} = \mu\left(\frac{\partial \bar{u}_i}{\partial x_j} + \frac{\partial \bar{u}_j}{\partial x_i} - \frac{2}{3}\frac{\partial \bar{u}_\alpha}{\partial x_\alpha}\delta_{ij}\right). \tag{11.2.10}$$

In Eqs. (11.2.7) to (11.2.10) we have left the molecular transport terms in the primitive, time-averaged form because Favre averaging does not help to simplify them. In high Reynolds number flows, these terms are usually small and can be neglected anyway, as mentioned earlier. This assumption, however, fails for flows near solid walls where effects of molecular viscosity are important. It could also become questionable for flows with highly diffusive species such as hydrogen, and in the high-temperature flame region where the turbulent Reynolds number can be significantly reduced.

An inspection of Eqs. (11.2.6) to (11.2.10) shows that the statistically averaged equations do not involve only the averaged quantities—additional terms describing exchanges between the fluctuating quantities are generated from the inertial terms. These additional, turbulent flux terms involve the Reynolds stress for $\bar{\rho}\widetilde{u''_i u''_j}$ and Reynolds flux for $\bar{\rho}\widetilde{u''_i h^{s\prime\prime}}$ and $\bar{\rho}\widetilde{u''_i Y''}$. Since there are now more unknowns than equations, the system of equations is not closed. Consequently, additional independent relations, which are necessarily intuitive and approximate in nature, are needed for a complete solution. This constitutes the closure problem in turbulence modeling. For turbulent combustion, closure is also needed for the averaged reaction rate term, $\bar{w}$, that appears in the reactive scalar equations (11.2.8) and (11.2.9). The highly nonlinear nature of this term makes its evaluation based on statistics very difficult. We shall first study the closure methods for nonreacting turbulent flows, and then discuss the reaction rate closure.

11.2.2.1. Gradient Transport Models: Since turbulence is diffusive and dissipative in nature, it is reasonable to draw analogy between the turbulent transport of a flow property and the corresponding molecular diffusive transport, which is proportional to the gradient of that property. As such, the gradient transport approach assumes that the turbulent transport term of a property is proportional to the negative of the gradient of the mean value of that property, with the proportionality constant being a corresponding turbulent transport coefficient. This is readily demonstrated by expressing the Reynolds flux term for the reaction scalar ϕ as

$$-\bar{\rho}\widetilde{u''_i\phi''} = \frac{\mu_T}{\sigma_\phi}\frac{\partial\tilde{\phi}}{\partial x_i}, \tag{11.2.11}$$

where μ_T is the turbulent viscosity coefficient, and σ_ϕ a turbulent Prandtl or Schmidt number. Contrary to their counterparts for the molecular transport, these turbulent transport parameters are properties of the flow, and as such must be independently specified.

To prescribe an expression for the Reynolds stress term $\bar{\rho}\widetilde{u_i''u_j''}$, we first note that the transport should not involve the symmetrical terms $\bar{\rho}\widetilde{u_i''u_i''}$ as momentum transport in one direction is caused by its gradient in the direction normal to it. If we further define the turbulent kinetic energy (per unit mass) based on Favre-averaged quantities as

$$\tilde{k} = \frac{1}{2}\widetilde{u_\alpha''u_\alpha''}, \tag{11.2.12}$$

then we can express the Reynolds stress terms as

$$-\left(\bar{\rho}\widetilde{u_i''u_j''} - \frac{2}{3}\bar{\rho}\tilde{k}\delta_{ij}\right) = \mu_T\left(\frac{\partial \tilde{u}_i}{\partial x_j} + \frac{\partial \tilde{u}_j}{\partial x_i}\right) - \frac{2}{3}\mu_T\frac{\partial \tilde{u}_\alpha}{\partial x_\alpha}\delta_{ij}. \tag{11.2.13}$$

The need for the second term on the LHS of Eq. (11.2.13) can be demonstrated by summing over all i and requiring that the resulting expression be balanced. When $\bar{\rho}\widetilde{u_i''u_j''}$ is used in the Favre averaged equation (11.2.7), this term can be absorbed in the pressure term as $(\partial/\partial x_i)(\bar{p} + 2\bar{\rho}\tilde{k}/3)$.

The assumption of gradient transport relegates closure to the specifications of μ_T and σ_ϕ. Two models have been developed towards specifying μ_T, namely the simplest, Prandtl's mixing length model, and the more descriptive k–ϵ model. Much less study has been conducted on the modeling of σ_ϕ, which is frequently treated as an empirical constant.

Thus, based on the concept of diffusive transport, we have

$$\nu_T = \frac{\mu_T}{\bar{\rho}} \sim u^*\ell^*, \tag{11.2.14}$$

where ν_T is a turbulent kinematic viscosity, and u^* and ℓ^* some velocity and length scales of the flow. Drawing analogy with molecular diffusion, Prandtl proposed the mixing length hypothesis by assuming that ℓ^* is some mixing length ℓ_m over which the turbulent transport is effected, and that

$$u^* \approx \ell_m\left|\frac{\partial \tilde{u}}{\partial y}\right| \tag{11.2.15}$$

for two-dimensional flows. Consequently,

$$\nu_T \approx \ell_m^2\left|\frac{\partial \tilde{u}}{\partial y}\right|. \tag{11.2.16}$$

For a general flow $|\partial\tilde{u}/\partial y|$ is to be evaluated on the basis of the mean rate of strain.

In this model ℓ_m still needs to be specified. For example, for channel or duct flows ℓ_m can be taken to be proportional to the channel height or the duct diameter. For flow near a solid wall, Prandtl assumed that ℓ_m is proportional to the distance from the wall.

The mixing length hypothesis is simple to use and can be predictive if ℓ_m is properly prescribed. However, it has two major limitations. First, ν_T, and, hence, turbulent transport, vanish whenever the gradient of $\tilde{u}$ vanishes, for example, along the axis of a jet. Second, it is inadequate to describe complex flows such as those with recirculation, for which a relevant ℓ_m cannot be readily identified.

A more descriptive approach that at present enjoys considerable usage is the k–ϵ model, also known as the two-equation model. The basic concept here is to describe the two fundamental parameters governing turbulent gradient transport, u^* and ℓ^* or their equivalents, through differential equations instead of algebraically as in the mixing length model. Specifically, we first express

$$u^* \approx \tilde{k}^{1/2}, \qquad \ell^* \approx \frac{\tilde{k}^{3/2}}{\tilde{\epsilon}}, \tag{11.2.17}$$

where $\tilde{\epsilon}$ is the viscous dissipation rate of the turbulent kinetic energy $\tilde{k}$, defined as

$$\tilde{\epsilon} = \nu_T \left\langle \frac{\partial u''_\alpha \partial u''_\alpha}{\partial x_\beta \partial x_\beta} \right\rangle, \tag{11.2.18}$$

and the meaning of the symbol $< \ >$ is identical to that of the superscript $\sim$, indicating Favre averaging. Consequently, from Eq. (11.2.14) we can write

$$\nu_T = C_\mu \frac{\tilde{k}^2}{\tilde{\epsilon}}, \tag{11.2.19}$$

where C_μ is a proportionality constant. Both experimental and DNS studies on simple flows have shown that $\nu_T \tilde{\epsilon}/\tilde{k}^2$ is indeed approximately a constant, having a numerical value that is close to 0.09.

Thus instead of specifying u^* and ℓ^* in order to determine ν_T, the task now is to determine $\tilde{k}$ and $\tilde{\epsilon}$. The approach is to derive from the Navier–Stokes equations differential conservation equations governing their evolution. Extra terms are generated upon averaging, and are subsequently modeled to effect closure. The advantage here is that the spatial variation of ν_T can be described through the corresponding variations of $\tilde{k}$ and $\tilde{\epsilon}$ as a result of their transport, production, and dissipation.

To derive an equation for $\tilde{k}$, we first write Eq. (11.2.2) for i and j. These equations are then respectively cross multiplied by u''_j and u''_i, Favre averaged, and added. We then set $i = j$ for the resulting equation and sum over i, yielding

$$\begin{aligned} \frac{\tilde{D}\tilde{k}}{\tilde{D}t} = &- \bar{\rho}\widetilde{u''_\alpha u''_\beta}\frac{\partial \tilde{u}_\beta}{\partial x_\alpha} - \frac{\partial}{\partial x_\alpha}\left(\bar{\rho}\widetilde{u''_\alpha k''}\right) - \overline{u''_\alpha}\left(\frac{\partial \bar{p}}{\partial x_\alpha} - g_\alpha\right) \\ &- \frac{\partial}{\partial x_\alpha}\left(\widetilde{p' u''_\alpha}\right) + \overline{p'\frac{\partial u''_\alpha}{\partial x_\alpha}} - \bar{\rho}\tilde{\epsilon}. \end{aligned} \tag{11.2.20}$$

In writing Eq. (11.2.20) we have neglected molecular transport on the assumption of high Reynolds number flows.

To proceed with the modeling of the various terms on the RHS of Eq. (11.2.20), we shall first drop the third and fifth terms which vanish for constant-density flows because $\overline{u'_i} \equiv 0$ and $\partial u'_\alpha/\partial x_\alpha \equiv 0$ from continuity. Since modeling of these two terms for variable-density flows is still not agreed upon, they are neglected here on the assumption that their influence on ν_T and, hence, the flow through variable density might be small.

The Reynolds stress $\bar{\rho}\widetilde{u_i''u_j''}$, appearing as the first term on the RHS and representing turbulence generation, will be simply modeled in the manner of Eq. (11.2.13). The second term involves the transport of k'' through the flow fluctuation u_i'', and, hence, can be modeled as the gradient transport of $\tilde{k}$. This also holds for the transport of the pressure work, $p'u_i''$. Consequently it is reasonable to collectively model these two terms as

$$-(\bar{\rho}\widetilde{u_\alpha''k''} + \widetilde{p'u_\alpha''}) = \frac{\mu_T}{\sigma_k}\frac{\partial \tilde{k}}{\partial x_\alpha}, \tag{11.2.21}$$

where $\sigma_k = 1.00$ is recommended.

Summarizing, the k-equation is given by

$$\frac{\tilde{D}\tilde{k}}{\tilde{D}t} = \frac{\partial}{\partial x_\alpha}\left(\frac{\mu_T}{\sigma_k}\frac{\partial \tilde{k}}{\partial x_\alpha}\right) + P_k - \bar{\rho}\tilde{\epsilon}, \tag{11.2.22}$$

where

$$P_k = \mu_T\left(\frac{\partial \tilde{u}_\alpha}{\partial x_\beta} + \frac{\partial \tilde{u}_\beta}{\partial x_\alpha}\right)\frac{\partial \tilde{u}_\alpha}{\partial x_\beta} - \frac{2}{3}\left(\mu_T\frac{\partial \tilde{u}_\beta}{\partial x_\beta} + \bar{\rho}\tilde{k}\right)\frac{\partial \tilde{u}_\alpha}{\partial x_\alpha}. \tag{11.2.23}$$

The basic nature of the ϵ-equation is quite different from that of the k-equation. The k-equation can be considered to be an exact one given the gradient transport assumption and the turbulent viscosity hypothesis. While a procedure similar to that for the k-equation can be performed to derive an equation that describes the transport of $\tilde{\epsilon}$, the various source terms on the RHS are still grouped into the three global terms representing gradient diffusion, production, and destruction. Furthermore, each of these terms requires an empirically determined coefficient. As such, the ϵ-equation is best viewed as entirely empirical (Pope 2000), given by

$$\frac{\tilde{D}\tilde{\epsilon}}{\tilde{D}t} = \frac{\partial}{\partial x_\alpha}\left(\frac{\mu_T}{\sigma_\epsilon}\frac{\partial \tilde{\epsilon}}{\partial x_\alpha}\right) + C_{\epsilon 1}\frac{\tilde{\epsilon}}{\tilde{k}}P_k - C_{\epsilon 2}\bar{\rho}\frac{\tilde{\epsilon}^2}{\tilde{k}}, \tag{11.2.24}$$

where $\sigma_\epsilon = 1.3$, $C_{\epsilon 1} = 1.44$, and $C_{\epsilon 2} = 1.92$.

The k–ϵ model is perhaps the simplest, complete model to describe turbulent flows, and as such is incorporated in many commercial CFD codes for extensive ranges of applications. However, while its performance is acceptable for simple flows, the predictions can be qualitatively incorrect for complex flows. Frequently the constant coefficients need to be adjusted to achieve "predictability," which is questionable from both fundamental and practical viewpoints. The inaccuracies of the model mostly come from the gradient transport assumption and the ϵ-equation.

11.2.2.2. Reynolds Stress Models: In the gradient transport models the Reynolds stress and flux are indirectly determined through the turbulent viscosity coefficient subjected to, and therefore limited by, the assumption of gradient transport. It is then logical to develop differential equations that would describe the transport of these quantities directly. Compared to the gradient transport models, modeling and closure are now delayed to the next level in the hierarchy of equations generated by taking moments of the Navier–Stokes equations. As such, the Reynolds stress models yield improved description of turbulent transport.

To obtain a differential equation for the Reynolds stress $\bar{\rho}\widetilde{u_i''u_j''}$, we follow the same manipulation as that for the derivation of the k-equation, Eq. (11.2.20), except the last steps of setting $i = j$ and summing over i are omitted. This yields

$$\frac{\tilde{D}}{\tilde{D}t}(\widetilde{u_i''u_j''}) = -\left(\bar{\rho}\widetilde{u_\alpha''u_j''}\frac{\partial\tilde{u}_i}{\partial x_\alpha} + \bar{\rho}\widetilde{u_\alpha''u_i''}\frac{\partial\tilde{u}_j}{\partial x_\alpha}\right) - \frac{\partial}{\partial x_\alpha}(\bar{\rho} < u_\alpha''u_i''u_j'' >)$$
$$-\left(\overline{u_j''\frac{\partial p}{\partial x_i}} + \overline{u_i''\frac{\partial p}{\partial x_j}}\right) + \overline{u_j''}g_i + \overline{u_i''}g_j + \overline{u_j''\frac{\partial\sigma_{i\alpha}}{\partial x_\alpha}} + \overline{u_i''\frac{\partial\sigma_{j\alpha}}{\partial x_\alpha}}. \quad (11.2.25)$$

On the RHS of Eq. (11.2.25), the first two terms represent the effects of interaction between fluctuations and mean velocity gradients. The third term is the triple correlation of velocity fluctuations, which requires further modeling approximation. The fourth and fifth terms in the parenthesis involve the pressure gradient, $\overline{u_i''(\partial p/\partial x_j)}$. Each of these terms can be broken up into two terms through $p = \bar{p} + p'$. The first of these two terms, $\overline{u_i''}(\partial\bar{p}/\partial x_j)$, does not exist in constant density flows because for them $\overline{u_i'} \equiv 0$. It represents the coupling between density inhomogeneities and the mean pressure gradient, and was shown (Bray et al. 1981) to account for turbulence production in premixed turbulent flames.

Similarly, to obtain a differential equation for the Reynolds flux $\bar{\rho}\widetilde{u''\phi''}$, we multiply Eq. (11.2.2) by ϕ'', multiply, say, Eq. (11.2.4) (with $Y \equiv \phi$) with the index j by u_i, average the resulting expressions, and then add them. This results in the expression

$$\frac{\tilde{D}}{\tilde{D}t}\left(\widetilde{u_i''\phi''}\right) = -\left(\bar{\rho}\widetilde{u_\alpha''\phi''}\frac{\partial\tilde{u}_i}{\partial x_\alpha} + \bar{\rho}\widetilde{u_\alpha u_i}\frac{\partial\tilde{\phi}}{\partial x_\alpha}\right) - \frac{\partial}{\partial x_\alpha}(\bar{\rho} < u_\alpha''u_i''\phi'' >)$$
$$-\overline{\phi''\frac{\partial p}{\partial x_i}} + \overline{\phi''}g_i + \overline{u_i''w} + \mu\overline{\frac{\partial u_i}{\partial x_\alpha}\frac{\partial\phi}{\partial x_\alpha}}. \quad (11.2.26)$$

Meaning of the individual terms in Eq. (11.2.26) can be interpreted in the same manner as that for Eq. (11.2.25).

In second moment modeling, we frequently also need to know the variance of the reaction scalars, $\widetilde{\phi''^2}$. An equation describing its variation can be obtained by multiplying Eq. (11.2.4) by ϕ'' and averaging,

$$\frac{\tilde{D}}{\tilde{D}t}\widetilde{\phi''^2} = 2\bar{\rho}\widetilde{u_\alpha''\phi''}\frac{\partial\tilde{\phi}}{\partial x_\alpha} - \frac{\partial}{\partial x_\alpha}\left(\bar{\rho}\widetilde{u_\alpha''\phi''^2}\right) + \overline{\phi''w} - \bar{\rho}\mu\overline{\left(\frac{\partial\phi}{\partial x_\alpha}\right)^2}. \quad (11.2.27)$$

11.2.3. Large Eddy Simulation

While the RANS approach has been used successfully, its inherent limitation is that the solution is a statistical mean and, hence, is not sufficient to capture highly transient phenomena or detailed structures of the turbulent flow. However, application of

DNS is still cost-prohibitive. Large eddy simulation (LES) finds a compromise in between, by resolving the large-scale structure while modeling the dissipative small-scale processes. The basic concept stems from Kolmogorov's theory in that, while the large-scale flow motion is geometry and problem specific, at sufficiently high Reynolds numbers the small-scale eddies are more canonical in nature, and their primary role is to dissipate the turbulent kinetic energy. Therefore the important large-scale characteristics are solved accurately while the "subgrid" effects below the numerical resolution are modeled.

To implement LES, all the conservation equations are spatially filtered with a filter of size Δ, which is in general equivalent to the grid size (hence called the "grid filter") of the LES simulation. Let the filtered flow variable, say u, be denoted by an overbar

$$\bar{u}(\mathbf{x}, t) = \int_D G(\mathbf{x}, \mathbf{r}) u(\mathbf{x} - \mathbf{r}, t) d\mathbf{r}, \tag{11.2.28}$$

where the integral is over the entire flow domain D and G is some filtering function satisfying

$$\int_D G(\mathbf{x}, \mathbf{r}) d\mathbf{r} = 1. \tag{11.2.29}$$

Typically, G is the product of three filter functions in each of the space dimensions $x_i, i = 1, 2, 3$. For example, the Gaussian filter

$$G_i(x_i, r_i) = \frac{\sqrt{6}}{\sqrt{\pi}\Delta_i} \exp\left[-6(x_i - r_i)^2/\Delta_i^2\right], \quad i = 1, 2, 3, \tag{11.2.30}$$

or the sharp cutoff filter defined as

$$G_i(x_i, r_i) = \frac{\sin[\pi(x_i - r_i)/\Delta_i]}{\pi(x_i - r_i)}, \quad i = 1, 2, 3, \tag{11.2.31}$$

may be used in each coordinate direction.

Applying the filter in Eq. (11.2.28) to the momentum equation (11.2.2) yields

$$\frac{\tilde{D}\tilde{u}_i}{\tilde{D}t} = -\frac{\partial \bar{p}}{\partial x_i} + \frac{\partial}{\partial x_\alpha}[\bar{\sigma}_{i\alpha} - \bar{\rho}(\widetilde{u_i u_\alpha} - \tilde{u}_i \tilde{u}_\alpha)] + \bar{\rho} g_i, \tag{11.2.32}$$

which looks very similar to the RANS equation (11.2.7). While the formalism of the filtered equation (11.2.32) is based on the filtering of Eq. (11.2.28), it is noted that such a filtering operation is not needed in actual computation because it is the filtered variables that are solved at the grid level. The main difference between Eq. (11.2.32) and the RANS counterpart (11.2.7) is the subgrid-scale stress term

$$\tau_{ij} = \bar{\rho}(\widetilde{u_i u_j} - \tilde{u}_i \tilde{u}_j), \tag{11.2.33}$$

which physically represents turbulent dissipation at the unresolved scale. Since $\widetilde{u_i u_j}$ cannot be determined from the solutions of the filtered equation, this term needs to be modeled. One of the most common approaches is the Smagorinsky model:

$$\tau_{ij} - \frac{1}{3}\delta_{ij}\tau_{kk} = -\nu_t \left(\frac{\partial \tilde{u}_i}{\partial x_j} + \frac{\partial \tilde{u}_j}{\partial x_i}\right) = -2\nu_t \tilde{S}_{ij}, \tag{11.2.34}$$

which is analogous to the gradient transport model. By dimensional argument analogous to that of the mixing length theory, the subgrid-scale viscosity, ν_t, has been modeled as

$$\nu_t = C_s \Delta^2 \left|\tilde{S}\right|, \tag{11.2.35}$$

where C_s, called the Smagorinsky coefficient, is to be determined.

In the Smagorinsky model (Smagorinsky 1963), C_s is predetermined as a constant number, typically 0.3 or less, depending on the flow configuration. In the past, having such an arbitrary constant renders the LES concept to be yet another engineering approximation, albeit much less so than the RANS approach. This has been overcome by the development of the dynamic subgrid model (Germano et al. 1991) that determines the Smagorinsky constant as a part of the integration procedure. The model starts from the scale invariance in the inertial subrange of turbulence spectrum. If the turbulent Reynolds number is sufficiently large and the grid resolution (Δ) properly falls into the inertial subrange, then the amount of the subgrid kinetic energy can be extrapolated from the shape of the spectrum that has a larger scale than Δ. This can be achieved by filtering the grid-level momentum equation (11.2.32) with a "test filter" $\hat{\Delta}$, which is a multiple of the original grid size Δ, typically twice as large. Applying the test filter to Eq. (11.2.36) results in the unresolved stress term

$$T_{ij} = \bar{\rho}\left(\widehat{\widetilde{u_i u_j}} - \hat{\tilde{u}}_i \hat{\tilde{u}}_j\right). \tag{11.2.36}$$

By further applying the test filter to (11.2.33), and then equating the term $\widehat{\widetilde{u_i u_j}}$ from Eqs. (11.2.33) and (11.2.36), we obtain a mathematical identity

$$\widehat{\tilde{u}_i \tilde{u}_j} - \hat{\tilde{u}}_i \hat{\tilde{u}}_j = T_{ij} - \hat{\tau}_{ij}, \tag{11.2.37}$$

which is called the Germano identity. Substituting the Smagorinsky model Eqs. (11.2.34) and (11.2.35) into Eq. (11.2.37) allows the determination of the Smagorinsky constant because it now depends on only $\tilde{u}_i$ and its gradients. It is now computed dynamically and is in general a function of space and time. Similar dynamic procedures can be formulated for other scalar transport equations in modeling subgrid scalar dissipation. It may be noted that there are many subtle issues involved in properly implementing the dynamic procedure without causing numerical instability or preventing the subgrid model from being affected by numerical errors.

11.2.4. Probability Density Functions

The gradient transport and Reynolds stress models are also called first and second moment methods respectively because the mean velocity $\bar{u}_i$ and the Reynolds stresses $\widetilde{u_i'' u_j''}$ are simply the first and second moments of the probability density function of the velocity, $P(\mathbf{u}; \mathbf{x}, t)$. Thus it is logical to attempt to develop a transport equation for P and determine P through it. Once P is known, then the Reynolds stresses can be readily evaluated. By the same token, by knowing the joint *pdf* $P_{\mathbf{u}\phi}(\mathbf{u}, \phi; \mathbf{x}, t)$ of $\mathbf{u}(\mathbf{x}, t)$ and a reaction scalar $\phi(\mathbf{x}, t)$, the Reynolds flux for ϕ, $\widetilde{u_i'' \phi''}$, can be evaluated.

Analogous to the Boltzmann equation for the velocity distribution function in the kinetic theory of gases, transport equations for $P(\mathbf{u}; \mathbf{x}, t)$ and $P_{\mathbf{u}\phi}(\mathbf{u}, \phi; \mathbf{x}, t)$ can be developed in terms of temporal, spatial, velocity, and the reaction scalar variations (Pope 1985, 1990). Thus for $P(\mathbf{u}; \mathbf{x}, t)$ we have

$$\rho\frac{\partial P}{\partial t} + \rho\mathbf{u}\cdot\nabla_{\mathbf{x}}P + (\rho\mathbf{g} - \nabla\bar{p})\cdot\nabla_{\mathbf{u}}P = \nabla_{\mathbf{u}}\cdot[< -\nabla\cdot\sigma + \nabla p'|\mathbf{u} > P], \quad (11.2.38)$$

where $\nabla_{\mathbf{u}}$ is the divergence operator with respect to the velocity vector, and the force factor associated with the $\nabla_{\mathbf{u}}P$ term is simply that due to gravity and the mean pressure gradient.

Similarly for $P_{\mathbf{u}\phi}(\mathbf{u}, \phi; \mathbf{x}, t)$, we have

$$\rho\frac{\partial P_{\mathbf{u},\phi}}{\partial t} + \rho\mathbf{u}\cdot\nabla_{\mathbf{x}}P_{\mathbf{u},\phi} + (\rho\mathbf{g} - \nabla\bar{p})\cdot\nabla_{\mathbf{u}}P_{\mathbf{u},\phi} + \frac{\partial(wP_{\mathbf{u},\phi})}{\partial\phi}$$
$$= \nabla_{\mathbf{u}}\cdot[< -\nabla\cdot\sigma + \nabla p'|\mathbf{u},\phi > P_{\mathbf{u},\phi}] - \frac{\partial}{\partial\phi}[< \nabla\cdot(\rho D\nabla\phi)|\mathbf{u},\phi > P_{\mathbf{u},\phi}]. \quad (11.2.39)$$

It is significant to note that all the terms on the LHS of Eqs. (11.2.38) and (11.2.39), including the inertial terms, are closed and, hence, do not require modeling. The source terms on the RHS, representing effects due to viscous stress, pressure fluctuation, and molecular diffusion, require modeling in the manner discussed previously. Hence the PDF method is still inherently approximate.

The primary merit of the PDF method is that aspects of turbulent transport are in closed form, hence avoiding the assumption of gradient transport. Furthermore, a probability density function contains much more information than the corresponding means and correlations, and, therefore, is potentially more useful. The difficulty lies in the large number of dimensions involved in characterizing the *pdf*; it being seven $(\mathbf{u}, \mathbf{x}, t)$ for single-component flows and many more for reacting flows because the enthalpy and concentrations of all species are independent variables that characterize the *pdf*. Thus it is a challenging task not only to acquire the *pdf* experimentally, but also to obtain the PDF solutions computationally.

11.2.5. Closure of the Reaction Rate Term

In previous sections, the various turbulence modeling approaches are discussed for nonreacting flows. While modeling nonreacting turbulent flows is already a challenging task, modeling reacting flows imparts even more significant challenges in the closure of the highly nonlinear chemical source terms. For example, in the context of the RANS model, solution of the Favre-averaged Eqs. (11.2.8) to (11.2.9) also requires an evaluation of the averaged reaction term, $\bar{w}$, which cannot be evaluated simply in terms of the average thermodynamic properties. That is,

$$\bar{w} \neq w(\bar{p}, \bar{Y}_i, \bar{T}, \ldots).$$

To demonstrate this point, let us consider only the term for the reaction rate constant, $k(T) = Be^{-T_a/T}$, in the reaction rate expression w; the symbol $k(T)$ here should not be confused with the turbulent kinetic energy k. Expanding $k(T)$ with $T = \bar{T} + T'$ for $T' \ll \bar{T}$, we have

$$k(T) = Be^{-T_a/\bar{T}}\left[1 + \left(\frac{T_a}{\bar{T}^2}\right)T' + \frac{1}{2}\left(\frac{T_a}{\bar{T}}\right)^2\left(\frac{T'}{\bar{T}}\right)^2 + \cdots\right],$$

which when averaged becomes

$$\bar{k} = Be^{-T_a/\bar{T}}\left[1 + \frac{1}{2}\left(\frac{T_a}{\bar{T}}\right)^2\left(\frac{\overline{T'T'}}{\bar{T}^2}\right) + \cdots\right]. \tag{11.2.40}$$

Thus in order for the average reaction rate constant $\bar{k}$ to depend only on the average temperature $\bar{T}$, such that $\bar{k} \approx Be^{-T_a/\bar{T}}$, we must require

$$\frac{(\overline{T'T'})^{1/2}}{\bar{T}} \ll \frac{\bar{T}}{T_a}. \tag{11.2.41}$$

However, since $T_a \gg \bar{T}$ for most combustion problems, (11.2.41) holds only for very small values of temperature fluctuations. Thus a straightforward application of averaging on the reaction term is not valid and alternative approaches for evaluation are needed.

One of the earlier attempts to provide a reaction term closure is to assume that the reaction rate is controlled by the mixing process only. The eddy-breakup (EBU) model, originally suggested by Spalding, sets the average reaction rate to be proportional to the variance of the mass fraction of the fuel and a characteristic mixing time determined from the turbulence characteristics, namely k/ϵ, hence

$$\bar{w}_F = \bar{\rho}C_{\text{EBU}}\left(\frac{\epsilon}{k}\right)\left(\overline{Y_F''^2}\right)^{1/2}, \tag{11.2.42}$$

where C_{EBU} is an empirical constant to be adjusted to fit the experimental data. The original EBU model has also been modified to the eddy dissipation model in order to accommodate both premixed and nonpremixed combustion by taking the minimum of the three production/consumption rates of the fuel, oxidizer, and product:

$$\bar{w}_F = \bar{\rho}A\left(\frac{\epsilon}{k}\right)\tilde{Y}_F, \qquad \bar{w}_O = \bar{\rho}A\left(\frac{\epsilon}{k}\right)\frac{\tilde{Y}_O}{\sigma_O},$$

$$\bar{w}_P = \bar{\rho}AB\left(\frac{\epsilon}{k}\right)\frac{\tilde{Y}_P}{1+\sigma_O}, \tag{11.2.43}$$

where σ_O is the stoichiometric oxidizer to fuel mass ratio, and A and B are $O(1)$ model constants. While these models have been widely used in many engineering calculations, their applicability and validity have been limited to specific combustion regimes.

A first-principle approach to the closure of the reaction terms that is applicable to general reacting flows is extremely difficult due to the fundamental differences between various combustion conditions. Therefore, a more feasible option is to identify and characterize distinct combustion regimes on the important physical parameters, and to develop methods that are effective in limited regimes.

A useful concept in the classification of combustion regimes of both premixed and nonpremixed flames is that of the flamelet, for which the laminar flame structure is confined to thin surfaces embedded within turbulent eddies such that the flame basically experiences a laminar flow. Under such situations, we can write

$$w(\mathbf{x}) = w_L \Sigma(\mathbf{x}), \tag{11.2.44}$$

where w_L is the reaction rate per unit area of the flamelet surface, and Σ the flamelet surface area per unit volume, called the flame surface density. An alternate expression for Eq. (11.2.44), based on counting the number of flamelets that cross a given location per unit time, is

$$w(\mathbf{x}) = w_L \nu(\mathbf{x}), \tag{11.2.45}$$

where w_L is now the reaction rate per flamelet crossing, and $\nu(\mathbf{x})$ the crossing frequency. Obviously $\Sigma(\mathbf{x})$ and $\nu(\mathbf{x})$ need to be modeled.

In the next two sections we shall study premixed and nonpremixed flames, respectively, discussing the various regimes characterizing their behavior and some modeling approaches to analyze them.

11.3. PREMIXED TURBULENT COMBUSTION

11.3.1. Regimes of Combustion Modes

Having identified the characteristic turbulent flow time and length scales, we can now compare these scales with those of chemical reaction and laminar flames in order to distinguish the various possible modes of premixed turbulent combustion. From Section 7.2, we have the characteristic length and time scales of the laminar flame as

$$\ell_L = \frac{\nu}{s_L}, \qquad \tau_L = \frac{\nu}{(s_L)^2}, \tag{11.3.1}$$

where $\ell_L = \ell_D^o$, $s_L = s_u^o$ and $\tau_L = \tau_D^o$ are the symbols conventionally used in the turbulent combustion literature for the laminar flame thickness, flame speed, and flame time respectively. We have also used ν in place of D as the characteristic diffusivity. It then follows that the turbulent Reynolds number based on the integral scale, Re_o, is related to the laminar flame quantities as

$$Re_o = \frac{u'_o \ell_o}{\nu} = \frac{u'_o}{s_L}\frac{\ell_o}{\ell_L}. \tag{11.3.2}$$

We are interested in assessing if a laminar flame structure can exist in a turbulent flow. The relevant parameter for such an assessment is the Karlovitz number, Ka, in that the laminar flame structure would be destroyed when the characteristic flow time is shorter than the characteristic flame time. Furthermore, since the smallest turbulence scale is the Kolmogorov scale, it is appropriate to base Ka on properties of the Kolmogorov eddy. Thus we can define a turbulent Karlovitz number as

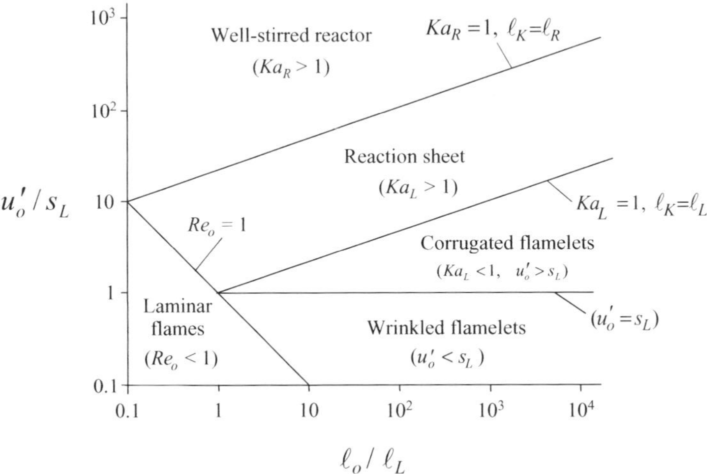

Figure 11.3.1. Regime diagram for premixed turbulent combustion (Peters 2000).

$Ka_L \approx \tau_L/\tau_K$. Furthermore, since $\tau_L = \nu/s_L^2 = \ell_L^2/\nu$ from (11.3.1), and $\tau_K = \nu/(u'_K)^2 = \ell_K^2/\nu$ from (11.1.22), we have

$$Ka_L = \frac{\tau_L}{\tau_K} = \left(\frac{\ell_L}{\ell_K}\right)^2 = \left(\frac{u'_K}{s_L}\right)^2. \tag{11.3.3}$$

Furthermore, from (11.1.25) we have $u'^3_o/\ell_o \approx u'^3_K/\ell_K$. Using these three relations, we have

$$\frac{u'_o}{s_L} = Ka_L^{2/3}\left(\frac{\ell_o}{\ell_L}\right)^{1/3}. \tag{11.3.4}$$

Note that from Eqs. (11.1.23) and (11.3.3) we have $\ell_o/\ell_L = Re_o^{3/4} Ka_L^{-1/2}$, and from Eqs. (11.1.24) and (11.3.3) we obtain $\tau_o/\tau_L = Re_o^{1/2} Ka_L^{-1}$, which are the relations used previously when discussing the computational burdens involved in the DNS of reacting turbulent flows.

In defining Ka_L we have referenced the flame thickness to the Kolmogorov scale. A more refined indication of the presence of chemical reactivity within a Kolmogorov eddy is to reference the reaction zone thickness, ℓ_R, to ℓ_K. Since $\ell_R \sim \ell_L/Ze$, we can then define a Karlovitz number based on ℓ_R as

$$Ka_R = \frac{\tau_R}{\tau_K} = \left(\frac{\ell_R}{\ell_K}\right)^2 = \left(\frac{\ell_R}{\ell_L}\right)^2\left(\frac{\ell_L}{\ell_K}\right)^2 = Ze^{-2} Ka_L. \tag{11.3.5}$$

Figure 11.3.1 plots $\log(u'_o/s_L)$ versus $\log(\ell_o/\ell_L)$ for the three relations (11.3.2), (11.3.4), and (11.3.5), with the specific transition values $Re_o = Ka_L = Ka_R = 1$ and a typical $Ze = 10$. These transition boundaries have slopes of -1, $\frac{1}{3}$, and $\frac{1}{3}$. A boundary for $u'_o/s_L = 1$ is also indicated. These four boundaries identify five burning regimes, with the following characteristics.

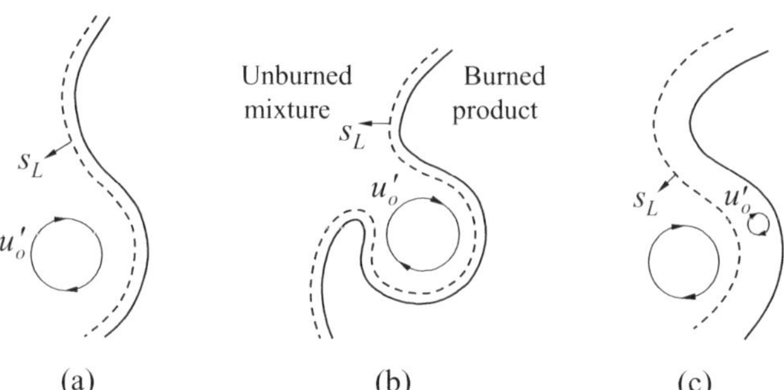

Figure 11.3.2. (a) Weak flame-vortex interaction ($u'_o < s_L$) resulting in a wrinkled flamelet. (b) Strong flame-vortex interaction ($u'_o > s_L$) resulting in a corrugated flamelet. (c) Strong flame-vortex interaction with the smaller eddies penetrating into and broadening the preheat zone of the flame (Peters 2000).

Laminar Flame Regime ($Re_o < 1$)*:* In this regime the turbulence intensity is weak and the turbulence scale is small. The flow is laminar and there is minimum extent of flame wrinkling.

Wrinkled Flamelet Regime ($Re_o > 1, u'_o/s_L < 1$)*:* Since $u'_o/s_L < 1$ and hence $Ka_L < 1$, the flame thickness is much smaller than the Kolmogorov scale. As such, the fundamental flame element retains the laminar flame structure within the turbulent flow field, hence the name laminar flamelet. Since u'_o can be interpreted as the turnover velocity of the large eddies, $u'_o < s_L$ further implies that the flamelet surface is only slightly wrinkled as it passes through these eddies (Figure 11.3.2a).

Corrugated Flamelet Regime ($Re > 1, Ka_L < 1, u'_o/s_L > 1$)*:* Since $Ka_L < 1$, the flame element still retains its laminar flame structure. However, since $u'_o > s_L$, the flamelet becomes highly convoluted upon traversing the eddy (Figure 11.3.2b), with the extent of distortion being of the same order as the size of the eddy and folding of the flamelet is expected. The characteristic eddy size that separates the behaviors of wrinkled and corrugated flames can be assessed by equating the turnover velocity with the laminar flame speed. By calling this eddy size as the Gibson scale, ℓ_G, and from the general relation (11.1.25), we have

$$\frac{\ell_G}{\ell_o} \approx \left(\frac{s_L}{u'_o}\right)^3. \tag{11.3.6}$$

It is reasonable to expect that folding of the flamelet can lead to pockets of unburned and burned mixtures. The unburned pocket will burn out by itself as the enclosing flame propagates inward, provided it does not extinguish due to curvature-induced stretch effects. The burned pocket, however, will grow as the enclosing flame propagates outward. Such a growth will be limited by the continuous interaction with eddies of size ℓ_G, indicating that there is a preference for the formation of burned pockets of size ℓ_G.

Reaction-Sheet Regime ($Re_o > 1$, $Ka_L > 1$, $Ka_R < 1$): The lower boundary of this regime, $Ka_L = 1$, implies $\ell_K \approx \ell_L$ from Eq. (11.3.3). Thus in this regime, although the flame still behaves as a flamelet for the large eddies, the smaller eddies can now penetrate into the preheat zone of the flame structure and thereby enhance the heat and mass transfer rates. The flame is broadened as a consequence (Figure 11.3.2c). The reaction sheet, being thinner than the Kolmogorov scale, $\ell_R^o < \ell_K$, is however only wrinkled, with its structure unaffected by the eddy motion.

Well-Stirred Reactor Regime ($Re_o > 1$, $Ka_R > 1$): In this regime the Kolmogorov eddies are smaller than the reaction zone thickness and as such can penetrate into the reaction zone structure. This facilitates diffusion, and, hence, heat transfer rate to the preheat zone, leading to a precipitous drop in the flame temperature and consequently extinction of the flame. The entire flow now behaves like a well-stirred reactor, without any distinct local structure.

The above classification of the flame regimes is based on comparison of characteristic length and time scales. The boundaries, however, can be significantly modified by considering additional physics. For example, the discussion on wrinkling and corrugation was conducted without considering the significant change in density across the flame. In reality, since the normal flow velocity is greatly increased due to thermal expansion, while the tangential velocity is continuous across the flame, the original vortex structure can be substantially modified downstream of the flame. Thus the efficiency of rolling up a flame by a vortex could be smaller than anticipated.

The impingement of a vortex on a flamelet represents a disturbance to the flamelet, no matter how weak is the vortex (Pan et al. 2002). Such a disturbance could then lead to the development of flamefront hydrodynamic and diffusional-thermal instabilities. For the latter, flame wrinkling can be either facilitated or retarded depending on the mixture Lewis number. Furthermore, since triggering of the diffusional-thermal instability, and the eventual establishment of both the hydrodynamic and diffusional cells, are length-scale dependent, the propensity to develop wrinkles also depends on the characteristic sizes of the eddies.

The discussion has also assumed that the flamelet is stationary, being passively distorted by the vortex. However, as we have learned from our studies on stretched flames in Chapter 10, a freely propagating flame adjusts its location in response to the upstream motion and therefore would resist wrinkling and possibly also extinction. Furthermore, it has been demonstrated that rapid fluctuations in the upstream motion reduces the sensitivity of the flame response, including extinction.

In view of these considerations, the various boundaries shown in Figure 11.3.1, except that of $Re_o = 1$, should be viewed as only tentative, pending further study. We mention in passing that a plot similar to that of Figure 11.3.1, but in slightly different coordinates, was first proposed by Borghi (1988), and is known as the Borghi diagram.

11.3.2. Turbulent Burning Velocities

Similar to the interest in the determination of laminar burning velocities for premixed flames propagation in laminar flows, there is both fundamental and practical interest to describe and determine the corresponding turbulent burning velocities in turbulent flows. The problem, however, is far more complex and less well defined in that while the laminar burning velocity is strictly a function of the diffusive–reactive properties of the mixture, the turbulent burning velocity in addition also depends on the turbulence characteristics of the flow as well as their coupling with the subsequent combustion processes occurring within the flame.

Experimental determination of turbulent burning velocities have adopted four major techniques, namely the Bunsen flame, the rod-stabilized flame, the stagnation or counterflow flame, and the expanding spherical flame. The first three methods involve stationary flames, with turbulence generated upstream by the use of screens, grids, or perforated plates. For the Bunsen and rod-stabilized flames, the turbulence decays as the flow approaches the flame and the true "upstream" turbulence intensity needs to be specified. For the stagnation flow, the adverse pressure gradient in the streamwise direction tends to retard the decay such that with judicious selection of a global strain rate, fairly constant turbulence intensity can be maintained (Cho et al. 1988). For the rod-stabilized flame, additional turbulence can also be generated in the form of the vortices produced as the flow passes over the rod. Because of the globally stationary nature of these three flames, these are also the configurations through which the turbulent flame structure is measured and studied. For the expanding spherical flame, turbulence is generated by several fans oppositely located within the combustion bomb.

Figure 11.3.3 (Kobayashi et al. 1996) shows typical data of the measured turbulent burning velocity, s_T/s_L, as a function of u'_o/s_L for constant pressure. It is seen that, with increasing turbulence intensity, s_T/s_L monotonically increases with increasing turbulence intensity, though with a gradually decreasing slope. This is known as the bending effect.

Various expressions have been derived and proposed for the turbulent burning velocity, mostly phenomenological in nature. We discuss in the following some of these expressions.

11.3.2.1. Reaction Sheet versus Flamelet Descriptions: Damköhler first recognized that, depending on whether the turbulence scale is smaller or larger than the laminar flame thickness, the turbulent flame propagation modes are fundamentally different, as are the situations corresponding to the reaction sheet and wrinkled flamelet regimes. Specifically, when the turbulence scale is smaller than the laminar flame thickness, the turbulent eddies simply modify the transport process between the

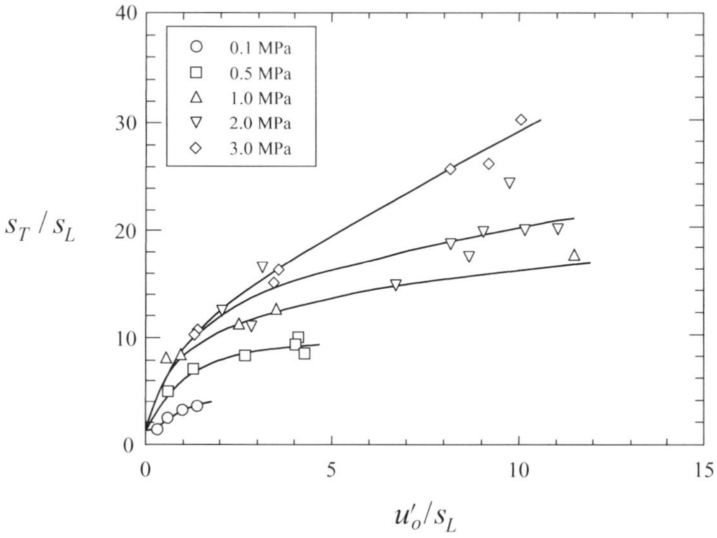

Figure 11.3.3. Experimental turbulent burning velocity as function of turbulence intensity and pressure, for $\phi = 0.9$ methane–air mixtures (Kobayashi et al. 1996).

reaction sheet and the unburned gas. Thus analogous to the laminar flame result of $s_L \sim D^{1/2}$, we can express the turbulent burning velocity as

$$s_T \sim \sqrt{D_T}, \tag{11.3.7}$$

where D and D_T are the molecular and turbulent diffusivities respectively. Consequently,

$$\frac{s_T}{s_L} = \sqrt{\frac{D_T}{D}}. \tag{11.3.8}$$

Furthermore, since $D_T \sim \nu_T \sim u'_o \ell_o$, while $D \sim \nu$, we have

$$\frac{s_T}{s_L} = \sqrt{\frac{u'_o \ell_o}{\nu}} = \sqrt{Re_o}. \tag{11.3.9}$$

When the turbulence scale is larger than the flame thickness, we are in the laminar flamelet regime. Wrinkling of the flame increases its surface area, and hence its total burning rate, such that a turbulent burning velocity can be defined according to (Figure 11.3.4)

$$s_T A = s_L A_T, \qquad \text{or} \qquad \frac{s_T}{s_L} = \frac{A_T}{A}, \tag{11.3.10}$$

where A_T is the total surface area of the wrinkled laminar flame and A the area of the approach flow. Thus the determination of s_T is reduced to an evaluation of the area ratio A_T/A. The following descriptions are all based on this concept, which was first discussed in Section 10.2.3.

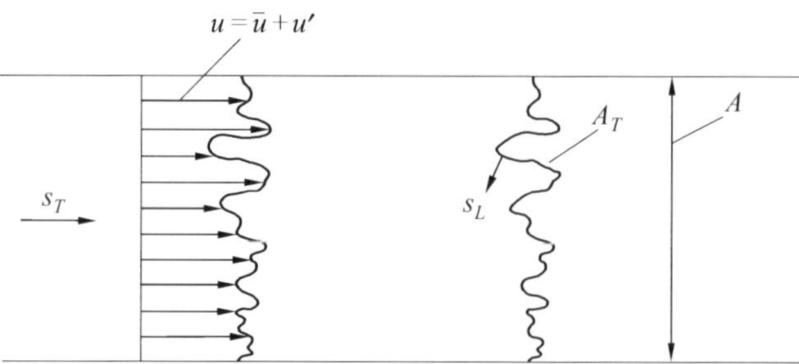

Figure 11.3.4. Definition of the turbulent burning velocity for wrinkled flamelets (Peters 2000).

11.3.2.2. Vector Description: In Section 10.2.3 we used the G-equation to show the dependence of the burning rate on the extent of wrinkling, Eq. (10.2.23), for flames that are not folded. To relate the flame geometry to the flow dynamics, from the flame geometry triangle in Figure 11.3.5 we have

$$\frac{s_T}{s_L} = \frac{A_T}{A} = \frac{\sqrt{\delta x^2 + \delta y^2}}{\delta x} = \sqrt{1 + \left(\frac{\delta y}{\delta x}\right)^2} = \sqrt{1 + \tan^2\theta}. \qquad (11.3.11)$$

For the triangle on flow dynamics, the velocity component normal to the flame surface is s_L while that tangential to it is u'_o, which represents the influence of stretch exerted by the turbulent eddy along the flame surface. Thus

$$\tan\theta = \frac{u'_o}{s_L}, \qquad (11.3.12)$$

which, when inserted into Eq. (11.3.11), yields

$$\frac{s_T}{s_L} = \sqrt{1 + \left(\frac{u'_o}{s_L}\right)^2}. \qquad (11.3.13)$$

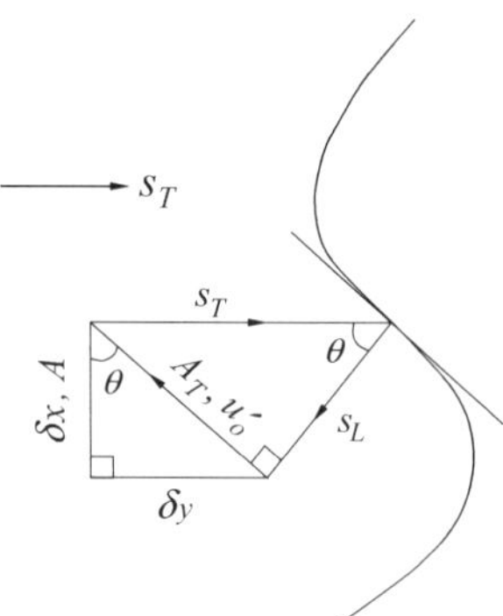

Figure 11.3.5. Vector diagram showing the triangle derivation of the turbulent burning velocity (Williams 1985).

In the limits of weak and strong turbulence, Eq. (11.3.13) respectively becomes

$$\frac{s_T}{s_L} \approx 1 + \frac{1}{2}\left(\frac{u'_o}{s_L}\right)^2, \qquad \text{for } u'_o/s_L \ll 1, \tag{11.3.14}$$

$$\frac{s_T}{s_L} \approx \frac{u'_o}{s_L}, \qquad \text{for } u'_o/s_L \gg 1, \tag{11.3.15}$$

which show quadratic and linear variations respectively. In particular, in the strong turbulence limit Eq. (11.3.15) is simply

$$s_T \approx u'_o, \tag{11.3.16}$$

which shows that the laminar flamelet loses its influence in that the flame surface is passively convected by the turbulent eddies. The turbulent burning rate is then just given by the turbulence intensity. This linear behavior at high turbulence intensities, however, is contrary to the observed bending effect shown in Figure 11.3.3.

Other derivations based on the representation of the various velocity vectors through the triangle relation have been developed, resulting in expressions similar to those identified above.

11.3.2.3. Fractal Description: The simple, triangle description allows only one scale of wrinkling. Since turbulence has a cascade of scales, it is reasonable to expect that surface wrinkling should also exhibit a range of scales. To determine the area of the rough surface of a wrinkled flame, the concept of statistical geometry known as fractals has been applied (Gouldin 1987). Fractals are geometrical objects such as curves, surfaces, volumes, and higher-dimensional bodies that have rugged boundaries and obey certain self-similarity behavior. To appreciate the concept of ruggedness, let us consider the following example.

The circumference of a circle can be determined by first inscribing a polygon of N sides within it and then summing over the length γ of the N sides. As we decrease the measuring length γ and, hence, increase N, the circumference of the polygon becomes closer to that of the circle and eventually approaches it, which has a finite value. The circle is therefore a smooth object. However, for a rugged boundary, such a limit is never approached because no matter how small γ is, there is always a ruggedness whose length is smaller. In fact, as $\gamma \to 0$, the length of the boundary approaches infinity.

For ruggedness that exhibits similarity with the measuring scale γ, the degree of ruggedness can be quantified by a fractal dimension, D, defined as

$$N\gamma^D = 1, \tag{11.3.17}$$

where N is the number of units in the measurement. Thus if we apply Eq. (11.3.17) to the nonfractal objects of a straight line, a square, and a cube, all with sides of unit length, and if we use $\gamma = 1/n$ as our measuring scale, then $N = n$, n^2 and n^3 for the straight line, square, and cube. These objects would respectively have dimensions of $D = 1, 2$, and 3, thereby satisfying Eq. (11.3.17). It is therefore reasonable to define

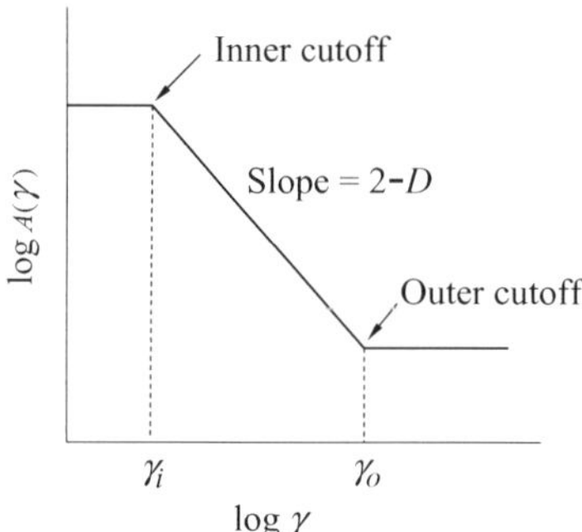

Figure 11.3.6. Self-similar representation of a fractal surface.

the dimension of a fractal object by a noninteger D, as the power of γ through Eq. (11.3.17), given by

$$D = -\frac{\log N}{\log \gamma}. \qquad (11.3.18)$$

Thus if the object is indeed a fractal, and hence exhibits self-similarity, a plot of $\log N$ versus $\log \gamma$ should yield a straight line, with its slope being $-D$.

The area of a wrinkled surface is therefore the sum of N units, each of which has an area of γ^2, as

$$A \sim N\gamma^2 \sim \gamma^{2-D}, \qquad (11.3.19)$$

where (11.3.17) has been used. This relation is shown in Figure 11.3.6. Since the minimum dimension of a wrinkled surface is 2 when the surface is not rugged, and the maximum dimension is 3 when the surface fully fills a volume, the fractal dimension of a wrinkled surface should be between 2 and 3. By the same reasoning the length of a fractal line is $\sim \gamma^{1-D}$, with a fractal dimension between 1 and 2.

The similarity behavior of a fractal phenomenon is expected to break down for sufficiently large and small values of γ, called the outer and inner cutoffs, γ_{outer} and γ_{inner}, respectively (Figure 11.3.6). For a turbulent flow it is reasonable to identify these two cutoffs as the integral and Kolmogorov scales, respectively. By further associating the integral scale eddies with the cross-sectional area of the turbulent flow, A, and the Kolmogorov scale eddies with the flame surface area A_T, we have

$$\frac{s_T}{s_L} = \frac{A_T}{A} = \left(\frac{\gamma_{\text{inner}}}{\gamma_{\text{outer}}}\right)^{(2-D)} = \left(\frac{\ell_K}{\ell_o}\right)^{(2-D)}. \qquad (11.3.20)$$

Using Eq. (11.1.23) which relates ℓ_o/ℓ_K to Re_o, we obtain

$$\frac{s_T}{s_L} = Re_o^{3(D-2)/4}. \qquad (11.3.21)$$

Further studies (Peters 1986; Kerstein 1988a) have suggested that the inner cutoff should be the Gibson scale at which the connected laminar flame structure could be destroyed through flame corrugation and local extinction. Using ℓ_G as γ_{inner} in Eq. (11.3.20) and Eq. (11.3.6), we have

$$\frac{s_T}{s_L} = \left(\frac{\ell_o}{\ell_G}\right)^{(D-2)} = \left(\frac{u'_o}{s_L}\right)^{3(D-2)}, \tag{11.3.22}$$

which exhibits dependence on the laminar burning velocity.

Measurements of turbulent flows and turbulent flames have shown that D varies between 2.31 and 2.36. Using these values in Eq. (11.3.22) shows that the exponent $3(D-2) \approx 1$ such that $s_T \approx u'_o$, the high turbulent intensity limit of the vector description given by Eq. (11.3.16).

11.3.2.4. Dynamic Evolution Description: In this formulation we consider the evolution of a flame surface as it is entrained by a turbulent flow. The surface will have its area stretched as it moves through the flow. Thus in the Lagrangian frame, the evolution of the flame surface area A can be described by the turbulent analogue of the laminar flame stretch equation (10.2.25),

$$\frac{1}{A}\frac{dA}{dt} = \kappa_T, \tag{11.3.23}$$

where κ_T is the characteristic turbulent strain rate. Integrating Eq. (11.3.23) from $t = 0$ to τ_T for $A(0)$ to $A(\tau_T)$ as the flame traverses the turbulent flow, we have

$$\frac{A(\tau)}{A(0)} = e^{\kappa_T \tau_T}. \tag{11.3.24}$$

Identifying the turbulent stretch rate with the turnover time of an integral eddy, we have $\kappa_T \sim u'_o/\ell_o$. Furthermore, the turbulent flame time is $\tau_T \sim \nu_T/s_T^2 \sim u'_o\ell_o/s_T^2$. Substituting these relations into Eq. (11.3.24), and using the definition $s_T A(0) = s_L A(\tau_T)$, we obtain

$$\begin{aligned} \frac{s_T}{s_L} = \frac{A(\tau_T)}{A(0)} &= \exp[(u'_o/s_T)^2] \\ &= \exp[(u'_o/s_L)^2/(s_T/s_L)^2], \end{aligned} \tag{11.3.25}$$

which can be alternately expressed as

$$\left(\frac{s_T}{s_L}\right)^2 \ln\left(\frac{s_T}{s_L}\right)^2 = 2\left(\frac{u'_o}{s_L}\right)^2. \tag{11.3.26}$$

Equation (11.3.26) shows the proper bending behavior as u'_o/s_L increases. A similar approach was developed by Kerstein (1988b), assuming an exponential growth of the flame surface. Also note the similarity of the functional form of Eq. (11.3.26) to Eqs. (8.4.9) and (10.4.28) for loss- and stretch-affected s_L, although the forcing term on the RHS of Eq. (11.3.26) is always positive.

11.3.2.5. Renormalization Theories: Recognizing that processes occurring in turbulent flows involve wide spectra of spatial-temporal scales, renormalization methods

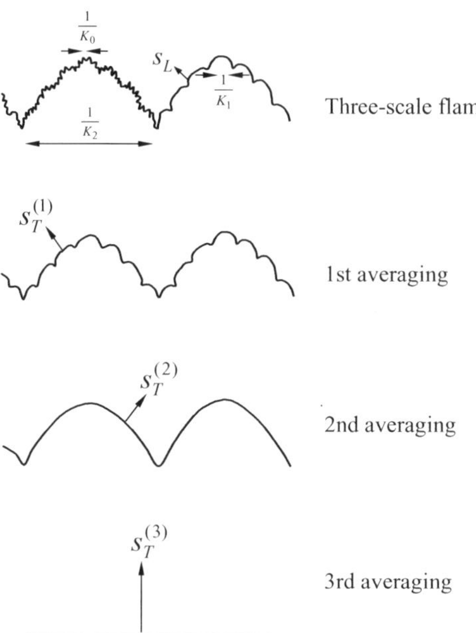

Figure 11.3.7. Concept of cascade renormalization theory used in deriving the turbulent burning velocity through successive averaging of flame wrinkles of progressively larger scales (Sivashinsky 1988).

have been applied to the evaluation of various turbulent properties such as the turbulent transport coefficients. The basic concept is the successive averaging over gradually increasing scales. Sivashinsky (1988), following Yakhot (1988), developed a cascade renormalization theory of turbulent burning velocities. In this approach, the continuous spectrum of a turbulent flow is first replaced by a cascade of eddies of widely separated scales. The relations obtained for these eddies are then extrapolated to the original continuous system. Figure 11.3.7 shows a conceptual representation of the successive averaging over flame wrinkles of progressively larger scales, leading to the derivation of a turbulent burning velocity at each scale of averaging until wrinkles of all scales are averaged with the corresponding identification of the final, global turbulent burning velocity.

It is rather interesting that the renormalization theories of both Yakhot (1988) and Sivashinsky (1988) yield an expression for the turbulent burning velocity that is the same as Eq. (11.3.26). Kerstein (1988b) provided some heuristic arguments that relate the concepts of the two approaches.

11.3.3. Flamelet Modeling

We shall demonstrate some useful concepts in the flamelet modeling of premixed turbulent flames through the Bray–Moss–Libby (BML) theory (Bray 1980), which is perhaps the first rational formulation of such an endeavor. The model basically

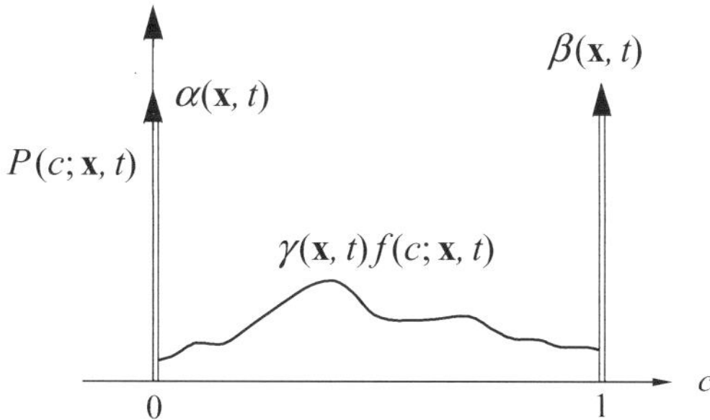

Figure 11.3.8. The probability density function of the progress variable $P(c;\mathbf{x},t)$ (Bray 1980).

tracks the statistics of a progress variable c for a global reaction, with $c(\mathbf{x},t)$ being the fraction of the products generated at $\mathbf{x}$ and time t,

$$c = \frac{Y_P}{Y_{P,b}}, \tag{11.3.27}$$

as already defined in Eq. (5.4.55), where $Y_{P,b}$ is the mass fraction of the product. Thus in terms of c, we have

$$\frac{T}{T_u} = 1 + \hat{q}_c c, \tag{11.3.28}$$

$$\frac{\rho}{\rho_u} = \frac{T_u}{T} = \frac{1}{1+\hat{q}_c c}, \tag{11.3.29}$$

where $\hat{q}_c = (q_c Y_u)/(c_p T_u) = (1-\alpha)/\alpha$, and $\alpha = \rho_b/\rho_u$.

Consider closure of the first moment equation for c, which is simply Eq. (11.2.9), with $\tilde{Y}$ replaced by $\tilde{c}$, and with $Y_{P,b}$ absorbed in $\bar{w}$,

$$\bar{\rho}\frac{\partial \tilde{c}}{\partial t} + \bar{\rho}\tilde{u}_\alpha \frac{\partial \tilde{c}}{\partial x_\alpha} = -\frac{\partial}{\partial x_\alpha}\left(\bar{\rho}\widetilde{u''_\alpha c''}\right) + \bar{w}, \tag{11.3.30}$$

where we have neglected the molecular diffusion term. We aim to express $\bar{\rho}$, $\tilde{u}_\alpha$, the Reynolds flux $\widetilde{u''_\alpha c''}$, and the average reaction rate $\bar{w}$ in terms of $\tilde{c}$.

We first note that by averaging Eq. (11.3.29) in the form $(\rho/\rho_u)(1+\hat{q}_c c) = 1$, say through the use of a *pdf*, $P(c;\mathbf{x},t)$, we readily obtain an explicit expression for $\bar{\rho}$ as

$$\frac{\bar{\rho}}{\rho_u} = \frac{1}{1+\hat{q}_c\tilde{c}}. \tag{11.3.31}$$

The basic premise of the BML model is that in a premixed turbulent flame the product exists in one of three states: a completely unreacted state with $c \equiv 0$, a completely reacted state with $c \equiv 1$, and an intermediate state with various extents of reactedness. Thus the probability of finding the product at $(\mathbf{x},t)$ is described by a *pdf* (Figure 11.3.8),

$$P(c;\mathbf{x},t) = \alpha(\mathbf{x},t)\delta(c) + \beta(\mathbf{x},t)\delta(1-c) + \gamma(\mathbf{x},t)f(c;\mathbf{x},t), \tag{11.3.32}$$

where $\alpha(\mathbf{x}, t)$ and $\beta(\mathbf{x}, t)$ are respectively the strengths of the delta functions $\delta(c)$ and $\delta(1 - c)$, and $f(c; \mathbf{x}, t)$ is the distribution of c in the reacting state. The function $\alpha(\mathbf{x}, t)$ is to be distinguished from the thermal expansion parameter $\alpha = \rho_b/\rho_u$ and the summation index α. From the normalization requirement of $\int_0^1 f(c; \mathbf{x}, t)dc \equiv 1$, we have

$$\alpha(\mathbf{x}, t) + \beta(\mathbf{x}, t) + \gamma(\mathbf{x}, t) = 1. \tag{11.3.33}$$

If we now assume that the flame is very thin, then the probability of finding c in the states of either 0 or 1 is much greater than the states in between. Thus in the flamelet regime $\gamma(\mathbf{x}, t)$ is much smaller than $\alpha(\mathbf{x}, t)$ and $\beta(\mathbf{x}, t)$, which are $O(1)$ quantities. Consequently we expect that the thermodynamic aspects of the flow are mainly affected by $\alpha(\mathbf{x}, t)$ and $\beta(\mathbf{x}, t)$, while the average reaction rate, $\bar{w}$, is solely controlled by $\gamma(\mathbf{x}, t) f(c; \mathbf{x}, t)$ because w vanishes for $c = 0$ and 1.

Thus by assuming $\gamma \ll 1$, the complete statistical description of c can be determined. For example, by evaluating the relation $\bar{\rho}\tilde{c} = \int_0^1 \rho c P(c; \mathbf{x}, t)dc$ using Eq. (11.3.32), we have

$$\alpha(\mathbf{x}, t) = \frac{1 - \tilde{c}}{1 + \hat{q}_c \tilde{c}}, \qquad \beta(\mathbf{x}, t) = \frac{(1 + \hat{q}_c)\tilde{c}}{1 + \hat{q}_c \tilde{c}}. \tag{11.3.34}$$

Other correlations involving $\tilde{c}$ can be similarly derived, such as

$$\tilde{c} = \beta(\mathbf{x}, t), \qquad \widetilde{c''^2} = \tilde{c}(1 - \tilde{c}), \qquad \widetilde{c''^3} = \tilde{c}(1 - \tilde{c})(1 - 2\tilde{c}). \tag{11.3.35}$$

To derive expressions for the averaged quantities of the turbulent fluxes, we utilize the joint *pdf* of c and any velocity component, say u, as

$$\begin{aligned} P(u, c; \mathbf{x}, t) = {} & \alpha(\mathbf{x}, t)\delta(c)P(u, 0; \mathbf{x}, t) + \beta(\mathbf{x}, t)\delta(1 - c)P(u, 1; \mathbf{x}, t) \\ & + \gamma(\mathbf{x}, t)P(u, c; \mathbf{x}, t), \end{aligned} \tag{11.3.36}$$

where $P(u, 0; \mathbf{x}, t)$ and $P(u, 1; \mathbf{x}, t)$ are the conditional *pdf*s representing the distribution of u within the reactants and products respectively. Using $P(u, c; \mathbf{x}, t)$ with $\gamma = 0$, we have

$$\tilde{u}(\mathbf{x}, t) = (1 - \tilde{c})\bar{u}_u(\mathbf{x}, t) + \tilde{c}\bar{u}_b(\mathbf{x}, t), \tag{11.3.37}$$

$$\widetilde{u''c''} = \frac{\overline{\rho(u - \tilde{u})(c - \tilde{c})}}{\bar{\rho}} = \tilde{c}(1 - \tilde{c})(\bar{u}_b - \bar{u}_u). \tag{11.3.38}$$

Equation (11.3.38) reveals an interesting property of the turbulent transport. That is, in a globally steady, planar turbulent flame, due to thermal expansion the mean velocity increases from $\bar{u}_u$ to $\bar{u}_b$ as c increases from 0 to 1. Consequently $\widetilde{u''c''} > 0$. This, however, contradicts the notion of gradient transport,

$$\widetilde{u''c''} = -D_T \frac{\partial \tilde{c}}{\partial x} < 0, \tag{11.3.39}$$

because $\partial \tilde{c}/\partial x > 0$. This phenomenon, called counter-gradient diffusion (Libby & Bray 1981), has been confirmed by extensive computational and experimental studies.

Equations (11.3.31), (11.3.37), and (11.3.38) provide the expressions for $\bar{\rho}$, $\tilde{u}$, and $\widetilde{u''c''}$ in terms of $\tilde{c}$ needed for the transport aspects of Eq. (11.3.30). Evaluation of the average reaction rate,

$$\bar{w}(\mathbf{x}) = \int_0^1 w(c;\mathbf{x},t)P(c;\mathbf{x},t)dc = \gamma(\mathbf{x},t)\int_0^1 w(c;\mathbf{x},t)f(c;\mathbf{x},t)dc, \quad (11.3.40)$$

however, requires the specification of the distribution function $f(c;\mathbf{x},t)$ and therefore is subjected to the uncertainty of modeling (Bray 1980). Alternatively, $\bar{w}$ can be evaluated based on the notions of flame surface density and flame crossing. For example, the flame surface density expression, Eq. (11.2.44), can be expressed as

$$\bar{w} = \rho_u s_L I_o \Sigma, \quad (11.3.41)$$

where I_o describes the effect of stretch on the laminar flame speed s_L, and Σ is to be modeled. In terms of flamelet crossing, we can write

$$\bar{w} = \rho_u s_L I_o \frac{\widetilde{c''^2}}{\ell_{\text{crossing}}} = \rho_u s_L I_o \frac{\tilde{c}(1-\tilde{c})}{\ell_{\text{crossing}}}, \quad (11.3.42)$$

where we have used Eq. (11.3.35), and ℓ_{crossing} is the characteristic length scale of either a laminar flame or isolated pockets of reactants and products. Phenomenologically, the factor $\tilde{c}(1-\tilde{c})$ represents the probability of detecting a flame crossing and $s_L I_o/\ell_{\text{crossing}}$ is the frequency of such crossings.

11.4. NONPREMIXED TURBULENT COMBUSTION

11.4.1. Regimes of Combustion Modes

As in premixed combustion, we shall discuss various combustion modes in nonpremixed turbulent combustion. The essential description should still be based on the characteristic time and length scales, although a nonpremixed flame does not have a propagation velocity and thus identifying a relevant characteristic time scale is not straightforward. Furthermore, the preheat zone of the nonpremixed flame is purely determined by the convective–diffusive transport caused by the turbulent flow field, and thus is little affected by the chemical reaction. In terms of the mixture fraction variable discussed in Chapters 5, 6, and 9, the transport layer thickness is expressed as

$$\ell_L = \frac{1}{|\nabla Z|_{\text{st}}} \approx \sqrt{\frac{\nu}{\chi_{\text{st}}}}, \quad (11.4.1)$$

where $\chi_{st} = 2\nu|\nabla Z|^2_{st}$ is the scalar dissipation rate, defined in Eq. (9.4.15), evaluated at the stoichiometric mixture fraction. Here we omit the factor 2 in the conversion for convenience. The characteristic time scale for transport is then given by

$$\tau_L = \frac{\ell_L^2}{\nu} = \frac{1}{\chi_{st}}. \tag{11.4.2}$$

The other characteristic time scale is that of the chemical reaction, denoted as τ_c. In contrast to the premixed flame where the balance between reaction and transport has to be made under all conditions, for nonpremixed combustion τ_L and τ_c do not have to be the same. In fact, the flame Damköhler number,

$$Da_L = \frac{\tau_L}{\tau_c} = \frac{1}{\chi_{st}\tau_c}, \tag{11.4.3}$$

is an important parameter that represents the relative chemical strength of the flame. If $Da_L \approx 1$, then the residence time within the reaction zone is not long enough to sustain combustion, and thus the flame is prone to extinction.

The above scaling argument can be approximately extended to turbulent combustion. In the spectrum of turbulent eddies, we expect that the Kolmogorov eddies have the shortest turnover time and are most effective in the transport process in the preheat zone (Poinsot & Veynante 2005). Therefore,

$$\ell_L \approx \ell_K, \tag{11.4.4}$$

$$\tau_L \approx \tau_K, \tag{11.4.5}$$

and we can derive the relation between Da_o and Re_o as

$$Da_o = \frac{\tau_o}{\tau_c} = \frac{\tau_o}{\tau_K}\frac{\tau_K}{\tau_c} \approx \frac{\tau_o}{\tau_K}\frac{\tau_L}{\tau_c} = \sqrt{Re_o}\,Da_L. \tag{11.4.6}$$

If $Da_L < 1$, then some reaction zones will be extinguished by the large scalar dissipation rate induced by the smaller eddies near the Kolmogorov scale. Therefore, $Da_L = 1$ sets the boundary between the reaction sheet regime and the broken sheet regime. Hence,

$$Da_o = \sqrt{Re_o} \tag{11.4.7}$$

determines the criterion for the reaction sheet limit. This is shown in Figure 11.4.1. As in Figure 11.3.1, for $Re_o < 1$ molecular diffusion dominates over the effect of turbulent eddies such that laminar flames are formed. For the weaker chemical reaction case of $Da_o < 1$, eddies at all scales can cause a sufficient amount of scalar dissipation to induce flame quenching, and thus no reaction sheet is expected throughout the entire combustion process. This can be referred to as the distributed reaction regime.

As in the premixed combustion case, caution is needed when referring to the regime diagram because boundaries can be modified by considering density change and transient effects.

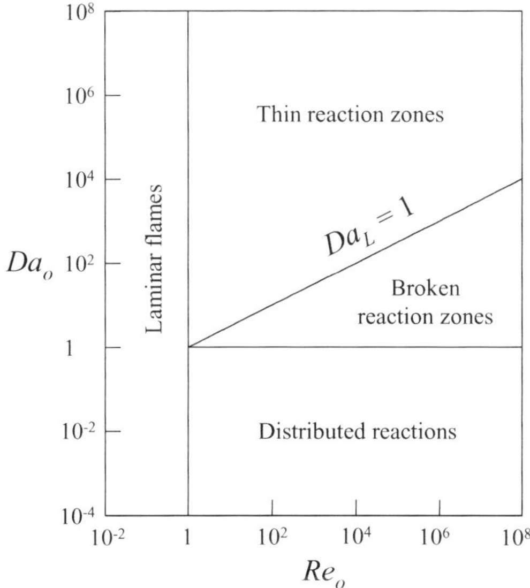

Figure 11.4.1. Regime diagram for nonpremixed turbulent combustion.

11.4.2. Mixture Fraction Modeling

This approach was originally formulated by Bilger (1980). There are basically three components in the formulation. The first component recognizes that description of nonpremixed flames can be facilitated through the canonical description based on the mixture fraction, Z. Specifically, when the diffusivities of all reaction scalars are the same, all conserved scalars including those of the total (chemical plus thermal) enthalpy and the element mass fractions are linear functions of Z. Consequently, if reaction rates are sufficiently rapid such that chemical equilibrium is maintained everywhere, then the local, instantaneous values of all state variables can be determined once we are given the local, instantaneous mixture fraction and, hence, the corresponding pressure, enthalpy, and element mass fractions.

The second component of the formulation is the determination of the statistical averages of these state variables. This is accomplished through the use of a *pdf* for Z, $P(Z)$, such that, for example, the average density is given by $\bar{\rho} = \int_0^1 \rho(Z; \mathbf{x}, t) P(Z; \mathbf{x}, t) dZ$.

There are two approaches to determine $P(Z; \mathbf{x}, t)$. The first is to solve for P differentially, through the probability density function approach of Section 11.2.4. This, however, suffers from uncertainties in the closure approximation as well as extensive computation time. The second approach, which appears to be more fruitful, is to assign a general class of shapes for the *pdf*, characterized by a minimum number of parameters. These parameters are then determined by taking appropriate moments of Z.

Figure 11.4.2 shows sketches of the anticipated shapes of the *pdf* for various types of flows (Bilger 1980). Of particular interest is the delta-function-like behavior at the

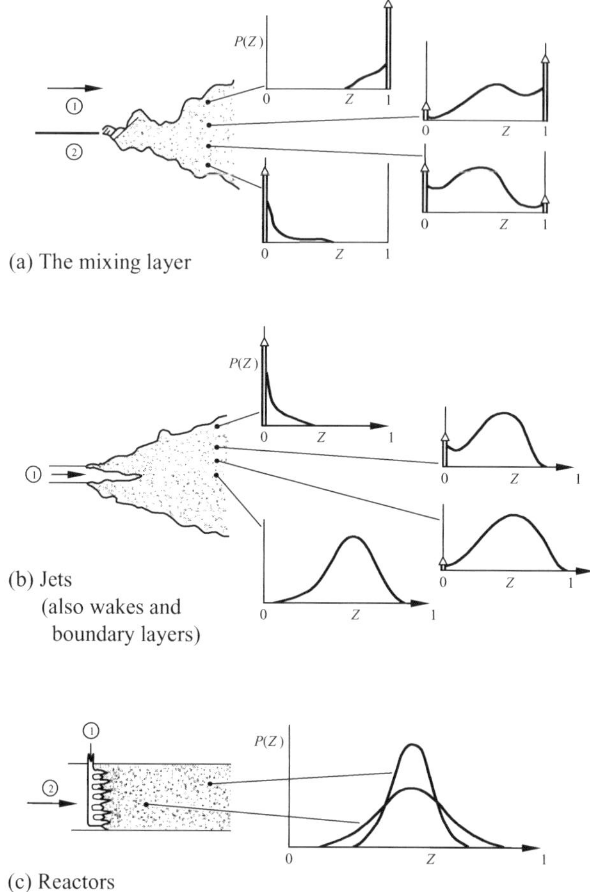

Figure 11.4.2. Probability density function forms for a conserved scalar in various types of flows (Bilger 1980).

edges of mixing layers, jets, and wakes because of intermittency, which is an inherent feature of nonpremixed flows. The function, which offers sufficient flexibility in shape while being characterized by only a small number of parameters, is the beta function, given by

$$\tilde{P}(Z;\mathbf{x},t) = Z^{a-1}(1-Z)^{b-1}\frac{\Gamma(a+b)}{\Gamma(a)\Gamma(b)}, \tag{11.4.8}$$

where a and b are nonnegative constants and Γ is the gamma function. Note that $\tilde{P}(Z;\mathbf{x},t)$ in Eq. (11.4.8) is Favre-designated, thereby indicating that the density term associated in Favre averaging is implicitly absorbed in the function. $\tilde{P}(Z;\mathbf{x},t)$ has therefore been called the Favre *pdf*.

Using $\tilde{P}(Z;\mathbf{x},t)$, it can then be readily shown that

$$a = \tilde{Z}c, \qquad b = (1-Z)c, \tag{11.4.9}$$

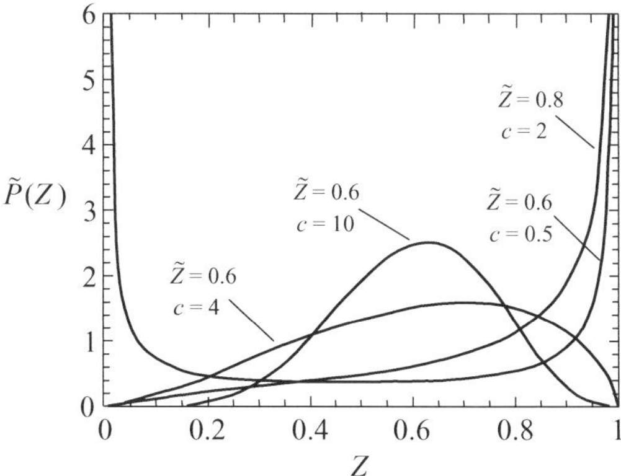

Figure 11.4.3. Shapes of the beta-function *pdf* for different parameters $\tilde{Z}$ and c (Peters 2000).

where

$$c = \frac{\tilde{Z}(1-\tilde{Z})}{\widetilde{Z''^2}} - 1 \geq 0, \tag{11.4.10}$$

which is not to be confused with the progress variable studied in the last section. Figure 11.4.3 plots this Favre *pdf* for various values of $\tilde{Z}$ and c. It is seen that the distribution function approaches that of a Gaussian for large values of c and, hence, small values of the variance $\widetilde{Z''^2}$. Furthermore, a singularity is developed at $\tilde{Z}=0$ for $a<1$ and at $\tilde{Z}=1$ for $b<1$, thereby providing sufficient flexibility to describe the intermittency behavior.

The third component of the formulation is the determination of $\widetilde{Z''}$ and $\widetilde{Z''^2}$. Recognizing that the conservation equations for the mixture fraction is simply those for the conserved scalars, as discussed in Chapter 5, the Favre-averaged equation for Z is

$$\bar{\rho}\frac{\partial \tilde{Z}}{\partial t} + \bar{\rho}\tilde{u}_\alpha \frac{\partial \tilde{Z}}{\partial x_\alpha} = -\frac{\partial}{\partial x_\alpha}\left(\bar{\rho}\widetilde{u''_\alpha Z''}\right), \tag{11.4.11}$$

where we have again neglected the molecular diffusion term. To derive an equation for the Favre variance $\widetilde{Z''^2}$, we multiply the equation for $\tilde{Z}$ by Z'' and average, yielding

$$\bar{\rho}\frac{\partial \widetilde{Z''^2}}{\partial t} + \bar{\rho}\tilde{u}_\alpha \frac{\partial \widetilde{Z''^2}}{\partial x_\alpha} = -\frac{\partial}{\partial x_\alpha}\left(\bar{\rho}\widetilde{u''_\alpha Z''^2}\right) - 2\bar{\rho}\widetilde{u''_\alpha Z''}\frac{\partial \tilde{Z}}{\partial x_\alpha} - 2\rho D\overline{\frac{\partial Z''}{\partial x_\alpha}\frac{\partial Z''}{\partial x_\alpha}}. \tag{11.4.12}$$

To effect closure of Eqs. (11.4.11) and (11.4.12), the Reynolds flux terms can be simply modeled through gradient transport,

$$\widetilde{u''_\alpha Z''} = -D_T\frac{\partial \tilde{Z}}{\partial x_\alpha}, \tag{11.4.13}$$

$$\widetilde{u''_\alpha Z''^2} = -D_T\frac{\partial \widetilde{Z''^2}}{\partial x_\alpha}. \tag{11.4.14}$$

The last term on the RHS of Eq. (11.4.12) is $\bar{\rho}\tilde{\chi}$, where $\tilde{\chi}$ is the Favre-averaged scalar dissipation rate. It can be simply modeled as

$$\tilde{\chi} = C_\chi \frac{\tilde{\epsilon}}{\tilde{k}} \widetilde{Z''^2}, \tag{11.4.15}$$

where C_χ is a constant, and $\tilde{\epsilon}$ and $\tilde{k}$ can be determined in the manner discussed previously. The problem is completely defined at this stage.

PROBLEMS*

1. Assume that experimental studies of turbulent-jet methane–air diffusion flames have been performed under the following sets of conditions of jet exit diameter D and average jet exit velocity U:

Experiment	D (mm)	U (m/s)
A	1	10
B	5	30
C	100	5

For each experiment, estimate at the duct exit and at the position of the maximum average rate of heat release: (a) the turbulence Reynolds number based on the integral scale, (b) the average rate of dissipation of turbulent kinetic energy, (c) the Kolmogorov microscale, (d) the Taylor microscale, (e) the Kolmogorov time, and (f) the time for convection of a fluid element over a Kolmogorov scale in the laboratory frame.

Supplemental information:
(1) The fluctuating velocity at the exit is usually taken as 10% of the mean flow velocity.
(2) For turbulent jets, $u'_{rms}\ell_o \approx const.$
(3) The integral scale ℓ_o is taken as the jet width, which is related to the axial distance of this position, namely the flame length L, to the jet exit, see pp. 202–203 (Peters, 2000).
(4) At the position of the maximum average rate of heat release, the convection velocity of the fluid element is $U(D/L)$.

2. For premixed turbulent combustion in a propane–air mixture at atmospheric pressure and room temperature, having a laminar flame speed of 0.5 m/s, estimate the regime of turbulent combustion (weak-turbulence, distributed-reaction, etc.) for each of the following situations:

(a) The flame is stabilized just downstream from a grid having a mesh spacing of 2 mm in a wind tunnel with an average velocity of 0.6 m/s at the grid.

* Problems 1–5 are courtesy of F. A. Williams.

(b) The flame is held in a burner duct of circular cross-section, 1 m in diameter, by a transverse cylindrical rod 0.1 m in diameter, in a flow with an average velocity of 50 m/s.

(c) The flame is propagating through a cloud 10 m high in the open atmosphere, horizontally, with an ambient wind of 5 m/s.

(d) Combustion occurs in a spherical chamber 0.1 m in diameter having many inlet jets 2 mm in diameter with jet velocities of 100 m/s.

In solving this problem, take the fluctuating velocity to be 10% of the exit velocity, except for Case (a) for which use 3% due to the grid. Take $Ze = 10$.

3. Estimate the fraction of material mixed to molecular scales for an unignited ethylene jet in air at an axial distance of 0.2 m from the jet exit if the exit diameter is 5 mm and the exit velocity is 50 m/s.

4. For a methane–air diffusion flame at normal atmospheric pressure, with reactants initially at 300 K, calculate the average temperature at a point in the turbulent flame brush where the probability density function for the mixture fraction is $P(Z) = 6Z(1 - Z)$, assuming the limit of large Damköhler number. For the methane–air diffusion flame, the stoichiometric mixture fraction is 0.054 and its flame temperature is 2300 K.

5. Show that the beta-function probability-density function

$$P(Z) = \frac{\Gamma(a+b)}{\Gamma(a)\Gamma(b)} Z^{a-1}(1 - Z)^{b-1}$$

approaches a Gaussian in the limit of large values of a and b for very small fluctuations $Z' = Z - \bar{Z}$ about the mean value $\bar{Z}$.

6. Derive the various relations in Eqs. (11.3.34), (11.3.35), and (11.3.38).

12 Combustion in Boundary-Layer Flows

In many practical situations of interest to combustion, high-speed gas flow prevails. Examples are flame stabilization by bluff bodies within the combustion chamber of a gas turbine, accidental or intentional explosion of a combustible by a hot metal particle or projectile, thermal protection of reentry vehicles by ablative heat shields, and the burning of solid and liquid surfaces in an oxidizing gas stream.

When such a high-speed flow is adjacent to either a solid surface or another flow with slower velocity, a transition region exists. Across this region, the flow velocity, and possibly also temperature and concentration, will change from their respective freestream values to either satisfy the boundary conditions required at the solid surface or approach the freestream values of the slower flow. For fluids with small viscosity μ, the transition region is thin and the normal gradient across it, $\partial u/\partial y$, is large such that despite the small μ, the shear stress, $\tau = \mu \partial u/\partial y$, may assume large values. Thus if the characteristic dimensions over which properties change appreciably in the x- and y-directions are ℓ and δ respectively, then the existence of a boundary layer is implied by the condition $\delta/\ell \ll 1$. Furthermore, since for gases the diffusive transport processes of heat, mass, and momentum occur at comparable rates, we expect that the boundary-layer thicknesses for these three processes also should not differ too much from each other. Finally, since it is within the boundary layer where significant changes in the flow properties occur, it is reasonable to expect that rapid variation in the chemical reaction rates, for example, those responsible for the abrupt occurrence of ignition, will also take place within the boundary layer rather than outside of it.

Compared to some of the simpler, one-dimensional problems we have studied, analysis of chemically reacting boundary layers offers two additional complexities. The first problem is the need to analyze momentum transport in addition to the transport of heat and mass. The second is that boundary-layer flows are at least two dimensional because the dominant diffusive transport is normal to the main flow, and are therefore governed by parabolic partial differential equations. Thus one major activity with boundary-layer research is to seek similarity solutions such that instead of depending on, say, both the x- and y-coordinates, the flow of interest varies with

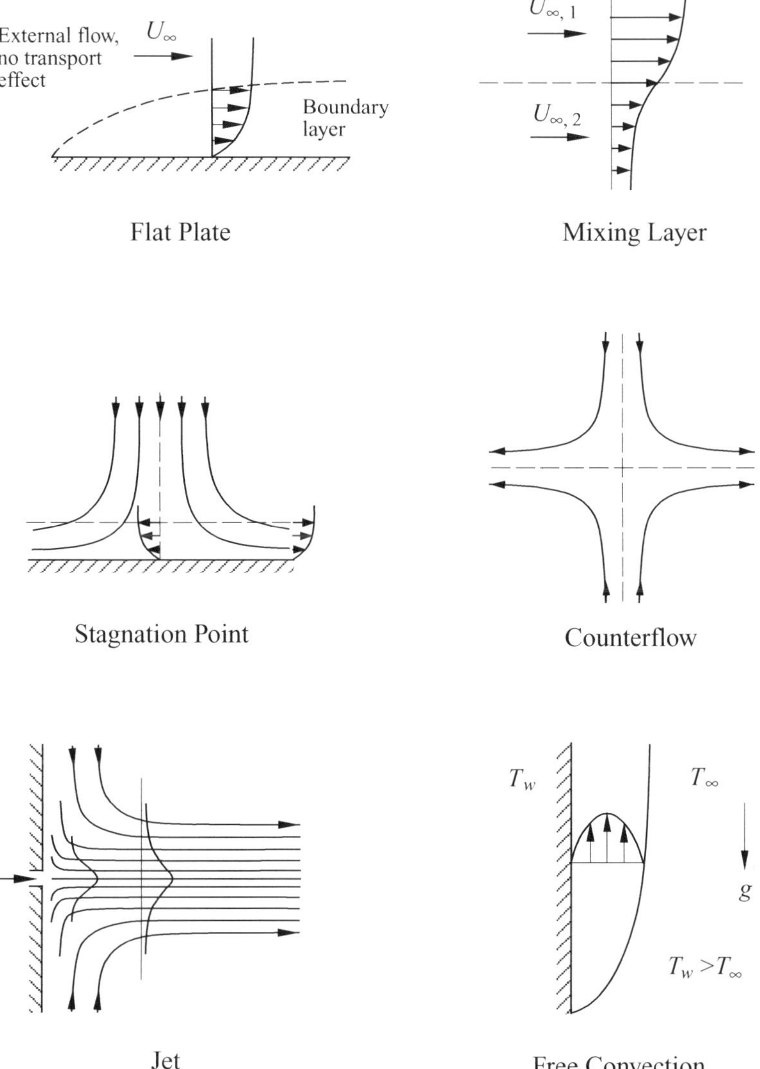

Figure 12.1.1. Various types of boundary- and mixing-layer flows.

only a scaled coordinate, say $\eta = \eta(x, y)$, which depends on both x and y. If such a solution exists, then the flow is governed by ordinary differential equations whose solutions are much simpler. Physically, the existence of a similarity solution implies that the variations of the flow properties in the x- and y-directions scale with each other in a definite relation given by η. Indeed, for boundary-layer flows without chemical reactions, many similarity solutions have been found. The requirement for similarity, however, is more stringent for chemically reacting boundary-layer flows, as will be shown later.

In this chapter, we shall restrict our discussion to steady, two-dimensional boundary layers; some of the examples are shown in Figure 12.1.1. The main purpose here is to demonstrate the general properties of chemically reacting boundary-layer flows and the analytical techniques developed in tackling them. We shall first study subsonic, forced convection flows. For this, we first derive the governing equations and address the question of similarity, in Section 12.1. We shall then present a number of boundary-layer flows, each illustrating some aspects of the flow that are of fundamental and practical interest. The first example, discussed in Section 12.2, is the flame-sheet nonpremixed burning of an ablating solid fuel in an oxidizing gas stream. Here because of the absence of finite-rate reactions, the governing equations are essentially those of the corresponding chemically inert case such that flow similarity is preserved. The next three problems, presented in Section 12.3, all involve the ignition of a premixed combustible. The first problem is in the stagnation flow against a hot surface, for which similarity exists. The second and third problems are ignition along a hot flat plate and by a parallel hot inert stream, for which similarity is weakly violated. The degrees of deviation from similarity, however, are different for these two cases.

In Section 12.4, we shall study boundary-layer flows involving jets, covering the topics of flame heights and the phenomena of flame stabilization and blowout in nonpremixed flows within the context of triple flames, first discussed in Sections 8.6.3 and 8.6.4.

Then, in Sections 12.5 and 12.6, we shall study two special kinds of boundary-layer flows, namely supersonic flows for which the conversion of the kinetic energy of the freestream to the thermal energy in the boundary layer, through viscous slowdown of the flow, can substantially affect the combustion process, and natural convection flows for which the boundary layer is generated through the temperature difference in the fluid medium in the presence of a body force such as gravity.

There is an extensive literature on boundary-layer theory, particularly those by Schlichting et al. (1999) for a comprehensive treatment of nonreactive boundary layers; Stewartson (1964) for compressible flows; Dorrance (1962) and Hayes and Probstein (1959) for supersonic flows; and Chung (1965) and Williams (1985) for chemically reacting flows.

12.1. CONSIDERATIONS OF STEADY TWO-DIMENSIONAL BOUNDARY-LAYER FLOWS

12.1.1. Governing Equations

A steady two-dimensional boundary-layer flow is shown schematically in Figure 12.1.2. The continuity equation and the conservation equations for momentum, energy, and species are respectively given by (Williams 1985):

Continuity:

$$\frac{\partial(\rho u)}{\partial x} + \frac{\partial(\rho v)}{\partial y} = 0 \tag{12.1.1}$$

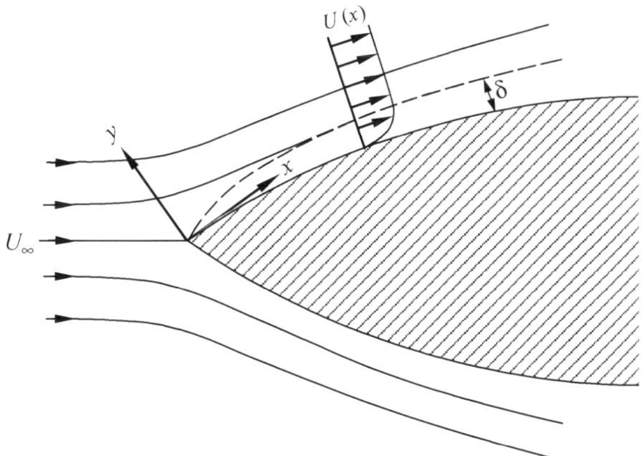

Figure 12.1.2. General representation of a forced-convection two-dimensional boundary-layer flow (Williams 1985).

x-Momentum:

$$\underbrace{u\frac{\partial u}{\partial x}+v\frac{\partial u}{\partial y}}_{(U_\infty^2/\ell)}=\underbrace{\frac{1}{\rho}\left\{-\frac{\partial}{\partial x}\left[\frac{2}{3}\mu\left(\frac{\partial u}{\partial x}+\frac{\partial v}{\partial y}\right)\right]+\frac{\partial}{\partial x}\left(2\mu\frac{\partial u}{\partial x}\right)+\frac{\partial}{\partial y}\left(\mu\frac{\partial v}{\partial x}\right)\right\}}_{Re_\ell^{-1}(U_\infty^2/\ell)}$$

$$+\underbrace{\frac{1}{\rho}\frac{\partial}{\partial y}\left(\mu\frac{\partial u}{\partial y}\right)}_{Re_\ell^{-1}(\ell/\delta)^2(U_\infty^2/\ell)}-\frac{1}{\rho}\frac{\partial p}{\partial x} \tag{12.1.2}$$

y-Momentum:

$$\underbrace{u\frac{\partial v}{\partial x}+v\frac{\partial v}{\partial y}}_{(\delta/\ell)(U_\infty^2/\ell)}=\underbrace{\frac{1}{\rho}\left\{-\frac{\partial}{\partial y}\left[\frac{2}{3}\mu\left(\frac{\partial u}{\partial x}+\frac{\partial v}{\partial y}\right)\right]+\frac{\partial}{\partial y}\left(2\mu\frac{\partial v}{\partial y}\right)+\frac{\partial}{\partial x}\left(\mu\frac{\partial u}{\partial y}\right)\right\}}_{Re_\ell^{-1}(\ell/\delta)(U_\infty^2/\ell)}$$

$$+\underbrace{\frac{1}{\rho}\frac{\partial}{\partial x}\left(\mu\frac{\partial v}{\partial x}\right)}_{Re_\ell^{-1}(\delta/\ell)(U_\infty^2/\ell)}-\frac{1}{\rho}\frac{\partial p}{\partial y} \tag{12.1.3}$$

Scalar Conservation:

$$\underbrace{u\frac{\partial(\tilde{T},\tilde{Y}_i)}{\partial x}+v\frac{\partial(\tilde{T},\tilde{Y}_i)}{\partial y}}_{(U_\infty/\ell)}=\underbrace{\frac{1}{\rho}\frac{\partial}{\partial x}\left[\rho D\frac{\partial(\tilde{T},\tilde{Y}_i)}{\partial x}\right]}_{Re_\ell^{-1}(U_\infty/\ell)}+\underbrace{\frac{1}{\rho}\frac{\partial}{\partial y}\left[\rho D\frac{\partial(\tilde{T},\tilde{Y}_i)}{\partial y}\right]}_{Re_\ell^{-1}(\ell/\delta)^2(U_\infty/\ell)}+\frac{1}{\rho}w_{(\tilde{T},\tilde{Y}_i)}, \tag{12.1.4}$$

where $w_{(\tilde{T},\tilde{Y}_i)}$ is the reaction rate pertinent to $(\tilde{T},\tilde{Y}_i)$, and we have assumed $\kappa=0$ and $Le=1$.

To examine the relative magnitudes of the various terms in the above equations, we first note that since

$$\frac{\partial}{\partial x} \sim \frac{1}{\ell} \quad \text{and} \quad \frac{\partial}{\partial y} \sim \frac{1}{\delta}, \tag{12.1.5}$$

the continuity equation implies that

$$v \sim u\frac{\delta}{\ell}. \tag{12.1.6}$$

Thus if we scale u, v, x, y, ρ, and μ by U_∞, $U_\infty(\delta/\ell)$, ℓ, δ, ρ_∞, and μ_∞ respectively, then the various terms in Eqs. (12.1.2)–(12.1.4) would have their respective magnitudes as indicated, where $Re_\ell = \rho_\infty U_\infty \ell/\mu_\infty$ is the system Reynolds number and U_∞ is some reference freestream velocity. Let us now examine Eq. (12.1.2) for the x-momentum in detail. Invoking the boundary-layer approximation

$$\frac{\delta}{\ell} \ll 1, \tag{12.1.7}$$

it is seen that the term $\rho^{-1}\partial(\mu\partial u/\partial y)/\partial y$ is larger than the rest of the viscous terms by a factor $(\ell/\delta)^2$. Therefore all terms of the order $Re_\ell^{-1}(U_\infty/\ell)$ can be neglected. Furthermore, since viscous effects must be important within the boundary layer, this remaining viscous term must be of the same order as the convection term, implying $Re_\ell^{-1}(\ell/\delta)^2 \sim O(1)$, or

$$Re_\ell \sim \left(\frac{\ell}{\delta}\right)^2 \gg 1. \tag{12.1.8}$$

Thus $Re_\ell \gg 1$ is required for boundary-layer flows. No statement is made about the pressure gradient term, which varies for different flows.

For the y-momentum conservation, we note that the term $\rho^{-1}\partial(\mu\partial v/\partial x)/\partial x$ is smaller than the rest of the viscous terms by a factor of Re_ℓ. Furthermore, the convective terms and the remaining viscous terms in Eq. (12.1.3) are smaller than the leading terms in Eq. (12.1.2) by a factor δ/ℓ. This also holds for the pressure gradient term, $\partial p/\partial y$, because it must balance with the rest of the terms in Eq. (12.1.3). Thus Eq. (12.1.3) can be neglected, implying that there is very small pressure variation across the boundary layer.

Momentum conservation is therefore simplified to

$$\rho u\frac{\partial u}{\partial x} + \rho v\frac{\partial u}{\partial y} - \frac{\partial}{\partial y}\left(\mu\frac{\partial u}{\partial y}\right) = -\frac{\partial p}{\partial x}. \tag{12.1.9}$$

Since in the external flow

$$\rho_\infty U\frac{dU}{dx} = -\frac{dp}{dx}, \tag{12.1.10}$$

where $U(x)$ is the external flow velocity, Eq. (12.1.9) can also be written as

$$\rho u\frac{\partial u}{\partial x} + \rho v\frac{\partial u}{\partial y} - \frac{\partial}{\partial y}\left(\mu\frac{\partial u}{\partial y}\right) = \rho_\infty U\frac{dU}{dx}. \tag{12.1.11}$$

Finally, for the scalar conservation, using $Sc = \mu/\rho D \sim 1$, it is seen that the diffusion term in the streamwise, x-direction can also be neglected.

Summarizing, the governing equations have now been simplified to the following:

Continuity:

$$\frac{\partial(\rho u)}{\partial x} + \frac{\partial(\rho v)}{\partial y} = 0 \tag{12.1.1}$$

x-Momentum:

$$L_b(u) = \rho_\infty U(x)\frac{dU(x)}{dx} \tag{12.1.12}$$

Energy:

$$L_b(\tilde{T}) = -w_F \tag{12.1.13}$$

Species:

$$L_b(\tilde{Y}_i) = w_F, \tag{12.1.14}$$

where the operator $L_b(\cdot)$ is

$$L_b(\cdot) = \left[\rho u\frac{\partial}{\partial x} + \rho v\frac{\partial}{\partial y} - \frac{\partial}{\partial y}\left(\mu\frac{\partial}{\partial y}\right)\right](\cdot). \tag{12.1.15}$$

It is noted that the transport operator, $L_b(\cdot)$, is the same for u, $\tilde{T}$, and $\tilde{Y}_i$; the only differences in Eqs. (12.1.12)–(12.1.14) are the source terms. Furthermore, the primary consequence of the boundary-layer assumption is the neglect of the streamwise diffusion terms. As such, the system is not capable of exhibiting any effect due to streamwise diffusion, especially those associated with preheating of the mixture and back diffusion of the radicals.

12.1.2. Transformation to Boundary-Layer Variables

Equations (12.1.1) and (12.1.12)–(12.1.14) show that the flow is described by a set of coupled parabolic partial differential equations. It is useful to first investigate if a similarity solution exists. To do so let us first transform the governing equations to the conventional variable-density boundary-layer variables through the Howarth–Dorodnitzyn transformation (Howarth 1938, 1948; Dorodnitzyn 1942).

By defining a stream function $\psi(x, y)$ as

$$\rho u = \frac{\partial\psi}{\partial y} \quad \text{and} \quad \rho v = -\frac{\partial\psi}{\partial x}, \tag{12.1.16}$$

the continuity equation is automatically satisfied. Next define the streamwise and transverse independent variables as

$$s = \rho_\infty\mu_\infty\int_0^x U(x')dx' \tag{12.1.17}$$

$$\eta = \frac{U(x)}{\sqrt{2s}} \int_0^y \rho(x, y')dy', \tag{12.1.18}$$

and a boundary-layer stream function $f(s, \eta)$ as

$$f(s, \eta) = \frac{\psi(x, y)}{\sqrt{2s}}. \tag{12.1.19}$$

Note that ψ and s have the dimensions (g/cm-s) and (g/cm-s)^2 respectively, while η and f are nondimensional.

In terms of the above variables, we have

$$\begin{aligned}
\frac{\partial}{\partial x} &= \frac{\partial s}{\partial x}\frac{\partial}{\partial s} + \frac{\partial \eta}{\partial x}\frac{\partial}{\partial \eta} = \rho_\infty \mu_\infty U \frac{\partial}{\partial s} + \frac{\partial \eta}{\partial x}\frac{\partial}{\partial \eta} \\
\frac{\partial}{\partial y} &= \frac{\partial s}{\partial y}\frac{\partial}{\partial s} + \frac{\partial \eta}{\partial y}\frac{\partial}{\partial \eta} = \frac{U\rho}{\sqrt{2s}}\frac{\partial}{\partial \eta} \\
\frac{\partial}{\partial y}\left(\mu \frac{\partial}{\partial y}\right) &= \frac{U^2 \rho}{2s}\frac{\partial}{\partial \eta}\left(\rho\mu \frac{\partial}{\partial \eta}\right) = \frac{\rho_\infty \mu_\infty U^2 \rho}{2s}\frac{\partial^2}{\partial \eta^2},
\end{aligned} \tag{12.1.20}$$

where we have made the Chapman–Rubesin assumption of $\rho\mu = \rho_\infty \mu_\infty =$ constant, which also implies $\rho^2 D =$ constant with the $Sc = 1$ assumption. Furthermore, since

$$\rho u = \frac{\partial \psi}{\partial y} = U\rho \frac{\partial f}{\partial \eta} \tag{12.1.21}$$

$$\begin{aligned}
\rho v &= -\frac{\partial \psi}{\partial x} = -\left(\frac{\partial s}{\partial x}\frac{\partial}{\partial s} + \frac{\partial \eta}{\partial x}\frac{\partial}{\partial \eta}\right)\sqrt{2s}\, f \\
&= -\frac{\rho_\infty \mu_\infty U}{\sqrt{2s}}\left(2s\frac{\partial f}{\partial s} + f\right) - \sqrt{2s}\frac{\partial \eta}{\partial x}\frac{\partial f}{\partial \eta},
\end{aligned} \tag{12.1.22}$$

the operator $L_b(\cdot)$ becomes

$$\begin{aligned}
L_b &\equiv \rho u \frac{\partial}{\partial x} + \rho v \frac{\partial}{\partial y} - \frac{\partial}{\partial y}\left(\mu \frac{\partial}{\partial y}\right) \\
&= U\rho \frac{\partial f}{\partial \eta}\left(\rho_\infty \mu_\infty U \frac{\partial}{\partial s} + \frac{\partial \eta}{\partial x}\frac{\partial}{\partial \eta}\right) \\
&\quad - \frac{\rho_\infty \mu_\infty U^2 \rho}{2s}\left(2s\frac{\partial f}{\partial s} + f\right)\frac{\partial}{\partial \eta} - U\rho \frac{\partial f}{\partial \eta}\frac{\partial \eta}{\partial x}\frac{\partial}{\partial \eta} \\
&\quad - \frac{\rho_\infty \mu_\infty U^2 \rho}{2s}\frac{\partial^2}{\partial \eta^2} \\
&= -\frac{\rho_\infty \mu_\infty U^2 \rho}{2s}\left(\frac{\partial^2}{\partial \eta^2} + f\frac{\partial}{\partial \eta} + 2s\frac{\partial f}{\partial s}\frac{\partial}{\partial \eta} - 2s\frac{\partial f}{\partial \eta}\frac{\partial}{\partial s}\right).
\end{aligned} \tag{12.1.23}$$

Applying Eq. (12.1.23) to Eqs. (12.1.12) through (12.1.14) yields

$$\frac{\partial^2 u}{\partial \eta^2} + f\frac{\partial u}{\partial \eta} + 2s\frac{\partial f}{\partial s}\frac{\partial u}{\partial \eta} - 2s\frac{\partial f}{\partial \eta}\frac{\partial u}{\partial s} = -\left(\frac{2s}{\mu_\infty U}\right)\frac{1}{\rho}\frac{dU}{dx} \tag{12.1.24}$$

$$\frac{\partial^2 \tilde{T}}{\partial \eta^2} + f\frac{\partial \tilde{T}}{\partial \eta} + 2s\frac{\partial f}{\partial s}\frac{\partial \tilde{T}}{\partial \eta} - 2s\frac{\partial f}{\partial \eta}\frac{\partial \tilde{T}}{\partial s} = \left(\frac{2s}{\rho_\infty \mu_\infty U^2}\right)\frac{w_F}{\rho} \tag{12.1.25}$$

$$\frac{\partial^2 \tilde{Y}_i}{\partial \eta^2} + f\frac{\partial \tilde{Y}_i}{\partial \eta} + 2s\frac{\partial f}{\partial s}\frac{\partial \tilde{Y}_i}{\partial \eta} - 2s\frac{\partial f}{\partial \eta}\frac{\partial \tilde{Y}_i}{\partial s} = -\left(\frac{2s}{\rho_\infty \mu_\infty U^2}\right)\frac{w_F}{\rho}. \tag{12.1.26}$$

The momentum equation can be expressed in an alternate form by noting that since $u = U\partial f/\partial \eta$, we have from Eq. (12.1.12)

$$\begin{aligned}
&\rho u\frac{\partial u}{\partial x} + \rho v\frac{\partial u}{\partial y} - \frac{\partial}{\partial y}\left(\mu\frac{\partial u}{\partial y}\right)\\
&\quad = U\left[\rho u\frac{\partial}{\partial x} + \rho v\frac{\partial}{\partial y} - \frac{\partial}{\partial y}\left(\mu\frac{\partial}{\partial y}\right)\right]\frac{\partial f}{\partial \eta} + \rho u\frac{dU}{dx}\frac{\partial f}{\partial \eta}\\
&\quad = -\frac{\rho_\infty \mu_\infty U^3 \rho}{2s}\left(\frac{\partial^3 f}{\partial \eta^3} + f\frac{\partial^2 f}{\partial \eta^2} + 2s\frac{\partial f}{\partial s}\frac{\partial^2 f}{\partial \eta^2} - 2s\frac{\partial f}{\partial \eta}\frac{\partial^2 f}{\partial s \partial \eta}\right)\\
&\qquad + \rho U\frac{dU}{dx}\left(\frac{\partial f}{\partial \eta}\right)^2 = \rho_\infty U\frac{dU}{dx},
\end{aligned} \tag{12.1.27}$$

which upon rearrangement yields

$$\begin{aligned}
&\frac{\partial^3 f}{\partial \eta^3} + f\frac{\partial^2 f}{\partial \eta^2} + 2s\frac{\partial f}{\partial s}\frac{\partial^2 f}{\partial \eta^2} - 2s\frac{\partial f}{\partial \eta}\frac{\partial^2 f}{\partial s \partial \eta}\\
&\quad = -\left(\frac{2s}{\rho_\infty \mu_\infty U^2}\right)\left[\frac{\rho_\infty}{\rho} - \left(\frac{\partial f}{\partial \eta}\right)^2\right]\frac{dU}{dx}.
\end{aligned} \tag{12.1.28}$$

Equations (12.1.24) or (12.1.28), together with (12.1.25) and (12.1.26) constitute the final governing equations expressed in boundary-layer variables. The presence of the density term, ρ_∞/ρ, couples the energy and species equations to the momentum equation, while variation of the flow field, described by the momentum equation, influences energy and species transport through the stream function f.

The $\rho\mu$ = constant assumption, or equivalently $\rho^2 D$ = constant assumption, is a natural outcome in the boundary-layer similarity analysis, and is to be contrasted with the ρD = constant assumption associated with the one-dimensional and quasi-one-dimensional flows studied earlier. Since the kinetic theory of gases shows that $D \sim T^\alpha$, with $\alpha = 1.5$ for the hard-sphere model, $\rho^2 D$ = constant implies a stronger temperature dependence of D while ρD = constant implies a weaker dependence. Thus the adoption of either assumption is equally accurate, or equally inaccurate. Therefore, caution should be exercised when comparing results obtained with different assumptions.

12.1.3. Discussion on Similarity

The coefficients for the source terms involving w_F and dU/dx in Eqs. (12.1.24) to (12.1.28) depend explicitly on the streamwise variable s. This implies that the flow variables should depend on both s and η, instead of on η alone, and hence are intrinsically nonsimilar. Thus a minimum requirement for similarity is to suppress the s-dependence on the RHS of these equations representing the respective forcing functions of the various transport processes. Several special situations exist for which such a requirement can be met.

For a chemically inert flow, $w_F \equiv 0$. Then a necessary condition for the flow to be similar such that all properties vary only with η, that is, $f = f(\eta)$, $\tilde{T} = \tilde{T}(\eta)$, $\tilde{Y}_i = \tilde{Y}_i(\eta)$, and so forth, is

$$\frac{2s}{\rho_\infty \mu_\infty U(x)^2}\frac{dU(x)}{dx} = \text{constant} \tag{12.1.29}$$

in Eq. (12.1.28). If this were not so, then even if we try to force $f = f(\eta)$ into its LHS, an explicit dependence on s, or x, will still appear on the RHS, indicating the inconsistency of the similarity assumption. It can then be easily shown that Eq. (12.1.29) is satisfied for flows described by

$$U(x) \sim x^m, \tag{12.1.30}$$

where m is a constant. These are the Falkner–Skan flows.

The requirement of similarity for a chemically reacting flow, $w_F \neq 0$, is more stringent in that not only Eq. (12.1.29) has to be satisfied, we require in addition that

$$\frac{2s}{\rho_\infty \mu_\infty U(x)^2} = \text{constant}, \tag{12.1.31}$$

such that there is also no explicit dependence on s on the RHS of Eqs. (12.1.25) and (12.1.26). The only flow that simultaneously satisfies Eqs. (12.1.29) and (12.1.31) is the stagnation flow given by

$$U(x) \sim x. \tag{12.1.32}$$

Thus expressing $U(x)$ as

$$U(x) = ax, \tag{12.1.33}$$

where a is the constant velocity gradient at the stagnation point, we have

$$\frac{2s}{\rho_\infty \mu_\infty U(x)^2} = \frac{1}{a}. \tag{12.1.34}$$

Consequently Eqs. (12.1.28), (12.1.25), and (12.1.26) are now ordinary differential equations given by

$$\frac{d^3 f}{d\eta^3} + f\frac{d^2 f}{d\eta^2} = -\left[\frac{\rho_\infty}{\rho} - \left(\frac{df}{d\eta}\right)^2\right] \tag{12.1.35}$$

$$\frac{d^2 \tilde{Y}_i}{d\eta^2} + f\frac{d\tilde{Y}_i}{d\eta} = -\frac{w_F}{a\rho} \tag{12.1.36}$$

$$\frac{d^2 \tilde{T}}{d\eta^2} + f\frac{d\tilde{T}}{d\eta} = \frac{w_F}{a\rho}. \tag{12.1.37}$$

It is also significant to note that since $U \sim x$, Eq. (12.1.18) shows that $\eta \sim y$. Consequently f, $\tilde{Y}_i$, and T all vary only with y, implying that isosurfaces of these quantities are parallel to the stagnation surface.

Another possibility of simplification is to decouple the momentum equation from the energy and species equations, and independently seek a similarity solution for f. There are two ways to suppress the density term, ρ_∞/ρ, which effects the coupling. The first is to simply assume that density is constant such that

$$\frac{\rho_\infty}{\rho} = 1. \tag{12.1.38}$$

Then together with Eq. (12.1.30), or

$$U = ax^m, \tag{12.1.39}$$

where a is now the proportionality constant, the momentum equation (12.1.28) in its similarity form is

$$\frac{d^3 f}{d\eta^3} + f\frac{d^2 f}{d\eta^2} = -\frac{2m}{m+1}\left[1 - \left(\frac{df}{d\eta}\right)^2\right]. \tag{12.1.40}$$

The second case is for $dU/dx = 0$, or

$$U(x) = \text{constant}. \tag{12.1.41}$$

Then Eq. (12.1.28) in its similarity form is

$$\frac{d^3 f}{d\eta^3} + f\frac{d^2 f}{d\eta^2} = 0, \tag{12.1.42}$$

while the density is still treated as varying. Equation (12.1.41) describes the flat-plate and mixing-layer flows and Eq. (12.1.42) is the well-known Blasius equation.

The above decoupling also offers an additional observation. Since now f is similar, the conserved scalar, $\beta_i = \tilde{Y}_i + \tilde{T}$, obtained by eliminating the reaction term from Eqs. (12.1.25) and (12.1.26), may also be similar, being governed by

$$\frac{d^2 \beta_i}{d\eta^2} + f\frac{d\beta_i}{d\eta} = 0, \tag{12.1.43}$$

whose solution is

$$\beta_i(\eta) = c_{i,1} + c_{i,2}\int^{\eta}\left[\exp\left(-\int^{\eta'} f(\eta'')d\eta''\right)\right]d\eta', \tag{12.1.44}$$

where $c_{i,1}$ and $c_{i,2}$ are the integration constants. For the Blasius equation, the integration in Eq. (12.1.44) can be readily effected, yielding

$$\beta_i(\eta) = c_{i,1} + c_{i,2}\frac{df}{d\eta}, \tag{12.1.45}$$

which shows that $\beta_i(\eta)$ varies linearly with the velocity $df/d\eta$. This linear relation between β_i and $df/d\eta$ can also be written, without derivation, by simply observing that

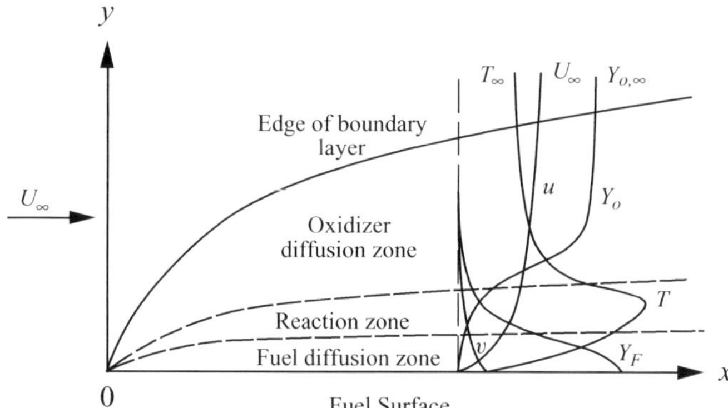

Figure 12.2.1. Velocity and scalar profiles of a nonpremixed flame established within an ablating flat-plate boundary-layer flow (Williams 1985).

they are governed by the same differential operators in Eqs. (12.1.42) and (12.1.43). In fact, by observing from Eqs. (12.1.12) to (12.1.14) that

$$L_b(u) = 0 \quad \text{and} \quad L_b(\beta_i) = 0, \tag{12.1.46}$$

we can readily write

$$\beta_i(x, y) = c_{i,1} + c_{i,3}u(x, y), \tag{12.1.47}$$

which is simply Eq. (12.1.45). The existence of such a linear relation is the consequences of the equal diffusivity assumption for mass, momentum and heat, and the absence of inhomogeneous source/sink terms in the transport processes.

Therefore for both decoupled cases one can first solve for $f(\eta)$ from Eqs. (12.1.40) or (12.1.42), then $\beta_i(\eta)$ from Eq. (12.1.45), and finally $\tilde{T}(s, \eta)$ from Eq. (12.1.25). The solution procedure is greatly facilitated this way.

Finally, it should be cautioned that our discussion on similarity is based only on the form of the differential equations. It is obvious that similarity also depends on the boundary conditions that, in turn, depend on the particular physical situation of interest. Therefore the influence of the boundary conditions on similarity should be examined for each individual case studied. Obviously similarity would not exist if a nonsimilar boundary condition is imposed, for example, an arbitrarily varying surface temperature along a flat plate.

12.2. NONPREMIXED BURNING OF AN ABLATING SURFACE

Figure 12.2.1 illustrates the physical situation of interest (Emmons 1956). Here a semi-infinite flat plate is placed parallel to a gaseous oxidizing stream with a uniform velocity $U(x) = U_\infty$. The fuel plate gasifies and releases fuel vapor into the boundary-layer where it reacts with the oxidizer, resulting in a nonpremixed flame. The practical objective is to determine the burning rate of the fuel, the drag force acting on its surface, and the flame temperature and location.

We shall restrict our study to the reaction-sheet limit such that there is no leakage of the reactants through the flame. This implies that the oxidizer concentration at the plate surface is zero. Furthermore, because of the high thermal conductivity of the condensed phase, we shall also assume that the surface temperature is a constant. Since the reaction-sheet assumption implies that the flow is essentially inert, and because of the constant boundary conditions for Y_O and T, the flow is self-similar.

Our previous discussion shows that for this flow the coupling function β_i is linearly related to the normalized velocity, $f'(\eta)$, as given by Eq. (12.1.45), where $(\cdot)' \equiv d/d\eta$. To evaluate the integration constants, we apply the following boundary conditions:

$$\text{At } \eta = 0: \quad f'(0) = 0, \quad \beta_{O,s} = \tilde{T}_s, \quad \beta_{F,s} = \tilde{Y}_{F,s} + \tilde{T}_s \tag{12.2.1}$$

$$\text{At } \eta = \infty: \quad f'(\infty) = 1, \quad \beta_{O,\infty} = \tilde{Y}_{O,\infty} + \tilde{T}_\infty, \quad \beta_{F,\infty} = \tilde{T}_\infty, \tag{12.2.2}$$

where we have invoked the no-slip boundary condition at the surface.

Thus applying Eqs. (12.2.1) and (12.2.2) to Eq. (12.1.45), we have

$$\tilde{Y}_O + \tilde{T} = \tilde{T}_s + (\tilde{Y}_{O,\infty} + \tilde{T}_\infty - \tilde{T}_s) f' \tag{12.2.3}$$

$$\tilde{Y}_F + \tilde{T} = (\tilde{Y}_{F,s} + \tilde{T}_s) + (-\tilde{Y}_{F,s} + \tilde{T}_\infty - \tilde{T}_s) f'. \tag{12.2.4}$$

Evaluating Eqs. (12.2.3) and (12.2.4) at the flame, where $\tilde{Y}_O(\eta_f) = \tilde{Y}_F(\eta_f) = 0$, we can solve for the flame location and temperature as

$$f'(\eta_f) = \frac{\tilde{Y}_{F,s}}{\tilde{Y}_{F,s} + \tilde{Y}_{O,\infty}} = \Phi^* \tag{12.2.5}$$

$$(\tilde{T}_f - \tilde{T}_s) = (\tilde{Y}_{O,\infty} + \tilde{T}_\infty - \tilde{T}_s)\Phi^*, \tag{12.2.6}$$

where $\Phi^* = \phi^*/(1 + \phi^*) = Z_f$ is simply the stoichiometric mixture fraction identified in Section 5.5 and Chapter 6, and $\tilde{Y}_{F,s}$ can be determined by using a surface gasification law, such as the Clausius–Clapeyron relation, which depends on the surface temperature. It is seen that the flame location η_f is given implicitly by the function $f'(\eta_f)$. Indeed, if $f'(\eta_f)$ were considered as the flame location in $f'(\eta)$ space, then the result is identical to that of the chambered flame treated in Section 6.1.

All of the above results depend on $f(\eta)$, which is to be determined from the Blasius equation (12.1.42),

$$f''' + ff'' = 0, \tag{12.2.7}$$

subject to

$$f'(0) = 0, \quad f'(\infty) = 1, \tag{12.2.8}$$

and a third boundary condition obtained from energy balance at the surface,

$$\left(\lambda \frac{\partial T}{\partial y}\right)_0 = (\rho v)_0 q_v, \tag{12.2.9}$$

where q_v is the heat of gasification per unit mass of fuel. If there is also heat transfer either from the surface to the plate interior or through surface radiation, then a factor

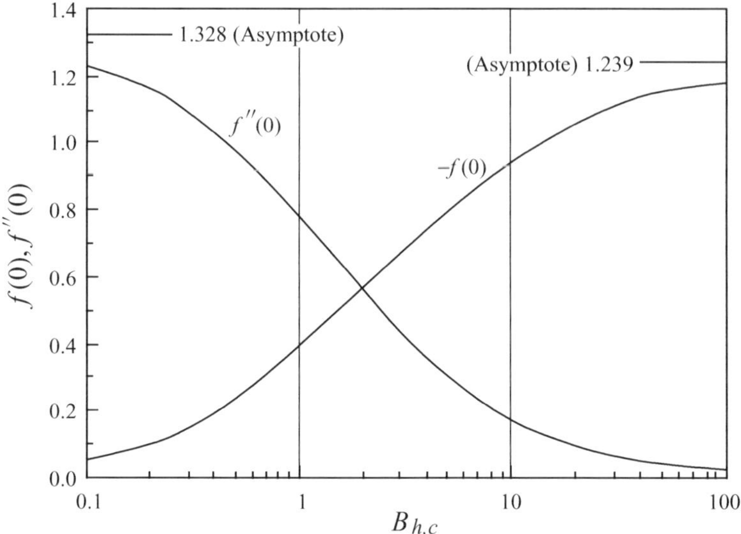

Figure 12.2.2. Solution for the nonpremixed burning of an ablating surface.

q_ℓ can be added to q_v to represent the heat transfer per unit mass of fuel gasified. Substituting Eqs. (12.1.22) and (12.2.3) into Eq. (12.2.9), and noting that $\tilde{Y}_O \equiv 0$ in the inner region to the flame and $\tilde{T} = c_p T/q_c$, we have

$$B_{h,c} f''(0) = -f(0), \tag{12.2.10}$$

where

$$B_{h,c} = \frac{(\tilde{T}_\infty - \tilde{T}_s) + \tilde{Y}_{O,\infty}}{\tilde{q}_v} \tag{12.2.11}$$

is the heat-transfer number identified previously in Eq. (6.4.31) for the droplet combustion problem. This illustrates the physical similarity between the two problems and the fundamental significance of the heat transfer number.

With the additional boundary condition Eq. (12.2.10), Eq. (12.2.7) can be solved by using $B_{h,c}$ as a parameter. The numerical solutions of $f(0)$ and $f''(0)$ are shown in Figure 12.2.2. It is seen that, with increasing $B_{h,c}$, the surface gasification rate represented by $-f(0)$ increases while the surface shear stress represented by $f''(0)$ decreases, which is physically reasonable.

A nondimensional local mass burning rate can be defined as

$$\tilde{m} = \frac{m}{\rho_\infty U_\infty} = \frac{(\rho v)_0}{\rho_\infty U_\infty} = -\frac{f(0)}{\sqrt{2Re_x}}, \tag{12.2.12}$$

where $Re_x = \rho_\infty U_\infty x/\mu_\infty$. Equation (12.2.12) shows that the local mass burning rate varies with $1/\sqrt{x}$. For a plate of length ℓ, an average burning rate can be defined as

$$\tilde{m}_{\text{av}} = \frac{\int_0^\ell \tilde{m} dx}{\int_0^\ell dx} = -\sqrt{2/Re_\ell}\, f(0). \tag{12.2.13}$$

Similarly, a nondimensional local shear stress acting at the plate surface can be defined as

$$\tilde{\tau}_s = \frac{\tau_s}{\rho_\infty U_\infty^2/2} = \frac{(\mu \partial u/\partial y)_0}{(\rho_\infty U_\infty^2/2)} = \sqrt{2/Re_x}\, f''(0), \tag{12.2.14}$$

and an average shear stress as

$$\tilde{\tau}_{s,\mathrm{av}} = \frac{\int_0^\ell \tilde{\tau}_s dx}{\int_0^\ell dx} = 2\sqrt{2/Re_\ell}\, f''(0), \tag{12.2.15}$$

which again indicates that $\tilde{\tau}_s$ varies with $x^{-1/2}$. Figure 12.2.2 shows that the viscous drag on the plate can be significantly reduced with surface blowing due to the outward mass transfer. In particular, the boundary layer can be blown off from the surface with a sufficiently strong blowing, leading to the formation of a viscous, mixing layer in the flow interior.

12.3. IGNITION OF A PREMIXED COMBUSTIBLE

In this section, we shall study the ignition of a cold combustible through boundary-layer heat transfer under three different extents of flow nonsimilarity. The problem is structurally analogous to that studied in Section 8.2 on the ignitability of a stagnant combustible mixture by a hot surface, with the additional presence of a flow.

12.3.1. Ignition at the Stagnation Point

We are interested in the condition to achieve ignition by a cold premixed combustible when it impinges onto a stagnation surface held at a constant temperature T_s (Law 1978a). The flow is self-similar and is governed by Eqs. (12.1.35) to (12.1.37).

Following the convection-free formulation of Section 8.2.5, we define

$$\xi = W(\eta)/W(\infty), \tag{12.3.1}$$

where

$$W(\eta) = \int_0^\eta \exp\left[-\int_0^{\eta'} f(\eta'')d\eta''\right] d\eta' \tag{12.3.2}$$

such that Eqs. (12.1.36) and (12.1.37) are transformed to

$$\frac{d^2\tilde{T}}{d\xi^2} = -\frac{d^2\tilde{Y}_i}{d\xi^2} = -\left[\frac{W(\infty)}{W'(\eta)}\right]^2 \frac{\rho_s}{\rho} Da_C \tilde{Y}_O \tilde{Y}_F e^{-\tilde{T}_a/\tilde{T}}, \tag{12.3.3}$$

with $Da_C = B_C/(\rho_s a)$. The boundary conditions are:

$$\text{At surface:}\quad \xi = 0\ (\eta = 0):\quad \tilde{T} = \tilde{T}_s,\quad d\tilde{Y}_i/d\xi = 0, \tag{12.3.4}$$

$$\text{At freestream:}\quad \xi = 1\ (\eta = \infty):\quad \tilde{T} = \tilde{T}_\infty,\quad \tilde{Y}_i = \tilde{Y}_{i,\infty}. \tag{12.3.5}$$

The transformed Eqs. (12.3.3) to (12.3.5) are basically the same as those for the hot plate problem of Section 8.2, differing only in the preexponential terms.

It is then obvious that, if we carry through the same derivation, the two problems can be made identical if we define a reduced Damköhler number for the present flow

$$\Delta = [W(\infty)]^2 \left(\frac{2\epsilon\, Da_C}{\beta^2}\right) \tilde{Y}_{O,\infty}\tilde{Y}_{F,\infty}e^{-\tilde{T}_a/\tilde{T}_s}, \tag{12.3.6}$$

which is to be compared with Eq. (8.2.7), where $\beta = \tilde{T}_s - \tilde{T}_\infty = O(1)$ and $\tilde{Y}_{F,\infty} \equiv 1$.

Comparing Eqs. (8.2.7) and (12.3.6), and recognizing that $\lambda/c_p \sim \rho D = \rho_s D_s$, it is seen that the two Damköhler numbers differ only by the factors ℓ_o^2/D_s in Eq. (8.2.7) and $[W(\infty)]^2/a$ in Eq. (12.3.6). Both these factors have the dimension of time, representing the characteristic physical time of the flow in the outer, frozen, region. When properly reduced by the multiplicative factor ϵ in the definition of Δ, they represent the physical time available for chemical reaction to take place in the inner, reactive, region. For the hot plate case the factor (ℓ_o^2/D_s) clearly indicates the diffusive nature of the flow, whereas for the stagnation flow the presence of the velocity gradient a through Da_C shows that the characteristic physical time is based on the intensity of the convection flow. However, since diffusion balances convection in the outer region, the characteristic convection time is also indicative of the characteristic diffusion time.

Having identified Δ, and drawing analogy with the relation (8.2.16), it can then be stated that ignition at the stagnation point is possible if

$$\Delta \geq 1 \tag{12.3.7}$$

is satisfied.

Finally, it may be noted that while apparently we have been able to identify the ignition criterion without solving the momentum equation, the effect of the flow is built into the factor

$$W(\infty) = \int_0^\infty \exp\left[-\int_0^\eta f(\eta')d\eta'\right]d\eta.$$

Thus solution for $f(\eta)$ from Eq. (12.1.35) is still needed. It can however be easily shown that to the present degree of accuracy, $W(\infty)$ depends only on the frozen solution through the factor (ρ_∞/ρ_s), but not on the chemical reaction rate. Thus a universal plot of $W(\infty)$ versus the temperature ratio (T_s/T_∞) can be computationally generated and used for all situations.

12.3.2. Ignition along a Flat Plate

In this problem (Law & Law 1979), we have a premixed combustible flowing over a parallel hot flat plate held at a constant temperature T_s. It is of interest to determine the minimum distance to achieve ignition.

Since the momentum equation is decoupled from the energy and species equations, the stream function $f(\eta)$ is just the well-known Blasius solution for the flat plate boundary-layer flow, and therefore can be considered to be known.

The individual variations of $\tilde{T}$ and $\tilde{Y}_i$ are governed by

$$\frac{\partial^2 \tilde{T}}{\partial \eta^2} + f\frac{\partial \tilde{T}}{\partial \eta} - 2xf'\frac{\partial \tilde{T}}{\partial x} = \left(\frac{2x}{U_\infty}\right)\frac{w_F}{\rho} \tag{12.3.8}$$

$$\frac{\partial^2 \tilde{Y}_i}{d\eta^2} + f\frac{\partial \tilde{Y}_i}{\partial \eta} - 2xf'\frac{\partial \tilde{Y}_i}{\partial x} = -\left(\frac{2x}{U_\infty}\right)\frac{w_F}{\rho}. \tag{12.3.9}$$

In writing Eqs. (12.3.8) and (12.3.9) we have used x instead of s because $s \sim x$ for constant $U(x) = U_\infty$.

For this problem we cannot assume that the coupling function β_i is similar and given by Eq. (12.1.45). This is because even though the surface temperature $\tilde{T}_s$ is constant and therefore similar, the surface concentration $\tilde{Y}_{i,s}$ does not need to be so because only its gradient is specified at the surface, that is, $(\partial \tilde{Y}_i/\partial y)_0 = 0$. Thus $\tilde{Y}_{i,s}$ is affected by chemical reaction and is expected to vary in the streamwise direction. Consequently the coupling function $\tilde{T}_s + \tilde{Y}_{i,s}$ does not exist at the surface.

Equations (12.3.8) and (12.3.9) show that the characteristic flow time is now (x/U_∞), and that the intensity of chemical reaction continuously increases with x as the mixture flows downstream. In fact the present problem is of the same nature as that of transient ignition discussed in Section 8.2.3 if we identify (x/U_∞) as a time-like coordinate. In Section 8.2.3, we have forced a similarity solution to the problem. We shall now justify such an approximation in the following.

To investigate the nature of similarity of the problem, let us again suppress the first-order transverse convection term by defining

$$\xi = f'(\eta). \tag{12.3.10}$$

Then Eq. (12.3.8) becomes

$$[f''(\eta)]^2\frac{\partial^2 \tilde{T}}{\partial \xi^2} - 2\xi x\frac{\partial \tilde{T}}{\partial x} = \left(\frac{2x}{U_\infty}\right)\frac{w_F}{\rho}. \tag{12.3.11}$$

Since flow nonsimilarity is caused by chemical reaction that is the most intense next to the hot surface, it is reasonable to expect that nonsimilarity is also the strongest there. To analyze the flow in this reactive region, from our experience with Section 8.2 we can define an inner variable $\chi = \beta\xi/\epsilon$, where $\beta = O(1)$ and $\epsilon \ll 1$. Thus transforming Eq. (12.3.11) into the χ coordinate, and noting that $f''(\eta) \simeq f''(0) = 0.4696$ next to the surface, we have

$$\frac{\partial^2 \tilde{T}_{\text{in}}}{\partial \chi^2} - \frac{2x}{[f''(0)]^2}\left(\frac{\epsilon}{\beta}\right)^3\frac{\partial \tilde{T}_{\text{in}}}{\partial x} = \left[\frac{\epsilon}{\beta f''(0)}\right]^2\left(\frac{2x}{U_\infty}\right)\left(\frac{w_F}{\rho}\right)_{\text{in}}. \tag{12.3.12}$$

Equation (12.3.12) shows that although $\tilde{T}$ is nonsimilar and varies with both χ and x, the influence by the streamwise convection term is extremely weak, being smaller than the $O(1)$ transverse diffusion term by a factor ϵ^3. This ϵ^3 factor results from the $O(\epsilon^2)$ convectional variation in the streamwise direction, and from the small, $O(\epsilon)$

velocity, f', in the reaction layer next to the plate. Thus by neglecting this $O(\epsilon^3)$ term, we have

$$\frac{d^2\tilde{T}_{\text{in}}}{d\chi^2} \approx \left[\frac{\epsilon}{\beta f''(0)}\right]^2 \left(\frac{2x}{U_\infty}\right)\left(\frac{w_F}{\rho}\right)_{\text{in}}. \tag{12.3.13}$$

Such an approximation is known as that of local similarity, in which the streamwise variation is sufficiently weak such that history effect can be neglected and its influence is therefore manifested parametrically rather than differentially. In the present case, the parameter is simply the coefficient x multiplying the reaction rate w_F. The mathematical simplification is substantial because the system is now governed by ordinary, instead of partial, differential equations.

Let us next consider the broad, diffusive-convective outer region. Here we cannot assume the flow is similar even if it is chemically nonreactive. The reason being that the outer flow is now not in direct contact with the flat plate, whose constant temperature would have rendered the flow field similar. Rather, the inner boundary of the outer flow, as $\xi \to 0$, is the outer boundary of the inner flow, as $\chi \to \infty$. Since the inner flow is not similar, the outer flow must necessarily be affected through its boundary with the inner flow, as $\xi \to 0$, and therefore should be also nonsimilar. In other words, the outer flow now experiences a "wall" temperature that continuously increases with x.

While the outer flow is not similar, it is nevertheless reasonable to expect that it is locally similar because the location at which similarity is most severely violated, $\xi \to 0$, is still locally similar, being described by the inner solution. This point can be proven through a more rigorous analysis.

Since both the inner and outer regions are locally similar, we can assume local similarity everywhere. Consequently Eq. (12.3.11) becomes

$$\frac{d^2\tilde{T}}{d\xi^2} \approx \frac{1}{[f''(\eta)]^2}\left(\frac{2x}{U_\infty}\right)\frac{w_F}{\rho}, \tag{12.3.14}$$

with the boundary conditions $\tilde{T}(\xi = 0) = \tilde{T}_s$ and $\tilde{T}(\xi = 1) = \tilde{T}_\infty$, and with $f''(\eta) \approx f''(0)$ because the reaction term is effective only in the inner region near $\eta \to 0$. Analogous discussion can be conducted for the concentration field, $\tilde{Y}_i$, to show that it is also locally similar. Thus our final governing equations again resemble those for both the stagnant hot-plate and the stagnation flow cases. Drawing direct analogy with these cases, it can be stated that ignition is expected to occur when the streamwise distance x_I satisfies the general relation $\Delta \geq 1$, or

$$\frac{2x_I}{U_\infty[f''(0)]^2}\left(\frac{2\epsilon\, Da_C}{\beta^2\rho_s}\right)\tilde{Y}_{O,\infty}\tilde{Y}_{F,\infty}e^{-\tilde{T}_a/\tilde{T}_s} \geq 1. \tag{12.3.15}$$

The local similarity assumption is expected to break down as the ignition location is approached. Here the rapid variation in the flow properties in the streamwise direction facilitates streamwise transport. However, since under such a situation the streamwise gradient can be so steep that even streamwise diffusion cannot be neglected, the basic boundary-layer assumption breaks down as well. This implies

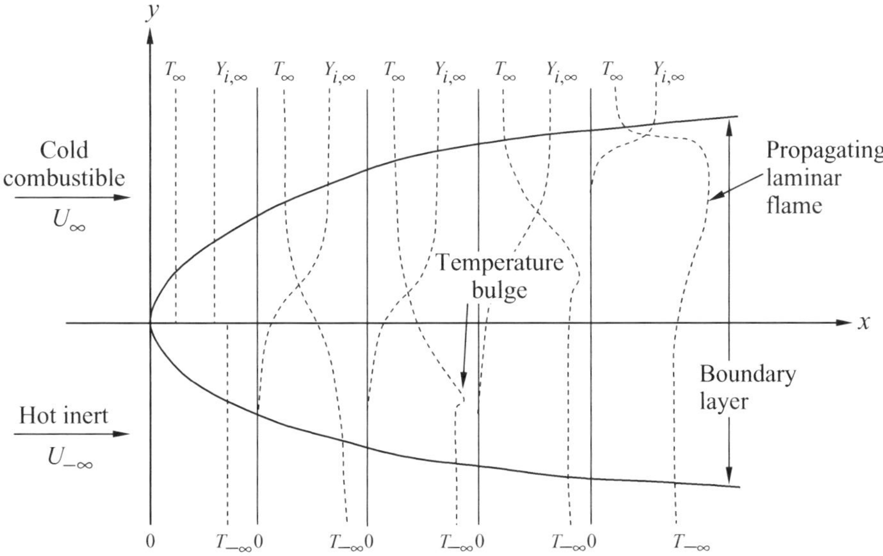

Figure 12.3.1. Development of the temperature and species profiles during the ignition of a premixed combustible within the laminar mixing layer (Williams 1985).

that the flow field is elliptic in nature around the ignition point and, therefore, should be treated as such.

12.3.3. Ignition in the Mixing Layer

We now study ignition in the laminar mixing-layer flow (Marble & Adamson 1954; Law & Law 1982a) that, although still locally similar, has a different scaling as compared to that of the flat plate.

The problem of interest is shown in Figure 12.3.1. Here a slow, hot, inert gas stream with velocity $U_{-\infty}$ and temperature $T_{-\infty}$ meets a parallel, fast, cold combustible gas stream with velocity U_{∞}, temperature T_{∞}, and reactant mass fractions $Y_{i,\infty}$ at $x = 0$. A mixing layer is subsequently developed through which thermal energy is transferred from the hot to the cold stream while the reactants are transferred from the cold to the hot stream. Chemical reaction is initiated within the mixing layer, with its intensity depending on the local temperature and species concentrations. Since chemical reaction is facilitated with increasing temperature and reactant concentrations, ignition is favored to occur near the hot boundary because of the high temperature, and near the cold boundary because of the high concentrations. However, because of the temperature-sensitive Arrhenius factor, ignition should occur near the hot boundary except for reactions with very large reaction orders that are unrealistic based on our understanding of chemical kinetics. It may also be noted that whereas for the flat-plate case the reactant concentrations in the reaction region are nearly uniform and close to their freestream values, for the mixing layer the concentrations in the reaction region are low and vary across it in an essential manner. This variation has to be allowed and thereby increases the complexity of the problem.

Similar to the flat plate problem, the practical objective of the study is to predict the minimum distance to achieve ignition. The governing differential equations are the same as those for flat-plate ignition except now the boundary conditions are

$$f'(-\infty) = U_{-\infty}/U_\infty = \lambda, \qquad f(0) = 0, \quad f'(\infty) = 1, \tag{12.3.16}$$

$$\tilde{T}(x, -\infty) = \tilde{T}_{-\infty}, \qquad \tilde{T}(x, \infty) = \tilde{T}_\infty, \tag{12.3.17}$$

$$\tilde{Y}_i(x, -\infty) = 0, \qquad \tilde{Y}_i(x, \infty) = \tilde{Y}_{i,\infty}, \tag{12.3.18}$$

and a self-similar initial profile for $\tilde{T}$,

$$\tilde{T}(0, \eta) = \tilde{T}_{-\infty} - \beta\xi, \tag{12.3.19}$$

where $\beta = \tilde{T}_{-\infty} - \tilde{T}_\infty$ and

$$\xi(\eta) = \frac{f' - \lambda}{1 - \lambda} \tag{12.3.20}$$

is the transformed independent variable used to suppress the transverse convection term and place the hot and cold streams at $\xi = 0$ and 1 respectively. Therefore, in the ξ-coordinate Eq. (12.3.8) becomes

$$\left[\frac{f''}{(1-\lambda)}\right]^2 \frac{\partial^2 \tilde{T}}{\partial \xi^2} - 2x[\lambda + (1-\lambda)\xi]\frac{\partial \tilde{T}}{\partial x} = \left(\frac{2x}{U_\infty}\right)\frac{w_F}{\rho}. \tag{12.3.21}$$

The coupling function β_i for this problem has a similarity profile, being given by Eq. (12.1.45), because both the temperature and concentration are specified constants at the two freestream boundaries. Thus applying these boundary conditions, we have

$$\tilde{Y}_i + \tilde{T} = \tilde{T}_{-\infty} + (\tilde{Y}_{i,\infty} - \beta)\xi. \tag{12.3.22}$$

To investigate similarity, we examine Eq. (12.3.21) for its behavior in the reaction region where $\eta \to -\infty$, or $\xi \to 0$. Since $f''(-\infty) = 0$, we have to expand $f''(\eta)$ for $\eta \to -\infty$ in order to obtain the proper variation of the coefficient $[f''(\eta)/(1-\lambda)]^2$. To do this we first integrate the boundary condition $f'(-\infty) = \lambda$ to yield

$$f \sim \lambda\eta + c_1 \tag{12.3.23}$$

as $\eta \to -\infty$, where c_1 is a constant. Substituting (12.3.23) into the Blasius equation $f''' + ff'' = 0$ and integrating again, we have

$$f'' \sim c_2 e^{-z^2}, \tag{12.3.24}$$

where $z = (\lambda\eta + c_1)/\sqrt{2\lambda}$, and c_2 is another constant. Integrating (12.3.24) yields

$$f' \sim \lambda + c_2\sqrt{\pi/2\lambda}\,\mathrm{erfc}(z) \tag{12.3.25}$$

with $\text{erfc}(z) = (2/\sqrt{\pi})\int_{-\infty}^{z} e^{-t^2}dt$. But $\text{erfc}(z) \sim -e^{-z^2}/\sqrt{\pi}z$ as $z \to -\infty$, therefore

$$f' \sim \lambda - \frac{c_2}{\sqrt{2\lambda}}\frac{e^{-z^2}}{z}, \tag{12.3.26}$$

which can also be expressed in terms of ξ as

$$\xi \sim -\frac{c_2}{\sqrt{2\lambda}(1-\lambda)}\frac{e^{-z^2}}{z}. \tag{12.3.27}$$

Taking the logarithm of (12.3.27), and keeping only the leading term for $z \to -\infty$, we obtain the asymptotic relation between ξ and z:

$$\ln \xi \sim -z^2. \tag{12.3.28}$$

Using (12.3.24), (12.3.27), and (12.3.28), we then have

$$\left[\left(\frac{f''}{1-\lambda}\right)_{\eta \to -\infty}\right]^2 \sim \lambda(2z^2)\xi^2 \sim -\lambda\xi^2 \ln \xi^2. \tag{12.3.29}$$

Substituting (12.3.29) into Eq. (12.3.21), we have, for the inner region, which is located at $\eta \to -\infty$,

$$-\lambda(\xi^2 \ln \xi^2)\frac{\partial^2 \tilde{T}_{\text{in}}}{\partial \xi^2} - 2x[\lambda + (1-\lambda)\xi]\frac{\partial \tilde{T}_{\text{in}}}{\partial x} = \left(\frac{2x}{U_\infty}\right)\left(\frac{w_F}{\rho}\right)_{\text{in}}. \tag{12.3.30}$$

Let us now investigate the relative orders of magnitude of the diffusive and convective terms in Eq. (12.3.30). Following the procedure for the flat plate case, we can introduce a stretched inner variable $\chi \sim \xi/\epsilon$. Using χ in Eq. (12.3.30), and noting that $\ln \xi^2 \sim \ln \epsilon^2 + \ln \chi^2 \sim \ln \epsilon^2$ for $\chi = O(1)$ in the inner region, we obtain

$$\chi^2 \frac{\partial^2 \tilde{T}_{\text{in}}}{\partial \chi^2} + \frac{x}{\ln \epsilon}\frac{\partial \tilde{T}_{\text{in}}}{\partial x} = \left(\frac{-x}{U_{-\infty}\ln \epsilon}\right)\left(\frac{w_F}{\rho}\right)_{\text{in}}. \tag{12.3.31}$$

It is therefore clear that in the limit $\epsilon \to 0$, the streamwise convection term again becomes very small relative to the diffusion term such that Eq. (12.3.31) simplifies to the locally similar form

$$\chi^2 \frac{d^2 \tilde{T}_{\text{in}}}{d\chi^2} \approx \left(\frac{-x}{U_{-\infty}\ln \epsilon}\right)\left(\frac{w_F}{\rho}\right)_{\text{in}}. \tag{12.3.32}$$

Comparing Eq. (12.3.31) with Eq. (12.3.12) for the flat plate case, it is seen that local similarity is a weaker assumption for the mixing-layer flow than the flat-plate flow, in which the streamwise convection is $O(-1/\ln \epsilon)$ and $O(\epsilon^3)$ respectively.

It is important to recognize that this discussion on similarity and the asymptotic scaling of the reaction zone are conducted in ξ-coordinate, in which the boundary-layer parameters ($\tilde{T}$, $\tilde{Y}_i$, f') vary linearly and the influence of chemical reactivity is confined to the region of $\xi = O(\epsilon)$. Since $\xi \sim e^{-\eta^2}/\eta \sim e^{-\eta^2}$ for $\eta \to -\infty$, chemical reaction is expected to be confined to the region $O(\sqrt{|\ln \epsilon|})$ in the η-coordinate.

Equation (12.3.32) also reveals another important result in that the equation contains only $U_{-\infty}$, but not U_∞ because there is no explicit λ dependence. Furthermore,

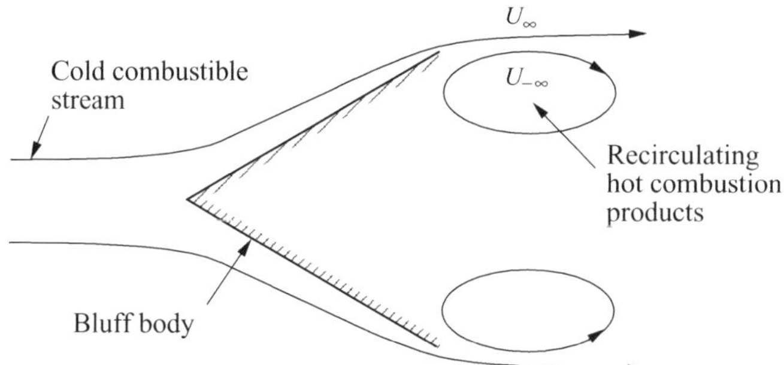

Figure 12.3.2. Schematic showing flame stabilization by bluff body in high-speed flows.

we expect that $\tilde{T}$ in the external, nonreactive region only minimally deviates from its frozen value $(\tilde{T}_{-\infty} - \beta\xi)$, which also does not depend on λ. Therefore it can be concluded that the ignition event is primarily dependent on the velocity of the hot inert stream, $U_{-\infty}$, but not on the velocity of the cold reactive stream, U_{∞}. This is again physically reasonable because ignition occurs close to the hot boundary that moves with $U_{-\infty}$. By the same reasoning, it can also be shown that if the upper stream is the cold mixture while the lower stream the hot inert, then in Eq. (12.3.32) $U_{-\infty}$ is simply replaced by U_{∞}.

Numerical solutions of the complete nonsimilar governing equations have also been obtained (Law & Law 1982a). The nonsimilar solution does not exhibit the nonlinear ignition turning-point behavior because it is evolutionary and each solution is well defined. Thus an alternate ignition criterion has to be used. Figure 12.3.1 shows the development of the temperature profile within the mixing layer. Initially the temperature across the mixing layer decreases monotonically from T_{∞} to $T_{-\infty}$. However, at a certain location, x_I, a temperature bulge appears and subsequently a laminar flame is developed which propagates into the combustible gas. Thus the first occurrence of a local temperature maximum,

$$\left(\frac{\partial \tilde{T}}{\partial \eta}\right)_{x_I, \eta_I} = 0, \tag{12.3.33}$$

can be used as an indication that ignition has been achieved.

12.3.4. Flame Stabilization and Blowoff in High-Speed Flows

Analysis and results of the mixing-layer ignition problem are frequently used to explain flame stabilization and blowoff in combustors with high volumetric flow rates such as the ramjets and afterburners. Here the high-speed, cold, combustible stream flows over a bluff body, as schematically shown in Figure 12.3.2. The bluff body generates a slowly moving recirculation zone of the combustion product, which serves as the hot stream to effect ignition. Thus if we identify the thermochemical state of the hot stream as the burned state of the cold combustible stream, then the reaction rate term in Eq. (12.3.32), w_F, is proportional to $(f^o)^2$, where f^o is the burning flux of the laminar flame identified in Section 7.2. Therefore the minimum distance to

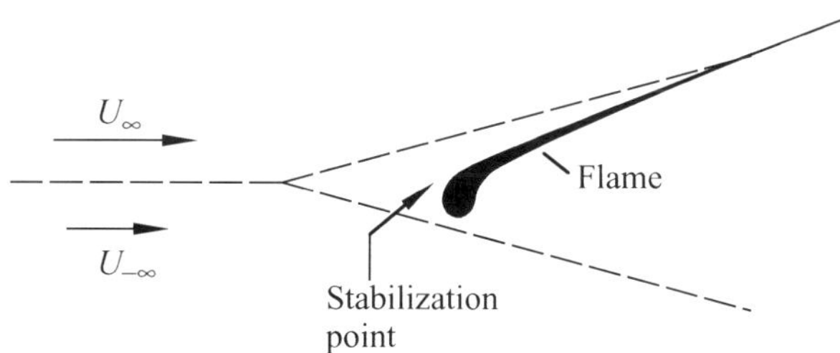

Figure 12.3.3. Schematic demonstrating that flame holding is a phenomenon of flame stabilization instead of ignition.

achieve ignition, x_I, is related to the dynamic and thermochemical aspects of the flow, $U_{-\infty}$ and f^o respectively, through

$$x_I \sim \frac{\rho_{-\infty} U_{-\infty}}{(f^o)^2} \tag{12.3.34}$$

as indicated by the RHS of Eq. (12.3.32). The parameter $\rho_{-\infty} U_{-\infty}$ can be further related to the mass flow rate $\rho_\infty U_\infty$ of the cold stream through an analysis of the dynamics of the recirculation zone, yielding a relation between the mass throughput of the combustible gas and the ignition distance. It is frequently inferred that if this ignition distance exceeds the length of the recirculation zone, then the flame is blown downstream and can no longer be stabilized in the combustor.

There is, however, a fundamental flaw associated with the above concept. That is, while ignition of the cold combustible is effected by its mixing with the hot inert stream of the recirculation zone of combustion products, these combustion products, however, do not exist before ignition is achieved and rigorous combustion established. Thus the fundamental phenomenon of bluff-body stabilized flame attachment and blowoff is not mixing-layer ignition. The problem of interest here is actually the stabilization of an existing flame in the mixing layer, with stabilization being achieved by the dynamic balance between the local flow velocity and flame velocity at the leading edge of the flame (Figure 12.3.3) in the manner discussed in Section 8.6. The fundamental mechanism of "ignition" of the cold combustible is very different here because it is now also influenced by the streamwise back diffusion of heat and radicals from the existing flame. The distance of flame stabilization would be much shorter than that for ignition.

We shall therefore study in the next section a problem of flame stabilization in a nonpremixed jet flow to demonstrate the basic concepts involved in analyzing such phenomena.

12.4. JET FLOWS

In many combustion devices the fuel is introduced into the oxidizing environment in the form of a jet, such as those associated with the Bunsen flame, the pilot flame in furnaces, fuel injection in Diesel engines, and the gas flare in oil fields. As the

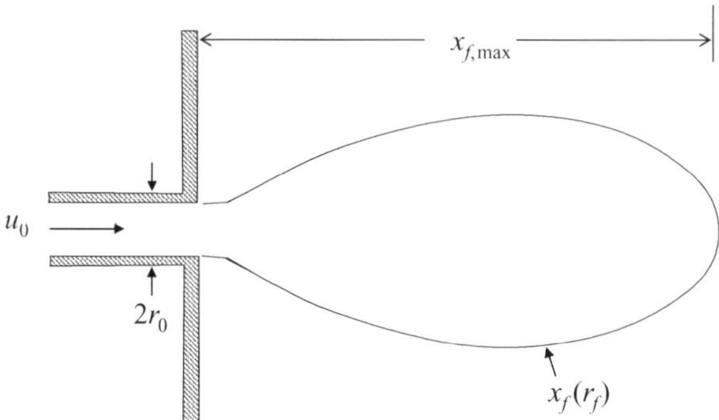

Figure 12.4.1. Schematic of the jet flame.

jet penetrates into the ambience, it entrains and mixes with the surrounding fluid through viscous action, while at the same time it loses its streamwise momentum, as shown in Figure 12.1.1. Upon ignition, a nonpremixed flame is established at a distance slightly downstream of the nozzle exit, being stabilized there through the mechanism described in Section 8.6. With continuous increase in the injection velocity, the flame will be first blown off of the nozzle port and eventually blown out altogether.

For phenomena involving jet flows we are frequently interested in two questions, namely what is the contour of the flame, shown schematically in Figure 12.4.1, especially its length because this affects the distribution of the heat generated, and the critical states of blowoff and blowout. Indeed, the study of the Burke–Schumann flame in Section 6.2 was motivated by the first question, although its highly idealized flow field, in which the momentums of the fuel and oxidizer flows are matched, can hardly be considered as representative of a jet flow. Therefore, in this section we shall analyze the geometry of the nonpremixed jet flame and the blowout of the lifted flame.

12.4.1. Similarity Solution

Let us first solve the flow field produced by issuing a uniform jet of density ρ_o and velocity u_o from, say, a circular nozzle of radius r_o, into a stagnant atmosphere. The governing equations are:

Continuity:

$$\frac{\partial(\rho u r)}{\partial x} + \frac{\partial(\rho v r)}{\partial r} = 0 \qquad (12.4.1)$$

x-Momentum:

$$\rho u r\frac{\partial u}{\partial x} + \rho v r\frac{\partial u}{\partial r} = \frac{\partial}{\partial r}\left(\mu r\frac{\partial u}{\partial r}\right), \qquad (12.4.2)$$

where (x, r) and (u, v) are the streamwise and radial coordinates and velocities respectively. The boundary conditions are

$$r = 0: \qquad \frac{\partial u}{\partial r} = 0, \qquad v = 0 \tag{12.4.3}$$

$$r \to \infty: \qquad \frac{\partial u}{\partial r} = 0, \qquad u = 0. \tag{12.4.4}$$

As usual, the continuity equation can be replaced by a stream function $\psi(x, r)$, defined as

$$\rho u r = \rho_o \frac{\partial \psi}{\partial r}, \qquad \rho v r = -\rho_o \frac{\partial \psi}{\partial x}. \tag{12.4.5}$$

Because of the finite dimension of the jet, the total amount of the momentum injected is conserved in the streamwise direction. This can be demonstrated by integrating Eq. (12.4.2) from $r = 0$ to ∞, using the boundary conditions of (12.4.3) and (12.4.4). The conserved momentum can therefore be defined at the nozzle exit, as

$$J = \int_0^\infty \rho u^2 (2\pi r) dr = \int_0^{r_o} \rho u^2 (2\pi r) dr = \pi r_o^2 \rho_o u_o^2. \tag{12.4.6}$$

The scaling of the boundary-layer variables with x can be identified by considering the simpler, incompressible situation. Thus letting x, η, and $f(\eta)$ assume the forms

$$\psi \sim x^p f(\eta), \qquad \eta \sim \frac{r}{x^n}, \tag{12.4.7}$$

we identify the dependence

$$u \sim x^{p-2n}, \quad \frac{\partial u}{\partial x} \sim x^{p-2n-1}, \quad \frac{\partial u}{\partial r} \sim x^{p-3n}, \quad \frac{1}{r}\frac{\partial}{\partial r}\left(r\frac{\partial u}{\partial r}\right) \sim x^{p-4n}. \tag{12.4.8}$$

Using the relations of (12.4.8) in Eq. (12.4.6) and requiring that the resulting expression be independent of x, and in Eq. (12.4.2) and requiring that the inertial and viscous terms balance, we obtain two relations for the exponents, p and n, from which they can be solved to yield $p = n = 1$.

Based on the above scaling with x, we introduce the boundary-layer variables for the compressible case as

$$\eta = \frac{\tilde{r}}{x} \tag{12.4.9}$$

$$\psi(x, y) = \frac{\mu_o x}{\rho_o} f(\eta), \tag{12.4.10}$$

which transform the governing equations to the incompressible form, where

$$\tilde{r}^2 = 2 \int_0^r \frac{\rho}{\rho_o} r dr. \tag{12.4.11}$$

Consequently, u, v, and Eq. (12.4.2) become

$$u = \frac{\mu_o}{\rho_o}\left(\frac{f'}{x\eta}\right) \tag{12.4.12}$$

$$\rho v r = -\mu_o\left(f + x f'\frac{\partial\eta}{\partial x}\right) \tag{12.4.13a}$$

$$= -\mu_o(f - \eta f') \quad \text{for incompressible flows} \tag{12.4.13b}$$

$$\frac{d}{d\eta}\left[C\eta\frac{d}{d\eta}\left(\frac{f'}{\eta}\right)\right] + \frac{d}{d\eta}\left(\frac{ff'}{\eta}\right) = 0, \tag{12.4.14}$$

where

$$C = \frac{\mu\rho r^2}{\mu_o\rho_o\bar{r}^2} \tag{12.4.15}$$

is the Chapman–Rubesin parameter. Applying the boundary conditions (12.4.3) and (12.4.4) to Eqs. (12.4.12) and (12.4.13a) yields

$$f(0) = 0, \quad f''(0) = 0, \quad f'(\infty) = 0. \tag{12.4.16}$$

Integrating Eq. (12.4.14) subject to (12.4.16), we have

$$f(\eta) = \frac{Ck\eta^2}{1 + \frac{k\eta^2}{4}}, \tag{12.4.17}$$

where k is the integration constant that can be determined by evaluating Eq. (12.4.6) using Eqs. (12.4.12) and (12.4.17),

$$k = \frac{3}{16\pi}\left(\frac{\rho_o J}{\mu_o^2 C^2}\right). \tag{12.4.18}$$

Finally, substituting $f(\eta)$ in Eqs. (12.4.12) and (12.4.13b) for u and v yields

$$u(x,\eta) = \frac{3}{8\pi}\left(\frac{J}{\mu_o C x}\right)\frac{1}{\left(1 + \frac{k\eta^2}{4}\right)^2}, \tag{12.4.19}$$

$$\rho v r = \frac{3}{16\pi}\left(\frac{\rho_o J}{\mu_o C}\right)\frac{\left(1 - \frac{k\eta^2}{4}\right)\eta^2}{\left(1 + \frac{k\eta^2}{4}\right)^2}, \tag{12.4.20}$$

with Eq. (12.4.20) applicable only for incompressible flows. Having determined the flow field, we now study the configuration of the flame situated within it.

12.4.2. Height of Nonpremixed Jet Flames

In 1949, Hottel and Hawthorne presented a classical paper on the mixing and combustion of nonpremixed turbulent jets. The results (Figure 12.4.2) show that when the jet velocity is low and the flow is laminar, the flame height increases with the

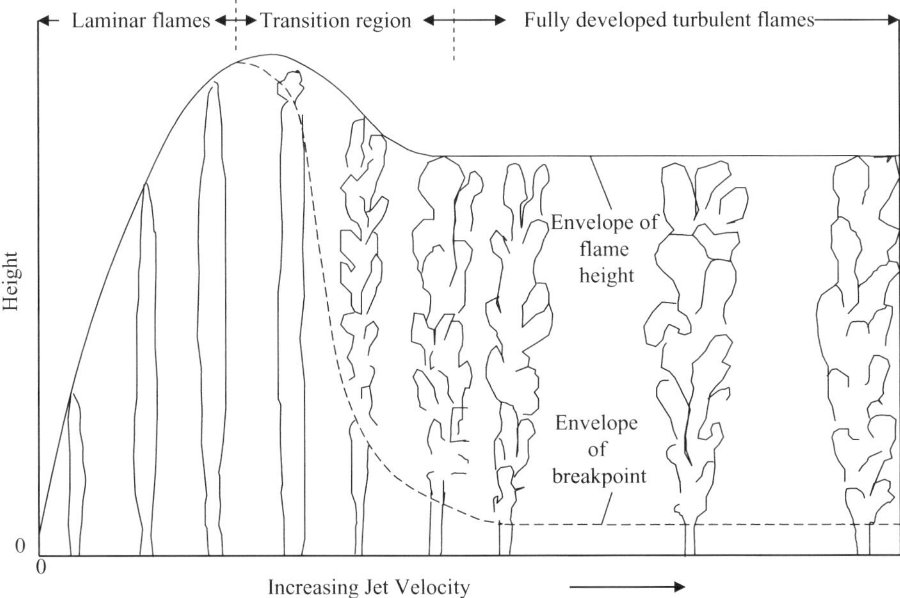

Figure 12.4.2. Schematic showing the experimentally observed flame height with increasing jet velocity (adapted from Hottel & Hawthorne 1949).

jet velocity until a maximum, at which instability is observed at the flame tip. With further increase of the jet velocity the flame becomes turbulent, and its height sharply decreases and, when scaled by the nozzle diameter, remains fairly constant thereafter. This phenomenon is a beautiful demonstration of the change in the controlling diffusion mechanism from laminar to turbulent, as we shall show subsequently.

The flow is still described by Eqs. (12.4.1) and (12.4.2), and the coupling function for fuel and oxidizer concentrations, $\beta_{F,O} = \tilde{Y}_F - \tilde{Y}_O$, is governed by

$$\rho u r \frac{\partial \beta_{F,O}}{\partial x} + \rho v r \frac{\partial \beta_{F,O}}{\partial r} = \frac{\partial}{\partial r}\left(\mu r \frac{\partial \beta_{F,O}}{\partial r}\right), \tag{12.4.21}$$

where we have assumed equal diffusivity and unity Schmidt number. The boundary conditions are the vanishing of the concentration gradients at $r = 0$, $\tilde{Y}_F(r = \infty) = 0$, and $\tilde{Y}_O(r = \infty) = \tilde{Y}_{O,\infty}$. The use of the fuel–oxidizer coupling function yields the flame geometry most expeditiously. A similar expression can be written for the coupling function for, say, the fuel concentration and enthalpy. This, however, is not necessary as its solution will only yield the flame temperature, which, as we know, is the adiabatic flame temperature for the nonpremixed system, as having been repeatedly demonstrated for other problems.

Comparing Eqs. (12.4.2) and (12.4.21), we see that u and β are governed by the same operator. Consequently a solution for $\beta_{F,O}$ is simply

$$\beta_{F,O} = \tilde{Y}_F - \tilde{Y}_O = c_1 u + c_2, \tag{12.4.22}$$

where the constants c_1 and c_2 can be readily determined by evaluating $\beta_{F,O}$ at the initial state of the jet where $Y_F = Y_{F,O}$, $u = u_o$, and at the ambient state where $Y_O = Y_{O,\infty}$ and $u = 0$. Thus we have

$$\tilde{Y}_F - \tilde{Y}_O = (\tilde{Y}_{F,O} + \tilde{Y}_{O,\infty})\frac{u}{u_o} - \tilde{Y}_{O,\infty}, \tag{12.4.23}$$

with u given by Eq. (12.4.19). Note that instead of the initial jet condition we can also use the statement for coupling function conservation obtained by multiplying Eq. (12.4.22) by $\rho u(2\pi r dr)$, and integrating the resulting expression from $r = 0$ to ∞ in the same manner as that led to the derivation of J in Eq. (12.4.6).

Evaluating at the flame surface where both $\tilde{Y}_F$ and $\tilde{Y}_O$ vanish readily yields its geometry,

$$x_f(\eta_f) = (1+\phi^*)\frac{3}{8\pi}\frac{J/u_o}{\mu_o C(1+\frac{k\eta_f^2}{4})^2}, \tag{12.4.24}$$

where $\phi^* = \tilde{Y}_{F,o}/\tilde{Y}_{O,\infty}$. Setting $\eta_f = 0$, we obtain the flame height as

$$x_{f,\max} = (1+\phi^*)\frac{3}{8\pi}\left(\frac{J/u_o}{\mu_o C}\right) = (1+\phi^*)\frac{3}{8\pi}\left(\frac{\pi r_o^2 \rho_o u_o}{\mu_o C}\right) = (1+\phi^*)\frac{3}{8\pi}\frac{Q}{\mu_o C}, \tag{12.4.25}$$

where $Q = \pi r_o^2 \rho_o u_o$ is the mass flow rate of the jet.

Equation (12.4.25) readily demonstrates the behavior of the flame height shown in Figure 12.4.2 for the laminar and turbulent flame regimes. Specifically, in the laminar regime the kinematic viscosity μ_o/ρ_o is a constant and as such $x_{f,\max}$ varies linearly with the jet exit velocity u_o, that is,

$$\text{Laminar flame regime:} \qquad x_{f,\max} \sim u_o. \tag{12.4.26}$$

However, in the turbulent flame regime the effects of turbulence can be approximately accounted for by using a turbulent kinematic viscosity ν_T for μ_o/ρ_o. Furthermore, if we use the simplest description, based on Prandtl's mixing length model, ν_T can be expressed as

$$\nu_T \sim u_o r_o$$

according to Eq. (11.2.14). Then Eq. (12.4.25) shows that

$$\text{Turbulent flame regime:} \qquad \frac{x_{f,\max}}{r_o} \sim \text{constant}. \tag{12.4.27}$$

The flame height behavior of Figure 12.4.2 is therefore satisfactorily explained.

12.4.3. Stabilization and Blowout of Lifted Flames

Next we consider the stabilization and blowout of lifted flames, first studied in Section 8.6.4. Figure 12.4.3 shows the photographic images of a jet-stabilized nonpremixed flame formed by issuing a propane jet into stagnant air at increasing flow rates (Chung & Lee 1991). As discussed previously, when the flow rate is small, the

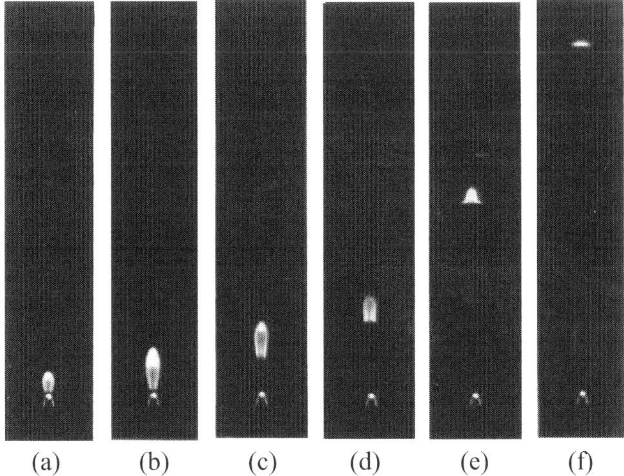

Figure 12.4.3. Photographic images of attached and lifted flames of propane jets in air. Tube diameter = 0.195 mm; Volumetric flow rate: 0.195, 17.3, 18.3, 20.3, 22.3, and 26.4 ml/min for (a)–(f) (Chung & Lee 1991).

flame is stabilized at the rim of the nozzle, as shown in Figures 12.4.3a and 12.4.3b. However, when the flow rate is sufficiently high, the flame is blown off, or lifted off, and becomes stabilized in a lifted state, as shown in the succeeding images. The flow at such a flow rate can be either laminar or turbulent, although we shall only consider the laminar situation here.

Figure 12.4.4 shows an enlarged view of the lifted flame together with a schematic of its global structure. Visually, the lifted flame has a blue-green inner cone, a blue-violet segment near the rim, and a faint blue segment between them. This observation then indicates that the inner segment is a rich premixed flame because it shows the green color due to C_2 emission, and the outer segment is a lean premixed flame because it shows the violet color due to OH emission. The blue color due to the CH emission is superimposed on them, and constitutes the middle diffusion flame. The lifted flame therefore has the structure of a triple

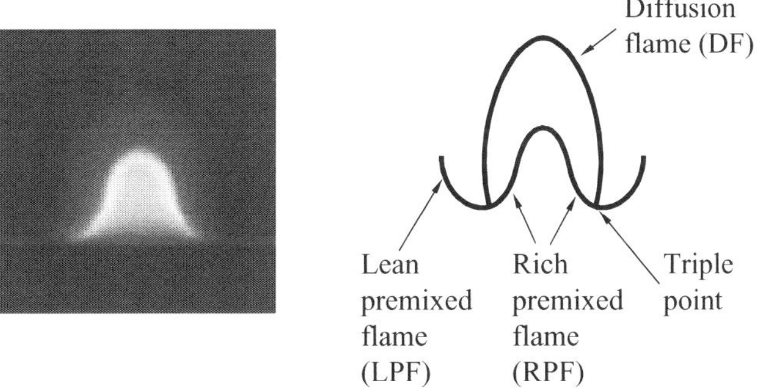

Figure 12.4.4. Photographic and schematic of the structure of the lifted tribrachial flame.

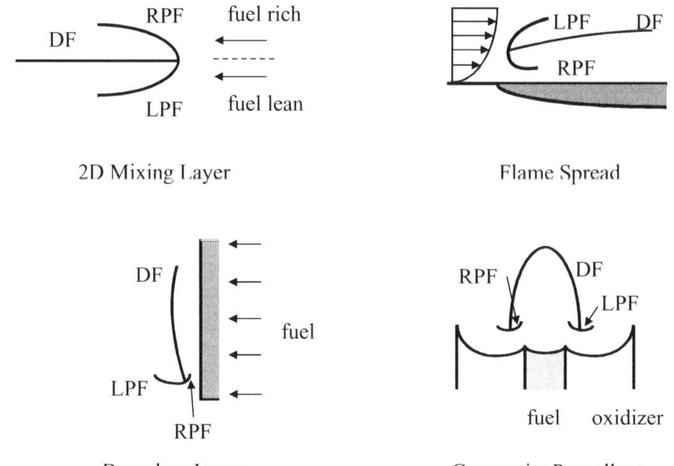

Figure 12.4.5. Several situations showing the presence of the tribrachial flame.

flame, which is also called a tribrachial or an edge flame, as previously discussed in Section 8.6 for the rim-stabilized nonpremixed flame. Thus the relevant phenomenon is that of the stabilization and blowout of a tribrachial flame situated in a free jet.

The stabilization, propagation, blowoff, and blowout of tribrachial flames are an essential component of the structure and response of nonpremixed flames in mixing-layer flows. For example, when we showed in Figure 12.2.1 the structure of the nonpremixed flame in the boundary layer of an ablating surface, the structure and processes occurring at the leading edge were not discussed. Clearly, with a sufficiently strong flow, the flame cannot be stabilized over the ablating surface and will be blown away. The flame component that controls the stabilization of the bulk nonpremixed flame is the tribrachial flame segment that constitutes the leading edge, imparting to it a premixed flame character with its inherent dependence on chemical kinetics. Figure 12.4.5 shows such a structure for various types of flows. The presence of tribrachial, or edge, flame segments is also an essential component of the structure of turbulent flames (Vervisch & Poinsot 1998).

In the following, we shall use the stabilization and blowout of the lifted flame as an example to demonstrate how such problems can be analyzed. Much of the understanding discussed was gained by Chung and coworkers (see, for example, Chung & Lee 1991; Lee & Chung 1997), and reviews on edge flames are given in Buckmaster (2002), and Chung (2007).

To analyze the stabilization mechanism, we first note from Figure 12.4.3 that the lifted flame is situated at a large distance from the nozzle exit, compared to the nozzle diameter, of the order of 10 cm for the situation shown. This then leads to the following two observations. First, the flow of interest is indeed that of the jet, being in the far field of the nozzle flow. Second, since the flame thickness is of the order of 1 mm, the bulk flow field is not affected by the flame. Consequently the stabilization is a passive process, with the flame being affected by the flow, and not vice versa as far as the upstream state of the flame is concerned. Solution of the problem is therefore decoupled in that all we need to do is to determine the flow field in the absence of reactions, and then use it to locate the stabilization point and the state of blowout.

The flow field is still governed by that derived in Section 12.3.1. In addition, we also need to solve for the distribution of the fuel concentration from

$$\rho u r \frac{\partial Y_F}{\partial x} + \rho v r \frac{\partial Y_F}{\partial r} = \frac{\partial}{\partial r}\left(\frac{\mu r}{Sc}\frac{\partial Y_F}{\partial r}\right), \tag{12.4.28}$$

subject to the boundary conditions $\partial Y_F/\partial r = 0$ at $r = 0$ and $Y_F = 0$ as $r \to \infty$. A general Schmidt number, Sc, is retained as it is a crucial parameter for the present problem. To solve for Y_F, we observe that the conservation equation for u and Y_F, Eqs. (12.4.2) and (12.4.28), bear close resemblance to each other, except for the presence of Sc in Eq. (12.4.28). We therefore seek a solution of the form

$$Y_F(x, \eta) = \frac{y_F(\eta)}{x}, \tag{12.4.29}$$

which has the same variation with x as that of u. Equation (12.4.28) is then transformed to

$$\frac{d}{d\eta}\left(\frac{C}{Sc}\eta\frac{dy_F}{d\eta}\right) + \frac{d}{d\eta}(f y_F) = 0, \tag{12.4.30}$$

which can be readily solved for y_F. By further applying the conservation requirement

$$I = \int_0^\infty \rho u Y_F(2\pi r)dr = \int_0^{r_o} \rho u Y_F(2\pi r)dr = \pi r_o^2 \rho_o u_o Y_{F,o} = (J/u_o)Y_{F,o} \tag{12.4.31}$$

we obtain

$$Y_F(x, \eta) = \frac{(1+2Sc)}{8\pi}\frac{(J/u_o)Y_{F,o}}{\mu_o C x}\frac{1}{\left(1+\frac{k\eta^2}{4}\right)^{2Sc}}. \tag{12.4.32}$$

We now discuss the stabilization criterion. Recognizing the basic structure of a tribrachial flame, it is reasonable to assume that the reactant concentrations around the triple point where the three flame segments meet should be close to stoichiometric, with $Y_F = Y_{F,\text{st}}$, and the burning is most intense. Consequently this would be the point of stabilization, which has a flame speed $s_{u,\text{st}}$ that should be close to the laminar flame speed for the stoichiometric mixture. Since at the point of stabilization the local flame speed should balance the local flow velocity, from Eq. (12.4.19) we have

$$s_{u,\text{st}}(x_L, \eta_L) = \frac{3}{8\pi}\left(\frac{J}{\mu_o C x_L}\right)\frac{1}{\left(1+\frac{k\eta_L^2}{4}\right)^2}, \tag{12.4.33}$$

while the fuel concentration there is stoichiometric, given by Eq. (12.4.32),

$$Y_{F,\text{st}}(x_L, \eta_L) = \frac{(1+2Sc)}{8\pi}\left(\frac{I}{\mu_o C x_L}\right)\frac{1}{\left(1+\frac{k\eta_L^2}{4}\right)^{2Sc}}, \tag{12.4.34}$$

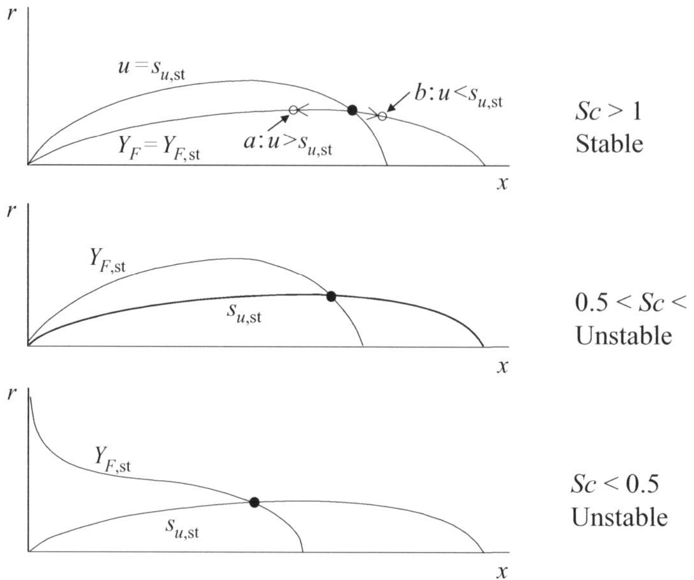

Figure 12.4.6. Phenomenological assessment of the dependence of the stability of lifted flame on Sc.

where the subscript L designates the lifted state. Solving for (x_L, η_L) from Eqs. (12.4.33) and (12.4.34) yields

$$\frac{\mu_o/\rho_o}{s_{u,\mathrm{st}}}\left(\frac{x_L}{r_o^2}\right) = \frac{1}{8C}\left[\frac{3^{Sc}}{(1+2Sc)}\frac{Y_{F,\mathrm{st}}}{Y_{F,o}}\right]^{\frac{1}{(Sc-1)}}\left(\frac{u_o}{s_{u,\mathrm{st}}}\right)^{\frac{(2Sc-1)}{(Sc-1)}} \tag{12.4.35}$$

$$\frac{k\eta_L^2}{4} = \left[\frac{3}{(2Sc+1)}\frac{Y_{F,\mathrm{st}}}{Y_{F,o}}\frac{u_o}{s_{u,\mathrm{st}}}\right]^{-\frac{1}{2(Sc-1)}} - 1. \tag{12.4.36}$$

Equation (12.4.35) shows that the liftoff height x_L increases with increasing flow velocity for $Sc > 1$ and $Sc < 0.5$, while it decreases for $0.5 < Sc < 1$, which is physically unrealistic. Therefore we can readily rule out the existence of stabilized lifted flames for the latter situation. In fact, all $Sc < 1$ systems can be ruled out based on stability considerations. To demonstrate this point, consider Figure 12.4.6 in which the isovelocity and isoconcentration lines based on Eqs. (12.4.33) and (12.4.34) are plotted, with their point of intersection being the stabilization point. Since Sc measures the rate of viscous diffusion relative to mass diffusion, and since diffusion occurs only in the transverse, η, direction, a faster diffusion process will lead to a correspondingly faster reduction of the transported quantity in the streamwise direction. Thus the isovelocity contour is wider and shorter than the isoconcentration contour for $Sc > 1$ mixtures, while the opposite holds for $Sc < 1$. Now first consider the $Sc > 1$ case of Figure 12.4.6, and momentarily displace a flame at the stabilizing point along the stoichiometric concentration line $Y_{F,\mathrm{st}}$ in, say, the upstream direction to point a. At this location the flame maintains its stoichiometric burning velocity $s_{u,\mathrm{st}}$ because

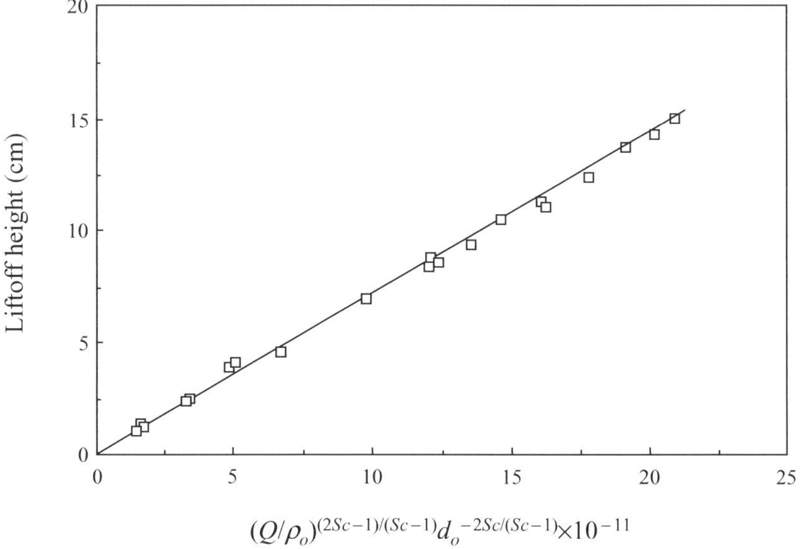

Figure 12.4.7. Experimental data demonstrating the validity of the derived relation between liftoff height, nozzle diameter (d_o), and flow rate (Q); d_o in mm, Q in ml/min; $Sc = 1.376$ for propane (Chung & Lee 1991).

$Y_{F,\mathrm{st}}$ is fixed. However, the streamwise flow velocity u here is larger than $s_{u,\mathrm{st}}$ because point a is located at a smaller r than that of the isovelocity line $u = s_{u,\mathrm{st}}$. Consequently the flame is pushed back to the original stabilized location. By the same token, a downstream displacement to point b will result in a smaller local velocity u relative to the stoichiometric flame speed, causing the flame to move back upstream. Thus $Sc > 1$ flames are dynamically stable. A similar argument for $Sc < 1$ mixtures shows that they are unstable.

Experimentally it has been found that propane and butane jets ($Sc > 1$) have stable lifted flames, while methane and ethane flames ($0.5 < Sc < 1$) are unstable in that they are blown out directly from the rim-stabilized state. For hydrogen flames ($Sc < 0.5$) the anchoring at the rim is so strong that blowoff cannot be achieved until the flow velocity is so high that the flow becomes turbulent.

Equation (12.4.35) shows that $x_L \sim \rho_o r_o^2 u_o^{(2Sc-1)/(Sc-1)}$, which can be alternately expressed in terms of the volumetric flow rate $Q \sim \rho_o r_o^2 u_o$ as

$$x_L \sim Q^{(2Sc-1)/(Sc-1)} r_o^{-2Sc/(Sc-1)}. \tag{12.4.37}$$

Figure 12.4.7 shows that such a relation indeed holds, where d_o is the nozzle diameter.

Equation (12.4.36) also shows that, in order for η_L to be real, we require

$$\frac{3}{(2Sc+1)}\left(\frac{Y_{F,\mathrm{st}}}{Y_{F,o}}\right)\frac{u_o}{s_{u,\mathrm{st}}} \leq 1 \quad \text{for} \quad Sc > 1. \tag{12.4.38}$$

Thus there exists a maximum velocity,

$$u_{o,BO} = \frac{(2Sc+1)s_{u,\mathrm{st}}}{3}\left(\frac{Y_{F,o}}{Y_{F,\mathrm{st}}}\right), \tag{12.4.39}$$

beyond which there is no stabilized solution. Consequently $u_{o,BO}$ can be identified as the velocity at blowout. It may be noted that $u_{o,BO}$ is only a function of Sc, $s_{u,\mathrm{st}}$, and $Y_{F,\mathrm{st}}$, and is independent of the nozzle diameter. This has been verified experimentally. Thus the proper parameter to characterize blowout is the nozzle exit velocity instead of the flow Reynolds number, $Re = \rho_o u_o d_o/\mu_o$, as was assumed in previous studies.

12.5. SUPERSONIC BOUNDARY-LAYER FLOWS

For propulsion devices that fly at supersonic to hypersonic speeds, with Mach number ranging from 3 to 7–8 for liquid hydrocarbon fuels and higher speeds for hydrogen, the flow within the combustion chamber can also be supersonic. Such high speeds then imply that the residence time available for combustion is extremely short, of the order of a few msec. Thus the issues of fuel–air mixing, ignition, flame-holding, complete combustion, and viscous wall heating assume particular significance in the design of these supersonic propulsion devices.

Two considerations are especially relevant in the study of supersonic boundary-layer flows. The first is the presence of shock and detonation waves when the supersonic freestream is oblique to the surface or interface along which the boundary layer would be otherwise generated. Thus an analysis of the boundary-layer flow must necessarily consider the presence and, hence, influence of these waves. The second is the substantial amount of conversion of the kinetic energy of the flow to thermal energy as the flow is viscously slowed down within the boundary layer. The associated heat release can significantly alter the combustion process within the boundary layer.

In this section, we shall study the influence due to the thermal energy recovery as a result of viscous slowdown. To avoid the presence of shocks and detonation waves, we shall therefore consider only supersonic flow over a flat plate. Specifically, we shall study two problems, namely the supersonic analogues of the nonpremixed burning of an ablating surface and the premixed ignition along a flat plate, as analyzed in Section 12.2 and Section 12.3.2 respectively.

Following the same derivation procedure that led to the governing equations for the steady two-dimensional boundary-layer flows for low-speed flows treated in Section 12.1, it can be readily shown that for high-speed flows the inclusion of the $\mathbf{P}:(\nabla\mathbf{v})$ term in Eq. (5.2.4) results in the dimensional energy conservation equation the extra source terms, $u\,dp/dx + \mu(\partial u/\partial y)^2$, subject to the boundary-layer approximation. By further assuming that the Prandtl number is unity, the governing equations for the supersonic flat-plate boundary-layer flow are given by

$$f''' + ff'' = 0 \tag{12.5.1}$$

$$\frac{\partial^2 \tilde{T}}{\partial \eta^2} + f\frac{\partial \tilde{T}}{\partial \eta} - 2xf'\frac{\partial \tilde{T}}{\partial x} = \frac{2x}{U_\infty}\frac{w_F}{\rho} - 2\tilde{H}(f'')^2 \tag{12.5.2}$$

$$\frac{\partial^2 \tilde{Y}_i}{\partial \eta^2} + f\frac{\partial \tilde{Y}_i}{\partial \eta} - 2xf'\frac{\partial \tilde{Y}_i}{\partial x} = -\left(\frac{2x}{U_\infty}\right)\frac{w_F}{\rho}, \tag{12.5.3}$$

where

$$\tilde{H} = \frac{(\gamma - 1)}{2}\tilde{T}_\infty M_\infty^2 \tag{12.5.4}$$

is the temperature recovery parameter, which vanishes for $M_\infty \to 0$, $\gamma = c_p/c_v$ is the specific heat ratio, and Eqs. (12.5.1) and (12.5.3) are just Eqs. (12.2.7) and (12.3.9) respectively.

12.5.1. Nonpremixed Burning of an Ablating Surface

Since the physical problem is identical to that of Section 12.2, the boundary conditions are still given by Eqs. (12.2.1) and (12.2.2). Forming the coupling function $\beta_i = \tilde{Y}_i + \tilde{T}$ from Eqs. (12.5.2) and (12.5.3), and noting that it is self-similar because the temperature recovery term does not depend on x, we have

$$\frac{d^2 \beta_i}{d\eta^2} + f\frac{d\beta_i}{d\eta} = -2\tilde{H}(f'')^2. \tag{12.5.5}$$

Transforming Eq. (12.5.5) to the $\xi = f'(\eta) = u/U_\infty$ coordinate, we have

$$\frac{d^2 \beta_i}{d\xi^2} = -2\tilde{H}, \tag{12.5.6}$$

which can be readily solved to yield

$$\beta_i = \beta_{i,s} + [(\beta_{i,\infty} - \beta_{i,s}) + \tilde{H}]\xi - \tilde{H}\xi^2. \tag{12.5.7}$$

Evaluating β_O and β_F at the reaction sheet, the flame location and temperature are then given by

$$f'(\eta_f) = \frac{\tilde{Y}_{F,s}}{\tilde{Y}_{F,s} + \tilde{Y}_{O,\infty}} = \Phi^* \tag{12.2.5}$$

$$(\tilde{T}_f - \tilde{T}_s) = [(\tilde{Y}_{O,\infty} + \tilde{T}_\infty - \tilde{T}_s) + \tilde{H}]f'(\eta_f) - \tilde{H}[f'(\eta_f)]^2. \tag{12.5.8}$$

Thus temperature recovery affects the flame temperature directly through Eq. (12.5.8), and the flame location indirectly through the stream function f.

It can also be shown that the solution for $f(\eta)$ is again given by that at the subsonic, $M_\infty = 0$ limit, except now the heat transfer number is given by

$$B_{h,c} = \frac{(\tilde{T}_\infty - \tilde{T}_s) + \tilde{Y}_{O,\infty} + \tilde{H}}{\tilde{q}_v}. \tag{12.5.9}$$

This result is physically reasonable in that temperature recovery provides an additional heat source in effecting gasification.

12.5.2. Ignition along a Flat Plate

Next we study the ignition of a supersonic combustible flow along a flat plate (Im, Bechtold & Law 1993). This is the analog of the subsonic situation analyzed in Section 12.3.2, except now ignition is expected to be facilitated by temperature recovery. Indeed, if temperature recovery was not accounted for in the supersonic combustor, then any reasonable estimate of the ignition distance will indicate that it is too long to be practical.

The influence of temperature recovery obviously increases with increasing M_∞. For a moderately supersonic flow over a heated, isothermal plate, we expect that the presence of temperature recovery in the boundary layer would reduce the extent of heat loss from the plate to the freestream, thereby facilitating ignition. Thus starting from a low subsonic M_∞, increasing M_∞ would initially lead to a corresponding increase in the ignition distance simply because the flow is faster. However, by continuously increasing M_∞, the extent of temperature recovery would gradually increase and could eventually reverse the increasing trend in the ignition distance. In fact, for sufficiently large Mach numbers and if the plate temperature is not too high, temperature recovery can be so strong as to cause the development of a temperature bulge within the boundary layer. In this case the ignition kernel develops within the boundary layer instead of being next to the plate, and loses heat to both the plate and the ambience. The plate now becomes a heat sink instead of the primary source of ignition.

We shall study only the moderate M_∞ situation for which ignition is still initiated at the plate. The problem, although quite simple, does exhibit the reversal in trend in the ignition distance due to temperature recovery. Following the same discussion that led to Eq. (12.3.14) for the low subsonic case, but now including the temperature recovery effect, we have

$$\frac{d^2\tilde{T}}{d\xi^2} \approx \frac{1}{(f'')^2}\left(\frac{2x}{U_\infty}\right)\frac{w_F}{\rho} - 2\tilde{H}. \tag{12.5.10}$$

The frozen solution is given by

$$\tilde{T}_0(\xi) = \tilde{T}_s - (\beta - \tilde{H})\xi - \tilde{H}\xi^2, \tag{12.5.11}$$

which shows that, in the frozen limit, the plate is subadiabatic relative to the adjacent gas for $\beta > \tilde{H}$, and superadiabatic otherwise.

An analysis similar to that of Section 12.3.2 can be conducted for the criterion on the ignition distance. The result, however, can also be readily obtained with the following observation. The heat transfer parameter β, identified to be $(\partial\tilde{T}_0/\partial\tilde{n})_0$ for the general $M_\infty = 0$ ignition situation treated in Section 8.2, is now $(\beta - \tilde{H})$,

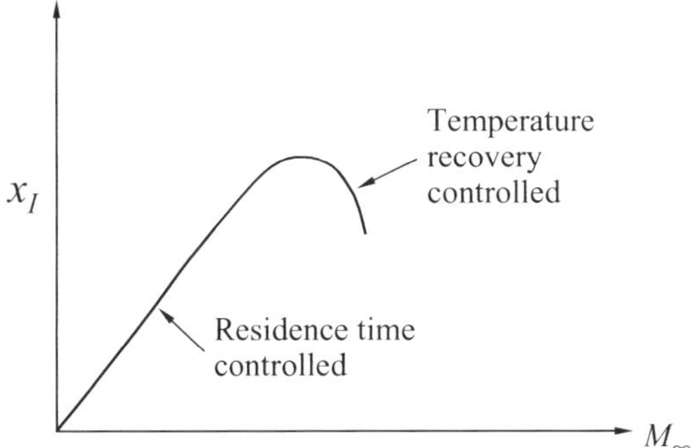

Figure 12.5.1. Dependence of ignition distance on Mach number in a supersonic flat-plate boundary-layer flow, demonstrating the concept of the ignition source being either the hot plate or through temperature recovery.

indicating the reduction in the heat loss rate from the plate due to the presence of temperature recovery. Furthermore, in the reaction zone next to the plate, the volumetric, heat generation term $2\tilde{H}$ in Eq. (12.5.10) is expected to be small relative to the diffusion and reaction terms. The problem is therefore asymptotically identical to the subsonic case, provided β is replaced by $(\beta - \tilde{H})$. From Eq. (12.3.15), we can therefore write down the relation governing the ignition distance x_I as

$$\frac{2x_I}{U_\infty[f''(0)]^2}\left[\frac{2\epsilon Da_C}{(\beta-\tilde{H})^2\rho_s}\right]\tilde{Y}_{O,\infty}\tilde{Y}_{F,\infty}e^{-\tilde{T}_a/\tilde{T}_s} \geq 1. \qquad (12.5.12)$$

Equation (12.5.12) demonstrates the nonmonotonic variation of x_I with M_∞ as

$$x_I \sim M_\infty(\beta-\tilde{H})^2 = M_\infty\left[\beta - \frac{(\gamma-1)}{2}\tilde{T}_\infty M_\infty^2\right]^2. \qquad (12.5.13)$$

Thus as M_∞ increases from zero, x_I first increases linearly with M_∞. The temperature recovery term, however, becomes progressively important with increasing M_∞, such that x_I eventually decreases with M_∞ after reaching a maximum at $M_\infty = \{(2\beta/5)/[(\gamma-1)\tilde{T}_\infty]\}^{1/2}$, with the corresponding maximum ignition distance given by $x_{I,\max} \sim (4\beta/5)^2 M_\infty$. Figure 12.5.1 illustrates such a variation. The rapid rate with which x_I decreases with M_∞ after the attainment of the maximum x_I is to be noted.

12.6. NATURAL CONVECTION BOUNDARY-LAYER FLOWS

Buoyancy is omnipresent in practically all combustion processes on Earth. Since it is a volumetric process, involving rising of the hot combustion products relative to the cold environment, the influence increases with increasing dimension of the combustion phenomenon. A relevant example is flame spreading over walls, ceilings, and floors in room fires.

Consider the natural convection flow adjacent to a vertical heated plate in a relatively cold, quiescent environment, as shown in Figure 12.1.1. The less dense, heated gas next to the plate tends to rise due to buoyancy, which in turn causes the

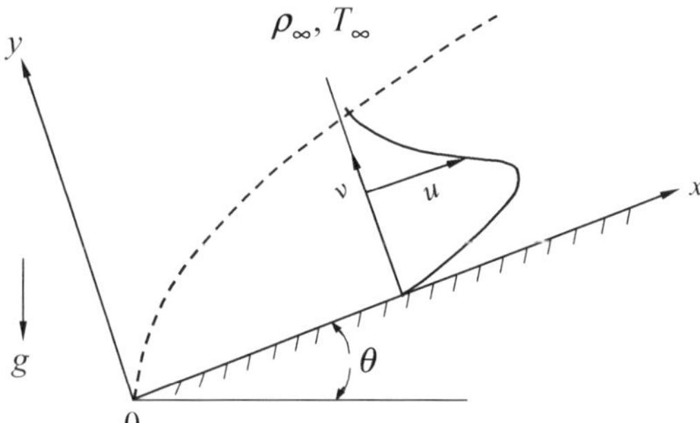

Figure 12.6.1. General representation of a natural convection two-dimensional boundary-layer flow.

environment gas to move toward the plate to replenish the rising gas. The intensity of such a natural convection flow can be characterized by a Grashof number defined as

$$Gr_\ell = \frac{g\ell^3}{\nu_\infty^2}\left(\frac{\rho_\infty - \rho_s}{\rho_\infty}\right) \approx \frac{g\ell^3}{\nu_\infty^2}\left(\frac{T_s - T_\infty}{T_\infty}\right)$$

based on a characteristic length of the system, ℓ. For sufficiently large values of Gr_ℓ, a boundary layer is developed over the plate. The magnitude of Gr_ℓ relevant to the development of such a boundary layer can be assessed by relating it to the Reynolds number Re_ℓ for the forced-flow boundary layer. That is, since $g\ell$ has the dimension of (velocity)^2, the term $g\ell^3/\nu_\infty^2$ in Gr_ℓ can be interpreted as Re_ℓ^2. We therefore expect that it requires a much larger numerical value of Gr_ℓ to develop the natural convection boundary layer than that of Re_ℓ for the forced convection boundary layer. Furthermore, since the forced-convection boundary-layer thickness varies as $\delta \sim Re_\ell^{1/2}$, while $Gr_\ell \sim Re_\ell^2$, we expect that the natural convection boundary layer should vary as $\delta \sim Gr_\ell^{1/4}$. This has been found to be largely correct provided that the plate is not nearly horizontal. The reason being the anticipated dominance of the body force component along the plate as compared to that normal to the plate. When the plate is near horizontal, however, the scaling and, hence, the nature of similarity change to $\delta \sim Gr_\ell^{2/5}$(Stewartson 1958). This change in scaling is rather interesting and is the central theme of this section (Umemura, Nam & Law 1990). Application of the results to specific combustion situations is fairly straightforward.

The problem of interest is shown in Figure 12.6.1, in which a flat plate of temperature T_s is inclined to the horizontal plane at an angle θ, with the gravity vector $\mathbf{g} = g\mathbf{j}$ pointed downward. We choose the x and y coordinates to be parallel and normal to the plate. The environment is stationary and has a temperature $T_\infty < T_s$. The conservation equations for mass, x-momentum, y-momentum, and energy for the two-dimensional natural convection boundary layer are

$$\frac{\partial(\rho u)}{\partial x} + \frac{\partial(\rho v)}{\partial y} = 0 \qquad (12.1.1)$$

$$\rho u \frac{\partial u}{\partial x} + \rho v \frac{\partial u}{\partial y} = -\frac{\partial p}{\partial x} + \mu \frac{\partial^2 u}{\partial y^2} + g(\rho_\infty - \rho) \sin\theta \tag{12.6.1}$$

$$0 = -\frac{\partial p}{\partial y} + g(\rho_\infty - \rho) \cos\theta \tag{12.6.2}$$

$$\rho u \frac{\partial T}{\partial x} + \rho v \frac{\partial T}{\partial y} = (\lambda / c_p) \frac{\partial^2 T}{\partial y^2} + q_c w, \tag{12.6.3}$$

subject to the boundary conditions

$$u = v = 0, \quad T = T_s \quad \text{at} \quad y = 0 \tag{12.6.4}$$

$$u = 0, \quad T = T_\infty \quad \text{in the ambience,} \tag{12.6.5}$$

where we have assumed μ = constant, which is a more convenient assumption for this problem than that of $\rho\mu$ = constant invoked in Section 12.1. The last terms in Eqs. (12.6.1) and (12.6.2) represent the components of the buoyancy force in the x and y directions respectively. In particular, Eq. (12.6.2) shows that $\partial p/\partial y > 0$, which is a consequence of decreasing temperature and, hence, increasing density as y increases.

To appreciate the different manner through which the normal and tangential components of the buoyancy force drive the fluid along the plate, we integrate Eq. (12.6.2) with $p(y = \infty) = 0$ to obtain

$$p(x, y) = -g \cos\theta \int_y^\infty (\rho_\infty - \rho) dy. \tag{12.6.6}$$

Substituting Eq. (12.6.6) into Eq. (12.6.1) yields

$$\rho u \frac{\partial u}{\partial x} + \rho v \frac{\partial u}{\partial y} = g \cos\theta \frac{\partial}{\partial x} \int_y^\infty (\rho_\infty - \rho) dy + \mu \frac{\partial^2 u}{\partial y^2} + g(\rho_\infty - \rho) \sin\theta. \tag{12.6.7}$$

Evaluating Eqs. (12.6.6) and (12.6.7) at the plate surface, $y = 0$, where the effects of both buoyancy force components become maximum, we obtain

$$p(x, 0) = -g \cos\theta \int_0^\infty (\rho_\infty - \rho) dy \tag{12.6.8}$$

$$-\mu \left(\frac{\partial^2 u}{\partial y^2} \right)_{y=0} = g \cos\theta \frac{d}{dx} \int_0^\infty (\rho_\infty - \rho) dy + g(\rho_\infty - \rho) \sin\theta. \tag{12.6.9}$$

The physical meaning of Eqs. (12.6.8) and (12.6.9) is the following. First, Eq. (12.6.8) indicates that, relative to $p(y = \infty) = 0$, the normal buoyant force acting per unit distance of x induces a negative pressure at the plate surface. Furthermore, since we expect $\int_0^\infty (\rho_\infty - \rho) dy \sim (\rho_\infty - \rho_s)\delta(x)$, where $\delta(x)$ is the boundary-layer thickness, the magnitude of this negative pressure will continuously increase downstream because of the monotonic increase of $\delta(x)$ and consequently the total amount of the buoyant mass. This normal pressure then induces a streamwise pressure gradient force, $\partial p/\partial x < 0$, as represented by the first term on the RHS of Eq. (12.6.9).

The second RHS term represents the streamwise buoyant force, which is a constant. Thus both components of the buoyant force contribute to the streamwise motion—indirectly through the normal component and directly through the tangential component. These two buoyant forces are balanced by the shear stress at the plate surface. Additionally, since the boundary-layer thickness $\delta(x)$ is expected to vary with x^m, with $0 < m < 1$ for proper behavior of its development, $d\delta(x)/dx$ would diverge at the leading edge ($x = 0$) but decrease to zero as $x \to \infty$. Consequently the influence of the normal buoyant force for the streamwise motion decreases with increasing x while the influence of the tangential buoyant force is a constant, which implies that the tangential motion is dominated by the normal force near the leading edge and by the tangential force further downstream. In other words, the boundary-layer flow is horizontal-plate-like near the leading edge and vertical-plate-like in the downstream. As such, the flow behavior depends on the inclination angle θ as well as the streamwise distance x.

We now transform the governing equations to the boundary-layer variables and then separately identify the similarity solution for the vertical ($\theta = \pi/2$) and horizontal ($\theta = 0$) plates. The boundary-layer variables are defined as

$$s = \left(Gr_x \frac{\sin^5\theta}{\cos^4\theta}\right)^{1/3} \sim x, \quad \eta = \frac{1}{\delta(s)}\int_0^y \frac{\rho}{\rho_\infty}dy,$$

$$\psi = \rho_\infty \nu \left(\frac{\cos\theta}{\sin\theta}\right) s^n f(s,\eta), \tag{12.6.10}$$

where

$$Gr_x = \frac{g(T_w - T_\infty)x^3}{\nu_\infty^2 T_\infty}, \quad \delta(s) = \left[\frac{\nu_\infty^2 T_\infty \cos\theta}{g(T_s - T_\infty)\sin^2\theta}\right]^{1/3} s^m, \tag{12.6.11}$$

and ψ is defined through $\rho u = \partial\psi/\partial y$ and $\rho v = -\partial\psi/\partial x$. We let δ and ψ vary with arbitrary powers of s, namely m and n respectively. These values are to be determined through the quest for similarity in general as well as in the limits of $\theta = 0$ and $\pi/2$, which respectively correspond to $s = 0$ and ∞ for fixed Gr_x and, hence, x.

Applying the transformation (12.6.10) and (12.6.11) to Eq. (12.6.7) in the same manner as that for the forced convection flow treated in Section 12.1, after some algebra we obtain

$$f''' - s^{m+n-1}\left[(n-m)(f')^2 - nff''\right] + ms^{5m-2}\left(\int_\eta^\infty \hat{T}d\eta + \eta\hat{T}\right) + s^{4m-1}\hat{T}$$

$$= s^{m+n}\left(f'\frac{\partial f'}{\partial s} - f''\frac{\partial f}{\partial s} - s^{3m-2n}\int_\eta^\infty \frac{\partial\hat{T}}{\partial s}d\eta\right), \tag{12.6.12}$$

where $\hat{T} = (T - T_\infty)/(T_s - T_\infty)$ and the transformation has eliminated the explicit dependence of the solution on θ.

To determine m and n, we first require that Eq. (12.6.12) degenerates to the similarity form for both the $\theta = 0$ and $\pi/2$ limits. In such limits, $(\partial f/\partial s) \equiv 0$, $\partial f'/\partial s \equiv 0$, and $(\partial \hat{T}/\partial s) \equiv 0$ such that the RHS of Eq. (12.6.12) vanishes. Furthermore, all terms on the LHS must also be independent of s. The requirement on the second term implies that $m + n - 1 \equiv 0$, or

$$n = 1 - m. \tag{12.6.13}$$

Thus Eq. (12.6.12) simplifies to

$$\begin{aligned} f''' - (1-2m)(f')^2 + (1-m) f f'' + m s^{5m-2} \left(\int_\eta^\infty \hat{T} d\eta + \eta \hat{T} \right) + s^{4m-1} \hat{T} \\ = s \left(f' \frac{\partial f'}{\partial s} - f'' \frac{\partial f}{\partial s} - s^{5m-2} \int_\eta^\infty \frac{\partial \hat{T}}{\partial s} d\eta \right). \end{aligned} \tag{12.6.14}$$

We now specialize to the two limiting cases of similarity flows for which the RHS of Eq. (12.6.14) vanishes, and examine the behavior of the terms on the LHS. Thus for the vertical plate, $\theta = \pi/2$ and $s \to \infty$, the fourth term vanishes if we make the fifth term to be independent of s by setting $4m - 1 = 0$ such that $s^{5m-2} = s^{-3/4} \to 0$. Consequently, to achieve similarity we require

$$m = \frac{1}{4} \quad \text{for} \quad \theta = \frac{\pi}{2}, \tag{12.6.15}$$

and Eq. (12.6.14) becomes

$$f''' - \frac{1}{2}(f')^2 + \frac{3}{4} f f'' + \hat{T} = 0. \tag{12.6.16}$$

By the same reasoning, since $s \to 0$ for the horizontal plate, the fifth term vanishes if we make the fourth term to be independent of s by setting $5m - 2 = 0$, such that $s^{4m-1} = s^{3/5} \to 0$. Thus similarity holds for

$$m = \frac{2}{5} \quad \text{for} \quad \theta = 0, \tag{12.6.17}$$

and Eq. (12.6.14) becomes

$$f''' - \frac{1}{5}(f')^2 + \frac{3}{5} f f'' + \frac{2}{5} \left(\int_\eta^\infty \hat{T} d\eta + \eta \hat{T} \right) = 0. \tag{12.6.18}$$

For the energy conservation equation (12.6.3), applying the transformation of (12.6.10) and (12.6.11), with $n = 1 - m$ and allowing for a general Prandtl number, Pr, we have

$$\begin{aligned} \frac{\partial^2 \hat{T}}{\partial \eta^2} + Pr(1-m) f \frac{\partial \hat{T}}{\partial \eta} - Prs f' \frac{\partial \hat{T}}{\partial s} - Prs \frac{\partial f}{\partial s} \\ = -s^{2m}(T_s - T_\infty) \left(\frac{c_p}{\lambda} \right) \left[\frac{\nu_\infty^2 T_\infty \cos\theta}{(T_s - T_\infty)\sin\theta} \right]^{2/3} \left(\frac{\rho_\infty}{\rho} \right)^2 q_c w. \end{aligned} \tag{12.6.19}$$

It is then apparent that in the presence of finite-rate chemistry the streamwise influence of the reaction rate will continuously render $\hat{T}$ to be nonsimilar. Furthermore, since $\hat{T}$ is coupled to the momentum equation, f is accordingly affected and is therefore nonsimilar as well. Thus the problem can be actually inherently nonsimilar, even in the limits of vertical and horizontal plates. However, as we have shown previously, the problem can be rendered similar for nonpremixed reaction-sheet combustion, and locally similar for weakly reactive situations.

Two additional points merit mentioning before closing. First, in conventional nonreactive heat and mass transfer problems the temperature variation in the flow is not large and the fluid density is frequently assumed to be constant. The density differential, which is responsible for the buoyant force, is then replaced by the temperature differential through the use of the coefficient of thermal expansion

$$\beta = \frac{1}{v}\left(\frac{\partial v}{\partial T}\right)_p = -\frac{1}{\rho}\left(\frac{\partial \rho}{\partial T}\right)_p \approx -\frac{1}{\rho}\left(\frac{\rho - \rho_\infty}{T - T_\infty}\right), \tag{12.6.20}$$

where v is the specific volume. The buoyancy force $g(\rho_\infty - \rho)$ in Eqs. (12.6.1) and (12.6.2) is then replaced by $g\beta\rho(T - T_\infty)$ with the density treated as constant. This is known as the Boussinesq approximation. In the present derivation the conversion of the density differential to the temperature differential was effected by using the isobaric approximation, $(\rho - \rho_\infty)/\rho \approx (T_\infty - T)/T_\infty$, which is not restricted to small temperature or density differences.

Second, following the convention in nonreactive heat transfer, our Grashof number is based on the temperature (or density) difference $(T_s - T_\infty)$. This is not an appropriate representation of the source of the driving force when a flame is present, as for example in the case of nonpremixed burning involving an ablating surface. The appropriate temperature difference used in assessing Gr should be $(T_f - T_\infty) \approx (T_{\rm ad} - T_\infty)$. Indeed, since T_s is frequently quite close to T_∞, use of $(T_s - T_\infty)$ can result in much smaller values of Gr.

PROBLEMS

1. While extensive theories on boundary-layer flows have been developed for incompressible flows, the flows of interest to combustion are characterized by variable density. This difficulty however can be circumvented by using the Howarth–Dorodnitzyn transformation, which reduces the equations for variable density flows to those of constant density flows. Take the simple flat-plate boundary flows for illustration. The continuity and momentum conservation equations are

$$\frac{\partial(\rho u)}{\partial x} + \frac{\partial(\rho v)}{\partial y} = 0, \tag{12.P.1}$$

$$\rho u\frac{\partial u}{\partial x} + \rho v\frac{\partial u}{\partial y} = \frac{\partial}{\partial y}\left(\mu\frac{\partial u}{\partial y}\right). \tag{12.P.2}$$

Show that by using the variables

$$\xi = \rho\mu x, \quad z = \int_0^y \rho dy, \quad w = \frac{1}{\rho\mu}\left(\rho v + u\int_0^y\left(\frac{\partial\rho}{\partial x}\right)_y dy\right),$$

with $\rho\mu$ assumed to be constant, the variable-density governing equations are transformed to the constant density form:

$$\frac{\partial u}{\partial \xi} + \frac{\partial w}{\partial z} = 0, \tag{12.P.3}$$

$$u\frac{\partial u}{\partial \xi} + w\frac{\partial u}{\partial z} = \frac{\partial^2 u}{\partial z^2}. \tag{12.P.4}$$

2. The various boundary-layer solutions studied in the text were all directly obtained from the governing differential equations. An alternate, approximate solution procedure is the integral analysis of Karman-Pohlhausen. The method integrates, say, the momentum equation, with an assumed velocity profile through the boundary layer, and solves for such global properties as functions of the momentum boundary-layer thickness. We shall demonstrate this method for the ablating flat-plate boundary layer of Section 12.2. For simplicity we shall use the incompressible form of the governing equations, recognizing that they can be obtained from the compressible form through the Howarth–Dorodnitzyn transformation, shown in Problem 1. Hence, we have

$$\frac{\partial u}{\partial x} + \frac{\partial v}{\partial y} = 0, \tag{12.P.5}$$

$$u\frac{\partial u}{\partial x} + v\frac{\partial u}{\partial y} = \nu\frac{\partial^2 u}{\partial y^2}, \tag{12.P.6}$$

subject to the boundary conditions

$$u(0) = 0, \quad v(0) = v_0, \quad u(\delta) = U_\infty, \tag{12.P.7}$$

where v_0 is the velocity of the gasifying species leaving the surface, and $\delta(x)$ is the boundary-layer thickness.

(a) With v given by integrating Eq. (12.P.5) from $y = 0$ to $\delta(x)$, show that a further integration of Eq. (12.P.6) yields

$$v_0 U_\infty - \frac{\partial}{\partial x}\int_0^\delta (U_\infty - u)u\,dy = -\nu\left(\frac{\partial u}{\partial y}\right)_0. \tag{12.P.8}$$

(b) Next, by defining the boundary-layer variables, $\eta = y/\delta(x)$, $u/U_\infty = \phi(\eta)$, and letting the surface blowing velocity vary inversely with $\delta(x)$, as $v_0 = \nu\beta/\delta(x)$, where β can be considered as a blowing rate coefficient, show that Eq. (12.P.8) becomes, after a further integration in x,

$$[\delta(x)]^2 = \frac{2\nu}{U_\infty I}[\beta + \phi'(0)]x, \tag{12.P.9}$$

where $I = \int_0^1 (\phi - \phi^2)d\eta$.

(c) Assume a fourth-degree polynomial in η for $\phi(\eta)$, which satisfies the profile requirements,

$$\phi(0) = 0, \quad \phi(1) = 1, \quad \phi'(1) = 0, \quad \phi''(1) = 0, \quad \beta\phi'(0) = \phi''(0), \tag{12.P.10}$$

where the last relation is obtained by evaluating Eq. (12.P.5) at the surface, and show that

$$\phi(\eta) = \frac{12}{(6+\beta)}\left[\eta + \beta\frac{\eta^2}{2} - (3+2\beta)\frac{\eta^3}{3} + (2+\beta)\frac{\eta^4}{4}\right]. \tag{12.P.11}$$

(d) Now let us finally let the system react in the manner of the nonpremixed burning problem of Section 12.2. Determine the mass burning flux and the shear stress at the surface, $(\rho v)_0$ and $(\mu \partial u/dy)_0$, respectively.

3. Rework the problem of Section 8.2.3 on the ignition of an unconfined mixture by a hot plate, up to the stage where you have shown that the problem is locally similar.

4. Use the jet flow system of Section 12.4 to analyze the ignition of a stagnant combustible of temperature T_∞ and reactant mass fractions $Y_{i,\infty}$ by a hot, inert jet of radius r_o and initial temperature T_o and velocity u_o. Assume $Sc = 1$ for simplicity. This problem is somewhat similar to that of the mixing layer treated in Section 12.3.3, except now the ignition source, that is, the jet, is of finite dimension and, hence, during the induction period is continuously cooled down through entrainment of the ambient gas. Just derive the final structure equation for the inner region, with the appropriate boundary conditions. Discuss the issues of similarity and jet cooling.

5. Set up and solve the problem of the nonpremixed burning of a vertical ablating plate subjected to natural convection (Kosdon, Williams & Buman 1969) in the same manner as that for the forced convection problem of Section 12.2.

13 Combustion in Two-Phase Flows

In many combustion applications the fuel is originally present as either liquid or solid. In order to facilitate mixing and the overall burning rate, as pointed out in Section 6.4.1, the condensed fuel is frequently first atomized or pulverized, and then sprayed or dispersed in the combustion chamber. Consequently, in these devices combustion actually takes place in a two-phase medium, consisting of the dispersed fuel droplets or particles in a primarily oxidizing gas.

A description of two-phase combustion consists of three components, namely the gasification and dynamics of individual and groups of droplets (or particles); a statistical characterization of the spray; and the collective interaction of these droplets with the bulk gaseous medium through the description of the two-phase flow in terms of heat, mass, and momentum transfer. The first component was introduced in Section 6.4 through the d^2-law of droplet vaporization and burning, and will be extensively studied in this chapter for both droplets and particles. Specifically, we shall first discuss the general phenomenology of droplet combustion with and without external convection, and the experimental methodologies commonly used in investigating droplet combustion. We shall then study the combustion of single-component droplets by relaxing the various assumptions associated with the d^2-law, and hence examine effects of droplet heating, fuel vapor accumulation, and variable transport properties that were briefly mentioned in Section 6.4.4. We shall also relax the assumptions of gas-phase quasi-steadiness, stagnant environment, and solitary droplet by discussing effects due to gas-phase transient diffusion, external convection, and droplet interaction, respectively. The influence of finite-rate kinetics through droplet ignition and extinction, and the phenomenon of droplet collision, will be discussed. We shall then study multicomponent droplet combustion because most practical liquid fuels are blends of many components. The gasification mechanisms of miscible fuel blends and of alcohols, emulsions, slurries, and reactive liquids will be covered.

Next we shall discuss carbon combustion for its relevance in the burning of organic matters such as coal particles, and metal particle combustion for its relevance in explosion hazards, propellant combustion, and the combustion synthesis of materials.

Following the extensive discussion on droplet and particle processes, we shall then study spray combustion, which includes spray statistics, the conservation equations, and problems on spray vaporization and flame propagation.

Reviews on droplet, particle, and spray combustion are given in Faeth (1977; 1996), Law (1982, 1998), Law and Law (1993), and Sirignano (1999), and pertinent review literature on other topics will be cited where appropriate.

13.1. GENERAL CONSIDERATIONS OF DROPLET COMBUSTION

13.1.1. Phenomenology

For a droplet undergoing combustion in a stagnant, gravity-free environment, both forced and natural convection are absent. This results in a spherically symmetric combustion configuration, as shown in Figure 6.4.3. Here the liquid fuel gasifies and forms a layer rich in fuel vapor next to the droplet surface. The fuel vapor is subsequently transported outward toward the reaction zone, which is deficient in the fuel concentration. Similarly the ambient oxidizer gas is transported inward. In the reaction zone fuel and oxidizer mix and react, liberating heat and combustion products that are transported both inward and outward. At the droplet surface the inwardly transported heat is utilized for further gasification to sustain combustion and also in heating up the droplet interior. Because of spherical symmetry, all transport processes occur only in the radial direction. In the gas phase both diffusion and Stefan convection exist, while in the liquid phase only diffusion is possible. The continuously regressing droplet surface, however, constitutes a passive mode of liquid-phase convection responsible in exposing the droplet interior to the gas medium. That is, in the reference frame attached to the regressing droplet surface, the state of the droplet interior will experience a net convective velocity as the liquid element "flows" toward the surface.

In realistic situations a relative nonradial velocity usually exists between the droplet and the ambience. This can be caused either by the inertia the droplet acquired during spraying, or by its slower response, compared with that of the gas, to changes in the flow velocity and configuration. Even in a stagnant environment, buoyancy alone can induce a relative velocity.

When the relative velocity is small, the flow configuration is slightly distorted from spherical symmetry, as shown in Figure 13.1.1a. Here the gas-phase processes should be qualitatively similar to the spherically symmetric situation. The mathematical analysis, however, is significantly more complicated because of the additional dependence on the nonradial coordinates. Furthermore, the existence of nonradial convection exerts a shear stress at the droplet surface and thereby generates a recirculatory motion within the droplet interior. Therefore in the liquid phase there is now also convective transport in addition to diffusion. Finally, since gas-phase transport rates are enhanced with this additional nonradial convection, the burning rate is also expected to be increased.

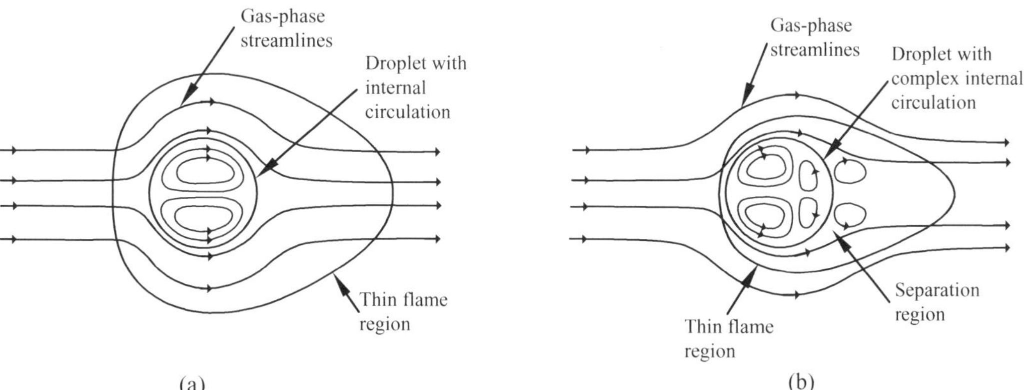

Figure 13.1.1. Schematic showing droplet combustion configurations in the presence of externally imposed convection: (a) Weak convection that induces internal circulation; (b) strong convection, leading to separation and hence a complex recirculation pattern.

With further increase in the flow velocity, whose intensity can be represented by the appropriate Reynolds and Grashof numbers, separation occurs at the droplet surface (Figure 13.1.1b). This obviously further complicates the flow configuration and analysis.

Finally, with very intense convection, the flame surrounding the droplet can be "blown off." The droplet now undergoes pure vaporization with a significantly reduced gasification rate. The fuel vapor generated at the droplet surface is then swept leeward where it mixes with the oxidizing gas to a certain extent. Combustion is sometimes possible in the wake region of the droplet, resulting in a wake flame. The phenomenon of blow-off of the enveloping nonpremixed flame is essentially an extinction event, caused by extinction of the flame in the front stagnation region.

Figures 13.1.2 and 13.1.3 show photographs of the various flame configurations discussed above. Figure 13.1.2 was obtained by burning a fuel droplet suspended at

Figure 13.1.2. Photographic image of an almost spherically symmetric droplet combustion configuration, obtained for a suspended droplet in reduced pressure and hence buoyancy.

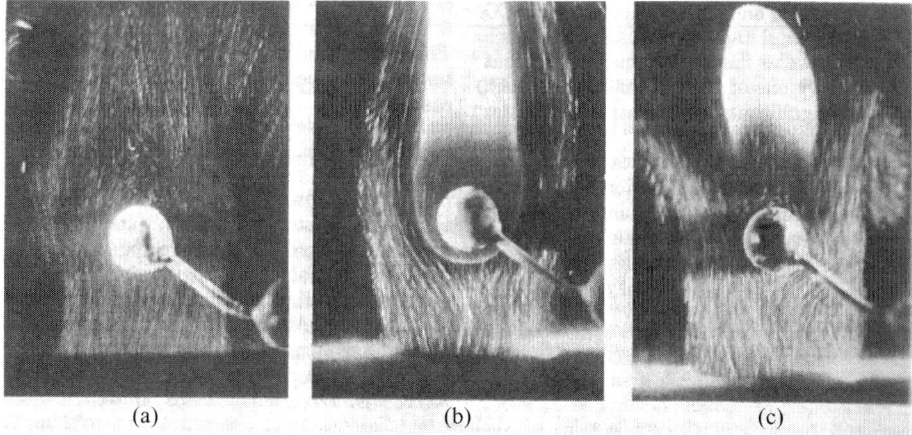

Figure 13.1.3. Particle track photographs. (a) Sphere without flame, $Re = 92$; (b) sphere with envelope flame, $Re = 92$; (c) sphere with wake flame, $Re = 152$ (Gollahalli & Brzustowski 1973).

the end of a quartz fiber in a quiescent environment of reduced buoyancy, yielding a nearly spherical flame. Figure 13.1.3 shows burning under highly convective situations resulting from blowing air below a porous sphere wetted with liquid fuel (Gollahalli & Brzustowski 1973). It is seen that, at a moderate blowing rate (Figure 13.1.3b), the flame completely envelopes the fuel sphere. However, at a higher blowing rate, the flame segment in the forward portion of the sphere appears to be "extinguished" (Figure 13.1.3c), and that combustion now takes place only in the wake region.

It is worthwhile to take this opportunity to mention some potential pitfalls when interpreting the photographic image of a flame. The wake "flames" in Figures 13.1.3b and 13.1.3c are partly images of soot emission, which is visually bright yellow. However, real flames of hydrocarbon fuel oxidation are frequently bluish in color whose intensity also progressively weakens as the extinction state is approached. Thus when observing or photographing a flame that is partly bright yellow and partly faint blue, the blue region may not be readily detected. Thus special care is needed in imaging sooty flames or flames with large variations of luminosity.

The bulk parameters of interest in droplet combustion studies are the droplet gasification rate, the location and temperature of the flame, the droplet drag and dynamics, ignition and extinction limits, and the extent of pollutant formation such as those of NO_x and soot. The droplet diameters of interest are typically between 10 and 100 μm. Gasification of smaller droplets can be considered to be instantaneous during the period between spray injection and active combustion, while larger droplets either tend to breakup or cannot achieve complete gasification during the available residence time. Droplet breakup also limits the maximum droplet Reynolds number to about 100 for typical surface tension values of hydrocarbon fuels. It is frequently desirable for the droplets within a spray to have a distribution of sizes and velocities in order to achieve an optimum spatial distribution through penetration, as well as a controlled rate of gasification and thereby chemical heat release.

13.1.2. Experimental Considerations

Experimental studies of droplet combustion have employed the following methods: (a) a single droplet suspended at the end of a thin quartz fiber (Godsave 1953; Goldsmith & Penner 1954); (b) a freely falling single droplet or droplet stream (Sangiovanni & Labowsky 1982; Lasheras, Fernandez-Pello & Dryer 1979; Wang, Liu & Law 1984); (c) a porous sphere with liquid fuel fed to its interior at such a rate that the surface is just wetted to support combustion (Wise, Lorell & Wood 1955; Agoston, Wise & Rosser 1957). Their merits and limitations are briefly discussed in the following.

The suspended droplet experiment can be easily set up and performed. Ignition is achieved by either a single pulse or continuous spark discharge. Furthermore, since the droplet is stationary, detailed cinemicrophotography can be taken of its burning sequence. Because of the thicknesses of the suspension fiber and its thickened end, it is difficult to suspend a droplet much smaller than 1 mm in diameter, which is much larger than typical droplet sizes within sprays. This should not be of serious concern if the size-dependence of the phenomenon of interest is known. However, the suspension fiber also distorts the droplet shape from spherical; the distortion is especially severe toward the end of the droplet lifetime when the droplet size becomes comparable to the suspension fiber and its thickened end. There is also heat transfer from the flame to the suspension fiber at the point where they intercept. The amount of heat conducted away from the droplet represents a loss, while the amount conducted toward the droplet can enhance the gasification rate because heat transfer through the fiber is more efficient than that through the gas medium between the flame and the droplet surface. It has been shown that interference from distortion and heat loss can be considered to be unimportant during much of the droplet lifetime for fiber diameters less than 100 μm.

The suspension technique is limited to fuels that are relatively nonvolatile, because otherwise substantial vaporization would have occurred during the period involved with suspending the droplet, charging the chamber with the proper environment, and applying the ignition stimulus. The problem is particularly severe for multicomponent fuels whose composition can be altered, from the prepared value, by an extent that is not known because of preferential vaporization of components with higher volatilities.

Free droplet experiments offer the advantages of small size, noninterference from a suspension fiber, the capability of using volatile fuels, and in situ sampling of the droplet composition. A notable technique of generating a stream of uniform-sized droplets with controllable spacing and minimum convection is that of ink-jet printing, which involves squeezing out droplets from a nozzled tube pressed against a piezoelectric transducer (Wang, Liu & Law 1984). Thus by applying pulses of voltage to the transducer, it is deformed and thereby generates a pressure wave that forces a droplet of given volume out of the nozzle. The regularity of the droplet stream produced permits detailed photography of the droplet size by using stroboscopy, which yields the instantaneous droplet burning rate. The technique is particularly attractive because of its "droplet-on-demand" feature, controllable through electronic circuitry.

The porous sphere experiment is a truly steady-state one because of the fixed "droplet" size and liquid temperature, and therefore most closely conforms with the steady-state gas-phase assumption of the d^2-law. This experiment also allows detailed probing of the flame structure. Its main drawbacks are the excessively large size and the preclusion of observing certain transient phenomena that are inherently present during droplet combustion.

All combustion experiments conducted under the influence of gravity are complicated by buoyancy (Law & Faeth 1994; Ross 2001). For droplet combustion the effects are manifested in two ways. First, the burning rate is increased because of the enhanced transport rates. Second, the flame shape is usually so severely distorted from spherical symmetry that it is not meaningful to identify a flame "diameter." The distortion increases with the droplet size, and therefore is particularly serious for experiments using suspended droplet and porous sphere techniques. Finally, since the intensity of buoyancy depends on the instantaneous droplet size, the above effects are also expected to be transient in nature.

Two techniques have been employed to minimize or eliminate buoyancy. Single droplet experiments have been conducted in gravity-free environments such as a freely falling chamber (Kumagai & Isoda 1957; Okajima & Kumagai 1975), or onboard an aircraft executing a parabolic trajectory, or in a space-based laboratory (Dietrich et al. 1996). Here a fuel droplet can be created by suspending a liquid mass between two capillary fibers that are then impulsively pulled apart. Upon ignition, the subsequent combustion sequence can be studied through cinephotography. This is probably the most desired technique to study spherically symmetric droplet combustion, although the experimental set-up and procedure can be quite complex and costly. An alternate, relatively simple, technique at minimizing buoyancy is to conduct the experiment under low pressure (Law, Chung & Srinivasan 1980; Sung, Zhu & Law 1998). The principle being that since the spherically symmetrical droplet combustion is diffusion controlled and therefore basically pressure-independent, reducing the chamber pressure therefore diminishes buoyancy, due to reduced density and hence density difference, without affecting the basic diffusion-limited combustion mechanism. Figure 13.1.2 shows that the resulting flame can achieve a high degree of sphericity and concentricity. There is, however, a lower limit in pressure below which finite-rate kinetic effects become important. This can be circumvented by enriching the combustion environment with oxygen, which also reduces the flame size and, hence, the extent of buoyancy.

Buoyancy effects generally are not strong for earth-bound experiments involving freely falling droplets generated by the ink-jet printing technique because of their relatively small sizes typically between 50 to 300 μm.

13.2. SINGLE-COMPONENT DROPLET COMBUSTION

13.2.1. Droplet Heating

We have shown that during the steady vaporization and combustion of a pure-component droplet the droplet temperature assumes a unique value for a given system. This temperature is usually much higher than the initial droplet temperature at the instant of injection or ignition. Therefore there must exist a transient droplet heating period during which the heat transferred to the droplet surface is used for both gasification as well as droplet heating, causing a reduction in the droplet gasification rate.

Since droplet heating involves the change of a liquid property, and therefore occurs over a longer period than that of gas-phase transport, gas-phase quasi-steadiness can still be assumed. Thus the only modification of the gas-phase solution such as the d^2-law is to substitute q_v by an effective latent heat of gasification, $q_{v,\mathrm{eff}}$, defined as

$$m_v q_{v,\mathrm{eff}} = m_v q_v + \left(4\pi r^2 \lambda_\ell \frac{\partial T}{\partial r}\right)_{r_s^-}, \tag{13.2.1}$$

where the heat conduction term represents the amount of heat transferred to the droplet interior to effect droplet heating, λ_ℓ is the liquid-phase thermal conductivity coefficient, and the subscript ℓ designates the liquid phase. Evaluation of this term requires knowledge of the droplet temperature distribution $T(r, t)$. In the absence of internal recirculatory motion, the unsteady heat transfer process within the droplet is simply given by the spherically symmetric heat conduction equation (Law & Sirignano 1977),

$$\frac{\partial T}{\partial t} = \frac{1}{r^2}\frac{\partial}{\partial r}\left(\alpha_{h,\ell} r^2 \frac{\partial T}{\partial r}\right), \tag{13.2.2}$$

subject to

$$T(r; t = 0) = T_o(r), \quad \left(\frac{\partial T}{\partial r}\right)_{r=0} = 0, \tag{13.2.3}$$

and Eq. (13.2.1), where $\alpha_{h,\ell} = \lambda_\ell/(c_{p,\ell}\rho_\ell)$ is the thermal diffusion coefficient of the liquid. Coupling between the gas- and liquid-phase solutions is achieved by using $q_{v,\mathrm{eff}}$ in place of q_v in Eq. (6.4.31). Numerical computation is needed for the solution.

There are three sources of unsteadiness in the above equations, namely the unsteady conduction term in Eq. (13.2.2), the continuously regressing droplet surface $r_s(t)$, and the continuously changing surface temperature $T_s(t)$. To simplify the analysis, frequently the droplet temperature is assumed to be spatially uniform but temporally varying (Law 1976). Then energy conservation at the droplet surface is simply

$$m_v q_{v,\mathrm{eff}} = m_v q_v + \left(\frac{4}{3}\pi r_s^3 \rho_\ell c_{p,\ell}\right)\frac{dT_s}{dt}, \tag{13.2.4}$$

from which $T_s(t)$ can be determined.

The above two models represent extreme rates of liquid-phase transport. Equation (13.2.2) only allows heat diffusion, which is always present, and is therefore the slowest limit. On the other hand Eq. (13.2.4) implies that the conductivity is infinitely large such that spatial variations are perpetually uniformized. Thus it represents the fastest possible limit. These two models are respectively referred to as the diffusion limit and the infinite conductivity limit; the latter is also conventionally called the batch distillation limit because it is analogous to the chemical process of distillation.

The parameter indicating the relative dominance of the diffusion limit versus the distillation limit behavior is the gasification Peclet number (Makino & Law 1988a), which can be defined as the ratio of the droplet surface regression rate to the liquid-phase thermal diffusivity,

$$Pe_h = \frac{K}{\alpha_{h,\ell}},$$

where K can be either K_v or K_c, for vaporization or burning respectively. Thus for situations involving either very high liquid-phase thermal diffusion or very slow gasification rates, that is, when $Pe_h \ll 1$, the distillation limit with uniform droplet temperature is favored. By the same reasoning, the diffusion limit with droplet temperature stratification is favored when $Pe_h \gg 1$.

Liquid-phase transport rates can also be facilitated in the presence of convection in the form of internal circulation. The increase, however, cannot be too large because temperature uniformization is effected only through diffusion that takes place in directions normal to the streamlines. It has been shown (Lara-Urbanejo & Sirignano 1981) that, in the limit of infinite recirculation rate, the center of the vortex core is located at $\frac{2}{3}r_s$, which gives the maximum reduction in the characteristic dimension for diffusive heating.

In Figures 13.2.1, 13.2.2, and 13.2.3, predictions from both models are shown for an octane droplet burning in the standard atmosphere, with an initial droplet temperature of 300 K. Figure 13.2.1 shows the variations of the surface and center temperatures with a nondimensional time $\tilde{t} = [(\rho_g D_g)/(\rho_\ell r_{s,o}^2)]t$. It is seen that in the diffusion limit the surface temperature initially increases rapidly while the core region slowly starts to heat up. This heating may or may not persist throughout the droplet lifetime, depending on the liquid-phase thermal diffusivity. In the distillation limit the increase in the uniform droplet temperature essentially follows that of the surface temperature in the diffusion limit. The increase is initially slower because of the additional heat needed for the core region, although heating of the complete droplet is finished earlier.

Figure 13.2.2 shows the temporal variations of $q_{v,\mathrm{eff}}/q_v$ and $\tilde{m}_v$. Since $q_{v,\mathrm{eff}}/q_v \to 1$ in the limit of vanishing droplet heating, $(q_{v,\mathrm{eff}}/q_v) - 1 > 1$ and < 1 should respectively represent periods dominated by droplet heating and droplet gasification. It is thus seen that intense droplet heating, and therefore rapid increase in the gasification rate $\tilde{m}_v$, can be considered to be over in less than 10 percent of the droplet lifetime. It may also be noted that the burning process is already gasification dominated before

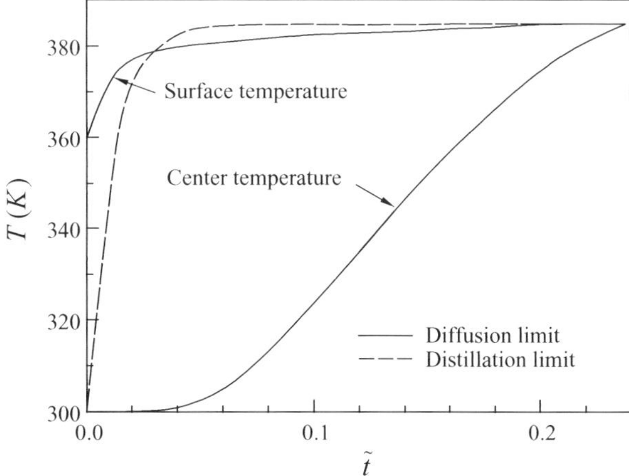

Figure 13.2.1. Temporal variation of the surface and center temperatures of an octane droplet after ignition, demonstrating the rapidity in the heating up of the droplet surface layer in both the diffusion and distillation limits.

the core region is appreciably heated. This demonstrates the important concept that because the surface layer, which consists of substantial amount of the droplet mass, is initially heated at approximately similar rates for the two limits, heating of the much lighter inner core constitutes only small perturbation to the total heat budget at the

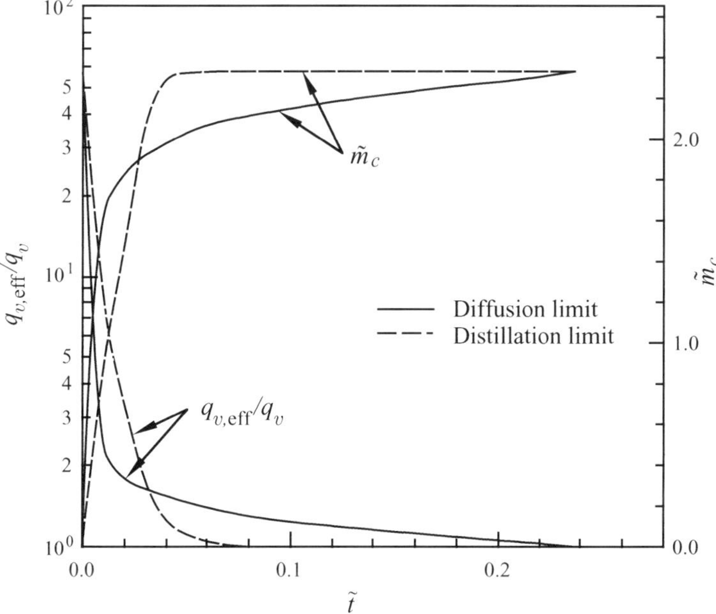

Figure 13.2.2. Temporal variation of the burning rate constant and normalized effective latent heat of vaporization of the octane droplet of Figure 13.2.1, demonstrating: (a) the sequential nature of active droplet heating and gasification; and (b) active droplet heating spans only the initial period of the droplet lifetime.

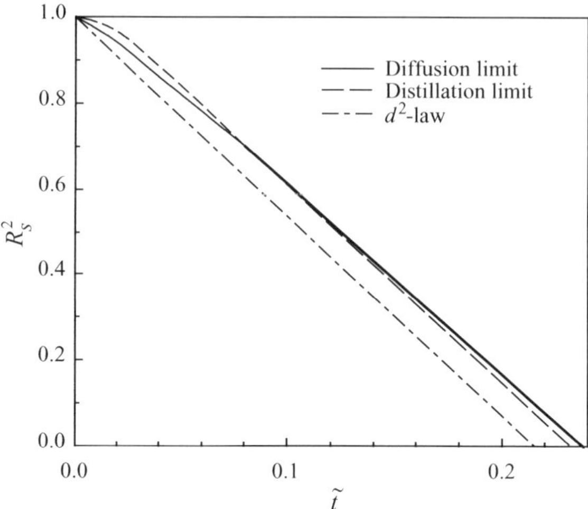

Figure 13.2.3. Comparison of the size history of the octane droplet of Figure 13.2.1, predicted by various assumptions regarding droplet heating.

surface, whether it takes place simultaneously with, or subsequent to, the surface heating process.

Figure 13.2.3 shows that once the intense heating period is over, R_s^2 regresses quite linearly with time, as would be expected, where $R_s = r_s/r_{s,o}$. Furthermore, the total burning times are remarkably close not only between the two limits of heat transfer, but also when compared with the d^2-law result.

Summarizing, the following conclusions can be made regarding droplet heating for a single-component fuel in a constant environment, whose pressure is also sufficiently below the critical pressure. First, active droplet heating and gasification occur somewhat sequentially, with the former mostly over in the initial 5–10 percent of the droplet lifetime depending on the fuel volatility and the initial droplet temperature. The fact that these two processes occur somewhat sequentially is also physically reasonable because active droplet heating takes place when the droplet temperature is low. A low droplet temperature implies a low fuel vapor concentration at the surface and consequently a slower gasification rate. As the droplet temperature is increased close to the final value, the droplet heating rate must slow down while the gasification rate increases because of the higher fuel vapor concentration at the droplet surface.

Droplet heating proceeds rapidly and therefore only slightly prolongs the total gasification time of the droplet. Furthermore, since the bulk gas-phase combustion characteristics are insensitive to the detailed heat transfer processes within the droplet, for convenience they can be simulated by using the distillation limit. The droplet temperature distribution, however, may remain nonuniform and temporally varying during much of the droplet lifetime. Therefore studies of those combustion characteristics that do depend on the droplet temperature distribution should employ accurate description of the internal transfer process. Processes that are likely to take place during the initial transient heating period, such as droplet ignition, also require an accurate description of droplet heating.

13.2.2. Fuel Vapor Accumulation

In addition to droplet heating, another important transient process occurring during droplet combustion is fuel vapor accumulation in the region between the droplet and the flame (Law, Chung & Srinivasan 1980). Its significance can be appreciated by considering the instant immediately after ignition. At that instant the amount of fuel vapor present in the inner region to the flame should be of the same order as the amount present in the droplet vicinity before ignition. Since this amount is very small because of the initially low droplet temperature, the flame initially must also lie close to the droplet surface. With subsequent gasification the fuel vapor present in the inner region increases and larger flames can be supported. Therefore only part of the fuel vaporized during this period is consumed at the flame, the rest is being accumulated in the inner region as the flame expands. This amount may or may not be totally consumed as the flame collapses toward the droplet during the latter part of its lifetime.

The existence of this accumulation process implies that mass conservation is actually violated in the d^2-law. Rather, overall mass conservation for the fuel vapor should read

Gasification rate at droplet surface (m_v)
= Consumption rate at flame (m_c)
+ Accumulation/depletion rate in the inner region,

which can be expressed as

$$m_v = m_c + \frac{d}{dt}\int_{r_s(t)}^{r_f(t)} (Y_F\rho)\,4\pi r^2 dr. \tag{13.2.5}$$

The last term is absent in Eq. (6.4.38) for the d^2-law.

It is reasonable to expect that the neglected mass accumulation term is significant because although the gas density is low compared to the liquid density, the flame size can be substantial such that volume effect dominates. Furthermore, the fuel vapor is accumulated at the expense of the finite droplet mass, based on which the droplet mass lifetime is computed. Indeed it can be easily demonstrated that the amount of fuel vapor present, as given by results of the d^2-law, is of the same order as the droplet mass, which is obviously incorrect.

While fuel vapor accumulation is a transient process, from overall mass conservation the accumulation rate should be of the same order as the droplet surface regression rate, and therefore is much slower than the gas-phase transport rates. Hence we expect that gas-phase quasi-steadiness still holds. Thus the d^2-law formulation can be modified to allow for the accumulation effect by distinguishing m_v and m_c, and relating them through Eq. (13.2.5).

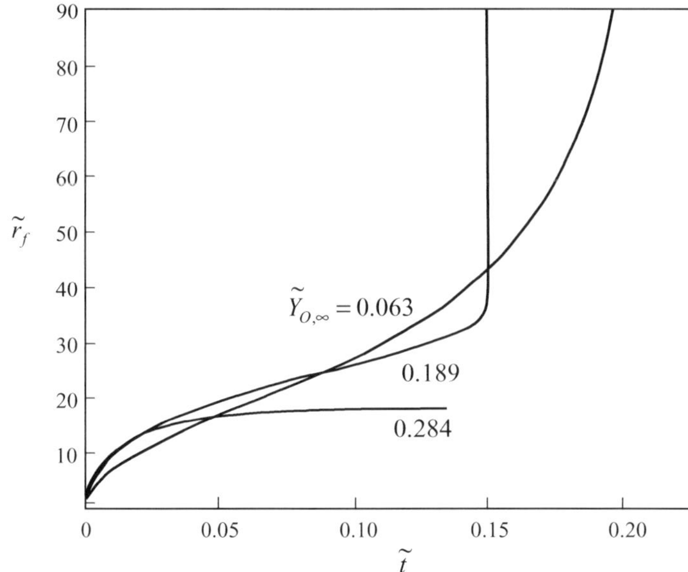

Figure 13.2.4. Temporal variation of the flamefront standoff ratio, $\tilde{r}_f$, for a heptane droplet burning in 300 K atmosphere, demonstrating that because of fuel vapor accumulation effects $\tilde{r}_f$ continuously increases with time in low-$\tilde{Y}_{O,\infty}$ environments but levels off in high-$\tilde{Y}_{O,\infty}$ environments.

Results from the above formulation completely substantiate the experimental observation presented in Section 6.4.4. Specifically, it is found that, in the absence of droplet heating, the droplet surface area regresses almost linearly with time, in the manner of d^2-law variation, except for a short initial period of slightly faster rate. The flamefront standoff ratio, $\tilde{r}_f$, however, increases after ignition (Figure 13.2.4). In a low oxidizer environment the increase is without bound, whereas in a richer environment it approaches a constant value, which is different from that of the d^2-law result. The actual flame radius (Figure 13.2.5), $R_f = r_f/r_{s,o}$, first increases and then decreases for the relatively rich oxidizer environments of $\tilde{Y}_{O,\infty} = 0.189$ and 0.284, but increases without bound for the relatively oxidizer-lean, air environment of $\tilde{Y}_{O,\infty} = 0.063$. These results are reasonable because the flame is smaller in an oxidizer-rich environment such that less fuel vapor is needed for accumulation.

Results further show that droplet heating dominates the droplet size history during the initial period of gasification, while fuel vapor accumulation dominates the flame size history throughout the droplet lifetime, as would be expected. The fuel consumption rate, m_c, is initially smaller than the fuel gasification rate, m_v (Figure 13.2.6). However, the opposite holds in the latter part of the droplet lifetime, implying that the initially accumulated fuel is now being depleted as the flame collapses inward such that the fuel consumption rate exceeds the gasification rate. In an oxidizer-rich environment a steady rate of depletion may be attained. However, this steady-state value is not the d^2-law solution, $m_c = m_v$, which occurs only at one instant in the droplet lifetime. Therefore, depending on the ambient oxygen concentration and hence the flame size, the fuel vapor accumulated initially may not be

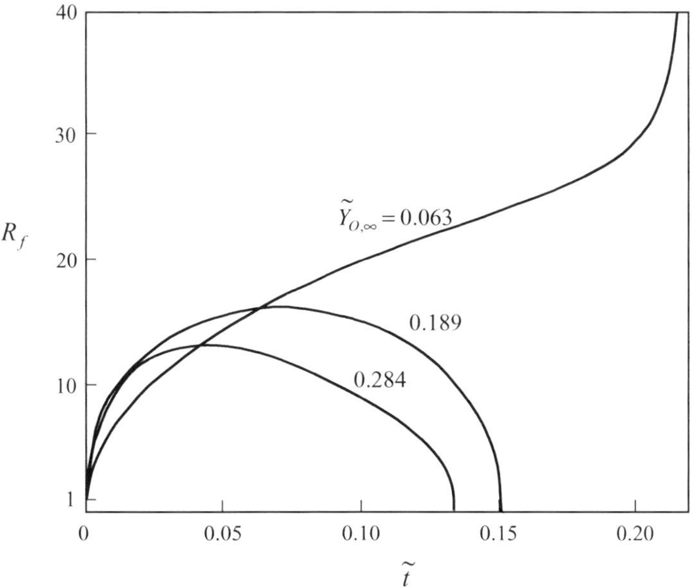

Figure 13.2.5. Temporal variation of the nondimensional flame radius, R_f, for the heptane droplet of Figure 13.2.4, demonstrating fuel vapor accumulation effects.

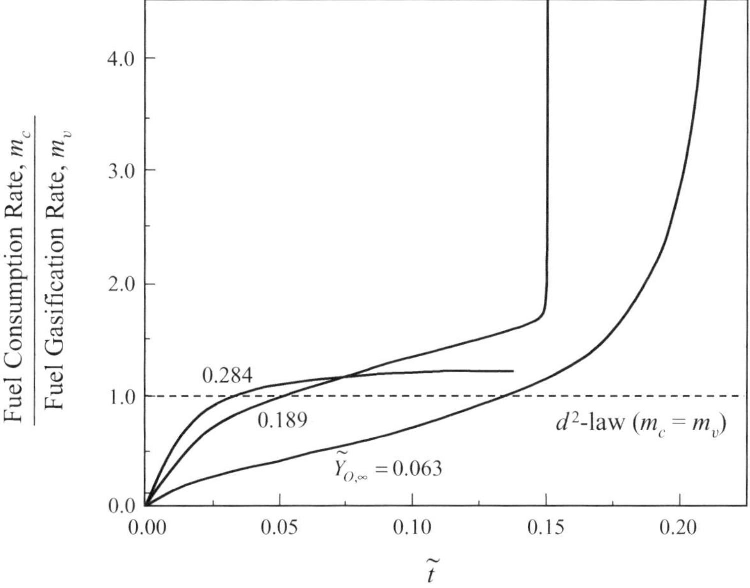

Figure 13.2.6. Temporal variation of the fractional fuel-consumption rate for the heptane droplet of Figure 13.2.4, demonstrating fuel vapor accumulation effects.

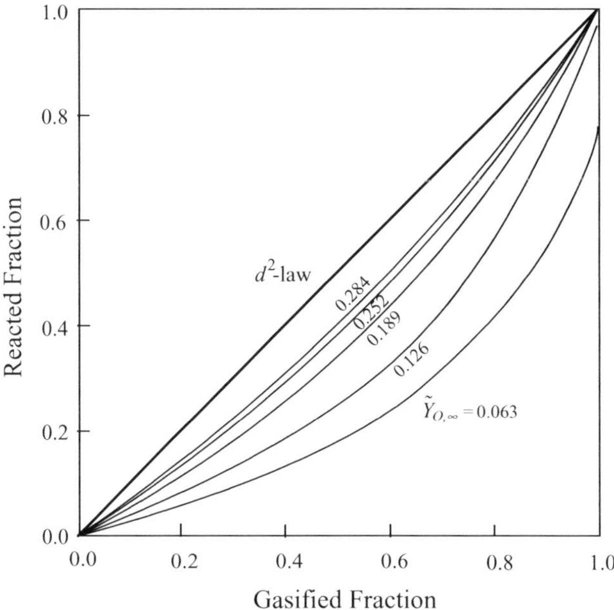

Figure 13.2.7. For the heptane droplet of Figure 13.2.4, demonstrating the concept that because of fuel vapor accumulation, the fractional amount of fuel reacted at any instant is always less than the fractional amount of fuel gasified.

totally consumed upon complete droplet vaporization (Figure 13.2.7). Consequently, an important practical implication of the fuel vapor accumulation process is that since the fuel gasification rate is not equal to the fuel consumption rate, adoption of the d^2-law in spray modeling may result in erroneous estimates of the instantaneous chemical heat release rate of the spray.

13.2.3. Variable Property Effects

Since diffusion is the dominant process in nonpremixed flames, and since the molecular weights of typical liquid hydrocarbons are substantially greater than that of air, the assumption of constant transport properties including unity Lewis number can have significant quantitative influence on the calculated flame characteristics. Using the reaction-sheet formulation of Section 6.1.2, the d^2-law can be reformulated allowing for constant, but different, values of c_p, λ, and ρD in the inner and outer regions to the flame, respectively designated by subscripts 1 and 2 (Law & Law 1976, 1977). This yields the following expressions for m_c, $\tilde{r}_f$, and T_f:

$$\frac{m_c}{4\pi r_s} = \ln\left\{\left[1 + \frac{c_{p,1}(T_f - T_s)}{q_v}\right]^{(\lambda_1/c_{p,1})}\left(1 + \tilde{Y}_{O,\infty}\right)^{(\rho D)_2}\right\} \tag{13.2.6}$$

$$\tilde{r}_f = 1 + \frac{(\lambda_1/c_{p,1})}{(\rho D)_2}\frac{\ln\left[1 + c_{p,1}(T_f - T_s)/q_v\right]}{\ln(1 + \tilde{Y}_{O,\infty})} \tag{13.2.7}$$

$$q_c = (c_{p,2}T_f - c_{p,1}T_s + q_v) + \frac{c_{p,2}(T_f - T_\infty)}{\left[(1 + \tilde{Y}_{O,\infty})^{1/Le_2} - 1\right]}. \tag{13.2.8}$$

Equation (13.2.8) shows that the flame temperature now depends on the transport process through Le_2, and consequently is not the adiabatic flame temperature. In particular since $\tilde{Y}_{O,\infty} \ll 1$ for practical hydrocarbon combustion in air, we have

$$\left(1+\tilde{Y}_{O,\infty}\right)^{1/Le_2} - 1 \simeq \frac{\tilde{Y}_{O,\infty}}{Le_2},$$

which shows that the effect of Le_2 is simply a modification of the oxidizer concentration experienced by the flame by a factor Le_2^{-1}. Therefore for $Le_2 > 1$, the oxidizer concentration is effectively reduced. The flame temperature is also reduced from the adiabatic flame temperature, which is obtained with $Le_2 = 1$. This is reasonable because now thermal diffusion is more efficient than mass diffusion, resulting in a relatively faster rate of heat transfer away from the flame. The converse holds for $Le_2 < 1$.

The expressions for m_c and $\tilde{r}_f$ show that the relevant Lewis number here is a mixed one, given by

$$Le_{\text{eff}} = \frac{\lambda_1/c_{p,1}}{(\rho D)_2},$$

which demonstrates that the dominant processes affecting the droplet mass burning rate and flame size are thermal conduction in the inner region and oxygen diffusion in the outer region. As such, the individual Lewis numbers, Le_1 and Le_2, are actually irrelevant in determining m_c and $\tilde{r}_f$. Furthermore, it has also been estimated that for light hydrocarbon fuels (e.g., heptane) burning in air, Le_{eff} is between $\frac{1}{3}$ to $\frac{1}{2}$. Therefore when it is used in Eq. (13.2.7), a much smaller $\tilde{r}_f$ results when compared to the one determined by using the d^2-law expression, Eq. (6.4.29), derived by assuming unity Le throughout. The predicted $\tilde{r}_f$ is then in the right range of the experimentally observed values.

The transport coefficients also need to be evaluated at a reference temperature T_{ref} in each of the inner and outer regions. Two such reference temperatures have been recommended, namely the simple arithmetic mean (Law & Williams 1972),

$$T_{\text{ref},1} = \frac{1}{2}(T_s + T_f), \qquad T_{\text{ref},2} = \frac{1}{2}(T_f + T_\infty), \tag{13.2.9}$$

and the $\frac{1}{3}$-rule modified to allow for the existence of the flame sheet and the temperature profiles across it,

$$T_{\text{ref},1} = \frac{1}{3}(T_s + 2T_f), \qquad T_{\text{ref},2} = \frac{1}{3}(T_f + 2T_\infty). \tag{13.2.10}$$

Similar mixing rules can be applied to the estimation of representative compositions when accounting for their influence on transport coefficients.

13.2.4. Gas-Phase Transient Diffusion and High-Pressure Combustion

We have just shown that the two major transient processes occurring during droplet combustion, namely droplet heating and fuel vapor accumulation, can all be satisfactorily described on the basis of gas-phase quasi-steadiness. We now study the characteristics of gas-phase transient diffusion and its influence on droplet combustion (Rosner & Chang 1973; Waldman 1975; Crespo & Liñán 1975; Matalon & Law 1983, 1985).

First we note that the quasi-steady assumption is expected to break down in regions far away from the droplet where the flow velocity is extremely slow such that the characteristic diffusion time is of the same order as that for surface regression. The location r_∞ at which this assumption breaks down can be estimated by equating the diffusion time at r_∞ with the regression time at r_s,

$$\frac{r_\infty^2}{D_g} \sim \frac{r_s^2}{K}. \tag{13.2.11}$$

Since $K/D_g \sim \rho_g/\rho_\ell$ from (6.4.41), we have

$$\frac{r_\infty}{r_s} \sim \left(\frac{\rho_\ell}{\rho_g}\right)^{1/2}. \tag{13.2.12}$$

In particular, for near- or super-critical combustion, $\rho_\ell/\rho_g = O(1)$. Then the assumption is invalid everywhere.

In order to unambiguously identify the extent of departure from the d^2-law caused by gas-phase transient diffusion, it is necessary to suppress droplet heating and fuel vapor accumulation. Analytically this can be achieved by respectively setting $q_{v,\text{eff}} = q_v$ and using the quasi-steady d^2-law results as the initial concentration and temperature profiles. Results from a perturbation analysis (Matalon & Law 1983), using $\delta \approx [\tilde{m}(\rho_g/\rho_\ell)]^{1/2}$ as the small parameter of expansion, where $\tilde{m}$ corresponds to either pure vaporization or reaction-sheet combustion as the case may be, shows (Figure 13.2.8) that the effect of transient diffusion, as represented by finite values of δ, is to decrease the droplet gasification rate. This result is physically reasonable because, with transient diffusion, the gas-phase processes cease to be infinitely responsive as compared to the droplet surface regression rate. The reduction, however, is quite small for sufficiently subcritical pressures. It can therefore be concluded that gas-phase quasi-steadiness is an adequate and useful assumption for the modeling of subcritical heterogeneous combustion systems, provided $\delta \ll 1$.

For $\delta \approx 1$ situations involving high-pressure, near-critical and supercritical combustion, gas-phase transient diffusion is inherently important (Bellan 2000; Yang 2000). From practical considerations, internal combustion engines operate under elevated pressures frequently in excess of the thermodynamic critical pressure of the liquid fuel. For example, while the critical pressures of diesel fuels are of the order of 20 atm, the pressure within diesel engines can range from 40 atm at the end of the compression stroke to twice that value at the peak combustion pressure. At such elevated pressures various aspects of low-pressure droplet combustion need to be revised. The elevation of the liquid boiling point lengthens the droplet heating period but reduces the subsequent latent heat of gasification, while the simultaneous reduction in the surface tension promotes droplet deformation and internal motion. Gas-phase compressibility and dissolution of gas into the liquid also become important. At and beyond the critical state the distinction between

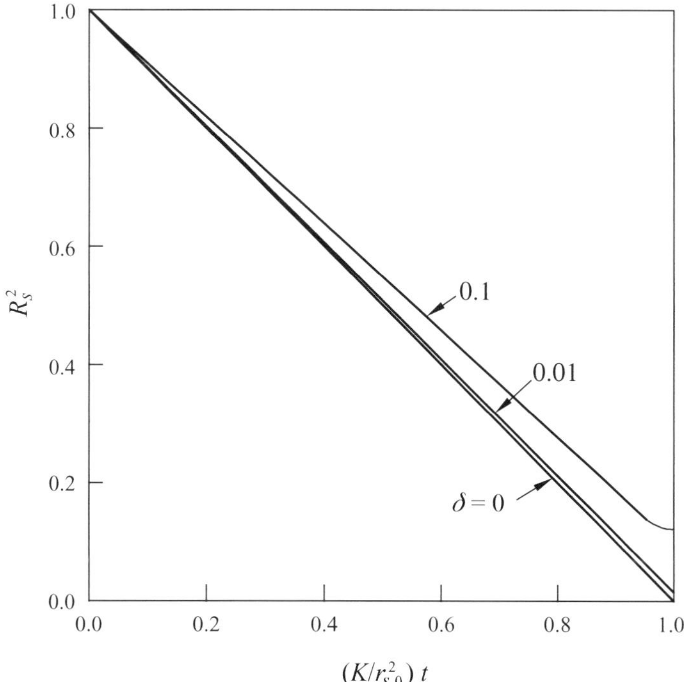

Figure 13.2.8. Calculated droplet size history with various extents of gas-phase transient diffusion as represented by $\delta \sim [\tilde{m}(\rho_g/\rho_\ell)]^{1/2}$.

gas and liquid vanishes and the phenomena of interest cease to be that of droplet combustion.

Figure 13.2.9 shows the burning times of n-octane droplets obtained under microgravity and normal gravity for subcritical and supercritical ambient pressures, with the critical pressure of octane being $p_{cr} = 2.52\ MP_a$ (Sato et al. 1990). It is seen that, up to the critical pressure, increasing the system pressure increases the droplet burning rate and, hence, decreases its lifetime. This is due to the reduced latent heat of vaporization. However, for pressures greater than the critical, the burning rate decreases with increasing pressure as indicated by the increase in the droplet lifetime. Since the problem of interest here is primarily that of the transient mixing between the fuel "vapor" sphere and the ambient "gas," the mixing and hence burning rates for unit mass now primarily vary with the mass diffusivity that is reduced with increasing pressure in this inherently transient regime.

Figure 13.2.9 also shows that the burning rate under normal gravity is substantially faster than that under microgravity. This is due to the augmented transfer rate in the presence of buoyancy.

13.2.5. Convection Effects and Droplet Dynamics

The heat and mass transfer processes of the spherically symmetric droplet combustion discussed so far are strongly influenced by external convection and droplet dynamics. Conversely, the droplet dynamics are also affected by droplet combustion.

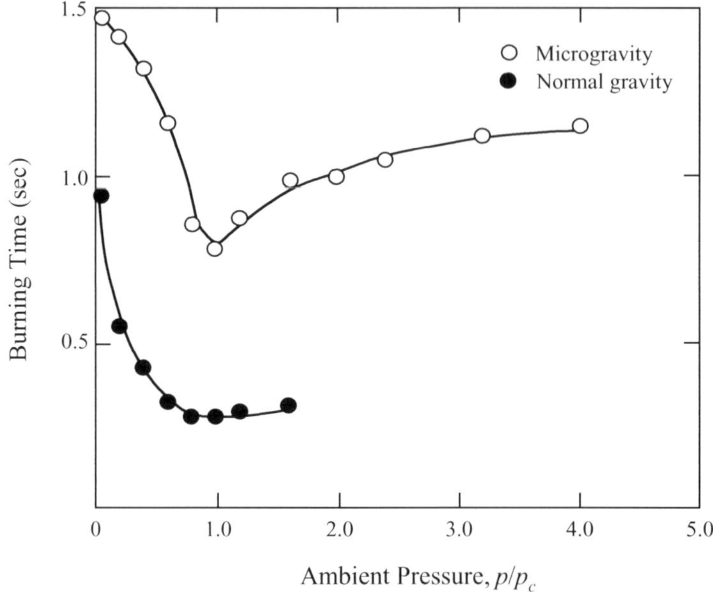

Figure 13.2.9. Experimentally determined droplet burning lifetime as a function of pressure under normal gravity and microgravity conditions (Sato et al. 1990).

During vigorous burning, external convection generally increases the droplet gasification rate. Various semiempirical correlations have been proposed for the augmentation in the gasification rate; a recommended one for forced convection (Faeth 1977) is

$$\frac{K}{K^o} = 1 + \frac{0.276 Re^{1/2} Pr^{1/3}}{(1 + 1.232/Re\, Pr^{4/3})^{1/2}}, \tag{13.2.13}$$

where K^o is the gasification rate constant from the spherically symmetric limit. Equation (13.2.13) is applicable to both pure vaporization and combustion, and gives the correct limits for low and high Re situations up to $Re = 1{,}800$. Specifically, for low Re flows, the correction varies with Re in accordance with the Stokes flow limit, while for high Re flows, the correction varies with $Re^{1/2}$ as in boundary-layer flows.

The corresponding correlations for K due to natural convection are (Law & Williams 1972)

$$\frac{K}{K^o} = 1 + 0.52 Gr^{1/2}, \quad Gr < O(1) \tag{13.2.14a}$$

$$= 1 + 0.85 Gr^{1/4}, \quad Gr > O(10), \tag{13.2.14b}$$

where Gr is the Grashof number. The $Gr^{1/2}$ and $Gr^{1/4}$ variations are in agreement with analyses for low-Gr creeping flows and high-Gr boundary-layer flows respectively. Figure 13.2.10 shows experimental data, from a variety of sources, demonstrating the transition from low- to high-Gr regimes.

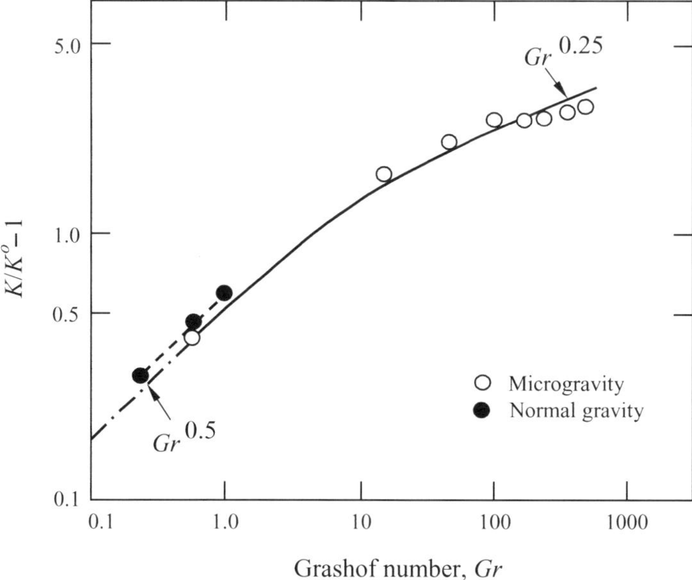

Figure 13.2.10. Dependence of the normalized burning rate constant, K/K^o, on the Grashof number, showing variation of the Grashof number scaling.

There exists an upper limit in the convection intensity above which the flame is "blown off" the droplet. This phenomenon is basically that of extinction in the forward stagnation region.

Vaporization and combustion can in turn influence the drag experienced by the droplet and thereby its dynamics in several essential ways. The outward mass flux at the droplet surface reduces friction drag but increases pressure drag, especially due to early separation. The existence of interfacial velocity, especially after the droplet has been substantially heated up, may delay separation and reduce both the friction and pressure drags. Furthermore, the significant elevation of temperature in both the gas and liquid phases, and the increase in the gas density because of the presence of the high-molecular-weight fuel vapor, can all greatly influence the fluid properties and thereby estimation of the drag coefficient. Since the droplet size continuously decreases, the drag force experienced by the droplet is inherently transient, varying over the same time scale as the droplet gasification process.

The effect of surface mass transfer on the drag experienced by an evaporating porous sphere has been studied both experimentally and computationally (Yuen & Chen 1976; Renksizbulut & Yuen 1983). It was concluded that while mass transfer reduces friction drag significantly, the pressure drag is increased by an almost equal amount. The net effect is that the standard drag curve for solid spheres can be used for evaporating droplets (Figure 13.2.11), provided that the density is that of the freestream and the viscosity of the vapor mixture is evaluated at the $\frac{1}{3}$-reference state. Comparable studies involving burning droplets or spheres with higher surface mass fluxes have not been conducted.

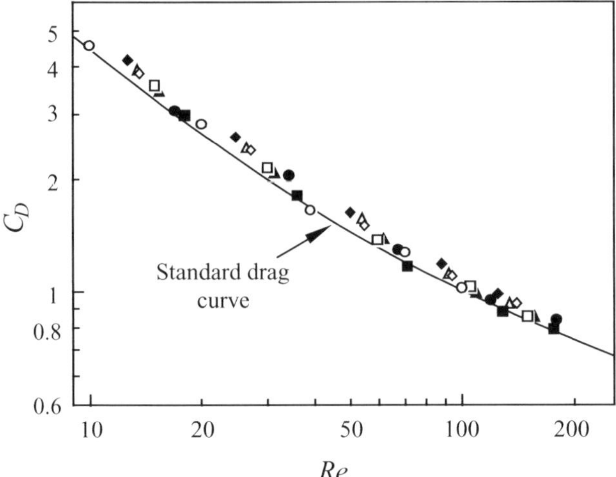

Figure 13.2.11. Comparison of the calculated drag coefficient with the standard drag curve for vaporizing droplets, demonstrating the approximate cancellation of the pressure drag and friction drag such that the standard drag curve can still be used (Renksizbulut & Yuen 1983).

The final effect of external convection is the generation of circulatory motion within the droplet interior, and thereby enhancement of the internal heat and mass transport rates. However, as mentioned earlier, rapid internal circulation cannot cause perpetual spatial uniformization throughout the droplet. In particular, the net effect of rapid internal circulation is to reduce the diffusion distance by a factor of three and thereby shorten the diffusion time by an order of magnitude. However, for large Reynolds number flows the burning time is reduced by $O(Re^{1/2})$, as shown by Eq. (13.2.13). Therefore for $Re = O(10^2)$, the burning time is also reduced by an order of magnitude. Since reductions in the characteristic diffusion and burning times are of the same order even in such a highly convective limit, it is then reasonable to expect that significant concentration nonuniformity should prevail throughout the droplet lifetime.

13.2.6. Droplet Interaction

Our discussion so far has been limited to an isolated droplet situated in an unbounded environment. In the practical situation within a spray, any given droplet is surrounded by and thereby interacts with the rest of the droplet ensemble. The intensity obviously depends on how closely the droplets are spaced.

To assess the effects of droplet–droplet interaction, studies have been conducted on the vaporization and combustion of only a few interacting droplets. Theoretical analysis of the problem is necessarily complicated because of the three-dimensional nature of the configuration. For a steady situation without external flow, continuity and coupling function conservation are given by

$$\nabla \cdot (\rho \mathbf{v}) = 0 \tag{13.2.15}$$

$$\nabla \cdot [\rho \mathbf{v} \beta_i - (\lambda / c_p) \nabla \beta_i] = 0. \tag{13.2.16}$$

A special solution of Eqs. (13.2.15) and (13.2.16) can be obtained (Labowsky 1978; Umemura 1981) by defining a velocity potential ϕ satisfying

$$\nabla\phi = \rho\mathbf{v}, \tag{13.2.17}$$

such that Eq. (13.2.15) is simply the Laplace equation

$$\nabla^2\phi = 0. \tag{13.2.18}$$

Furthermore, with $\rho\mathbf{v}$ defined by Eq. (13.2.17), it can be easily shown that

$$\beta_i = c_{1,i} + c_{2,i}\exp[\phi/(\lambda/c_p)] \tag{13.2.19}$$

satisfies Eq. (13.2.16), where $c_{1,i}$ and $c_{2,i}$ are constants to be determined by applying the appropriate boundary conditions.

Equation (13.2.18) can be rigorously solved for the binary droplet system by using the bispherical coordinate (Twardus & Brzustowski 1977; Brzustowski et al. 1979; Umemura 1981; Umemura, Ogawa & Oshima 1981). However, numerical solution is needed for systems with more droplets (Labowsky 1978). The extent of computation can become prohibitively tedious and therefore impractical as the number of droplets increases. It seems that in order to incorporate interaction effects into spray modeling, either statistical methods or the concept of screening distance (Samson et al. 1978, Samson, Bedeaux & Deutch 1978) are needed in order to limit the extent of computation.

The major conclusions drawn from the analytical results of the binary droplet system are that interaction reduces the droplet gasification rate from its isolated value because of the competition for oxygen, and that d^2-law does not hold because of the continuous increase in the separation distance between the droplet surfaces as gasification proceeds. There is also a minimum droplet separation distance below which the individual flames will merge.

Figure 13.2.12a shows a burning sequence of three horizontally aligned, suspended droplets burning interactively, with buoyancy being minimized by using a low-pressure environment (Miyasaka & Law 1981). It is seen that initially a single, merged flame exists because of the proximity of the droplets. As burning progresses, however, flame separation occurs as the droplet surfaces recede from each other. Since the center droplet suffers stronger oxygen competition as compared to the edge droplets, it has a larger flame and a slower burning rate, which causes it to burn out last.

Figure 13.2.12b shows a corresponding burning sequence conducted under atmospheric pressure and thereby increased extent of buoyancy. It is seen that the center droplet now has a faster burning rate and therefore burns out first. The reason is that the edge droplets now help in inducing the buoyant flow, which increases the oxygen supply rate.

The above concepts are quantitatively demonstrated in Figure 13.2.13, in which the normalized burning rate constant, $K_c(t)/K_{c,\text{isolated}}$, is plotted against the normalized separation distance, $L/d_s(t)$, for a two-droplet array, where L is the fixed separation distance between the two droplet centers. It is clear that both the theoretical curve

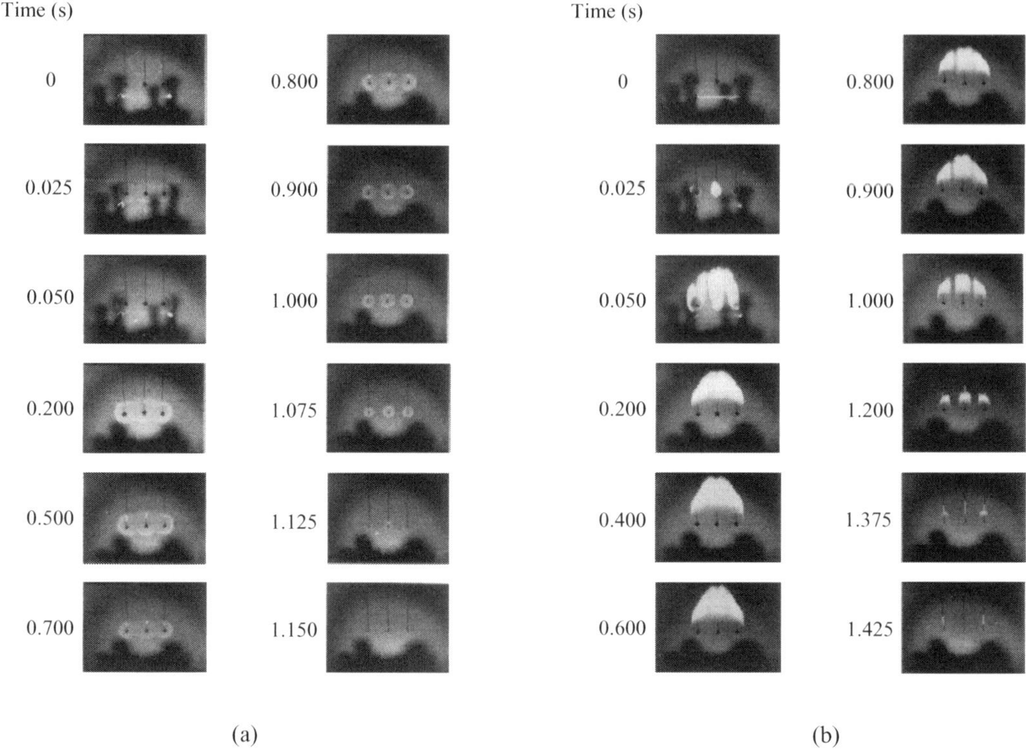

Figure 13.2.12. Burning sequence of a three-droplet array with (a) minimum and (b) strong buoyancy. The center droplet in (a) has a larger flame because it has less oxygen supply; it burns out last while the center droplet in (b) burns out first.

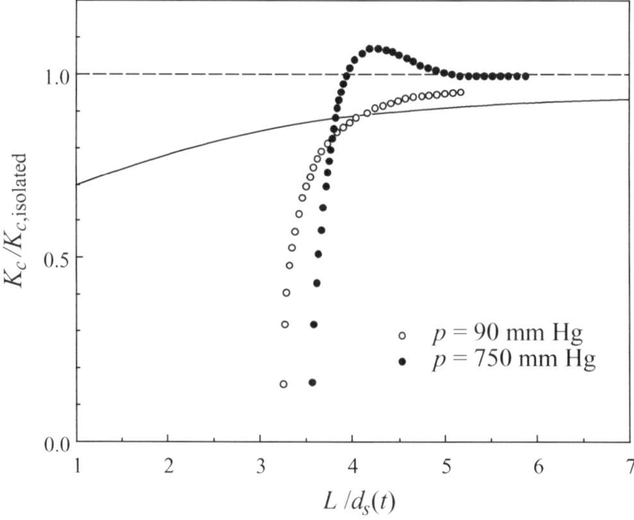

Figure 13.2.13. Normalized burning rate constant for a two-droplet array in the presence of droplet interaction and buoyancy; theoretical result is for a two-droplet array undergoing quasi-steady burning without buoyancy.

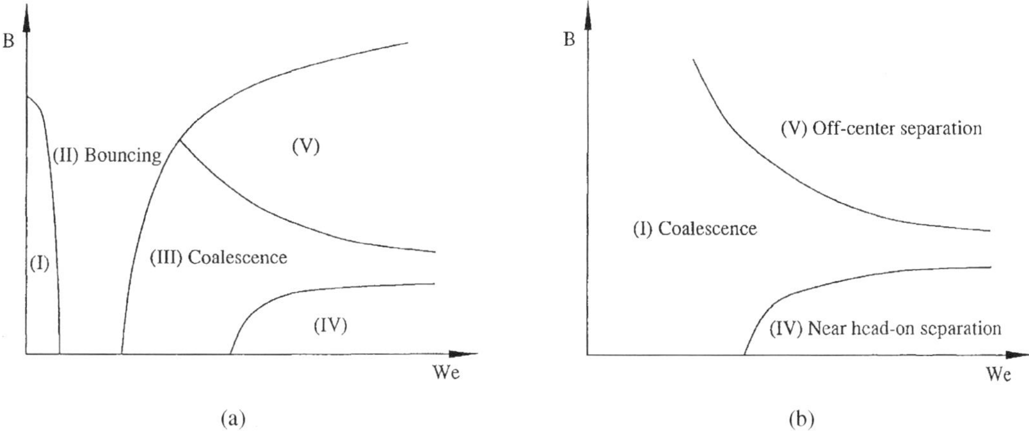

Figure 13.2.14. Collision regime diagram for: (a) hydrocarbon droplets at atmospheric pressure and water droplet at elevated pressures, (b) water droplet at atmospheric pressure and hydrocarbon droplets at reduced pressures.

and the low pressure experimental data show that with increasing L/d_s as burning progresses, K_c continuously increases because of the diminishing interaction effect. The theoretical curve, however, shows a stronger long range effect than the experimental data. This is likely caused by the constant property assumption in the theory, which results in larger flame sizes and thereby stronger interaction effects.

With increasing pressure and thereby buoyancy, the experimental data show a local burning rate maxima at an optimum separation distance (Feedoseeva 1973; Miyasaka & Law 1981), in accordance with the discussion on the burning sequence of Figure 13.2.12b.

13.2.7. Dynamics of Droplet Collision

Since the region immediately downstream of a spray injector has a high concentration of droplets, it is reasonable to expect that some of these droplets can collide and, depending on the outcome of the collision, can substantially modify the subsequent development of the spray. Droplet collision is also of interest to the study of rain drop formation and the modeling of nuclear fusion.

Experimental studies of binary droplet collision have revealed a rich variety of collision outcomes, which can be represented in a B–We regime diagram shown in Figure 13.2.14 for a typical hydrocarbon, say an n-alkane, and water (Jiang, Umemura & Law 1992; Qian & Law 1997). Here $We = d_s \rho_\ell U^2/\sigma$ is the liquid Weber number, $B = \chi/d_s$ the impact parameter, d_s the droplet diameter, U the relative velocity of the droplets, σ the surface tension of the liquid, and χ the perpendicular distance between the two lines passing through the droplet centers with direction parallel to the droplet relative velocity. Thus $B = 0$ and 1 respectively designate head-on and grazing collisions. Figures 13.2.15 and 13.2.16 show the time-resolved images of the various outcomes for head-on and off-center collisions.

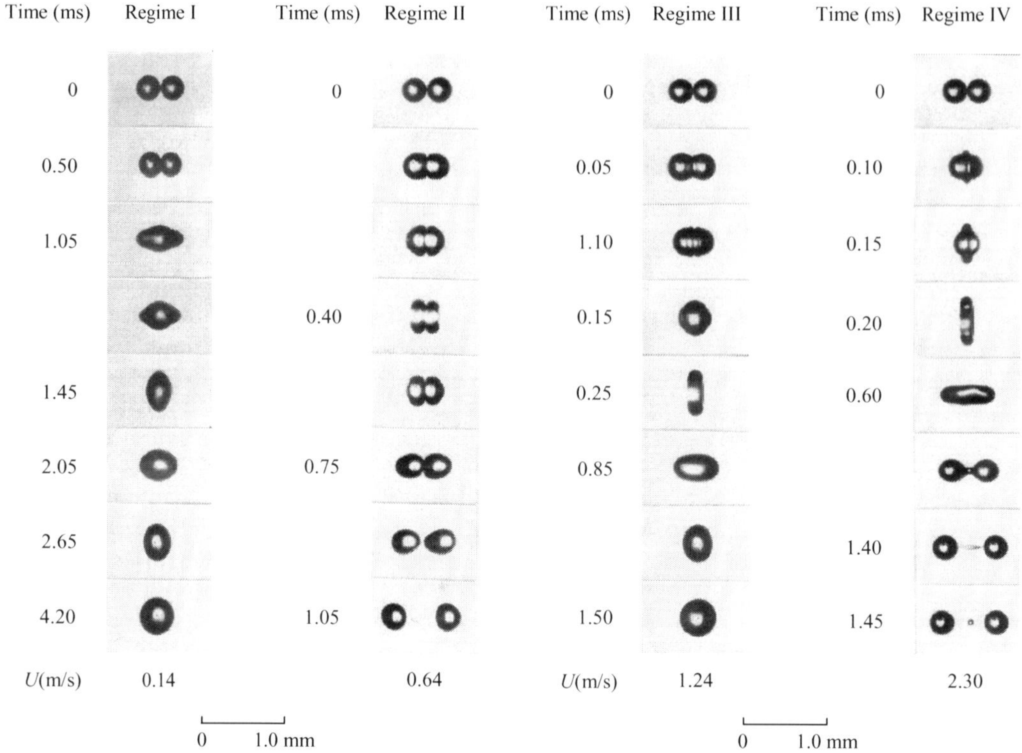

Figure 13.2.15. Photographic images showing representative head-on collision in Regimes I–IV.

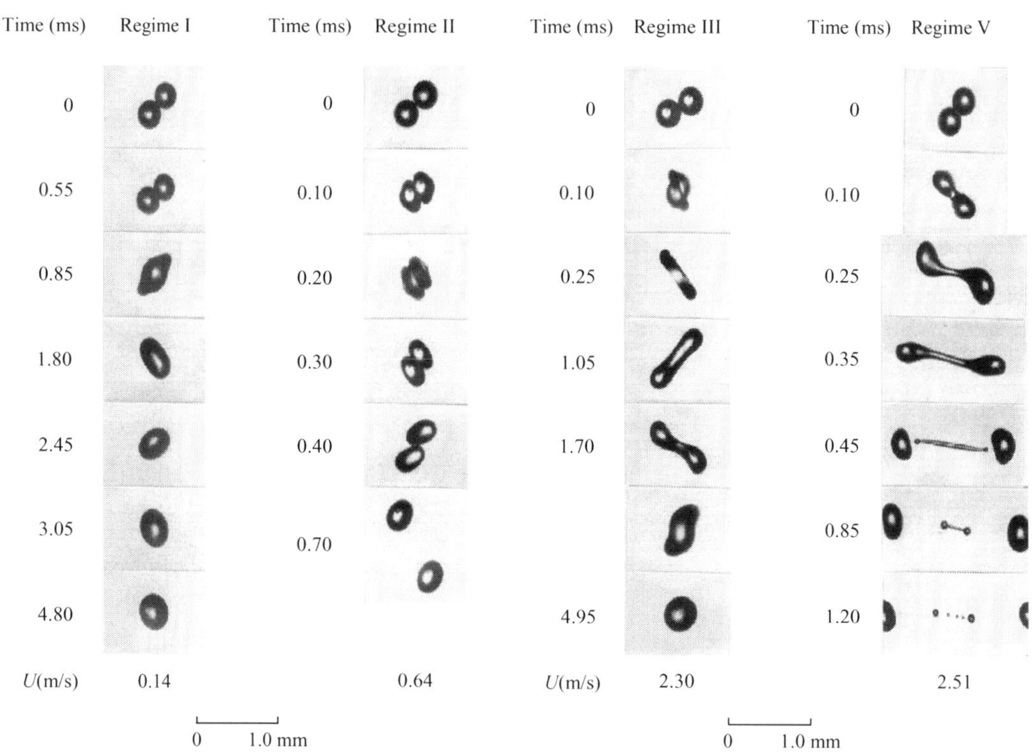

Figure 13.2.16. Photographic images showing representative off-center collision in Regimes I, II, III, and V.

Let us first consider the head-on collision events for hydrocarbon droplets at 1 atm. It is seen that when the collision inertia is small (Regime I), with small We, the droplets will first merge upon contact, and then oscillate for a few periods before they settle down to the final, merged droplet of the combined mass. Since the droplets have negligible kinetic energy before collision for the present low We situation, the initial total energy of the system is the surface tension energy of the two droplets. Furthermore, since this value is larger than that of the droplet consisting of the combined mass, the initial shape of the combined mass cannot be spherical. Referring to the image at 1.05 ms for Regime I in Figure 13.2.15, we see that the larger curvature and hence surface tension force at the rim of the merged mass creates an inward motion, causing it to contract. This motion over-shoots, resulting in the shape at 1.45 ms. At this stage, the kinetic energy of the internal motion is mostly converted to the surface tension energy again, with the remaining energy having been viscously dissipated. This exchange between the kinetic energy of the internal motion and surface tension energy provides the mechanism for oscillation, while viscous damping allows the combined mass to eventually assume the spherical shape.

By increasing the impact inertia and hence We (Regime II), Figure 13.2.15 shows that the droplets are substantially squashed upon impact, although a distinct interface still exists between the droplets as evidenced by the presence of a cusp at the edge of the interface. Thus coalescence does not occur in this regime and the droplets subsequently bounce off.

A further increase in the collision energy again results in permanent coalescence, in Regime III, and is characterized by the total deformation of the merged mass to the shape of a dimpled disc. Finally, in Regime IV, the collision energy is so large that the surface energy of the coalesced mass is not sufficient to contain the liquid in a closed surface. Thus after the initial coalescence, contraction of the deformed mass causes it to split. This is effected through the pinching off of the connecting ligament, resulting in the formation of a satellite droplet.

For off-center collisions (Figure 13.2.16), the behavior is qualitatively similar to the corresponding head-on situations, except a rotational motion is now imparted to the droplets in contact. Regime V, however, is distinctive in that the collision is grazing and highly energetic. Thus shearing action dominates and there is very little rotational motion. Satellite droplets of several generations can be formed as the merged mass splits apart.

The nonmonotonic transition behavior between Regimes, I, II, and III is quite interesting. The factor that controls the possibility of droplet bouncing is whether the colliding droplets can squeeze out the gas film between them or whether the inter-droplet pressure build-up is sufficiently large to repel them. To verify this concept, experiments have been conducted by decreasing either the chamber pressure or the molecular weight of the gas. The reduction of mass associated with the inter-droplet gas film indeed inhibits bouncing and promotes coalescence. Consequently, Regime II is mostly or totally eliminated for such cases, resulting in direct transition from Regime I to Regime III.

For the collision of water droplets, because the surface tension and viscosity are different from those of the hydrocarbons, the collision outcome at 1 atm is actually described by Figure 13.2.14b, without Regime II. Bouncing is recovered only at higher pressures, with the outcome described by Figure 13.2.14a.

13.2.8. Ignition and Extinction Criteria

When studying spray combustion, it is frequently necessary to assess the instant at which droplets ignite or extinguish. Such an assessment can be readily performed by evaluating appropriate ignition and extinction Damköhler numbers based on the S-curve analysis for nonpremixed flames discussed in Chapter 9. Here, we shall outline the key steps and give the criteria (Law 1975).

For a one-step overall, second-order reaction, energy conservation is given by

$$\frac{\tilde{m}}{\tilde{r}^2}\frac{d\tilde{T}}{d\tilde{r}} - \frac{1}{\tilde{r}^2}\frac{d}{d\tilde{r}}\left(\tilde{r}^2\frac{d\tilde{T}}{d\tilde{r}}\right) = Da_C\tilde{Y}_O\tilde{Y}_F e^{-\tilde{T}_a/\tilde{T}}, \tag{13.2.20}$$

where $Da_C = (B_C r_s^2)/(\lambda/c_p)$ is the collision Damköhler number and other quantities are defined in Section 6.4 for droplet combustion. The analysis can be facilitated by drawing analogy with the nonpremixed chambered flame of Chapter 9. This can be accomplished by using the convection-free formulation through the change of variable

$$\xi = 1 - e^{-\tilde{m}/\tilde{r}}, \tag{13.2.21}$$

such that $\tilde{r} = (1, \infty)$ correspond to $\xi = (\xi_s, 0)$. This reduces Eq. (13.2.20) to

$$\frac{d^2\tilde{T}}{d\xi^2} = -Da_C\left[\frac{\tilde{m}^2}{(1-\xi)^2[\ln(1-\xi)]^4}\right]\tilde{Y}_O\tilde{Y}_F e^{-\tilde{T}_a/\tilde{T}}, \tag{13.2.22}$$

where the coupling functions $\tilde{Y}_i + \tilde{T}$ are given by Eq. (6.4.24) and (6.4.25),

$$\tilde{T} + \tilde{Y}_O = \tilde{Y}_{O,\infty}(1-\xi) - \beta_v\xi + \tilde{T}_\infty \tag{13.2.23}$$

$$\tilde{T} + \tilde{Y}_F = (1-\beta_v)\xi + \tilde{T}_\infty, \tag{13.2.24}$$

with $\beta_v = (\tilde{T}_\infty - \tilde{T}_s) + \tilde{q}_v$, and the boundary conditions are

$$\tilde{T}(0) = \tilde{T}_\infty, \tag{13.2.25}$$

$$\tilde{T}(\xi_s) = \tilde{T}_s, \tag{13.2.26}$$

$$\left(\frac{d\tilde{T}}{d\xi}\right)_{\xi_s} = -\tilde{q}_v e^{\tilde{m}}. \tag{13.2.27}$$

The extra boundary condition for the second-order Eq. (13.2.22) allows the determination of $\tilde{m}$.

Ignition is expected to be initiated in a narrow region next to the hot ambience, which has the highest temperature. Following Section 9.3.1, the structure equation for the temperature perturbation in the reaction zone is

$$\frac{d^2\theta}{d\chi^2} = -\Delta\frac{(\chi - \theta)}{\chi^4}e^{(\theta - \beta_v\chi)} \tag{13.2.28}$$

$$\theta(0) = 0, \tag{13.2.29}$$

$$\left(\frac{d\theta}{d\chi}\right)_\infty = 0, \tag{13.2.30}$$

where

$$\Delta = \left(\frac{\tilde{m}_{v,0}}{\tilde{T}_\infty^2/\tilde{T}_a}\right)^2 Da_C Y_{O,\infty}e^{-\tilde{T}_a/\tilde{T}_\infty} \tag{13.2.31}$$

is the reduced Damköhler number and $m_{v,0}$ the droplet vaporization rate in the frozen limit. The only difference between Eq. (13.2.28) with Eq. (9.3.2) is the extra χ^4 term as a consequence of the present spherical geometry. A solution of Eqs. (13.2.28) to (13.2.30) yields the ignition turning point and consequently the ignition Damköhler number, $\Delta(\beta_v)$. Although an analytic solution for $\Delta_I(\beta_v)$ cannot be found, a semi-analytical expression (Makino 1991)

$$\Delta_I(\beta_v) = 0.5\left[\frac{4}{e(1-\beta_v)}\right]^4 - 0.6\left[\frac{4}{e(1-\beta_v)}\right]^2 \tag{13.2.32}$$

has been derived that correlates the numerical solution well. Thus a droplet is expected to ignite when $\Delta > \Delta_I(\beta_v)$ is satisfied.

Equation (13.2.32) shows that ignition is not possible for $\Delta_I \to \infty$ as $\beta_v \to 1$. Physically, the condition $\beta_v > 1$ corresponds to $c_p(T_\infty - T_s) > (q_c - q_v)$, which implies that the ambient temperature is higher than the adiabatic flame temperature. The system is therefore superadiabatic and ignition is irrelevant.

The extinction analysis completely parallels that in Section 9.3.4. Furthermore, since the thin reaction zone is now situated within the bulk of the flow field, the structure equation for the perturbed temperature is identical to Eqs. (9.3.23) through (9.3.25), with the reduced Damköhler number defined as

$$\delta = \frac{4\left(\tilde{T}_f^2/\tilde{T}_a\right)^3\tilde{m}_{c,0}^2}{[\ln(1+\tilde{Y}_{O,\infty})]^4}Da_C e^{-\tilde{T}_a/\tilde{T}_f}, \tag{13.2.33}$$

where $\tilde{m}_{c,0}$ is the droplet burning rate in the flame-sheet limit. Consequently, the extinction Damköhler number δ_E is again given by Eq. (9.3.28), with

$$\gamma = 1 - \frac{2(1-\beta_v)}{1+\tilde{Y}_{O,\infty}}. \tag{13.2.34}$$

13.3. MULTICOMPONENT DROPLET COMBUSTION

Earlier studies on droplet combustion were mostly based on pure fuels. Multicomponent effects were not considered to be serious for the reasons that commercial fuel blends were highly refined and lie within narrow specification ranges, and that requirements of combustor efficiency and emission were generally not stringent.

However, recent developments in engine design and fuel formulation indicate that multicomponent effects will become progressively more important in the utilization of liquid fuels. Combustion processes within engines will be more tightly controlled to further improve efficiency and reduce emissions. The synthetic fuels derived from coal, tar sand, and oil shale will have more complex composition as well as higher and wider boiling point ranges. There also exists considerable interest in the utilization of such hybrid fuels as water–oil emulsions, alcohol–oil solutions, and coal–water and coal–oil slurries. The widely different physical and chemical properties of the constituents of these hybrid fuels render it essential to consider multicomponent effects in their gasification behavior.

To understand heterogeneous multicomponent fuel combustion, either as a droplet or in some other form (e.g., pool burning), the following three factors need to be considered because they directly control the temporal variations of the relative gasification rates as well as the concentration and spatial distributions of the individual components within the droplet: (a) The relative concentrations and volatilities of the liquid constituents. (b) The miscibility of the liquid constituents and the ideality of the mixture. These affect the phase change and thereby the surface vapor pressure characteristics. (c) The rate of liquid-phase mass diffusion, the rate of droplet surface regression, and the intensity of motion within the droplet. These influence the rate with which the liquid components can be brought to the surface at which gasification takes place.

In the following we shall discuss the gasification mechanisms of various types of multicomponent mixtures.

13.3.1. Miscible Mixtures

Liquid-phase mass diffusion is a crucial process in the gasification of miscible mixtures. To appreciate this point, we first recognize that no matter how volatile a liquid element is, it cannot gasify unless it is exposed at the droplet surface. As discussed previously, exposure can be achieved through either the passive mode of surface regression or the active modes of diffusion and internal circulation. However, liquid-phase mass diffusion is an extremely slow process; its rate being one to two orders slower than those of surface regression and liquid-phase thermal diffusion during combustion or rapid vaporization. Therefore, with it being the dominant active mode of transport, it is reasonable to expect that the liquid element in the core of the droplet will be "trapped" during most of the droplet lifetime. Under this situation the relative volatilities of the individual components obviously cannot be the only factor in effecting gasification.

Solution of the gas-phase transport processes, with the assumption that reaction between the N fuel species and the oxidizer still occurs at a single flame sheet, yields the same expressions (Law & Law 1982b) for $\tilde{m}_c$, $\tilde{r}_f$, and $\tilde{T}_f$ as in Eqs. (6.4.28) to (6.4.31), provided we identify

$$q_v = \Sigma\epsilon_i q_{v,i}, \quad q_c = \Sigma\epsilon_i q_{c,i}, \quad \sigma_O = \Sigma\epsilon_i\sigma_{O,i}, \quad m_c = \Sigma m_{c,i},$$

where $\epsilon_i = \tilde{m}_{c,i}/\Sigma\tilde{m}_{c,i}$ is the fractional mass burning rate, which can be shown to be

$$\epsilon_i = Y_{i,s}/\Sigma Y_{j,s}, \tag{13.3.1}$$

where the indexes i and j refer only to the fuel species.

For the single-component d^2-law, the gas-phase solution essentially furnishes the major parameters of interest, namely $\tilde{m}_c$, $\tilde{r}_f$, and $\tilde{T}_f$. For the multicomponent case, however, the gas-phase solution intimately depends on the liquid-phase transport processes. For example, droplet heating is now an essential process in that, in addition to the need to heat up the initially cold droplet, the continuous change in the droplet surface composition also causes a corresponding continuous change in the attainable droplet temperature. The droplet heating effect can be accounted for by the use of $q_{v,\text{eff}}$ as discussed previously.

To account for the temporal variation of the droplet composition profile, liquid-phase mass diffusion in the spherically symmetric limit is described by (Law 1978b)

$$\frac{\partial Y_{\ell,i}}{\partial t} = \frac{1}{r^2}\frac{\partial}{\partial r}\left(D_\ell r^2 \frac{\partial Y_{\ell,i}}{\partial r}\right), \tag{13.3.2}$$

subject to

$$Y_{\ell,i}(r, t = 0) = Y_{\ell,i,o}(r), \tag{13.3.3}$$

$$\left(\frac{\partial Y_{\ell,i}}{\partial r}\right)_{r=0} = 0, \tag{13.3.4}$$

$$m_v Y_{\ell,i,s} - \left(4\pi r^2 \rho_\ell D_\ell \frac{\partial Y_{\ell,i}}{\partial r}\right)_{r=r_s} = m_{v,i}, \tag{13.3.5}$$

where $Y_{\ell,i}$ is the mass fraction of the ith component in the liquid phase. Note that we have assumed a single D_ℓ, and that for liquids $Le_\ell = \alpha_{h,\ell}/D_\ell \gg 1$. Furthermore, D_ℓ varies sensitively with temperature.

The last piece of specification needed is a relation for the surface vapor pressure. The simplest situation is when the mixture is ideal, obeying Raoult's law (Guggenheim 1952),

$$p_{i,s} = X_{\ell,i,s} p_{i,s,\text{pure}}, \tag{13.3.6}$$

which simply states that the vapor pressure of the ith component at the surface is its vapor pressure when it is pure, weighted by its molar fraction in the liquid at the interface; $p_{i,s,\text{pure}}$ can be given by, say, the Clausius–Clapeyron relation, while $p_s = \sum p_{i,s}$.

Analogous to the discussion on the Peclet number Pe_h for droplet heating, we can define a Peclet number for mass diffusion,

$$Pe_m = \frac{K}{D_\ell},$$

which is the ratio of the gasification rate constant to the liquid-phase mass diffusivity, where K is a characteristic gasification rate of the mixture. The droplet gasification behavior is therefore influenced by two parameters, namely Pe_h and Pe_m, which are related through

$$\frac{Pe_m}{Pe_h} = \frac{\alpha_{h,\ell}}{D_\ell} = Le_\ell.$$

In the limit of $Pe_m \to \infty$, which is likely caused by $D_\ell \to 0$, as in a solid, instead of $K \to \infty$, the internal composition of the droplet remains frozen and the gasification rate of the individual species would be proportional to its original quantity within the droplet at a location that is subsequently exposed by the regression surface. This mode of gasification is referred to as the onion skin model. In the opposite limit of $Pe_m \to 0$, which is likely caused by $K \to 0$ instead of $D_\ell \to \infty$, the internal composition is perpetually uniformized and the individual gasification rate is controlled by the volatility differentials among the individual species. This mode of gasification is referred to as the batch distillation model, similar to the consideration for droplet heating.

For small but finite values of Pe_m, droplet gasification takes on a boundary-layer characteristics (Law 1978b). To demonstrate such a behavior, we first note that for the vigorous combustion of a hydrocarbon droplet in air, $K = O(10^{-3} - 10^{-2}$ cm^2/sec). Since the thermal diffusivities of liquid hydrocarbons have the same order of magnitude as K, while $D_\ell = O(10^{-5} - 10^{-4}$ cm^2/sec), we have $Pe_h = O(1)$ and $Le_\ell = O(Pe_m^{-1}) = O(10 - 10^3) \gg 1$. The very slow value of the liquid-phase mass diffusivity is the controlling factor in the gasification behavior of multicomponent droplets.

To demonstrate this point, Figure 13.3.1 shows the development of the concentration profile of the volatile component of a 50–50 percent octane–decane droplet vaporizing in a 2,000 K environment, for $Le_\ell = O(Pe_m^{-1}) = 30$; where $M = 1 - R_s^3$ is the amount of droplet mass gasified and, hence, provides an indication of the progress in the droplet lifetime. These profiles show the attainment of a thin concentration boundary layer at the droplet surface, through which the concentrations adjust from their respective surface values to those in the inner core. Figures 13.3.2 and 13.3.3 show the temporal variations of the surface and center values of the volatile molar fraction and the temperature, as functions of Le_ℓ. These results reveal the following mechanism for strongly diffusion-limited multicomponent droplet combustion.

As droplet gasification is initiated, the surface concentration of the more volatile component decreases because of its relative volatility; the larger the Le_ℓ the more rapid the reduction. This preferential depletion can be initially supported by the volatile components in the surface layer because even though diffusion is slow,

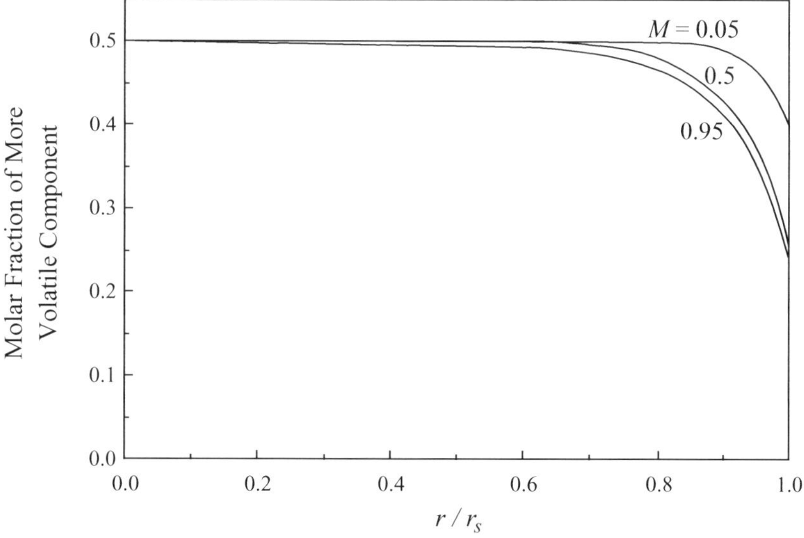

Figure 13.3.1. Development of the concentration profiles within the droplet at different stages of the droplet lifetime, $M = 1 - R_s^3$, showing establishment of the concentration boundary layer at the surface and its persistence throughout most of the droplet lifetime.

the diffusion distance is short. As such, during this period the droplet heating process is primarily controlled by the volatile component, while the droplet temperature is also closer to its boiling point. However, when the volatile concentration in this surface layer is reduced to a sufficiently low level, its continued preferential gasification would require supply from the inner core, implying that the volatile supply rate will

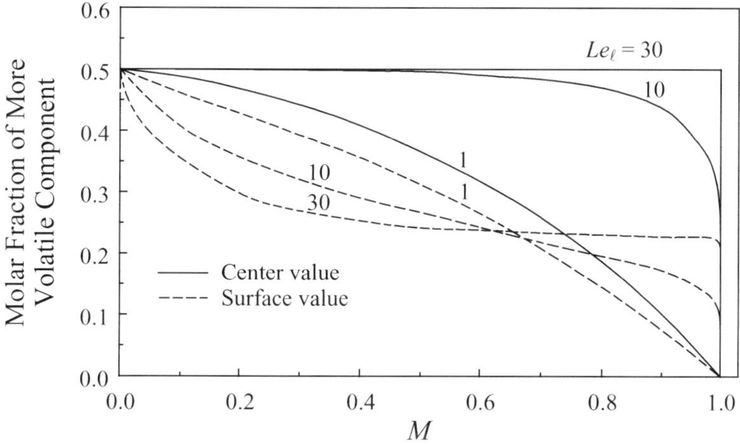

Figure 13.3.2. Concentration history of the volatile component at the surface and center locations of the droplet demonstrating, for large Le, the initial transition involving changes in the surface layer, a subsequent nearly steady-state behavior, and a final, extremely short period of volatility-dominated gasification. With decreasing Le_ℓ and, hence, less diffusional resistance, the transitions become less distinct.

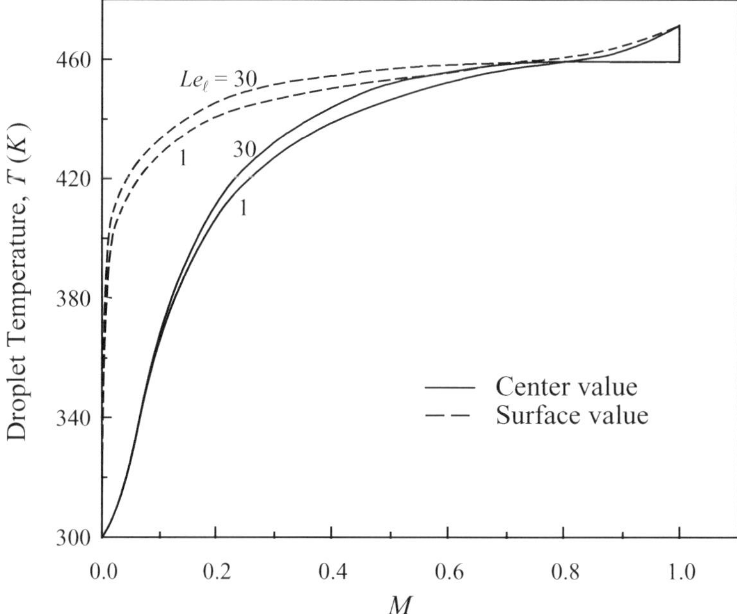

Figure 13.3.3. Temperature history for the same case as Figure 13.3.2.

slow down because of the much greater diffusional resistance. The reduced volatile supply rate now allows the less volatile component at the surface to participate more actively in gasification. The droplet temperature then starts to increase because of the higher boiling point of the less volatile component. For very large values of Le_ℓ, a diffusion-limited steady state is soon reached within the droplet, being characterized by approximately uniform and constant temperature, as well as constant concentration profiles. Species diffusion also balances surface regression such that the various species are transported to, and exposed at, the surface at approximately constant rates. The concentration at the droplet center remains practically at its initial value.

Finally, as the droplet size is reduced to be comparable to the thickness of the concentration boundary layer, then mass diffusion ceases to be the rate-controlling factor and volatility differential again exerts its dominance. The volatile component is then quickly depleted from the entire droplet, causing a rapid increase in the droplet temperature to approach the boiling point of the less volatile component. Thus as gasification is about to be completed the droplet basically consists of only the less volatile component. Figures 13.3.2 and 13.3.3 show the rapid changes that take place at the end of the droplet lifetime, especially for large values of Le_ℓ. The strength of diffusional resistance, however, is weakened with decreasing Le_ℓ. In particular, for $Le_\ell = 1$ the droplet concentration profile varies continuously throughout its lifetime.

The strongly diffusion-limited gasification mechanism described above is therefore qualitatively different from the earlier concept based on batch distillation, which requires that the droplet composition and temperature be perpetually uniformized such that volatility differential is the only controlling mechanism. In the distillation

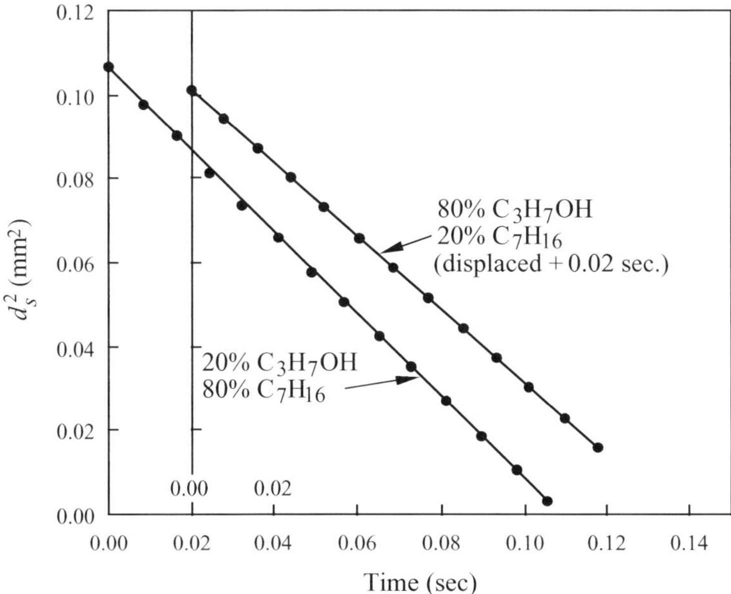

Figure 13.3.4. Experimental d_s^2-t plot for a heptane–propanol droplet. The nearly constant regression rate of each individual mixture indicates that the droplet gasifies like a pure component one because the volatilities of its components are almost the same.

limit (Law 1976b) the components would gasify in an approximately sequential fashion according to their relative volatilities. As mentioned earlier, the situation under which the batch distillation behavior prevails is when the droplet undergoes slow vaporization such that $Pe_h \ll 1$ and $Pe_m \ll 1$.

Experiments have been conducted to determine the instantaneous droplet size, flame size, and droplet composition of freely falling multicomponent droplets (Wang, Liu & Law 1984; Randolph, Makino & Law 1988). Figure 13.3.4 shows experimental plots of $d_s^2(t)$ for mixtures of heptane and propanol. Since these two compounds have almost the same normal boiling points (about 100°C), effects due to their volatility differential and thereby preferential gasification and diffusional resistance are eliminated such that a given mixture should behave like a single species. This point is substantiated by the linearity of the $d_s^2(t)$ plots in Figure 13.3.4. The slope of each curve, which yields the burning rate constant K_c, thus depends on the specific initial composition of the mixture. Furthermore, from overall mass conservation we expect that the fractional mass gasification rate of the ith species must be equal to its initial mass fraction in the mixture, or

$$\epsilon_i = Y_{\ell,i,o}. \tag{13.3.7}$$

Thus, the heat transfer number for combustion, $B_{h,c}$, can be directly written from Eq. (6.4.31) as

$$B_{h,c} = \frac{c_p(T_\infty - T_s) + (Y_{O,\infty}/\Sigma Y_{\ell,i,o}\sigma_{O,i})\Sigma Y_{\ell,i,o}q_{c,i}}{\Sigma Y_{\ell,i,o}q_{v,i}}. \tag{13.3.8}$$

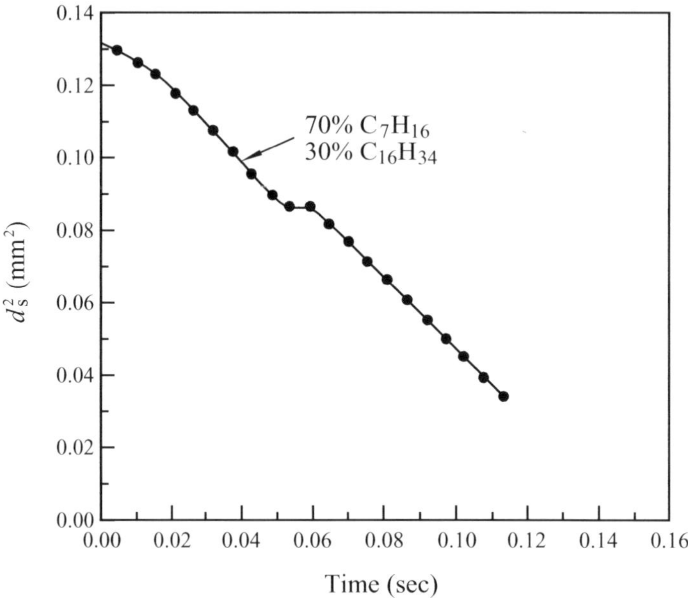

Figure 13.3.5. Experimental d_s^2-t plot for a heptane–hexadecane droplet, demonstrating the three-stage behavior during the gasification of a two-component droplet with vastly different volatilities.

Since an accurate knowledge of T_s is unimportant in the evaluation of $B_{h,c}$, Eq. (13.3.8) allows a direct assessment of the burning rate constant K_c during steady-state combustion.

If the volatilities of the two components are quite different, then both volatility differential and diffusional resistance are important factors and the resulting behavior can be qualitatively different. Figure 13.3.5 shows a three-staged behavior for the $d_s^2(t)$ plot of a 70 percent-C_7H_{16}–30 percent-$C_{16}H_{34}$ droplet. Such a behavior is in agreement with our previous discussion of the theoretical results. That is, the first stage involves the preferential gasification of the more volatile component and establishment of the concentration boundary layer. The second stage, represented by the flat portion of the d_s^2-t plot, indicates transitional droplet heating as the less volatile component starts to actively participate in the gasification process. Since active droplet heating implies reduced gasification rate, and since the flame size is proportional to the gasification rate, we would expect that the flame size and intensity will both decrease during this period. This flame shrinkage phenomenon is demonstrated in Figure 13.3.6, which is a time-integrated flame streak of a freely falling burning droplet stream, showing the sudden reduction and the subsequent resumption of the flame size and intensity during this second droplet heating period. Figure 13.3.7 quantifies the extent and history of the flame size by showing that after the initial rise in $\tilde{r}_f$ due to the diminishment of the initial droplet heating, it subsequently decreases during this second droplet heating period.

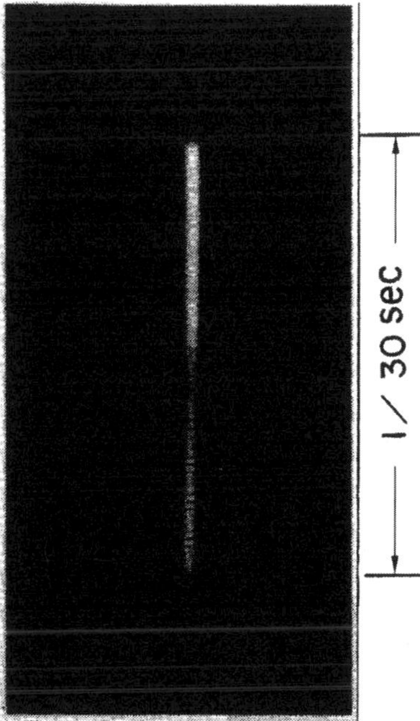

Figure 13.3.6. Flame streak of a freely falling burning droplet stream, showing the phenomenon of flame shrinkage due to transitional droplet heating as the droplet surface layer becomes more concentrated with the less volatile component.

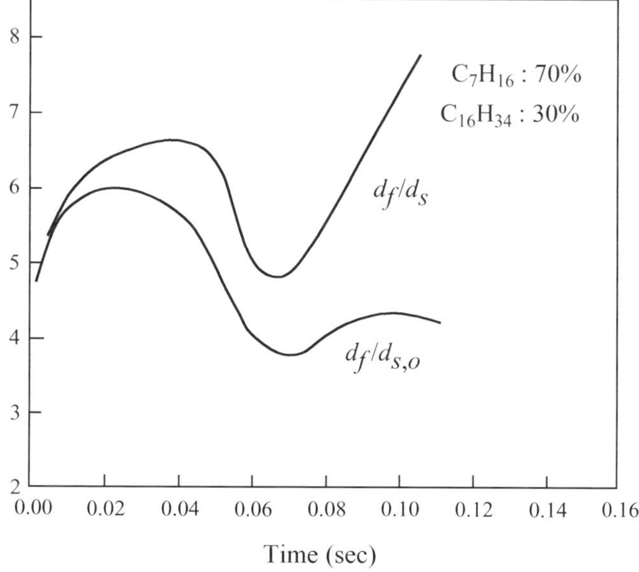

Figure 13.3.7. Flame size variations demonstrating the phenomenon of flame shrinkage.

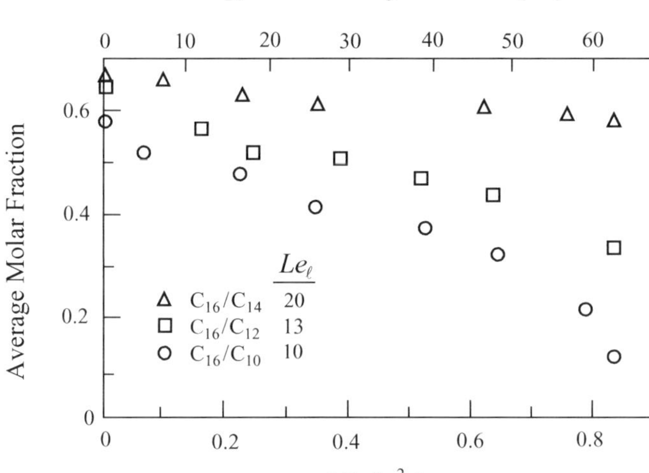

Figure 13.3.8. Experimental result on the temporal variation of the molar fraction of the more volatile component in a two-component alkane droplet, demonstrating the weakening of diffusional resistance with decreasing liquid-phase Lewis number.

The third stage of the d_s^2-t plot in Figure 13.3.5 shows a linear behavior, indicating the possible attainment of diffusion-limited quasi-steady state. This, however, turns out not to be the case. Figure 13.3.8 shows that the experimentally determined temporal variation of the average molar fraction of the more volatile component in a bicomponent droplet, where $(1 - R_s^2) \approx K_c t / r_{s,o}^2$ when the d^2-law holds rigorously. It is seen that the volatile component is neither rapidly depleted as required by the batch distillation mechanism, nor does it attain a constant value as based on the diffusion-limited mechanism with an infinitely thin concentration boundary layer at the surface. The departure from the diffusion-limited behavior becomes more severe with increasing volatility differential between the components, as would be expected. An estimate of the Le_ℓ of the mixtures shows that they are around 10–20, which are not sufficiently large for the diffusion-limited behavior to rigorously hold. Indeed, estimates of conventional hydrocarbon mixtures show that their Lewis numbers fall within this range, and, hence, cannot be considered to be limitingly large. Thus while the combustion of multicomponent droplets is diffusion controlled, the diffusional resistance is not sufficiently strong to yield the quasi-steady behavior, especially for mixtures with large volatility differentials. The fact that the third stage of Figure 13.3.5 is linear is because once vigorous burning is established, the burning rates of different alkanes are actually quite close to each other.

Finally, we note that even though batch distillation is not the correct gasification mechanism for multicomponent droplets, it is nevertheless a good approximation under two situations. The first is during the early part of the droplet lifetime when the more volatile component in the surface layer is preferentially gasified. Since this layer consists of substantial amount of the initial droplet mass because of volume effect, the influence of this volatility-controlled, surface layer gasification period can be quite significant. That is, much of the volatile components are preferentially gasified relatively early in the droplet lifetime in spite of the low value of the liquid-phase diffusion coefficient. The second situation is for relatively slow vaporization such that more time is available for liquid-phase mass diffusion to be effective.

13.3.2. Microexplosion Phenomenon

An interesting event that may occur during multicomponent droplet combustion is its sudden fragmentation, frequently violently. The basic mechanism responsible for this microexplosion event for miscible multicomponent mixtures is the diffusional entrapment of the volatile components in the droplet's inner core as just discussed. That is, after establishment of the concentration boundary layer at the droplet surface (Figure 13.3.1), the droplet temperature attains a high value because it is controlled by the more abundant, higher-boiling-point component at the surface. On the other hand the droplet interior has relatively higher concentration of the more volatile, lower-boiling-point component. Thus it is possible that the liquid elements in the droplet interior can be heated beyond the local boiling point and thereby possess substantial amount of superheat. According to thermodynamics there is a maximum limit on the amount of superheat a liquid can accumulate. Therefore if the droplet temperature is sufficiently high such that this limit is reached, then the liquid element will homogeneously nucleate and gasify, leading to intense internal pressure build-up and thereby the catastrophic fragmentation of the droplet. Experimentally, microexplosion has been frequently observed (Lasheras, Fernandez-Pello & Dryer 1980; Wang, Liu & Law 1984; Wang & Law, 1985). Figure 13.3.9 is the flame streak of a stream of freely falling droplets terminated by microexplosion. Since the droplet size is typically much smaller than the flame size, while Figure 13.3.10 shows that the "explosion ball" is in turn much larger than the dimension of the flame streak, the intensity of such a microexplosion event is quite evident.

Theoretical assessment (Law 1978b) of the potential occurrence of microexplosion can be obtained by first calculating the temperature and species distributions within the droplet. Homogeneous nucleation will initiate at a location r where the temperature $T(r, t)$ exceeds the local concentration-weighted limit of superheat, $T_L[X_{\ell,i}(r, t)]$, which is a thermodynamic property of the mixture. Empirically it has been found (Blander & Katz 1975) that the limits of superheat of many liquids are about 90 percent of their respective critical temperatures.

Theoretical studies of microexplosion show three distinctive properties. First, it can occur only if the volatilities of the components are sufficiently different and their initial concentrations lie within an optimum range. The reason being that microexplosion requires the nonvolatile components to drive up the droplet temperature and the volatile components to facilitate internal nucleation. Second, since the droplet center has the highest concentration of the volatile components while the droplet surface has the highest temperature, homogeneous nucleation should initiate somewhere between these two locations. Third, the occurrence of microexplosion is

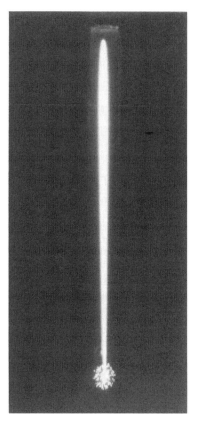

Figure 13.3.9. Flame streak of a freely falling droplet stream, showing the phenomenon of droplet microexplosion.

facilitated with increasing pressure. This is because while the droplet temperature increases substantially with increasing pressure as a result of the elevation of the boiling point of the mixture at the surface, the homogeneous nucleation temperature is almost insensitive to pressure variations when the pressure is not close to that of the critical state.

These three distinctive properties have all been experimentally verified. In particular it was shown that the optimum composition of a two-component mixture for enhanced microexplosion is around 50–50 percent, and that nucleation is initiated close to the droplet center and thereby has the maximum effect in shattering the droplet.

The phenomenon of microexplosion offers interesting potential in optimizing charge preparation in liquid-fueled combustors. For example, present designs of spray systems emphasize producing optimum droplet size distributions such that the droplets are both large enough for penetration into the combustor interior, but also small enough for rapid gasification. However, if microexplosion can be controlled to occur after penetration is achieved, then rapid gasification does not need to be a primary concern in designing spray systems. In this manner large-scale mixing can be

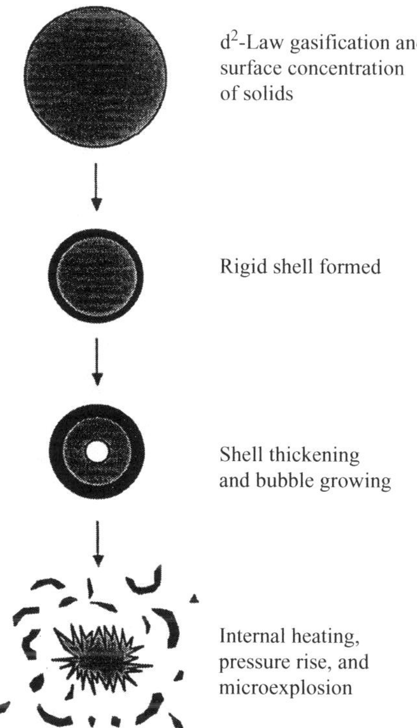

Figure 13.3.10. Schematic showing the gasification and shell formation mechanism of a slurry droplet.

achieved through spraying and penetration, with larger droplets, which is followed by instant gasification and local mixing through the onset of microexplosion.

Microexplosion may also improve the utilization of synthetic and less-refined fuels, which generally have higher boiling point ranges. Thus, with microexplosion, the fuel volatility becomes less crucial in effecting complete gasification within the combustor.

13.3.3. Emulsions and Slurries

Unlike a miscible mixture, which constitutes a single liquid phase, emulsions and slurries are multiphase mixtures. The most widely studied emulsions for combustion applications are water-in-oil emulsions in which water microdroplets are dispersed and stabilized in an oil with the addition of a small quantity of chemical surfactants (Lasheras, Fernandez-Pello & Dryer 1979; Wang & Law 1985; Randolph & Law 1988; Chung & Kim 1990). Water addition generally does not exceed 20 percent for smooth engine operations. The oils used usually have high boiling points around those of diesel and heavier oils. Water–oil emulsions have been tested in a variety of combustors including diesel engines, gas turbines, furnaces, and boilers. The results generally indicate reduction in soot and NO_x emissions, slight increase in emissions of CO and unburned hydrocarbons, and minimal change in the combustion efficiency. No beneficial effect has been observed for volatile fuels such as gasoline. Attempts to make stable methanol–oil emulsions have not been successful because of the extreme hygroscopic nature of methanol, causing its solution with water to readily phase separate from oil upon absorption of atmospheric moisture.

Thermodynamically, according to the phase rule a water–oil emulsion is considered to consist of two liquid phases whose vapor pressures are independent of their relative liquid concentrations, provided they can freely vaporize. This is to be contrasted with a miscible mixture whose partial vapor pressures of the individual components depend on the liquid surface composition according to, say, Raoult's law. Furthermore, while the components in a miscible mixture can freely diffuse, diffusion of the dispersed-phase water microdroplets in the continuous oil phase is negligible. Thus concentration of the water microdroplets within an emulsion droplet can be considered to be frozen during gasification. If the oil itself is multicomponent, diffusion of course still takes place among the miscible components.

The frozen nature of the emulsion droplet composition implies that liquid gasification is described by an "onion skin" model in that the fractional gasification rates of the water and (pure) oil components correspond to their respective initial mass fractions (Law 1977a). Consequently, only the abundant component, which is usually the oil, can gasify freely and hence maintain its equilibrium vapor pressure at the droplet surface. The vapor pressure of the less abundant component, water, is then limited by its concentration in the liquid. The attainable droplet temperature is therefore limited by the boiling point of oil, which is much higher than that of water. Consequently, the embedded water microdroplets can be readily heated to its limit of superheat and subsequently microexplode.

Microexplosion of emulsions also occurs more readily and with greater intensity as compared to droplets of miscible mixtures. The reason is that the tendency of the more volatile component in the miscible mixture to nucleate is "diluted" by the less volatile component because they are mixed at the molecular level. Furthermore, continuous and rapid bubble expansion requires a correspondingly rapid rate of supply of the volatile component to the nucleation site which, however, is not favored in the presence of strong diffusional resistance. In the case of water–oil emulsions, the water microdroplets are themselves quite large, of the order of a few microns for macroemulsions. Therefore they contain sufficient mass for instantaneous and sustained conversion into gas to facilitate microexplosion.

Emulsions of water with a heavy oil such as jet fuel have also been found to be fire safe in that, upon spillage, it is very difficult for them either to be ignited or to sustain steady burning. The reason (Law 1981) being that, since both water and oil can now freely vaporize, the liquid temperature is controlled by the boiling point of water, which is lower than that of the oil. This therefore suppresses the vapor pressure of the oil to levels below its flammability limit.

Studies on slurry combustion have been motivated by two quite different applications. Coal–oil and coal–water slurries consisting of 200–400 mesh (74–37 μm) coal particles have been used in furnaces and boilers as a means of direct coal utilization in conventional liquid-fueled burners, while slurries consisting of micronized particles (e.g., boron and carbon) in jet fuels have been formulated as high-energy propellants for tactical use.

Studies have shown that during slurry droplet gasification, the total droplet lifetime consists of a relatively short initial period, during which the liquid fuel vaporizes while the suspended particles agglomerate, followed by a very long period of agglomerate burning (Lee & Law 1991). Under most situations the agglomerate is hollow, with a porous shell structure. This behavior is described by the following mechanism (Takahashi, Dryer & Williams 1986; Antaki & Williams 1986; Lee & Law 1991), depicted in Figure 13.3.10. During the initial stage the slurry droplet gasifies as if it were a pure liquid, with a gasification rate equal to that of the liquid. The continuously regressing droplet surface concentrates the particles in the surface layer until a porous, rigid shell is formed, typically when the shell is two to four particles thick and the shell porosity is about 0.5. Gasification subsequently takes place at the outer surface of the shell, whose diameter remains fixed. The shell, however, steadily thickens through growth at its inner surface. A continuously growing gas bubble is also formed in the droplet interior because of liquid depletion. During this period the droplet gasification rate is constant because the (outer) surface of gasification is fixed. Thus, from volume conservation consideration, the diameters of the inner surface of the shell, as well as the bubble, must vary cubically with time. This mode of gasification has thus been referred to as the d^3-law. Based on the above understanding, the gasification history of a slurry droplet can be described until complete liquid depletion.

Microexplosion occurs readily for slurry droplets because the solid particles facilitate heterogeneous nucleation. The intensity of microexplosion, however, is not strong because the extent of superheating at the onset of nucleation is less for heterogeneous nucleation than for homogeneous nucleation.

13.3.4. Alcohols and Reactive Liquid Propellants

The droplet gasification phenomena studied so far involve heat transfer from the gas to the droplet surface to effect liquid vaporization. Since the rate of heat transfer not only is limited by diffusion, but is actually also retarded by the outwardly directed Stefan flow, any attempt to increase the inwardly directed heat diffusion rate will lead to a correspondingly higher Stefan flow because of the increase in the gasification rate. This in turn reduces the diffusive heat transfer rate to the surface.

Droplet gasification, however, can be greatly facilitated if a heat source is present either at the droplet surface or in its interior because it can be directly utilized for liquid vaporization. Its significance can be readily appreciated by examining the effect of adding a given amount of heat, either to the gas or to the liquid, on the heat transfer number for droplet gasification, $B_{h,c}$ or $B_{h,v}$, given by Eqs. (6.4.31) and (6.4.8) respectively. Since heat addition to the gas increases the heat transfer to the droplet as given by the numerator of B, while its addition to the liquid reduces the latent heat of vaporization as given by its denominator, it can be readily appreciated

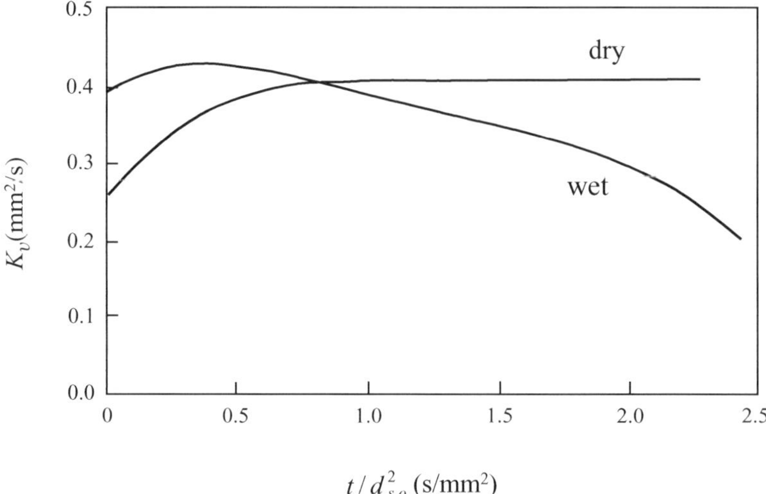

Figure 13.3.11. Experimental result on the temporal variation of the vaporization rate constant of a methanol droplet in dry and humid environments, demonstrating that water vapor condensation at the droplet surface can initially increase the gasification rate through the condensation heat release, but will eventually slow down the gasification rate as the droplet becomes highly concentrated with water.

that, for many situations, direct heat addition to the liquid would result in a greater increase in B. In the limit when the amount of heat addition is close to the latent heat of vaporization, the denominator of B could approach zero, implying that B and, hence, the droplet gasification rate would become very large. We discuss in the following two situations in which the droplet gasification rate is affected by heat addition to the liquid.

Methanol and ethanol have been used as alternate fuels. It has been found that their droplet gasification rate can be substantially increased through condensation of the water vapor from the environment (Lee & Law 1992). That is, since the boiling points of methanol and ethanol are lower than that of water, and since they are also water soluble, water vapor from a humid environment could condense onto and subsequently dissolve into the relatively cold droplet. The condensation heat release could then be used by the alcohol for its own vaporization, thereby facilitating its gasification rate. Furthermore, since the latent heat of vaporization for water (9.73 kcal/mole) is comparable to that for the alcohols (8.42 and 9.40 kcal/mole for methanol and ethanol, respectively), effects due to the exchange between the water condensation heat release and alcohol vaporization heat requirement can be quite substantial. This is an interesting concept in that gasification is effected, and enhanced, by utilizing the thermal as well as the moisture content of the environment. The same consideration can be extended to droplet burning because the water vapor generated at the flame can again diffuse to the droplet surface and condense.

Figure 13.3.11 shows the temporal change in the gasification rate constant, K_v, of a methanol droplet undergoing vaporization in dry and humid environments.

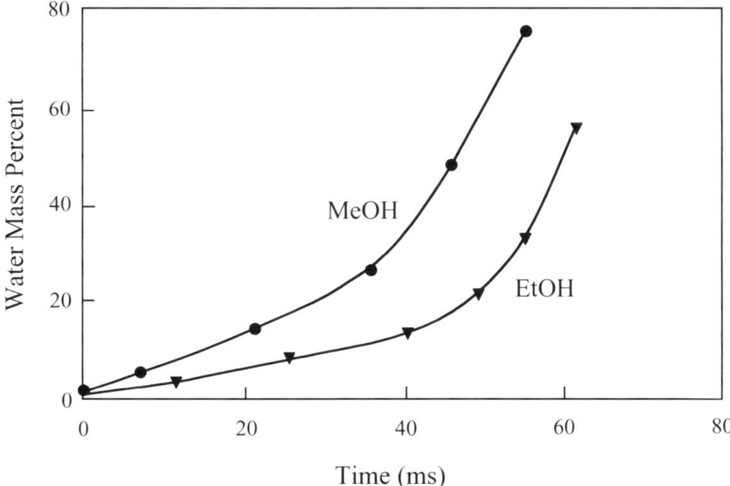

Figure 13.3.12. Experimental result on the temporal variation of the water content in the droplet for burning methanol (MeOH) and ethanol (EtOH) droplets, demonstrating that flame-generated water vapor can condense at the droplet surface and subsequently dissolve into the droplet interior.

It is seen that during the initial stage of gasification, K_v is substantially increased in the presence of water condensation. However, as the water uptake increases, K_v subsequently decreases because the droplet now has a higher water content. Figure 13.3.12 shows the increase in the water content in methanol and ethanol droplets undergoing combustion, with the last datum taken just prior to extinction. It is significant to note the extensive amount of water present within the droplet prior to extinction, indicating the strong propensity for these alcohols to sustain combustion even with substantial amount of water dissolution.

The second example is the gasification of a class of high-energy liquid propellants called organic diazides (Law 1998). These compounds have the generic structure of N_3–R–N_3, where N_3 is an azido group and R an organic functional group. The azido group decomposes at about 170°C, and releases about 50 kcal/mole of heat. Since the boiling points of these diazides are usually higher than their decomposition temperatures, the compounds will decompose in the liquid phase as the droplet is heated up. The decomposition heat release then facilitates droplet gasification. Figure 13.3.13 compares the gasification rate constants of alkyl diazides and *n*-alkanes, for both the purely vaporizing and burning cases. It is seen that, while the alkanes typically gasify with K around 1 mm^2/s, K can be significantly higher for the diazides, especially for the light compounds. For example, for diazidopropane, K_c is about 7 mm^2/s. Such a large increase in K can only be realized with liquid-phase heat generation. Furthermore, with the continuous heat generation, the droplet will eventually be superheated and microexplode. This further reduces the time to achieve complete droplet gasification. Figure 13.3.13 shows that the propensity to microexplode is significantly advanced for the heavier compounds.

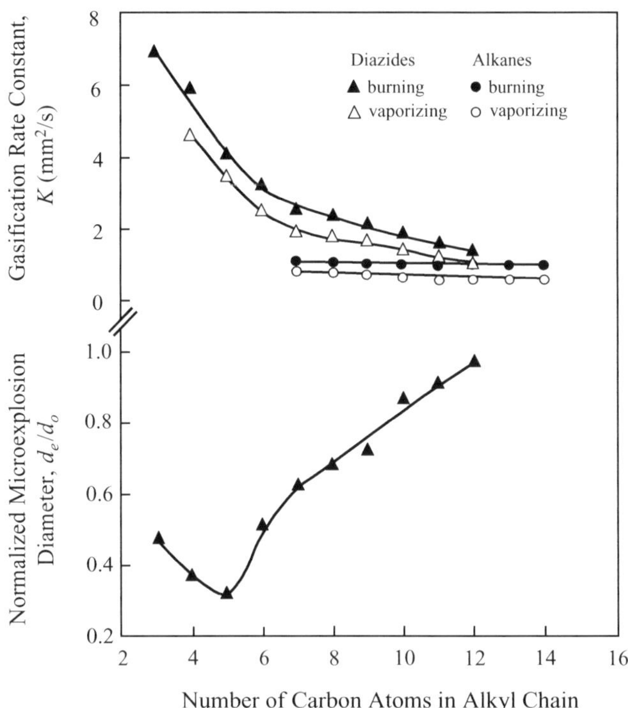

Figure 13.3.13. Gasification rate constants and microexplosion sizes of various n-alkanes and diazido-alkanes undergoing either pure vaporization or combustion. The very high gasification rates of the light diazides and strong microexplosion propensity of heavy diazides are to be noted.

13.4. CARBON PARTICLE COMBUSTION

13.4.1. Phenomenology

When a small piece of organic solid such as coal is introduced into the hot environment of, say, a furnace, the following sequence of events takes place. At first the particle is dark for a short period, during which drying of the fuel and pyrolysis of its volatile matter take place, leading to the emission of a combustible fuel vapor from the particle. If the furnace temperature is sufficiently high, then the fuel vapor is ignited, resulting in either an envelope or wake flame in the same manner as droplet burning. During this period the particle's surface temperature generally does not exceed 1,000 K. The gas-phase flame lasts for a while and then gradually disappears as the bulk of the volatile matter is pyrolyzed and reacted. This pyrolysis period is usually very short, spanning about 10 percent of the total particle lifetime. The solid pyrolysis itself is a highly complicated kinetic process. Depending on the coal type, the particle size, the heating rate and the furnace temperature, the amount and nature of the pyrolyzed gas, as well as the characteristics of the remaining particle, usually called coke or char, can be quite different. The char is porous and is made up of carbon and mineral components. For most chars the carbon content ranges from 55 to 97 percent while the heating value comprises 60 to 95 percent of the total calorific value of the combustible mass.

After pyrolysis of the solid, active burning of carbon in the char is initiated. The particle temperature rapidly increases and can exceed the furnace temperature. Carbon is gasified over the external and internal surfaces of the porous char while the gasified product can further react with air in a gas-phase flame. This carbon gasification period spans the remaining 90 percent of the particle lifetime. Upon complete gasification an ash residue, consisting mostly of mineral oxides, is left behind.

From the above discussion it is clear that an essential element in modeling coal combustion is the kinetics of carbon combustion, which constitutes most of the particle lifetime. Compared with the combustion of liquid hydrocarbon fuels, carbon combustion has the following distinctively different features: (a) The boiling points of conventional hydrocarbon liquids seldom exceed 700–800 K, while the sublimation temperature of carbon is in excess of 4,000 K. Thus carbon is very nonvolatile. (b) The high sublimation temperature implies that carbon can be heated to very high temperatures. When it is also actively reacting, its temperature can easily exceed the combustor temperature as just mentioned. (c) The continuous increase in the particle temperature is eventually arrested through radiative heat loss. Thus the particle temperature seldom exceeds 2,500–3,000 K and therefore is substantially below its sublimation temperature. (d) Because of the low particle temperature relative to the sublimation temperature, the carbon burning rate is expected to be less than those of liquid hydrocarbons, which can be heated to close to their boiling points. (e) Low volatility and high surface temperature indicate that surface reactions are important. (f) The existence of surface and gas-phase reactions implies the need for reaction schemes that describe the coupling between them.

In the next section, the kinetics of carbon oxidation is discussed, followed by an analysis of the combustion of a carbon particle.

13.4.2. Global Kinetics of Carbon Oxidation

The major surface reactions for carbon oxidation are

$$\mathrm{C} + \mathrm{O_2} \rightarrow \mathrm{CO_2} \tag{C1}$$

$$2\mathrm{C} + \mathrm{O_2} \rightarrow 2\mathrm{CO} \tag{C2}$$

$$\mathrm{C} + \mathrm{CO_2} \rightarrow 2\mathrm{CO} \tag{C3}$$

$$\mathrm{C} + \mathrm{H_2O} \rightarrow \mathrm{CO} + \mathrm{H_2}. \tag{C4}$$

Between the two oxidation reactions of (C1) and (C2), CO formation is the preferred route at higher temperatures. In particular, for particle temperatures higher than 1,000 K, the relative contribution from (C1) can be considered to be negligible. Thus reaction (C2) will be referred to as the C–O_2 reaction.

Comparing (C2) and (C3) as alternate routes of CO production in the presence of both O_2 and CO_2, the C–O_2 reaction is the preferred CO production route at low carbon temperatures. It is initiated around 600 K and is saturated, proceeding infinitely fast relative to diffusion, around 1,200 K. The C–CO_2 reaction of (C3) is the high temperature route, which is initiated around 1,600 K and becomes saturated around 2,500 K. It is of particular significance because CO_2 is the product of the gas-phase, water-catalyzed oxidation of CO,

$$2CO + O_2 \rightarrow 2CO_2, \tag{C5}$$

which will be referred to as the $CO–O_2$ reaction. Thus the $C–CO_2$ and $CO–O_2$ reactions form a loop.

The $C–H_2O$ reaction, (C4), is used to generate CO and H_2 from coal as gaseous fuel. It is important when the combustion environment consists of an appreciable amount of water.

13.4.3. Analysis

We now study the oxidation of a spherical carbon particle in an environment consisting of oxygen, carbon dioxide, and an inert, say, nitrogen (Libby & Blake 1979; Makino & Law 1988b; Makino 1992). A trace amount of water vapor is implicitly assumed to exist to facilitate the wet-CO oxidation.

The three major reactions here are the surface $C–O_2$ and $C–CO_2$ reactions and the gas-phase $CO–O_2$ reaction. For ease of referencing they are listed together in the following:

$$\begin{aligned} \text{C–O}_2 \textit{ surface reaction}: &\quad 2C + O_2 \rightarrow 2CO \\ \text{C–CO}_2 \textit{ surface reaction}: &\quad C + CO_2 \rightarrow 2CO \\ \text{CO–O}_2 \textit{ gas-phase reaction}: &\quad 2CO + O_2 \rightarrow 2CO_2 \end{aligned}$$

The combustion process depends critically on whether the $CO–O_2$ gas-phase reaction is activated. If it is not (Figure 13.4.1a), then the O_2 in the ambience can readily reach the carbon particle surface to participate in the $C–O_2$ reaction. Activation of the surface $C–CO_2$ reaction depends on whether the environment contains any CO_2. However, if the gas-phase $CO–O_2$ reaction is activated (Figure 13.4.1b), then the existence of the gas-phase flame cuts off most of the supply of oxygen to the surface such that the surface $C–O_2$ reaction is suppressed. At the same time, the CO_2 generated at the flame activates the surface $C–CO_2$ reaction.

To be more specific, let us consider an initially cold particle ($\sim$ 800–1,000 K) and a cold environment ($\sim$ 1,000 K) without any CO_2. Since the initial temperature of a char particle should approximately correspond to the final temperature of the pyrolyzing coal, it is reasonable to consider 800–1000 K as the lowest temperature range of the char. Under this situation only the $C–O_2$ reaction is effective, producing CO which is dispersed to the ambience as shown in Figure 13.4.1a.

As the particle heats up, the $C–CO_2$ reaction should be initiated except there is no CO_2 in the gas. However, since the particle is now quite hot, the gas-phase $CO–O_2$ reaction is initiated, resulting in the flame configuration of Figure 13.4.1b. Here the ambient oxygen reacts with the CO generated from the surface $C–CO_2$ reaction, and produces CO_2 which sustains this surface reaction. It is of interest to note that in the present case the ignition source for the gas-phase reaction can be either the hot particle or the hot ambience, while in the droplet case it is the hot ambience because

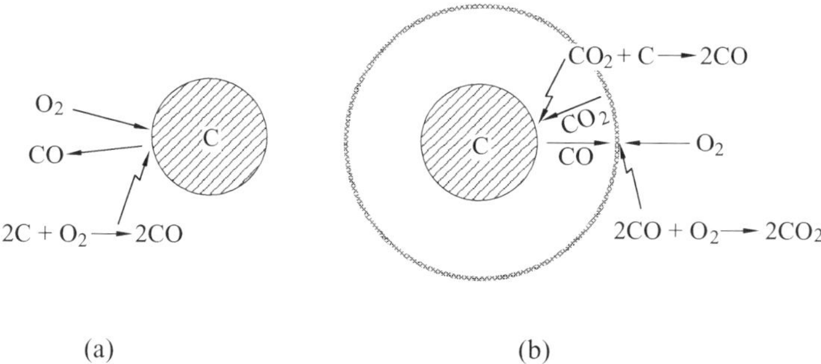

Figure 13.4.1. Possible burning configurations of carbon: (a) frozen gas-phase reaction, (b) detached flame-sheet burning.

the attainable droplet temperature is limited by the boiling point of the liquid and, hence, is too low to effect ignition. In contrast, the attainable temperature of the carbon particle can be quite high, being limited by radiative heat loss, which in turn depends on the particle size.

In the problem to be analyzed we shall assume that diffusion of the various species to and from the carbon surface is uninhibited, implying that there is either no ash layer or, if there were one, it is sufficiently porous and hence has very low diffusional resistance. It is further assumed that surface reactions take place only at the particle surface in that there is no species diffusion into the pores of the particle causing internal burning. It is noted that internal burning can be quite significant because of the large surface area associated with the porous structure, especially when the particle temperature is low such that the gaseous reactants can readily diffuse into the pores without suffering much depletion of its concentration through reaction in the particle surface region.

The quasi-steady gas-phase processes will be formulated and solved allowing for finite rates of the three major reactions and for a given surface temperature T_s, which can be separately determined through overall particle energy balance including, say, particle heating and radiative heat loss.

Using the coupling function formulation, the quasi-steady gas-phase heat and mass conservation equations are

$$\tilde{L}(\tilde{Y}_{CO} + \tilde{Y}_{CO_2}) = \tilde{L}(\tilde{Y}_{O_2} + \tilde{Y}_{CO_2}) = \tilde{L}(\tilde{Y}_{CO_2} - \tilde{T}) = 0 \qquad (13.4.1)$$

$$\tilde{L}(\tilde{T}) = \tilde{w}_g, \qquad (13.4.2)$$

where

$$\tilde{L}(\cdot) = \left[\frac{\tilde{m}}{\tilde{r}^2}\frac{d}{d\tilde{r}} - \frac{1}{\tilde{r}^2}\frac{d}{d\tilde{r}}\left(\tilde{r}^2\frac{d}{d\tilde{r}}\right)\right](\cdot)$$

is the spherically symmetric convective–diffusive operator, $\tilde{m} = m/(4\pi\rho D r_s)$, $\tilde{r} = r/r_s$, $\tilde{T} = c_p T/(q_c\sigma_{CO}\delta)$, $\tilde{Y}_i = Y_i/(\sigma_i\delta)$, q_c is the heat release per unit mass of CO consumed, σ_i is the stoichiometric mass ratio of the ith species to CO according to the CO–O$_2$ reaction, $\delta = W_{\rm CO}/W_{\rm C}$, $\tilde{w}_g$ is its nondimensional gas-phase reaction rate whose specification does not need to concern us here, and ρD is taken as a constant.

The boundary conditions for Eqs. (13.4.1) and (13.4.2) are

$$\begin{aligned} &\tilde{r} = 1: \quad \tilde{T} = \tilde{T}_s, \quad \tilde{Y}_i = \tilde{Y}_{i,s}, \quad i = \mathrm{CO, O_2, CO_2} \\ &\tilde{r} = \infty: \quad \tilde{T} = \tilde{T}_\infty, \quad \tilde{Y}_i = \tilde{Y}_{i,\infty}, \quad \tilde{Y}_{\rm CO} = 0, \quad i = \mathrm{O_2, CO_2}, \end{aligned} \tag{13.4.3}$$

which are to be supplemented with the conservation relations at the particle surface.

Let us first consider surface mass conservation for CO, which states that the difference between the bulk convective and diffusive transport is the net flow rate of CO,

$$mY_{\rm CO,s} - 4\pi r_s^2 \rho D \left(\frac{dY_{\rm CO}}{dr}\right)_{r_s} = m_{\rm CO}. \tag{13.4.4}$$

Since $m_{\rm CO} = m_{\rm CO,1} + m_{\rm CO,2}$, where $m_{\rm CO,1}$ and $m_{\rm CO,2}$ are the production rates of CO from the C–O$_2$ and C–CO$_2$ reactions respectively, with the subscripts 1,2 designating these two reactions, and since the reaction rates of CO and C are related through $m_{\rm CO,1} = (W_{\rm CO}/W_{\rm C})m_{\rm C,1}$ and $m_{\rm CO,2} = 2(W_{\rm CO}/W_{\rm C})m_{\rm C,2}$, Eq. (13.4.4) can be written as

$$\tilde{m}\tilde{Y}_{\rm CO,s} - \left(\frac{d\tilde{Y}_{\rm CO}}{d\tilde{r}}\right)_1 = \tilde{m}_{\rm C,1} + 2\tilde{m}_{\rm C,2}, \tag{13.4.5}$$

where $\tilde{Y}_{\rm CO} = (W_{\rm C}/W_{\rm CO})Y_{\rm CO}$. Similarly, the boundary conditions for $\tilde{Y}_{\rm O_2}$ and $\tilde{Y}_{\rm CO_2}$ are

$$\tilde{m}\tilde{Y}_{\rm O_2,s} - \left(\frac{d\tilde{Y}_{\rm O_2}}{d\tilde{r}}\right)_1 = -\tilde{m}_{\rm C,1}, \tag{13.4.6}$$

$$\tilde{m}\tilde{Y}_{\rm CO_2,s} - \left(\frac{d\tilde{Y}_{\rm CO_2}}{d\tilde{r}}\right)_1 = -\tilde{m}_{\rm C,2}. \tag{13.4.7}$$

The flow rate $m_{\rm C,1}$ is related to the reaction rate per unit area $w_{\rm C,1}$ through $m_{\rm C,1} = -4\pi r_s^2 w_{\rm C,1}$, where $w_{\rm C,1} = -2W_{\rm C}B_1 c_{\rm O_2,s}\exp(-E_{a,1}/R^oT_s)$ and B_1 is the frequency factor for the surface C–O$_2$ reaction. Similar expressions can be written for $m_{\rm C,2}$ and $w_{\rm C,2}$. We therefore have

$$\tilde{m}_{\rm C,1} = \tilde{Y}_{\rm O_2,s}k_1, \quad \tilde{m}_{\rm C,2} = \tilde{Y}_{\rm CO_2,s}k_2, \tag{13.4.8}$$

where

$$k_1 = \left(\frac{B_1 r_s}{\rho D}\right)e^{-\tilde{T}_{a,1}/\tilde{T}_s}, \quad k_2 = \left(\frac{B_2 r_s}{\rho D}\right)e^{-\tilde{T}_{a,2}/\tilde{T}_s},$$

are the nondimensional surface reaction rate constants, and B_1 and B_2 are the density-weighted preexponential factors. The particle burning rate is thus

$$\tilde{m} = \tilde{m}_{\rm C} = \tilde{m}_{\rm C,1} + \tilde{m}_{\rm C,2}. \tag{13.4.9}$$

Solving Eq. (13.4.1) subject to the boundary conditions, it can be shown that the various coupling functions are given by

$$\tilde{Y}_{\rm CO} + \tilde{Y}_{\rm CO_2} = \frac{(\tilde{Y}_{\rm CO_2,\infty} + \beta) + (\tilde{Y}_{\rm CO_2,\infty} - 1)\beta\xi}{1+\beta} \tag{13.4.10}$$

$$\tilde{Y}_{O_2} + \tilde{Y}_{CO_2} = \frac{(\tilde{Y}_{O_2,\infty} + \tilde{Y}_{CO_2,\infty} - \beta) + (\tilde{Y}_{O_2,\infty} + \tilde{Y}_{CO_2,\infty} + 1)\beta\xi}{1+\beta} \tag{13.4.11}$$

$$\tilde{Y}_{CO_2} - \tilde{T} = -\tilde{T}_s + (\tilde{Y}_{CO_2,\infty} - \tilde{T}_\infty + \tilde{T}_s)\xi + (1-\xi)\left(\frac{\tilde{Y}_{O_2,\infty} + \tilde{Y}_{CO_2,\infty} - \beta}{1+\beta} - \tilde{Y}_{O_2,s}\right), \tag{13.4.12}$$

where

$$\xi = \frac{e^{-\tilde{m}/\tilde{r}} - e^{-\tilde{m}}}{1 - e^{-\tilde{m}}}, \qquad \beta = e^{\tilde{m}} - 1.$$

The problem is therefore reduced to solving the energy equation (13.4.2) subject to the boundary conditions $\tilde{T}(1) = \tilde{T}_s$ and $\tilde{T}(\infty) = \tilde{T}_\infty$, using the coupling functions given by Eqs. (13.4.10)–(13.4.12) and $\tilde{m}$ given by Eq. (13.4.9). The solution including states of ignition and extinction can be obtained either computationally or through asymptotic analysis of the gas-phase reaction. The system behavior, however, can be bracketed by the following limiting solutions.

13.4.4. Limiting Solutions

All possible combustion states of the system are bounded by the limiting situations that the gas-phase reaction is either completely frozen or occurs infinitely fast, while the surface reactions are allowed to proceed at finite rates. Three possible burning modes can be envisioned, as follows.

Frozen Limit: Here we have $\tilde{L}(\tilde{T}) = 0$ in Eq. (13.4.2), which readily yields

$$\tilde{T} = \tilde{T}_s + (\tilde{T}_\infty - \tilde{T}_s)\xi. \tag{13.4.13}$$

Evaluating the coupling functions $\tilde{Y}_{O_2} + \tilde{Y}_{CO_2}$ and $\tilde{Y}_{CO_2} - \tilde{T}$ and their derivatives at the surface yields $\tilde{Y}_{O_2,s}$ and $\tilde{Y}_{CO_2,s}$, which when substituted into Eq. (13.4.9) yields the mass burning rate as a function of the particle temperature through k_1 and k_2,

$$\tilde{m} = \left(\frac{\tilde{Y}_{O_2,\infty}}{1+\beta+k_1\beta/\tilde{m}}\right)k_1 + \left(\frac{\tilde{Y}_{CO_2,\infty}}{1+\beta+k_2\beta/\tilde{m}}\right)k_2. \tag{13.4.14}$$

Detached Flame Limit: When the gas-phase reaction is infinitely fast, we obtain a detached flame, characterized by

$$\tilde{Y}_{O_2}(1 \le \tilde{r} \le \tilde{r}_f) = \tilde{Y}_{CO}(\tilde{r} \le \tilde{r}_f \le \infty) = 0.$$

Thus by using the coupling functions of Eqs. (13.4.10) to (13.4.12), we obtain

$$\tilde{m} = \left(\frac{\tilde{Y}_{O_2,\infty} + \tilde{Y}_{CO_2,\infty} - \beta}{1+\beta}\right)k_2 \tag{13.4.15}$$

$$\tilde{r}_f = \frac{\tilde{m}}{\ln(1 + \tilde{Y}_{O_2,\infty}/2)} \tag{13.4.16}$$

$$\tilde{T}_f = \tilde{T}_s + (\tilde{Y}_{O_2,\infty} + \tilde{T}_\infty - \tilde{T}_s)\frac{(2\beta - \tilde{Y}_{O_2,\infty})}{(2 + Y_{O_2,\infty})\beta}. \tag{13.4.17}$$

In this detached flame limit the burning rate $\tilde{m}$ still depends on the finite reaction rate of the surface C–CO_2 reaction. The surface C–O_2 reaction is however suppressed because there is no oxygen leakage through the flame.

Attached Flame Limit: With $\tilde{m}$ continuously decreasing in the detached flame limit, a value will be reached at which $\tilde{r}_f = 1$, implying that the flame is now contiguous to the surface. With further reduction in $\tilde{m}$, the combustion evolves into one with an attached reaction sheet. In this case the gas-phase reaction is still infinitely fast such that there is no leakage of CO into the gas phase. The rate of generation of CO through the surface C–CO_2 reaction, however, is not as fast as the situation in the detached reaction-sheet limit as to totally consume the oxygen at the surface. The presence of oxygen at the surface, then, also activates the surface C–O_2 reaction. Thus with $\tilde{r}_f = 1$, $\tilde{T}_f = \tilde{T}_s$, $\tilde{Y}_{CO,s} = 0$ but $\tilde{Y}_{O_2,s} \neq 0$, it can be shown by using Eqs. (13.4.10) and (13.4.11) that in this limit the mass burning rate is given by

$$\tilde{m} = \left(\frac{\tilde{Y}_{O_2,\infty} - 2\beta}{1+\beta}\right) k_1 + \left(\frac{\tilde{Y}_{CO_2,\infty} + \beta}{1+\beta}\right) k_2. \tag{13.4.18}$$

Diffusion Limit: The three limits identified above all pertain to the limiting behavior of the gas-phase reaction, with one or both of the surface reactions activated with finite rates. We now consider situations limited by the gas-phase transport.

We first consider the situation when the surface C–O_2 and C–CO_2 reactions both occur infinitely fast. This implies that $\tilde{Y}_{O_2,s} = \tilde{Y}_{CO_2,s} = 0$, which when applied to the coupling function Eq. (13.4.11) readily yields

$$\beta_{\max} = \tilde{Y}_{O_2,\infty} + \tilde{Y}_{CO_2,\infty}, \tag{13.4.19}$$

and thereby the burning rate

$$\tilde{m}_{\max} = \ln(1 + \beta_{\max}). \tag{13.4.20}$$

This is the maximum burning rate attainable for given ambient O_2 and CO_2 concentrations, with the carbon consumption rate limited by diffusion. The second situation occurs when the gas-phase CO–O_2 and the surface C–CO_2 reactions proceed infinitely fast. Then we still require $\tilde{Y}_{O_2,s} = 0$ and $\tilde{Y}_{CO_2,s} = 0$, which again yields $\beta_{\max}$ given by Eq. (13.4.20). Thus the diffusion-limited behavior can be attained by requiring only two of the three reactions to proceed infinitely fast. It should also be pointed out that $\tilde{m}_{\max}$ here pertains to only the mass gasification rate of carbon at the surface without being specific to the identity of the final product, which still depends on the nature of the gas-phase reaction. Thus the combustion product is mainly CO when the gas-phase reaction rate is slow, while it becomes CO_2 when it is fast.

We also note that since $m = 4\pi \rho D r_s \tilde{m}$, the fact that $\tilde{m}_{\max}$ is a constant implies that the mass burning rate m varies with the instantaneous particle radius r_s, which is just the diffusion-limited d^2-law result, as should be the case.

Surface Reaction Limit: In this limit $k_1 \ll 1$ and $k_2 \ll 1$. Then the mass burning rate for the frozen and attached flame limits degenerate to

$$\tilde{m} = \tilde{Y}_{O_2,\infty} k_1 + \tilde{Y}_{CO_2,\infty} k_2, \qquad (13.4.21)$$

while that for the detached flame limit becomes

$$\tilde{m} = (\tilde{Y}_{O_2,\infty} + \tilde{Y}_{CO_2,\infty}) k_2. \qquad (13.4.22)$$

The above expressions were obtained by assuming, realistically, that burning under the present situation is necessarily very slow, that is, $\tilde{m} \approx \beta \ll 1$.

It is of interest to note that Eqs. (13.4.21) and (13.4.22) show that $\tilde{m} \sim k$ in this limit. Since $m \sim r_s \tilde{m}$ and $k \sim r_s$, we have $m \sim r_s^2$. Furthermore, recognizing that $m \sim dr_s^3/dt \sim r_s^2 dr_s/dt$, we obtain the result that

$$\frac{dr_s}{dt} = \text{constant}, \qquad (13.4.23)$$

which can be considered as a d-law for particle burning controlled by surface reactions. It is important to mention again that this result is only for a solid carbon particle, without any internal pores. Since particles of char are usually porous, and if burning is kinetically controlled, then diffusion of the gas into the pores is efficient, resulting in reactions over the surface of the internal pores. Because of the much larger surface area associated with the pores, the gasification rate of the carbon can be greatly facilitated. At the same time, however, the size of the char particle itself does not change much, rendering the d-law inappropriate. This internal burning mode could eventually lead to rupturing of the pore structure and, hence, fragmentation of the particle.

Small Burning Rate Limit: Because of the nonvolatile nature of carbon gasification, its burning rates are typically small compared to those of hydrocarbon droplets, regardless of the intensity of the surface reactions. Consequently the relation $\tilde{m} \approx \beta \ll 1$ can be applied in general. Taking such a limit, the burning rates for the various limiting modes of the gas-phase reaction identified above can be explicitly expressed as:

Frozen Limit:

$$\tilde{m} \approx Y_{O_2,\infty}\left(\frac{k_1}{1+k_1}\right) + \tilde{Y}_{CO_2,\infty}\left(\frac{k_2}{1+k_2}\right) \qquad (13.4.24)$$

Detached Flame Limit:

$$\tilde{m} \approx (\tilde{Y}_{O_2,\infty} + \tilde{Y}_{CO_2,\infty})\left(\frac{k_2}{1+k_2}\right) \qquad (13.4.25)$$

Attached Flame Limit:

$$\tilde{m} \approx \frac{\tilde{Y}_{O_2,\infty} k_1 + \tilde{Y}_{CO_2,\infty} k_2}{1 + 2k_1 - k_2}. \qquad (13.4.26)$$

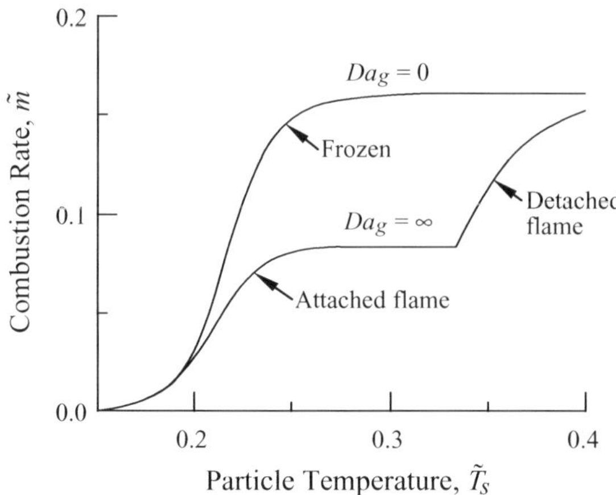

Figure 13.4.2. Limiting solutions of carbon particle combustion, bracketing all feasible combustion states ($Br_s/D_\infty = 10^8$, $\tilde{Y}_{CO_2,\infty} = 0$).

The previous expressions can be cast into a form that frequently appears in the literature on carbon burning. Use Eq. (13.4.25) for illustration. With $m = 4\pi\rho Dr_s\tilde{m}$, $k_2 = (B_2 r_s/\rho D)\exp(-T_a/T_s)$, Eq. (13.4.25) can be rearranged to assume the form

$$\frac{1}{m} = \frac{1}{h_{\text{diff}}} + \frac{1}{h_{\text{kin}}}, \tag{13.4.27}$$

where

$$h_{\text{diff}} = 4\pi\rho Dr_s\beta_{\max}, \quad h_{\text{kin}} = 4\pi r_s^2\beta_{\max}B_2 e^{-T_a/T_s} \tag{13.4.28}$$

are respectively the diffusion and surface reaction rates. Equation (13.4.27) shows in a transparent manner that the particle burning rate is diffusion limited when $h_{\text{kin}} \to \infty$, and is kinetically limited when $h_{\text{diff}} \to \infty$. Furthermore, because of their different dependence on r_s, the burning becomes diffusion controlled with increasing particle size, and kinetically controlled for smaller particles.

Having identified the various limiting modes of burning, we show in Figure 13.4.2 a representative plot of the combustion rate $\tilde{m}$ as a function of the particle temperature $\tilde{T}_s$ for the three limiting situations of frozen flow, attached flame, and detached flame for the gas-phase reaction. The curves are generated with typical kinetic values of carbon oxidation, with $B_1 = B_2 = B$, $Br_s/D_\infty = 10^8$, and $\tilde{Y}_{CO_2,\infty} = 0$. Combustion responses with finite values of gas-phase reactivity are bounded by these limit curves.

It is seen that in the limit of frozen gas-phase reactivity and the absence of CO_2 in the ambience, the only surface reaction occurring is that of C–O_2. This reaction becomes fully activated around $\tilde{T}_s = 0.25$, causing the curve to flatten out. In the limit of infinite gas-phase reactivity, the C–O_2 reaction is again activated first. This reaction, however, produces CO which subsequently reacts with the incoming O_2 to yield CO_2. Thus the CO–O_2 reaction, which is always activated in the present limit, reduces the amount of O_2 that can reach the surface and thereby reduces the intensity of the C–O_2 reaction. This explains the lower combustion rate here. The plateau regime is caused by the competition between the C–O_2 and CO–O_2 reactions. The combustion rate starts to increase again only when the flame detaches from the surface, implying that the CO–O_2 reaction is now being actively enhanced. Both the frozen and infinite gas-phase reactivity results asymptote to the maximum burning rate identified in Eq. (13.4.20).

In summary, from the present problem we not only have learned the mechanism of carbon combustion, but we have again appreciated the possible intricate couplings between different reaction steps in a multistep reaction scheme. For example, for the same particle temperature, the presence of gas-phase reaction can actually slow down the particle gasification rate. Furthermore, the existence of the two surface reactions can be competitive in that an increase in the C–CO_2 reaction, due to flame-generated CO_2, can lead to a corresponding decrease in the C–O_2 reaction.

13.5. METAL PARTICLE COMBUSTION

Although metals are generally not considered to be flammable materials because of their high ignition temperatures, they burn extremely vigorously once ignition is achieved. There are several practical interests related to metal particle combustion. Accidental fires and explosions in mine galleries are frequently caused by the flying incendiary (i.e., burning) metal particles which are abraded off drilling machines. The explosion of fuel tanks of automobiles and aircraft can also be caused by these particles abraded off the tank wall when it is either accidentally ruptured through collision or intentionally pierced by munition. Furthermore, metal particles have been added to both solid and liquid propellants in order to increase their energy density. Recently, there has also been considerable interest in the combustion synthesis of novel materials (Merzhanov 1990; Makino 2001) such as titanium nitride, molybdenum silicide, and zirconium carbide for their mechanical, thermal, and electrical properties. In one of these methods, called the self-propagating high-temperature synthesis (SHS), powders of two constituents (e.g., zirconium and carbon) are pressed to form a solid compact. By igniting the compact at one end, a combustion wave supported by the reaction heat release between the constituents subsequently propagates through the compact, forming the synthesized material as the combustion product. This process basically does not require external heating for synthesis, is self-purifying because the high flame temperature usually drives out the impurities, and has the potential of preshaping the product. It is clear that the combustion of metals is an essential element in SHS.

An important difference between the combustion characteristics of hydrocarbon droplets and carbon particles just studied, and of metal particles, is that the combustion products of the former, H_2O and CO_2, are gaseous, while those of the latter are frequently either solid or liquid under the prevailing temperatures (Law 1973; Williams 1997). Indeed, it has been suggested that since most metal oxides are refractory in nature, the heat of gasification of the oxide frequently can be so substantial that the flame temperature can be assumed to be at the boiling point of the oxide, as long as some condensed oxide is formed in the flame zone (Glassman 1960). The motion of the condensed oxides in the flame zone is also of particular interest in that, since they are not subjected to concentration diffusion, they are either convected or thermophoretically diffused. Consequently, depending on the directions of the Stefan flow in the inner and outer regions to the flame, the net convection can cause these condensed products to be either dispersed to the ambience, or convected toward the droplet surface, or stagnantly trapped at the flame.

The fact that the melting points of most metals are lower than those of their respective oxides implies that the oxides are present in condensed phase at the surface of the molten metal particle. These condensed oxides can be either brought in from the flame, or formed when the vapor product condenses at the surface, or are originally present over the particle surface. Indeed, all metals exposed to air have a thin film of oxide coating over its surface. The presence of these condensed oxides at the particle surface can affect particle combustion in an essential way. The influence may not be too serious if the particle temperature is higher than the melting point of the oxide, and if the molten oxide can contract under surface tension. The bulk of the molten metal is then exposed to the ambience and, hence, can undergo either gasification or oxidation. If, however, surface tension does not permit contraction, then the metal particle will be covered by the condensed oxide. In such a case reaction is still possible if the molten metal can diffuse through the molten oxide layer to react with the ambient air, or the ambient air can diffuse through the layer to react with the metal (Glassman, Williams & Antaki 1984). Either way, the reaction time is expected to be significantly slowed down due to this diffusional resistance. If the oxide coating is in solid state but is porous, then diffusion of either air or metal through the pores is still possible, leading to oxidation. However, if the solid coating is nonpermeable, then contact between metal and air is broken and oxidation could be inhibited.

The importance of oxide coating can be best illustrated for the oxidation of boron, which theoretically is an attractive propellant additive because of its large values of heat content on both mass and volume bases. However, because of the oxide coating, its successful utilization has not been realized. Specifically, it has been found that ignition of boron can only be achieved when the ambient temperature exceeds the boiling point of the oxide, at which state the original oxide coating is vaporized, exposing boron for oxidation.

Recent studies have found that surface oxides can dissolved into the particle interior, forming a complex metal-oxide system (Molodetsky, Dreizin & Law 1996). Furthermore, nitrogen in the air can also participate in surface reaction and dissolution, leading to interesting phase change processes within the particle interior. Indeed, in materials synthesis the oxidizer does not need to be oxygen. For example, the reaction between metals and nitrogen or hydrogen leads to the formation of nitrides and hydrides respectively. The potential combinations between the chemical elements and their stoichiometry are numerous, and the phase diagrams for these systems can be quite complex.

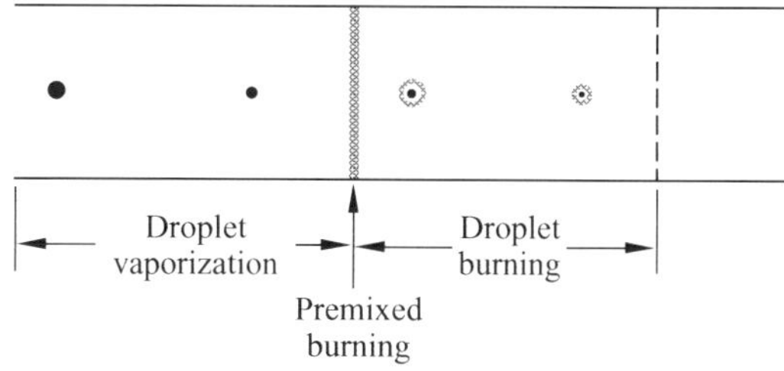

Figure 13.6.1. Schematic of one-dimensional spray combustion.

13.6. PHENOMENOLOGY OF SPRAY COMBUSTION

13.6.1. One-Dimensional, Planar, Spray Flames

We now begin our study on spray combustion. In order to gain a phenomenological understanding of the various processes of interest, let us first suppress jet mixing by considering flame propagation within a steady, one-dimensional, well-mixed spray, as shown in Figure 13.6.1. Thus, far upstream of the flame, we have a two-phase mixture consisting of fuel droplets, fuel vapor, and oxidizing and inert gases. Since the overall fuel–oxidizer ratio during combustion should be close to stoichiometric, and since some of the fuel is initially present in the liquid phase, it is reasonable to assume that the upstream gaseous mixture must be lean in the fuel vapor concentration.

Thus as the mixture approaches the flame in the flame-stationary frame, the droplets are heated up, their vaporization rate increases, and they become more ignitable. The gas medium is also heated up and, having been enriched with fuel vapor, becomes more stoichiometric and thereby also more reactive. It is conceivable that ignition of either the droplets or within the bulk gas can occur first. Ignition of the other mode should follow rapidly due to the extensive heat release from the initiating mode. The subsequent flame structure is expected to consist of a thin premixed flame reaction zone followed by a broader heat release region supported by the combustion of droplets either singly or in groups. The droplets will then either burn to completion if the mixture is overall fuel lean, or extinguish upon near complete depletion of the oxidizing gas if it is overall fuel rich.

There exists a minimum droplet size below which droplet burning does not need to be considered. This is because a sufficiently small droplet will be completely vaporized in crossing the preheat zone of the flame. To estimate this size, we note that the characteristic time for traversing the preheat zone is

$$\tau_D = \frac{\ell_D^o}{s_u^o} = \frac{\lambda/c_p\rho_g}{(s_u^o)^2}, \tag{13.6.1}$$

while from Eq. (6.4.16) the droplet vaporization time is given by

$$\tau_v = \frac{d_{s,o}^2}{4K_v} = \frac{d_{s,o}^2\rho_\ell}{8(\lambda/c_p)\ln(1+B_v)}. \tag{13.6.2}$$

Thus equating $\tau_D = \tau_v$, the minimum droplet size is

$$d^2_{s,o,\min} = 8\frac{\rho_g}{\rho_\ell}\left(\frac{\lambda/c_p}{f^o}\right)^2 \ln(1+B_v), \tag{13.6.3}$$

where f^o is the laminar burning flux of the flame, and the subscript o designates the initial state. If we take $B_v = 0.5$, $\rho_g/\rho_\ell = 10^{-3}$, $\lambda/c_p\rho_g = 10^0$ cm^2/sec, and $s_u^o = 40$ cm/sec for an atmospheric, stoichiometric alkane–air flame, then we get $d_{s,o,\min} \simeq 10\,\mu$m. This gives an order of magnitude estimate of the smallest droplet size below which droplet burning is not important for atmospheric flames. If we further recall that $(f^o)^2 \sim p^n$, then Eq. (13.6.3) shows that $d^2_{s,o,\min} \sim p^{(1-n)}$ for sufficiently subcritical situations. Since the droplet vaporization process is pressure insensitive, this dependence is totally due to changes in the flame thickness and thereby the droplet residence time in crossing the flame.

It is also of interest to assess the extent of droplet interaction straightly due to the proximity of their flames. If we take a uniform fuel–air mixture with the fuel completely existing in the form of droplets of equal size, then the fuel–air mixture ratio is

$$F/A \approx \frac{\frac{1}{6}\pi d_s^3 \rho_\ell}{\frac{1}{6}\pi d_g^3 \rho_g} = \frac{\rho_\ell}{\rho_g}\left(\frac{d_s}{d_g}\right)^3, \tag{13.6.4}$$

where d_g is an average inter-droplet distance. For near-stoichiometric combustion of practical hydrocarbon fuels, the fuel-to-air mass ratio is $F/A \simeq 0.05$. Thus $d_g/d_s \simeq 25$ for combustion under atmospheric pressure with $\rho_g/\rho_\ell = 10^{-3}$. This value is further reduced within internal combustion engines because of the higher pressure and therefore the higher density ρ_g. For example, with a compression ratio of 15, we get $d_g/d_s \simeq 15$. Therefore if we take the droplet flamefront standoff ratio, d_f/d_s, to be around 5 to 10, it is then obvious that the droplet flames can be quite close to each other within a spray when droplet burning is the dominant combustion mode.

13.6.2. Spray Jet Flames

An intrinsic weakness of the one-dimensional spray analysis is the assumption that all of the oxidizing gas is already present within the spray interior. In realistic situations, the fuel is frequently sprayed into the hot oxidizing gas such that initially there is very little or practically no oxidizer present within the spray. Since the hot oxidizing gas has to be entrained into the jet interior for reaction to be possible, the spray combustion characteristics then depend critically on the relative rates of droplet vaporization versus oxidizer entrainment. The situation is shown in Figure 13.6.2.

Specifically, it is reasonable to expect that if the entrainment rate is sufficiently fast, then the spray interior is rapidly heated up and enriched with oxygen. These favor droplet ignition. Furthermore, once the envelope flames are established around the droplets, either singly or in groups, then the fuel vapor that is subsequently generated through droplet vaporization will be consumed at these flames and therefore cannot

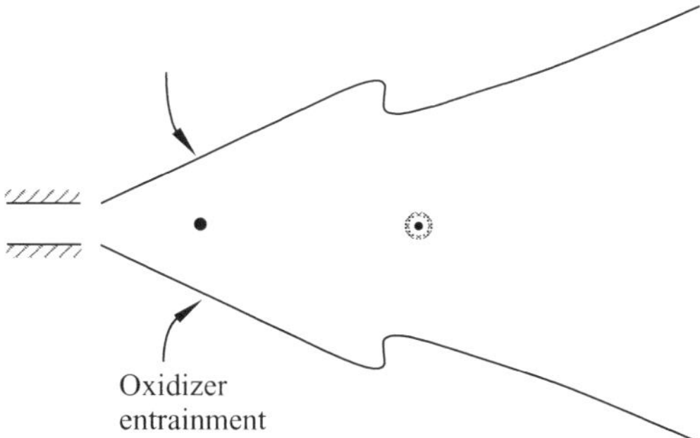

Figure 13.6.2. Schematic of spray jet combustion.

reach the bulk gaseous medium of the spray. Thus the droplet combustion mode will persist once established.

However, if the droplet vaporization rate is sufficiently fast or the entrainment rate sufficiently slow, then the spray interior will be rapidly cooled and further enriched with the fuel vapor. Both these factors inhibit droplet combustion. It is then likely that combustion will take place at the spray sheath, which will cut off further oxygen supply to the spray interior needed for droplet burning. Thus the jet sheath combustion mode will persist once established.

For coal dust flames, since the coal particles are quite nonvolatile as compared with oil droplets, oxidizer entrainment is expected to dominate, hence favoring the particle burning mode.

13.6.3. Cloud and Dense Spray Combustion

The above discussion demonstrates that single droplet combustion is not favored to occur if the droplets are sufficiently close to each other and/or the spray interior is sufficiently oxidizer lean. Under such situations the spray burns either with a single flame surrounding it or with clusters of flames enveloping groups of droplets. Such a combustion mode has been termed cloud combustion or group combustion (Chiu et al. 1978; Tishkoff 1979; Chiu, Kim & Croke 1983; Annamalai & Ryan 1992, 1993). Chiu and coworkers have proposed the use of a group combustion number, $G \sim N(d_{s,o}/\ell)$, to represent the relative tendencies for either single droplet combustion or group combustion, where $N = n\ell^3$ is the total number of droplets, n the droplet number density, and ℓ a characteristic dimension of the spray. It is clear that group combustion is favored for large G while single droplet combustion is favored for small G.

The definition of G is strictly geometrical in nature and does not include effects due to droplet vaporization and the heat transfer rate to the droplet cloud. In a sense it is similar to the idea leading to the derivation of Eq. (13.6.4). Nevertheless we expect that the trend in the cloud behavior with increasing or decreasing G should still hold. Thus for a spherical cloud four combustion modes can be identified, as shown in

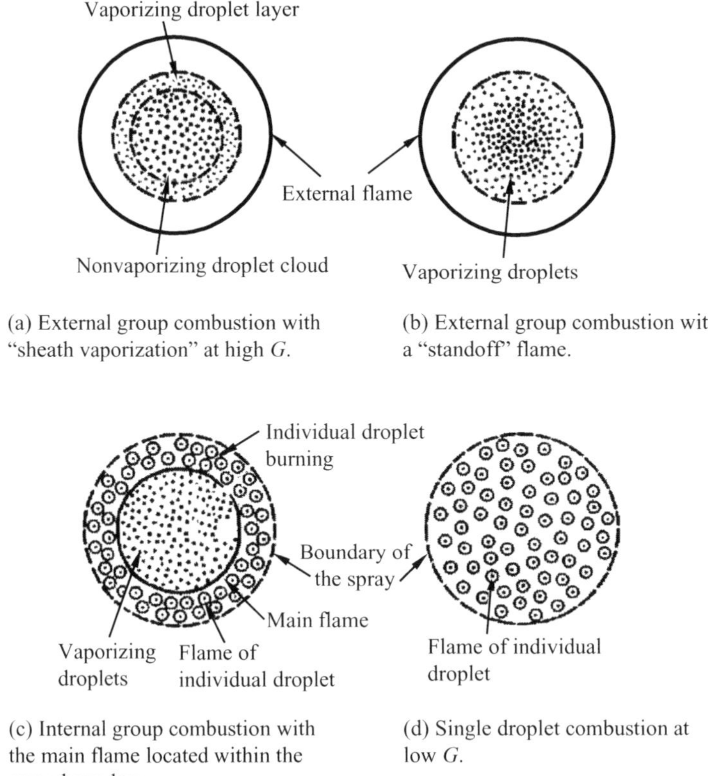

Figure 13.6.3. Schematic showing the four group combustion modes of a spherical droplet cloud with decreasing G from (a) to (d) (Chiu, Kim & Croke 1983).

Figure 13.6.3. Specifically, Figure 13.6.3a shows that for very high values of G, the inner core of the droplet cloud is fully saturated, with vaporization taking place only within a thin layer at the outer edge of the cloud. The vaporized fuel then diffuses outward and reacts with the inwardly diffusing oxidizer gas at a cloud diffusion flame. By continuously reducing G, this vaporizing droplet layer becomes progressively thicker until all the droplets in the cloud participate in vaporization, as shown in Figure 13.6.3b. With further reduction in G, the group diffusion flame moves into the cloud, resulting in an internal combustion mode, in which individual droplet combustion occurs outside the group diffusion flame while droplet vaporization occurs inside it. Finally, for very small values of G, individual droplet combustion prevails.

Figure 13.6.4 is the analog of Figure 13.6.3 for the spray jet configuration. It is seen that the continuous penetration of the jet is equivalent to a gradual lowering of G in terms of the spray combustion mode.

While further development of the group combustion concept is needed to enable a quantitative assessment of the prevailing burning mode of a spray, it is reasonable to expect that the region immediate downstream of the spray injection and formation processes must be sufficiently cold and fuel rich that individual droplet combustion is not possible. Processes occurring within this dense spray region are crucial

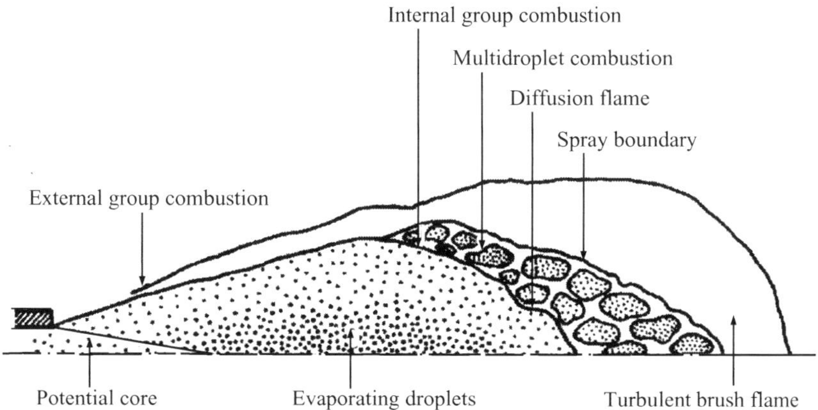

Figure 13.6.4. Schematic showing the group combustion concept of a spray jet (Chiu et al. 1978).

because it is precisely here the spray flame is likely to be anchored. In addition to droplet vaporization, droplet collision, coalescence, and shattering are also expected to be important in this region because of the high droplet velocity acquired through injection and the proximity between the droplets.

From the above discussion we have identified several important issues when analyzing spray combustion. These include the need to assess the ignitability of the droplet versus the spray sheath, the relative extents of heterogeneous droplet combustion versus homogeneous spray sheath combustion, and the relative extents of diffusional burning around droplets and the spray sheath as versus premixed burning taking place within the gaseous medium of the spray interior. These issues, which are largely unanswered at present, directly influence such practical concerns as flame stabilization, the fuel burning rate, and the extent of pollutant generation as in the case of NO_X formation, which is favored for nonpremixed burning because of its high flame temperature.

In the following we shall first discuss the statistics of sprays and the governing equations for two-phase flows. We shall then solve two problems involving one-dimensional sprays, one is on spray vaporization and the other on spray combustion supported only by droplet burning. These two problems demonstrate certain aspects of spray modeling as well as some useful physical insights.

13.7. FORMULATION OF SPRAY COMBUSTION

13.7.1. Spray Statistics

13.7.1.1. Spray Distribution Function: In the kinetic theory of gases, a velocity distribution function $f(\mathbf{x}, \mathbf{u}, t)$ is defined such that $f(\mathbf{x}, \mathbf{u}, t)d\mathbf{x}d\mathbf{u}$ represents the probable number of molecules lying within the spatial range $\mathbf{x}$ and $\mathbf{x} + d\mathbf{x}$, with velocities in the range $\mathbf{u}$ and $\mathbf{u} + d\mathbf{u}$, at time t. Based on the same concept, a spray distribution function $f(r_s, \mathbf{x}, \mathbf{u}, t)$ has been defined (Williams 1985) such that

$$f(r_s, \mathbf{x}, \mathbf{u}, t)dr_s d\mathbf{x}d\mathbf{u}$$

represents the number of droplets with radius within the range r_s and $r_s + dr_s$, located within the spatial range $\mathbf{x}$ and $\mathbf{x} + d\mathbf{x}$, and having velocities within the range $\mathbf{u}$ and $\mathbf{u} + d\mathbf{u}$, at time t.

The need to introduce r_s in the spray distribution function is obvious in that, unlike the point masses assumed in the kinetic theory of gases, the "particles" in the spray have finite dimension and their variation is a central concern in our study. The use of a single parameter, the radius r_s, to represent the structure of the droplet assumes that it is spherical in shape. This assumption is considered adequate because droplets in practical sprays are invariably very small such that the strong surface tension force promotes sphericity. The parameter indicating the tendency of the droplet to deviate from spherical symmetry and eventually break up is the gas-phase Weber number, $We = 2r_s\rho_g|\mathbf{u} - \mathbf{v}|^2/\sigma$, which is the ratio of the dynamic force to the surface tension force, where $|\mathbf{u} - \mathbf{v}|$ is the velocity difference between the droplet velocity $\mathbf{u}$ and the gas velocity $\mathbf{v}$, and ρ_g the gas density. For $We \gtrsim 20$, droplets tend to deform and break up.

For solid particles such as those of coal, spherical symmetry obviously does not exist, at least not before massive pyrolysis has occurred. In such cases it is sometimes possible to define an equivalent particle radius, provided that the different dimensions of the particle do not deviate too much from each other.

13.7.1.2. Spray Equation: A spray equation describing the change of spray distribution function $f(r_s, \mathbf{x}, \mathbf{u}, t)$ can be readily derived in the same manner as that of the control volume derivation for the conservation of mass, momentum, and energy discussed in Section 5.1, accounting for the fact that the control volume is now a seven-dimensional space in $(r_s, \mathbf{x}, \mathbf{u})$. Its derivation is also analogous to that of the Boltzmann equation governing the evolution of the velocity distribution function (Hirschfelder, Curtiss & Bird 1954). Thus we have (Williams 1985)

$$\frac{\partial f}{\partial t} + \frac{\partial}{\partial r_s}(\dot{r}_s f) + \nabla_{\mathbf{x}} \cdot (\mathbf{u} f) + \nabla_{\mathbf{u}} \cdot (\mathbf{g} f) = S, \tag{13.7.1}$$

where $\dot{r}_s = dr_s/dt$, $\mathbf{g}$ is the force per unit mass, and S is the source term. Note that $\mathbf{g}$ here includes body forces as well as surface forces such as the drag force. The function $\dot{r}_s$ is supplied from studies on droplet vaporization and combustion. When d^2-law holds, we simply have $\dot{r}_s \sim 1/r_s$.

The source term S represents the creation and destruction of droplets. As examples, droplets can be created by nucleation in a highly saturated vapor region, by aerodynamic shattering or stripping of liquid jets and droplets with sufficiently large Weber numbers, or by droplet microexplosion discussed earlier. On the other hand, droplets can be "destroyed" through coalescence either with other droplets, especially in the dense regions of the spray, or with the walls of the combustor. All these processes are not well understood and quantified, thereby making prescription of the source term difficult. For example, as we have seen, collision between two droplets does not always result in coalescence. The same applies to droplets impacting a wall, which is usually wetted by the deposition of previous droplets.

The spray distribution function and the associated spray equation enjoy far less fundamental significance and utility than the velocity distribution function of the Boltzmann equation. For example, the Boltzmann equation yields the Maxwell velocity distribution in the limit of thermodynamic equilibrium, which is an excellent assumption for most problems on gasdynamics including combustion. However, the velocity component of the spray distribution function is highly system dependent, being sensitive functions of the atomization process and the subsequent droplet dynamics.

13.7.1.3. Droplet Size Distribution Functions: Since the lifetime of a droplet varies quadratically with its initial size, the burning rate of a spray also depends sensitively on the initial size of its droplets, which usually varies over wide ranges. For example, droplet sizes within gas turbines typically range in diameters from 10 to 100 μm.

The distribution of the initial droplet size is invariably empirically determined. Because of the discrete nature of the experimental measurement, the distribution is usually represented as a histogram. By continuously reducing the measurement interval of the droplet size, a smooth function, $G(r_s)$, called the droplet size distribution function, can be determined. Frequently it is necessary to truncate $G(r_s)$ at a maximum droplet radius $r_{s,\max}$. The need to do so is due to the fact that $G(r_{s,\max})$ usually is not vanishingly small. Because of volume effect the tail end of the distribution function can represent a disproportionately large amount of the liquid mass contained within the spray. Thus it is important to have an accurate accounting of the droplet population within this large size regime.

Once $G(r_s)$ is known, other spray properties can be defined. For example the integrals

$$\int_0^{r_s} G(r_s)dr_s \quad \text{and} \quad \frac{4\pi}{3}\int_0^{r_s} r_s^3 G(r_s)dr_s,$$

respectively represent the total number and volume of the droplets having radii smaller than r_s.

A useful parameter in spray studies is the Sauter Mean Diameter (SMD), defined in discrete and continuous forms as

$$2\frac{\sum n_i r_{s,i}^3}{\sum n_i r_{s,i}^2} \quad \text{and} \quad 2\frac{\int_0^{r_{s,\max}} r_s^3 G(r_s)dr_s}{\int_0^{r_{s,\max}} r_s^2 G(r_s)dr_s} \tag{13.7.2}$$

respectively, where $r_{s,i}$ is the droplet radius in the ith-size class that has n_i droplets, and $r_{s,\max}$ is sometimes replaced by infinity here. SMD, which is the ratio of the total droplet volume to the total droplet surface area, is an appropriate representation of the characteristic droplet size because it relates the total amount of liquid fuel that needs to be gasified to the total surface area available for gasification.

A function that well describes the measured droplet size distribution for a variety of sprays is the generalized Rosin–Rammler distribution function, given by

$$G(r_s) = b(r_s)^t \exp[-a(r_s)^s], \tag{13.7.3}$$

where a, b, s, and t are the constant fitting parameters. The special case of $t = 2$ is the Nukiyama–Tanasawa distribution, $t = s - 4$ is the Rosin–Rammler distribution, and $s = 1$ is the chi-square distribution.

The fact that Eq. (13.7.3) can satisfactorily describe the monomodal size distribution of droplets and particles is not surprising because the four adjustable constants constitute the minimum number of parameters needed to describe the four basic characteristics of such a distribution, namely the total number of droplets, the average droplet size, the standard deviation, and the skewness, which are respectively given by

$$n = \int_0^\infty G dr_s, \qquad < r_s > = \frac{1}{n}\int_0^\infty r_s G dr_s,$$

$$\sigma = \frac{1}{n}\int_0^\infty (r_s - < r_s >)^2 G dr_s, \qquad s\sigma^3 = \frac{1}{n}\int_0^\infty (r_s - < r_s >)^3 G dr_s.$$

13.7.2. Conservation Equations

In a two-phase mixture, because of the volume occupied by the droplets, it is necessary to distinguish the fluid density ρ_f, which is the mass of the gas per unit volume of physical space, from the gas density ρ_g. They are related through

$$\rho_f/\rho_g = 1 - \int\int \left(\frac{4}{3}\pi r_s^3\right) f dr_s d\mathbf{u}. \tag{13.7.4}$$

The conservation equations are given by the following.

Overall Continuity for Gas:

$$\frac{\partial \rho_f}{\partial t} + \nabla_{\mathbf{x}} \cdot (\rho_f \mathbf{v}) = -\int\int \rho_\ell \left(4\pi r_s^2\right) \dot{r}_s f dr_s d\mathbf{u}, \tag{13.7.5}$$

where the source term on the RHS represents the reduction in the liquid mass due to droplet gasification, causing the variation in ρ_f.

Conservation of Momentum:

$$\rho_f \frac{\partial \mathbf{v}}{\partial t} + \rho_f \mathbf{v} \cdot \nabla_{\mathbf{x}} \mathbf{v} = -(\nabla_{\mathbf{x}} \cdot \mathbf{P}) + \rho_f \sum_{i=1}^{N} Y_i f_i - \int\int \rho_\ell \left(\frac{4}{3}\pi r_s^3\right) \mathbf{g} f dr_s d\mathbf{u}$$
$$+ \int\int \rho_\ell (4\pi r_s^2) \dot{r}_s (\mathbf{u} - \mathbf{v}) f dr_s d\mathbf{u}, \tag{13.7.6}$$

where f_i is the force per unit mass acting on species i in the gas. Furthermore, the first source term, involving $\mathbf{g}$, represents force per unit volume exerted on the droplets by the gas, while the second source term, involving $\dot{r}_s$, represents the momentum transferred to the gas by the vaporizing mass. For droplets in sprays, the dominant force contributing to **g** is that due to drag, given by

$$\mathbf{g} = \frac{3}{8}\frac{\rho_g}{\rho_\ell}\frac{|\mathbf{v}-\mathbf{u}|(\mathbf{v}-\mathbf{u})}{r_s}C_D, \tag{13.7.7}$$

where C_D is the drag coefficient, which is a function of the Reynolds number, $Re = 2r_s\rho_g|\mathbf{u}-\mathbf{v}|/\mu_g$, as well as the mass transfer number B_m, where μ_g is the gas viscosity.

Conservation of Energy:

$$\begin{aligned}&\frac{\partial}{\partial t}\left[\rho_f\left(h_f+\frac{v^2}{2}\right)\right]+\nabla_{\mathbf{x}}\cdot\left[\rho_f\mathbf{v}\left(h_f+\frac{v^2}{2}\right)\right]\\&=-\nabla_{\mathbf{x}}\cdot\mathbf{q}-\mathbf{P}:(\nabla\mathbf{v})+\frac{\partial p}{\partial t}+\rho_f\sum_{i=1}^{N}Y_i\mathbf{f}_i\cdot(\mathbf{v}+\mathbf{V}_i)\\&-\int\int\rho_\ell\left(\frac{4}{3}\pi r_s^3\right)(\mathbf{g}\cdot\mathbf{u})\,f\,dr_s d\mathbf{u}-\int\int\rho_\ell\left(4\pi r_s^2\right)\dot{r}_s\left(h+\frac{u^2}{2}\right)f\,dr_s d\mathbf{u},\end{aligned} \tag{13.7.8}$$

where h_f is the total enthalpy per unit mass of the gas, and the meanings of the two source terms are analogous to those in momentum conservation.

Conservation of Species:

$$\frac{\partial}{\partial t}(\rho_f Y_i)+\nabla_{\mathbf{x}}\cdot[\rho_f(\mathbf{v}+\mathbf{V}_i)Y_i]=w_i-\int\int\rho_\ell\left(4\pi r_s^2\right)\dot{r}_s\Omega_i\,f\,dr_s d\mathbf{u}, \tag{13.7.9}$$

where Ω_i is the radial mass flux of species i. Thus in a chemically reacting flow, creation and destruction of chemical species can arise either from reactions in the bulk gaseous medium and/or from droplet vaporization or combustion.

13.8. ADIABATIC SPRAY VAPORIZATION

We consider the vaporization and transport of an adiabatic spray in a quasi-one-dimensional chamber with cross-sectional area $A(x)$ (Law 1977b). For simplicity we shall assume that the spray is monodisperse in that all the droplets have the same size, and there is no velocity lag between the droplets and the gas.

In Section 6.4.2 we have studied the general vaporization characteristics of an isolated droplet in an unbounded environment of constant temperature and concentration of the vaporizing species, say T_∞ and Y_∞. In a spray interior the ambience the droplet perceives is actually the inter-droplet spacing whose properties are represented by those of the bulk gaseous medium, say T_g and Y_g. If the droplets are sufficiently far apart, we can approximately identify T_∞ and Y_∞ by T_g and Y_g respectively. However, unlike the case of an isolated droplet, the collective vaporization of the droplets will continuously chill the gas medium and also enrich it with the vaporizing species, thereby causing T_g to decrease and Y_g to increase. It is therefore conceivable that for a spray that is initially quite cold and/or fuel rich, a state of saturation can be reached such that vaporization is completely terminated. A practical example is the vaporization and transport of the spray in the region immediately downstream of the spray injector.

From Section 6.4.2 we have shown that the nondimensional droplet vaporization rate $\tilde{m} = m/[4\pi(\lambda/c_p)r_s]$ and the vapor concentration at the droplet surface Y_s are respectively given by

$$\tilde{m} = \ln(1 + B_v) \tag{13.8.1}$$

$$Y_s = \frac{B_v + Y_g}{1 + B_v}, \tag{13.8.2}$$

where

$$B_{h,v} = \frac{c_p(T_g - T_s)}{q_v + q_\ell} \tag{13.8.3}$$

is the heat transfer number for vaporization, T_s the droplet temperature, and q_ℓ the amount of heat used to heat up the droplet interior per unit mass of the droplet vaporized.

In the above relations, T_g, Y_g, and T_s continuously change as vaporization proceeds, and are to be determined through the conservation relations of energy and vapor for the entire mixture,

$$(1 + \eta Y_{\ell,o})c_p T_g + (1 - \eta)Y_{\ell,o}(c_p T_s - q_v) = c_p T_{g,o} + Y_{\ell,o}(c_p T_{s,o} - q_v) \tag{13.8.4}$$

$$Y_g = \frac{Y_{g,o} + \eta Y_{\ell,o}}{1 + \eta Y_{\ell,o}}, \tag{13.8.5}$$

where $Y_{\ell,o}$ and $Y_{g,o}$ are the initial mass fractions of the liquid and vapor in the mixture, $\eta = 1 - R_s^3$ the fractional amount of liquid vaporized and can be termed the evaporation efficiency of the mixture, $R_s = r_s/r_{s,o}$, and we have also assumed the same c_p for liquid and gas.

Equations (13.8.4) and (13.8.5) allow us to assess a priori the ability of the spray to achieve complete vaporization. Since vaporization occurs only for $T_g > T_s$ and $Y_g < Y_s$, where Y_s is the vapor concentration at the surface, at complete spray vaporization, $\eta = 1$, we must have

$$T_g^* = \frac{T_{g,o} + Y_{\ell,o}(T_{s,o} - q_v/c_p)}{1 + Y_{\ell,o}} > T_s^* \tag{13.8.6}$$

$$Y_g^* = \frac{Y_{g,o} + Y_{\ell,o}}{1 + Y_{\ell,o}} < Y_s^*, \tag{13.8.7}$$

where the superscript * designates the completely vaporized state. Furthermore, since T_s and Y_s are related through the Clausius–Clapeyron vapor pressure relation,

$$Y_s = Y_s(T_s), \tag{13.8.8}$$

such that $Y_s^* = Y_s(T_s^*)$ at $\eta = 1$, the inequalities of (13.8.6) and (13.8.7) imply that

$$Y_g^* < Y_s(T_g^*) \tag{13.8.9}$$

must hold for complete vaporization to be possible.

To trace through the spray vaporization history, we need to know quite accurately the variation of the droplet temperature T_s because the droplet vaporization rate depends sensitively on T_s, especially for slow rates of vaporization. It is, however, reasonable to anticipate that, except for an initial transient period, T_s may not vary much during vaporization. This is due to the fact that T_s, which is just the wet-bulb temperature, decreases with decreasing ambient temperature but increases with increasing ambient vapor concentration, as shown in Figure 6.3.2. Consequently, the net effect of a decreasing T_g and increasing Y_g on T_s, as vaporization proceeds, can be small.

Thus if we assume that the droplets attain an equilibrium temperature $T_{s,e}$, then using Eqs. (13.8.4) and (13.8.5) it can be readily shown that Y_s and B of Eqs. (13.8.2) and (13.8.3) are given by

$$Y_{s,e} = \frac{Y_{g,o} + B_{v,e}}{1 + B_{v,e}}, \tag{13.8.10}$$

$$1 + B_v = \frac{1 + B_{v,e}}{1 + \eta Y_{\ell,o}} \tag{13.8.11}$$

where

$$B_{v,e} = (\tilde{T}_{g,o} - \tilde{T}_{s,e}) + (\tilde{T}_{s,o} - \tilde{T}_{s,e})Y_{\ell,o}, \tag{13.8.12}$$

and $\tilde{T} = c_p T / q_v$. It is significant to note that since $Y_{s,e}$ is a constant, independent of η, an equilibrium droplet temperature $T_{s,e}$, as originally assumed, indeed exists. $T_{s,e}$ can be solved by equating Eq. (13.8.10) to the Clausius–Clapeyron relation Eq. (13.8.8). It is important to emphasize that this equilibrium droplet temperature, attained in an adiabatic environment of continuously changing temperature and vapor concentration, is not the wet-bulb temperature attained in a constant environment.

Since vaporization terminates with $B_v = 0$, an alternate criterion for complete spray vaporization ($\eta = 1$) is given by Eq. (13.8.11) as

$$B_{v,e} > Y_{\ell,o}. \tag{13.8.13}$$

If Eq. (13.8.13) is not satisfied, vaporization will terminate with $B_v = 0$ and, hence,

$$\eta_e = B_{v,e} / Y_{\ell,o}, \tag{13.8.14}$$

which corresponds to a final unvaporized droplet size

$$R_{s,e} = (1 - B_{v,e} / Y_{\ell,o})^{1/3}. \tag{13.8.15}$$

To study the transport of the spray, we have, from d^2-law,

$$\frac{dr_s^2}{dt} = -\frac{2(\lambda/c_p)}{\rho_\ell}\ln(1+B_v). \tag{13.8.16}$$

Since there is no velocity lag between the droplets and the gas, we can express $d/dt = ud/dx$, such that Eq. (13.8.16) becomes

$$\tilde{u}R_s\frac{dR_s}{d\tilde{x}} = \ln(1+B_v), \tag{13.8.17}$$

where $\tilde{u} = u/u_o$ and $\tilde{x} = [(\lambda/c_p)/(u_o r_{s,o}^2\rho_\ell)]x$ are the nondimensional velocity and distance respectively.

If the transport occurs in a quasi-one-dimensional channel with slowly varying cross-sectional area $A(x)$, and if we assume that the density of the mixture does not change much, then continuity gives

$$\tilde{u}\tilde{A} = 1, \tag{13.8.18}$$

where $\tilde{A} = A/A_0$. Substituting Eq. (13.8.18) into Eq. (13.8.17) and integrating, we obtain

$$C(\tilde{x}) = \int_{R_s}^{1}\frac{R_s}{\ln(1+B_v)}dR_s, \tag{13.8.19}$$

where we have defined a chamber function

$$C(\tilde{x}) = \int_0^{\tilde{x}}\tilde{A}(\tilde{x})d\tilde{x}, \tag{13.8.20}$$

which is known once $\tilde{A}(x)$ is given. For a chamber of constant cross-sectional area, $C(\tilde{x}) = \tilde{x}$.

Equation (13.8.19) can be numerically integrated by using Eq. (13.8.11),

$$C(\tilde{x}) = \int_{R_s}^{1}\frac{R_s}{\ln[(1+B_{v,e})/(1+\eta Y_{\ell,o})]}dR_s. \tag{13.8.21}$$

If we further assume that the spray is dilute, with $Y_{\ell,o} \ll 1$, then Eq. (13.8.21) can be readily integrated to yield

$$C(\tilde{x}) = \frac{(1+\gamma)\left(1-R_s^2\right) - 0.4\gamma\left(1-R_s^5\right)}{2\tilde{m}_e}, \tag{13.8.22}$$

where $\tilde{m}_e = \ln(1+B_{v,e})$ and $\gamma = Y_{\ell,o}/\tilde{m}_e$. Equation (13.8.22) explicitly relates the droplet size with the distance along the evaporator. The minimum chamber length $\tilde{x}^*$ needed to achieve complete spray vaporization, $R_s = 0$, is then given by

$$C(\tilde{x}^*) = \frac{1+0.6\gamma}{2\tilde{m}_e}. \tag{13.8.23}$$

For a chamber with constant cross-sectional area, the minimum chamber length in dimensional form is

$$x^* = \frac{u_o r_{s,o}^2\rho_\ell}{2(\lambda/c_p)\tilde{m}_e}\left(1+0.6\frac{Y_{\ell,o}}{\tilde{m}_e}\right), \tag{13.8.24}$$

which shows that $x^* \sim r_{s,o}^2$. The importance of minimizing the chamber length by using small droplets is again demonstrated.

13.9. HETEROGENEOUS LAMINAR FLAMES

Analogous to the standard, one-dimensional laminar flame propagation in a premixed gaseous combustible, which is characterized by the laminar flame speed s_u^o, a corresponding flame speed can also be defined for the propagation of a combustion wave in a two-phase mixture. The propagation is supported by homogeneous burning in the bulk gaseous medium and possibly also nonpremixed burning around the droplets. The relative extent of homogeneous versus heterogeneous burning depends on the initial liquid loading, initial droplet size, and the liquid volatility. For mixtures whose initial heterogeneity is low, for example if the initial droplet size is less than 10 μm as estimated previously, the flame basically behaves as a gaseous flame. In this section, we study the other limit in which the initial mixture does not have any fuel vapor and droplet prevaporization before reaching the flame front is also totally suppressed. At the bulk flamefront the droplets are instantly ignited and subsequently burn according to the d^2-law, but with the burning rate dependent on the instantaneous values of the temperature and oxygen concentration of the gas downstream of the flamefront.

A heterogeneous flame speed can be estimated in the manner of the homogeneous premixed flame of Section 7.2. Thus if $(1+\sigma)$ is the mass of the mixture reacted per unit mass of fuel burned, where σ is the oxidizer-to-fuel mass ratio, then the mixture reaction rate can be estimated as

$$w \approx (1+\sigma)\int_0^\infty \rho_\ell\left(4\pi r_s^2\right)(-\dot{r}_s)G dr_s. \tag{13.9.1}$$

Since

$$\dot{r}_s = \frac{dr_s}{dt} = \frac{1}{2r_s}\frac{dr_s^2}{dt} = -\frac{K_c/2}{r_s}, \tag{13.9.2}$$

where $K_c = 2[(\lambda/c_p)/\rho_\ell]\ln(1+B_{h,c})$ as given by Eq. (6.4.41), and if we assume a monodisperse spray with a uniform droplet size r_s, then

$$w \approx 2\pi(1+\sigma)n_o r_s \rho_\ell K_c, \tag{13.9.3}$$

which shows that w decreases as the droplet size diminishes. Thus using this w as the characteristic reaction rate in the expression for the laminar burning flux, $f^2 = (\rho_o u_o)^2 = (\lambda/c_p)w$, with $r_s \to r_{s,o}$ as the characteristic droplet size, we have

$$f^2 = 2\pi(1+\sigma)n_o r_{s,o}\rho_\ell(\lambda/c_p)K_c \tag{13.9.4a}$$

$$= 4\pi(1+\sigma)n_o r_{s,o}(\lambda/c_p)^2\ln(1+B_{h,c}). \tag{13.9.4b}$$

Furthermore, since $n_o\rho_\ell\left(\frac{4}{3}\pi r_{s,o}^3\right) =$ constant is the initial mass of the liquid, $n_o \sim (r_{s,o})^{-3}$, and we have

$$f \sim \frac{\lambda/c_p}{r_{s,o}}. \tag{13.9.5}$$

Thus the heterogeneous burning flux is inversely proportional to $r_{s,o}$, and is proportional to λ/c_p, which demonstrates the basic diffusive nature of burning.

We shall next demonstrate how the problem can be solved more rigorously (Williams 1985).

13.9.1. Gas-Phase Flames

With the assumptions of monodispersity and no velocity lag between the gas and droplets, we have

$$f = \rho u = \rho_o u_o, \tag{13.9.6}$$

$$nu = n_o u_o, \tag{13.9.7}$$

which show that $n/\rho = n_o/\rho_o$ is a conserved quantity. We shall also use the mass fraction of the gas $Z = \rho_f/\rho$ as the quantity to indicate the extent of heterogeneity; $Z = 1$ for a completely gaseous mixture. Thus, with

$$\rho = \rho_f + n\rho_\ell\left(\frac{4}{3}\pi r_s^3\right), \tag{13.9.8}$$

we can divide Eq. (13.9.8) by ρ to get

$$1 = Z + \frac{n\rho_\ell}{\rho}\left(\frac{4}{3}\pi r_s^3\right) = Z_o + \frac{n_o\rho_\ell}{\rho_o}\left(\frac{4}{3}\pi r_{s,o}^3\right). \tag{13.9.9}$$

Equation (13.9.9) shows that

$$\frac{1-Z}{1-Z_o} = \left(\frac{r_s}{r_{s,o}}\right)^3 = R_s^3, \tag{13.9.10}$$

from which we can also obtain the relation

$$\frac{dZ}{dx} = \frac{2\pi\rho_\ell u_o n_o r_{s,o}}{f} K_c\left(\frac{1-Z}{1-Z_o}\right)^{1/3}, \tag{13.9.11}$$

where we have used the d^2-law relation

$$\frac{dr_s}{dt} = \frac{dx}{dt}\frac{dr_s}{dx} = u\frac{dr_s}{dx} = -u\frac{K_c/2}{r_s}, \tag{13.9.12}$$

with

$$K_c = \frac{2(\lambda/c_p)}{\rho_\ell}\ln\left[1 + \frac{c_p(T_g - T_s) + q_c Y_{O,g}/\sigma}{q_v}\right]. \tag{13.9.13}$$

Equation (13.9.13) shows that K_c depends on the temperature T_g and oxygen concentration $Y_{O,g}$ in the spray interior. The variable u in Eq. (13.9.11) can be related to the thermodynamic variables T and Z as

$$\frac{u}{u_o} = \frac{\rho_o}{\rho} = \frac{\rho_{f,o}}{\rho_f}\frac{Z}{Z_o} \approx \frac{\rho_{g,o}}{\rho_g}\frac{Z}{Z_o} = \frac{T_g}{T_{g,o}}\frac{Z}{Z_o}, \tag{13.9.14}$$

where we have approximated $\rho_f = \rho_g$ and have assumed that the average molecular weight remains the same.

For species conservation, we have from Eq. (13.7.9)

$$\frac{d}{dx}\left(\rho_f u Y_{g,i} - \rho_f D\frac{dY_{g,i}}{dx}\right) = -\Omega_i \frac{d(\rho_\ell u)}{dx}, \tag{13.9.15}$$

where $\Omega_F = 0, \Omega_O = -\sigma$, and we have used Fick's law of diffusion, Eq. (5.2.16). Integrating Eq. (13.9.15), using the relations $\rho_f u = fZ$, $\rho_\ell u = f(1 - Z)$, and evaluating the integration constant at the upstream state, we have

$$\frac{\rho_f D}{f}\frac{dY_{g,i}}{dx} = Z(Y_{g,i} - \Omega_i) - Z_o(Y_{g,i,o} - \Omega_i). \tag{13.9.16}$$

Finally, energy conservation is governed by

$$\frac{d}{dx}(\rho_f u h_g + \rho_\ell u h_\ell + q_x) = 0. \tag{13.9.17}$$

To evaluate Eq. (13.9.17), we note that the enthalpies of the gas and liquid are respectively

$$h_g = \sum_{i=1}^{N} Y_{g,i} h_{g,i}, \tag{13.9.18}$$

$$h_\ell = h_\ell^o + c_p(T_\ell - T^o), \tag{13.9.19}$$

where

$$h_{g,i} = h_{g,i}^o + c_p(T_g - T^o), \tag{13.9.20}$$

h_ℓ^o is related to $h_{F,i}^o$ through the latent heat of gasification q_v,

$$h_\ell^o = h_{g,F}^o - q_v, \tag{13.9.21}$$

and we have assumed the same c_p for the gas and liquid. Furthermore, the heat flux vector, given by Eq. (5.2.7), is

$$q_x = -\lambda\frac{dT_g}{dx} - \rho_f D\sum_{i=1}^{N} h_{g,i}^o \frac{dY_{g,i}}{dx}. \tag{13.9.22}$$

Thus the integrated form of Eq. (13.9.17) is

$$\sum_{i=1}^{N} h^o_{g,i} Z Y_{g,i} - Z\left(h^o_{g,F} - q_v\right) + c_p(T_\ell - T_{\ell,o}) + Z c_p(T_g - T_\ell)$$

$$\frac{\lambda}{f}\frac{dT_g}{dx} \quad \frac{\rho_f D}{f} \sum_{i=1}^{N} h^o_{g,i} \frac{dY_{g,i}}{dx}$$

$$= \sum_{i=1}^{N} h^o_{g,i} Z_o Y_{g,i,o} - Z_o\left(h^o_{g,F} - q_v\right) + Z_o c_p(T_{g,o} - T_{\ell,o}), \qquad (13.9.23)$$

where the integration constant is evaluated at the upstream state where dT_g/dx and $dY_{g,i}/dx$ vanish.

Substituting Eq. (13.9.16) into Eq. (13.9.23) yields

$$\frac{\lambda}{f}\frac{dT_g}{dx} = Z(c_p T_g - q') - Z_o(c_p T_{g,o} - q'_o) + c_p(T_\ell - T_{\ell,o}), \qquad (13.9.24)$$

where

$$q' = q_c - q_v + c_p T_\ell, \qquad (13.9.25)$$

in which $q_c = h^o_{g,F} - \sum h^o_{g,i}\Omega_i$ is the heat of reaction per unit mass of fuel.

Two observations can be made at this stage. First, evaluating Eq. (13.9.24) at the downstream boundary, at which $dT_g/dx = 0$ and $Z = 1$, we obtain the downstream, adiabatic flame temperature,

$$T_{g,\infty} = Z_o T_o + (1 - Z_o) q'/c_p$$
$$= T_o + (1 - Z_o)(q_c - q_v)/c_p, \qquad (13.9.26)$$

where we have assumed for simplicity that $T_{\ell,o} = T_{g,o} = T_o$.

The second point is that the functional forms of Eq. (13.9.16) and (13.9.24) are very similar to each other. Indeed, if we assume unity Lewis number, or $\lambda/c_p = \rho_f D$, then T_g and $Y_{g,i}$ are linearly related through $T_g = a_i + b_i Y_{g,i}$. Substituting this expression into Eq. (13.9.16), and comparing the resulting coefficients with those of Eq. (13.9.24) yields the coefficients a_i and b_i. Thus we have

$$Y_{g,i} = \Omega_i + (Y_{g,i,o} - \Omega_i)\left(\frac{q' - c_p T_g}{q' - c_p T_o}\right). \qquad (13.9.27)$$

Since $Y_{g,i}$ is known once T_g is determined, we only need to solve for T_g from Eq. (13.9.24) and Z from Eq. (13.9.11). The solution can be further facilitated by dividing Eq. (13.9.24) by Eq. (13.9.11), yielding

$$\frac{dT_g}{dZ} = \frac{Z(c_p T_g - q') - Z_o(c_p T_o - q')}{F(T_g, Z)}, \qquad (13.9.28)$$

where

$$F(T_g, Z) = \frac{4\pi\lambda^2 n_o r_{s,o}}{c_p f^2}\left(\frac{Z_o}{Z}\right)\left(\frac{T_o}{T_g}\right)\left(\frac{1 - Z}{1 - Z_o}\right)^{1/3} \ln\left[1 + \frac{c_p(T_g - T_o) + q_c Y_{O,g}/\sigma}{q_v}\right],$$

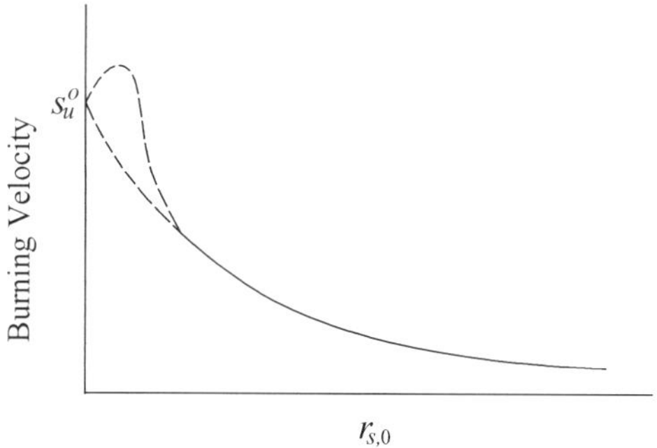

Figure 13.9.1. Schematic showing the dependence of the burning velocity on the initial droplet size (- heterogeneous limit; - - - two possible transitional behavior).

from which $T_g(Z)$ can be solved subject to $T_g = T_o$ at $Z = Z_o$ and $(dT_g/dZ) \to 0$ as $Z \to 1$. In analogy with the laminar premixed flame propagation, the solution here also yields a burning flux eigenvalue from which the burning flux f can be determined.

A plot of the heterogeneous flame speed u_o versus $r_{s,o}$ (Figure 13.9.1) shows the monotonically increasing trend as $r_{s,o}$ decreases, as indicated from the phenomenological result of Eq. (13.9.5). The fact that u_o varies inversely with the initial droplet size is entirely reasonable based on the d^2-law consideration. On the other hand, the present analysis must breakdown as $r_{s,o} \to 0$, at which homogeneous burning prevails with $u_o \to s_u^o$ and allowing for the presence of the latent heat of vaporization. At present it is not clear whether the approach to s_u^o is monotonic or nonmonotonic (Figure 13.9.1). In the latter case a maximum s_u^o could exist for an optimum droplet size because the stoichiometric, nonpremixed droplet burning should have a faster burning rate.

13.9.2. Condensed-Phase Flames

Next we shall discuss a rather interesting, and for a while puzzling, phenomenon in flame propagation observed in self-propagating high-temperature synthesis (SHS) of materials (Makino & Law 2001; Makino 2001). Let us consider the synthesis of a compound AB from mixed and compacted powders of a high-melting-point material A and a relatively low-melting-point material B, such as carbon and titanium respectively (Figure 13.9.2). As the compact is heated by the approaching flame front, the particles of B would melt, forming a suspension of particles of A in molten B. Reaction between A and B subsequently takes place at the surface of A. Because of the very high temperature involved in materials synthesis, the reaction is usually very fast and hence diffusion controlled. The product AB can be either in the solid phase or dissolve in the molten B and crystallize later when it is cooled down.

While the phenomenon of interest is that of heterogeneous, diffusion-controlled nonpremixed burning, most theoretical studies and interpretation of experimental results have been based on the premixed flame propagation mode (Merzhanov 1990).

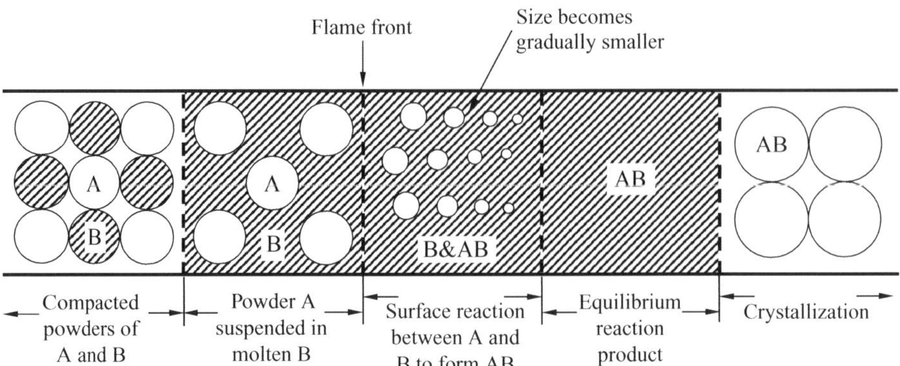

Figure 13.9.2. Schematic showing flame structure in the self-propagating high-temperature synthesis process (SHS).

Indeed, these results seem to support the notion of premixed burning in terms of the Arrhenius sensitivity of the propagation velocity and the extinction state on the adiabatic flame temperature, and the presence of pulsating and spinning flame fronts characteristic of large Lewis number mixtures. In fact, as mentioned in Section 10.9.3, early studies on pulsating instabilities suggested that they are observable only in condensed-phase flames because of the very small mass diffusivity, and hence large Lewis number, required for their onset.

To resolve this inconsistency in observation and interpretation, we compare the phenomenological laminar burning fluxes for the gaseous and heterogeneous flames, which are respectively given by Eqs. (7.2.8) and (13.9.4a),

$$f_{\text{gas}}^2 \sim (\lambda/c_p)w \sim (\lambda/c_p)e^{-T_a/T_{\text{ad}}}, \tag{13.9.29}$$

$$f_{\text{SHS}}^2 \sim (\lambda/c_p)K_c. \tag{13.9.30}$$

Furthermore, similar to the derivation of the d^2-law, it can be easily shown that the burning rate constant for a spherical particle A reacting at an infinitely fast rate with B at its surface is given by

$$K_c = (2\rho_{\text{B}} D/\rho_{\text{A}})\ln(1 + Y_{\text{B}}/\sigma_O). \tag{13.9.31}$$

The crucial point to recognize here is that liquid-phase mass diffusivity varies sensitively with temperature. Thus if we represent such a variation in an Arrhenius manner, with a correspondingly large activation temperature, say T_d (Bechtold & Margolis 1992), we can write

$$D \sim e^{-T_d/T}. \tag{13.9.32}$$

Using these results in Eq. (13.9.30), we have

$$f_{\text{SHS}}^2 \sim (\lambda/c_p)e^{-T_d/T_{\text{ad}}}. \tag{13.9.33}$$

Comparing Eq. (13.9.33) with Eq. (13.9.29), a complete correspondence in the temperature dependence is observed. It is therefore clear that the Arrhenius sensitivity of the chemical reactions is now mimicked by that of the liquid-phase mass diffusivity, rendering the diffusion-controlled heterogeneous flame to exhibit phenomena commonly associated with premixed flames.

PROBLEMS

1. Solve the droplet combustion problem by using the flame-sheet formulation. Do not invoke the $Le = 1$ assumption by using $\lambda_1, \lambda_2, c_{p,1}, c_{p,2}, (\rho D)_1, (\rho D)_2$ to designate the various transport coefficients, where the subscripts 1,2 respectively represent the inner and outer regions to the flame. First show that the first integrals of the governing equations yield

$$\left.\begin{aligned} m_c Y_F - 4\pi r^2 (\rho D)_1 \frac{dY_F}{dr} &= m_c \\ m_c c_{p,1}(T - T_s) - 4\pi r^2 \lambda_1 \frac{dT}{dr} &= -m_c q_v \end{aligned}\right\} r_s < r < r_f$$

$$\left.\begin{aligned} m_c Y_O - 4\pi r^2 (\rho D)_2 \frac{dY_O}{dr} &= -\sigma_O m_c \\ m_c c_{p,2}(T - T_s) - 4\pi r^2 \lambda_2 \frac{dT}{dr} &= -m_c[(q_v - q_c) - (c_{p,1} - c_{p,2})T_s] \end{aligned}\right\} r_f < r < \infty$$

where $m_c = 4\pi r^2 \rho u$. Discuss the physical meaning of the various integration constants. Use these equations to solve for $m_c, \tilde{r}_f$ and T_f as given by Eqs. (13.2.6) to (13.2.8).

2. We now examine in more detail the adequacy of the gas-phase quasi-steady assumption in droplet vaporization and combustion (Matalon & Law 1983, 1985). The problem is that of a droplet of radius r_s and temperature T_s undergoing gasification with a finite surface regression rate, dr_s/dt, in a stagnant environment with temperature T_∞. Assume $Le = 1$ and only work with the pure vaporization case as the implication is the same for burning.

(a) Write down the gas-phase governing equations for the problem.

(b) Nondimensionalize the governing equations based on the following scaling: initial droplet radius $r_{s,o}$ for length, ρ_∞ for density, T_∞ for temperature, $c_p T_\infty$ for energy, $\lambda/(c_p \rho_\infty r_{s,o})$ for velocity, and $r_{s,o}^2 \rho_\ell c_p/\lambda$ for time associated with the gas-phase diffusion processes. You will notice that through such a scaling the ratio $\delta^2 = \rho_\infty/\rho_\ell$ emerges as a system parameter. You can now observe that the condition under which the gas-phase transport can be assumed to be quasi-steady, as far as the governing equations are concerned, is for $\delta \ll 1$. The equations then degenerate to those used in our analysis of droplet vaporization and combustion, first encountered in Section 6.4.

(c) In addition to the result of (b), we also need to assure that the boundary conditions at the regressing surface can also be approximated as quasi-steady. This can be demonstrated by using the general jump relations derived in

Section 5.1.5. Derive these jump relations across the liquid–gas interface and make the appropriate observation. For example, the nondimensional continuity relation at the interface is

$$(1 - \delta^2 \rho)\frac{dr_s}{dt} = -\rho u,$$

which degenerates to Eq. (6.4.11) for $\delta \ll 1$.

3. Derive the d^3-law for the expanding vapor bubble within the agglomerated shell during slurry droplet gasification. Specifically, show that

$$\frac{dr_g^3}{dt} = \frac{3}{2} R_s K_c,$$

where r_g is the instantaneous radius of the vapor bubble, R_s the outer radius of the agglomerated shell, and K_c the droplet burning rate constant.

4. A monopropellant is a chemical that is capable to undergo exothermic reactions without the need for another reagent. The reaction can be either decomposition in nature or the monopropellant consists of both fuel and oxidizer components. Conceptually we can simply consider it as a premixture.

Thus consider a monopropellant droplet of radius r_s and constant temperature T_s that undergoes flame-sheet combustion in an inert environment of temperature T_∞. The monopropellant has a laminar burning flux f^o and adiabatic flame temperature T_{ad}. If the burning is adiabatic such that $T_\infty = T_{\text{ad}}$, determine the nondimensional droplet burning rate $\tilde{m}$ and the flamefront standoff ratio $\tilde{r}_f$. Show that droplet burning follows a d-law for fast reaction rates, and a d^2-law for slow reaction rates.

5. A spherical metal particle burns in an oxidizing environment with an infinitely fast surface reaction. Show that its burning rate constant is given by Eq. (13.9.31).

6. Some metals are excellent rocket fuels because of their large heat of combustion on either a mass or volume basis. Compared with hydrocarbon combustion, however, metal combustion is complicated by the possibility that the metal oxide formed frequently exist in condensed phase. For example, Glassman (1960) hypothesized that because of the large latent heat of vaporization of the metal oxide, the flame temperature T_f is suppressed to the boiling point of the oxide, say T_f^o, as long as condensed and vaporized metal oxide co-exist at the flame.

Formulate a model for metal droplet combustion with the following assumptions: (a) the flame temperature is at $T_f = T_f^o$; (b) a fraction θ of the product formed at the flame is vaporized with the rest $(1 - \theta)$ in condensed phase; (c) the condensed oxide can still diffuse; (d) the droplet is at the boiling point of the metal. Solve for the mass burning rate $\tilde{m}_c$, the flamefront standoff ratio $\tilde{r}_f$, and θ. Let $q_{c,1}$ and $q_{c,2}$ respectively designate the heat of combustion of the metal with oxygen to form the oxide vapor, and the condensation heat release as the oxide condenses. Assume $Le = 1$.

7. Another feature of metal particle combustion is the potential condensation of the oxide vapor at the particle surface, releasing the condensation heat that can be directly used by the metal particle for gasification. A similar situation can exist for the combustion of a methanol or ethanol droplet in which the water vapor produced at the flame diffuses back to the droplet surface where it condenses.

Rework the droplet combustion problem, but allow for the complete condensation of the product vapor at the droplet surface. Let $q_{v,F}$ and $q_{v,P}$ be the heats of vaporization of the fuel and the product respectively. Assume $Le = 1$ and T_s given.

14 Combustion in Supersonic Flows

In almost all the combustion problems studied so far, the flows are sufficiently subsonic such that $M \ll 1$, where M is the Mach number. There are, however, situations in which reactions take place in flows whose velocities can be sufficiently high such that they either are close to sonic or are supersonic. Examples are combustion within supersonic ramjet (scramjet) engines and the initiation and propagation of blast waves.

In nonreactive fluid mechanics such high-speed flows are called compressible flows because density now varies appreciably with the flow velocity. Such a density variation is to be distinguished from that caused by the large amount of heat release in reactive, low subsonic flows studied in previous chapters. It is therefore important to recognize that density can still vary significantly in an aerodynamically incompressible, low subsonic flow due to heat release. When it is actually assumed to be constant for such a flow, either due to the smallness of the heat release or for analytical expediency, then the flow is said to be one of constant density.

There are several fundamental differences between high-speed flows and low subsonic flows. First, the isobaric assumption of Section 5.2.4 ceases to hold. Second, the kinetic energy of the flow is now appreciable as compared to the chemical energy and frequently needs to be considered. Indeed, we have already encountered this issue in Section 12.5 on supersonic boundary-layer flows. Third, while diffusion is an essential process in low subsonic flows, convection frequently dominates over diffusion in high-speed flows except for situations involving steep gradients such as those within boundary layers. Since diffusion is represented by second-order differentials, which are the highest-order terms in the conservation equations, its insignificance in high-speed flows then completely changes the characteristics of the flow, from being elliptic or parabolic to hyperbolic. Physically, changes in the flow properties are now effected through either Mach lines or shock/detonation waves, which respectively propagate at sonic and supersonic speeds.

The rapidity with which events take place in supersonic combustion offers significant challenges in practical situations. For example, high Mach number supersonic aeropropulsion engines have only a few milliseconds of residence time for mixing and reaction to be completed, and it is very difficult to quench an explosion wave once it is initiated.

The contents of this chapter are divided into two major classes of supersonic flows. The first involves weakly perturbed flows with the disturbances propagating at the sonic speed. Here we shall first discuss the general influence of chemical reactions on sound wave propagation, and their limiting behavior in chemically frozen and equilibrium flows. We shall then analyze, in Sections 14.2 and 14.3, several problems involving the propagation of such weak disturbances. In Section 14.4 we present the method of characteristics, which is frequently used for the solution of hyperbolic partial differential equations. Materials presented in Sections 14.1 to 14.3 follow closely those in Vincenti and Kruger (1965) and Williams (1985).

In the second part of this chapter, we study the structure and dynamics of detonation waves, which embody shock waves in their structure and hence propagate at supersonic speeds. Here we shall first discuss the basic, one-dimensional wave propagation and structure, and then the observed three-dimensional, unsteady structure, in Sections 14.5 and 14.6 respectively. In Section 14.7 the propagation and decay of a strong blast wave in a nonreactive environment, initiated by an intense energy source, are analyzed. In Sections 14.8 and 14.9 we study the initiation of detonations, directly through energy deposition, and indirectly through transition from states of slow burning to detonation.

Further exposition on the fundamental aspects of the subject matters can be found in Clarke and McChesney (1964), Vincenti and Kruger (1965), and Williams (1985) for the first part; and Landau and Lifshitz (1959), Zel'dovich and Kompaneets (1960), Fickett and Davis (1979), Toong (1983), Strehlow (1984), Lee (1984, 2001), Stewart (1998), Clavin (2000), and Kailasanath (2003) for the second part.

14.1. FROZEN AND EQUILIBRIUM FLOWS

14.1.1. Governing Equations for Nondiffusive Flows

Assuming that all diffusive transport is negligible as compared to convective transport, and that body forces and radiative transport are absent, the conservation equations (5.2.1) to (5.2.4) for high-speed flows are given by

$$\frac{\partial \rho}{\partial t} + \nabla \cdot (\rho \mathbf{v}) = 0 \tag{14.1.1}$$

$$\rho \frac{D\mathbf{v}}{Dt} = -\nabla p \tag{14.1.2}$$

$$\rho \frac{Dh}{Dt} = \frac{Dp}{Dt} \tag{14.1.3}$$

$$\rho \frac{DY_i}{Dt} = w_i, \quad i = 1, 2, \ldots, N, \tag{14.1.4}$$

where $D/Dt = \partial/\partial t + \mathbf{v} \cdot \nabla$ is the material derivative, and we have used the enthalpy $h = e + p/\rho$ instead of the internal energy e for energy conservation. It is seen that by neglecting the second-order diffusive transport terms, the conservation equations degenerate to a system of first-order nonlinear partial differential equations.

Equations (14.1.1) to (14.1.4) are supplemented by the equations of state,

$$h = h(p, \rho, Y_i), \tag{14.1.5}$$

$$T = T(p, \rho, Y_i), \tag{14.1.6}$$

constituting $5 + N$ equations to solve for the $5 + N$ unknowns p, ρ, T, h, Y_i, and $\mathbf{v}$.

14.1.2. Entropy Production

The nonequilibrium process of chemical reaction generates entropy. Expressing Eq. (1.2.7) in terms of intensive quantities (per unit mass), we have

$$Tds = dh - (1/\rho)dp - \sum_{i=1}^{N}(\bar{\mu}_i/W_i)dY_i, \tag{14.1.7}$$

where dY_i/W_i is dN_i per unit mass. In terms of material derivatives, Eq. (14.1.7) can be written as

$$T\frac{Ds}{Dt} = \frac{Dh}{Dt} - \frac{1}{\rho}\frac{Dp}{Dt} - \sum_{i=1}^{N}(\bar{\mu}_i/W_i)\frac{DY_i}{Dt}. \tag{14.1.8}$$

Substituting Eqs. (14.1.3) and (14.1.4) into Eq. (14.1.8) yields

$$\frac{Ds}{Dt} = -\frac{1}{\rho T}\sum_{i=1}^{N}(\bar{\mu}_i/W_i)w_i. \tag{14.1.9}$$

Equation (14.1.9) shows that the entropy of a fluid element is constant, with

$$\frac{Ds}{Dt} = 0 \tag{14.1.10}$$

in the two limiting situations of frozen and equilibrium flows, defined by:

Frozen Flow:

$$w_i = 0 \tag{14.1.11}$$

Equilibrium Flow:

$$\sum_{i=1}^{N}(\bar{\mu}_i/W_i)w_i \sim \sum_{i=1}^{N}(\bar{\mu}_i/W_i)dY_i \sim \sum_{i=1}^{N}\bar{\mu}_i dN_i = 0. \tag{14.1.12}$$

The last relation in Eq. (14.1.12) is simply the definition of chemical equilibrium, Eq. (1.2.12).

14.1.3. Speed of Sound

An important parameter indicating the extent of compressibility of a high-speed flow is the speed of sound, defined as $a^2 = (\partial p/\partial \rho)_s$, for a nonreacting gas. For a reacting gas, the state of reactedness also needs to be considered. We now investigate the limiting situations of the speeds of sound in frozen and equilibrium flows.

From $h = h(p, \rho, Y_i)$, we have

$$dh = \left(\frac{\partial h}{\partial p}\right)_{\rho, Y_i} dp + \left(\frac{\partial h}{\partial \rho}\right)_{p, Y_i} d\rho + \sum_{i=1}^{N} \left(\frac{\partial h}{\partial Y_i}\right)_{p, \rho, Y_{j(j \neq i)}} dY_i. \quad (14.1.13)$$

Substituting dh into Eq. (14.1.7), we have

$$Tds = \left[\left(\frac{\partial h}{\partial p}\right)_{\rho, Y_i} - \frac{1}{\rho}\right] dp + \left(\frac{\partial h}{\partial \rho}\right)_{p, Y_i} d\rho + \sum_{i=1}^{N} \left[\left(\frac{\partial h}{\partial Y_i}\right)_{p, \rho, Y_{j,(j \neq i)}} + \left(\frac{\bar{\mu}_i}{W_i}\right)\right] dY_i. \quad (14.1.14)$$

Consequently, for sound propagation in a frozen flow, $ds \equiv 0$ and $dY_i \equiv 0$, such that Eq. (14.1.14) yields

$$a_f^2 = \left(\frac{\partial p}{\partial \rho}\right)_{s, Y_i} = -\frac{(\partial h/\partial \rho)_{p, Y_i}}{(\partial h/\partial p)_{\rho, Y_i} - (1/\rho)}. \quad (14.1.15)$$

For an equilibrium flow, $ds \equiv 0$ and $\Sigma(\bar{\mu}_i / W_i) dY_i \equiv 0$. Furthermore, we have an additional constraint that the equilibrium concentrations are functions of two of the state variables, say, p and ρ,

$$Y_{i,e} = Y_{i,e}(p, \rho), \quad (14.1.16)$$

from which we have

$$dY_{i,e} = \left(\frac{\partial Y_{i,e}}{\partial p}\right)_{\rho} dp + \left(\frac{\partial Y_{i,e}}{\partial \rho}\right)_{p} d\rho. \quad (14.1.17)$$

Substituting Eq. (14.1.17) into Eq. (14.1.14), with $dY_i = dY_{i,e}$, yields

$$a_e^2 = \left(\frac{\partial p}{\partial \rho}\right)_{s, Y_{i,e}} = -\frac{(\partial h/\partial \rho)_{p, Y_i} + \sum_{i=1}^{N} (\partial h/\partial Y_i)_{p, \rho, Y_{j(j \neq i)}} (\partial Y_{i,e}/\partial \rho)_p}{(\partial h/\partial p)_{\rho, Y_i} - (1/\rho) + \sum_{i=1}^{N} (\partial h/\partial Y_i)_{p, \rho, Y_{j(j \neq i)}} (\partial Y_{i,e}/\partial p)_\rho}. \quad (14.1.18)$$

An alternate expression can be derived for a_e^2 by relating it to a_f^2. Specifically, by expressing (14.1.16) as $p = p(\rho, Y_i = Y_{i,e})$, we have

$$\left(\frac{\partial p}{\partial \rho}\right)_{s, Y_i = Y_{i,e}} = \left(\frac{\partial p}{\partial \rho}\right)_{s, Y_i} + \sum_{i=1}^{N} \left(\frac{\partial p}{\partial Y_i}\right)_{s, \rho, Y_{j(j \neq i)}} \left(\frac{\partial Y_i}{\partial \rho}\right)_{s, Y_i = Y_{i,e}}, \quad (14.1.19)$$

which is simply

$$a_e^2 = a_f^2 + d, \quad (14.1.20)$$

with d being the second term on the RHS of Eq. (14.1.19). It can be demonstrated from general thermodynamic considerations that $a_f > a_e$. Mechanistically, it is reasonable to expect that an equilibrium medium has a greater "capacitance" to absorb the energy of the propagating sound wave, rendering a slower a_e. By the same token, a frozen medium is more "rigid," hence possessing a higher a_f.

14.1.4. Acoustic Equations

We now study the propagation of a small disturbance in an otherwise equilibrium flow. For simplicity, we shall use the one-step, two-reactant equation, $A \rightleftharpoons B$, such that the chemical nature of the flow can be represented by a single variable Y, the mass fraction of A. Furthermore, for clarity in presenting the basic concepts, we shall represent Eq. (14.1.4) by

$$\frac{DY}{Dt} \approx -\frac{Y - Y_e}{\tau_o}, \tag{14.1.21}$$

which represents the relaxation nature of the present phenomena. Here

$$\tau_o = \frac{\rho_o}{[(\partial w/\partial Y)_{p,T}]_o} \tag{14.1.22}$$

is a characteristic reaction time at the initial, undisturbed, equilibrium state o, at which $Y_o = Y_{e,o}$.

Next we introduce small amplitude disturbances in the forms of

$$\begin{aligned} &p = p_o + p', \qquad &&\rho = \rho_o + \rho', \qquad &&T = T_o + T', \\ &h = h_o + h', &&Y = Y_{e,o} + Y', &&Y_e = Y_{e,o} + Y_e', \\ & &&\mathbf{v} = \mathbf{v}', && \end{aligned}$$

where we have let the medium be initially stationary, with $\mathbf{v}_o \equiv 0$. Substituting the above quantities into Eqs. (14.1.1), (14.1.2), (14.1.3), and (14.1.21), expanding and keeping the leading-order terms, we obtain

$$\frac{\partial \rho'}{\partial t} + \rho_o \nabla \cdot \mathbf{v}' = 0 \tag{14.1.23}$$

$$\rho_o \frac{\partial \mathbf{v}'}{\partial t} = -\nabla p' \tag{14.1.24}$$

$$\rho_o \frac{\partial h'}{\partial t} = \frac{\partial p'}{\partial t} \tag{14.1.25}$$

$$\tau_o \frac{\partial Y'}{\partial t} = -(Y' - Y_e'). \tag{14.1.26}$$

Differentiating Eq. (14.1.26) with respect to t, we have

$$\tau_o \frac{\partial}{\partial t}\left(\frac{\partial Y'}{\partial t}\right) + \frac{\partial}{\partial t}(Y' - Y_e') = 0. \tag{14.1.27}$$

From Eq. (14.1.13), we have

$$\left(\frac{\partial h}{\partial Y}\right)_{p,\rho} \frac{\partial Y'}{\partial t} = \frac{\partial h'}{\partial t} - \left(\frac{\partial h}{\partial p}\right)_{\rho,Y} \frac{\partial p'}{\partial t} - \left(\frac{\partial h}{\partial \rho}\right)_{p,Y} \frac{\partial \rho'}{\partial t}. \tag{14.1.28}$$

Substituting $\partial p'/\partial t$ and $\partial h'/\partial t$ respectively from Eqs. (14.1.23) and (14.1.25) into Eq. (14.1.28), and using the definition of a_f^2 given by Eq. (14.1.15), we obtain

$$\frac{\partial Y'}{\partial t} = \frac{(\partial h/\partial \rho)_{p,Y}}{(\partial h/\partial Y)_{p,\rho}} \left(\frac{1}{a_{f,o}^2}\frac{\partial p'}{\partial t} + \rho_o \nabla \cdot \mathbf{v}'\right). \tag{14.1.29}$$

Furthermore, differentiating Eq. (14.1.17) and using Eq. (14.1.23) yield

$$\frac{\partial Y_e'}{\partial t} = \left(\frac{\partial Y_e}{\partial p}\right)_\rho \frac{\partial p'}{\partial t} - \left(\frac{\partial Y_e}{\partial \rho}\right)_p \rho_o \nabla \cdot \mathbf{v}'. \tag{14.1.30}$$

Substituting Eqs. (14.1.29) and (14.1.30) into Eq. (14.1.27), and defining a stream function ϕ as

$$p' = -\rho_o \frac{\partial \phi}{\partial t}, \qquad \mathbf{v}' = \nabla \phi, \tag{14.1.31}$$

which satisfies Eq. (14.1.24), we obtain the acoustic wave equation

$$\tilde{\tau}_o \frac{\partial}{\partial t}\left(\frac{1}{a_{f,o}^2}\frac{\partial^2 \phi}{\partial t^2} - \nabla^2 \phi\right) + \left(\frac{1}{a_{e,o}^2}\frac{\partial^2 \phi}{\partial t^2} - \nabla^2 \phi\right) = 0, \tag{14.1.32}$$

where

$$\tilde{\tau}_o = \left\{\left[1 + \frac{(\partial h/\partial Y)_{p,\rho}(\partial Y_e/\partial \rho)_p}{(\partial h/\partial \rho)_{p,Y}}\right]_o\right\}^{-1} \tau_o. \tag{14.1.33}$$

Let us now discuss the wave equation (14.1.32), together with the rate equation (14.1.26). In the limit of frozen flow, $\tau_o \to \infty$, we have

$$\frac{\partial}{\partial t}\left(\frac{1}{a_{f,o}^2}\frac{\partial^2 \phi}{\partial t^2} - \nabla^2 \phi\right) = 0,$$

which implies

$$\frac{1}{a_{f,o}^2}\frac{\partial^2 \phi}{\partial t^2} - \nabla^2 \phi = fn(\mathbf{x}). \tag{14.1.34}$$

If the disturbance ϕ vanishes at sometime throughout the flow, we can set $fn(\mathbf{x}) = 0$. Wave propagation is then described by the classical wave equation,

$$\frac{1}{a_{f,o}^2}\frac{\partial^2 \phi}{\partial t^2} - \nabla^2 \phi = 0, \tag{14.1.35}$$

characterized by the frozen sound speed, as it should be. Furthermore, Eq. (14.1.26) shows that in this limit $\partial Y'/\partial t \to 0$ so that its RHS is bounded. This then implies $Y' \to$ constant $= 0$.

In the limit of equilibrium flow, $\tau_o \to 0$. Equation (14.1.32) shows that

$$\frac{1}{a_{e,o}^2}\frac{\partial^2 \phi}{\partial t^2} - \nabla^2 \phi = 0, \tag{14.1.36}$$

which is characterized by the equilibrium sound speed. Inspecting Eq. (14.1.26), we see that as $\tau_o \to 0$, we must have $Y' \to Y_e'$ in order for $\partial Y'/\partial t$ to remain finite. The results are therefore consistent. Furthermore, in this limit the order of the general wave equation (14.1.32) drops from three to two in the time coordinate, indicating the potential breakdown of the solution at some specific t.

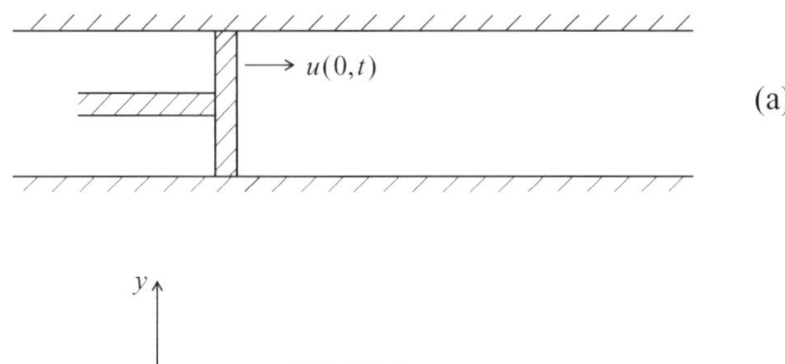

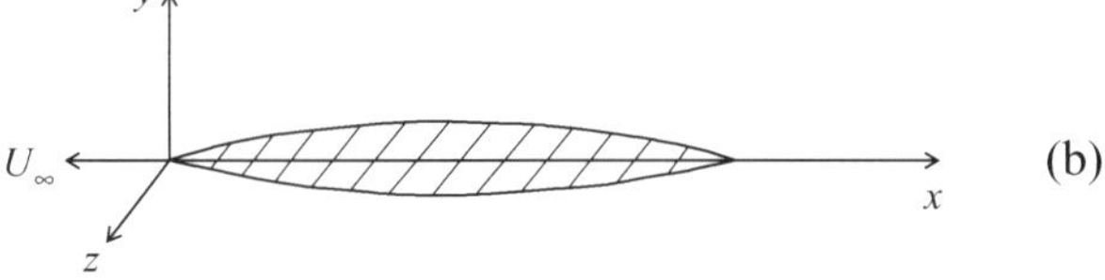

Figure 14.2.1. Schematics of the (a) sinusoidally oscillating piston problem, and (b) slender body problem.

14.2. DYNAMICS OF WEAKLY PERTURBED FLOWS

In the last section we identified the properties of frozen and equilibrium flows, which respectively represent the slowest and fastest limits of chemical reactivity. In a flowing system the extent of the progress of chemical reactivity, and consequently the system response, is to be measured by comparing the characteristic chemical time, designated by $\tilde{\tau}_o$ in Eq. (14.1.33), with a characteristic flow time. Furthermore, entropy is constant in frozen and equilibrium flows, each of which is characterized by its own speed of sound and, hence, wave shape. It is, however, reasonable to anticipate that in the presence of chemical reaction, which is a nonequilibrium process, entropy is produced and the flow is dissipative and distorting. We now study two problems to demonstrate these concepts.

14.2.1. One-Dimensional Propagation of Acoustic Waves

Here we analyze the simplest possible mode of acoustic wave propagation, namely one involving planar waves generated by the motion of a piston in a constant area duct of infinite extent, as shown in Figure 14.2.1a. By oscillating the piston harmonically with an angular frequency ω and a small amplitude $U(0)$ about $x = 0$, acoustic waves are generated and propagate to the right of the piston. The velocity of the piston is described by the real part of

$$u(0, t) = Re\{U(0)e^{iwt}\}, \tag{14.2.1}$$

where $Re\{\cdot\}$ designates the real part. Since this boundary condition is applied to the velocity $u' = \partial\phi/\partial x$ instead of the stream function ϕ, we differentiate Eq. (14.1.32) by x to yield, in one dimension,

$$k\frac{\partial}{\partial \tilde{t}}\left(\frac{\partial^2 \tilde{u}}{\partial \tilde{t}^2} - \frac{\partial^2 \tilde{u}}{\partial \tilde{x}^2}\right) + \left(\alpha\frac{\partial^2 \tilde{u}}{\partial \tilde{t}^2} - \frac{\partial^2 \tilde{u}}{\partial \tilde{x}^2}\right) = 0, \tag{14.2.2}$$

where we have written $u = u'$ for simplicity, and have also introduced the nondimensional quantities

$$\tilde{x} = \frac{\omega}{a_{f,o}}x, \qquad \tilde{t} = \omega t, \qquad \tilde{u} = \frac{u}{U(0)},$$

$$k = \omega\tilde{\tau}_o, \qquad \alpha = \frac{a_{f,o}^2}{a_{e,o}^2} > 1.$$

Following standard approach, we assume a harmonic solution for Eq. (14.2.2), given by

$$\tilde{u}(\tilde{x}, \tilde{t}) = Re\{f(\tilde{x})e^{i\tilde{t}}\}, \tag{14.2.3}$$

where $f(\tilde{x})$ is a complex function. Substituting (14.2.3) into Eq. (14.2.2), we get

$$\frac{d^2 f}{d\tilde{x}^2} + \left(\frac{\alpha + ik}{1 + ik}\right) f = 0, \tag{14.2.4}$$

with the boundary condition $f(0) = 1$. The solution of Eq. (14.2.4) is straightforward, given in general by

$$f = Ae^{c^- \tilde{x}} + Be^{c^+ \tilde{x}}, \tag{14.2.5}$$

where c^- and c^+ are the two roots of

$$c^2 = -\frac{\alpha + ik}{1 + ik},$$

which can be explicitly expressed in terms of their respective real and imaginary parts through algebraic manipulation. The results then show that the B solution represents waves propagating to the left, and hence is discarded. Furthermore, the boundary condition $f(0) = 1$ yields $A = 1$. Thus the final solution for Eq. (14.2.3) is

$$\tilde{u}(\tilde{x}, \tilde{t}) = e^{-\delta\tilde{x}} \cos(\tilde{t} - \lambda\tilde{x}), \tag{14.2.6}$$

where

$$(\delta, \lambda) = \left\{\frac{1}{2(1 + k^2)}\left[\mp(\alpha + k^2) + \sqrt{(1 + k^2)(\alpha^2 + k^2)}\right]\right\}^{1/2}, \tag{14.2.7}$$

with δ and λ respectively correspond to the negative and positive signs. In dimensional form Eq. (14.2.6) becomes

$$u(x, t) = U(0) \exp\left(-\frac{\omega\delta}{a_{f,o}}x\right) \cos\left[\omega\left(t - \frac{\lambda}{a_{f,o}}x\right)\right]. \tag{14.2.8}$$

Equation (14.2.8) shows that the disturbance propagates with a phase velocity $a_{f,o}/\lambda$ and is being attenuated with a damping constant $\omega\delta/a_{f,o}$, in units of inverse length. Both these quantities depend on k and α, which in turn depend on the relaxation time $\tilde{\tau}_o$ and hence the nonequilibrium reaction process. The dependence of phase velocity on the wave frequency is known as sound dispersion. The amplitude attenuation represents sound absorption.

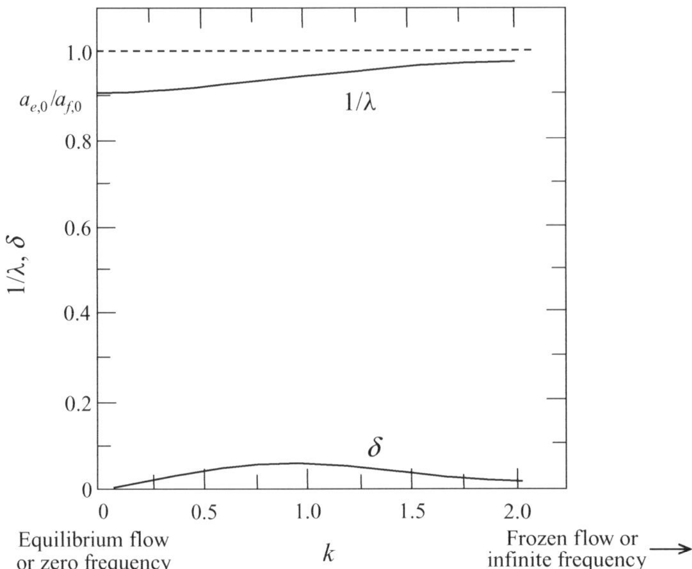

Figure 14.2.2. Variation of δ and $1/\lambda$ with k for $a_{f,0}/a_{e,0} = 1.1$ (Vincenti & Kruger, 1965).

In the limit of equilibrium flow, $\tilde{\tau}_o \to 0$ and therefore $k \to 0$. Thus $\delta \to 0$ and $\lambda \to \sqrt{\alpha} = a_{f,o}/a_{e,o}$. Therefore the wave propagates with the equilibrium speed of sound and without change in shape. Similarly, in the limit of frozen flow, $\tilde{\tau}_o \to \infty$ and $\delta \to 0$, $\lambda \to 1$. The wave then propagates with the frozen speed of sound, but again without change in shape.

Figure 14.2.2 shows the dependence of δ and $1/\lambda$ on k. It is seen that with increasing k, which can be interpreted as either increasing reaction time or decreasing flow time (increasing frequency), the speed of sound increases from $a_{e,o}$ to $a_{f,o}$ as reaction becomes progressively slower relative to the time available for it to approach equilibrium.

The damping constant, $\omega\delta/a_{f,o}$, approaches 0 as $\omega \to 0$ and $(k \to 0, \delta \to 0)$. The opposite limit, however, is not readily apparent because as $\omega \to \infty$, $\delta \to 0$. An expansion of δ^2 for $1/k^2 \ll 1$ yields $\delta^2 \approx [(\alpha - 1)/2k]^2$ such that

$$\frac{\omega\delta}{a_{f,o}} \longrightarrow \frac{1}{2a_{f,o}\tilde{\tau}_o}\left(\frac{a_{f,o}^2}{a_{e,o}^2} - 1\right). \tag{14.2.9}$$

Thus damping is finite for high-frequency flows. Physically, in low-frequency flows, near equilibrium is maintained and the rate of entropy production is small. For high-frequency flows, deviation from equilibrium is significant and the rate of entropy production can be substantial.

The problem of impulsively imparting and sustaining a constant velocity U to the piston at $t = 0$, with the subsequent propagation and evolution of the wave shape, has also been analyzed using Laplace transform. Figure 14.2.3 shows the velocity profile with increasing time. It is seen that for small time, with $t \ll \tau_o$, the wave

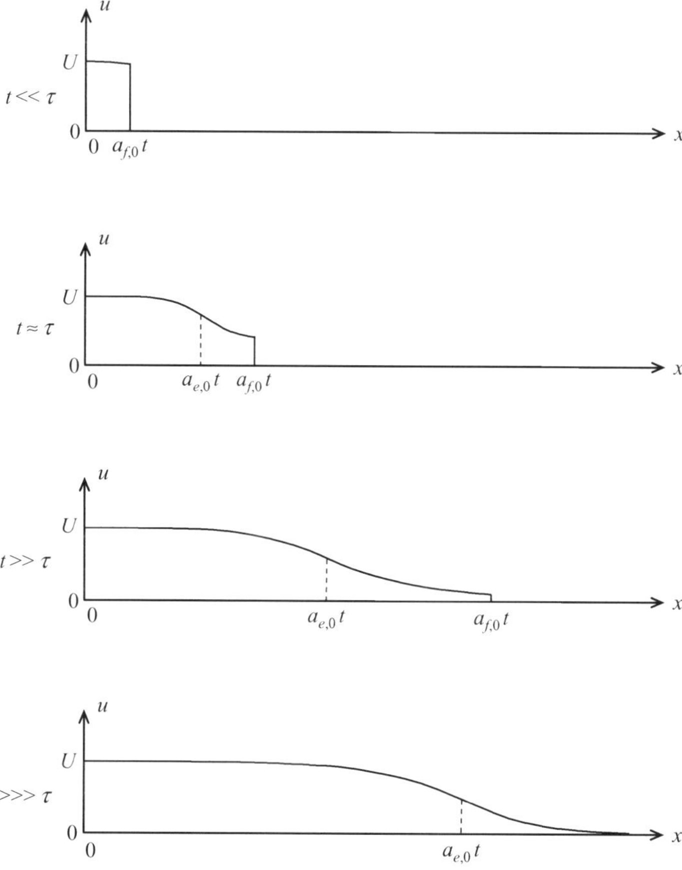

Figure 14.2.3. Schematic of the one-dimensional, unsteady propagation of a sound pulse into a reacting gas (Williams 1985).

propagates with the piston velocity U at the frozen speed of sound. However, for $t \gg \tau_o$, propagation of the wave is led by a region of equilibrium that continuously spreads out.

14.2.2. Uniform Flow over Slender Bodies

We now consider steady flow over a slender body, which causes small disturbances in the flow. The body has a constant velocity U_∞ parallel to x, the axis in the slender direction (Figure 14.2.1b). Transforming Eq. (14.1.32) to a coordinate system attached to the body defined by $\hat{x} = x + U_\infty t$, $\hat{y} = y$, $\hat{z} = z$, $\hat{t} = t$, we have

$$\tilde{\tau}_\infty \left(\frac{\partial}{\partial \hat{t}} + U_\infty \frac{\partial}{\partial \hat{x}} \right) \left[\frac{1}{a_{f,\infty}^2} \left(\frac{\partial}{\partial \hat{t}} + U_\infty \frac{\partial}{\partial \hat{x}} \right)^2 - \hat{\nabla}^2 \right] \phi + \left[\frac{1}{a_{e,\infty}^2} \left(\frac{\partial}{\partial \hat{t}} + U_\infty \frac{\partial}{\partial \hat{x}} \right)^2 - \hat{\nabla}^2 \right] \phi = 0. \tag{14.2.10}$$

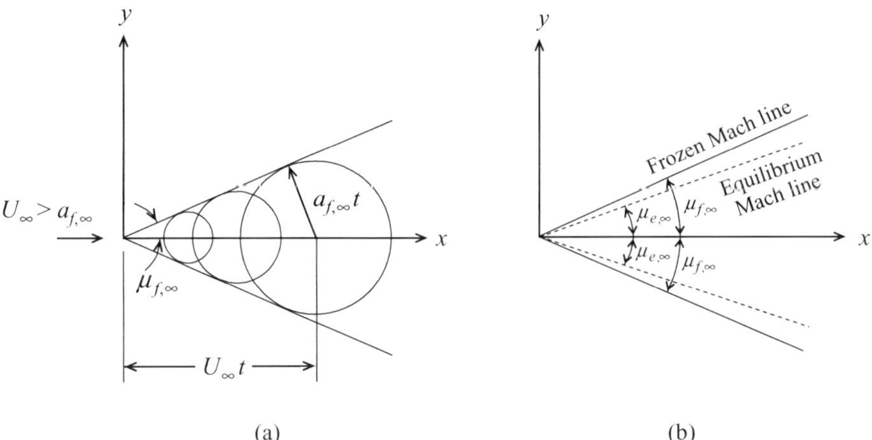

Figure 14.2.4. Regions of influence in nonequilibrium and equilibrium flows (Vincenti & Kruger 1965).

For steady flows, $\partial/\partial\hat{t} \equiv 0$ and Eq. (14.2.10) becomes

$$\tilde{\tau}_\infty U_\infty \frac{\partial}{\partial \hat{x}}\left[\left(M_{f,\infty}^2 - 1\right)\frac{\partial^2\phi}{\partial \hat{x}^2} - \frac{\partial^2\phi}{\partial \hat{y}^2} - \frac{\partial^2\phi}{\partial \hat{z}^2}\right] + \left[\left(M_{e,\infty}^2 - 1\right)\frac{\partial^2\phi}{\partial \hat{x}^2} - \frac{\partial^2\phi}{\partial \hat{y}^2} - \frac{\partial^2\phi}{\partial \hat{z}^2}\right] = 0, \tag{14.2.11}$$

where $M_{f,\infty} = U_\infty/a_{f,\infty}$ and $M_{e,\infty} = U_\infty/a_{e,\infty}$ are respectively the frozen and equilibrium Mach numbers evaluated at the ambience. The parameter $\tilde{\tau}_\infty U_\infty$ is now a relaxation length.

Since $M_{e,\infty} > M_{f,\infty}$, the flow changes from elliptic to hyperbolic when $M_{f,\infty} > 1$, for $\tilde{\tau}_\infty \neq 0$ because the first term in Eq. (14.2.11) is of higher order than the second term, and thus controls the nature of the solution. For $\tilde{\tau}_\infty \equiv 0$, the transition occurs when $M_{e,\infty} > 1$.

For supersonic flows, as long as $\tilde{\tau}_\infty \neq 0$, an infinitesimal disturbance travels with the frozen speed of sound, $a_{f,\infty}$, because it is larger than the equilibrium speed of sound, $a_{e,\infty}$. Together with the gas motion U_∞, a Mach cone of half-apex angle

$$\mu_{f,\infty} = \tan^{-1}\left[\frac{a_{f,\infty}t}{\sqrt{(U_\infty t)^2 - (a_{f,\infty}t)^2}}\right] = \tan^{-1}\frac{1}{\sqrt{M_{f,\infty}^2 - 1}} \tag{14.2.12}$$

is swept out, as shown in Figure 14.2.4. For $\tilde{\tau}_\infty \equiv 0$, the Mach angle is given by the equilibrium Mach number as

$$\mu_{e,\infty} = \tan^{-1}\frac{1}{\sqrt{M_{e,\infty}^2 - 1}}. \tag{14.2.13}$$

Since $a_{f,\infty} > a_{e,\infty}$, the frozen Mach angle is greater than the equilibrium value. The change from $M_{f,\infty}$ to $M_{e,\infty}$ is again abrupt, as $\tilde{\tau}_\infty$ drops to zero.

14.3. STEADY, QUASI-ONE-DIMENSIONAL FLOWS

14.3.1. Nonlinear Flows

Now we consider the characteristics of steady, quasi-one-dimensional nonlinear flows. Since there is no diffusional transport, the governing equations for momentum, heat, and mass are basically those of the one-dimensional flow. The effect of multidimensionality is accounted for by replacing Eq. (14.1.1) by overall mass conservation,

$$\rho u A = \text{constant}, \tag{14.3.1}$$

such that

$$\frac{d\rho}{\rho} + \frac{du}{u} + \frac{dA}{A} = 0, \tag{14.3.2}$$

where $A = A(x)$ is the cross-sectional area of the flow. Equations (14.1.2) to (14.1.4) remain unchanged, and are given by

$$\rho u du + dp = 0 \tag{14.3.3}$$

$$u du + dh = 0 \tag{14.3.4}$$

$$\rho u \frac{dY_i}{dx} = w_i, \quad i = 1, 2, \ldots, N. \tag{14.3.5}$$

Thus for given $A(x)$, Eqs. (14.3.2) to (14.3.5), together with the equation of state (14.1.5), constitute $4 + N$ equations to solve for the $4 + N$ unknowns $\rho, u, p, h,$ and Y_i.

From Eq. (14.3.3), we can write

$$u du = -\frac{dp}{\rho} = -\frac{dp}{d\rho}\frac{d\rho}{\rho},$$

which, when substituted into Eq. (14.3.2), yields

$$\frac{du}{u} = \frac{dA/A}{\frac{u^2}{(dp/d\rho)} - 1}. \tag{14.3.6}$$

Thus if entropy is constant along the flow, as in frozen and equilibrium flows, then $(dp/d\rho)_s = a^2$ and we retrieve the well-known result for adiabatic, inviscid flows,

$$\frac{du}{u} = \frac{dA/A}{M^2 - 1}, \tag{14.3.7}$$

showing that in order to accelerate a flow from subsonic to supersonic, or decelerate it from supersonic to subsonic, we must have $M = 1$ at the throat of a converging–diverging channel where $dA \equiv 0$. For a general, nonequilibrium flow, Eq. (14.3.7) does not hold and $u(x)$ is determined from a complete evaluation of the flow.

To consider the nonequilibrium situations, we eliminate $d\rho, dp,$ and dh from Eq. (14.1.13) by using Eqs. (14.3.2)–(14.3.4). Furthermore, by using the definition of the frozen speed of sound, Eq. (14.1.15), we obtain

$$\frac{du}{u} = \frac{\frac{dA}{A} - \left[\rho\left(\frac{\partial h}{\partial \rho}\right)_{p,Y_i}\right]^{-1} \sum_{i=1}^{N}\left(\frac{\partial h}{\partial Y_i}\right)_{p,\rho,Y_{j(j\neq i)}} dY_i}{M_f^2 - 1}. \tag{14.3.8}$$

By substituting $dY_i \sim w_i$ from Eq. (14.3.5) into Eq. (14.3.8), and for given reaction rate relations, the resulting expression can be numerically integrated for a nonequilibrium channel flow of given $A(x)$. Equation (14.3.8) shows that for a continuously accelerating flow, $du > 0$, the requirement that the numerator and denominator of its RHS must vanish at $M_f = 1$, so that the RHS remains finite, implies that the state $M_f = 1$ is displaced from the throat. The displacement is usually to the downstream of the throat, as the nonequilibrium term in the numerator is positive under most situations (Vincenti & Kruger, 1965).

Studies of nonequilibrium nozzle flows are of practical interest in the freezing of recombination reactions in rocket nozzles. Here products from the combustion chamber are highly dissociated because of the high temperature. Thus in order to achieve the maximum thrust, it is desirable to maintain the flow in the nozzle to be close to equilibrium so as to capture the heat release through recombination. An expansion process that is too rapid can prematurely freeze the recombination reactions and thereby withhold the recombination heat release for thrust.

14.3.2. Linearized Nozzle Flows

Now we study the linearized limit of steady-state, quasi-one-dimensional flows characterized by gradual area change. Thus if we start with the generalized equation for the steady, quasi-one-dimensional flow, and linearize them in the same manner as that for the wave equation, it can be shown that, for a flow that is in equilibrium at an initial state $x = x_o$, the velocity perturbation $u' = u - u_o$ corresponding to an area perturbation $A' = A - A_o$ is described by

$$\tilde{\tau} u_o \frac{d}{dx}\left[\left(M_{f,o}^2 - 1\right)\frac{du'}{dx} - \frac{u_o}{A_o}\frac{dA'}{dx}\right] + \left[\left(M_{e,o}^2 - 1\right)\frac{du'}{dx} - \frac{u_o}{A_o}\frac{dA'}{dx}\right] = 0. \quad (14.3.9)$$

Comparing Eq. (14.3.9) with the linearized two-dimensional flow described by Eq. (14.2.11), with $u' = \partial\phi/\partial x$ and $v' = \partial\phi/\partial y$,

$$\tilde{\tau}_o u_o \frac{\partial}{\partial x}\left[\left(M_{f,o}^2 - 1\right)\frac{\partial u'}{\partial x} - \frac{\partial v'}{\partial y}\right] + \left[\left(M_{e,o}^2 - 1\right)\frac{\partial u'}{\partial x} - \frac{\partial v'}{\partial y}\right] = 0, \quad (14.3.10)$$

it is clear that by treating a two-dimensional flow as a quasi-one-dimensional flow with a slowly varying area $A(x)$, we are making the approximation

$$\frac{\partial v'}{\partial y} \approx \frac{u_o}{A_o}\frac{dA'}{dx}$$

in the linearized limit. The advantage gained here is that we now only need to solve ordinary instead of partial differential equations.

Integrating Eq. (14.3.9) once, and applying the initial conditions that at $x = 0$, $u' = A' \equiv 0$ and $du'/dx = dA'/dx \equiv 0$, we have

$$\tilde{\tau}_o u_o \left[\left(M_{f,o}^2 - 1\right)\frac{du'}{dx} - \frac{u_o}{A_o}\frac{dA'}{dx}\right] + \left[\left(M_{e,o}^2 - 1\right)u' - \frac{u_o}{A_o}A'\right] = 0. \quad (14.3.11)$$

Equation (14.3.11) can be readily integrated, yielding

$$u'(x) = \frac{u_o}{(M_{f,o}^2 - 1)A_o}\left\{A'(x) - \left[\left(\frac{M_{e,o}^2 - 1}{M_{f,o}^2 - 1}\right) - 1\right] I(x)\right\}, \tag{14.3.12}$$

where

$$I(x) = \int_o^x A'(\eta)\exp\left[-\left(\frac{Me_{e,o}^2 - 1}{M_{f,o}^2 - 1}\right)\frac{(x-\eta)}{\tilde{\tau}_o u_o}\right]\frac{d\eta}{\tilde{\tau}_o u_o}. \tag{14.3.13}$$

Equation (14.3.12) shows that $u'(x)$ depends not only on the local value of $A'(x)$, as given by the first term, but also on the history of the flow, including the area variation, through the integral term $I(x)$. Thus a nonequilibrium flow has memory.

In the limit of frozen flow, $\tilde{\tau}_o \to \infty$ and $I(x) \to 0$, Eq. (14.3.12) becomes

$$u'(x) = \frac{u_o}{(M_{f,o}^2 - 1)}\frac{A'(x)}{A_o}, \quad \text{as } \tilde{\tau}_o \to \infty. \tag{14.3.14}$$

In the equilibrium limit of $\tilde{\tau}_o \to 0$, the integral of $I(x)$ is indeterminate. An inspection of $I(x)$, however, shows that for small $\tilde{\tau}_o$, the exponential term is essentially zero except when $\eta \to x$. Therefore we can change the integration variable from η to a new $O(1)$ variable $\chi = (x-\eta)/(\tilde{\tau}_o u_o)$, with $\tilde{\tau}_o u_o \ll 1$. Consequently $A'(\eta) \approx A'(x) + O(\tilde{\tau}_o u_o)$ and the integration can be readily carried out, yielding $I(x) \approx \left[(M_{f,o}^2 - 1)/(M_{e,o}^2 - 1)\right]A'(x)$ and

$$u'(x) = \frac{u_o}{(M_{e,o}^2 - 1)}\frac{A'(x)}{A_o}, \quad \text{as } \tilde{\tau}_o \to 0, \tag{14.3.15}$$

as the case should be. Thus in both the frozen and equilibrium limits $u'(x)$ depends only on the value of the local cross-sectional area.

14.4. METHOD OF CHARACTERISTICS

So far we have demonstrated that supersonic flows are governed by nonlinear hyperbolic differential equations. Analytical solutions can be obtained for specialized situations either for weakly perturbed flows for which these equations can be linearized, or for quasi-one-dimensional flows for which spatial variation occurs only in one direction. In this section, we consider flows that are quasi-linear in that the coefficients of the differentials are at most functions of the dependent variables but not their derivatives. Within such flows, surfaces can be identified over which certain flow properties either assume constant values or vary only over the surface but not in the direction normal to it. The procedure, called the method of characteristics, involves first identifying these characteristic surfaces, and then obtaining the description of the appropriate flow properties over them.

In the following we shall first present the general procedure of the method of characteristics, restricting nevertheless to first-order equations with two independent variables. The characteristic surfaces then degenerate to characteristic curves, with property variation described by ordinary differential equations. We shall then apply this procedure to the identification of the characteristics of two familiar problems, namely an unsteady one-dimensional flow and a steady two-dimensional flow. Further discussion on this topic can be found in Courant and Friedrichs (1948).

14.4.1. General Procedure for Two Independent Variables

The system of interest consists of n first order quasi-linear hyperbolic partial differential equations, represented by the index i, that describe the variations of n dependent variables $u_k(x, y)$, $k = 1, 2, \ldots, n$, with the two independent variables (x, y),

$$L_i(u_1, u_2, \ldots, u_n) = \sum_{k=1}^{n}\left(a_{i,k}\frac{\partial u_k}{\partial x} + b_{i,k}\frac{\partial u_k}{\partial y}\right) + c_i = 0, \quad i = 1, 2, \ldots, n. \quad (14.4.1)$$

The coefficients $a_{i,k}$, $b_{i,k}$, and c_i are functions of x, y, and u_k only. We aim to identify a characteristic direction $\sigma(x, y)$ such that the flow varies only along σ, that is, $u_k(x, y) \to u_k(\sigma)$.

Multiplying Eq. (14.4.1) by λ_i and summing over i, we have

$$\begin{aligned}
\sum_{i=1}^{n}\lambda_i L_i(u_1, u_2, \ldots, u_n) &= \sum_{i=1}^{n}\left(\lambda_i\sum_{k=1}^{n}a_{i,k}\frac{\partial u_k}{\partial x} + \lambda_i\sum_{k=1}^{n}b_{i,k}\frac{\partial u_k}{\partial y}\right) + \sum_{i=1}^{n}\lambda_i c_i \\
&= \sum_{k=1}^{n}\left[\left(\sum_{i=1}^{n}\lambda_i a_{i,k}\right)\frac{\partial u_k}{\partial x} + \left(\sum_{i=1}^{n}\lambda_i b_{i,k}\right)\frac{\partial u_k}{\partial y}\right] + \sum_{i=1}^{n}\lambda_i c_i \\
&= \sum_{k=1}^{n}\left(\sum_{i=1}^{n}\lambda_i a_{i,k}\right)\left[\frac{\partial u_k}{\partial x} + \left(\frac{\sum_{i=1}^{n}\lambda_i b_{i,k}}{\sum_{i=1}^{n}\lambda_i a_{i,k}}\right)\frac{\partial u_k}{\partial y}\right] + \sum_{i=1}^{n}\lambda_i c_i = 0.
\end{aligned} \quad (14.4.2)$$

We now seek to express each of the kth term in the square bracket as a function that varies only along one independent variable, σ. Noting that

$$\begin{aligned}
\frac{du_k}{d\sigma} &= \frac{\partial x}{\partial \sigma}\frac{\partial u_k}{\partial x} + \frac{\partial y}{\partial \sigma}\frac{\partial u_k}{\partial y} \\
&= \frac{\partial x}{\partial \sigma}\left[\frac{\partial u_k}{\partial x} + \left(\frac{\partial y/\partial \sigma}{\partial x/\partial \sigma}\right)\frac{\partial u_k}{\partial y}\right],
\end{aligned} \quad (14.4.3)$$

and by comparing the terms within the square brackets of Eqs. (14.4.2) and (14.4.3), it is clear that a characteristic direction

$$\zeta = \frac{dy}{dx} = \frac{\partial y/\partial \sigma}{\partial x/\partial \sigma} = \frac{\sum_{i=1}^{n}\lambda_i b_{i,k}}{\sum_{i=1}^{n}\lambda_i a_{i,k}}, \quad k = 1, 2, \ldots, n \quad (14.4.4)$$

can be defined for every kth term in Eq. (14.4.2). Thus substituting Eq. (14.4.3) into Eq. (14.4.2) and using the definition of ζ yields

$$\sum_{k=1}^{n}\left(\sum_{i=1}^{n}\lambda_i a_{i,k}\right)\left(\frac{du_k}{d\sigma}\right) + \left(\sum_{i=1}^{n}\lambda_i c_i\right)\frac{dx}{d\sigma} = \sum_{i=1}^{n}\lambda_i\left[\left(\sum_{k=1}^{n}a_{i,k}\frac{du_k}{d\sigma}\right) + c_i\frac{dx}{d\sigma}\right] = 0. \quad (14.4.5)$$

Equation (14.4.5) can be alternately expressed as

$$\sum_{i=1}^{n} \lambda_i \left[\left(\sum_{k=1}^{n} b_{i,k} \frac{du_k}{d\sigma} \right) + c_i \frac{dy}{d\sigma} \right] = 0 \tag{14.4.6}$$

by using the last equality in Eq. (14.4.4).

Equation (14.4.4) yields a system of n homogeneous equations for the unknowns λ_i, given by

$$\sum_{i=1}^{n} \lambda_i (a_{i,k}\zeta - b_{i,k}) = 0, \quad k = 1, 2, \ldots, n. \tag{14.4.7}$$

In order for Eq. (14.4.7) to have nontrivial solutions, the determinant of its coefficient matrix must vanish,

$$|a_{k,i}\zeta - b_{k,i}| \equiv 0, \tag{14.4.8}$$

which yields an nth-order polynomial for ζ from which n roots can be solved. If all the roots are real, then the system is said to be totally hyperbolic. These n roots then correspond to the n characteristic directions defined by

$$\zeta_l = \left(\frac{dy}{dx} \right)_l, \quad l = 1, 2, \ldots, n. \tag{14.4.9}$$

Having determined the characteristics, the original system of partial differential equations, (14.4.1), can be transformed to an equivalent system of ordinary differential equations. Thus by using any $(n-1)$ of the equations from (14.4.7), and either Eq. (14.4.5) or Eq. (14.4.6), a new system of n homogeneous equations is formed. Vanishing of the determinant of its coefficient matrix then yields the needed equations governing the variation of u_k with σ along ζ. Using different ζ_l leads to the governing equations for the specific characteristic.

When the number of equations that needs to be solved is two, the algebraic complexity of the solution is greatly reduced. Specifically, (14.4.8) can now be expressed as

$$\begin{vmatrix} a_{1,1}\zeta - b_{1,1} & a_{2,1}\zeta - b_{2,1} \\ a_{1,2}\zeta - b_{1,2} & a_{2,2}\zeta - b_{2,2} \end{vmatrix} \equiv 0, \tag{14.4.10}$$

which yields a quadratic equation in ζ, from which the two roots, say $\zeta^{\pm}$, can be solved. The requirement that $\zeta^{\pm}$ be real could impose certain constraints on the system.

To determine the governing equations for u_k along σ, we use the first equation in (14.4.7) and Eq. (14.4.5). Vanishing of the determinant of its coefficient matrix,

$$\begin{vmatrix} a_{1,1}\zeta - b_{1,1} & a_{2,1}\zeta - b_{2,1} \\ a_{1,1}\frac{du_1}{d\sigma} + a_{1,2}\frac{du_2}{d\sigma} + c_1\frac{dx}{d\sigma} & a_{2,1}\frac{du_1}{d\sigma} + a_{2,2}\frac{du_2}{d\sigma} + c_2\frac{dx}{d\sigma} \end{vmatrix} \equiv 0, \tag{14.4.11}$$

yields a first order linear ODE for u_k, from which the particular equations valid for the separate $\zeta^{\pm}$ can be determined. These equations are called compatibility relations.

As examples, we shall apply the method of characteristics to two problems, namely those involving unsteady, one-dimensional, frozen, isentropic flows, and steady, two-dimensional, chemically reacting flows.

14.4.2. Unsteady, One-Dimensional, Frozen, Isentropic Flows

We use this very simple case to demonstrate the method of characteristics. The continuity and momentum equations are respectively given by

$$\frac{\partial \rho}{\partial t} + \rho \frac{\partial u}{\partial x} + u \frac{\partial \rho}{\partial x} = 0, \tag{14.4.12}$$

$$\rho \frac{\partial u}{\partial t} + \rho u \frac{\partial u}{\partial x} + \frac{\partial p}{\partial x} = 0. \tag{14.4.13}$$

There are three dependent variables, namely u, ρ, and p. For an isentropic flow, $p = p(\rho)$ such that

$$dp = \left(\frac{\partial p}{\partial \rho}\right)_s d\rho = a^2 d\rho. \tag{14.4.14}$$

Consequently

$$\frac{Dp}{Dt} = a^2 \frac{D\rho}{Dt}. \tag{14.4.15}$$

Since Eq. (14.4.15) holds on a particle trajectory, a characteristic of the flow is simply the particle line defined by

$$\zeta^0 = \left(\frac{dx}{dt}\right)^0 = u. \tag{14.4.16}$$

To determine the other characteristics, we identify $u_1 = u$ and $u_2 = p$. Using Eq. (14.4.14), Eqs. (14.4.13) and (14.4.12) can be expressed in the forms of

$$\frac{\partial u}{\partial t} + 0 \cdot \frac{\partial p}{\partial t} + u \frac{\partial u}{\partial x} + \frac{1}{\rho} \frac{\partial p}{\partial x} + 0 = 0 \quad (i = 1), \tag{14.4.17}$$

$$0 \cdot \frac{\partial u}{\partial t} + \frac{1}{\rho a} \frac{\partial p}{\partial t} + a \frac{\partial u}{\partial x} + \frac{u}{\rho a} \frac{\partial p}{\partial x} + 0 = 0 \quad (i = 2). \tag{14.4.18}$$

Comparing Eqs. (14.4.17) and (14.4.18) with Eq. (14.4.1), we have

$$\begin{array}{lllll} a_{1,1} = 1, & a_{1,2} = 0, & b_{1,1} = u, & b_{1,2} = 1/\rho, & c_1 = 0, \\ a_{2,1} = 0, & a_{2,2} = 1/(\rho a), & b_{2,1} = a, & b_{2,2} = u/(\rho a), & c_2 = 0. \end{array} \tag{14.4.19}$$

Using these coefficients, Eq. (14.4.10) becomes

$$\begin{vmatrix} \zeta - u & -a \\ -(1/\rho) & \zeta/(\rho a) - u/(\rho a) \end{vmatrix} = (\rho a)^{-1} \left[(\zeta - u)^2 - a^2\right] = 0, \tag{14.4.20}$$

which yields two roots, and, hence, two characteristic directions,

$$\zeta^{\pm} = \left(\frac{dx}{dt}\right)^{\pm} = u \pm a. \tag{14.4.21}$$

The compatibility relations are given by evaluating Eq. (14.4.11) using the coefficients (14.4.19),

$$\begin{vmatrix} \zeta - u & -a \\ \frac{du}{d\sigma} & (1/\rho a)\frac{dp}{d\sigma} \end{vmatrix} = a\frac{du}{d\sigma} + \frac{(\zeta - u)}{\rho a}\frac{dp}{d\sigma} = 0. \tag{14.4.22}$$

Substituting $\zeta^{\pm}$ given by Eq. (14.4.21) in Eq. (14.4.22) yields

$$\frac{du}{d\sigma} \pm \frac{1}{\rho a}\frac{dp}{d\sigma} = 0 \qquad \text{on} \qquad \zeta^{\pm} = u \pm a. \tag{14.4.23}$$

For an isotropic flow with constant γ, $a = (\gamma p/\rho)^{1/2}$. Substituting a into Eqs. (14.4.23) and integrating, we obtain the relation

$$J_{\pm} = \frac{2a}{\gamma - 1} \pm u, \tag{14.4.24}$$

which are constants along the characteristics $\zeta^{\pm}$. The quantities $J_{\pm}$ are called the Riemann invariants of the flow.

14.4.3. Steady Two-Dimensional Flows

The governing equations are given by Eqs. (14.1.1)–(14.1.5),

$$\nabla \cdot (\rho \mathbf{v}) = \rho \nabla \cdot \mathbf{v} + \mathbf{v} \cdot \nabla \rho = 0 \tag{14.4.25}$$

$$\rho \mathbf{v} \cdot \nabla u = -\frac{\partial p}{\partial x} \tag{14.4.26}$$

$$\rho \mathbf{v} \cdot \nabla v = -\frac{\partial p}{\partial y} \tag{14.4.27}$$

$$\rho \mathbf{v} \cdot \nabla \left(h + \tfrac{1}{2}q^2\right) = 0 \tag{14.4.28}$$

$$\rho \mathbf{v} \cdot \nabla Y_i = w_i, \quad i = 1, 2, \ldots, N \tag{14.4.29}$$

$$h = h(p, \rho, Y_i) \tag{14.1.5}$$

where $q = \left(u^2 + v^2\right)^{1/2}$ is the magnitude of the flow velocity. Furthermore, Eq. (14.4.28), obtained from Eqs. (14.1.2) and (14.1.3), replaces Eq. (14.1.3), and Eqs. (14.4.26) and (14.4.27) are the x- and y-momentum equations. Equations (14.1.5) and (14.4.25) to (14.4.29) constitute $5 + N$ equations to solve for the $5 + N$ unknowns, p, ρ, h, Y_i, u, and v.

Equations (14.4.28) and (14.4.29) readily show that the streamline is a characteristic of the flow, as will be further demonstrated later. Equation (14.4.28) also shows that the quantity,

$$h + \tfrac{1}{2}q^2 = \text{constant}, \tag{14.4.30}$$

holds along the streamline.

The task of finding the characteristics is now reduced to working with Eqs. (14.4.25) to (14.4.27). Since the streamline has already been identified as a characteristic, a natural coordinate for the flow should be one consisting of the streamline direction and the direction normal to it. Designating the new coordinate by (s, n), and the angle between the flow direction and the x-axis by $\theta = \tan^{-1}(v/u)$, we have the relations

$$ds = \cos\theta dx + \sin\theta dy, \qquad dn = -\sin\theta dx + \cos\theta dy, \tag{14.4.31}$$

$$u = q\cos\theta, \qquad v = q\sin\theta. \tag{14.4.32}$$

To transform the governing equations from the (x, y) coordinate to the (s, n) coordinate, we have

$$\begin{aligned}\frac{\partial}{\partial x} &= \frac{\partial s}{\partial x}\frac{\partial}{\partial s} + \frac{\partial n}{\partial x}\frac{\partial}{\partial n} = \cos\theta\frac{\partial}{\partial s} - \sin\theta\frac{\partial}{\partial n},\\ \frac{\partial}{\partial y} &= \frac{\partial s}{\partial y}\frac{\partial}{\partial s} + \frac{\partial n}{\partial y}\frac{\partial}{\partial n} = \sin\theta\frac{\partial}{\partial s} + \cos\theta\frac{\partial}{\partial n}.\end{aligned} \tag{14.4.33}$$

Using Eqs. (14.4.31) to (14.4.33), the convection operator $\mathbf{v}\cdot\nabla$ becomes

$$\mathbf{v}\cdot\nabla = u\frac{\partial}{\partial x} + v\frac{\partial}{\partial y} = q\frac{\partial}{\partial s}. \tag{14.4.34}$$

Applying Eq. (14.4.34) to Eqs. (14.4.28) and (14.4.29), we have

$$\frac{d}{ds}\left(h + \frac{1}{2}q^2\right) = 0, \tag{14.4.35}$$

$$q\frac{dY_i}{ds} = w_i/\rho, \quad i = 1, 2, \ldots, N, \tag{14.4.36}$$

which demonstrate clearly the sole dependence of $\left(h + \frac{1}{2}q^2\right)$ and Y_i on s.

To transform the continuity equation (14.4.25), the $\mathbf{v}\cdot\nabla\rho$ term is simply $q(\partial\rho/\partial s)$ by use of Eq. (14.4.34). Furthermore, the factor $\nabla\cdot\mathbf{v}$ is $(\partial q/\partial s) + q(\partial\theta/\partial n)$. Thus Eq. (14.4.25) becomes

$$\frac{1}{\rho}\frac{\partial\rho}{\partial s} + \frac{1}{q}\frac{\partial q}{\partial s} + \frac{\partial\theta}{\partial n} = 0. \tag{14.4.37}$$

The momentum equations (14.4.26) and (14.4.27) are transformed by using Eqs. (14.4.32) to (14.4.34), yielding

$$\rho q\cos\theta\frac{\partial q}{\partial s} - \rho q^2\sin\theta\frac{\partial\theta}{\partial s} = -\cos\theta\frac{\partial p}{\partial s} + \sin\theta\frac{\partial p}{\partial n}, \tag{14.4.38}$$

$$\rho q\sin\theta\frac{\partial q}{\partial s} + \rho q^2\cos\theta\frac{\partial\theta}{\partial s} = -\sin\theta\frac{\partial p}{\partial s} - \cos\theta\frac{\partial p}{\partial n}. \tag{14.4.39}$$

Multiply Eqs. (14.4.38) and (14.4.39) by $\cos\theta$ and $\sin\theta$, respectively, and add the resulting expressions to yield

$$\rho q\frac{\partial q}{\partial s} + \frac{\partial p}{\partial s} = 0. \tag{14.4.40}$$

Similarly, multiply Eqs. (14.4.38) and (14.4.39) by $\sin\theta$ and $\cos\theta$, respectively, and subtract the resulting expressions to yield

$$\rho q^2\frac{\partial\theta}{\partial s} + \frac{\partial p}{\partial n} = 0. \tag{14.4.41}$$

We see that Eq. (14.4.40) is another relation describing changes along the streamline. The problem is now reduced to solving only Eqs. (14.4.37) and (14.4.41). However, in order to apply the method of characteristics, we need to reduce the flow variables in Eq. (14.4.37) to only two, say p and θ in accordance with Eq. (14.4.41). To accomplish this, we first differentiate $h(p, \rho, Y_i)$ with respect to the streamline direction s as

$$\frac{\partial h}{\partial s} = \left(\frac{\partial h}{\partial p}\right)_{\rho, Y_i}\left(\frac{\partial p}{\partial s}\right) + \left(\frac{\partial h}{\partial \rho}\right)_{p, Y_i}\left(\frac{\partial \rho}{\partial s}\right) + \sum_{i=1}^{N}\left(\frac{\partial h}{\partial Y_i}\right)_{p,\rho,Y_{j(j\neq i)}}\left(\frac{\partial Y_i}{\partial s}\right). \quad (14.4.42)$$

We can now replace $\partial h/\partial s$, $\partial \rho/\partial s$, and $\partial Y_i/\partial s$ by Eqs. (14.4.35), (14.4.37), and (14.4.36) respectively. By further using Eq. (14.4.40), Eq. (14.4.42) becomes

$$\left(\frac{q^2}{a_f^2} - 1\right)\frac{\partial p}{\partial s} + \rho q^2 \frac{\partial \theta}{\partial n} - q\sum_{i=1}^{N}\alpha_i(w_i/\rho) = 0, \quad (14.4.43)$$

where $\alpha_i = (\partial h/\partial Y_i)_{p,\rho,Y_{j(j\neq i)}}/(\partial h/\partial \rho)_{p,Y_i}$.

Equations (14.4.41) and (14.4.43) are the two equations to solve for the characteristics, with (s, η) being the two directions and p, θ the two functions. Thus we have

$$\begin{array}{lllll} a_{1,1} = 0, & a_{1,2} = \rho q^2, & b_{1,1} = 1, & b_{1,2} = 0, & c_1 = 0, \\ a_{2,1} = \left(M_f^2 - 1\right), & a_{2,2} = 0, & b_{2,1} = 0, & b_{2,2} = \rho q^2, & c_2 = -q\sum_{i=1}^{N}\alpha_i(w_i/\rho), \end{array} \quad (14.4.44)$$

where $M_f^2 = q^2/a_f^2$ and ζ is defined as dn/ds. Substituting the coefficients in (14.4.44) into Eq. (14.4.10), we have

$$\begin{vmatrix} -1 & (M_f^2 - 1)\zeta \\ \rho q^2\zeta & -\rho q^2 \end{vmatrix} = \rho q^2[1 - (M_f^2 - 1)\zeta^2] = 0, \quad (14.4.45)$$

from which we obtain the two characteristics as

$$\zeta^{\pm} = \left(\frac{dn}{ds}\right)^{\pm} = \pm\frac{1}{(M_f^2 - 1)^{1/2}}. \quad (14.4.46)$$

Using Eq. (14.4.11), we have

$$\begin{vmatrix} -1 & (M_f^2 - 1)\zeta \\ \rho q^2 \frac{d\theta}{d\sigma} & (M_f^2 - 1)\frac{dp}{d\sigma} - q\sum_{i=1}^{N}\alpha_i(w_i/\rho)\frac{ds}{d\sigma} \end{vmatrix} = 0. \quad (14.4.47)$$

Solving Eq. (14.4.47) yields the compatibility relations

$$\left(M_f^2 - 1\right)^{1/2}\frac{dp}{d\sigma} \pm \rho q^2 \frac{d\theta}{d\sigma} + \frac{q}{(M_f^2 - 1)^{1/2}}\sum_{i=1}^{N}\alpha_i(w_i/\rho)\frac{ds}{d\sigma} = 0,$$

$$\text{on } \zeta^{\pm} = \pm\frac{1}{(M_f^2 - 1)^{1/2}}. \quad (14.4.48)$$

14.5. STEADY ONE-DIMENSIONAL DETONATIONS

In Section 7.1, we have identified the existence of subsonic, deflagration waves and supersonic, detonation waves through the Rankine–Hugoniot relations. The structure and propagation of the deflagration waves have been studied in detail in previous chapters. We shall now study the characteristics of detonation waves.

The Rankine–Hugoniot relations show that, in crossing a detonation wave, the pressure and density increase while the velocity decreases. Since these relations are only concerned with the upstream and downstream equilibrium states that bound the wave, they do not provide any information regarding the nonequilibrium structure of the wave. Furthermore, for a given detonation wave velocity, the problem is not uniquely determined in that there is a strong solution and a weak solution. The strong wave has larger changes in the flow quantities and its downstream flow is subsonic in the wave-stationary frame, while the weak wave has relatively smaller changes in the flow quantities and the downstream flow is supersonic. In addition, there also exists a Chapman–Jouguet (CJ) wave, corresponding to the tangency point of the Rayleigh line and the Hugoniot curve, for which the wave velocity is well defined and its downstream flow is sonic. It was stated that only the strong solution is physically realistic, and that in most situations detonation propagates with the CJ velocity, which is the weakest of the strong detonations.

In this section, we shall further study the characteristics of the one-dimensional detonation wave propagation, first without and then considering the wave structure. We shall also provide more precise descriptions of the various propagation modes, including effects of loss and nonideal processes.

14.5.1. Chapman–Jouguet Detonations

Chapman–Jouguet detonations are the only detonation waves that are well defined within the context of the Rankine–Hugoniot relations based on mass, momentum, and energy conservation. Furthermore, they are dynamically stable as compared to strong detonation waves. This stability arises from the consideration that detonations, being compressive waves, are always followed by rarefaction waves due to mass conservation in a given physical space. Since the downstream flow of strong detonations is subsonic, these rarefaction waves will penetrate into the detonation wave structure and continuously weaken it until the flow becomes sonic, which corresponds to the CJ propagation. The resulting sonic plane then acts as an information barrier that isolates the detonation wave structure from further attenuation, and hence enables the CJ detonation to be self-sustained.

As physical examples of the preference for CJ propagation, let us first consider propagation of the planar detonation in a tube with a closed end, with the wave moving away from it. The boundary conditions in the laboratory frame are that the gas velocities are zero both ahead of the detonation wave and at the closed end. Since the gas has acquired a velocity upon passing the detonation, this velocity must diminish and eventually vanish at the closed end. Such a transition is accomplished through rarefaction waves. Alternatively, we can argue that since the gas is highly compressed upon passing through the wave, it must be sufficiently expanded before reaching the closed end so that a void is not created. Thus if the flow downstream of the detonation wave is subsonic relative to the detonation wave, that is, the detonation is stronger than the CJ wave, these rarefaction waves will intrude into and weaken the wave structure until the flow becomes sonic at its downstream boundary, as just mentioned. The gas velocity downstream of this boundary will be continuously reduced by the rarefaction waves until it vanishes in the laboratory frame.

A similar reasoning can be extended to a detonation wave propagating from the open end of a tube. Here the far downstream pressure is the same as the upstream pressure, and hence is lower than that immediately downstream of the detonation wave. Consequently, for a subsonic downstream, rarefaction waves generated from the open end will propagate into the wave structure to reduce the pressure until the CJ state is attained. The overall detonation structure is therefore similar to that identified for the closed end, except the downstream boundary of the rarefaction region is determined by the attainment of the ambient pressure.

The final example is the spherically expanding detonation, which is the three-dimensional analog of the one-dimensional propagation from the closed end. Since the velocities at the center and upstream of the detonation are both zero, we can therefore again argue for the preference for CJ propagation.

Explicit expressions were derived in Section 7.1 for the CJ wave velocity and the downstream state. It can be readily shown that, in the limit of a strong wave, that is, $M_u \gg 1$ or $\hat{q}_c \gg 1$, the CJ detonation velocity, D_{CJ}, and the downstream state are given by

$$D_{\mathrm{CJ}} = \sqrt{2(\gamma^2 - 1)q_c}, \quad \frac{p_{\mathrm{CJ}}}{\rho_u D_{\mathrm{CJ}}^2} = \frac{v_{\mathrm{CJ}}}{D_{\mathrm{CJ}}} = \frac{1}{\gamma + 1}, \quad \frac{a_{\mathrm{CJ}}}{D_{\mathrm{CJ}}} = \frac{\gamma}{\gamma + 1}. \tag{14.5.1}$$

Note that for the rest of this chapter we shall use the symbols u and v to designate flow velocities measured in the wave-stationary and laboratory frames respectively; the latter symbol is not to be confused with that used for the specific volume in the Rankine–Hugoniot relations or the y-velocity in the description of two-dimensional flows. Furthermore, we have also omitted the subscript b to the downstream properties p_{CJ}, v_{CJ}, and a_{CJ} because the implication is obvious.

14.5.2. Overdriven Detonations

Conceptually, a strong detonation can be maintained by having it supported by a piston whose velocity is independently controlled, thereby providing the downstream boundary condition of the wave. Thus if the piston velocity is larger than the downstream velocity of the corresponding CJ wave, then the downstream velocity of such a wave can remain subsonic, resulting in a strong detonation. Such an overdriven detonation is not self-sustaining because it relies on the piston motion to maintain the rear fluid pressure. It is also highly unstable in that any small change in the piston velocity will generate either compression or rarefaction waves that will modify the detonation velocity.

14.5.3. Taylor Expansion Waves

Knowing the CJ velocity, it is straightforward to determine the one-dimensional flow field that is established behind a CJ wave, which is followed by a piston of given velocity. This flow field is of practical interest in assessing the mechanical effects of detonation on confining structures or the flow field established by propagating detonation waves in propulsive devices such as the pulse detonation engine. Figure 14.5.1 illustrates the flow fields for four situations of different piston velocities that we shall study in some detail, recognizing that the piston velocity governs the structure of the flow field. Specifically, case (a) is for the situation of a piston moving at exactly the velocity of the burned products immediately behind the detonation wave, v_{CJ}. Consequently, the pressure and flow velocity remain constant behind the detonation wave. Faster piston velocities will lead to overdriven detonations. Cases (b) and (c) are for piston velocities lower than v_{CJ}. In these cases, an unsteady expansion is developed behind the detonation wave, which decelerates the particle velocity so as to match the piston velocity. Case (c) is for zero piston velocity, which corresponds to the special situation of the detonation propagating from the closed end of a tube. Case (d) corresponds to a receding piston representing the expansion front of the detonation products propagating into a medium of low pressure, as would occur in the case of detonation initiation at the open end of a tube.

Analytical solutions for the different flow fields (Taylor 1950a) can be obtained by the method of characteristics. We assume the CJ detonation is initiated at the piston surface at ($x = 0$, $t = 0$). Since there is no reference time or length scale involved, the flow field is self-similar. Therefore we seek the dependence of the flow velocity and thermodynamic profiles on the similarity variable x/ta_{CJ}, where the downstream sonic speed at the CJ state is chosen as a convenient reference velocity. The boundary conditions are prescribed along the detonation path given by $x/ta_{\mathrm{CJ}} = D_{\mathrm{CJ}}/a_{\mathrm{CJ}} = (v_{\mathrm{CJ}} + a_{\mathrm{CJ}})/a_{\mathrm{CJ}}$, where the CJ state prevails ($p = p_{\mathrm{CJ}}, a = a_{\mathrm{CJ}},\ \rho = \rho_{\mathrm{CJ}}$). Along the trajectory of the piston surface given by $x/ta_{\mathrm{CJ}} = v_p/a_{\mathrm{CJ}}$, the fluid velocity equals that of the piston, that is, $v = v_p$.

Assuming the flow to be isentropic, with a constant γ, the problem is simply that analyzed in Section 14.4.2. The solution yields the Riemann invariants

$$J_{\pm} = \frac{2a}{\gamma - 1} \pm v = \text{constant}, \tag{14.4.24}$$

which are conserved along the two families of characteristics ζ^{+} and ζ^{-} given by

$$\frac{dx}{dt} = v \pm a. \tag{14.4.21}$$

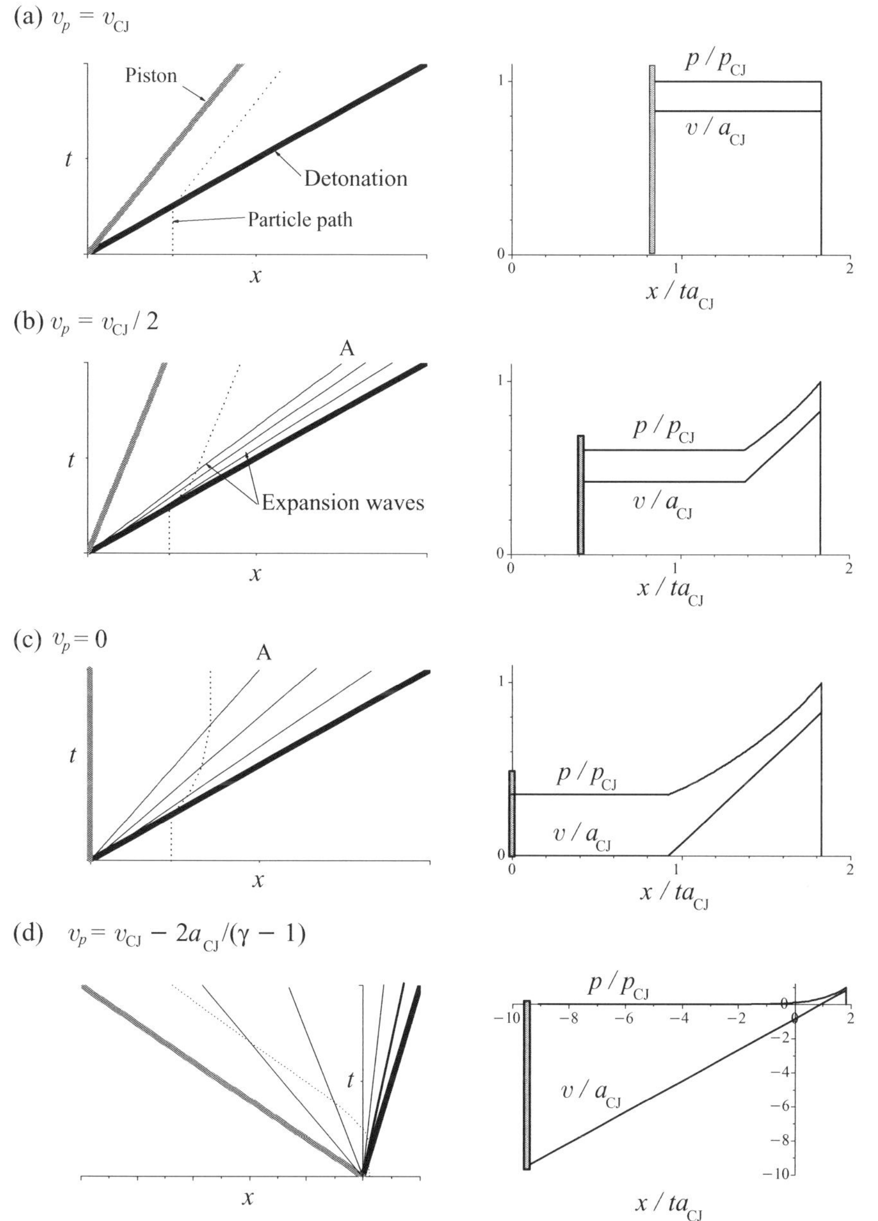

Figure 14.5.1. The structure of the Taylor expansion waves between a CJ detonation and a constant velocity piston for four different piston velocities. First column illustrates the detonation path, piston path, particle path, and expansion waves in a space-time diagram. The second column provides the pressure and particle velocity distributions in terms of the similarity variable x/ta_{CJ}.

Since all ζ^- characteristics pass through the CJ state, all Riemann invariants J_- are constant and have the same value, yielding a general relation between the sonic speed and particle velocity for any interior point,

$$J_- = \frac{2a}{\gamma - 1} - v = \frac{2a_{CJ}}{\gamma - 1} - v_{CJ}. \tag{14.5.2}$$

We note here that Eq. (14.5.2) can be directly used to determine the piston velocity corresponding to the free expansion illustrated in Figure 14.5.1d, as discussed above. Since the ζ^- characteristics passing through the CJ point also extend into the vanishing pressure medium (i.e., $a/a_{\mathrm{CJ}} = 0$), we arrive at the boundary condition

$$v_p = -\frac{2a_{\mathrm{CJ}}}{\gamma - 1} + v_{\mathrm{CJ}} \tag{14.5.3}$$

for case (d).

Further recognizing that in general the solution is self-similar in x/t and the detonation propagates along a straight path $x/t = D_{\mathrm{CJ}}$ with constant state, there is a simple wave solution such that the ζ^+ characteristics are all straight lines given by

$$\frac{x}{t} = v + a. \tag{14.5.4}$$

Equation (14.5.4) thus sets the dependence of the profiles on the similarity variable. Combined with Eq. (14.5.2) we get the dependence of the particle velocity on x/t:

$$\frac{v}{a_{\mathrm{CJ}}} = \left(\frac{\gamma - 1}{\gamma + 1}\right)\frac{D_{\mathrm{CJ}}}{a_{\mathrm{CJ}}} - 1 + \frac{2}{(\gamma + 1)}\frac{x}{ta_{\mathrm{CJ}}}. \tag{14.5.5}$$

The solution given by Eq. (14.5.5) is shown in Figure 14.5.1 for the different cases considered, where we have assumed that the CJ state is given by the strong wave approximation, Eq. (14.5.1). The velocity decays to the piston velocity at an interior point, hence a fluid element is first subjected to gas expansion, followed by a region of constant state.

It is of interest to obtain the solution for the trajectory of the boundary separating the constant and varying regions of particle velocities. This occurs when $v = v_p$, from which we obtain the trajectory of the trailing edge of the expansion wave, A, shown in Figure 14.5.1, given by

$$\left(\frac{x}{ta_{\mathrm{CJ}}}\right)_{\mathrm{A}} = \frac{(\gamma + 1)}{2}\frac{v_p}{a_{\mathrm{CJ}}} - \frac{(\gamma - 1)}{2}\frac{D_{\mathrm{CJ}}}{a_{\mathrm{CJ}}} + \frac{\gamma + 1}{2}. \tag{14.5.6}$$

It can be verified that for case (d), the piston trajectory coincides with that given by Eq. (14.5.6). The gas therefore undergoes a continuous expansion when it is expanded to zero pressure.

Once the particle velocity is known, the distribution of sonic speed is readily obtained from Eq. (14.5.2). Since the flow is isentropic, the pressure and density profiles are given by the isentropic relations for a perfect gas,

$$\left(\frac{p}{p_{\mathrm{CJ}}}\right)^{(\gamma-1)/2\gamma} = \left(\frac{\rho}{\rho_{\mathrm{CJ}}}\right)^{(\gamma-1)/2} = \frac{a}{a_{\mathrm{CJ}}}. \tag{14.5.7}$$

The pressure profiles are also shown in Figure 14.5.1 for the four cases considered. The constant pressure at the piston surface, equal to that at the weak discontinuity A, is given by letting $v = v_p$ in Eq. (14.5.2) and with Eq. (14.5.7), yielding

$$\frac{p_p}{p_{\mathrm{CJ}}} = \left[1 - \frac{(\gamma - 1)}{2}\frac{(v_{\mathrm{CJ}} - v_p)}{a_{\mathrm{CJ}}}\right]^{2\gamma/(\gamma-1)}. \tag{14.5.8}$$

For case (c) corresponding to the closed wall condition of zero velocity piston, and assuming the detonation state is given by Eq. (14.5.1), the pressure at the wall is simply given by

$$\frac{p_P}{p_{\mathrm{CJ}}} = \left(\frac{\gamma+1}{2\gamma}\right)^{2\gamma/(\gamma-1)}. \tag{14.5.9}$$

This so-called pressure "plateau" value is generally of substantial interest in assessing the impulse generated by a detonation on a flat surface.

14.5.4. ZND Structure of Detonation Waves

A description of the structure of the steadily propagating detonation wave was independently proposed by Zel'dovich (1940), von Neumann (1942), and Döring (1943), and, hence, is referred to as the ZND structure. The structure consists of an infinitely thin leading shock that compresses the reactant gas to high pressure and temperature. The state immediately downstream of this shock is called the Neumann state, designated by the subscript N. Due to the high velocity of the flow, diffusive transport is neglected and the shocked gas undergoes a convected thermal explosion. The compression waves emanating from the reacting gas then maintains the leading shock at constant strength. For CJ propagation, the structure is terminated when the gas reaches the sonic velocity in the detonation wave frame. For the idealized adiabatic one-dimensional planar propagation, this sonic state is attained at downstream infinity where chemical equilibrium is attained, provided the reaction mechanism does not attain a state where the main rate of energy release falls to zero before global equilibrium—an aspect of the eigenvalue nature of the detonation structure that will be discussed in the next section.

The path taken by a fluid element in the ZND structure from the initial state i to the final CJ state can be followed in the Rankine–Hugoniot pressure–volume, p-ρ^{-1} phase diagram, shown in Figure 14.5.2. It is seen that, instead of the direct upward transition from i to CJ along the Rayleigh line, the flow actually reaches the final state CJ from the initial state i by a discontinuous transition from i to N_{CJ} across a shock, and then $N_{\mathrm{CJ}} \to$ CJ along the Rayleigh line.

Parameters characterizing the Neumann state for a perfect gas with constant γ are given by the normal shock jump relations (Liepmann & Roshko 1957):

$$M_N^2 = \frac{2+(\gamma-1)M_u^2}{2\gamma M_u^2-(\gamma-1)} \tag{14.5.10}$$

$$\frac{\rho_N}{\rho_u} = \frac{(\gamma+1)M_u^2}{(\gamma-1)M_u^2+2} \tag{14.5.11}$$

$$\frac{p_N}{p_u} = 1+\frac{2\gamma}{\gamma+1}(M_u^2-1) \tag{14.5.12}$$

$$\frac{T_N}{T_u} = \frac{a_N^2}{a_u^2} = 1+\frac{2(\gamma-1)}{(\gamma+1)^2}\frac{(\gamma M_u^2+1)}{M_u^2}(M_u^2-1). \tag{14.5.13}$$

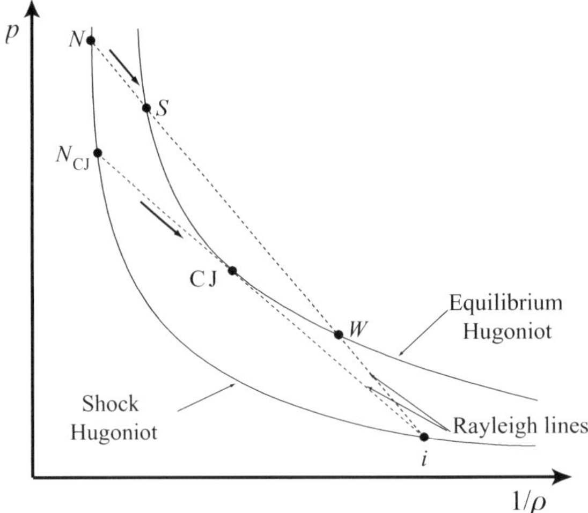

Figure 14.5.2. Rankine–Hugoniot pressure–volume diagram representing the evolution of a fluid element state traversing the reaction zone structure of overdriven and CJ detonations.

These relations assume particularly simple expressions for strong shocks, for which $M_u \gg 1$:

$$\rho_N \approx \frac{\gamma+1}{\gamma-1}\rho_u, \qquad p_N \approx \frac{2}{\gamma+1}\rho_u u_u^2,$$

$$T_N \approx \frac{2\gamma}{(\gamma+1)^2}\frac{u_u^2}{c_p}, \qquad M_N \approx \sqrt{\frac{\gamma-1}{2\gamma}},$$

$$u_N = \frac{\rho_u}{\rho_N}u_u \approx \frac{\gamma-1}{\gamma+1}u_u, \qquad v_N = u_u - u_N \approx \frac{2}{\gamma+1}u_u. \tag{14.5.14}$$

With the detonation velocity $D_{CJ} = u_u$ given by the equilibrium CJ theory, and using the Neumann state as the initial condition, the detonation structure can be obtained by integrating the steady, one-dimensional conservation equations for the flow, together with the chemical kinetic rate equations, until equilibrium is achieved asymptotically. Figure 14.5.3 shows the calculated profiles for the temperature, pressure, particle velocity (u), and heat release rate for a detonation wave in an initial mixture of stoichiometric methane and air at atmospheric conditions. It is seen that the wave structure consists of an almost thermally neutral induction zone, followed by rapid exothermicity and hence gas expansion.

Because of the near thermally neutral character of the induction zone, the various thermodynamic and gasdynamic state variables do not change appreciably, and the induction stage can be approximated by a convected thermal explosion. The induction length obtained from the ZND model can thus be approximated by

$$\ell_{ig} = u_N \tau_{ig}(p_N, T_N), \tag{14.5.15}$$

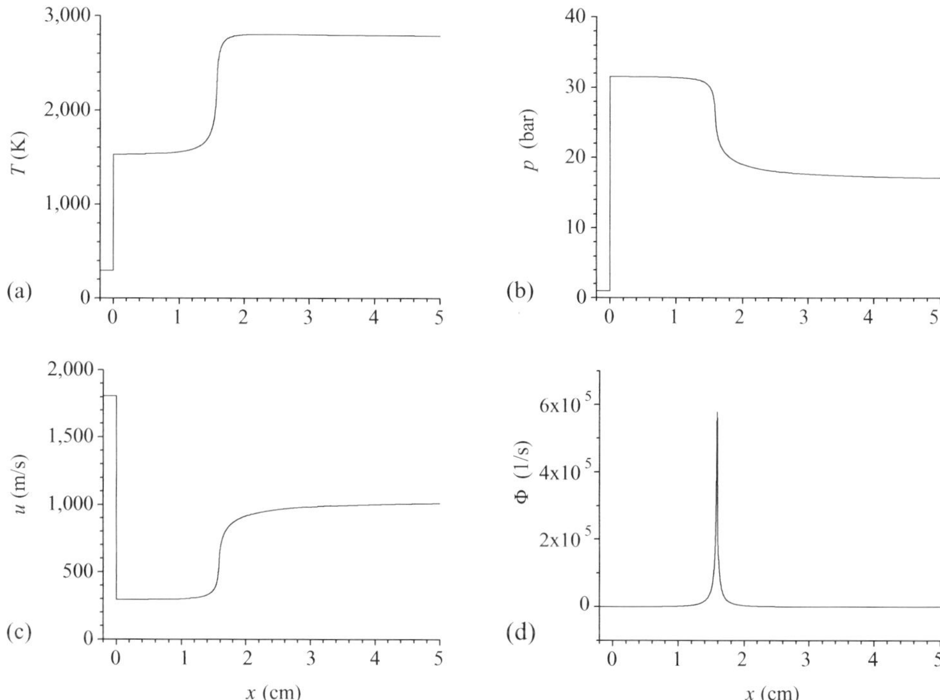

Figure 14.5.3. Calculated ZND structure of a CJ detonation in a stoichiometric methane–air mixture initially at atmospheric conditions: (a) temperature, (b) pressure, (c) gas velocity and (d) rate of energy release are plotted in terms of distance behind the leading shock.

where τ_{ig} is the ignition delay time for the mixture evaluated at the Neumann state of p_N and T_N, given by

$$\tau_{ig}(p_N, T_N) \sim \rho_N/w(p_N, T_N) \sim \rho_N/\left(p_N^n e^{-T_a/T_N}\right). \qquad (14.5.16)$$

The ignition delay time is the link to chemistry for detonations and can be simply evaluated in the manner of the thermal explosion discussed in Section 8.1.

The structure of an overdriven detonation supported by a piston can be obtained in a similar fashion. In particular, the flow now reaches the final state (S) corresponding to the strong detonation from the initial state i by the path $i \to N \to S$, as shown in Figure 14.5.2.

The ZND structure for the overdriven detonation can also be used to rule out the existence of weak detonations, which terminate at point W. Since the gas always proceeds down the Rayleigh line from the Neumann state, state S is first achieved. A transition from state S to state W would involve an entropy-violating expansion shock, which is physically not possible.

The solution for the structure of an overdriven detonation can be similarly obtained by integrating the governing equations with the initial, Neumann state being stronger than that of the CJ state. The ratio of the square of the detonation velocity to that of the ideal CJ value defines the overdrive factor F. Figure 14.5.4 shows the

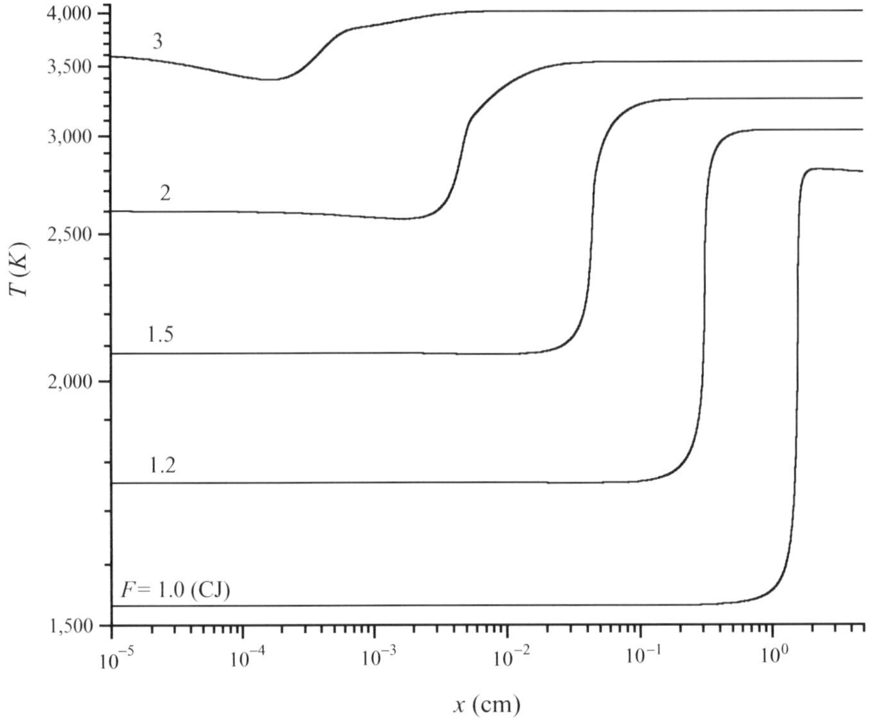

Figure 14.5.4. Calculated ZND structure of overdriven stoichiometric methane–air detonations initially at atmospheric conditions. F denotes the overdrive parameter, the ratio of the square of the detonation velocity to that of the ideal CJ detonation velocity.

ZND structure of overdriven stoichiometric methane–air detonations with various overdrive factors. It is seen that, with increasing overdrive, the shock is stronger and the ignition delay time is consequently much shorter. The sequence of overdriven profiles also illustrates well the Arrhenius dependence of the global induction kinetics on the Neumann state. Furthermore, for high values of the overdrive, say $F = 3$, the shock temperature is so high that products remain dissociated and the flow is mainly endothermic during the induction period before thermal runaway.

14.5.5. Eigenvalue Structure of Quasi-One-Dimensional Detonations

The ZND model presented above is highly idealized. In reality, there are many effects that can compromise the strict one dimensionality and adiabaticity of the wave. For example, as we shall discuss later, the ZND model is unstable to multi-dimensional perturbations and as such exhibits spatial and temporal fluctuations. The departure from strict one dimensionality in partially confined or unconfined detonations also leads to divergence of the streamlines behind the detonation front, which can act as lateral losses of mass and momentum. Furthermore, for detonations propagating in narrow tubes, the front can be affected by frictional and heat loss, and there can also be volumetric heat and momentum loss through radiation and particle suspension. If these processes are weak, they can be modeled as source terms in the governing equations for the steady one-dimensional ZND structure. We shall now show that the inclusion of such effects fundamentally changes the wave structure, rendering it to assume an eigenvalue character.

In the fixed reference frame, the one-dimensional conservation equations for mass, momentum, and energy with their respective source terms m, f, and g can be written as

$$\frac{\partial \rho}{\partial t} + \frac{\partial(\rho v)}{\partial x} = m \tag{14.5.17}$$

$$\frac{\partial(\rho v)}{\partial t} + \frac{\partial(p + \rho v^2)}{\partial x} = f \tag{14.5.18}$$

$$\frac{\partial \rho(h + \frac{1}{2}v^2)}{\partial t} + \frac{\partial \rho v(h + \frac{1}{2}v^2)}{\partial x} - \frac{\partial p}{\partial t} = g, \tag{14.5.19}$$

where for simplicity we assume the gas to be perfect and the total enthalpy given by

$$h = \frac{\gamma}{(\gamma - 1)}\frac{p}{\rho} + \lambda q_c, \tag{14.5.20}$$

with λ being a reaction-progress variable such that $\lambda = 0$ denotes chemical equilibrium. For a one-step overall reaction λ can be simply identified as the normalized mass fraction of the reactant.

Applying the transformation

$$\begin{aligned} &\chi = x_s(t) - x, \qquad \tau = t, \\ &\frac{\partial}{\partial t} = \frac{\partial}{\partial \tau} + D\frac{\partial}{\partial x}, \qquad \frac{\partial}{\partial x} = -\frac{\partial}{\partial \chi}, \\ &u = D - v, \end{aligned} \tag{14.5.21}$$

the above equations can be transformed into the reference frame moving with the steady detonation wave, with their steady-state being

$$\frac{d(\rho u)}{d\chi} = m \tag{14.5.22}$$

$$\frac{d(p + \rho u^2)}{d\chi} = Dm - f \tag{14.5.23}$$

$$\frac{d\rho u(h + \frac{1}{2}u^2)}{d\chi} = D^2 m - \tfrac{1}{2}Df + g. \tag{14.5.24}$$

The state at the end of the reaction zone, where the flow is sonic, cannot be obtained based on the CJ criterion for equilibrium, which requires $m = f = g \equiv 0$. Instead, it is to be determined from integration of the governing equations for the structure between the Neumann state and the sonic plane. This can be demonstrated by rearranging the governing equations to the following form for, say, u:

$$\frac{du}{d\chi} = \frac{\Phi}{1 - M^2}, \tag{14.5.25}$$

where

$$\Phi = \{(\gamma - 1)\left(g - \rho u q_c \frac{d\lambda}{d\chi}\right) + m[(\gamma - 1)\left(\tfrac{1}{2}D^2 - (h + \tfrac{1}{2}u^2)\right) + a^2 - \gamma u(D - u)]$$
$$+ f[\gamma u - (\gamma - 1)D]\}/(\rho a^2) \quad (14.5.26)$$

is called the thermicity of the system. Similar expressions can be derived for $d\rho/d\chi$ and $dp/d\chi$, with their respective denominators for the RHS containing the $(1 - M^2)$ term.

By inspection, the solution to Eq. (14.5.25) is singular at the sonic surface, where $M = 1$ and the denominator of the RHS vanishes. Thus a unique solution for the reaction zone structure is obtained by the simultaneous requirement of $\Phi = 0$. This is the generalized CJ criterion (Fickett & Davis 1979), which is almost always encountered in the presence of source terms in the governing equations. Clearly, at the sonic surface, the requirement $\Phi = 0$ implies either chemical equilibrium, $d\lambda/d\chi = 0$, and all sources have decayed to zero, or a balance between the different source terms m, f, g, and $d\lambda/d\chi$. The generalized CJ criterion is satisfied for only specific values of the detonation wave velocity, D, which therefore becomes the "eigenvalue."

In deriving the above results we have not specified the nature of the reaction and the distinction between frozen and equilibrium sound speeds. If the reaction is irreversible, then such a distinction is irrelevant. However, if the reaction is reversible, then the relation $a^2 = \gamma p/\rho$ that we have used in deriving Eq. (14.5.25) would correspond to the frozen sound speed. This is consistent with the eigenvalue nature of the problem in that vanishing of the denominator at the sonic state, in the presence of loss, defines the eigenvalue of the detonation velocity, with the sonic plane located at a finite distance downstream of the shock. On the other hand, in the absence of loss as in CJ propagation, this generalized CJ criterion no longer applies. The time to reach the sonic plane then becomes infinitely long and the equilibrium sound speed is the relevant sound speed. In this case, the denominator of Eq. (14.5.25) does not vanish when the numerator vanishes at the equilibrium state (Sharpe 2000).

The eigenvalue nature of the wave will be further studied in Section 14.8.2 through the problem of the propagation of a spherical detonation wave, where the effect of flow divergence is examined in the context of detonation initiation.

14.6. UNSTEADY THREE-DIMENSIONAL DETONATIONS

While much work has been devoted to the simple steady-state, one-dimensional planar configuration for detonations, and much insight has been gained on the limiting forms for the structure in the presence of loss, it is known since the early 1960s that this structure is highly unstable and is prone to result in transient three-dimensional structures. Dissipative and turbulent processes have also been found to be important. Furthermore, although detonation velocities are predicted well by the CJ values for fairly reactive mixtures, large quantitative departures are frequently found between experimental values and predictions based on the idealized one-dimensional model for the detonation dynamic parameters related to limit situations such as initiation requirements, detonation limits, and diffraction criteria (Lee 1984; Radulescu & Lee 2002).

In the following we shall first demonstrate that detonations are inherently unstable by discussing the pulsating instability of a one-dimensional detonation. We then discuss the complex three-dimensional structure of detonations characterized by the interaction of triple-shock units.

14.6.1. Pulsating Instability of the ZND Structure

We first analyze the sensitivity of the one-dimensional ZND model to perturbations in terms of the reaction and the flow. Let us momentarily introduce a disturbance to a detonation such that its propagation speed $D_{CJ} = u_u$ is, say, increased. The strengthening of the shock then increases the post-shock temperature and pressure, which leads to an increase in the reaction rate and, hence, a decrease in the induction length. At the same time, an increase in the shock strength would increase the downstream flow velocity and, hence, the induction length. If the effect of these two competing factors results in a reduction in the ignition length, then the net heat release rate is increased. This generates a pressure pulse that would propagate upstream acoustically, leading to a further strengthening of the shock. The system is therefore unstable. The converse holds if the ignition length is increased.

The propensity to destabilize can therefore be characterized by the sensitivity of the induction length ℓ_{ig} to changes in the detonation velocity D_{CJ} in that instability is expected for

$$\frac{d\ln\ell_{ig}}{d\ln D_{CJ}} < 0. \tag{14.6.1}$$

Criterion (14.6.1) can be readily evaluated by using Eqs. (14.5.15) and (14.5.16) for ℓ_{ig}, the strong shock relations of (14.5.14), and the strong detonation relations of (14.5.1), with

$$u_N \sim D_{CJ}, \; \rho_N \sim \text{constant}, \; p_N \sim D_{CJ}^2,$$

$$T_N \approx \frac{2\gamma(\gamma-1)}{(\gamma+1)^2}\frac{D_{CJ}^2}{c_p}, \quad D_{CJ}^2 = 2(\gamma^2-1)q_c.$$

This yields

$$\frac{8\gamma(\gamma-1)}{(\gamma+1)}\frac{E_a}{R^o T_N}\frac{q_c}{c_p T_N} + 2n > 1, \tag{14.6.2}$$

as the criterion for inherent instability. It shows the dual influence of the Arrhenius temperature sensitivity and chemical heat release on the propensity to destabilize, being promoted with larger values of E_a and q_c. With the exception of situations with $\gamma \to 1$ and $n \ll 1$, (14.6.2) is invariably satisfied for combustion systems. Alternate, rigorous stability analyses of the ZND structure (Lee & Stewart 1990; Clavin & Williams 2002) yield similar implications.

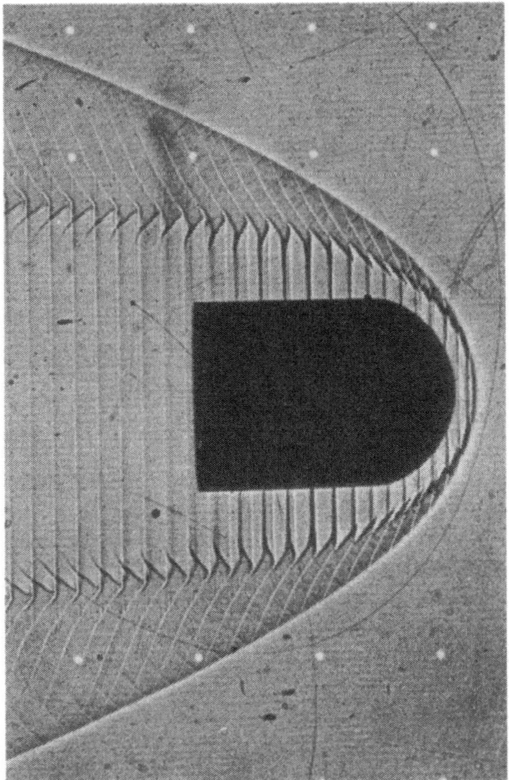

Figure 14.6.1. Schlieren photograph of a blunt axisymmetric projectile moving into a dilute hydrogen–oxygen mixture at a velocity close to the CJ detonation velocity, showing that the shock-induced reaction zone is unstable to perturbations in the bow shock, leading to the saw tooth shape in the interfaces separating reacted and unreacted gases with very long induction delay times (Lehr 1972).

Figure 14.6.1 shows a schlieren photograph (Lehr 1972) of the one-dimensional instability induced by the bow shock generated by firing a blunt supersonic projectile into a hydrogen–oxygen–inert mixture. High frequency periodic oscillations are clearly exhibited in the pattern of the reaction zones and are consequences of the wave interactions along the streamline of the projectile (Alpert & Toong 1972), as shown in the space-time diagram in Figure 14.6.2. The instability occurs primarily along the stagnation streamline, where the reactions are fast due to the high temperatures. The density discontinuities generated during the quasi-one-dimensional pulsating instability, clearly identifiable in the highly sensitive laser schlieren photograph, are convected downstream to a region of colder flow. The mechanism leading to the oscillation can be demonstrated by first letting the strength of the shock be momentarily increased by a disturbance at $t = t_0$. Due to the increase in temperature at the shock, a particle shocked after the disturbance, at $t = t_0^+$, reacts at $t = t_1$, before the fluid element shocked before the disturbance, at $t = t_0^-$, reacts at $t = t_2$. The earlier heat release in the reaction zone at t_1 sends compression waves to the

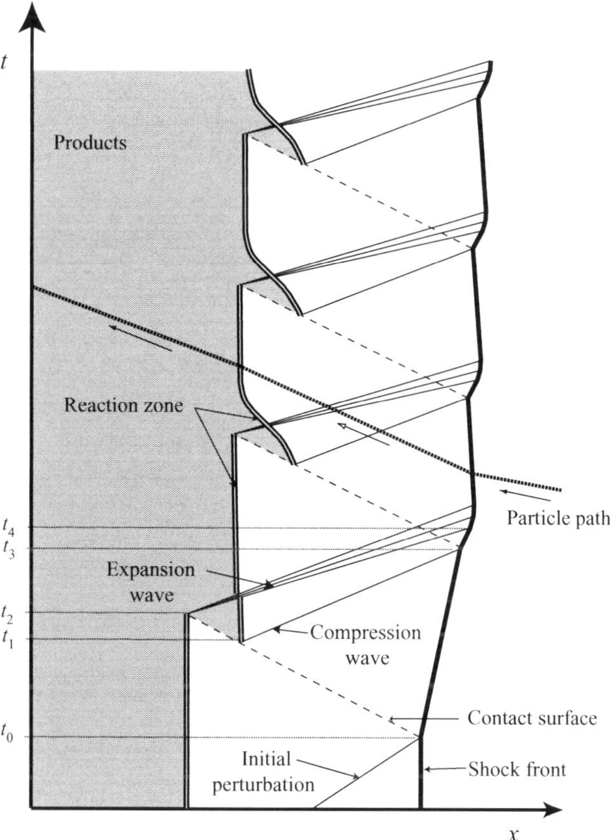

Figure 14.6.2. Space-time diagram illustrating the mechanism producing weak one-dimensional periodic instabilities in shock-induced reactions (adapted from Alpert & Toong 1972).

shock, hence further strengthening it at $t = t_3$. However, when the undisturbed particle reacts at $t = t_2$, the lower pressure in the undisturbed product is communicated to the leading shock as expansion waves, which reach the shock at $t = t_4$. At about the same time, the increased shock strength at t_3 has again led to an increase in the heat release rate in the reaction zone, which has sent a new compression wave that arrives and hence strengthens the shock again. The positive and negative forcing through compressions and expansions reaching the leading shock from the reaction zone thus give rise to the observed cyclic pattern of Figure 14.6.1, with the frequency of the pulsation controlled by the induction plus the acoustic times. This phenomenon is the simplest mode of oscillation, and is usually encountered near stability boundaries.

14.6.2. Triple-Shock Structure

The detonation structure is also spatially unstable. Similar to the discontinuous pulsating detonation front discussed above, the three-dimensional structure of the detonation front is discontinuous and unsteady due to the transverse waves that propagate

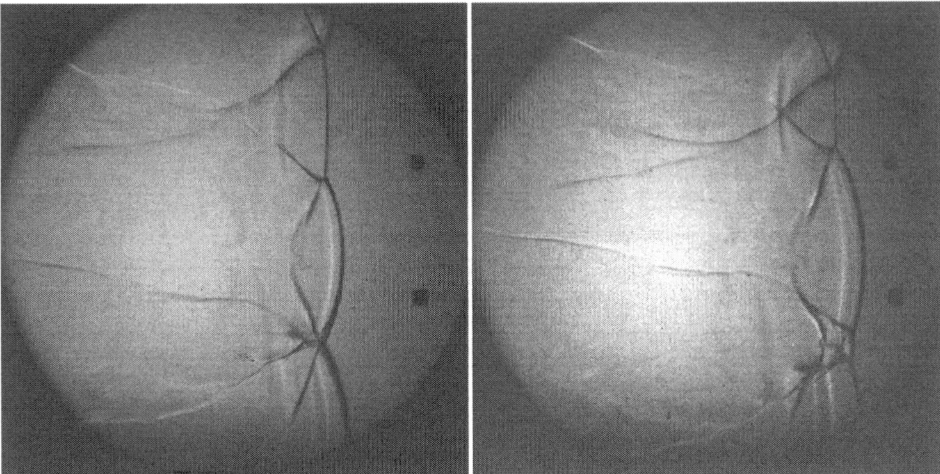

Figure 14.6.3. Schlieren photographs obtained in a thin channel illustrating the triple-shock structures evolving on the surface of a weakly unstable detonation; the initial mixture is $2H_2 + O_2 + 85$ percent Ar at 20 kPa initial pressure (Austin 2003).

obliquely to the front. An example of the evolution of the frontal structure of a weakly unstable detonation wave is shown in Figure 14.6.3 (Austin 2003). The shadowgraph clearly illustrates that the front is neither planar nor smooth. Shock–shock intersections occur, resulting in triple-shock, Mach reflection patterns that evolve with time and interact with each other. In this section, we shall therefore discuss the structure of these triple-shock interactions. They have significant effects on the local reaction rates because the mixture is shocked to different extents when it passes through the different components of a triple-shock structure.

Figure 14.6.4 shows an interferogram of a triple-shock system (Law & Glass 1971) obtained by diffracting a normal incident shock I of Mach number M_I by a sharp compressive wedge. When the wedge angle is sufficiently large, the diffraction results in a regular reflection pattern, consisting of the incident shock and a trailing reflected shock. However, when the wedge angle is smaller than a critical value, regular reflection is not possible. The diffraction instead results in a Mach reflection pattern, consisting of a Mach stem, M, which protrudes ahead of the incident shock, and a reflected shock R, which trails the incident shock, as shown schematically in Fig. 14.6.5. These three shocks meet at a triple point T, which travels in the laboratory frame with an angle θ_I relative to I. Thus in the shock-stationary frame the upstream flow approaches I with a Mach number $M_1 = M_I/\sin\theta_I$ at an angle θ_I. Since there is no characteristic dimension in the system, the entire triple-shock configuration remains similar as it translates along the triple-point trajectory.

Each of I, R, and M is oriented obliquely to its respective upstream flow. Thus the fundamental element of a triple-shock interaction is the deflection of a flow by an oblique shock. Taking the incident shock as an example, it is seen that the incident

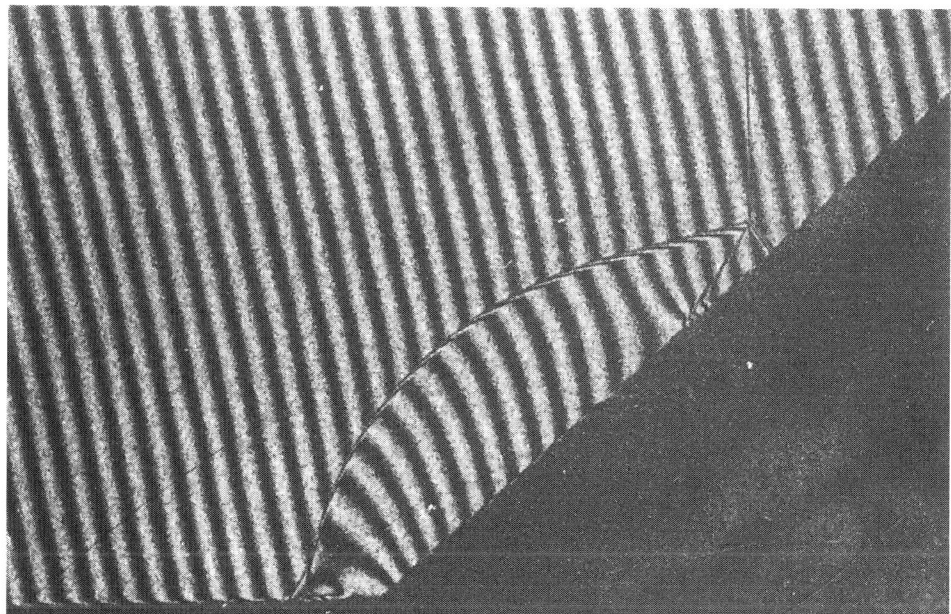

Figure 14.6.4. An interferogram showing the triple-shock configuration resulting from the diffraction of a normal shock by a wedge; the incident shock Mach number is 2.1, the initial temperature and pressure are 298.8 K and 55.6 torr respectively, and the gas is O_2 (Law & Glass 1971).

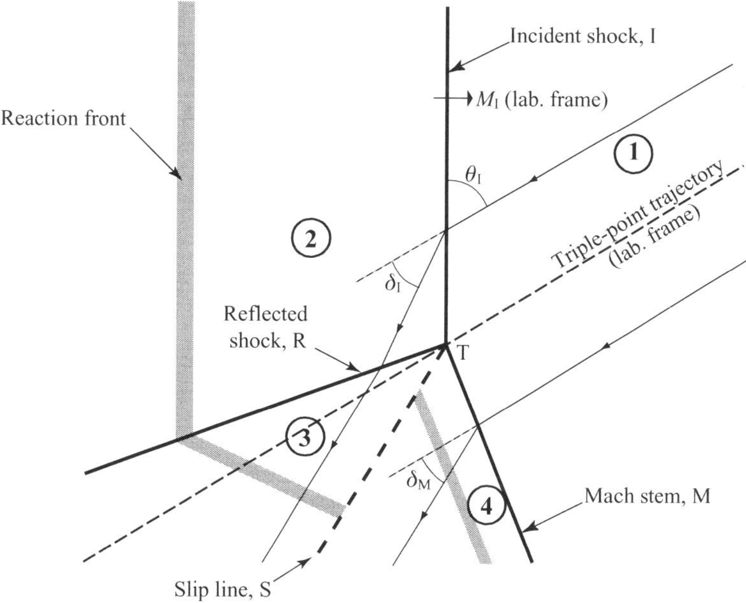

Figure 14.6.5. Schematic showing the triple-shock configuration.

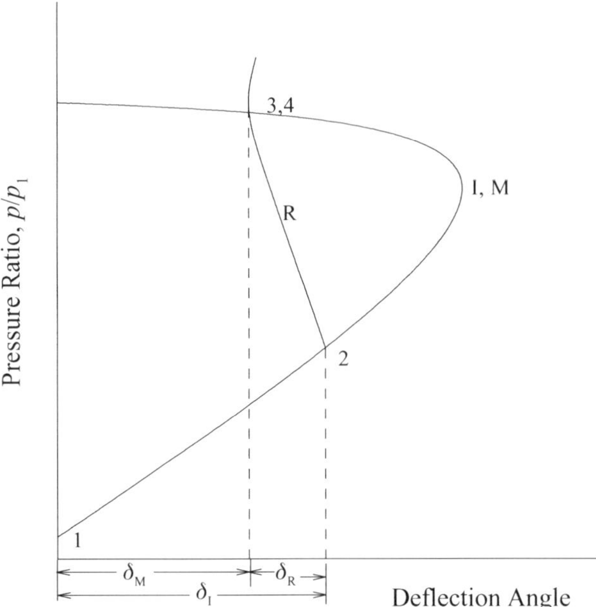

Figure 14.6.6. Schematic of the shock polar for the Mach reflection solution.

flow is deflected by an angle δ_I in crossing I, its Mach number changes from M_1 to M_2, and its pressure jumps from p_1 to p_2. For an ideal gas with constant specific heat the state downstream of this incident shock is given by

$$M_2^2 = \frac{2+(\gamma-1)M_1^2}{2\gamma M_1^2\sin^2\theta_I-(\gamma-1)} + \frac{2M_1^2\cos^2\theta_I}{2+(\gamma-1)M_1^2\sin^2\theta_I} \tag{14.6.3}$$

$$\cot\delta_I = \tan\theta\left[\frac{(\gamma+1)M_I^2}{2(M_I^2\sin^2\theta_I-1)}-1\right] \tag{14.6.4}$$

$$\frac{p_2}{p_1} = \frac{2\gamma M_I^2\sin^2\theta_I-(\gamma-1)}{(\gamma+1)}. \tag{14.6.5}$$

For a given M_1, a plot of p_2/p_1 versus the deflection angle δ_I yields the shock polar I as shown in Figure 14.6.6. It is seen that there exists a maximum δ_I beyond which there is no solution, and that for smaller values of δ_I there are two solution branches. Furthermore, it can also be shown that the downstream flow is subsonic for the upper branch ($M_2 < 1$) and supersonic for the lower branch ($M_2 > 1$), except for a small regime around the turning point on the lower branch for which the flow is subsonic. Thus the shock strength, as represented by the pressure ratio p_2/p_1, is weaker for the lower branch and stronger for the upper branch.

Now consider the interaction of I with the Mach stem M. Since these two shocks are oriented at different angles to the oncoming flow, the only possibility that dynamic equilibrium can be achieved for the downstream flow is through the triple-shock configuration shown in Figures 14.6.4 and 14.6.5. This involves the modification of the orientation of M as well as the presence of the reflected shock R such that the flows in regions 3 and 4 have the same direction and pressure. The temperature, density, and flow speed are however still different in these two regions. The surface separating them is called a slipstream S, which is dynamically unstable and tends to roll up into a vortex.

This arrangement can be further clarified by the shock polar analysis of Figure 14.6.6, noting that the polar I now also represents all possible orientations of the Mach stem M because both I and M have the same upstream flow Mach number M_1. Thus we can mark off state 2 corresponding to the incident shock angle θ_I on the shock polar (I, M). From state 2 we can erect a second polar corresponding to the Mach number M_2 in region 2. The intersection point, which gives state 3 on the R shock polar and state 4 on the (I, M) polar, assures the attainment of dynamic equilibrium in regions 3 and 4 because of the equality of pressure and deflection angle. We also note that while state 3 is supersonic, state 4 is always on the subsonic branch of the (I, M) polar. The subsonic nature of state 4 then allows pressure waves downstream of M to reach it so that its orientation and strength can be adjusted to produce the overall triple-shock structure. Furthermore, since the incident shock has a smaller angle of incidence than that of the Mach shock, or $\theta_I < \theta_M$, M is a stronger shock than I.

When the flow is reactive, there will be a broad region of induction followed by a narrow region of rapid heat release behind each of the incident, Mach, and reflected shocks, as schematically shown in Figure 14.6.5. Since M is stronger than I, the induction period is shorter behind M. This is an important feature in the propagation of multidimensional detonation waves, to be discussed next.

14.6.3. Triple-Shock Interactions

Figure 14.6.7 shows the etchings left on smoke foils placed on the walls of a detonation tube upon the passage of a detonation wave. These etchings are the trajectories traced out by the triple points as they undergo repeated collisions. The cell spacing is typically one to two orders larger than the chemical induction length, ℓ_{ig}. With special experimental conditions, the cellular structure can be very regular (Figure 14.6.7a), although in general the patterns observed are quite irregular (Figure 14.6.7b) and often display substructures superimposed on the larger scale cellular structure (Figure 14.6.7c).

To analyze the collision dynamics of the triple-shock systems, Figure 14.6.8 shows the trajectories of a pair of identical triple-shock units undergoing collision (Toong 1983; Strehlow 1984). It is seen that since the TI shock has a smaller angle of incidence than that of the TM shock, measured relative to the triple-point trajectory, TI is clearly the incident shock, TM the Mach shock, and TR the reflected shock. After collision the triple points T′ move away from each other, and the incident angle of the T′I′ shock is now smaller than that of the T′M′ shock. Thus T′I′ is now the incident shock while T′M′ the Mach shock, with T′R′ being the reflected shock.

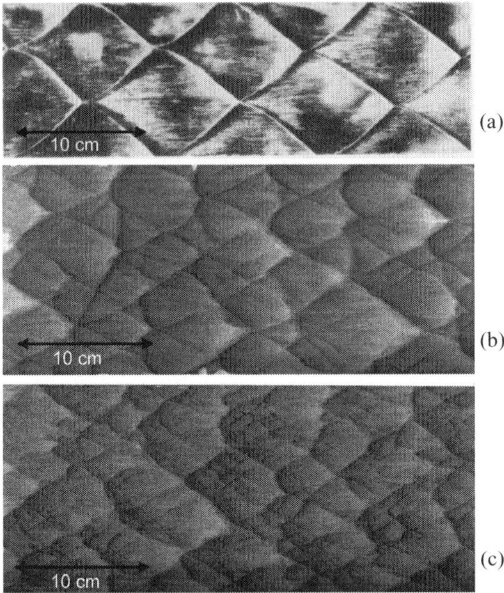

Figure 14.6.7. Detonation cellular structure recorded by the etchings of triple points on plates covered with soot for (a) a regular cellular structure ($2H_2 + O_2 + 70$ percent Ar at 9.3 kPa initial pressure), (b) irregular cellular structure ($2H_2 + O_2 + 60$ percent N_2, 20 kPa initial pressure) and (c) irregular cellular structure with substructure ($C_3H_8 + 5O_2 + 60$ percent N_2 at 20 kPa initial pressure); Figure 14.6.7a from Strehlow (1968) and Figures 14.6.7b and 14.6.7c from Austin (2003).

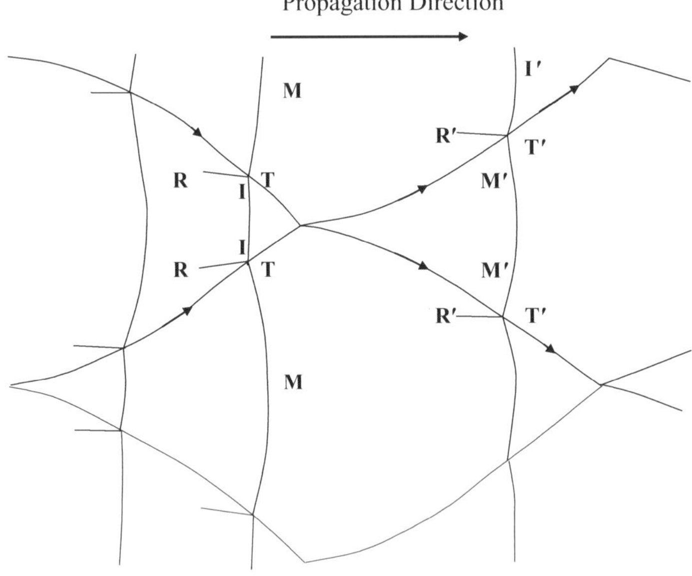

Figure 14.6.8. Schematic showing the collision between two triple-shock units.

The next point to note is that in spite of the different designations, the TM and T′I′ shocks are actually the same shock because it is not affected by the collision. Thus in terms of the shock strength we can write TM = T′I′. Further recognizing that the Mach shock is always stronger than the incident shock, the relative strengths of the various shocks before and after collision can be ordered according to

$$\mathrm{TI} < \mathrm{TM} = \mathrm{T'I'} < \mathrm{T'M'}. \tag{14.6.6}$$

Thus by changing into T′M′, the TI shock is repeatedly strengthened upon collisions. The strengthening is particularly pronounced in the presence of chemical reactions, which are readily initiated downstream of the Mach shock as compared to the incident shock, and compensates for the decay of the detonation wave due to the various loss or weakening mechanisms between collisions. This is the mechanism that has been suggested to be responsible for the sustenance of detonation waves in multi-dimensions, although recent studies (Austin, Pintgen & Shepherd 2005; Radulescu, Law & Sharpe 2005; Radulescu et al. 2005) have shown a more complex process, involving turbulent mixing and diffusion. Some aspects of these processes are briefly discussed next.

14.6.4. The Complex Structure

The picture that emerges is that shock–shock interactions occurring on the detonation front introduce periodic variations in the strength of the leading shock. In between triple-shock collisions, it has been determined that the velocity of the leading shock changes continuously from approximately 1.6 to 0.7 of the average velocity, which is close to the CJ velocity. The substantial variation in shock strength and temperature behind the shock therefore significantly affects the reaction rate and thereby the shock structure. For example, overdriving a wave to conditions generally encountered in the beginning of the cell cycle can reduce the ignition delay by four orders of magnitude. Conversely, reaction rates can also drop substantially during the decay of the leading shock to strengths below the average, causing a large fraction of the gas to accumulate unreacted behind the detonation front. Figure 14.6.9 shows the structure of such a typical detonation. The schlieren photograph clearly identifies the two interacting triple-shock units as well as the density interfaces separating the reacted gas from the unreacted gas. It is seen that the reaction rate behind the protruding, Mach stems is very high, with the reaction fronts being very close to the shock front. Detailed studies from computation and experiments using chemiluminescence showed that the center portion of the shock is the weaker incident shock behind which the reaction rate is much slower than the average, and the gas accumulates unreacted in tongue-shaped structures. After the triple-point collisions, these tongues are subsequently separated from the front and convected downstream as unreacted pockets. These pockets then react, possibly through turbulent mixing with the hot reacted gas, leading to wave structures more complicated than described through the well-defined scenario of triple-shock interactions.

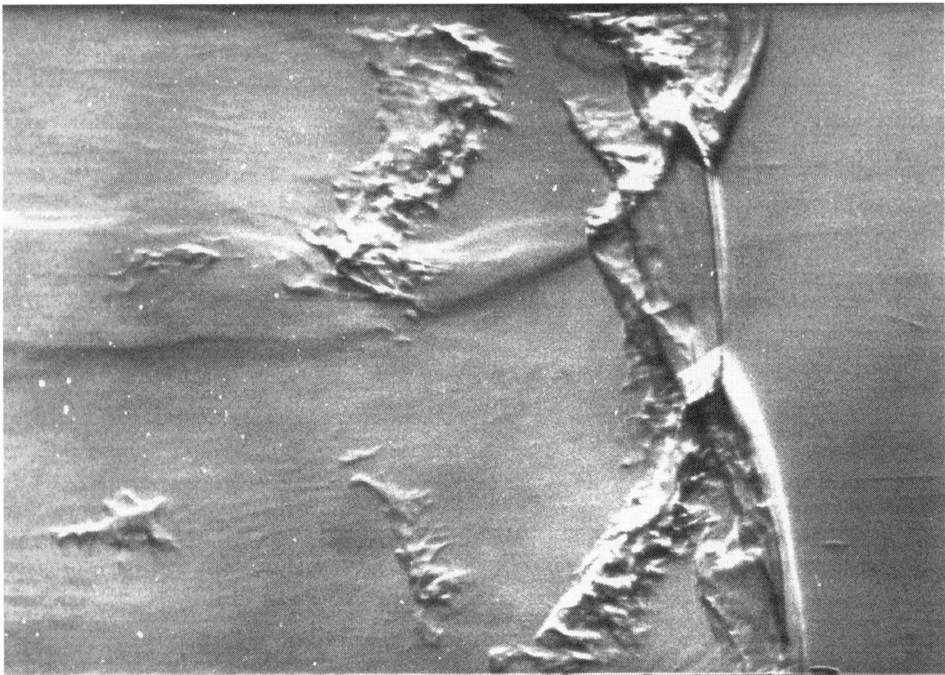

Figure 14.6.9. Schlieren photograph of the unstable cellular reaction zone structure of a $CH_4 + 2O_2$ detonation wave with 3.4 kPa initial pressure. Due to the high sensitivity of the reaction rates to the leading shock strength, the reaction rates behind the incident shock (central part of the front) are significantly lower than the reaction rates behind the Mach stems (top and bottom bulges), leading to large portions of unreacted gases convected behind the front (Radulescu, Law & Sharpe 2005).

14.7. PROPAGATION OF STRONG BLAST WAVES

There are many situations in which the deposition of a finite amount of energy at a sufficiently rapid rate in a compressible medium can lead to the formation of a spherical shock wave propagating away from the source. Examples are the setting off of a chemical or nuclear bomb in mid-air, and the accidental explosion of a combustible through spark discharge. If the environment is inert, then the intensity of the blast wave will gradually decay as it expands. However, if the environment is reactive, the blast wave can initiate chemical reactions in its downstream. If the reactions are sufficiently fast and can be coupled to the dynamics of the blast wave, a detonation wave can subsequently be initiated. In this section, we shall study the former situation, involving the propagation and decay of a blast wave in an inert environment. The situation when the environment is reactive will be studied in the next section, in which we shall derive the criterion governing the success or failure in the initiation of a detonation wave by a blast wave. It is also emphasized that although the present problem does not involve chemical reactions, it is of central interest to combustion because blast waves are frequently initiated by the rapid, localized release of chemical energy.

The problem was first solved independently by Taylor (1950b), von Neumann (1941), and Sedov (1946). Thus at time $t = 0$ a large quantity of energy E is instantaneously released at a certain point, $r = 0$, in a quiescent nonreactive gas. A strong spherical shock wave is generated and subsequently propagates outward with progressively reduced intensity. We are interested in determining the trajectory of the shock radius, $R_s(t)$, and the development of the flow field behind it, during the initial stage when the shock remains strong such that the contribution of the initial internal energy of the gas to the total energy engulfed by the blast is negligible.

Referring to the strong shock relations in (14.5.14), and by using the subscripts 1 and 2 in place of u and N respectively, it is seen that in this limit the density ratio across the shock, ρ_2/ρ_1, approaches the constant $(\gamma + 1)/(\gamma - 1)$, and the upstream pressure p_1 can be neglected relative to the downstream pressure p_2 because $p_2/p_1 \sim M_1^2 \gg 1$. Thus the problem is completely defined by two parameters, namely E and ρ_1. Since the dimensions of these two quantities are respectively ML^2/T^2 and M/L^3, it is not possible to form either a characteristic length or a characteristic time. The problem is therefore of a self-similar nature.

From the dimensions of E, ρ_1, r, and t, the only dimensionless similarity independent variable that can be formed is

$$\xi = \frac{r}{(E/\rho_1)^{1/5} t^{2/5}}. \tag{14.7.1}$$

Equation (14.7.1) shows that $r \sim t^{2/5}$. In particular, for the location of the shock surface, R_s, we have

$$R_s(t) \sim t^{2/5}. \tag{14.7.2}$$

Consequently the shock velocity $u_1 = D$ is given by

$$D(t) = \frac{dR_s(t)}{dt} = \frac{2}{5}\frac{R_s}{t} \sim t^{-3/5}, \tag{14.7.3}$$

which shows that the blast velocity diminishes with time as $t^{-3/5}$. A celebrated verification of the above result is given by the observation of the first atomic explosion in New Mexico in 1945 (Taylor 1950c), as shown in Figure 14.7.1.

Knowing $u_1(t)$, the strong shock relations of (14.5.14) then imply that immediately behind the blast wave ρ_2 remains constant in time while v_2 and p_2 diminish as $t^{-3/5}$ and $t^{-6/5}$ respectively.

The complete solution for this problem can be found in the texts by Landau and Lifshitz (1959) and Barenblatt (1996). We shall, however, present an approximate, but simpler and physically more illuminating solution given by Chernyi in 1957 and discussed in Zel'dovich and Raizer (1966). The analysis assumes that the gas compressed by the strong explosion wave is concentrated in a thin layer behind the shock front. Within this layer the density and velocity remain constant at their respective values, ρ_2 and v_2, immediately behind the shock. The thickness of this layer, δ,

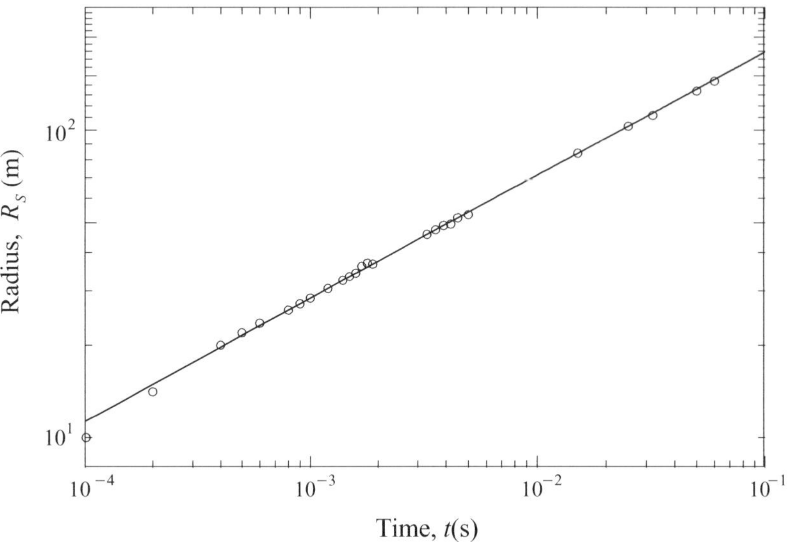

Figure 14.7.1. Experimentally measured radii of the blast wave from the first atomic bomb, showing their similarity variation according to $R_s \sim t^{2/5}$ (adapted from Taylor 1950c).

can be estimated by equating the mass engulfed by the blast wave, m, to that in the layer, as

$$m = \frac{4\pi}{3} R_s^3 \rho_1 = 4\pi R_s^2 \delta \rho_2, \quad \text{or} \quad \frac{\delta}{R_s} = \frac{1}{3}\frac{\rho_1}{\rho_2} \approx \frac{1}{3}\frac{\gamma - 1}{\gamma + 1}. \tag{14.7.4}$$

As an example, for $\gamma = 1.4$ we have $\delta / R_s = 0.0556$, which is a rather small number.

Next we assume that the motion of this thin layer is sustained by the pressure p_c within the inner core of the expanding sphere that acts on the interior surface of the layer. Thus even we have concentrated the entire mass in this layer, we are also implicitly assuming the presence of a small amount of mass in the inner core in order to effect the core pressure. Then, from Newton's law of motion, we have

$$\frac{d(mv_2)}{dt} = 4\pi R_s^2 p_c. \tag{14.7.5}$$

Expressing p_c as a fraction α of p_2, $p_c = \alpha p_2$, and since from strong shock relations we have $v_2 \approx 2D/(\gamma + 1)$ and $p_2 \approx 2\rho_1 D^2/(\gamma + 1)$, Eq. (14.7.5) becomes

$$\frac{1}{3}\frac{d(R_s^3 D)}{dt} = \alpha R_s^2 D^2. \tag{14.7.6}$$

Using $d/dt = (dR_s/dt)d/dR_s = D d/dR_s$, Eq. (14.7.6) can be integrated to yield

$$D = c R_s^{3(\alpha - 1)}, \tag{14.7.7}$$

where c is an integration constant. Further integrating Eq. (14.7.7) with $D = dR_s/dt$ yields the history of the blast wave,

$$R_s = [(4 - 3\alpha)ct]^{1/(4-3\alpha)}, \tag{14.7.8}$$

and subsequently that of its velocity.

The above results are given in terms of the constants α and c, which can be determined through energy conservation in that the initial energy, E, is a constant and is equal to the kinetic energy of the layer, $mv_2^2/2$, plus the internal energy of the core, given by

$$\frac{4\pi}{3}R_s^3\rho_c c_v T_c = \frac{4\pi}{3}R_s^3\frac{p_c}{(\gamma - 1)}. \tag{14.7.9}$$

Using Eq. (14.7.7) and the strong shock relations for v_2 and $p_2 = p_c/\alpha$, energy conservation is given by

$$E = \frac{4\pi}{3}\rho_1 c^2\left[\frac{2}{(\gamma+1)^2} + \frac{2\alpha}{(\gamma^2 - 1)}\right]R_s^{3(2\alpha-1)}. \tag{14.7.10}$$

Since E is a constant and hence independent of R_s, the exponent to R_s must vanish, implying that

$$\alpha = \frac{1}{2}, \tag{14.7.11}$$

and

$$c = \left[\frac{3}{4\pi}\frac{(\gamma - 1)(\gamma + 1)^2}{(3\gamma - 1)}\right]^{1/2}\left(\frac{E}{\rho_1}\right)^{1/2}. \tag{14.7.12}$$

Using $\alpha = \frac{1}{2}$ in Eqs. (14.7.7) and (14.7.8), we retrieve the similitude relations such as $R_s \sim t^{2/5}$ and $D \sim R_s^{-3/2} \sim t^{-3/5}$ identified earlier.

Substituting c into Eq. (14.7.7) yields the relation between the radius and velocity of the blast wave for a given E,

$$E = \left[\frac{4\pi}{3}\frac{(3\gamma - 1)}{(\gamma - 1)(\gamma + 1)^2}\right]\rho_1 D^2 R_s^3. \tag{14.7.13}$$

This relation will be used in the next section on the minimum energy needed for direct detonation initiation. For $\gamma = 1.4$, the coefficient term within the square bracket in Eq. (14.7.13) assumes the numerical value of 5.82, which is quite close to the value 5.31 obtained from the rigorous solution.

Finally, we note that since $p_2 \sim D^2$ and $D \sim c^{2/5} \sim E^{1/5}$ from Eqs. (14.7.7), (14.7.8), and (14.7.12), the overpressure varies with the energy release as $p_2 \sim E^{2/5}$.

Figure 14.7.2 shows the variations of v/v_2, p/p_2, and ρ/ρ_2 as functions of r/R_s, obtained from the rigorous solution. It is seen that while v/v_2 varies somewhat linearly downstream of the blast wave, both p/p_2 and ρ/ρ_2 decrease rapidly, with p/p_2 approaching a finite $O(1)$ value and ρ/ρ_2 decreasing to almost zero. Consequently, the temperature would increase to very large values in the opposite manner as the

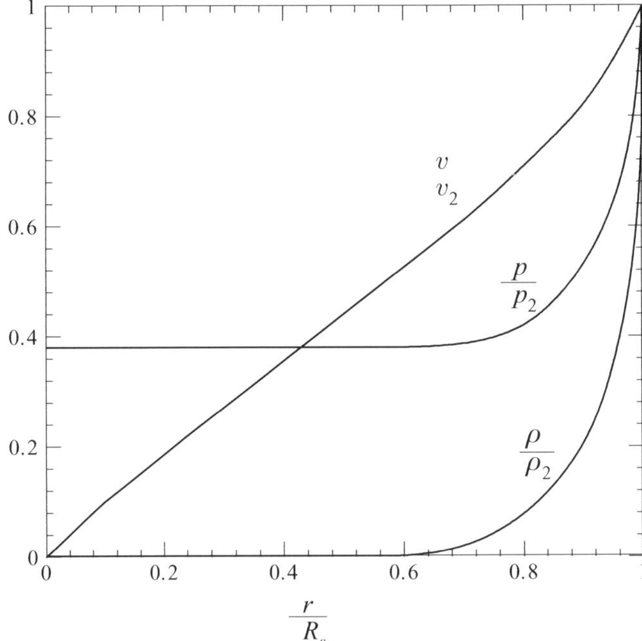

Figure 14.7.2. Theoretical profiles of flow velocity, pressure, and density behind a strong blast wave, for $\gamma = 1.4$.

decrease of the density, and almost all the gas resides in a thin layer immediately downstream of the shock. These results justify the assumption made in obtaining the approximate solution. In reality, such large temperature and density gradients near the core of the explosion are very unstable and are rapidly damped by viscous effects. The extreme low density of the bulk of the spherical wave also explains the highly buoyant nature of such strong blast waves.

14.8. DIRECT DETONATION INITIATION

The need to predict the initiation of detonation waves is perhaps one of the key fundamental and practical issues in the study of detonations. Experimentally, there are various ways through which a detonation can be initiated, with each mechanism embodying a different facet of the initiation phenomenon. Categorically, a detonation wave can be initiated either directly, through a rapid deposition of energy, or indirectly where a detonative state is achieved through the acceleration of a series of progressively more reactive states. In this section, we shall study the process of direct initiation, using the problem of blast wave initiation as an example.

14.8.1. The Zel'dovich Criterion

In blast wave initiation, if the reaction rate induced by the blast wave is sufficiently fast, a sonic surface is established behind the detonation wave and isolates the detonation structure from the downstream rarefaction waves, leading to self-sustained propagation. Zel'dovich et al. (1957) first proposed that, for successful initiation, the radius of the blast wave should be at least equal to the induction length of the CJ wave, ℓ_{ig}, when the velocity of the blast wave has decayed to that of the CJ wave, D_{CJ}. Substituting these conditions in Eq. (14.7.13) for the blast front decay, but generalizing the result to the planar, cylindrical, and spherical geometries, for $j = 0, 1, 2$, yields an explicit relation for the critical energy:

$$E_{j,cr} = k_j \rho_1 D_{CJ}^2 \ell_{ig}^{j+1}, \tag{14.8.1}$$

where $k_j = 1, 3.93, 5.31$ for the three geometries, for $\gamma = 1.4$. Equation (14.8.1) gives the important relationship that the critical energy for a spherical blast wave varies with the cube of the induction length of the detonation wave.

Surprisingly, the above result underpredicts the experimentally determined critical energy by a huge amount—approximately six to eight orders of magnitude. If instead of using the ZND induction length, we substitute the real thickness of the detonation wave structure, which is of the order of the dimension of a cell, the criterion still underestimates the experimentally determined critical energy in the spherical geometry, but yields reasonable agreement for the planar initiation. These considerations then suggest that the Zel'dovich criterion still captures the correct physics, but only when the front is not curved. It thus appears that significantly more energy is required to establish a self-sustained curved detonation front. This brings us back to the discussion of the eigenvalue structure of detonations, and particularly to the detonation structure in the presence of mass divergence after a flow has crossed a curved shock that is convex relative to the upstream flow, as would occur in the outwardly propagating spherical or cylindrical waves. The divergence of the streamlines introduces attenuating effects on the front, and hence inhibits the ease of initiation of these waves. This mechanism is demonstrated in the following.

14.8.2. Curvature-Induced Quenching Limit

We now analyze the structure and propagation of a curved detonation wave. The governing equations for a spherically propagating detonation in the laboratory coordinate are given by

$$\frac{\partial \rho}{\partial t} + \frac{\partial(\rho v)}{\partial r} + \frac{2\rho v}{r} = 0 \tag{14.8.2}$$

$$\frac{\partial(\rho v)}{\partial t} + \frac{\partial(\rho v^2 + p)}{\partial r} + \frac{2\rho v^2}{r} = 0 \tag{14.8.3}$$

$$\frac{\partial \rho(h + \frac{1}{2}v^2)}{\partial t} + \frac{\partial \rho u(h + \frac{1}{2}v^2)}{\partial r} + \frac{2}{r}\rho u(h + \tfrac{1}{2}v^2) - \frac{\partial p}{\partial t} = 0, \tag{14.8.4}$$

for mass, momentum, and energy conservation, where h is the total enthalpy of the gas, given by Eq. (14.5.20), and

$$(h_b - h_u) = -q_c + c_p(T_b - T_u) = -q_c + \left(\frac{\gamma}{\gamma - 1}\right)\left(\frac{p_b}{\rho_b} - \frac{p_u}{\rho_u}\right). \tag{14.8.5}$$

Transforming Eqs. (14.8.2) to (14.8.4) to the coordinate attached to the shock front, which has a radius R_s and propagates with a velocity D, using

$$\tau = t, \qquad \chi = R_s - r, \qquad u = D - v,$$

we have

$$\frac{\partial \rho}{\partial \tau} + \frac{\partial(\rho u)}{\partial \chi} + \frac{2\rho(D-u)}{(R_s - \chi)} = 0 \tag{14.8.6}$$

$$\frac{\partial(\rho u)}{\partial \tau} + \frac{\partial(\rho u^2 + p)}{\partial \chi} + \frac{2\rho u(D-u)}{(R_s - \chi)} - \rho\frac{dD}{d\tau} = 0 \tag{14.8.7}$$

$$\rho\frac{\partial(h + \frac{1}{2}u^2)}{\partial \tau} + \rho u\frac{\partial(h + \frac{1}{2}u^2)}{\partial \chi} - \frac{\partial p}{\partial \tau} - \rho u\frac{dD}{d\tau} = 0. \tag{14.8.8}$$

If we assume that the flow is quasi-steady and the detonation structure is quasi-planar such that R_s is much larger than the detonation structure ($R_s \gg \chi$), whose length scale is the induction length ℓ_{ig}, then Eqs. (14.8.6)–(14.8.8) simplify to

$$\frac{d(\rho u)}{d\chi} = -\frac{2\rho(D-u)}{R_s} \tag{14.8.9}$$

$$\frac{d(\rho u^2 + p)}{d\chi} = -\frac{2\rho u(D-u)}{R_s} \tag{14.8.10}$$

$$\frac{d(h + \frac{1}{2}u^2)}{d\chi} = 0. \tag{14.8.11}$$

It is instructive to see that Eqs. (14.8.9)–(14.8.11) take the generic form of the eigenvalue structure, Eqs. (14.5.22)–(14.5.24), discussed in Section 14.5.5, with "source" terms in the mass and momentum equations. In fact, Eq. (14.5.25) is retrieved by defining the thermicity as

$$\Phi = -\frac{2(D-u)}{R_s} - \frac{(\gamma-1)u}{a^2}q_c\frac{d\lambda}{d\chi}. \tag{14.8.12}$$

The above expression clearly demonstrates the effect of curvature on the detonation structure. In particular, the term $2(D-u)/R_s$ can be interpreted as the stretch rate experienced by the expanding spherical wave, having the dimension of 1/s. This positive stretch causes flow divergence, and hence slows down the subsonic flow downstream of the shock. At the same time chemical heat release through $d\lambda/d\chi(<0)$ tends to increase the flow velocity. Consequently, if the stretch rate is not large, then a sonic state can be attained, resulting in a steadily propagating detonation wave. However, if the curvature effect is too strong, then a sonic state cannot be reached and detonation initiation fails.

A formal asymptotic analysis (Yao & Stewart 1995) would require the additional conservation equation for the reactant concentration. We shall, however, solve this problem by following the approach of He and Clavin (1994) based on the square-wave model, in which chemical reaction is suppressed during the induction period, and is instantaneously completed at the end of this period. Consequently the chemical effect on the structure is simply manifested through the induction length, given by Eq. (14.5.15). The solution with this approach is algebraically simpler and physically more transparent.

We first integrate Eqs. (14.8.9)–(14.8.11) across the detonation structure, using

$$\begin{aligned} \chi = 0: &\quad u = u_u, \quad p = p_u, \quad \rho = \rho_u, \quad h = h_u. \\ \chi = \ell_{\text{ig}}: &\quad u = u_b, \quad p = p_b, \quad \rho = \rho_b, \quad h = h_b. \end{aligned} \tag{14.8.13}$$

This yields

$$\hat{\rho}_b \hat{u}_b = \hat{D} - 2 \int_0^{\hat{\ell}_{\text{ig}}} \hat{\rho}(\hat{D} - \hat{u}) d\hat{\chi} \tag{14.8.14}$$

$$\gamma \hat{\rho}_b \hat{u}_b^2 + \hat{p}_b = (\gamma \hat{D}^2 - 1) - 2\gamma \int_0^{\hat{\ell}_{\text{ig}}} \hat{\rho}\hat{u}(\hat{D} - \hat{u}) d\hat{\chi} \tag{14.8.15}$$

$$\left(\frac{\gamma}{\gamma - 1}\right) \frac{\hat{p}_b}{\hat{\rho}_b} + \frac{1}{2}\gamma \hat{u}_b^2 = \frac{\gamma}{\gamma - 1} + \frac{1}{2}\gamma \hat{D}^2 + \hat{q}_c, \tag{14.8.16}$$

where the nondimensional quantities are defined as $\hat{\rho} = \rho/\rho_u$, $\hat{p} = p/p_u$, $\hat{\chi} = \chi/R_s$, $\hat{\ell}_{\text{ig}} = \ell_{\text{ig}}/R_s$, $\hat{u} = u/a_u$, $\hat{q}_c = q_c\gamma/[c_p(\gamma - 1)T_u]$. Note that $\hat{D}$ is the Mach number of the detonation propagation velocity.

The terms containing the two integrals in Eqs. (14.8.14) and (14.8.15) represent effects of curvature. Since we have assumed weak curvature, they are perturbation terms and as such can be evaluated by using the one-dimensional planar result $\hat{\rho}\hat{u} \approx \hat{\rho}_u \hat{D}_{\text{CJ}} = \hat{D}_{\text{CJ}}$, yielding

$$2 \int_0^{\hat{\ell}_{\text{ig}}} \hat{\rho}(\hat{D} - \hat{u}) d\hat{\chi} \approx 2\hat{D}_{\text{CJ}} \int_0^{\hat{\ell}_{\text{ig}}} (\hat{\rho} - 1) d\hat{\chi} \approx 2\hat{D}_{\text{CJ}}(\hat{\rho}_{N,\text{CJ}} - 1)\hat{\ell}_{\text{ig}} \approx \frac{4}{\gamma - 1} \hat{D}_{\text{CJ}} \hat{\ell}_{\text{ig}} \tag{14.8.17}$$

$$2 \int_0^{\hat{\ell}_{\text{ig}}} \hat{\rho}\hat{u}(\hat{D} - \hat{u}) d\hat{\chi} \approx 2\hat{D}_{\text{CJ}}^2 \int_0^{\hat{\ell}_{\text{ig}}} \left(1 - \frac{1}{\hat{\rho}}\right) d\hat{\chi} \approx 2\hat{D}_{\text{CJ}}^2 \left(1 - \frac{1}{\hat{\rho}_{N,\text{CJ}}}\right) \hat{\ell}_{\text{ig}} \approx \frac{4}{\gamma + 1} \hat{D}_{\text{CJ}}^2 \hat{\ell}_{\text{ig}}. \tag{14.8.18}$$

In obtaining the last equalities in Eqs. (14.8.17) and (14.8.18), we have used the strong shock relation of $\hat{\rho}_{N,\text{CJ}} \approx (\gamma + 1)/(\gamma - 1)$. Equations (14.8.17) and (14.8.18) can be further developed by using $a_b^2 = \gamma p_b/\rho_b$ and the CJ condition that the flow in the burned state is sonic, that is, $u_b \approx a_b$, yielding $\hat{p}_b/\hat{\rho}_b = \hat{u}_b^2$. Consequently we have

$$(\gamma + 1)\hat{\rho}_b \hat{u}_b^2 = (\gamma \hat{D}^2 + 1) - \frac{4\gamma}{\gamma + 1} \hat{D}_{\text{CJ}}^2 \hat{\ell}_{\text{ig}} \tag{14.8.19}$$

$$\frac{\gamma + 1}{2(\gamma - 1)} \hat{u}_b^2 = \frac{1}{\gamma - 1} + \frac{1}{2}\hat{D}^2 + \frac{\hat{q}_c}{\gamma}. \tag{14.8.20}$$

Using Eqs. (14.8.16), (14.8.19), and (14.8.20), $\hat{\rho}_b$, $\hat{u}_b$, and $\hat{D}$ can be solved. In particular, $\hat{\rho}_b$ can be eliminated by dividing Eq. (14.8.19) by Eq. (14.8.16), resulting in an expression for $\hat{u}_b$, which can be eliminated by combining this expression with Eq. (14.8.20). By further assuming a large $\hat{D}$, an expression can be obtained for $\hat{D}$, as

$$\frac{\gamma^2}{2(\gamma^2-1)}\hat{D}^2\left(1+\frac{8}{\gamma^2-1}\hat{\ell}_{\text{ig}}\right)^2=\frac{1}{\gamma-1}+\frac{1}{2}\hat{D}^2+\frac{\hat{q}_c}{\gamma}. \quad (14.8.21)$$

Finally, expressing $\hat{D}$ as $\hat{D}=\hat{D}_{\text{CJ}}(1-\triangle)$, where $\triangle=(\hat{D}_{\text{CJ}}-\hat{D})/\hat{D}_{\text{CJ}}$ is the deficit of the detonation velocity from its CJ value, substituting it into Eq. (14.8.21), expanding for $\triangle \ll 1$ and $\hat{\ell}_{\text{ig}} \ll 1$, and collecting terms of equal order, we obtain the leading order solution for $\hat{D}_{\text{CJ}}$ as

$$\hat{D}_{\text{CJ}}=2(\gamma+1)+2(\gamma^2-1)\frac{\hat{q}_c}{\gamma}\rightarrow 2(\gamma^2-1)\frac{\hat{q}_c}{\gamma}\ \text{for large}\ \hat{q}_c. \quad (14.8.22)$$

This result agrees with that of (14.5.1), obtained by taking the limit of large $\hat{q}_c$ in Eq. (7.1.19) for CJ detonation, as should be the case. Results at the next order yields the relation for the change in the propagation velocity from the CJ value due to the curvature effect,

$$\triangle=\frac{8\gamma^2}{(\gamma^2-1)}\hat{\ell}_{\text{ig}}. \quad (14.8.23)$$

Equation (14.5.15) provides an independent statement on the influence of chemistry on the induction length. When referenced to the CJ state, the induction length is given by

$$\frac{\ell_{\text{ig}}}{\ell_{\text{ig,CJ}}}=\frac{\exp(T_a/T_N)}{\exp(T_a/T_{N,\text{CJ}})}\approx\exp\left[\beta\frac{(T_{N,\text{CJ}}-T_N)}{T_{N,\text{CJ}}}\right], \quad (14.8.24)$$

where $\beta=T_a/T_{N,\text{CJ}}$ is the relevant Arrhenius number for this problem. However, $T_N \sim \hat{D}^2$ in the limit of a strong shock, which in turn implies that

$$\frac{T_{N,\text{CJ}}-T_N}{T_{N,\text{CJ}}}=\frac{\hat{D}^2_{\text{CJ}}-\hat{D}^2}{\hat{D}^2_{\text{CJ}}}\approx 2\,\triangle\,. \quad (14.8.25)$$

Substituting Eq. (14.8.25) into Eq. (14.8.24) then yields

$$\frac{\ell_{\text{ig}}}{\ell_{\text{ig,CJ}}}\approx 2\beta\triangle. \quad (14.8.26)$$

Eliminating ℓ_{ig} from Eqs. (14.8.23) and (14.8.26) finally results

$$(2\beta\triangle)e^{-2\beta\triangle}=\beta\frac{16\gamma^2}{\gamma^2-1}\frac{\ell_{\text{ig,CJ}}}{R_s}. \quad (14.8.27)$$

Equation (14.8.27) thus provides the link between the detonation velocity and the radius of curvature of the front. Figure 14.8.1 shows the classical, dual-solution, turning point behavior of an inverse C-shaped curve for a fixed radius of curvature, where we have designated the RHS of Eq. (14.8.27) as K. Between the two solutions, the lower branch is the physically realistic one because it retrieves the CJ solution as

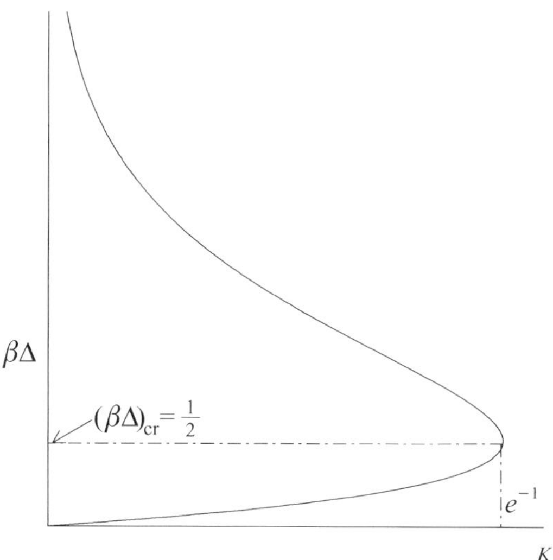

Figure 14.8.1. The dual-solution C-curves for curvature-induced detonation quenching.

the detonation radius approaches infinity and the detonation wave becomes planar. The upper branch is not physical because it requires $\beta\Delta \to \infty$ as $R_s \to \infty$, which contradicts the requirement that $\beta\Delta$ is an $O(1)$ quantity in the above expansion.

Thus starting from the CJ solution on the lower branch, the detonation velocity decreases with decreasing radius of curvature and, hence, increasing flow divergence. Propagation is possible as long as the sonic state can be attained. However, beyond the critical, turning point, flow divergence is of such a magnitude that a sonic surface cannot be attained. Consequently the rear expansion waves would penetrate the reaction zone and significantly attenuate the detonation, rendering steady propagation not possible. Recognizing that the parameter K indicates the severity of curvature, it is clear that at the turning point, designating the state of quenching, we have $K_{\mathrm{cr}} = e^{-1}$ and $(2\beta\Delta)_{\mathrm{cr}} = 1$, or

$$\Delta_{\mathrm{cr}} = \frac{1}{2\beta}, \tag{14.8.28}$$

and no solution exists for $K > K_{\mathrm{cr}}$.

Knowing K_{cr}, the critical detonation radius for sustained propagation, $R_{s,\mathrm{cr}}$, is then given by evaluating Eq. (14.8.23), as

$$\frac{R_{s,\mathrm{cr}}}{\ell_{\mathrm{ig,CJ}}} = \beta\frac{16e\gamma^2}{\gamma^2 - 1}, \tag{14.8.29}$$

which can be generalized to the cylindrical ($j = 1$) and spherical ($j = 2$) geometries by replacing the factor 16 on the RHS by $8j$.

The factor $16e\gamma^2/(\gamma^2 - 1)$ in Eq. (14.8.29) is a large number, assuming a value of about 90 for $\gamma = 1.4$. Furthermore, since β typically assumes values between 5 and 10, the critical radius for the successful initiation of detonation is 500–1,000 times that of the ignition length. This result is to be contrasted with that of the

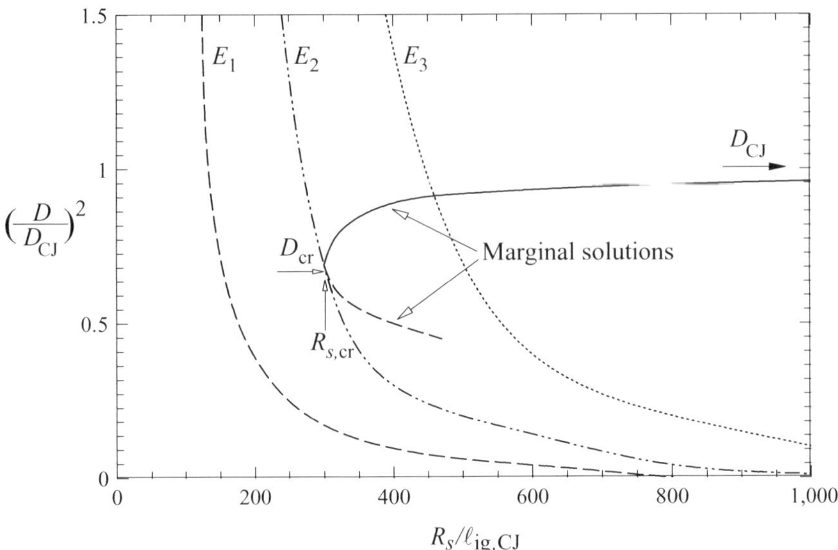

Figure 14.8.2. Combined plots of the decay of the strong blast wave solution and the detonation quenching C-curve, demonstrating the solution range for direct detonation initiation.

Zel'dovich criterion, which equates the critical radius with the ignition length, that is, $R_{s,\mathrm{cr}}/\ell_{\mathrm{ig,CJ}} = O(1)$. This exerts an enormous influence on the critical energy needed to initiate detonation, to be discussed next.

14.8.3. Curvature-Affected Initiation Limit

Knowing the rate of decay of the inert blast wave and the dependence of the detonation velocity on the detonation radius, the detonability of a blast wave initiated by the deposition of a certain amount of energy E can be readily assessed. Figure 14.8.2 shows solutions for the decay of the spherical blast waves for three values of the source energy, together with the quenching curve of Eq. (14.8.27). It is then clear that the blast wave with the source energy E_1 is beyond the turning point of the quenching curve and hence fails to initiate detonation, while that with the source energy E_3 intersects the upper branch of the quenching curve and as such is capable to initiate the detonation, at the critical radius $R_{s,\mathrm{cr}}$. Using this critical value, given by Eq. (14.8.29), for the blast wave radius R_s in the expression for the blast wave energy, Eq. (14.7.13) then readily yields the critical energy for initiation, for the general configuration j, as

$$E_{j,\mathrm{cr}} = \left(\frac{8je\beta\gamma^2}{\gamma^2-1}\right)^{j+1} k_j \rho_1 D_{\mathrm{CJ}}^2 \ell_{\mathrm{ig}}^{j+1}. \tag{14.8.30}$$

Comparing Eq. (14.8.30) with that based on the Zel'dovich criterion, Eq. (14.8.1), shows that the present expression is larger by a factor

$$\left(\frac{8je\beta\gamma^2}{\gamma^2-1}\right)^{j+1},$$

which has a numerical value of the order of 10^8 to 10^9 for the spherical case, and as such places the critical energy within the right range of observed values. The need to properly account for effects due to curvature and hence flow divergence is therefore demonstrated.

14.9. INDIRECT DETONATION INITIATION

There are two major mechanisms through which a detonation can be established through a progression of increased intensity of subsonic burning, namely synchronized initiation and deflagration-to-detonation transition (DDT). These are separately discussed in the following.

14.9.1. Synchronized Initiation

The concept of synchronized initiation can be appreciated by considering a mixture with a temperature gradient which is preconditioned to explode with different induction times along the gradient. If ignition is initiated at the high-temperature end of such a gradient such that compression waves are sent forward, and if the gradient is of such a nature that successive explosions along it occur upon the arrival of the compression waves generated upstream, then these compression waves are amplified by a mechanism which is essentially that of the Rayleigh's criterion, that is, energy sources are in phase with the acoustic waves. This mechanism was dubbed SWACER (Lee & Moen 1980) to describe initiation through "Shock Wave Amplification by Coherent Energy Release." It requires only the existence of a preconditioned induction time gradient field in the medium, which can be produced through, for example, ultraviolet irradiation to generate a gradient of radicals, or turbulent mixing to produce gradients of temperature and reactant concentrations.

To determine whether a spontaneous wave resulting from an induction time gradient, $g_{ig}(x)$, can lead to the formation of a detonation wave, it suffices to compare the speed of the spontaneous ignition wave, U_{spon}, which is simply given by $1/g_{ig}(x)$, with the sonic speed of the gas, a, and the detonation speed D. Based on these parameters, as well as the laminar flame speed, s_L, Zel'dovich (1980) proposed four different possibilities:

(a) $U_{spon} > D$: The reaction wave is so rapid that it resembles a constant volume explosion.
(b) $a < U_{spon} < D$: Transition to detonation through synchronized initiation.
(c) $s_L < U_{spon} < a$: The reaction wave propagates at nearly the spontaneous wave speed with small pressure change across it because the compression waves essentially run away.
(d) $U_{spon} < s_L$: Diffusion dominates, leading to the formation of the laminar flame.

The above consideration does not include the close coupling between gasdynamics and exothermicity via compression and expansion waves that would heat or cool the medium nonuniformly. Due to the extreme sensitivity of induction delay times to temperature variations, this gives rise to highly nonlinear processes that could modify the limits set by the simple induction time gradient consideration presented above.

14.9.2. Deflagration-to-Detonation Transition

Conceptually, the problem of DDT consists of describing the physical processes that are responsible for the acceleration of a flame, propagating via mass and heat transport at velocities lower than a few meters per second, to velocities three to four orders of magnitude higher. At such high velocities convective motion dominates over molecular transport and ignition is effected mainly by adiabatic shock compression. Thus we are interested in how the wave changes from a diffusive to a convective structure. Consider as an example the acceleration process obtained in a smooth tube. Here, after ignition of a flame at the closed end of the tube, the flame propagates into the fresh gas. Due to expansion of the burned gas, compression waves are generated ahead of the diffusion-dominated flame. These compression waves coalesce to form a shock, which compresses the initial gas to higher temperature and pressure. The acceleration of the leading shock preconditions the medium ahead of the flame following it by consecutively shortening the induction delay in successive fluid elements. With increasing velocity of this shock-flame complex, the induction delay time behind the shock can become in phase with the convective motion in the flame complex. Spontaneous fast flames can then develop, replacing the diffusive nature of the flame by a convective nature. This accelerating effect leads to the establishment of detonation waves when the velocity of the spontaneous wave becomes equal to that of the detonation.

The flame acceleration processes described above are rationalized on the basis of laminar flows. However, the smallness of molecular diffusivities, and thereby the laminar flame speeds, compared with the substantially higher velocity needed to sustain a detonation wave, make the evolution of the system from one mechanism to the other extremely inefficient. Indeed, predictions for the onset of detonations in such laminar scenarios would lead to distances on the order of several kilometers! This is contrary to the very short formation distance observed, say for the case of Figure 14.9.1 (Urtiew & Oppenheim 1966; Oppenheim 1985) in which the transition to detonation is accomplished within a distance of about a meter. The images not only show the coalescence of the compression waves but also the presence of complex flow. Detailed studies of such phenomena have revealed the importance of flame wrinkling and turbulent transport. Specifically, it was found that after the formation of a laminar flame, it can become cellularly unstable by the mechanisms discussed in Section 10.9. Furthermore, the flow ahead of the flame is accelerated by the compression waves from the burned products, causing the development of boundary layers along the walls and the flow to become turbulent overall. The presence of obstacles also promotes turbulization of the flow and transverse pressure waves. The subsequent flame interaction with the turbulent flow further wrinkles the flame surface, amplifies the flame burning rate, strengthens the compression waves ahead of the flame, and thereby facilitates the transition to detonation.

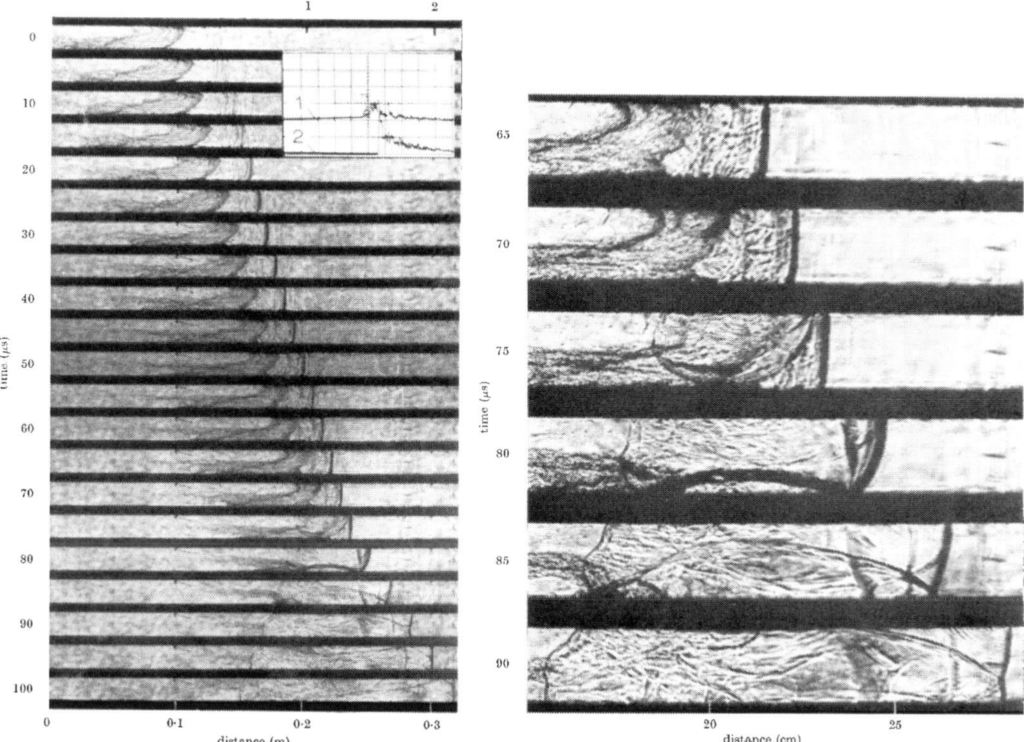

Figure 14.9.1. Sequence of schlieren photographs illustrating the transition of a turbulent deflagration to a detonation wave through the amplification of pressure waves originating near the turbulent flame brush, in $2H_2 + O_2$ and 0.78 bar initial pressure (Urtiew & Oppenheim 1966).

PROBLEMS

1. The coupling between unsteady compressible flows and chemical reactions, which deposit energy in the flow and serve as entropy sources, is best examined by transforming the governing conservation laws in characteristic form. Starting from the conservation laws for a perfect gas in one dimension, show that when the entropy is not constant, the characteristic equations are

$$\frac{D_\pm J_\pm}{Dt} = \frac{a}{\gamma R}\frac{D_\pm s}{Dt} + \left(\frac{\gamma - 1}{\gamma R}\right) a \frac{Ds}{Dt},$$

where $J_\pm$ are the Riemann variables defined by (14.4.24), s is the entropy, R is the gas constant, and the differentials are applied along the three family of characteristics ζ^0, ζ^+, and ζ^- given by Eqs. (14.4.16) and (14.4.21), that is,

$$\frac{D}{Dt} = \frac{\partial}{\partial t} + v\frac{\partial}{\partial x} \quad \text{and} \quad \frac{D_\pm}{Dt} = \frac{\partial}{\partial t} + (v \pm a)\frac{\partial}{\partial x}.$$

Clearly, for nonisentropic flows, the Riemann variables are no longer invariant. Their variation along the ζ^+ and ζ^- characteristic directions depends on the rate of entropy change along a particle path, as given by the rate of energy release following a reacting fluid element, and depends also on the "convected" entropy changes along the ζ^+ and ζ^- characteristic directions. Given the rate of energy release following a fluid element

$$T\frac{Ds}{Dt} = q_c w,$$

and appropriate initial conditions, the evolution of a given flow field can be integrated directly with the system of the three ordinary differential equations given above along the three characteristic directions (Foa 1960).

2. Derive the entropy increase across a reaction wave in a perfect gas mixture for fixed initial stagnation state. Show that the entropy at the CJ detonation state is greater than the entropy at the CJ deflagration state, and that the minimum entropy increase occurs for a constant-pressure combustion process. The entropy increase can be used to evaluate propulsion efficiencies in steady engines using different modes of combustion; the penalties growing with entropy increase. For engines operating with stationary combustion waves, where the entropy increase throughout the steady flow of the engine occurs mainly across the combustion wave, evaluation of the entropy generation should be referenced to the upstream stagnation state. Hence, for this type of engines, the analysis indicates that constant pressure combustion is most beneficial. On the other hand, the static reference state (Problem 7.3) is more appropriate for engines operating unsteadily, such as a pulsed device where the gas is ignited intermittently. Consequently, for a given static initial state, detonations appear to be more efficient than deflagrations based on entropy generation across the combustion wave. (Foa 1960; Wintenberger & Shepherd 2006)

3. The impact of a strong detonation wave at a rigid wall results in a reflected shock wave behind which the gas velocity is zero. Show that the pressure behind the reflected shock is given by

$$p = \frac{5\gamma + 1 + \sqrt{17\gamma^2 + 2\gamma + 1}}{2\gamma}(\gamma - 1)\rho_1 q_c,$$

where $\rho_1 q_c$ is the available energy per unit volume of the reactant gas. You may assume that the detonation wave is a gasdynamic discontinuity where the relations (14.5.1) for a strong detonation hold for the state behind a freely propagating detonation.

4. Consider a detonation initiated at the closed end of a tube and its subsequent reflection at the open end of the tube. Figure 14.P.1 illustrates the wave interactions. The reflected expansion wave originating at A interacts nonlinearly with the Taylor expansion wave until point B and subsequently propagates through a quiescent medium and reaches the closed wall at C.

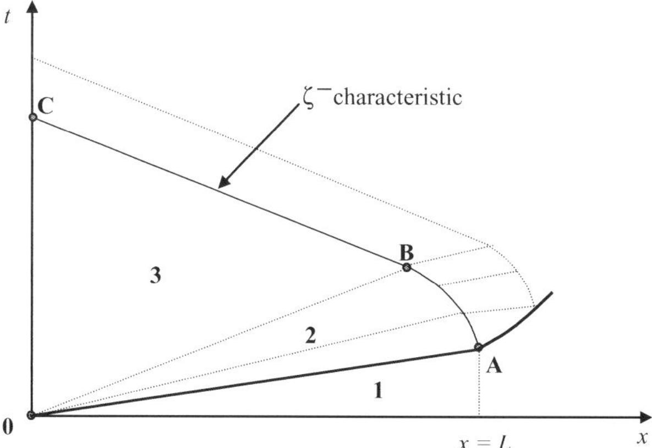

Figure 14.P.1. Wave diagram for a detonation wave reflection at the open end of a tube.

Since the head of the expansion wave propagates along a ζ^- characteristic, with the solution for the $v(\zeta)$ and $a(\zeta)$ profiles in the Taylor wave 2 given in Section 14.5.3, show that the trajectory AB satisfies the following first-order ordinary differential equation:

$$t\frac{d\xi}{dt} + 2\frac{\gamma-1}{\gamma+1}\left(\xi - \frac{D_{\mathrm{CJ}}}{a_{\mathrm{CJ}}} + \frac{\gamma+1}{\gamma-1}\right) = 0,$$

where the similarity variable ξ is defined by $\xi \equiv x/ta_{\mathrm{CJ}}$. Integrate this ODE with appropriate boundary conditions at A to determine the time interval $(t_{\mathrm{B}} - t_{\mathrm{A}})$. Assuming the detonation to be strong, show that the total time between the detonation initiation at the closed wall (t_0) and arrival of the reflected expansion wave back at the closed wall (t_{C}) is given by

$$\frac{2L}{D_{\mathrm{CJ}}}\left(\frac{\gamma+1}{2\gamma}\right)^{-\frac{\gamma+1}{2(\gamma-1)}}.$$

5. The shock-change equation relates the shock pressure to the pressure gradient, the reaction rate, and the sonic parameter behind the shock. Starting from the one-dimensional conservation equations, show that in the laboratory frame for a planar geometry, the rate of change of the shock pressure following the shock is given by

$$\left(\frac{dp}{dt}\right)_{\mathrm{shock}} \equiv \left(\frac{\partial}{\partial t} + D\frac{\partial}{\partial x}\right)p = \frac{\Phi - \frac{\rho a^2}{\rho_u(D-v_u)}\eta\frac{\partial p}{\partial x}}{1 + \rho_u(D-v_u)(1-\eta)^{-1}\left(\frac{dv}{dp}\right)_{\mathrm{Hugoniot}}},$$

where Φ is the thermicity given by $-(\gamma-1)q_c(D\lambda/Dt)$, $(dv/dp)_{\mathrm{Hugoniot}}$ a property of the material relating p and v along the shock Hugoniot, D the shock velocity, v_u the gas velocity ahead of the shock and the sonic parameter is given by

$$\eta \equiv 1 - \frac{(D-v)^2}{a^2}.$$

Clearly, for a nonreactive gas, a positive pressure gradient in the direction of shock propagation, which can be due to expansions, results in the shock decay. Show that for a steadily moving shock, the thermicity balances the expansion due to the pressure gradients. Also show that the above expression results in the master equation given by Eq. (14.5.25) if the shock is steady. (Fickett & Davis 1979)

6. Show that for a strong detonation ($p_u = 0$) in a perfect gas, the sonic condition ($u = a$) can be written uniquely in terms of the density as

$$\frac{\rho^*}{\rho_u} = \frac{\gamma + 1}{\gamma}.$$

Hint: for a strong detonation, the Rayleigh line is given by $p = -\rho_u^2 D^2 \left(\frac{1}{\rho} - \frac{1}{\rho_u}\right)$.

7. Consider the problem of a steady strong detonation with a constant volumetric heat loss q_L. Show that the energy equation

$$\rho u \frac{d}{dx}\left(c_p T + \lambda q_c + \frac{u^2}{2}\right) = -q_L$$

can be written uniquely in terms of the density, yielding

$$\frac{1}{\rho}\frac{d\rho}{dx} = \left(\frac{\gamma - 1}{\gamma}\right)\frac{\rho^2}{\rho_u^2}\frac{1}{D^2}\left(\frac{-q_c\frac{d\lambda}{dx} - \frac{q_L}{\rho_u D}}{\frac{\gamma+1}{\gamma} - \frac{\rho}{\rho_u}}\right).$$

If the reaction rate is given by an Arrhenius reaction of the form

$$\frac{d\lambda}{dt} = -k\lambda^n e^{-E_a/R^0 T},$$

show that at the sonic plane, the reaction is not complete ($\lambda^* \neq 0$) and the reactant mass fraction is given by

$$\lambda^* = \left[\frac{\gamma q_L}{(\gamma + 1)\rho_u q_c k}\right]^{1/n} e^{-E_a(\gamma+1)^2/(n\gamma D)}.$$

By integrating the energy equation between the shock and the sonic plane, show that the eigenvalue detonation velocity is given implicitly by

$$\frac{D_{\mathrm{CJ}}^2 - D^2}{D_{\mathrm{CJ}}^2} = \frac{q_L}{\rho_u D Q} x^*(D) + \lambda^*(D),$$

where x^* is the location of the sonic plane. Using the square wave approximation of Section 14.8, estimate $x^*(D)$ by assuming it corresponds to the ignition length l_{ig}. Assuming complete reaction at the sonic plane ($\lambda^* = 0$), determine the maximum value of heat loss permitting the existence of a steady detonation. (Zel'dovich & Kompaneets 1960)

8. Show that the period of the pulsation of the detonation in Figure 14.6.2 is given approximately by

$$\frac{2 + \sqrt{2(\gamma - 1)}}{3 - \gamma} t_{\mathrm{ig}},$$

if one assumes a square-wave strong detonation with weak pulsations. In practice, such oscillations are observed near the stability boundary of real detonations. As the pulsation amplitude grows, much stronger nonlinear pulsations are established with typical frequencies lower by an order of magnitude.

9. Generalize the strong blast solution of Eq. (14.7.1) for planar and cylindrical symmetries.

References

Agoston, G. A., Wise, H. & Rosser, W. A. 1957. Dynamic factors affecting the combustion of liquid spheres. *Proc. Combust. Inst.* **6**, 708–717.

Alpert, R. L. & Toong, T. Y. 1972. Periodicity in exothermic hypersonic flows about blunt projectiles. *Astronaut. Acta* **17**, 539–560.

Annamalai, K. & Ryan, W. 1992. Interactive processes in gasification and combustion. 1. Liquid-drop arrays and clouds. *Prog. Energy Combust. Sci.* **18**, 221–295.

Annamalai, K. & Ryan, W. 1993. Interactive processes in gasification and combustion. 2. Isolated carbon, coal and porus char particle. *Prog. Energy Combust. Sci.* **19**, 383–446.

Antaki, P. 1986. Transient processes in a rigid slurry droplet during liquid vaporization and combustion. *Combust. Sci. Technol.* **46**, 113–135.

Antaki, P. & Williams, F. A. 1986. Transient processes in a nonrigid slurry droplet during liquid vaporization and combustion. *Combust. Sci. Technol.* **49**, 289–296.

Ashurst, Wm. T. 1994. Modeling turbulent flame propagation. *Proc. Combust. Inst.* **25**, 1075–1089.

Austin, J. M. 2003. The role of instability in gaseous detonation. Ph.D. thesis, California Institute of Technology, Pasadena, CA.

Austin, J. M., Pintgen, F. & Shepherd, J. E. 2005. Reactions in highly unstable detonations. *Proc. Combust. Inst.* **30**, 1849–1857.

Barenblatt, G. I. 1996. *Scaling, Self Similarity, and Intermediate Asymptotics*. Cambridge.

Barin, I. 1989. *Thermochemical Data of Pure Substances*, Part 1. VCH Publishers.

Bechtold, J. K. & Margolis, S. B. 1992. The structure of supercritical flames with Arrhenius mass diffusion. *Combust. Sci. Technol.* **83**, 257–290.

Bechtold, J. K. & Matalon, M. 1987. Hydrodynamic and diffusion effects on the stability of spherically expanding flames. *Combust. Flame* **67**, 77–90.

Bechtold, J. K. & Matalon, M. 2001. The dependence of the Markstein length on stoichiometry. *Combust. Flame* **127**, 1906–1913.

Bellan, J. 2000. Supercritical (and subcritical) fluid behavior and modelling: Drops, streams, shear and mixing layers, jets, and sprays. *Prog. Energy Combust. Sci.* **26**, 329–366.

Benson, S. W. 1960. *The Foundations of Chemical Kinetics*. McGraw-Hill.

Benson, S. W. 1981. The kinetics and thermochemistry of chemical oxidation with application to combustion and flames. *Prog. Energy Combust. Sci.* **7**, 125–134.

Bilger, R. W. 1980. Turbulent flows with nonpremixed reactants. In *Turbulent Reacting Flows*, P. A. Libby & F. A. Williams (eds.). Springer-Verlag, pp. 65–113.

Blander, M. & Katz, J. L. 1975. Bubble nucleation in liquids. *AIChE J.* **21**, 833–848.

Borghi, R. 1988. Turbulent combustion modeling. *Prog. Energy Combust. Sci.* **14**, 245–292.

Borman, G. L. & Ragland, K. W. 1998. *Combustion Engineering*. McGraw-Hill.

Bosschaart, K. J. & De Goey L. P. H. 2003. Detailed analysis of the heat flux method for measuring burning velocities. *Combust. Flame* **132**, 170–180.

Botha, J. P. & Spalding, D. B. 1954. The laminar flame speed of propane–air mixtures with heat extraction from the flame. *Proc. R. Soc. Lond. A* **225**, 71–96.

Bowman, C. T. 1973. Kinetics of nitric oxide formation in combustion processes. *Proc. Combust. Inst.* **14**, 729–738.

Bowman, C. T. 1992. Control of combustion-generated nitrogen oxide emission: Technology driven by regulation. *Proc. Combust. Inst.* **24**, 859–878.

Bowman, C. T. 2000. Gas-phase reaction mechanisms for nitrogen oxide formation and removal in combustion. In *Pollutants from Combustion*, C. Vovelle (ed.). Kluwer, pp. 123–144.

Bradley, D. 1992. How fast we can burn? *Proc. Combust. Inst.* **24**, 247–262.

Bradley, J. 1979. *Flames and Combustion Phenomena*. Chapman and Hall.

Bray, K. N. C. 1980. Turbulent flows with premixed reactants. In *Turbulent Reacting Flows*. P. A. Libby & F. A. Williams (eds.). Springer-Verlag, pp. 115–183.

Bray, K. N. C. 1996. The challenge of turbulent combustion. *Proc. Combust. Inst.* **26**, 1–26.

Bray, K. N. C., Libby, P. A., Masuya, G. & Moss, J. B. 1981. Turbulence production in premixed turbulent flames. *Combust. Sci. Technol.* **25**, 127–140.

Brezinsky, K. 1986. The high-temperature oxidation of aromatic hydrocarbons. *Prog. Energy Combust. Sci.* **12**, 1–24.

Brown, G. L. & Roshko, A. 1974. On density effects and large structure in turbulent mixing layers. *J. Fluid Mech.* **64**, 775–816.

Brzustowski, T. A., Twardus, E. M., Wojcicki, S. & Sobiesiak, A. 1979. Interaction of two burning droplets of arbitrary size. *AIAA J.* **17**, 1234–1242.

Buckmaster, J. D. 2002. Edge flames. *Prog. Energy Combust. Sci.* **28**, 435–475.

Buckmaster, J. D., Clavin, P., Liñán, A., Matalon, M., Peters, N., Sivashinsky, G. I. & Williams, F. A., 2005. Combustion Theory and Modeling. *Proc. Combust. Inst.* **30**, 1–19.

Buckmaster, J. D. & Ludford, G. S. S. 1982. *Theory of Laminar Flames*. Cambridge.

Buckmaster, J. D. & Ludford, G. S. S. 1983. *Lectures on Mathematical Combustion, CBMS-NSF Regional Conference Series* in *Applied Mathematics* **43**. Society of Industrial and Applied Mathematics.

Burgess, M. J. & Wheeler, R. V. 1911. The lower limit of inflammation of mixtures of the paraffin hydrocarbons with air. *J. Chem. Soc.* (*London*) **99**, 2013–2030.

Burgoyne, J. H. & Weinberg, F. J. 1954. Determination of the distribution of some parameters across the combustion zone of a flat flame. *Proc. R. Soc. Lond. A* **224**, 286–308.

Burke, S. P. & Schumann, T. E. W. 1928. Diffusion Flames, *Proc. Combust. Inst.* **1–2**, 2–11.

Bush, W. B. & Fendell, F. E. 1970. Asymptotic analysis of laminar flame propagation for general Lewis number. *Combust. Sci. Technol.* **1**, 421–428.

Bychkov, V. V. & Liberman, M. A. 2000. Dynamics and stability of premixed flames. *Phys. Rep.* **325**, 116–237.

Calcote, H. F., Gregory C.A., Jr., Barnett, C.M. & Gilmer, R.B. 1952. Spark ignition: Effect of molecular structure. *Ind. Eng. Chem.* **44**, 2656–2659.

Calcote, H. F. & Pease, R. N. 1951. Electrical properties of flames. *Ind. Eng. Chem.* **43**, 2726–2731.

Candel, S. 2002. Combustion dynamics and control. *Proc. Combust. Inst.* **29**, 1–28.

Chao, B. H., Law, C. K. & Tien, J. S. 1991. Structure and extinction of diffusion flames with flame radiation. *Proc. Combust. Inst.* **23**, 523–531.

Chen, J. Y. 1988. A general procedure for constructing reduced reaction mechanisms with given independent relations. *Combust. Sci. Technol.* **57**, 89–94.

Chigier, N. 1981. *Energy, Combustion, and Environment*. McGraw-Hill.

Chiu, H. H., Ahluwalia, R. K., Koh, B. & Croke, E. J. 1978. Spray group combustion. *AIAA Paper* No. 78–75, 16th Aerospace Sciences Meeting, Huntsville, AL.

Chiu, H. H., Kim, H. Y. & Croke, E. J. 1983. Internal group combustion of liquid droplets. *Proc. Combust. Inst.* **19**, 971–980.

Cho, P., Law, C. K., Hertzberg, J. & Cheng, R. K. 1988. Structure and propagation of turbulent premixed flames stabilized in a stagnation flow. *Proc. Combust. Inst.* **21**, 1493–1499.

Christiansen, E. W., Sung, C. J. & Law, C. K. 1998. Pulsating instability in near-limit propagation of rich hydrogen/air flames. *Proc. Combust. Inst.* **27**, 555–562.

Christiansen, E. W., Law, C. K. & Sung, C. J. 2001. Steady and pulsating propagation and extinction of rich hydrogen/air flames at elevated pressures. *Combust. Flame* **124**, 35–49.

Chung, P. M. 1965. Chemically reacting nonequilibrium boundary layers. *Advances in Heat Transfer* **2**, 109–270.

Chung, S. H. 2007. Stabilization, propagation and instability of tribrachial triple flames. *Proc. Combust. Inst.* **31**, 877–892.

Chung, S. H. & Kim, J. S. 1990. An experiment on vaporization and microexplosion of emulsion fuel droplets on a hot surface. *Proc. Combust. Inst.* **23**, 1431–1435.

Chung, S. H. & Law, C. K. 1984a. Burke–Schumann flame with streamwise and preferential diffusion. *Combust. Sci. Technol.* **37**, 21–47.

Chung, S. H. & Law, C. K. 1984b. An invariant derivation of flame stretch. *Combust. Flame* **55**, 123–125.

Chung, S. H. & Law, C. K. 1988. An integral analysis of the structure and propagation of stretched premixed flames. *Combust. Flame* **72**, 325–336.

Chung, S. H. & Lee, B. J. 1991. On the characterization of laminar lifted flames in a nonpremixed jet. *Combust. Flame* **86**, 62–72.

Ciezki, H. K. & Adomeit, G. 1993. Shock-tube investigation of self-ignition of *n*-heptane–air mixtures under engine relevant conditions. *Combust. Flame* **93**, 421–433.

Clarke, J. F. & McChesney, M. 1964. *The Dynamics of Real Gases*. Butterworth.

Clavin, P. 1985. Dynamic behavior of premixed flame fronts in laminar and turbulent flows. *Prog. Energy Combust. Sci.* **11**, 1–59.

Clavin, P. 2000. Dynamics of combustion fronts in premixed gases: From flames to detonations. *Proc. Combust. Inst.* **28**, 569–585.

Clavin, P. & Williams, F. A. 1982. Effects of molecular diffusion and of thermal-expansion on the structure and dynamics of premixed flames in turbulent flows of large-scale and low intensity. *J. Fluid Mech.* **116**, 251–282.

Clavin, P. & Williams, F. A. 2002. Dynamics of planar gaseous detonations near Chapman–Jouguet conditions for small heat release. *Combust. Theory Modeling* **6**, 127–140.

Courant, R. & Friedrichs, K. O. 1948. *Supersonic Flow and Shock Waves*. Interscience.

Crespo, A. & Liñán, A. 1975. Unsteady effects in droplet evaporation and combustion. *Combust. Sci. Technol.* **11**, 9–18.

Dagaut, P., Pengloan, G. & Ristori, A. 2002. Oxidation, ignition, and combustion of toluene: Experimental and detailed chemical kinetic modeling. *Phys. Chem. Chem. Phys.* **4**, 1846–1854.

Darrieus, G. 1938. Propagation d'un front de flamme. Presented at *La Technique Moderne*, Paris, France in 1938, and at *Congres de Mechanique Appliquee* in 1945.

Davis, S. G. & Law, C. K. 1999. Determination of and fuel structure effects on laminar flame speeds of C_1 to C_8 hydrocarbons. *Combust. Sci. Techno.* **140**, 427–450.

Dean, A. M. 1985. Predictions of pressure and temperature effects upon radical addition and recombination reactions, *J. Phys. Chem.* **89**, 4600–4608.

De Goey, L. P. H., van Maaren, A. & Quax, R. M. 1993. Stabilization of adiabatic premixed laminar flames on a flat flame burner. *Combust. Sci. Technol.* **92**, 201–207.

Dietrich, D. L., Haggard, J. B., Jr., Dryer, F. L., Nayagam, V., Shaw, B. D. & Williams, F. A. 1996. Droplet combustion experiments in spacelab. *Proc. Combust. Inst.* **26**, 1201–1207.

Dixon-Lewis, G. 1990. Structure of laminar flames. *Proc. Combust. Inst.* **23**, 305–324.

Döring, W. 1943. On detonation processes in gases. *Ann. Phys.* **43**, 421–436.

Dorodnitzyn, A. 1942. *Dokl. Akad. Nauk SSSR* **34**, 213.

Dorrence, W. H. 1962. *Viscous Hypersonic Flow*. McGraw-Hill.

Dowdy, D. R., Smith, D. R., Taylor, S. C. & Williams, A. 1990. The use of expanding spherical flames to determine burning velocities and stretch effects in hydrogen–air mixtures. *Proc. Combust. Inst.* **23**, 325–332.

Eckbreth, A. C. 1996. *Laser Diagnostics for Combustion Temperature and Species*, 2nd ed. Gordon and Breach.

Egolfopoulos, F. N. & Campbell, C. S. 1996. Unsteady counterflowing strained diffusion flames: Diffusion-limited frequency response. *J. Fluid Mech.* **318**, 1–29.

Emmons, H. W. 1956. The film combustion of liquid fuel. *Z. Angew. Math. Mech.* **36**, 60–71.

Eng, J. A., Law, C. K. & Zhu, D. L. 1994. On burner-stabilized cylindrical premixed flames in microgravity. *Proc. Combust. Inst.* **25**, 1711–1718.

Eng, J. A., Zhu, D. L. & Law, C. K. 1995. On the structure, stabilization, and dual response of flat-burner flames. *Combust. Flame* **100**, 645–652.

Faeth, G. M. 1977. Current status of droplet and liquid combustion. *Prog. Energy Combust. Sci.* **3**, 191–224.

Faeth, G. M. 1996. Spray combustion phenomena. *Proc. Combust. Inst.* **26**, 1593–1612.

Feedoseeva, N. V. 1973. Interaction of two vigorously evaporating drops in the absence of combustion. *Adv. Aerosol Phys.* **3**, 27–34.

Fendell, F. E. 1965. Ignition and extinction in combustion of initially unmixed reactants. *J. Fluid Mech.* **21**, 281–303.

Fenimore, C. P. 1964. *Chemistry in Premixed Flames*. Pergamon.

Fenimore, C. P. 1971. Formation of nitric oxide in premixed hydrocarbon flames. *Proc. Combust. Inst.* **13**, 373–380.

Ferguson, C. R. & Keck, J. C. 1979. Stand-off distances on a flat flame burner. *Combust. Flame* **34**, 85–98.

Fickett, W. & Davis, W. C. 1979. *Detonation*. University of California Press.

Foa, J. V. 1960. *Elements of Flight Propulsion*. Wiley.

Fotache, C. G., Sung, C. J., Sun, C. J. & Law, C. K. 1998. Mild oxidation regimes and multiple criticality in nonpremixed hydrogen–air counterflow. *Combust. Flame* **112**, 457–471.

Fowler, R. H. & Guggenheim, E. A. 1939. *Statistical Thermodynamics*. MacMillan.

Frank-Kameneskii, D. A. 1969. *Diffusion and Heat Transfer in Chemical Kinetics*. Plenum Press.

Frenklach, M., Clary, D. W., Gardiner, W. C., Jr. & Stein, S. E. 1984. Detailed kinetic modeling of soot formation in shock-tube pyrolysis of acetylene. *Proc. Combust. Inst.* **20**, 887–901.

Frenklach, M. & Warnatz, J. 1987. Detailed modeling of PAH profiles in a sooting low-pressure acetylene flame. *Combust. Sci. Technol.* **51**, 265–283.

Fristrom, R. M. & Westenberg, A. A. 1965. *Flame Structure*. McGraw-Hill.

Gardiner, W. C., Jr. 1999. *Combustion Chemistry*. Springer-Verlag.

Gaydon, A. G. & Wolfhard, H. G. 1970. *Flames*. Chapman and Hall.

Germano, M., Piomelli, U., Moin, P. & Cabot, W. H. 1991. A dynamic subgrid scale eddy viscosity model. *Phys. Fluids* **A3**, 1760–1765.

Gilbert, R. G., Luther, K. & Troe, J. 1983. Theory of thermal unimolecular reactions in the fall-off range. II. Weak collision rate constants. *Ber. Bunsenges. Phys. Chem.* **87**, 169–177.

Gilbert, R. G. & Smith, S. C. 1990. *Theory of Unimolecular and Recombination Reactions*. Blackwell-Scientific, Oxford.

Glassman, I. 1960. Combustion of metals: Physical considerations. In ARS *Progress in Astronautics and Rocketry: Solid Propellant Rocket Research*, M. Summerfield (ed.). Academic, pp. 253–258.

Glassman, I. 1996. *Combustion*. Academic.

Glassman, I. & Law, C. K. 1991. Sensitivity of metal reactivity to gaseous impurities in oxygen environments. *Combust. Sci. Technol.* **80**, 151–157.

Glassman, I., Williams, F. A. & Antaki, P. 1984. A physical and chemical interpretation of boron particle combustion. *Proc. Combust. Inst.* **20**, 2057–2064.

Glasstone, S. 1958. *Elements of Physical Chemistry*. D. Van Nostrand.

Godsave, G. A. E. 1953. Studies of the combustion of drops in a fuel spray: The burning of single drops of fuel. *Proc. Combust. Inst.* **4**, 818–830.

Goldsmith, M. & Penner, S. S. 1954. On the burning of single drops of fuel in an oxidizing atmosphere. *Jet Propulsion* **24**, 245–251.

Goldstein, H. 1980. *Classical Mechanics*, 2nd ed. Addison-Wesley.

Gollahalli, S. R. & Brzustowski, T. A. 1973. Experimental studies on the flame structure in the wake of a burning droplet. *Proc. Combust. Inst.* **14**, 1333–1344.

Gordon, S. & McBride, B. J. 1971. Computer program for calculation of complex equilibrium compositions, rocket performance, incident and reflected shocks, and Chapman–Jouguet detonations. *NASA* SP-273.

Gouldin, F. C. 1987. An application of fractals to modeling premixed turbulent flame. *Combust. Flame* **68**, 249–266.

Goussis, D. A. & Lam, S. H. 1992. A study of homogeneous methanol oxidation kinetics using CSP. *Proc. Combust. Inst.* **22**, 113–120.

Gray, B. F. & Yang, C. H. 1969. On slow oxidation of hydrocarbon and cool flame. *J. Phys. Chem.* **73**, 3395–3406.

Gray, P. 1991. Chemistry and combustion. *Proc. Combust. Inst.* **23**, 1–19.

Griffith, J. F. & Barnard, J. A. 1995. *Flame and Combustion*, 3rd ed. CRC Press.
Guggenheim, E. A. 1952. *Mixtures*. Oxford.
Guggenheim, E. A. 1957. *Thermodynamics*. Interscience.
Hayes, W. D. & Probstein, R. F. 1959. *Hypersonic Flow Theory*. Academic.
Haynes, B. S. & Wagner, H. Gg. 1981. Soot Formation. *Prog. Energy Combust. Sci.* **7**, 229–273.
He, L. & Clavin, P. 1994. On the direct initiation of gaseous detonations by an energy source. *J. Fluid Mech.* **277**, 227–248.
Hinze, J. O. 1975. *Turbulence*, 2nd ed. McGraw-Hill.
Hirschfelder, J. O., Curtiss, C. F. & Bird, R. B. 1954. *Molecular Theory of Gases and Liquids*. John Wiley.
Hottel, H. C. & Hawthorne, W. R. 1949. Diffusion in laminar flame jets. *Proc. Combust. Inst.* **3**, 254–266.
Howarth, L. 1938. On the solution of the laminar boundary layer equations. *Proc. R. Soc. Lond. A* **164**, 547–579.
Howarth, L. 1948. Concerning the effect of compressibility on laminar boundary layers and their separation. *Proc. R. Soc. Lond. A* **194**, 16–42.
Im, H. G., Bechtold, J. K. & Law, C. K. 1993. Thermal ignition analysis in supersonic flat-plate boundary layers. *J. Fluid Mech.* **249**, 99–120.
Im, H. G. & Chen, J. H. 2000. Effects of flow transients on the burning velocity of laminar hydrogen–air premixed flames. *Proc. Combust. Inst.* **28**, 1833–1840.
Ishizuka, S. & Law, C. K. 1983. An experimental study of extinction and stability of stretched premixed flames. *Proc. Combust. Inst.* **19**, 327–335.
Janicka, J. & Sadiki, A. 2005. Large eddy simulation of turbulent combustion systems. *Proc. Combust. Inst.* **30**, 537–547.
Jiang, Y. J., Umemura, A. & Law, C. K. 1992. An experimental investigation on the collision behavior of hydrocarbon droplets. *J. Fluid Mech.* **234**, 171–190.
Joulin, G. & Clavin, P. 1979. Linear stability analysis of nonadiabatic flames—diffusional–thermal model. *Combust. Flame* **35**, 139–153.
Ju, Y. & Niioka, T. 1994. Reduced kinetic mechanism of ignition for nonpremixed hydrogen–air in a supersonic mixing layer. *Combust. Flame* **19**, 240–246.
Kailasanath, K. 2003. Recent developments in the research on pulse detonation engines. *AIAA J.* **41**, 145–159.
Kanury, A. M. 1975. *Introduction to Combustion Phenomena*. Gordon and Breach.
Kassoy, D. R. 1975. Theory of adiabatic, homogeneous explosion from initiation to completion. *Combust. Sci. Technol.* **10**, 27–35.
Kassoy, D. R. & Liñán, A. 1978. Influence of reactant consumption on critical conditions for homogeneous thermal explosion. *Quar. J. Mech. Appl. Math.* **31**, 99–112.
Kee, R. J., Coltrin, M. E. & Glarborg, P. 2003. *Chemical Reacting Flows: Theory and Practice*. John Wiley.
Kelley, A. P. & Law, C. K. 2009. Nonlinear effects in the extraction of laminar flame speeds from expanding spherical flames. *Combust. Flame* **156**, 1844–1851.
Kennedy, I. M. 1997. Models of soot formation and oxidation. *Prog. Energy Combust. Sci.* **23**, 95–132.
Kerstein, A. R. 1988a. Fractal dimension of turbulent premixed flames. *Combust. Sci. Technol.* **60**, 441–445.

Kerstein, A. R. 1988b. Simple derivation of Yakhot's turbulent premixed flame speed formula. *Combust. Sci. Technol.* **60**, 163–165.

Kerstein, A. R. 2002. Turbulence in combustion processes: Modeling challenges. *Proc. Combust. Inst.* **29**, 1763–1773.

Kerstein, A. R., Ashurst, W. T. & Williams, F. A. 1988. Field equation for interface propagation in an unsteady homogeneous flow field. *Phys. Rev. A* **37**, 2728–2731.

Kim, T. J., Yetter, R. A., & Dryer, F. L. 1994. New results on moist CO oxidation: High-pressure, high-temperature experiments and comprehensive kinetic modeling. *Proc. Combust. Inst.* **25**, 759–766.

Klimov, A. M. 1963. Laminar flames in a turbulent flow. *Zhur. Priki. Mekh. Tekhn. Fiz.* **3**, 49–58.

Kobayashi, H., Tamura, T., Maruta, K., Niioka, T. & Williams, F. A. 1996. Burning velocity of turbulent premixed flames in a high pressure environment. *Proc. Combust. Inst.* **26**, 389–396.

Kohse-Höinghaus, K., Barlow, R.S., Aldén, M. & Wolfrum, J. 2005. Combustion at the focus: Laser diagnostics and control. *Proc. Combust. Inst.* **30**, 89–123.

Kohse-Höinghaus, K. & Jeffries, J. B. (eds.) 2002. *Applied Combustion Diagnostics*. Taylor and Francis.

Kosdon, F. J., Williams, F. A. & Buman, C. 1969. Combustion of vertical cellulosic cylinders. *Proc. Combust. Inst.* **12**, 253–264.

Kreutz, T. G. & Law, C. K. 1996. Ignition in nonpremixed counterflowing hydrogen versus heated air: Computational study with detailed chemistry. *Combust. Flame* **104**, 157–175.

Kreutz, T. G. & Law, C. K. 1998. Ignition in nonpremixed counterflowing hydrogen versus heated air: Computational study with skeletal and reduced chemistry. *Combust. Flame* **114**, 436–456.

Kumagai, S. & Isoda, H. 1957. Combustion of fuel droplets in a falling chamber. *Proc. Combust. Inst.* **6**, 726–731.

Kuo, K. K. 2002. *Principles of Combustion*, 2nd ed. John Wiley.

Labowsky, M. J. 1978. A formalism for calculating the evaporation rates of rapidly evaporating interacting particles. *Combust. Sci. Technol.* **18**, 145–151.

Laidler, K. J. 1965. *Chemical Kinetics*, McGraw-Hill.

Lam, S. H. & Goussis, D. A. 1988. Understanding complex chemical kinetics with computational singular perturbation. *Proc. Combust. Inst.* **22**, 931–941.

Lam, S. H. 1993. Using CSP to understand complex chemical kinetics. *Combust. Sci. Technol.* **89**, 375–404.

Lam, S. H. 1994. The CSP method for simplifying kinetics. *Int. J. Chem. Kinet.* **26**, 461–486.

Landau, L. 1945. On the theory of slow combustion. *Acta Physicochimica URSS* **19**, 77–85.

Landau, L. D. & Lifshitz, E. M. 1959. *Fluid Mechanics*. Addison-Wesley.

Lara-Urbanejo, P. & Sirignano, W. A. 1981. Theory of transient multicomponent droplet vaporization in a convective field. *Proc. Combust. Inst.* **18**, 1365–1374.

Lasheras, J. C., Fernandez-Pello, A. C. & Dryer, F. L. 1979. Initial observations on the free droplet combustion characteristics of water-in-oil emulsions. *Combust. Sci. Technol.* **21**, 1–14.

Lasheras, J. C., Fernandez-Pello, A. C. & Dryer, F. L. 1980. Experimental observations on the disruptive combustion of free droplets of multicomponent fuels. *Combust. Sci. Technol.* **22**, 195–209.

Lasheras, J. C., Fernandez-Pello, A. C. & Dryer, F. L. 1981. On the disruptive burning of free droplets of alcohol/*n*-paraffin solutions and emulsions. *Proc. Combust. Inst.* **18**, 293–305.

Launder, B. E. & Spalding, D. B. 1972. *Mathematical Models of Turbulence*. Academic.

Law, C. K. 1973. A simplified theoretical model for the vapor-phase combustion of metal particles. *Combust. Sci. Technol.* **7**, 197–212.

Law, C. K. 1975. Asymptotic theory for ignition and extinction in droplet burning. *Combust. Flame* **24**, 89–98.

Law, C. K. 1976. Unsteady droplet combustion with droplet heating. *Combust. Flame* **26**, 17–22.

Law, C. K. 1977a. A model for the combustion of oil–water emulsion droplets. *Combust. Sci. Technol.* **17**, 29–38.

Law, C. K. 1977b. Adiabatic spray vaporization with droplet temperature transient. *Combust. Sci. Technol.* **15**, 65–73.

Law, C. K. 1978a. On the stagnation-point ignition of a premixed combustible. *Intl. J. Heat Mass Transfer* **21**, 1363–1368.

Law, C. K. 1978b. Internal boiling and superheating in vaporizing multicomponent droplets. *AIChE J.* **24**, 626–632.

Law, C. K. 1981. On the fire-resistant nature of oil–water emulsions. *Fuel* **60**, 998–999.

Law, C. K. 1982. Recent advances in droplet vaporization and combustion. *Prog. Energy Combust. Sci.* **8**, 171–201.

Law, C. K. 1989. Dynamics of stretched flames. *Proc. Combust. Inst.* **22**, 1381–1402.

Law, C. K. 1998. Droplet combustion of energetic liquid fuels. In *Propulsion Combustion: Fuels to Emission*, G. D. Roy (ed.). Taylor and Francis, pp. 63–92.

Law, C. K. 2006. Propagation, structure, and limit phenomena of laminar flames at elevated pressures. *Combust. Sci. Technol.* **178**, 335–360.

Law, C. K., Chao, B. H. & Umemura, A. 1992. On closure in activation energy asymptotics of premixed flames. *Combust. Sci. Technol.* **88**, 59–88.

Law, C. K., Cho, P., Mizomoto, M. & Yoshida, H. 1988. Flame curvature and preferential diffusion in the burning intensity of Bunsen flames. *Proc. Combust. Inst.* **21**, 1803–1809.

Law, C. K., Chung, S. H. & Srinivasan, N. 1980. Gas-phase quasi-steadiness and fuel vapor accumulation effects in droplet burning. *Combust. Flame* **38**, 173–198.

Law, C. K. & Egolfopoulos, F. N. 1992. A united chain-thermal theory of fundamental flammability limits. *Proc. Combust. Inst.* **24**, 137–144.

Law, C. K. & Faeth, G. M. 1994. Opportunities and challenges of combustion in microgravity. *Prog. Energy Combust. Sci.* **20**, 65–113.

Law, C. K. & Glass, I. I. 1971. Diffraction of strong shock waves by a sharp compressive corner. *Canadian Aeronautics and Space Institute Transactions* **4**, 2–12.

Law, C. K., Ishizuka, S. & Cho, P. 1982. On the opening of premixed Bunsen flame tips. *Combust. Sci. Technol.* **28**, 89–96.

Law, C. K. & Law, H. K. 1976. Quasi-steady diffusion flame theory with variable specific heats and transport coefficients. *Combust. Sci. Technol.* **12**, 207–216.

Law, C. K. & Law, H. K. 1977. Theory of quasi-steady one-dimensional diffusional combustion with variable properties including distinct binary diffusion coefficients. *Combust. Flame* **29**, 269–275.

Law, C. K. & Law, H. K. 1979. Thermal ignition analysis in boundary layer flows. *J. Fluid Mech.* **92**, 97–108.

Law, C. K. & Law, H. K. 1982a. A theoretical study of ignition in the laminar mixing layer. *J. Heat Transfer* **104**, 329–337.

Law, C. K. & Law, H. K. 1982b. A d^2-law for multicomponent droplet vaporization and combustion. *AIAA J.* **20**, 522–527.

Law, C. K. & Law, H. K. 1993. On the gasification mechanisms of multicomponent droplets. In *Modern Developments in Energy, Combustion, and Spectroscopy*, F. A. Williams, A. K. Oppenheim, D. B. Olfe & M. Lapp (eds.). Pergamon, pp. 29–48.

Law, C. K., Makino, A. & Lu, T. F. 2006. On the off-stoichiometric peaking of adiabatic flame temperature. *Combust. Flame* **145**, 808–819.

Law, C. K. & Sirignano, W. A. 1977. Unsteady droplet combustion with droplet heating–II: conduction limit. *Combust. Flame* **28**, 175–186.

Law, C. K. & Sung, C. J. 2000. Structure, aerodynamics, and geometry of premixed flamelets. *Prog. Energy Combust. Sci.* **26**, 459–505.

Law, C. K., Sung, C. J., Yu, G. & Axelbaum, R. L. 1994. On the structural sensitivity of purely strained planar premixed flame to strain rate variation. *Combust. Flame* **98**, 139–154.

Law, C. K. & Williams, F. A. 1972. Kinetics and convection in the combustion of alkane droplets. *Combust. Flame* **19**, 393–405.

Lawton, J. & Weinberg, F. J. 1969. *Electrical Aspects of Combustion*. Clarendon.

Lee, A. & Law, C. K. 1991. Gasification and shell characteristics in slurry droplet burning. *Combust. Flame* **85**, 77–93.

Lee, A. & Law, C. K. 1992. An experimental investigation on the vaporization and combustion of methanol and ethanol droplets. *Combust. Sci. Technol.* **86**, 253–265.

Lee, B. J. & Chung, S. H. 1997. Stabilization of lifted tribrachial flames in a laminar nonpremixed jet. *Combust. Flame* **109**, 163–172.

Lee, H. I. & Stewart, D. S. 1990. Calculation of linear detonation instability—one-dimensional instability of plane detonation. *J. Fluid Mech.* **216**, 103–132.

Lee, J. H. S. 1984. Dynamic parameters of gaseous detonations. *Ann. Rev. Fluid Mech.* **16**, 311–336.

Lee, J. H. S. 2001. Detonation waves in gaseous explosives. In *Handbook of Shock Waves*, G. Ben-Dor, O. Igra & T. Elperin (eds.). Academic, Chapter 17, Volume 3.

Lee, J. H. S. & Moen, I. O. 1980. The mechanism of transition from deflagration to detonation in vapor cloud explosion. *Prog. Energy Combust. Sci.* **6**, 359–389.

Lehr, H. F. 1972. Experiments on shock-induced combustion. *Astronaut. Acta* **17**, 589–597.

Lewis, B. & von Elbe, G. 1987. *Combustion, Flames, and Explosions of Gases*, 3rd ed. Academic.

Libby, P. A. & Blake, T. R. 1979. Theoretical study of burning carbon particles. *Combust. Flame* **36**, 139–169.

Libby, P. A. & Bray, K. N. C. 1981. Countergradient diffusion in premixed turbulent flames. *AIAA J.* **19**, 205–213.

Libby, P. A. & Williams, F. A. 1980. *Turbulent Reacting Flows*. Springer-Verlag.

Libby, P. A. & Williams, F. A. 1994. *Turbulent Reacting Flows*. Academic.

Lide, D. R. 1990–1991. *CRC Handbook of Chemistry and Physics*, 71st ed. CRC Press.

Liepmann, H. W. & Roshko, A. 1957. *Elements of Gasdynamics*. John Wiley.

Liñán, A. 1974. The asymptotic structure of counterflow diffusion flames for large activation energies. *Acta Astronautica* **1**, 1007–1039.

Lindstedt, R. P. 1988. Modeling of chemical complexities of flames. *Proc. Combust. Inst.* **27**, 269–285.

Lipatnikov, A. N. & Chomiak, J. 2002. Turbulent flame speed and thickness: Phenomenology, evaluation, and application in multidimensional simulation. *Prog. Energy Combust. Sci.* **28**, 1–74.

Lo Jacono, D., Papas, P., Matalon, M. & Monkewitz, P. A. 2005. An experimental realization of an unstrained, planar diffusion flame. *Proc. Combust. Inst.* **30**, 501–509.

Longwell, J. P. & Weiss, M. A. 1955. High temperature reaction rates in hydrocarbon combustion. *Ind. Eng. Chem.* **47**, 1634–1643.

Lovas, T., Nilsson, D. & Mauss, F. 2000. Automatic reduction procedure for chemical mechanisms applied to premixed methane–air flames. *Proc. Combust. Inst.* **28**, 1809–1815.

Lu, T. F. 2004. Comprehensive reduction and application of chemical reaction mechanisms. Ph.D. thesis, Princeton University.

Lu, T. F. & Law, C. K. 2005. A directed relation graph method for mechanism reduction. *Proc. Combust. Inst.* **30**, 1333–1341.

Lu, T. F. & Law, C. K. 2009. Toward accommodating realistic fuel chemistry in large-scale computations. *Prog. Energy Combust. Sci.* **35**, 192–215.

Lu, T. F., Ju, Y. & Law, C. K. 2001. Complex CSP for chemistry reduction and analysis. *Combust. Flame* **126**, 1445–1455.

Lu, T. F., Law, C. K. & Ju, Y. 2003. Some aspects of chemical kinetics in C–J detonation: Induction length analysis. *J. Propulsion Power* **19**, 901–907.

Maas, U. & Pope, S. B. 1992. Simplifying chemical kinetics: Intrinsic low dimensional manifolds in composition space. *Combust. Flame* **88**, 239–264.

Makino, A. 1991. Ignition criteria for a fuel droplet expressed in explicit form. *Combust. Sci. Technol.* **80**, 305–317.

Makino, A. 1992. An approximate explicit expression for the combustion rate of a small carbon particle. *Combust. Flame* **90**, 143–154.

Makino, A. 2001. Fundamental aspects of the heterogeneous flame in the self propagating high-temperature synthesis (SHS) process. *Prog. Energy Combust. Sci.* **27**, 1–74.

Makino, A. & Law, C. K. 1988a. On the controlling parameter in the gasification of multicomponent droplets. *Combust. Flame* **73**, 331–336.

Makino, A. & Law, C. K. 1988b. Quasi-steady and transient combustion of a carbon particle: Theory and experimental comparisons. *Proc. Combust. Inst.* **21**, 183–191.

Makino, A. & Law, C. K. 2001. On the correspondence between the homogeneous and heterogeneous theories of SHS, *Combust. Flame* **124**, 268–274.

Marble, F. E. & Adamson, T. C. 1954. Ignition and combustion in a laminar mixing zone. *Jet Propulsion* **24**, 85–94.

Markstein, G. H. 1964. *Nonsteady Flame Propagation*. Pergamon.

Maruta, K., Yoshida, M., Ju, Y. & Niioka, T. 1996. Experimental study on methane–air premixed flame extinction at small stretch rates in microgravity. *Proc. Combust. Inst.* **26**, 1283–1289.

Massias, A., Diamantis, D., Mastorakos, E. & Goussis, D. A. 1999a. An algorithm for the construction of global reduced mechanisms with CSP data. *Combust. Flame* **117**, 685–708.

Massias, A., Diamantis, D., Mastorakos, E. & Goussis, D. A. 1999b. Global reduced mechanisms for methane and hydrogen combustion with nitric oxide formation constructed with CSP data. *Combust. Theory Modeling* **3**, 233–257.

Matalon, M. 1983. On flame stretch. *Combust. Sci. Technol.* **31**, 169–181.

Matalon, M. & Law, C. K. 1983. Gas-phase transient diffusion in droplet vaporization and combustion. *Combust. Flame* **50**, 219–229.

Matalon, M. & Law, C. K. 1985. Gas-phase transient diffusion in droplet vaporization and combustion: Errata and extension. *Combust. Flame* **59**, 213–215.

Matalon, M. & Matkowsky, B. J. 1982. Flames as gas-dynamic discontinuity. *J. Fluid Mech.* **124**, 239–259.

Matkowsky, B. J. & Sivashinsky, G. I. 1979. Asymptotic derivation of two models in flame theory associated with the constant density approximation. *SIAM J. Appl. Math.* **37**, 686–699.

McMillen, D. F. & Golden, D. M. 1982. Hydrocarbon bond dissociation energies. *Annu. Rev. Phys. Chem.* **33**, 493–532.

Merzhanov, A. G. 1990. Self-propagating high-temperature synthesis: Twenty years of search and findings. In *Combustion and Plasma Synthesis of High-temperature Materials*, Z. A. Munir & J. B. Holt (eds.). VCH Publishers, pp. 1–53.

Merzhanov, A. G. 1994. Solid flames—discoveries, concepts, and horizons of cognition. *Combust. Sci. Technol.* **98**, 307–336.

Miller, J. A. 1996. Theory and modeling in combustion chemistry. *Proc. Combust. Inst.* **26**, 461–480.

Miller, J. A. & Bowman, C. T. 1989. Mechanism and modeling of nitrogen chemistry in combustion. *Prog. Energy Combust. Sci.* **4**, 287–338.

Miller, J. A., Pilling, M. J. & Troe, J. 2005. Unraveling combustion mechanisms through a quantitative understanding of elementary reactions. *Proc. Combust. Inst.* **30**, 43–88.

Miyasaka, K. & Law, C. K. 1981. Combustion of strongly interacting linear droplet arrays. *Proc. Combust. Inst.* **18**, 283–292.

Mizomoto, M., Asaka, Y., Ikai, S. & Law, C. K. 1985. Effects of preferential diffusion on the burning intensity of curved flames. *Proc. Combust. Inst.* **20**, 1933–1939.

Molodetsky, I. E., Dreizin, E. L. & Law, C. K. 1996. Evolution of particle temperature and internal composition for zirconium burning in air. *Proc. Combust. Inst.* **26**, 1919–1923.

Monchick, L. & Mason, E. A. 1961. Transport properties of polar gases. *J. Chem. Phys.* **35**, 1676–1697.

Monin, A. S. & Yaglom, A. M. 1965. *Statistical Fluid Mechanics: Mechanics of Turbulence*. MIT.

Moskaleva, L. V. & Lin, M. C. 2000. The spin-conserved reaction $CH + N_2 \rightarrow H + NCN$: A major pathway to prompt NO studied by quantum–statistical theory calculations and kinetic modeling of rate constant. *Proc. Combust. Inst.* **28**, 2393–2401.

Mueller, M. A., Kim, T. J., Yetter, R. A. & Dryer, F. L. 1999. Flow reactor studies and kinetic modeling of the H_2–O_2 reaction. *Intl. J. Chem. Kinetics* **31**, 113–125.

Okajima, S. & Kumagai, S. 1975. Further investigations of combustion of free droplets in a freely falling chamber including moving droplets. *Proc. Combust. Inst.* **15**, 401–417.

Okajima, S. & Kumagai, S. 1983. Experimental studies on combustion of fuel droplets in flowing air under zero- and high-gravity conditions. *Proc. Combust. Inst.* **19**, 1021–1027.

Oppenheim, A. K. 1985. Dynamic features of combustion. *Phil. Trans. R. Soc. Lond. A* **315**, 471–508.

Oran, E. S. & Boris, J. P. 2001. *Numerical Simulation of Reactive Flow*. Cambridge.

Pan, K. L., Qian, J., Law, C. K. & Shyy, W. 2002. The role of hydrodynamic instability in flame-vortex interaction. *Proc. Combust. Inst.* **29**, 1695–1704.

Pelce, P. & Clavin, P. 1982. Influence of hydrodynamic and diffusion upon the stability limits of laminar premixed flames. *J. Fluid Mech.* **124**, 219–237.

Penner, S. S. 1958. *Chemistry Problems in Jet Propulsion*. Pergamon.

Peters, N. 1985. Numerical and asymptotic analysis of systematically reduced reaction schemes for hydrocarbon flames. *Lecture Notes in Physics* **241**, 90–109.

Peters, N. 1986. Laminar flamelet concepts in turbulent combustion. *Proc. Combust. Inst.* **21**, 1231–1250.

Peters, N. 1991. Reducing mechanisms. *Lecture Notes in Physics* **384**, 48–67.

Peter, N. 1997. Kinetic foundation of thermal flame theory. *In Honor of Ya. B. Zel'dovich*. W. A. Sirignano, A. G. Merzhanov & L. De Luca (eds.). *Prog. Astronautics & Aeronautics* **173**, 73–94.

Peters, N. 2000. *Turbulent Combustion*. Cambridge.

Peters, N. & Kee, R. J. 1987. The computation of stretched laminar methane–air diffusion flames using a reduced four-step mechanism. *Combust. Flame* **68**, 17–29.

Peters, N. & Rogg, B. 1993. *Reduced Kinetic Mechanisms for Applications in Combustion Systems. Lecture Notes in Physics* **M15**.

Peters, N. & Williams, F. A. 1987. The asymptotic structure of stoichiometric methane–air flames. *Combust. Flame* **68**, 185–207.

Pilling, M. J. & Seakins, P. W. 1995. *Reaction Kinetics*. Oxford.

Piomelli, U. 1999. Large eddy simulation: Achievements and challenges. *Prog. Aerospace Science* **35**, 335–362.

Poinsot, T. 1996. Using direct numerical simulation to understand premixed turbulent combustion. *Proc. Combust. Inst.* **26**, 219–232.

Poinsot, T., Candel, S. & Trouvé, A. 1995. Application of direct numerical simulation to premixed turbulent combustion. *Prog. Energy Combust. Sci.* **21**, 531–576.

Poinsot, T., Echekki, T. & Mungal, M. G. 1992. A study of the laminar flame tip and implications for premixed turbulent combustion. *Combust. Sci. Technol.* **81**, 45–73.

Poinsot, T. & Veynante, D. 2005. *Theoretical and Numerical Combustion,* 2nd ed. Edwards.

Pope, S. B. 1985. PDF method for turbulent reactive flows. *Prog. Energy Combust. Sci.* **11**, 119–192.

Pope, S. B. 1990. Computations of turbulent combustion: Progress and challenges. *Proc. Combust. Inst.* **23**, 591–612.

Pope, S. B. 1997. Computationally efficient implementation of combustion chemistry using in situ adaptive tabulation. *Combust. Theory Modeling* **1**, 41–63.

Pope, S. B. 2000. *Turbulent Flows*. Cambridge.

Puri, I. K. 1993. *Environmental Implications of Combustion Processes*. CRC Press.

Qian, J. & Law, C. K. 1997. Regimes of coalescence and separation in droplet collision. *J. Fluid Mech.* **331**, 59–80.

Qin, Z., Lissianski, V., Yang, H., Gardiner, W. C. Jr., Davis, S. G. & Wang, H. 2000. Combustion chemistry of propane: A case study of detailed reaction mechanism. *Proc. Combust. Inst.* **28**, 1663–1669.

Radulescu, M. I., Law, C. K. & Sharpe, G. J. 2005. Structure of unstable gaseous detonation waves. *Phys. Fluids* **17**, 091105.

Radulescu, M. I. & Lee, J. H. S. 2002. The failure mechanism of gaseous detonations: experiments in porous wall tubes. *Combust. Flame* **131**, 29–46.

Radulescu, M. I., Sharpe, G. J., Lee, J. H. S., Kiyanda, C. B., Higgins, A. J. & Hanson, R. K. 2005. The ignition mechanism in irregular structure gaseous detonations. *Proc. Combust. Inst.* **30**, 1859–1867.

Randolph, A. L. & Law, C. K. 1988. Time-resolved gasification and sooting characteristics of droplets of alcohol–oil blends and water–oil emulsions. *Proc. Combust. Inst.* **21**, 1125–1130.

Randolph, A. L., Makino, A. & Law, C. K. 1988. Liquid-phase diffusional resistance in multicomponent droplet gasification. *Proc. Combust. Inst.* **21**, 601–608.

Reid, R. C., Prausnitz, J. M. & Sherwood, T. K. 1987. *The Properties of Gases and Liquids*, 3rd ed. McGraw-Hill.

Renksizbulut, M. & Yuen, M. C. 1983. Numerical study of droplet evaporation in a high-temperature stream. *J. Heat Transfer* **105**, 389–397.

Reynolds, W. C. 1986. The element potential for chemical equilibrium analysis: Implementation in the interactive program STANJAN. *Tech. Rept.* A-3391, Dept. of Mechanical Engineering, Stanford University.

Richter, H. & Howard, J. B. 2000. Formation of polycyclic aromatic hydrocarbons and their growth to soot—a review of chemical reaction pathways. *Prog. Energy Combust. Sci.* **26**, 565–608.

Robinson, P. J. & Holbrook, K. A. 1972. *Unimolecular Reactions*. Wiley.

Ronney, P. D. 1985. Effects of gravity on laminar premixed gas combustion. 2. Ignition and extinction phenomena. *Combust. Flame* **62**, 121–133.

Ronney, P. D. 1998. Understanding combustion processes through microgravity research. *Proc. Combust. Inst.* **27**, 2485–2506.

Ronney, P. D. & Wachman, H. Y. 1985. Effects of gravity on laminar premixed gas combustion. 1. Flammability limits and burning velocities. *Combust. Flame* **62**, 107–119.

Rosner, D. E. 1986. *Transport Processes in Chemically Reacting Flow Systems*. Butterworth.

Rosner, D. E. & Chang, W. S. 1973. Transient evaporation and combustion of a fuel droplet near its critical temperature. *Combust. Sci. Technol.* **7**, 145–158.

Ross, R. D. 2001. *Microgravity Combustion: Fire in Free Fall*. Academic.

Samson, R., Bedeaux, D. & Deutch, J. M. 1978. A simple model of fuel spray burning. II. Linear droplet streams. *Combust. Flame* **31**, 223–229.

Samson, R., Bedeaux, D., Saxton, M. J. & Deutch, J. M. 1978. A simple model of fuel spray burning. I. Random sprays. *Combust. Flame* **31**, 215–221.

Sangiovanni, J. J. & Labowsky, M. 1982. Burning times of linear fuel droplet arrays: A comparison of experiment and theory. *Combust. Flame* **47**, 15–30.

Sarofim, A. F. 1986. Radiative heat transfer in combustion, friend or foe? *Proc. Combust. Inst.* **21**, 1–23.

Sato, J., Tsue, M., Niwa, M. & Kono, M. 1990. Effects of natural convection on high-pressure droplet combustion. *Combust. Flame* **82**, 142–150.

Schlichting, H., Gersten, K., Krause, E., Mayes, K. & Oertel, H., Jr. 1999. *Boundary Layer Theory*, 8th ed. Springer-Verlag.

Sedov, L. I. 1946. Propagation of strong blast waves. *Prikl. Mat. Mech.* **10**, 241–250.

Semenov N. N. 1958. *Some Problems in Chemical Kinetics and Reactivity.* Vol. I. Princeton University Press.

Semenov N. N. 1959. *Some Problems in Chemical Kinetics and Reactivity.* Vol. II. Princeton University Press.

Seshadri, K. 1996. Multistep asymptotic analyses of flame structures. *Proc. Combust. Inst.* **26**, 831–846.

Seshadri, K. & Ilincic, N. 1995. The asymptotic structure of nonpremixed methane–air flames with oxidizer leakage of order unity. *Combust. Flame* **101**, 69–80.

Seshadri, K. & Peters, N. 1988. Asymptotic structure and extinction of methane—air diffusion flames. *Combust. Flame* **73**, 23–44.

Seshadri, K., Peters, N. & Williams, F. A. 1994. Asymptotic analysis of stoichiometric and lean hydrogen–air flames. *Combust. Flame* **96**, 407–427.

Seshadri, K. & Williams, F. A. 1978. Laminar flow between parallel plates with injection of a reactant at high Reynolds number. *Intl. J. Heat Mass Transfer* **21**, 251–253.

Seshadri, K. & Williams, F. A. 1994 Reduced chemical systems and their application in turbulent combustion. In *Turbulent Reacting Flows*, P. A. Libby & F. A. Williams (eds.). Academic.

Sharpe, G. J. 2000. The structure of planar and curved detonation waves with reversible reactions. *Phys. Fluids* **12**, 3007–3020.

Shy, S. S., Jang, R. H. & Ronney, P. D. 1996. Laboratory simulation of flamelet and distributed models for premixed turbulent combustion using aqueous autocatalytic reactions. *Combust. Sci. Technol.* **114**, 329–350.

Simmie, J. M. 2003. Detailed chemical kinetic models for the combustion of hydrocarbon fuels. *Prog. Energy Combust. Sci.* **29**, 599–634.

Sirignano, W. A. 1999. *Fluid Dynamics and Transport of Droplets and Sprays*. Cambridge.

Sivashinsky, G. I. 1976. Distorted flame front as a hydrodynamic discontinuity. *Acta Astronautica* **3**, 889–918.

Sivashinsky, G. I. 1977. Diffusional-thermal theory of cellular flames. *Combust. Sci. Technol.* **15**, 137–146.

Sivashinsky, G. I. 1983. Instabilities, pattern formation, and turbulence in flames. *Ann. Rev. Fluid Mech.* **15**, 179–199.

Sivashinsky, G. I. 1988. Cascade-renormalization theory of turbulent flame speed. *Combust. Sci. Technol.* **62**, 77–96.

Sivashinsky, G. I., Law, C. K. & Joulin, G. 1982. On stability of premixed flames in stagnation-point flow. *Combust. Sci. Technol.* **28**, 155–159.

Smagorinsky, J. 1963. General circulation experiments with the primitive equations: I. The basic equations. *Mon. Weather Rev.* **91**, 99–164.

Smith, F. A. & Pickering, S. F. 1928. Bunsen flames of unusual structure. *Proc. Combust. Inst.* **1–2**, 24–26.

Smooke, M. D. 1991. *Reduced Kinetic Mechanisms and Asymptotic Approximations for Methane-Air Flames. Lecture Notes in Physics* **384**.

Smooke, M. D., Puri, I.K. & Seshadri, K. 1986. A comparison between numerical calculations and experimental measurements of the structure of a counterflow diffusion flame burning diluted methane in diluted air. *Proc. Combust. Inst.* **21**, 1783–1792.

Sohrab, S. H. & Law, C. K. 1985. Influence of burner rim aerodynamics on polyhedral flames and flame stabilization. *Combust. Flame* **62**, 243–254.

Sohrab, S. H., Ye, Z. Y. & Law, C. K. 1985. An experimental investigation on flame interaction and the existence of negative flame speeds. *Proc. Combust. Inst.* **20**, 1957–1965.

Souders, M. & Eshbach, O. W. 1975. *Handbook of Engineering Fundamentals*, 3rd ed. Wiley.

Spalding, D. B. 1953. The burning of liquid fuels. *Proc. Combust. Inst.* **4**, 847–864.

Spalding, D. B. 1957. A theory of inflammability limits and flame-quenching. *Proc. R. Soc. Lond. A* **240**, 83–100.

Spalding, D. B. & Yumlu, V. S. 1959. Experimental demonstration of the existence of two flame speeds. *Combust. Flame* **3**, 553–556.

Stewart, D. S. 1998. The shock dynamics of multidimensional condensed and gas-phase detonations. *Proc. Combust. Inst.* **27**, 2189–2205.

Stewartson, K. 1958. On free convection from a horizontal plate. *Z. Angew. Math. Phys.* **9a**, 276–282.

Stewartson, K. 1964. *The Theory of Laminar Boundary Layers in Compressible Fluids*. Clarendon.

Strahle, W. C. 1993. *Introduction to Combustion*. Gordon and Breach.

Strehlow, R. 1968. Gas phase detonations: Recent developments. *Combust. Flame* **12**, 81–101.

Strehlow, R. A. 1984. *Fundamentals of Combustion*, McGraw-Hill.

Stull, D. R. & Prophet, H. 1971. *JANAF Thermochemical Tables*, 2nd ed. National Bureau of Standards, NSRDS-NBS37.

Sun, C. J. & Law, C. K. 1998. On the consumption of fuel pockets via inwardly propagating flames. *Proc. Combust. Inst.* **27**, 963–970.

Sun, C. J. & Law, C. K. 2000. On the nonlinear response of strongly stretched premixed flames. *Combust. Flame* **121**, 236–248.

Sun, C. J., Sung, C. J., He, L. & Law, C. K. 1999. Dynamics of weakly stretched flames: Quantitative description and extraction of global flame parameters. *Combust. Flame* **118**, 108–128.

Sun, C. J., Sung, C. J. & Law, C. K. 1994. On adiabatic stabilization and geometry of Bunsen flames, *Proc. Combust. Inst.* **25**, 1391–1398.

Sung, C. J. & Law, C. K. 1996. Extinction mechanisms of near-limit premixed flames and extended limits of flammability. *Proc. Combust. Inst.* **26**, 865–873.

Sung, C. J. & Law, C. K. 2000. Structural sensitivity, response, and extinction of diffusion and premixed flames in oscillating counterflow. *Combust. Flame* **123**, 375–388.

Sung, C. J., Law, C. K. & Chen, J. Y. 1998. An augmented reduced mechanism for methane oxidation with comprehensive global parametric validation. *Proc. Combust. Inst.* **27**, 295–304.

Sung, C. J., Liu, J. B. & Law, C. K. 1995. Structural response of counterflow diffusion flames to strain rate variations. *Combust. Flame* **102**, 481–492.

Sung, C. J., Makino, A. & Law, C. K. 2002. On stretch-affected pulsating instability in rich hydrogen–air flames: Asymptotic analysis and computation. *Combust. Flame* **128**, 422–434.

Sung, C. J., Yu, K. M. & Law, C. K. 1994. On the geometry and burning intensity of Bunsen flames. *Combust. Sci. Technol.* **100**, 245–270.

Sung, C. J., Zhu, D. L. & Law, C. K. 1998. On microbuoyancy spherical diffusion flames and a double luminous zone structure of the hydrogen–methane flame. *Proc. Combust. Inst.* **27**, 2559–2567.

Takahashi, F., Dryer, F. L. & Williams, F. A. 1986. Combustion behavior of free boron slurry droplets. *Proc. Combust. Inst.* **21**, 1983–1991.

Taylor, G. I. 1950a. The dynamics of the combustion products behind plane and spherical detonation front in explosives. *Proc. R. Soc. Lond. A* **200**, 235–247.

Taylor, G. I. 1950b. The formation of a blast wave by a very intense explosion. I. Theoretical discussion. *Proc. R. Soc. Lond. A* **201**, 159–174. (First published under the same title as Report RC-210, 27 June 1941, Civil Defense Research Committee.)

Taylor, G. I. 1950c. The formation of a blast wave by a very intense explosion. II. The atomic explosion of 1945. *Proc. R. Soc. Lond. A* **201**, 175–186.

Tennekes, H. & Lumley, J. L. 1972. *A First Course in Turbulence*. MIT Press.

Tien, C. L. & Lee, S. C. 1982. Flame radiation. *Prog. Energy Combust. Sci.* **8**, 41–59.

Tien, J. H. & Matalon, M. 1991. On the burning velocity of stretched flames. *Combust. Flame* **84**, 238–248.

Tishkoff, J. M. 1979. A model for the effect of droplet interactions on vaporization. *Intl. J. Heat Mass Trans.* **22**, 1407–1415.

Tomlin, A. S., Pilling, M. J., Turanyi, T., Merkin, J. H. & Brindley, J. 1992. Mechanism reduction for the oscillatory oxidation of hydrogen: Sensitivity and quasi-steady-state analyses. *Combust. Flame* **91**, 107–130.

Tomlin, A. S., Turanyi, T. & Pilling, M. J. 1997. Mathematical tools for the construction, investigation and reduction of combustion mechanisms. *Comprehensive Chemical Kinetics*, 293–437, Elsevier.

Toong, T. Y. 1983. *Combustion Dynamics: The Dynamics of Chemically Reacting Fluids*. McGraw-Hill.

Troe, J. 1979. Predictive possibilities of unimolecular rate theory. *J. Phys. Chem.* **83**, 114–126.

Troe, J. 1983. Theory of thermal unimolecular reactions in the fall-off range. I. Strong collision rate constants. *Ber. Bunsenges. Phys. Chem.* **87**, 161–169.

Tse, S. D., Zhu, D. L. & Law, C. K. 2004. An optically accessible high-pressure combustion apparatus. *Rev. Sci. Instruments* **75**, 233–239.

Tsuboi, T. & Wagner, H. Gg. 1975. Homogeneous thermal oxidation of methane in reflected shock waves. *Proc. Combust. Inst.* **15**, 883–890.

Tsuji, H. 1982. Counterflow diffusion flames. *Prog. Energy Combust. Sci.* **8**, 93–119.

Tsuji, H. & Yamaoka, I. 1983. Structure and extinction of near-limit flames in a stagnation flow. *Proc. Combust. Inst.* **19**, 1533–1540.

Turanyi, T. 1990. Reduction of large reaction mechanisms. *New J. Chem.* **14**, 795–803.

Turns, S. R. 2000. *An Introduction to Combustion: Concepts and Applications*. McGraw-Hill.

Twardus, E. M. & Brzustowski, T. A. 1977. The interaction between two burning fuel droplets. *Archiwum Termodynamiki I Spalania* **8**, 347–358.

Umemura, A. 1981. A unified theory of quasi-steady droplet combustion. *Proc. Combust. Inst.* **18**, 1355–1363.

Umemura, A., Nam, S. & Law, C. K. 1990. Natural convection boundary layer flow over a heated plate with arbitrary inclination. *J. Fluid Mech.* **219**, 571–584.

Umemura, A., Ogawa, S. & Oshima, N. 1981. Analysis of the interaction between two burning droplets. *Combust. Flame* **41**, 45–55.

Urtiew, P. A. & Oppenheim, A. K. 1966. Experimental observations of the transition to detonation in an explosive gas. *Proc. R. Soc. Lond. A* **295**, 13–28.

Vagelopoulos, C. M. & Egolfopoulos, F. N. 1998. Direct experimental determination of laminar flame speeds. *Proc. Combust. Inst.* **27**, 513–519.

Vervisch, L. & Poinsot, T. 1998. Direct numerical simulation of nonpremixed turbulent flames. *Ann. Rev. Fluid Mech.* **30**, 655–692.

Veynante, D. & Vervisch, L. 2002. Turbulent combustion modeling. *Prog. Energy Combust. Sci.* **28**, 193–266.

Vincenti, W. G. & Kruger, C. H., Jr. 1965. *Introduction to Physical Gas Dynamics*, Kreiger.

Viskanta, R. & Mengüc, M. P. 1987. Radiation heat transfer in combustion systems. *Prog. Energy Combust. Sci.* **13**, 97–160.

von Neumann, J. 1941. The point source solution. National Defence Research Committee, Div. B, Report AM 9. Reprinted in *The Collected Works of John von Neumann*, A. J. Taub (ed.), Pergamon, 1963.

von Neumann, J. 1942. Theory of detonation waves. O.S.R.D. Rept. No. 549. Reprinted in *The Collected Works of John von Neumann*, A. J. Taub (ed.), Pergamon, 1963.

Waldman, C. H. 1975. Theory of nonsteady state droplet combustion. *Proc. Combust. Inst.* **15**, 429–442.

Wang, C. H. & Law, C. K. 1985. Microexplosion of fuel droplets under high pressure. *Combust. Flame* **59**, 53–62.

Wang, C. H., Liu, X. Q. & Law, C. K. 1984. Combustion and microexplosion of freely falling multicomponent droplets. *Combust. Flame* **56**, 175–197.

Wang, H. & Frenklach, M. 1991. Detailed reduction of reaction mechanisms for flame modeling. *Combust. Flame*, **87**, 365–370.

Wang, W., Rogg, B. & Williams, F. A. 1993 Reduced kinetic mechanisms for wet CO flames. In *Reduced Kinetic Mechanisms for Application in Combustion Systems*, N. Peters & B. Rogg (eds.). Springer-Verlag, pp. 845–856.

Warnatz, J. 1981. The structure of laminar alkane-, alkene-, and acetylene flames. *Proc. Combust. Inst.* **18**, 369–384.

Warnatz, J., Maas, U. & Dibble, R. W. 2001. *Combustion: Physical and Chemical Fundamentals, Modeling and Simulation, Experiments, Pollutant Formation*. Springer-Verlag.

Westbrook, C. K. & Dryer, F. L. 1981. Chemical kinetics and modeling of combustion processes. *Proc. Combust. Inst.* **18**, 749–767.

Westbrook, C. K. & Dryer, F. L. 1984. Chemical kinetics modeling of hydrocarbon combustion. *Prog. Energy Combust. Sci.* **10**, 1–57.

Williams, F. A. 1975. A review of some theoretical considerations of turbulent flame structure. In *Analytical and Numerical Methods for Investigation of Flow Fields with Chemical Reactions, Especially Related to Combustion*, M. Barrére (ed.), AGARD Conf. Proc. No. 164.

Williams, F. A. 1985. *Combustion Theory*. Addison-Wesley.

Williams, F. A. 1992. The role of theory in combustion science. *Proc. Combust. Inst.* **24**, 1–17.

Williams, F. A. 1997. Some aspects of metal particle combustion. In *Physical and Chemical Aspects of Combustion*, F. L. Dryer & R. F. Sawyer (eds.). Gordon & Breach, pp. 267–288.

Williams, F. A. 2000. Progress in knowledge of flamelet structure and extinction. *Prog. Energy Combust. Sci.* **26**, 657–682.

Wintenberger, E. & Shepherd, J. E. 2006. The stagnation Hugonist analysis for steady combustion waves in propulsion system. *J. Propul. Power* **22**, 835–844.

Wise, H., Lorell, J. & Wood, B. J. 1955. The effects of chemical and physical parameters on the burning rates of a liquid droplet. *Proc. Combust. Inst.* **5**, 132–141.

Wohl, K., Kapp, N. M. & Gazley, C. 1949. The stability of open flames. *Proc. Combust. Inst.* **3**, 3–40.

Wolfrum, J. 1998. Lasers in combustion: From basic theory to practical devices. *Proc. Combust. Inst.* **27**, 1–41.

Wu, C. K. & Law, C. K. 1985. On the determination of laminar flame speeds from stretched flames. *Proc. Combust. Inst.* **20**, 1941–1949.

Yakhot, V. 1988. Propagation velocity of premixed turbulent flame. *Combust. Sci. Technol.* **60**, 191–214.

Yang, V. 2000. Modeling of supercritical vaporization: Mixing and combustion processes in liquid-fueled propulsion systems. *Proc. Combust. Inst.* **28**, 925–942.

Yao, J. & Stewart, D. S. 1995. On the normal detonation shock velocity-curvature relationships for materials with large activation energy. *Combust. Flame* **100**, 519–528.

Yuen, M. C. & Chen, L. W. 1976. On drag of evaporating liquid droplets. *Combust. Sci. Technol.* **14**, 147–154.

Zabetakis, M. G. 1965. Flammability characteristics of combustible gases and vapors. *U.S. Department of Mines Bulletin* **627**.

Zel'dovich, Y. B. 1940. On the theory of propagation of detonation in gaseous systems. *Zhur. Eksp. Teor. Fiz.* **10**, 542–568. (English translation: NACA TM 1261, 1950).

Zel'dovich, Y. B. 1946. The oxidation of nitrogen in combustion and explosions. *Acta Physicochem, USSR* **21**, 577–628.

Zel'dovich, Y. B. 1980. Regime classification of an exothermic reaction with nonuniform initial conditions. *Combust. Flame* **39**, 211–214.

Zel'dovich, Y. B., Barenblatt, G. I., Librovich, V. B. & Makhviladze, G. M. 1985. *The Mathematical Theory of Combustion and Explosions*. Consultants Bureau.

Zel'dovich, Y. B. & Frank-Kamenetskii, D. A. 1938. Thermal theory of flame propagation, *Zhur Fiz. Khim.* **12**, 100.

Zel'dovich, Y. B., Kogarko, S. M. & Semenov, N. N. 1957. An experimental investigation of spherical detonation in gases. *Sov. Phys. Tech. Phys.* **1**, 1689–1713.

Zel'dovich, Y. B. & Kompaneets, A. S. 1960, *Theory of Detonation*, Academic.

Zel'dovich, Y. B. & Raizer, Y. P. 1966. *Physics of Shock Waves and High-Temperature Hydrodynamic Phenomena*. Academic.

Zheng, X. L., Blouch, J. D., Zhu, D. L., Kreutz, T. G. & Law, C. K. 2002. Ignition of premixed hydrogen–air by heated counterflow. *Proc. Combust. Inst.* **29**, 1637–1643.

Zheng, X. L. & Law, C. K. 2004. Ignition of premixed hydrogen–air by heated counterflow under reduced and elevated pressures. *Combust. Flame* **136**, 168–179.

Author Index

Subject Index

Note: Page numbers followed by "t" indicate tables. Page numbers followed by "f" indicate figures.

Made in the USA
San Bernardino, CA
11 October 2013